Sensory Individuals

Sensory Individuals

Unimodal and Multimodal Perspectives

Edited by
Aleksandra Mroczko-Wąsowicz
and Rick Grush

OXFORD
UNIVERSITY PRESS

OXFORD
UNIVERSITY PRESS

Great Clarendon Street, Oxford, OX2 6DP,
United Kingdom

Oxford University Press is a department of the University of Oxford.
It furthers the University's objective of excellence in research, scholarship,
and education by publishing worldwide. Oxford is a registered trade mark of
Oxford University Press in the UK and in certain other countries

First Edition published in 2023

Published in the United States of America by Oxford University Press
198 Madison Avenue, New York, NY 10016, United States of America

British Library Cataloguing in Publication Data
Data available

Library of Congress Control Number: 2022952177

ISBN 978–0–19–886630–5

DOI: 10.1093/oso/9780198866305.001.0001

Printed and bound by
CPI Group (UK) Ltd, Croydon, CR0 4YY

For those who supported me when I was preparing this book.
Special thanks to Konstancja, Vincent, and Paweł.

Aleksandra Mroczko-Wąsowicz

Contents

Acknowledgements

Our warm thanks to all contributors and our editors at Oxford University Press.

Aleksandra Mroczko-Wąsowicz has received research support for the project 'Perceptual objects in the unimodal and multimodal settings' from the National Science Centre, Poland (grant 2019/35/B/HS1/04386) and the University of Warsaw, and Rick Grush from the University of California, San Diego (UCSD).

During her research stay at the Department of Philosophy, UCSD in winter 2019, supported by Taiwan's Ministry of Science and Technology and the National Yang-Ming University, Aleksandra Mroczko-Wąsowicz had a chance to refresh collaboration with Rick Grush that was initiated few years earlier in Germany within the interdisciplinary research network MIND Group led by Thomas Metzinger.

The two co-editors Aleksandra Mroczko-Wąsowicz and Rick Grush benefited from the inspiring atmosphere of the Philosophy of Perception Workshop organized by Jonathan Cohen in February 2019 at the Department of Philosophy, UCSD, featuring many of the world's leading philosophers of perception who later contributed to this collection.

About the editors

Aleksandra Mroczko-Wąsowicz is currently Associate Professor of Philosophy and Cognitive Science at the University of Warsaw. She earned her PhD in philosophy and neuroscience from the University of Mainz and the Max Planck Institute for Brain Research in Frankfurt am Main in 2011. Before coming to Warsaw, she was a Postdoctoral Fellow in the Department of Philosophy at Ruhr University Bochum and the University of Duesseldorf, and Associate Professor at the Institute of Philosophy of Mind and Cognition at the National Yang-Ming University in Taipei. Her research centres on empirically informed philosophy of perception and philosophy of cognitive neuroscience. Aleksandra Mroczko-Wąsowicz has worked on the phenomenon of synaesthesia, the unity of consciousness, modularity, cognitive enhancement, and interactions between perceptual modalities.

Rick Grush is currently Professor of Philosophy at UCSD. He received a joint doctorate in philosophy and cognitive science from UCSD in 1995. From his doctoral dissertation to the early 2000s, Rick Grush developed the emulation theory of perception, imagery, and motor control. He then extended this framework to account for spatial and temporal aspects of perception. His research centres on theories of spatial and temporal representation and theoretical cognitive neuroscience.

Contributors

Clare Batty Philosophy, University of Kentucky, USA

J. Gavin Bremner Psychology, Lancaster University, UK

Berit Brogaard Philosophy, University of Miami, USA

Alessandra Buccella Philosophy, Chapman University, USA

Fabrizio Calzavarini Philosophy, University of Turin, Italy

Mazviita Chirimuuta Philosophy, University of Edinburgh, UK

Jonathan Cohen Philosophy, University of California San Diego, USA

Yale E. Cohen Neuroscience, University of Pennsylvania, USA

Frédérique de Vignemont Philosophy, Institut Jean Nicod, ENS-PSL University-EHESS-CNRS, France

Ronald W. DiTullio Neuroscience, University of Pennsylvania, USA

Viyehni Fuchser Psychology, University of California Los Angeles, USA

Matthew Fulkerson Philosophy, University of California San Diego, USA

E. J. Green Philosophy, Massachusetts Institute of Technology, USA

Scott P. Johnson Psychology, University of California Los Angeles, USA

Philip J. Kellman Psychology, University of California Los Angeles, USA

Simon Lacey Neurology, Penn State University, USA

William G. Lycan Philosophy, University of Connecticut, USA

Mohan Matthen Philosophy, University of Toronto, Canada

Bence Nanay Philosophy, University of Antwerp, Belgium

Casey O'Callaghan Philosophy, Washington University St Louis, USA

Jonas K. Olofsson Psychology, Stockholm University, Sweden

Joan Danielle K. Ongchoco Psychology, Yale University, USA

Stephen Pierzchajlo Psychology, Stockholm University, Sweden

Jake Quilty-Dunn Philosophy, Washington University St Louis, USA

K. Sathian Neurology, Penn State University, USA

Brian J. Scholl Psychology, Yale University, USA

Błażej Skrzypulec Philosophy, Jagiellonian University, Poland

Alan M. Slater Psychology, University of Exeter, UK

Barry C. Smith Philosophy, School of Advanced Study, University of London, UK

Thomas Alrik Sørensen Neuroscience, Aalborg University, Denmark

Charles Spence Psychology, Oxford University, UK

Nastazja Stoch Philosophy, University of Warsaw, Poland

Alberto Voltolini Philosophy, University of Turin, Italy

Benjamin D. Young Philosophy/Neuroscience, University of Nevada, Reno, USA

Nick Young Philosophy, University of Genoa, Italy

Paulina Zguda Philosophy, Jagiellonian University, Poland

Abbreviations

ACC	anterior cingulate cortex
aIPS	anterior intraparietal sulcus
AL	anterolateral
AVI	anterior ventral insula
BOLD	blood-oxygen level-dependent
BRDF	bidirectional reflectance distribution function
CFS	continuous flash suppression
CS	central sulcus
2D	two-dimensional
3D	three-dimensional
EEG	electroencephalographic
EP	exploratory procedure
ERC	entorhinal cortex
FAE	face adaptation effect
FEF	frontal eye field
FFA	fusiform face area
FFOV	Functional Field of View
FG	fusiform gyrus
FIT	feature integration theory
fMRA	functional magnetic resonance adaptation
fMRI	functional magnetic resonance imaging
GradCPT	Gradual Continuous Performance Task
HFR	haptic face region
IAT	Implicit Association Test
IPS	intraparietal sulcus
LOC	lateral occipital complex
MDS	multidimensional scaling
MFI	multisensory feature integration
ML	medial lateral
MLE	maximum likelihood estimation
MOT	multiple object tracking
MRI	magnetic resonance imaging
MRSC	maximal regularity-supporting causation
MST	Molecular Structure Theory
MTL	medial temporal lobe
MVPA	multivoxel pattern analysis
ODC	object-directed capacity
OFC	orbitofrontal cortex
ORN	olfactory receptor neuron
OSIQ	Object and Spatial Imagery Questionnaire
OSPB	object-specific preview benefit
PCA	principal component analysis
PCS	post-central sulcus

PET	positron emission tomography
PO	perceptual object
RT	rostral–temporal
rTMS	repetitive transcranial magnetic stimulation
SSD	sensory substitution device
SSR	spectral surface reflectance
SUIS	Spontaneous Use of Imagery Survey
TMS	transcranial magnetic stimulation
TTC	time-to-collision
TVA	theory of visual attention
vlPFC	ventrolateral prefrontal cortex
VoE	violation of expectation
VVIQ	Vividness of Visual Imagery Questionnaire
VWM	visual working memory

Introduction

Sensory individuals: Contemporary perspectives on modality-specific and multimodal objecthood

Aleksandra Mroczko-Wąsowicz and Rick Grush

1 Sensory individuals—current state of the field

The collection *Sensory Individuals: Unimodal and Multimodal Perspectives* is intended for a wide interdisciplinary audience. It provides a convergent focus of research from a number of different subject areas, primarily integrating three methodological approaches: philosophy of perception, experimental psychology, and cognitive neurosciences. The collection is a core sample of the recent explosion of work in the interdisciplinary study of perception. Blending theoretical and empirical investigations on the origin and nature of human object perception and the involved interactions between different sense modalities, new and exciting ideas have been developed concerning what exactly it is that is perceived and how we perceive it.

What is the nature of the entities that we represent perceptually, and what is the nature of the attribution of sensory properties to such entities? In particular, given that perception is effected by a number of sense modalities, each with very different characteristics, how are we to understand the relationship between the entities represented perceptually and the contribution of the modality or modalities involved in the perception of these items? One such relationship concerns the kinds of entities we can perceive through individual sense modalities, how they vary, and how the contributions made by different modalities are related to one another. The contributing chapters explore various possibilities for such entities, including (but not limited to) material bodies, events, and mereological complexes. The ultimate goal is to understand how it is possible that, and the extent to which, multiple sense modalities can work together to construct multisensory objects and events, like a crackling campfire or an ice cream cone, which we can see, feel, smell, hear, or taste.

So while it is true that perceptual objects are units of perception, such examples highlight the fact that it is not clear what these units are. When sitting at the campfire, are the objects the crackling sound, the smokey smell, and the flickering flame? Or is it a higher-level entity with some number of 'parts' or aspects? The term 'object' is ambiguous and used differently, depending on the discipline, theoretical perspective, and underlying ontological assumptions (see Schwartz, 2001).

It is possible to distinguish a notion of perceptual object from a broader notion of *the objects of perception* (O'Callaghan, 2016; cf. Casati, 2015). Perceptual objects cover extramental, concrete objects—that is, stable three-dimensional (3D) entities and two-dimensional (2D) stimuli. They are 'items that bear perceptible features and have perceptible parts arranged to

Aleksandra Mroczko-Wąsowicz and Rick Grush, *Introduction* In: *Sensory Individuals.* Edited by: Aleksandra Mroczko-Wąsowicz and Rick Grush, Oxford University Press. © Oxford University Press 2023. DOI: 10.1093/oso/9780198866305.003.0001

form a unified whole' (O'Callaghan, 2016, pp. 1271–2). Prototypical perceptual objects are material objects, often described as everyday, physical, middle-sized dry goods (Casati, 2015, p. 393; O'Callaghan, 2008, p. 803; 2016, p. 1271). However, perceptual objects cannot also be equated with material objects due to the existence of perceptual objects such as ephemeral visibilia, for example, rainbows, clouds, holes, or shadows (O'Callaghan, 2008, pp. 816, 825; 2016, p. 1274). Apart from material bodies and other object-like individuals, perceptual objects include event-like individuals and scattered objects—that is, groupings that are unified structurally by causal regularities like flocks of birds. Tracking any of them requires a perceptual suite of object-directed capacities relative to a specific modality (Green, 2018; O'Callaghan, 2016). A broader notion of the objects of perception would include extramental entities, events, properties, property bundles, spatial locations, and boundaries (Casati, 2015, p. 393; Clark, 2004, p. 463; Schwartz, 2001, p. 5), as well as existent and non-existent objects, their parts, properties, and relations, thus intentional objects.

Any attempt to characterize perceptual objects should accommodate differences in object perception between various sense modalities. Ideally such a characterization would settle whether entities beyond ordinary material objects (i.e., non-prototypical visual objects, sounds, odours, or entities targeted by other non-visual sense modalities) can qualify as perceptual objects for a given modality, and if so, by what criteria. Formulating a plausible account of perceptual objects would require criteria for both object conception and object individuation that specify how certain entities are singled out from the environment. Accordingly, an account of perceptual objecthood has at least two aspects: within a theory of perceptual objects in general (e.g., Green, 2018; O'Callaghan, 2016) and as applied to the targets of specific sense modalities.

By way of illustration of these points, Stevenson (2014, p. 1362) introduces 'eight criteria that seem to characterize object-hood', which he also calls 'capacities for object processing'. These are binding of parts and being a default mode of perception which is analytically less costly than processing parts, properties, generic impressions, or superordinate categories. Objects are structured, more memorable than the background, and subject of figure-ground segregation. Perceiving them is learning- and memory-dependent, as well as perceptually invariant. Other researchers mention also simultaneous tracking of several objects among object-related perceptual capacities (Green, 2018; O'Callaghan, 2016; Scholl *et al.*, 2001). Some of those capacities of perceptual systems are present in specific sense modalities, but not in others.

Although we can perceive many perceptual objects with more than one sense (e.g., ordinary visual objects are often tactile objects as well), there are objects (special sensibles) that are presented only to one modality—such as light for vision, and sounds for audition. These objects are called unimodal perceptual objects or proper objects. Proper objects of a certain modality are its intentional objects and are typically inaccessible by other sense modalities.

Various sense modalities operate differently and reveal diverse aspects of object perception. There has been a significant disproportion in quantity between studies on vision (e.g., Alvarez and Franconeri, 2007; Brewer, 2007; Lycan, 2014; Scholl *et al.*, 2001; Siegel, 2006; Wolfe *et al.*, 2002) and those on other modalities (Batty, 2014; Lycan, 2018; Matthen, 2010; Millar, 2019). Such visuocentrism has resulted in little attention devoted to non-visual perceptual objects, and this imbalance has invited hasty generalization of findings from the perception of visual objects to the perception of objects in other modalities (Vernazzani *et al.*, 2021).

Of course, an insistence that we take seriously the diversity to be found in different modalities should not blind us to the obvious fact that there are some common characteristics among perceptual objects of the visual, auditory, and olfactory modalities. An example of this sort of unified modality-independent theory is O'Callaghan's (2008, pp. 825–6; 2016, p. 1276) account of structured, mereologically complex individuals. On this view, a perceptual object has to be a feature bearer whose qualitative parts collectively form a cohesive whole, and whose internal and external structures are determined by a given sense modality.

It has only been relatively recently that the multimodality of (object) perception has gained prominence as a research topic (Auvray and Spence, 2008; Briscoe, 2017; Kubovy and Schutz, 2010; O'Callaghan, 2014, 2016). Multimodal perceptual objects result from integrative interactions between different sense modalities known as multimodal binding (Macpherson, 2011; O'Callaghan, 2015). In contrast to unimodal binding, in which sensory properties of a single sense modality are typically understood to be predicated of a single individual entity, in multimodal binding, sensory properties processed by different sense modalities may be predicated, depending on a conception, to multiple parts of the perceptual object (constituents of the mereological complex)—for instance, when perceivers attribute visual, auditory, olfactory, and tactile properties to the same multimodal object such as a cat. Such properties presented in multimodal experiences are not merely presented as co-occurring, but usually seem to be unified within some structured complexes. In particular, multimodal experiences commonly present entities as jointly having characteristics processed by various sense modalities. For example, we are able to experience not only something that looks cat-like and something that makes a meowing sound, but also a single object which both looks like a cat and meows.

In addition, entities presented in perceptual experiences, like objects, events, or spatiotemporal regions, are often amodal in the sense that information about them and their properties is common across multiple sense modalities and may be presented by more than one modality (Nudds, 2014). For instance, the same spatial location may be visually represented as containing a colour and auditorily represented as comprising some noise, or the same shape, size, and texture are common across visual and tactile stimulation. Furthermore, in phenomena such as amodal completion, perceptual experiences can include something that is not given in direct sensory stimulation of any sense modality. This is why we perceive the world as made of coherent wholes, even though we are presented with partly occluded items. The above observations show that perceptual experiences are, in an important respect, organized around multimodal and amodal entities.

Perceptual objects are sometimes associated with sensory individuals (cf. O'Callaghan, 2016, p. 1276). A sensory individual on this view is something to which sensory properties are attributed (Cohen, 2004, pp. 471–3). It marks two distinctions. First, the notion differentiates properties attributed perceptually (e.g., shape, colour, pitch, timbre) from properties that may be attributed otherwise (e.g., the authorship of a given piece) (Nanay, 2017, pp. 9, 12–14). Second, it distinguishes 'features from that to which they are attributed' (O'Callaghan, 2008, p. 807). One motivation for the positing of sensory individuals was to explain why perceptual systems do not simply detect and extract features from the environment, but also predicate them to particulars (Jackson, 1977). This is the well-known feature integration/binding problem or the 'many properties' problem (Clark, 2000, 2004; Cohen, 2004; Treisman, 1996; cf. O'Callaghan, 2008, p. 807).

Different candidates have been proposed for sensory individuals in the literature. A standard view is to treat spatio-temporal regions, places, or locations as sensory individuals in this sense. Referring to the notion of feature-placing, where regions are subjects of sensory qualities, Clark (2000, p. 61; 2004, p. 463) proposed that sensory individuals (in his words, 'phenomenal individuals') may be understood as 'regions of spacetime, in or around the body of the sentient organism'. Similarly, O'Callaghan (2008, pp. 805, 807, 825) has assumed that spatial location, as the simplest individual, may be necessary for individuating more complex individuals (coherent collections of spatio-temporally bounded locations, parts, or surfaces), since co-instantiation or feature binding could be explained only if 'distinct locations are distinct sensory individuals that instantiate visual features'. Recently, Nanay (2022) has argued that spatio-temporal regions, not ordinary objects, are sensory individuals of early vision.

Another, much more prevalent, conception of sensory individuals assumes they are particulars to which properties are bound—an approach that assimilates sensory individuals to perceptual objects.[1] The accounts which identify perceptual objects with sensory individuals may differ. According to Cohen (2004, pp. 489–90), 'individual bearers of sensory features are best conceived as (visual) objects'. On this conception of the sensory individuals of vision, visual properties qualify token ordinary spatially extended and temporally persistent objects (e.g., Cohen, 2004; Matthen, 2004, 2005, pp. 277–82; Nanay, 2013, p. 51; Pylyshyn, 2007; Scholl, 2007; Spelke, 1990). An alternate view is to understand sensory individuals to be not ordinary material objects, but rather *events*. Accordingly, sensory individuals may be thought of as occurrences, happenings, or interactions in the environment to which our perceptual states attribute sensory properties (Nanay, 2017, pp. 12–13; O'Callaghan, 2008, p. 824).

It has been suggested that whether an account of sensory individuals is plausible or not depends on a sense modality (Nanay, 2022). Within this conception of sensory individuals, this is related to a question about which modalities have objects or whether what is perceptually presented should be characterized in terms of objects.

For this reason, sensory individuals specific to non-visual sense modalities have been postulated primarily in the context of auditory and olfactory modalities—that is, sounds as auditory sensory individuals (Cohen, 2010; Nudds, 2010; O'Callaghan, 2008, 2011) and odours as olfactory sensory individuals (Batty, 2014; Lycan, 1996, 2000; Young, 2016). Finally, since some chemical senses seem to not exhibit object perception (see Keller, 2016; Barwich, 2019 for olfaction; O'Callaghan, 2016 for flavour perception; Stevenson, 2014 for gustation, chemesthesis, and flavour perception in the mouth), it may be claimed that the only appropriate approach to those senses is 'no sensory individuals' account, because they represent uninstantiated properties, which are not attributed to anything.

2 Summary and overview of the contributions

The contributions in this volume are organized, in order of appearance and in the table of contents, into three groups. The first concerns conceptual and methodological issues; the

[1] Since perceptual objects bear perceptible features, this understanding of sensory individuals stands in contradiction to the definition of sensory individuals as featureless bearers of features or bare particulars.

second concerns unimodal phenomena; and the third concerns multimodal phenomena. This is a useful way to carve up the conceptual space, but it is not the only useful way. In this introduction, we will discuss the contributions as falling into groups seen from a different perspective, in hope that this will provide added value.

A great deal of work in sensory processing and perception concerns how sensory input can be segmented into elements that are likely to correspond to a single object. The problem arises because for much perception targeting the external environment, relevant objects do not have dedicated sensors. A Golgi tendon organ is a sense receptor that is mechanically linked to a single tendon and is sensitive to the tension on *that* tendon. Light-sensitive cells in the retina are not similarly tied to proprietary objects. The same retinal cell conveys information about the sky at one time, a car at another, and a tree branch at another. This means that the identity of the sensor by itself is insufficient to determine what (if any) object the information that sensor is conveying concerns. The perceptual system in such cases brings additional resources to bear, including: (1) how sets of sensory receptors are responding collectively, both synchronically and diachronically, and relatedly; and (2) top-down expectations about the sorts of objects likely to be responsible for sensory stimulation. Under (1) fall many of the Gestalt criteria. One result of such processing would seem to be an estimate of which groups of sensory receptors are currently providing information about the same thing. Under (2) fall resources such as the 'object concept', which, inter alia, adds expectations about the sources of sensory information that are likely to be causing such stimulation. The two are closely related, since the explanation for why many of the criteria under (1) are effective is that they are decently reliable low-level ways of getting a head start on the things that fall under (2)—for example, the manifolds bounded in the sensory field as *figures* (vs ground) are statistically likely to correspond to causally connected objects.

The contribution by *Kellman and Fuchser* argues that both of these resources are in play in vision as distinct stages. In the initial stage, contour links are posited on purely bottom-up geometric grounds. This involves the 'linking of visible contour fragments' that produces an intermediate representation between mere edge/junctions representation and an object/scene representation. A key aspect of this is a phenomenon Kellman and Fuchser's article sheds considerable light upon: *path detection*. At the second stage, top-down constraints are brought in to discern those linkages that provide for a scene analysis. Crucially, they argue that the second stage cannot posit linkages that are not proposed by earlier stages but does prune many of the provisional linkages out.

Vision, of course, is not the only modality in which a manifold of sensation needs to be processed into object representations. The contribution by *diTullio and Cohen* discusses the processes by which the auditory system moves from raw sensory input to a representation of 'auditory objects' such as the honk of a car. One part of this process is *auditory scene analysis*, the parsing of sensory input into groupings based on Gestalt principles. These support completion—a tone followed by a brief noise burst, followed by the same tone again, is perceived as a single continuous tone that persists through the noise burst. Much of this sort of work has been psychological, but the authors here delve into the neurophysiology, especially the role of the ventral auditory pathway, in supporting auditory scene analysis. In particular, they take a deep dive into the different kinds of processing that seems to be going on in different areas of the ventral auditory pathway, including low-level segmentation, auditory decisions and evaluations, and higher-level auditory object categorizations (e.g., a *sonata* or *the English word 'red'*).

Johnson, Bremner, and Slater's contribution discusses the relationship between the sorts of low-level constraints exemplified by Gestalt criteria and the higher-level aspects of perception involving knowledge of the sorts of entities perception concerns—for example, the 'object concept'. Specifically, the question is whether the latter is a distinct innate capacity of the perceptual system or a capacity that emerges from experience characterized in terms of the lower-level processes. The latter would amount to a 'perceptual account of origins of object concepts'. The key aspect of the object concept with which the authors are concerned is what Piaget called 'objectification'—roughly, the capacity to understand that the objects of perception persist through time in spatially continuous locations/paths throughout episodes of not being perceived. They discuss two standard approaches to explaining the emergence of this capacity: a Piagetian sensorimotor learning approach and a nativist approach. According to the former, object persistence is learnt on the basis of scaffolded sequence of stages of sensorimotor learning. According to the latter, it is the manifestation of an innate core concept. The authors argue for a third option—that the emergence of infants' appreciation of object persistence is 'specified directly by perceptual information'. Infants have a limited capacity to exploit this information, but these limitations lessen over time.

Ongchoco and Scholl make a fascinating intervention in this topic. The usual view (outlined a few paragraphs back) is that the lower-level processes operate primarily on given sensory inputs, and the higher-level processes are driven by knowledge of the domain that is perceived. Ongchoco and Scholl's contribution does not suggest any of this is wrong, but it does suggest that it might not be the complete picture. They discuss a common phenomenon where subjects report perceiving patterns emerging from neutral, but structured, stimuli—for example, seeing a square or cross emerge while looking at a uniform grid pattern, or hearing a rhythm emerge from a uniform series of audible clicks. The authors present evidence that these 'objects' have many features of genuine perceptual objects, including same-object attentional advantages. They describe how this scaffolded attention process produces 'bona fide object representations' not only in the spatial domain, but also in the temporal domain—for example, the hearing of rhythmic patterns when presented with a regularly temporally spaced sequence of stimuli of equal pitch, volume, etc. The authors provide evidence that these objects are not the result of imagery (as one might have expected) but seem more closely connected to attention.

The group of contributions just described addressed the issue of the relation between low-level sensory segmentation/grouping and higher-level processing involving something like a conception of the external entities that perception is attempting to track. The next group of contributions is oriented towards a detailed investigation of these latter conceptions. What are these perceptual targets, how are they conceived, and what role do these conceptions play in unimodal, multimodal, and perhaps even *amodal* processing?

Our perceptual processes arguably have many kinds of objects—sounds, shadows, pains, etc. But the prototype is, as *Matthen* points out, *material bodies*. Examples would be chairs, rocks, animals, and apples. Roughly, these are the worldly counterparts of the 'object concept' discussed by Johnson. Several topics converge here. One is why such things should be of particular importance for perceptual processes: *why are they the prototype?* Matthen argues that motile organisms—those able to move themselves around—have a vested interest in tracking what material objects are up to. Such objects are represented as possessing networks of stable internal and external causal characteristics that explain how the object might behave and how the organism might interact with it. The fact that such aspects of material bodies play a role in the content of perception is illustrated in what Matthen calls *generalized amodal completion*: perceptual processing that 'inserts features' into the content of perception that could not have an impact on the sense receptors, but do reveal these causal potentials. As he puts it: 'What I seem to see and hear is a single object that has unseen and unheard properties. My perceptual system autonomously inserts into perceptual awareness a multimodal, not a merely visual, subject for both the visual and auditory properties that I perceive. This common subject is a material object.'

This last point, the idea that material bodies are the common objects of perception in multiple modalities, is precisely the topic explored by *Lacey and Sathian's* contribution. Focusing on the interactions between haptic and visual perception, the authors argue for a conception of perception that provides 'largely modality-independent, task-based' representations. For example, while there are differences between haptic and visual object perception and recognition, the authors report that 'crossmodal recognition is view-independent even for unfamiliar objects that are highly similar ... whether visual study is followed by haptic test or vice versa, and whatever the axis of rotation', and that 'both visual and haptic within-modal view-independence rely on the same view-independent representation'. That is, each modality seems to be geared to the production of a representation that is modality- and viewpoint-independent. They then turn from psychological studies to neurophysiology, and in particular, growing evidence of multimodal involvement in cortical processing areas and streams traditionally conceived as visual. As they summarize their findings, the 'visual and haptic modalities are interconnected at almost every level of object processing, sharing common object representations in crossmodal and view-independent recognition, and common dimensions in imagery preferences'.

One aspect of the object concept is that of *persistence while unperceived*. That aspect was the focus of Johnson, Bremner, and Slater's contribution, and is part of Matthen's conception of material objects as a persisting causal ground of both sensed and unsensed properties and processes. But while the basic idea of such persistence is easy enough to convey, the details of what precisely is involved in such a conception and how a modality could contribute to it is more complicated. Taking groundbreaking conceptual work by Peter Strawson and Gareth Evans as their starting point, *Batty and Smith* look at the role that space plays in such a conception. Whether the scenario is one in which one entity occludes another—because it is spatially *between* the perceiver and the unperceived object—or one in which the perceived object is simply not in the vicinity, it seems that an understanding of spatial relations is deeply involved in the notion of unperceived persistence. The authors get at this by asking whether olfactory experience can support a notion of existence unperceived. That is, suppose a subject smells lavender at t_1, followed by a gap of smelling nothing at t_2, and then smelling lavender again at t_3. What would be required for olfactory experience to support a distinction between (1) it being the numerically same odour perceived at t_1 and t_3 and persisting unperceived through t_2, and (2) two numerically distinct, but qualitatively identical, lavender odours smelled at t_1 and t_3? The authors follow Strawson and Evans in connecting this question to spatial representation: the difference between (1) and (2) is underwritten by the ability to conceive of the odour as located, and as myself as being on a route that sometimes is, and at other times is not, at *that* location. They examine empirical work aimed at just this question, whether the brain can construct a 'map' of odour space that supports something like 'location' of experienced odours. They discuss empirical work aimed at showing that humans are able to construct a representation of, and navigate, '... a 2D odour landscape solely on the basis of olfactory inputs', an ability shown to employ the entorhinal cortex. Batty and Smith argue, however, that this work fails to do the job. Interpreting the behaviour of the subjects as 'navigating an odour landscape', as opposed to gaining a sensitivity to certain temporal regularities, requires building in assumptions about causal structure that are not available in olfactory experience alone, but involve a conception of 'a 3D space in which material objects (i.e., odour *sources*) are basic—a world in which, as embodied creatures, we jointly exercise many senses to navigate our environment'.

The conception of material objects as persisting things is not only connected to our ability to conceive of them as existing unperceived, but is also connected to details of how perception works when we are perceiving the object. *Green's* contribution asks whether the perception of persistence is multimodal (think, for example, of seeing a chair as you approach, and then when you turn to sit, you feel the chair but no longer see it). Green breaks this into two questions. One is whether the computations behind the perception of persistence make use of information from multiple modalities, and the second is whether the phenomenology of the perception of persistence can sometimes be multimodal. An example of the first case would be where a visual presentation could be interpreted as either two circles moving towards and passing through each other on a straight line, or meeting and bouncing. An audible click or a beep at the moment of overlap greatly increases the chances of the percept being interpreted as a bounce. Green takes this as good evidence for multimodal information playing a role in the computations behind persistence. The phenomenological question does not have as definitive an answer, Green argues. The question is whether there are ever cases where the phenomenal character of a perceptual experience of persistence cannot 'be fully factored into components each of which is uniquely associated with just one modality'. Green takes crossmodal apparent motion—where, for example, a sequence of linearly separated noise bursts and flashes is perceived as uniform motion—as one case study. Such cases might initially seem to be examples of phenomenal multimodal persistence, but Green points out that the phenomenology in such cases does not support the conclusion. Subjects interpret the percepts from different modalities as of distinct objects that stand in causal relations rather than as single persisting objects. Green also reviews other results that might be taken to suggest positive answers to the phenomenology question, and shows that they too are not as decisive as they appear. He concludes that more work on the topic is called for.

De Vignemont takes the topic of the perception of persistence to the next level by exploring the question of whether we can perceive the future—for example, the impending collision when you are about to be struck by a ball. The author points out that there is a phenomenal element to impending collisions (e.g., seeing a ball that will hit you in less than a second) that is absent in seeing something moving towards you that is still some distance away. But is this phenomenal difference a matter of some (anticipated) future object or event being an aspect of perceptual phenomenology? De Vignemont argues that, in fact, it is a kind of temporal amodal completion in play in these cases. The phenomenon raises questions. What mechanism is responsible for this completion? And what is the perceptual object in such a case? On the first question, de Vignemont admits that many will assume that the mechanisms are not perceptual, but rather imaginative—one is imagining the impending impact. But though initially appealing, he points out that this proposal ends up making little sense on closer scrutiny. On the second, he points out that the objects of perception in such cases might be dynamical temporally extended events. In virtue of being in sensory contact with part of an event, we can (in some cases) perceive the entire event, including stages that are in the future.

The set of contributions just discussed each explored one or more aspects of our perceptual access to material bodies as spatial, persisting grounds of causal efficacy. One of the roles of a perceptual object (broadly construed) is to serve as a principle of *unity*—this unity might be the binding of different features in one modality (binding colour with shape and texture, for example) or integrating features registered in multiple modalities. On many accounts, one of the functions of the perception of material bodies is that it is to such bodies that sensory features are *attributed*, and if multiple features are attributed to the same material body, this can account for how different sensory features are bound or integrated. Attribution of multiple properties and features to a single material body is one way of achieving unity from diversity. The next set of contributions focuses on these two issues: are there candidates for the objects of unimodal and multimodal perception other than material bodies that might serve as principles of unity? And are there candidates for kinds of unity other than attribution? As is clear from these articles, these two questions, while different (see discussion of Callaghan below), are very closely related.

Cohen describes what he calls the *convergence conception*, according to which binding is effected by 'the representation of features as convergently exemplified by one and the same object . . .'. He suggests that while there has been motivation for this view—that binding is a matter of attribution of features to a single individual entity—in unimodal cases, it may not generalize well to multimodal cases. For the case of audition, in particular, with the 'sensory individuals' (medium-disturbing events, compression wave structures) posited by most accounts as the point of convergence for unimodal binding, Cohen points out that the convergence conception seems inapplicable when it comes to intermodal cases such as audition and vision. We do not see compression waves, for example. Cohen then discusses two kinds of response to this worry. The first is to take the objects to be mereological complexes. (Note that Young and Nanay argue in this volume for precisely that view, but their motivation is not multimodal integration, but rather different possible kinds of objects within a single modality.) Cohen points out that this proposal does supply a single individual as the target of multiple modalities, but 'it does not vindicate the convergence conception of multimodal binding as perceptual attribution . . .', because relevant features are often attributed only to the parts, not to the complexes. Second, Cohen considers views that claim that such binding is via association—attribution is unimodal and to modality-specific individuals, and multimodal binding is a matter of association between such individuals. The core problem here is that association is cheap and the proposal 'overgenerates instances of binding'. Cohen frames his proposed solution as a synthesis of the best parts of each of these views. Multimodal binding does not involve predication to any single individual, but it does involve predication to multiple individuals, each of which is represented as a part of a mereological complex.

To see how these topics interact, consider *Young and Nanay's* lead example of audition, where standardly a distinction is made between *source* hearing (whose object would be the vibrating tuning fork, or an *interaction event* consisting of a tuning fork causing compression waves in the air) and *sound* hearing (whose object would be the sound of middle C). This approach distinguishes two kinds of hearing, each with its own kind of non-composite object. The authors show that this approach raises difficult questions. For instance, do we hear sources indirectly via hearing the sounds directly? And they offer an alternative view according to which perception can have, as its object, *composite individuals*. For instance, you can perceive a table as a composite consisting of a top and legs. They claim that there are several kinds of composites: spatial (a table), temporal (a spoken word with syllables), and causal (one ball hitting another). All can count as a 'mereologically complex composite sensory individual'. As applied to the case of hearing, the view provides the option of saying that *the* object of hearing is a causally composite individual consisting of parts, which can include interaction events, vibrations, and sounds. Note how this approach might interact with Green's point discussed above about crossmodal apparent motion not being an example of multimodal perception of persistence because the lights and sounds seem phenomenally to be causally related, as opposed to a single persisting individual. But if some sensory individuals are causal composites, then perhaps there is multimodal access to the persistence of a single (composite) object in such cases.

But despite the different motivations and applications, Cohen converges on a similar view to Young and Nanay's when he points out that such mereological complexes 'must be structured complexes, where the structure in question extends to part-whole relations, predicational relations, and possibly more'. Recall that Young and Nanay point out three kinds of structure: spatial, temporal, and causal. But while Young and Nanay treat this as a premise, Cohen argues explicitly for it. Mereological complexes are cheap, but if we are to appeal to them as part of a substantive account of integration, they must have some non-trivial psychological import.

O'Callaghan approaches the issue from a different direction. Regardless of whether one takes the entity to which features are bound or predicated to be material bodies or mereological complexes, a question can be raised concerning whether the resultant unification is perceptual or cognitive. Of course, I take the ice cream whose temperature and texture I feel with my lips to be the same ice cream whose flavour I am tasting. But is this identification perceptual or cognitive? Or to put it another way, can crossmodal identification be *perceptual*? O'Callaghan argues that it cannot—that crossmodal identification is always cognitive. The thesis is nuanced. Perceptual processing is sensitive to crossmodal identity, but does not *represent* the individuals perceived by one modality *as identical to* the individuals perceived by another modality. One part of the argument borrows material from philosophy of language, and argues that the perception of an object in different modalities amounts to distinct Fregean modes of presentation of the object, in that identity statements concerning such object representations can always be informative—a '. . . subject rationally can doubt whether what is seen is the same as what is touched'. O'Callaghan agrees that 'perception is differentially sensitive to situations in which one thing is perceived using distinct senses', a fact that presumably underwrites much of the conviction that crossmodal identification can be perceptual. The conclusion is applied even to proposals on which the objects of perception are mereological complexes. As O'Callaghan puts it, 'The theory of binding is one thing; a theory of perceptible objects is another.'

This last claim was made by O'Callaghan as part of an exploration of multimodal perception. But a similar claim is argued for in the context of unimodal binding by *Quilty-Dunn*, who identifies the problem in which he is interested as the *syntactic binding problem*: 'how visual representations package information in such a way that feature combinations are distinguished from non-co-extensive combinations that incorporate the same primitive features.' He claims that attribution of multiple features to a single object is one way to solve this, but not necessarily the only one. He argues there seems to be another mechanism that can do the job—what he calls the *coordination model*. The argument proceeds by pointing out that attribution has a discursive format (it is the *predication* of a property or feature to an object), but that there are clear examples of binding that are not discursive. His example is *icons*, which he points out are not discursive, but holistic. As he put it, 'the same part of the icon that corresponds to that part of the scene and represents one feature will do double-duty and represent the other as well'. Icons serve for Quilty-Dunn as an example to motivate the idea that some binding is non-discursive. He further motivates the idea by pointing out that attribution is standardly taken to require attention, but there seem to be kinds of binding in vision that are independent of attention, a fact that is masked by many of the experimental protocols used to study binding. His specific proposal is the *coordination model*. The idea is that the visual system can plot coordinates in quality space, where each dimension of the space corresponds to a set of mutually exclusive qualities (red vs yellow; square vs circular vs triangular, etc.) and the value of each dimension corresponds to the instantiated property. The binding effected by such coordinates is not discursive—none of the coordinates is predicated of any of the others. The point (3, 4) in a Cartesian plane does not involve predicating 3 of 4, or 4 of 3, for instance.

After explaining that multimodal feature integration occurs in spatial senses (e.g., vision and audition) and chemical senses (e.g., taste and smell), *Spence* describes his contribution as focused on the question of whether '... the same rules of multisensory feature integration (MFI) apply in the two cases'. This description undersells the complexity of principles and processes that Spence's discussion reveals. Spence points out that some approaches to MFI (regardless of modalities) are based on an assumption of non-redundant cues, and the binding problem as classically understood is in this camp—knowing that something is green vs red does not implicate it being square any more than circular. Other approaches, such as Bayesian, assume redundant cues, where a cue itself provides information that has implications for the probability of other cues. Spatial attention can play a role in both, but not always in a way that requires spatial coincidence (as assumed in the Feature Integration Theory). It can manifest in different ways in both spatial and chemical senses. Moreover, top-down knowledge of the sort appealed to in Bayesian approaches can play a role in both spatial and chemical senses, but not always. For example, vanilla smells sweet despite the pods being bitter, because of an association between vanilla and sweet foods such as ice cream. After discussing many influences on multimodal feature integration and other crossmodal effects, Spence concludes that there are both similarities and differences between the way in which MFI operates in the spatial and chemical senses. But, in fact, his discussion shows that even when attention is restricted to MFI between any modalities, there is a great deal of complexity and variation.

Just as Spence highlighted the complexity and variation in MFI, the contribution by *Mroczko-Wąsowicz, Stoch, and Zguda* brings out many kinds of complexity and variation in approaches to theories of perceptual objects. The authors point out that there are three sources of constraints on theories of object perception: neurophysiological/psychological, phenomenological, and ontological. While each brings different resources to the endeavour, the authors argue that ultimately a theory of perceptual objects need not be *directly* beholden to ontological constraints, but should be beholden to both processing (neurophysiological and psychological) as well as phenomenological constraints (though, of course, these latter constraints may have implications for the relevant ontology). Moreover, while the theory should be responsive to considerations at the levels of both sensory processing (bottom-up) and object-conception (top-down expectations, assumptions, and associations), it need not be limited to considerations arising from those levels. After examining a variety of proposals, these authors conclude that two extant theories in the literature illustrate the most promising proposals for meeting the various desiderata of a theory of perceptual objects. The first is O'Callaghan's appeal to mereologically complex individuals and the second is Green's appeal to object-directed capacities. While the two approaches differ, each finds a way to expand the domain of perceptual objects beyond the prototype of material bodies while avoiding the potential anything-goes danger of appealing to unstructured mereological entities.

The final group of contributions are all deep dives that resist inclusion into any of the above clusters. First up are two contributions, each of which examines a very special kind of perceptual object: *faces* and *pictures*, respectively. While faces are obviously material bodies, *Fulkerson* shows that they are very special material bodies as far as perceptual processing is concerned. Fulkerson argues that face perception has dedicated unimodal processing mechanisms not only for vision, but also for touch. The fusiform face area (FFA) has long been known to have mechanisms dedicated to visual face perception, but Fulkerson discusses results indicating the presence of a haptic face area (HFA) near the FFA. And in addition, face perception features an unusually rich set of multimodal interactions not only between vision and touch, but also between other modalities. While there is currently no evidence supporting specialized multimodal perceptual face perception mechanisms, there do appear to be rich *domain-specific* multimodal interactions between face perception and other modalities such as hearing (e.g., voices), and smell.

Calzavarini and Voltolini explore another very special kind of material body, *pictures*—that is, material bodies with 2D surface features that are perceived as representing a 3D scene. The authors point out that a key feature of visual pictures is that they are such that the perceptual system is able to apply grouping operations on the basis of the depicted 2D features so as to impose object grouping that are sensitive to a dimension other than the two that are depicted, namely depth. The driving question is whether this sort of perceptual capacity is exclusively visual, or if it can operate in the tactile modality as well. Calzavarini and Voltolini point out that extant results strongly suggest that such capacities are supramodal. Multiple lines of evidence suggest that congenitally blind subjects are able to process tangible pictures (line relief drawings) in a way that captures things like depth ambiguity (e.g., Necker cube). Direct neurophysiological evidence is not extant, but Calzavarini and Voltolini clearly specify what sort of experiments would confirm the hypothesis.

Each of the next four contributions delves into nuances of one or another particular *modality*, though in some cases, the issue is whether the system in question is best thought of as a single modality. *Young* explores a puzzle concerning olfaction. Orthonasal (inward airflow) olfaction and retronasal (outward airflow) olfaction are sensitive to the same stimuli and employ the same receptors. Nevertheless, there are core differences that raise the question of whether 'these distinct pathways yield two senses of smell . . ., one perceptual modality of smell despite the duality . . ., or perhaps two olfactory sensory systems, only one of which yielding the perceptual modality of smell'. After detailing a number of fascinating differences between the two kinds of olfaction (inter alia, differences in cortical processing, and also evidence that habituating an odour orthonasally does not habituate it retronasally!), Young concludes that orthonasal olfaction is distinct from retronasal olfaction, with the former constituting the sense of smell and the latter being a component of the perceptual modality of flavour.

Pierzchajlo and Olofsson also focus on olfaction, and on the sense of smell (orthonasal) proper. The issue is whether the sense of smell is best understood as a bottom-up process tuned primarily for the detection of novelty, or if rather the sense of smell can be understood, like other modalities, to involve a top-down cognitive element of something like object recognition. The authors point out that much of the impetus for the novelty detection case comes from specific paradigms that downplay multimodal interactions, but when paradigms that probe such interactions are explored, a case can be made that the sense of smell is, in fact, geared towards object identification, but is more dependent than other modalities on multimodal interactions. The conclusion is that '... olfactory processing is ... highly dependent on prior experiences and contextual predictions, making active use of crossmodal cues and memory traces to facilitate performance'.

Lycan examines the modality of touch in great detail, arguing that it is an outlier in a number of respects. Lycan's discussion reveals a great many idiosyncratic features of touch, more than can be summarized in quick order. But one of the more sustained aspects of his discussion concerns *layering*. Layering is a phenomenon where a 'perceptual state may represent one object or property by representing a more primitive or less ambitious one', as when, for example, 'you hear a bang, and by hearing the bang, you hear a collision'. Lycan argues that while layering can occur in any modality, it is especially rampant in touch. One reason is that nearly all objects perceived by touch are perceived through a more primitive perception of one's own skin: 'we represent hardness, vibration, wetness, and motion by representing patterns of pressure on the skin.'

Skrzypulec explores 'phenomenally multimodal pain experiences'—experiences that include not just a phenomenal element of pain proper, so to speak, but also something like a thermal (burn) or tactile (pinch) aspect. This raises the question of how to analyse such experiences. Skrzypulec considers both an *attribution* framework according to which such experiences are attributing two properties to some perceptual object, and a mereological approach that takes the experience to present some sort of whole with a 'pain' proper part and a tactile (or thermal, etc.) proper part. He argues that neither approach is promising in the case of pain, and suggests that a better approach would be to treat such experiences as a type of *crossmodal correspondence*. Skrzypulec suggests that all three major mechanisms (structural, statistical, and semantic) usually taken to support crossmodal correspondences are plausibly operable in the case of pain experience. For example, in the same way that bright lights and loud noises are taken to correspond due to a similarity in the mechanism (increased neural firing rate), so too the fact that an experience of pain and a thermal or tactile experience can arise from similar mechanisms can underwrite a crossmodal correspondence.

The final two contributions concern details of sensory qualities. *Buccella and Chirimuuta* explore the perceptible qualities of gloss and timbre. First, they detail how these qualities are not well explained by the 'detection framework'—that is, they do not seem to be a matter of a quality that is keyed to the detection of some feature of the perceived object itself. For instance, gloss is not, despite what one might have initially assumed, a sensory feature that tracks the way in which a surface reflects light. Rather, they argue that gloss, while a genuinely real property of visible objects, is a context- and perceiver-dependent property. As they put it: 'gloss is a property of the process consisting in a viewer perceptually interacting with an object embedded in a scene and bathed in ambient illumination.' The authors turn to timbre, and though the details of the argumentation are different, the conclusion is the same: timbre is best understood as a property of a sound source in a particular context and for a certain kind of perceiver. The results of this contribution are thus twofold: a fascinating dive into the nuances of two ubiquitous perceptible qualities, and a general argument for a way to understand what at least some perceptible qualities are.

The topic of the complicated nature of sensory features is continued in the contribution of *Brogaard and Sørensen*. They argue that perceptual categorization, even in cases seemingly as straightforward as colour, is in fact variable across different individuals. Such variability is a reflection of the fact that different individuals employ different 'strategies for solving the different tasks we encounter in our environment', a stance that obviously resonates with Buccella and Chirimuuta's interactionist approach. The authors discuss a number of categorization strategies and give empirical support that the categorization of sensory qualities depends upon the strategies employed. The authors discuss evidence that perceptual categorization, even as influenced by the colour categories used by a speaker's native language, makes impact 'at the lowest level of perceptual processing . . .'. They argue for a perceptual pluralism, according to which the categorization of even sensory features can be driven by a variety of factors, depending on context, history, and strategies.

Acknowledgements

Thanks to all contributors and our editors at Oxford University Press. The work was supported by the National Science Centre, Poland (A. M. W.'s grant 2019/35/B/HS1/04386).

References

Alvarez, G. A. and Franconeri, S. L. (2007). How many objects can you attentively track?: evidence for a resource-limited tracking mechanism. *Journal of Vision*, 7(13), 1–10.
Auvray, M. and Spence, C. 2008. The multisensory perception of flavor. *Consciousness and Cognition*, 17, 1016–31.
Barwich, A.-S. (2019). A critique of olfactory objects. *Frontiers in Psychology*, 10, 1337.
Batty, C. (2014). Olfactory objects. In Perception and its modalities, edited by D. Stokes, M. Matthen, and S. Biggs, 222–46. Oxford: Oxford University Press.
Brewer, B. (2007). Perception and its objects. *Philosophical Studies*, 132, 87–97.
Briscoe, R. E. (2017). Multisensory processing and perceptual consciousness: Part II. *Philosophy Compass*, 12(12), e12423.
Casati, R. (2015). Object perception. In *Oxford Handbook of Philosophy of Perception*, edited by M. Matthen, 393–404. Oxford: Oxford University Press.
Clark, A. (2000). *A Theory of Sentience*. Oxford: Oxford University Press.
Clark, A. (2004). Feature-placing and proto-objects. *Philosophical Psychology*, 17(4), 443–69.
Cohen, J. (2004). Objects, places, and perception. *Philosophical Psychology*, 17(4), 471–95.
Cohen, J. (2010). Sounds and temporality. *Oxford Studies in Metaphysics*, 5, 303–20.
Green, E. J. (2018). A theory of perceptual objects. *Philosophy and Phenomenological Research*, 99(3), 1–31.
Jackson, F. (1977). *Perception: A Representative Theory*. Cambridge: Cambridge University Press.
Keller, A. (2016). *Philosophy of Olfactory Perception*. New York, NY: Palgrave Macmillian.
Kubovy, M. and Schutz, M. (2010). Audio-visual objects. *Review of Philosophy and Psychology*, 1, 41–61.
Lycan, W. G. (1996). *Consciousness and Experience*. Cambridge, MA: MIT Press.
Lycan, W. G. (2000). The slighting of smell. In *Of Minds and Molecules: New Philosophical Perspectives on Chemistry*, edited by N. Bhushan and S. Rosenfeld, 273–89. Oxford: Oxford University Press.
Lycan, W. G. (2014). What does vision represent? In *Does Perception Have Content?*, edited by B. Brogaard, 311–28. Oxford: Oxford University Press.
Lycan, W. G. (2018). What does taste represent? *Australasian Journal of Philosophy*, 96(1), 28–37.
Macpherson, F. (2011). Cross-modal experiences. *Proceedings of the Aristotelian Society*, 111, 429–68.

Matthen, M. (2004). Features, places, and things: reflections on Austen Clark's theory of sentience. *Philosophical Psychology*, 17(4), 497–518.
Matthen, M. (2005). *Seeing, Doing, and Knowing: A Philosophical Theory of Sense Perception*. Oxford: Oxford University Press.
Matthen, M. (2010). On the diversity of auditory objects. *Review of Philosophy and Psychology*, 1, 63–89.
Millar, B. (2019). Smelling objects. *Synthese*, 196, 4279–303.
Nanay, B. (2013). *Between Perception and Action*. Oxford: Oxford University Press.
Nanay, B. (2017). Philosophy of perception: a road map with lots of bypass roads. In *Current Controversies in Philosophy of Perception*, edited by B. Nanay, 1–20. New York, NY: Routledge.
Nanay, B. (2022). What do we see in pictures? The sensory individuals of picture perception. *Philosophical Studies*, 179, 3729–46. Available from: https://doi.org/10.1007/s11098-022-01864-9
Nudds, M. (2010). What are auditory objects? *Review of Philosophy and Psychology*, 1, 105–22.
Nudds, M. (2014). Is audio-visual perception 'amodal' or 'crossmodal'? In *Perception and Its Modalities*, edited by S. Biggs, M. Matthen, and D. Stokes, 166–88. Oxford: Oxford University Press.
O'Callaghan, C. (2008). Object perception: vision and audition. *Philosophy Compass*, 3(4), 803–29.
O'Callaghan, C. (2011). Lessons from beyond vision. *Philosophical Studies*, 153, 143–60.
O'Callaghan, C. (2014). Not all perceptual experience is modality specific. In *Perception and Its Modalities*, edited by D. Stokes, M. Matthen, and S. Biggs, 133–63. Oxford: Oxford University Press.
O'Callaghan, C. (2015). The multisensory character of perception. *Journal of Philosophy*, 112(10), 551–69.
O'Callaghan, C. (2016). Objects for multisensory perception. *Philosophical Studies*, 173, 1269–89.
Pylyshyn, Z. W. (2007). *Things and Places: How the Mind Connects with the World*. Cambridge, MA: MIT Press.
Scholl, B. J. (2007). Object persistence in philosophy and psychology. *Mind and Language*, 22(5), 563–91.
Scholl, B. J., Pylyshyn, Z. W., and Feldman, J. (2001). What is a visual object? Evidence from target merging in multiple object tracking. *Cognition*, 80, 159–77.
Schwartz, R. (2001). The concept of 'an object' in perception and cognition. In *From Fragments to Objects: Segmentation and Grouping in Vision*, edited by T. F. Shipley and P. J. Kellman, 3–17. Amsterdam: Elsevier.
Siegel, S. (2006). How does visual phenomenology constrain object-seeing? *Australasian Journal of Philosophy*, 84(3), 429–41.
Spelke, E. S. (1990). Principles of object perception. *Cognitive Science*, 14, 29–56.
Stevenson, R. J. (2014). Object concepts in the chemical senses. *Cognitive Science*, 38, 1360–83.
Treisman, A. (1996). The binding problem. *Current Opinion in Neurobiology*, 6, 171–8.
Vernazzani, A., Skrzypulec, B., and Schlicht, T. (2021). Structure of perceptual objects: introduction to the synthese topical collection. *Synthese*, 199(1), 1819–30.
Wolfe, J. M., Oliva, A., Horowitz, T. S., Butcher, S. J., and Bompas, A. (2002). Segmentation of objects from backgrounds in visual search tasks. *Vision Research*, 42(28), 2985–3004.
Young, B. D. (2016). Smelling matter. *Philosophical Psychology*, 29(4), 520–34.

PART 1

CONCEPTUAL, DEVELOPMENTAL, AND METHODOLOGICAL ASPECTS OF STUDYING SENSORY INDIVIDUALS

1
Origins of object concepts in infancy: A perceptual account

Scott P. Johnson, J. Gavin Bremner, and Alan M. Slater

1 Introduction

When we encounter a visual scene, we quickly form an impression of its contents and make moment-to-moment decisions about actions appropriate to the context, due to rich *object concepts*. Visual scenes tend to be very complex, consisting of overlapping and adjacent surfaces with distinct shapes, colours, textures, and depths relative to the observer. Object concepts generally lead us to expect most objects to have a consistent, regular, solid shape, even though most things cannot be seen in their entirety from a single vantage point because of occlusion—objects farther away are hidden by nearer ones, and the far sides of individual objects are hidden by the nearer sides due to opacity. Observers' typical visual experience, of course, is not one of incomplete fragments of surfaces, but instead one of solid objects, and we infer object shapes from partial views and incomplete information. In addition, we have certain common-sense expectations about the objects we see, such as their predictability over time (Köster *et al.*, 2020). Most objects are expected to be continuous across space and time (with exceptions like clouds and bubbles) despite gaps in perception due to occlusion, and objects are perceived as separate from each other. We expect objects to have a certain mass and structure, and to be solid in three-dimensional (3D) space. Movements of objects likewise tend to be predictable—we know that inanimate objects do not move in the absence of some outside force, cannot be in two places at once (nor two objects in the same place), are susceptible to gravitational constraints, and do not vanish out of existence and then reappear. Interestingly, the visual environment that is seen 'directly' (i.e., the light that reaches the eye after being reflected by visible surfaces in the environment) changes with every head and eye movement, and many objects such as animals and people frequently may move out of sight and subsequently return into view (O'Regan and Noë, 2002). Nevertheless, our experience is not a world of transitory, intangible, disembodied shapes, but rather one of substance, volume, and depth.

Is infants' experience of the world also guided by object concepts—knowledge of solid, predictable objects that maintain their properties over time? Alternatively, might infants instead perceive the world as shapes and surfaces rather than as objects as such, lacking object concepts without extensive experience in viewing and handling objects, a certain level of brain or visual acuity development, or all three? This chapter describes empirical approaches to this issue with a focus on development of infants' perception of partly and fully occluded objects—a foundational perceptual skill for object concepts. Sections 2 and 3 describe

Scott P. Johnson, J. Gavin Bremner, and Alan M. Slater, *Origins of object concepts in infancy* In: *Sensory Individuals*.
Edited by: Aleksandra Mroczko-Wąsowicz and Rick Grush, Oxford University Press.
DOI: 10.1093/oso/9780198866305.003.0002

two theoretical approaches that have divergent views about the origins of object concepts. Sections 4 and 5 describe a number of studies of infants' perception of partly and fully occluded objects. These studies provide compelling evidence for a perceptual account of the origins of object concepts—namely, infants' object perception seems to develop over time as perceptual constraints are overcome. We finish the chapter with some concluding remarks.

2 Piaget's theory

Empirical approaches to the origins of object concepts originated in the twentieth century with the publication of a succession of books by the Swiss psychologist Jean Piaget (Piaget, 1952, 1954). As part of his extensive investigations of early cognitive development, Piaget introduced a series of tasks to his own children and carefully charted their behaviour. From these observations, he formed his 'constructivist' theory of the development of object concepts and other cognitive skills (Piaget, 1954). This theory proposes that object knowledge develops in tandem with knowledge of spatial relations, guided by the assumption that one cannot perceive or act on objects accurately without awareness of their positions in space relative to other objects. The foremost explanandum of the Piagetian theory was *objectification*, the knowledge of the self and external objects as distinct entities, spatially segregated, persisting across time and space, and obeying certain common-sense causal constraints (e.g., when one object strikes another, the second object often starts to move). Objectification was proposed to be founded in children's recognition of their own bodies as independent objects and their own movements as movements of objects through space, like movements of other objects (a transition from 'egocentric' to 'allocentric' reasoning). By the end of the sensorimotor period at 18–24 months, actions that were formerly carried out can be imagined, and actions become placed on the ongoing, observed series of surrounding events as each child constructs the reality of an independent time and space. At that point, it may not always be necessary to act upon objects to gain an understanding of their properties.

Objectification was proposed to be an outcome of coordination of actions. Initially, behaviours are simply repeated but then become organized into more complex sequences of multiple actions. At the same time, infants start to explore novelty by trying new behaviours and by using familiar actions with no clear expectation of the outcome. These behaviours are evident in everyday play activities, as when Piaget (1954) observed his daughter repeatedly hide and reveal a toy under a blanket. Such simple games lead the child to establish spatial relations among objects (e.g., above, below, and behind) largely by manual engagement. For example, infants who are learning to reach (at 4–6 months) have to determine which objects are within reach and which are not, thus appreciating depth relations among different objects and the self. Therefore, both direct experience (to learn when search is successful or not) and deduction (reasoning from general principles to specific instances) contribute to the developmental process.

Development of action systems, spatial concepts, and object concepts was organized into six sensorimotor stages. During stages 1 and 2 (up to about four months), infants exhibit rudimentary recognition memory, for example, seeking the mother's breast after losing contact and continuing to look in the direction of a person's exit from the room. These behaviours were not observed systematically, however. For Piaget, *active* search, initiated by the child, was a critical feature of emergent object concepts.

In infants, more active search behaviour emerges after four months and marks the beginning of 'true' objectification during sensorimotor stage 3. Piaget outlined five examples, in roughly chronological order. The first of these was *visual accommodation* to rapid movements, for example, when an infant follows a dropped object by looking down towards the floor (this behaviour becomes more systematic when the infant drops the object). A second example was *interrupted prehension*, an attempt to reacquire an object that was dropped or taken from the hand if it was out of sight momentarily and within easy reach. (This occurs earlier in developmental time than a search for a fully hidden object.) *Deferred circular reactions* describes repetitive gestures during object-oriented play that resume after being interrupted for some delay (necessitating memory of the object, the actions, and their context). *Reconstruction of an invisible whole from a visible fraction* is seen, for example, when the infant retrieves an object from a cover when only a part of the object is visible. Finally, there is *removal of obstacles preventing perception*, as when infants pull away a cover from the face during a game of peekaboo or withdraw a fully hidden toy from beneath a blanket. This behaviour marks the transition to sensorimotor stage 4.

During sensorimotor stage 4, beginning at about 8 months, infants will search actively for a fully hidden object. However, if multiple potential hiding locations are present, searches may be erroneous—that is, the infant might look in the wrong place. For example, if an experimenter hides an object in a location (call it location A) and the infant searches successfully, and then the experimenter hides it in another location (location B), infants may remove the obstacle at the *first* location where the object was hidden (location A) rather than at the correct, current location (Figure 1.1). This response is known as the *A-not-B error* (Piaget, 1954). These behaviours mark an awareness of, and the search for, a vanished object, but their erroneous nature indicates a fundamental limitation of the emerging object concept and a sign of a lack of true objectification. According to Piaget (1954), these behaviours during sensorimotor stage 4 suggest that the object is an extension of the infants' own behaviour and does not exist independently of it. The identity of objects is not preserved across perceptual contacts, and the object that appears at a specific hiding place is a function of the child's activity. There is no concept yet of continuity across time and space.

Objectification is completed across the next two sensorimotor stages as each infant solves the problems of multiple visible displacements and invisible displacements, such that infants search systematically at all potential hiding locations visited by the object. For Piaget, mature

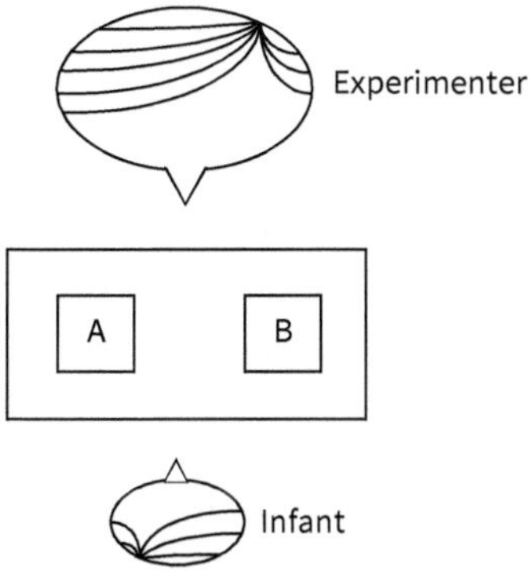

Figure 1.1 Scenario eliciting the A-not-B search error in infants, as discussed in Piaget (1954).

search revealed detachment of the object from the action, and knowledge that the infant's body itself is one object among many, brought into an allocentric system of organized objects and events—a complete object concept.

3 The nativist theory

In the 1960s, investigators began to develop ways of testing young infants' responses to partly hidden objects that did not rely on object search, and these experiments led to the view that knowledge of the physical world, including object concepts, is *innate*—that is, part of our nature, unlearned and not dependent on experience to cause them or to determine their function (Baillargeon, 2008; Spelke, 1990). The nativist theory of innate, unlearned object concepts is thus distinct from Piaget's view that sensorimotor experience is necessary for their development. According to this nativist view, infants possess 'core knowledge' and use this knowledge to reason about the events that they encounter. Some forms of this knowledge, such as the understanding that previously encountered objects continue to exist when out of view, are available too early to have been learned (Baillargeon *et al.*, 1985; Spelke *et al.*, 1992; Spelke and Kinzler, 2007). (Other kinds of object knowledge improve with experience, such as the prediction of the future locations of moving objects (Johnson and Shuwairi, 2009; von Hofsten *et al.*, 1998).) Particularly relevant to object concepts and a key aspect of innate knowledge is the principle of *object persistence* (Baillargeon, 2008)—that is, young infants are proposed to expect objects to exist continuously and to remain cohesive, in accordance with the unlearned principles of *continuity* (objects do not spontaneously appear and disappear) and *cohesion* (objects do not spontaneously fragment as they move).

These experiments rely on *visual preference* methods using techniques developed in the 1960s by Bower (1967) and based on observations reported by Fantz (1964). Fantz discovered that infants often lose interest in repetitive visual stimuli, and recover interest to novel stimuli. This decline in response tells us that infants can recognize previously encountered visual stimuli for brief intervals lasting a few minutes, an ability that is clearly required (but not sufficient) for mature object concepts. The visual preference methods were honed and standardized sufficiently such that habituation and recovery of interest have become a common, reliable means of investigating perceptual and cognitive abilities in infants from birth onwards (Bornstein, 1985). Some researchers devised a variant of this novelty-preference paradigm known as the *violation-of-expectation* (VoE) method, which relies on the assumption that infants will look preferentially in general at odd or unusual events. The looking time response does not require any kind of manual search, and thus it is more suitable for testing object concepts in young infants than behaviours that require manual activity and direct engagement with objects. However, an important difference relative to visual preference methods is that longer looking is driven not by simple perceptual novelty, but by knowledge that reality has been violated. In particular, a number of studies have reported that longer looking at events that are manipulated by the experimenters appear to break physical laws. In one example, an infant views two objects on a small stage (Figure 1.2A): a screen (that can rotate) and a box (Baillargeon, 1987; Baillargeon *et al.*, 1985). In a 'possible' version of the event (Figure 1.2B), the rotating screen stops at the (hidden) box. In an 'impossible' event (Figure 1.2C), however, the box rotates all the way back, through the space where the box should be (accomplished surreptitiously by an experimenter). Infants were reported to look longer at the impossible

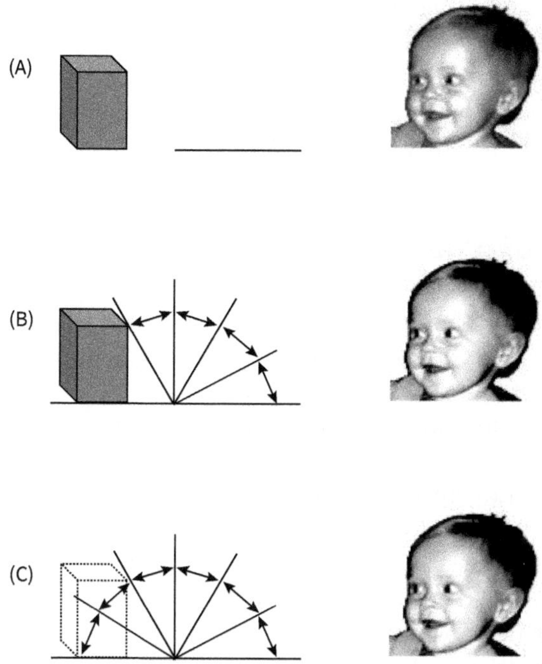

Figure 1.2 Objects and events shown to infants in studies using the violation-of-expectation method. (A) Infants view a box and a screen. (B) 'Possible' event. (C) 'Impossible' event.
Source data from Baillargeon (1987).

than at the possible event, taken as evidence that infants' expectations of object permanence and solidity were violated.

Studies that rely on the VoE measure have used experimental designs that attempt to rule out low-level perceptual bases for longer looking at the violation trial (see Cohen, 2004). Nevertheless, it can be difficult to identify the lower-level perceptual principles that may be driving infants' looking. Yet they are of paramount importance in understanding how infants respond to objects and their motions, as we show in the following sections of the chapter.

Much of the evidence on which claims about innate object concepts are based is from experiments with infants from 3 or 4 months of age upwards (Baillargeon, 2008). This leaves a gap of several months, starting at birth, during which infants gain experience viewing the world, and this contributes to developing object concepts. In the next two sections, we review evidence that considerable development occurs during the first 6 months after birth that likely forms a perceptual basis for some kinds of knowledge that nativists identify as innate.

This is not to deny the presence of innate ability (e.g., abilities to follow motion or to discriminate between new and old events). Our argument is that *perceptual* ability serves as a platform for *conceptual* knowledge to emerge later in infancy (we return to this argument later in the chapter). When an object disappears behind an occluder, several perceptual cues specify its continuity (Gibson, 1979; Michotte *et al.*, 1991). Young infants often seem to have limited ability to interpret these cues, and developments occur during the first 6 months after birth that result in robust perception of object unity and persistence across various forms of occlusion. This account is aligned with the theory of direct perception (Gibson,

1979) in assuming that the continued existence of objects is directly specified by perception. Specifically, the continued existence of an object is specified by the way in which it disappears behind an occluder. Thus, rather than support innate object concepts, this research reveals the conditions under which infants detect persistence and coherence of objects and perceptual completion in the events that they see.

4 Perception of object unity

Our starting point is the literature on infants' perception of *object unity*—perception of a partly hidden object as a continuous, complete surface behind an occluder. In the first published study on infants' perception of object unity, Bower (1967) devised a clever task to examine infant discrimination of the 'perceptual equivalence' of two visual stimuli. Bower used an operant conditioning procedure with 1-month-old infants. Sucking rate was the operant response, and the infant was required to suck on a pacifier for the visual display to appear. Infants were first exposed to a partly occluded triangle (Figure 1.3A), and their sucking rates were measured as new stimuli were introduced. Maintenance of sucking rates was taken as evidence that infants did not notice, or were unable to recognize, the change in stimulus, and reduced sucking rates were interpreted as evidence for perceptual discrimination of different versions of the triangle. Infants were reported to maintain sucking rates in response to a complete (unoccluded) triangle (Figure 1.3B), taken as evidence for perception of the training stimulus as a complete triangle behind the occluder. Presentation of triangle parts (separated by a gap) or other configurations (Figure 1.3C–E) resulted in reduced sucking rates. The decreased responses were viewed as evidence that the incomplete or altered forms were perceived by infants as different from the partly occluded triangle.

In later experiments, Kellman and Spelke (1983) investigated conditions under which older infants (4-month olds) would perceive a centre-occluded rod as continuous behind a box. After habituation to a partly occluded rod, infants looked longer at two rod parts than at a complete object, interpreted as a novelty preference and indicating perception of the rod parts' unity behind the occluder. Yet the effect was obtained only under certain conditions. Kellman and Spelke were unable to replicate the finding of perceptual completion in arrangements of stationary objects (Bower, 1967); instead, they reported that infants perceived object unity only when the rod parts underwent a common motion.

Kellman and Spelke (1983) tested infants' perception of object unity with a method of habituation, whereby 4-month olds viewed repeated presentations of a rod-and-box display until their looking times declined over trials. Infants were habituated either with a stimulus depicting a rod moving smoothly back and forth (Figure 1.4A) or with one in which the top part moved as the bottom part remained stationary (Figure 1.4B). Following habituation,

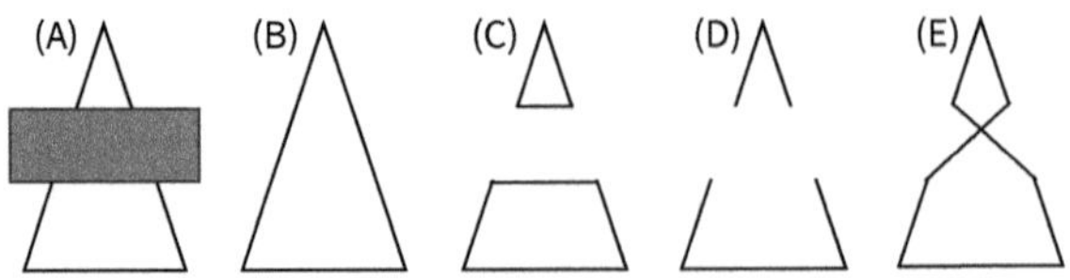

Figure 1.3 Training (A) and four test stimuli (B–E).
Source data from Bower (1967).

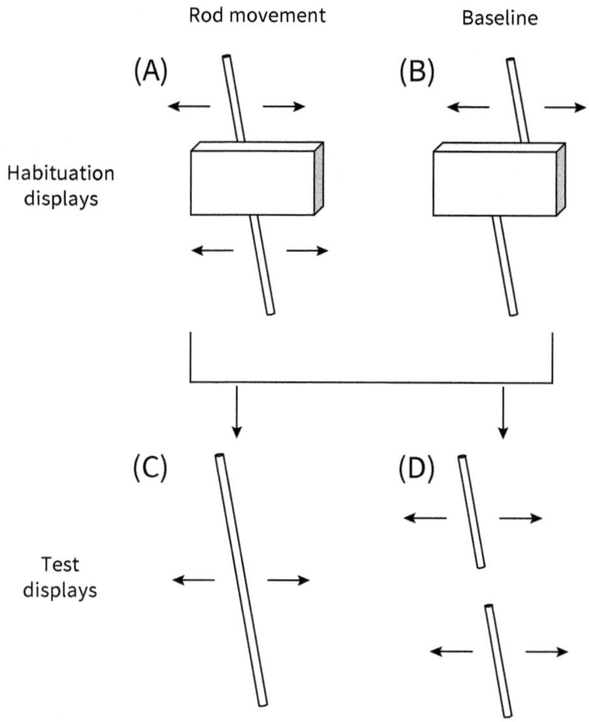

Figure 1.4 Habituation (top) and test stimuli (bottom), from Kellman and Spelke (1983). (A) Moving rod display. (B) Display to test for baseline test preferences. (C) Complete rod. (D) Broken rod.
Source data from Kellman and Spelke (1983).

infants in both groups viewed new moving rod test displays; one was a complete rod (Figure 1.4C) and the other consisted of two rod parts with a gap in the centre, a so-called 'broken rod' (Figure 1.4D). Infants who were habituated to the moving rod looked significantly longer at the broken rod (vs the complete rod), taken as evidence that the broken rod was considered relatively novel and the complete rod as familiar. Infants habituated to the alternate display showed no consistent preference—that is, there was no 'baseline' broken rod preference in the absence of exposure to an actual connected rod. Kellman and Spelke also tested 4-month olds' perceptual completion with displays of stationary objects, all of which led to null results.

The Kellman and Spelke (1983) experiments provided evidence for motion as a primary cue for infant's unity perception—in stimuli with a moving, partly hidden object, infants appeared to infer its unity, but not in stimuli with a stationary object. In addition, object motion progressively hides and reveals the background (also known as deletion and accretion of background texture) and so serves as depth cue, allowing the observer to perceive the relative distance of surfaces (Johnson and Aslin, 1996; Johnson and Náñez, 1995). Later experiments established that unity is only perceived when rod parts are aligned (Johnson, 2004; Johnson and Aslin, 1996) and is influenced by the Gestalt principle of 'figural goodness' (Johnson *et al.*, 2000).

Importantly, perceptual completion develops over time in infancy. When newborn infants were tested with rod-and-box displays similar to those depicted in Figure 1.4, the infants

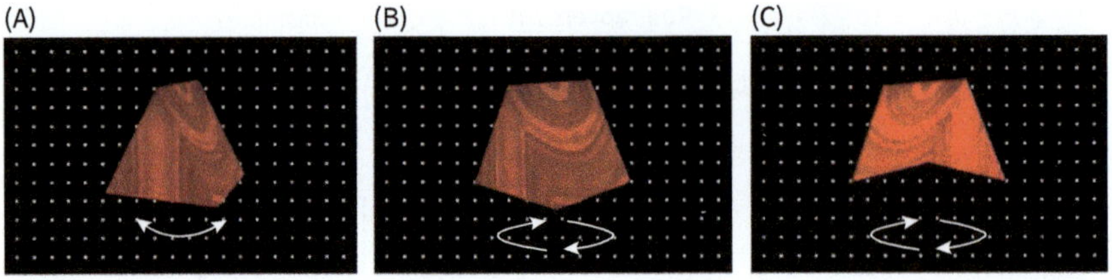

Figure 1.5 Stimuli from Soska and Johnson (2008). (A) Habituation display. (B) Complete object test display. (C) Hollow object test display.
Source data from Soska and Johnson (2008).

(observed within days after birth) showed test display preferences in the opposite direction relative to the 4-month olds observed by Kellman and Spelke (1983)—after habituation to a moving, centre-occluded rod, newborns looked longer at a complete vs a broken rod (Slater *et al.*, 1990; Slater *et al.*, 1994; Slater *et al.*, 1996). These results appear to suggest that newborn infants do not perceive object unity, but instead only visible surfaces (in line with the Piagetian theory). Interestingly, however, newborns have been shown to perceive unity in stimuli with apparent (stroboscopic) motion presentation conditions (Valenza *et al.*, 2006; Valenza and Bulf, 2011), though it remains unknown why apparent motion would be more likely than smooth motion to support perceptual completion. Perceptual completion becomes more robust by 2 months when the occluder is narrow or contains gaps in rod-and-box stimuli (Johnson, 2004; Johnson and Aslin, 1995).

Studies using more complex stimuli (rotating 3D objects; see Figure 1.5) revealed that perceptual completion continues to develop over the first year, such that by 6 months, infants appear to perceive an object as having a coherent 3D shape, even when seen from partial views (3D object completion; Soska and Johnson, 2008; see also Johnson *et al.*, 2003c). For example, after habituation to a wedge shape that was seen to rotate through 15° on the vertical axis, 6-month-old infants (but not 4-month olds) looked longer at a 'hollow' version of the shape when it rotated all the way through 360° (Figure 1.5C) vs a complete version (Figure 1.5B). In a follow-up study, Soska and Johnson (2013) found that a more complicated object (with more vertices and sides) was not perceived as a coherent 3D structure until 9.5 months. Thus, robustness of object unity perception increases over the early months and it is influenced by a range of perceptual variables.

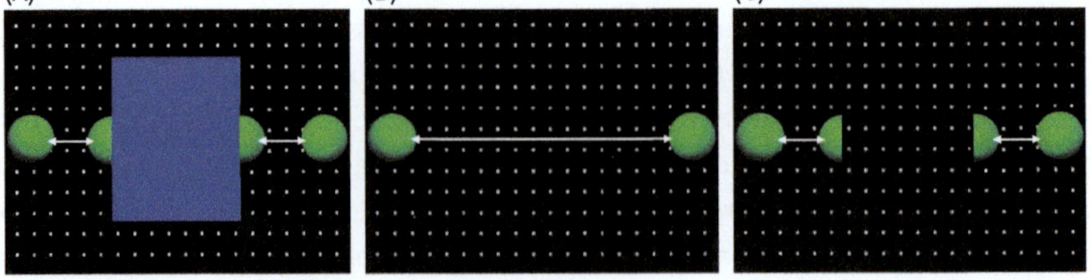

Figure 1.6 Stimuli from Johnson *et al.* (2003). (A) Habituation display. (B) Continuous trajectory test display. (C) Discontinuous trajectory test display.
Source data from Johnson *et al.* (2003).

5 Perception of trajectory continuity

Another area in which a perceptual account of infant object concepts appears appropriate is studies of perception of *trajectory continuity*—perception of a moving object as persisting and continuous when it passes temporarily behind an occluder. Johnson *et al.* (2003b) presented 4-month olds with a ball cyclically moving back and forth, disappearing behind a box in the central section of its trajectory, and reappearing (Figure 1.6A). In test trials, the occluder was removed, and the ball either moved continuously across the display (Figure 1.6B) or disappeared and reappeared where the edges of the occluder had been during habituation (Figure 1.6C), analogous to the continuous and broken rods described in the previous section. If infants perceived the ball's trajectory as continuous during habituation, they should look more at the discontinuous display as a novel event (Figure 1.6C), whereas if they perceived the trajectory as discontinuous, they should look more at the continuous test display (Figure 1.6B).

Interestingly, both outcomes occurred, depending on the age of the infants (Figure 1.7). Four-month olds looked longer at the continuous test display (Figure 1.6B), suggesting perception of the habituation trajectory as discontinuous, whereas 6-month olds showed the opposite preference, looking longer at the discontinuous stimulus (Figure 1.6C), implying they perceived the habituation trajectory as continuous. In a second experiment, Johnson *et al.* (2003b) presented 2- and 4-month olds with an occlusion event with a narrow occluding box, just wider than the ball, so that the moving ball was invisible for a very brief interval (667 ms). Under these conditions, 4-month olds' visual preference shifted and they looked longer at a discontinuous test display. The 2-month olds showed no preference. When the perceptual load was reduced to a minimum, therefore, 4-month olds seemed to have perceived a continuous trajectory, whereas there was no evidence that 2-month olds did. This indicates that (for 4-month olds at any rate), perception of trajectory continuity is dependent on occluder width, as it is for perception of object unity (Johnson, 2004; Johnson and Aslin, 1995). A further experiment revealed a more or less linear relation between occluder width and the direction of post-habituation preference, suggesting that the strength of the trajectory percept might range continuously from perception of discontinuity (at younger ages and with longer

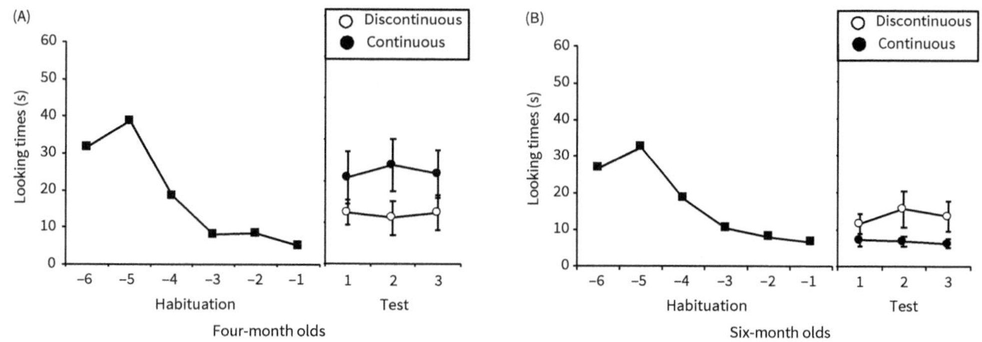

Figure 1.7 Looking time results from Johnson *et al.* (2003) to habituation and test stimuli. (A) Four-month olds. (B) Six-month olds. Error bars denote standard errors of the mean. Source data from Johnson *et al.* (2003).

occlusion durations) to perception of continuity, through an 'intermediate' state of no percept either way. This strongly suggests a perceptual basis for performance, and it is hard to conceptualize innate knowledge of object concepts that is moderated by age or perceptual variables in this way.

Converging evidence comes from work in which predictive eye tracking is used to measure perception of object persistence across occlusion. These studies show greater anticipatory tracking with age (Gredebäck and von Hofsten, 2004; Rosander and von Hofsten, 2004). Johnson *et al.* (2003a) found that 4-month olds showed little consistent predictive tracking when viewing the display depicted in Figure 1.6A, whereas 6-month olds showed significantly higher levels of prediction. In addition, 4-month olds' predictive tracking increased after they first viewed a continuous trajectory with no occluder (see also Johnson and Shuwairi, 2009). It does not seem to matter whether a two-dimensional (2D) or 3D display is used (Johnson *et al.*, 2012).

6 Perceptual constraints on perception of object continuity

In this section, we describe evidence from looking time studies with young infants (mainly 4-month olds) demonstrating perceptual constraints on infants' perception of object continuity. As noted, 4-month olds appeared to perceive trajectory continuity only when the occluder was narrow (Figure 1.8A). Bremner *et al.* (2005) asked whether this was because the object was out of sight for a short time or because it was out of sight across a short distance. They first increased the ball diameter until it was almost as wide as the wide occluder, in Johnson *et al.* (2003b) (Figure 1.8B), to result in a short temporal gap, but a large spatial gap in the trajectory. Four-month olds seemed to perceive trajectory continuity under these conditions, suggesting that time out of sight was the key factor. In another experiment, however, the researchers sped up a small ball while it was behind the wide occluder, or slowed it down behind a narrow occluder, and both of these manipulations resulted in continuity perception. Thus, 4-month olds perceived continuity when the perceptual load was reduced by limiting either the temporal or the spatial gap over which the trajectory had to be interpolated. Further studies revealed that a change in the object's trajectory, from a high to a low horizontal trajectory (Figure 1.8C) or from a falling to a rising diagonal trajectory (Figure 1.8D), seemed to prevent perception of continuity by 4-month olds (Bremner *et al.*, 2007). (In the case of a 'falling–rising' trajectory, this was true even if a surface was placed in the scene to provide an explanation for the change from a falling to a rising trajectory through occlusion (Figure 1.8E).) Even if the two visible portions of the trajectory were aligned across the occluder (Figure 1.8F), 4-month olds failed to perceive their continuity.

To further explore this phenomenon, Bremner *et al.* (2017) tested 4-month olds with stimuli that varied systematically in the relative orientations of occluder and trajectory. Infants perceived continuity when the object's trajectory was either horizontal or vertical and the occluder was either orthogonal or oblique to the background and the trajectory (Figure 1.9A–C), but not when the trajectory was oblique, no matter the orientation of the occluder (Figure 1.9D). It is possible that this striking limitation is due to difficulty in tracking oblique movements arising from the need to coordinate the extraocular muscles

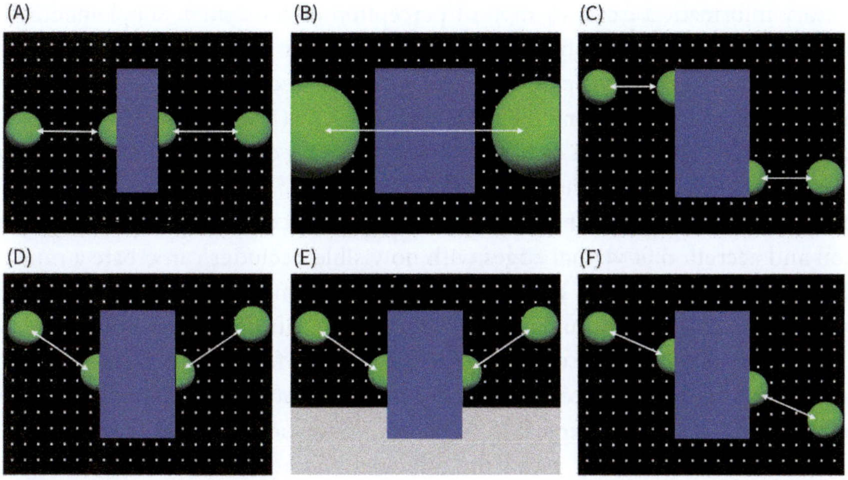

Figure 1.8 Habituation stimuli from Johnson *et al.* (2003) and Bremner *et al.* (2007). (A) Narrow occluder. (B) Large ball. (C) 'High–low' trajectory. (D) 'Falling–rising' trajectory. (E) 'Falling–rising' trajectory with surface. (F) Linear oblique trajectory.
Source data for A–B from Johnson *et al.* (2003); source data for C–F from Bremner *et al.* (2007).

that control the vertical and horizontal components of eye movements (Schiller, 1998). Infants also have difficulty in tracking circular object motions (Grönqvist *et al.*, 2006) and are less accurate in tracking unoccluded oblique vs horizontal or vertical object trajectories (Tham *et al.*, 2021).

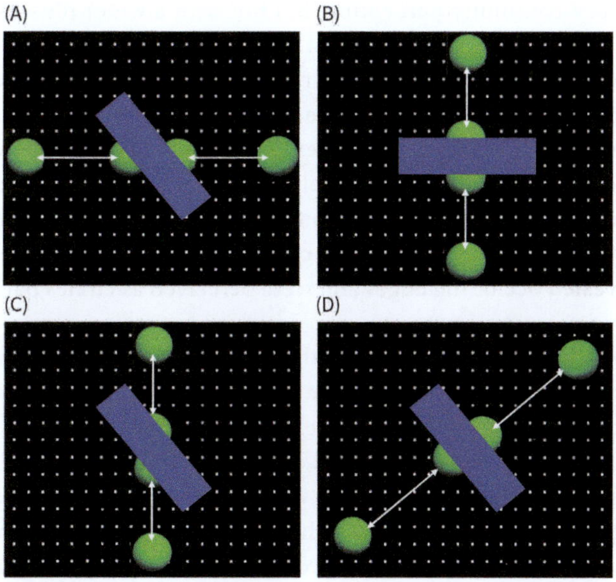

Figure 1.9 Habituation stimuli from Bremner *et al.* (2017). (A) Horizontal trajectory, oblique occluder. (B) Vertical trajectory, horizontal occluder. (C) Vertical trajectory, oblique occluder. (D) Oblique trajectory and oblique occluder.
Source data from Bremner *et al.*, 2017.

Auditory information can also support perception of continuity. Supplementing object movement during habituation trials with stereophonically presented sound that moved in tandem with the visible ball supported perception of trajectory continuity across occluder widths that yielded negative or null results when silent (Bremner *et al.*, 2012a). Moving sounds also enhanced 4-month olds' anticipatory tracking of objects that moved behind an occluder and re-emerged (Kirkham *et al.*, 2012).

A final question concerns the visual cues that specify an occlusion event. For adults, deletion and accretion at virtual edges with no visible occluder can create a *tunnel effect* (Burke, 1952; Michotte *et al.*, 1991)—the percept of a moving object disappearing and reappearing as if passing through an invisible slit in a surface. These are also important cues for infants. Bertenthal, Longo, and Kenny (2007) found that 5- to 9-month olds' predictive tracking was reduced if an object underwent instantaneous disappearance/reappearance or implosion/explosion at the occluder boundaries instead of the more natural deletion and accretion. Nevertheless, deletion and accretion are not sufficient by themselves to specify occlusion to infants; otherwise the discontinuous test displays used in the experiments described in this section would appear similar to the habituation display (as occlusion events), and hence attract low levels of looking. Furthermore, if infants perceived the discontinuous test display (e.g., Figure 1.6C) as an occlusion event, we would not have obtained the strong differences in preferences for the continuous or discontinuous test stimuli that we described.

To further investigate this issue, Bremner *et al.* (2012b) tested for 4-month-old infants' perception of a 'Kanizsa' figure (Kanizsa, 1979) as an occluding surface. The Kanizsa figure consisted of four so-called 'inducing elements', in this case partial circles arranged so as to appear as complete circles that were partly hidden by an invisible (illusory) surface. When the Kanizsa figure was narrow (Figure 1.10A), the 4-month olds provided evidence for trajectory continuity perception, but not with a wider illusory surface (Figure 1.10B), in parallel with earlier findings on the effects of occluder width (Johnson *et al.*, 2003b). Bremner *et al.* (2012b) then tested trajectory continuity perception with a visible edge (Figure 1.10C) and occlusion of the background texture (Figure 1.9D) but without the inducing elements. Neither of those cues were sufficient on their own, but presentation of both together (but still with no visible, luminance-defined surface) was sufficient to support perception of trajectory continuity. Taken together, these findings imply that 4-month olds treated the Kanizsa figure as one in which the inducing elements (if sufficiently close) revealed occluding edges and were perceived as circles partly hidden by the illusory surface.

These Kanizsa figure effects indicate that the precise locations of occlusion cues, such as the inducing elements that specify an illusory surface, need not be spatially contiguous with the path of object motion for infants to perceive trajectory continuity. This raises the question of whether accretion–deletion cues and occluding surface cues need to be congruent. This was addressed by Bremner *et al.* (2016) with stimuli that manipulated the relation between the ball's going in and out of view and the edges of the box. When 4-month-old infants viewed events in which accretion–deletion occurred at a point earlier (Figure 1.11A) or later (Figure 1.11B) in the trajectory than it should, or as if at an oblique occluding edge when the real occluding edges were vertical (Figure 1.11C), infants showed no signs of perceiving

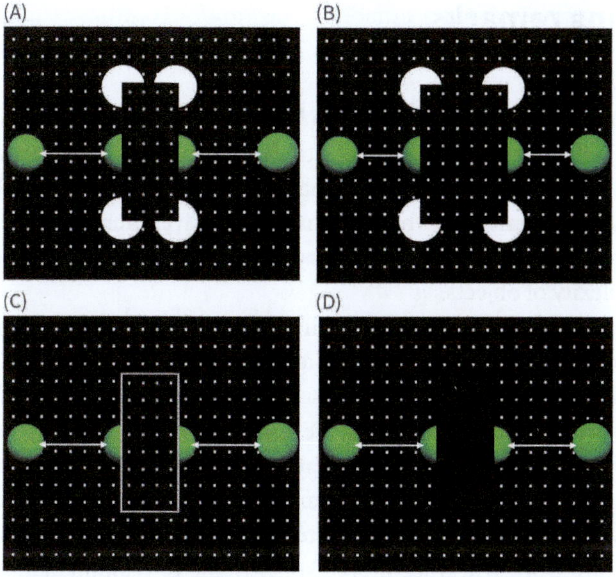

Figure 1.10 Habituation stimuli from Bremner *et al.* (2012b). (A) Narrow illusory surface. (B) Wide illusory surface. (C) Visible outline. (D) Occlusion of background texture.
Source data from Bremner *et al.* (2012b).

trajectory continuity. Notably, however, 4-month olds had little difficulty perceiving continuity when the occluding edges were oblique (e.g., Figure 1.8A, C), as long as the edges were congruent with the occluder (Bremner *et al.*, 2017). Taken together, the studies described in this section reveal the effects of trajectory orientation, auditory cues, and occlusion cues that have a strong influence on infants' ability to track hidden objects. These studies therefore provide convincing evidence for a perceptual basis for performance, as opposed to sensorimotor experience or unlearned core knowledge, the key principles of Piaget's theory and the nativist theory, respectively.

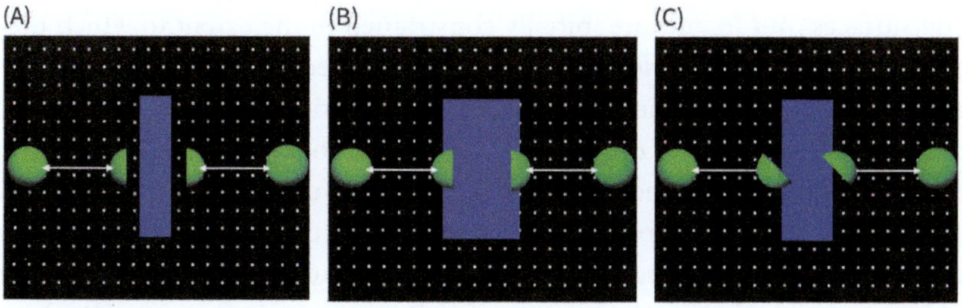

Figure 1.11 Habituation stimuli from Bremner *et al.* (2016). Accretion and deletion of the ball occurred (A) too early, (B) too late, or (C) at the wrong angle.
Source data from Bremner *et al.* (2016).

7 Concluding remarks

This body of work indicates a substantial number of perceptual constraints on young infants' detection of the unity of objects and the continuity of their trajectories that speak for a perceptual account of the origins of object permanence. The evidence recounted in this chapter suggests gradual development in which infants progressively achieve perceptual completion across larger spatial and temporal gaps, across an increasing range of trajectories, and with increasing complexity of objects.

Rather than advance object concepts as an innate principle through which events are interpreted (Spelke, 1990) or as the result of a lengthy process of manual engagement with objects alongside a detailed resolution of their relative spatial positions (Piaget, 1954), we propose instead that unity and persistence are specified by perceptual events such as deletion and accretion. The initial developmental question therefore is about infants' emerging ability to perceive completion of objects and their motions on the basis of these cues, in particular perceiving unity and persistence across shorter spatial and temporal gaps and in more complex objects. Future work will trace more explicitly the developmental sequence by which this occurs, perhaps with brain imaging or eye tracking methods. Approaches that can elucidate the 'developmental cascades' by which perceptual skills and learning are built on one another (Spencer *et al.*, 2009) are particularly likely to be fruitful. Our account predicts that with development, infants will need progressively fewer cues for perceptual completion (cf. Johnson, 1997).

Some nativist accounts (e.g., of language acquisition) have relied on pre-programmed developmental processes such as 'tuning' to the features of the immediate linguistic environment. The nativist account of object concepts, however, does not claim that the *process* of acquiring concepts of objects as solid and continuous is innate. Rather, it argues that core object *knowledge itself* (e.g., objects exist continuously across time and space) is innate and unlearned (Spelke, 1990; Spelke and Kinzler, 2007). (If it is just the processes that are innate, then Piaget was a nativist as well, given that processes of assimilation and accommodation were effectively built in.) Our perceptual account is more closely aligned with the Gibsonian theory (Gibson, 1950, 1979), according to which there is no need for representational processes to support perception. The structure of the world, including temporarily invisible parts, is specified directly by perceptual information. Our claim is that continuity of objects, while they are out of sight, is specified by the way in which they move out of sight behind occluding surfaces. But infants are initially constrained in the extent to which they can extract this information, and development is a progressive reduction in these constraints.

A limitation of our current views is that much of the evidence we presented comes from a visuo-motor analysis of object concepts, with its attendant focus on visual scene parsing and the amodal completion of visual objects and trajectories, and so our discussion is restricted to visual perception of rigid, inanimate, solid bodies. Yet if we think of object concepts more broadly, as noted earlier, comprising knowledge that inanimate objects do not move in the absence of some outside force, cannot be in two places at once (nor two objects in the same place), are susceptible to gravitational constraints, and do not vanish out of existence and then reappear, then a richer notion of material objects is needed. Moreover, animate objects have characteristics that go well beyond this description. These questions remain for future research.

To conclude, knowledge of at least some object concepts may have its base in the emergence of early perceptual capacities. As noted by Gibson (1979) and others (e.g., Cutting and Vishton, 1995), the principles of physical reality exist in information in the world. Origins of object concepts inhere in the improvement in the efficiency of perception of this information, and in constructing representations of object persistence from repeated exposures to it. In our view, the emergence of general representations of reality is only possible once the perceptual bases for these representations become stable across situations. A world in which objects persist under some circumstances, but not under others, does not present a stable basis for detection of the general principles upon which representations can be based. It is plausible that Piaget's constructivist account is relevant to the later development of representational processes. Although we claim that infants do not need to construct representations of physical reality, it is plausible that they have to construct relations between this reality and developing action systems, as infants learn how to use perceptual information to operate effectively in the world.

References

Baillargeon, R. (1987). Object permanence in 3.5- and 4.5-month-old infants. *Developmental Psychology*, 23, 655–64.

Baillargeon, R. (2008). Innate ideas revisited: for a principle of persistence in infants' physical reasoning. *Perspectives on Psychological Science*, 3, 2–13.

Baillargeon, R., Spelke, E. S., and Wasserman, S. (1985). Object permanence in 5-month-old infants. *Cognition*, 20, 191–208.

Bertenthal, B. J., Longo, M. R., and Kenny, S. (2007). Phenomenal permanence and the development of predictive tracking in infancy. *Child Development*, 78, 350–63.

Bornstein, M. H. (1985). Habituation of attention as a measure of visual information processing in human infants: summary, systemization, and synthesis. In *Measurement of Audition and Vision in the First Year of Postnatal Life: A Methodological Overview*, edited by G. Gottlieb and N. A. Krasnegor, 253–300. Norwood, NJ: Ablex.

Bower, T. G. R. (1967). Phenomenal identity and form perception in an infant. *Perception and Psychophysics*, 2, 74–6.

Bremner, J. G., Johnson, S. P., Slater, A. M., Mason, U., Cheshire, A., and Spring, J. (2007). Conditions for young infants' failure to perceive trajectory continuity. *Developmental Science*, 10, 613–24.

Bremner, J. G., Johnson, S. P., Slater, A. M., Mason, U., Foster, K., Cheshire, A., and Spring, J. (2005). Conditions for young infants' perception of object trajectories. *Child Development*, 76, 1029–43.

Bremner, J. G., Slater, A. M., Johnson, S. P., Mason, U. C., and Spring, J. (2012a). The effects of auditory information on 4-month-olds' perception of trajectory continuity. *Child Development*, 83, 954–64.

Bremner, J. G., Slater, A. M., Johnson, S. P., Mason, U., and Spring, J. (2012b). Illusory contour figures are perceived as occluding surfaces by four-month-old infants. *Developmental Psychology*, 48, 398–405.

Bremner, J. G., Slater, A. M., Mason, U. C., Spring, J., and Johnson, S. P. (2016). Perception of occlusion by young infants: must the occlusion event be congruent with the occluder? *Infant Behavior and Development*, 44, 240–8.

Bremner, J. G., Slater, A. M., Mason, U. C., Spring, J., and Johnson, S. P. (2017). Limits of object persistence: young infants perceive continuity of vertical and horizontal trajectories, but not 45-degree oblique trajectories. *Infancy*, 22, 303–22.

Burke, L. (1952). On the tunnel effect. *Quarterly Journal of Experimental Psychology*, 4, 121–38.

Cohen, L. B. (2004). Uses and misuses of habituation and related preference paradigms. *Infant and Child Development*, 13, 349–52.

Cutting, J. E. and Vishton, P. M. (1995). Perceiving layout: the integration, relative potency, and contextual use of different information about depth. In *Handbook of Perception and Cognition: Vol. 5. Perception of Space and Motion*, edited by W. Epstein and S. Rogers, 69–107. San Diego, CA: Academic Press.
Fantz, R. L. (1964). Visual experience in infants: decreased attention to familiar patterns relative to novel ones. *Science*, 146, 668–70.
Gibson, J. J. (1950). *The Perception of the Visual World*. Boston, MA: Houghton Mifflin.
Gibson, J. J. (1979). *The Ecological Approach to Visual Perception*. Boston, MA: Houghton Mifflin.
Gredebäck, G. and von Hofsten, C. (2004). Infants' evolving representations of object motion during occlusion: a longitudinal study of 6- to 12-month-old infants. *Infancy*, 6, 165–84.
Grönqvist, H., Gredebäck, G., and von Hofsten, C. (2006). Developmental asymmetries between horizontal and vertical tracking. *Vision Research*, 46, 1754–61.
Johnson, S. P. (1997). Young infants' perception of object unity: implications for development of attentional and cognitive skills. *Current Directions in Psychological Science*, 6, 5–11.
Johnson, S. P. (2004). Development of perceptual completion in infancy. *Psychological Science*, 15, 769–75.
Johnson, S. P., Amso, D., and Slemmer, J. A. (2003). Development of object concepts in infancy: evidence for early learning in an eye-tracking paradigm. *Proceedings of the National Academy of Sciences of the United States of America*, 100, 10568–73.
Johnson, S. P. and Aslin, R. N. (1995). Perception of object unity in 2-month-old infants. *Developmental Psychology*, 31, 739–45.
Johnson, S. P. and Aslin, R. N. (1996). Perception of object unity in young infants: the roles of motion, depth, and orientation. *Cognitive Development*, 11, 161–80.
Johnson, S. P., Bremner, J. G., Slater, A. M., and Mason, U. (2000). The role of good form in infants' perception of partly occluded objects. *Journal of Experimental Child Psychology*, 76, 1–25.
Johnson, S. P., Bremner, J. G., Slater, A. M., Mason, U. C., Foster, K., and Cheshire, A. (2003). Infants' perception of object trajectories. *Child Development*, 74, 94–108.
Johnson, S. P., Bremner, J. G., Slater, A. M., *et al.* (2012). Young infants' perception of the trajectories of two- and three-dimensional objects. *Journal of Experimental Child Psychology*, 113, 177–85.
Johnson, S. P., Cohen, L. B., Marks, K. H., and Johnson, K. L. (2003). Young infants' perception of object unity in rotation displays. *Infancy*, 4, 285–95.
Johnson, S. P. and Náñez, J. E. (1995). Young infants' perception of object unity in two-dimensional displays. *Infant Behavior and Development*, 18, 133–43.
Johnson, S. P. and Shuwairi, S. M. (2009). Learning and memory facilitate predictive tracking in 4-month-olds. *Journal of Experimental Child Psychology*, 102, 122–30.
Kanizsa, G. (1979). Organization in Vision. New York, NY: Praeger.
Kellman, P. J. and Spelke, E. S. (1983). Perception of partly occluded objects in infancy. *Cognitive Psychology*, 15, 483–524.
Kirkham, N. Z., Wagner, J., Swan, K., and Johnson, S. P. (2012). Sound support: intermodal information facilitates infants' perception of an occluded trajectory. *Infant Behavior and Development*, 35, 174–8.
Köster, M., Kayhan, E., Langeloh, M., and Hoehl, S. (2020). Making sense of the world: infant learning from a predictive processing perspective. *Perspectives on Psychological Science*, 15, 562–71.
Michotte, A., Thines, G., and Crabbe, G. (1991). Les compléments amodaux des structures perceptives [Amodal completion of perceptual structures]. Louvain: Publications Universitaires. Excerpted in: *Michotte's Experimental Phenomenology of Perception*, edited by G. Thines, A. Costall, and G. Butterworth, 140–67. Hillsdale, NJ: Erlbaum.
O'Regan, J. J. and Noë, A. (2002). A sensorimotor account of vision and visual consciousness. *Behavioral and Brain Sciences*, 24, 939–73.
Piaget, J. (1952). *The Origins of Intelligence in Children*. New York, NY: International Universities Press.
Piaget, J. (1954). *The Construction of Reality in the Child* (M. Cook, Trans.). New York, NY: Basic Books.
Rosander, K. and von Hofsten, C. (2004). Infants' emerging ability to represent occluded object motion. *Cognition*, 91, 1–22.
Schiller, P. H. (1998). The neural control of visually guided eye movements. In *Cognitive Neuroscience of Attention: A Developmental Perspective*, edited by J. E. Richards, 3–50. Mahwah, NJ: Erlbaum.

Slater, A., Johnson, S. P., Brown, E., and Badenoch, M. (1996). Newborn infants' perception of partly occluded objects. *Infant Behavior and Development*, 19, 145–8.

Slater, A., Johnson, S. P., Kellman, P. J., and Spelke, E. S. (1994). The role of three-dimensional depth cues in infants' perception of partly occluded objects. *Early Development and Parenting*, 3, 187–91.

Slater, A., Morison, V., Somers, M., Mattock, A., Brown, E., and Taylor, D. (1990). Newborn and older infants' perception of partly occluded objects. *Infant Behavior and Development*, 13, 33–49.

Soska, K. C. and Johnson, S. P. (2008). Development of 3D object completion in infancy. *Child Development*, 79, 1230–6.

Soska, K. C. and Johnson, S. P. (2013). Development of three-dimensional completion of complex objects. *Infancy*, 18, 325–44.

Spelke, E. S. (1990). Principles of object perception. *Cognitive Science*, 14, 29–56.

Spelke, E. S., Breinlinger, K., Macomber, J., and Jacobson, K. (1992). Origins of knowledge. *Psychological Review*, 99, 605–32.

Spelke, E. S. and Kinzler, K. D. (2007). Core knowledge. *Developmental Science*, *10*, 89–96.

Spencer, J. P., Blumberg, M. S., McMurray, B., Robinson, S. R., Samuelson, L. K., and Tomblin, J. B. (2009). Short arms and talking eggs: why we should no longer abide the nativist–empiricist debate. *Child Development Perspectives*, 3, 79–87.

Tham, D. S. Y., Rees, A., Bremner, J. G., Slater, A. M., and Johnson, S. P. (2021). Orientation effects in development of linear object tracking in early infancy. *Child Development*, 92, 324–34.

Valenza, E. and Bulf, H. (2011). Early development of object unity: evidence for perceptual completion in newborns. *Developmental Science*, 14, 799–808.

Valenza, E., Leo, I., Gava, L., and Simion, F. (2006). Perceptual completion in newborn human infants. *Child Development*, 77, 1810–21.

Von Hofsten, C., Vishton, P., Spelke, E. S., Feng, Q., and Rosander, K. (1998). Predictive action in infancy: tracking and reaching for moving objects. *Cognition*, 67, 255–85.

2
What makes something a perceptual object?

Aleksandra Mroczko-Wąsowicz, Nastazja Stoch, and Paulina Zguda

1 Introduction

Any account of perceptual objects (POs) has to explain what can be treated as an object in perception. Yet providing an adequate explanation is problematic from the start. Articulating such an explanation gives rise to delicate issues—for instance, the whole-part relation (If my bottle is an object, is its cap only a part of it and can it not constitute a separate object? Can a pile of books be considered a unitary object?); the status of fairly permanent non-solids (Can objecthood be ascribed to substances like water?); and the status of transient things (Are clouds or smoke objects?) or of extended continuous surfaces (Is a driveway an object?) (Russell, 2009, p. 81; Schwartz, 2001, p. 4; Snowdon, 2006, pp. 46–7). In other words, an account of POs must specify their scope, for not any perceptible entity is a PO.

Additionally, an adequate theory of POs faces several other, methodological and theoretical, challenges. Many theories have already been developed. This chapter evaluates them, based on the criteria presented in Section 2. In this section, we discuss the aspects of POs that pose potential problems for formulating a theory of POs. These potential problems allow us to assess a selection of leading proposals overviewed in Section 3. This results in our preference for the unified theories of POs, which seem to have more advantages than the other theories discussed.

2 Conceptual, methodological, and developmental aspects of research on perceptual objects

In this section, we outline potential problems for formulating an account of POs. These may stem from differences in methods dictated by disciplines and approaches to studying POs. Others may result from visuocentrism, which neglects the nature of objects targeted by sense modalities other than vision (Section 2.1). Moreover, the abundance of different kinds of 'objects' in the literature obscures a proper definition of a PO (Section 2.2). Also, there is a tendency to describe POs without specifying which perspective is assumed. This results in an oversimplified notion of POs (Section 2.3). Finally, object perception necessarily comes along with object conception (Section 2.4). Therefore, a comprehensive theory of POs should take these problems into account.

Aleksandra Mroczko-Wąsowicz, Nastazja Stoch, and Paulina Zguda, *What makes something a perceptual object?* In: *Sensory Individuals*. Edited by: Aleksandra Mroczko-Wąsowicz and Rick Grush, Oxford University Press.
DOI: 10.1093/oso/9780198866305.003.0003

2.1 Disciplines studying perceptual objects

How POs are defined depends on a given theoretical perspective, and that includes the discipline and specific research interests. But even within the same discipline, approaches may differ; for example, philosophers may focus either on the analysis of a conscious experience of what is perceived (Green, 2018; O'Callaghan, 2008, 2016) or on theoretical investigations reaching to the ontology of perceivable objects (Schwartz, 2001; Skrzypulec, 2018).

Overall, similar questions are raised across different disciplines concerning object perception. The main issue is the nature of POs—that is, how a stimulus array is segmented into individual objects, and how POs are distinguished from perceptual non-objects.

On the one hand, philosophers try to develop a conceptual framework that would allow us to grasp the nature of POs and their properties. Philosophers describe conditions under which something counts as a PO (see Section 3). In particular, they articulate criteria: capturing objects perceived in different modalities; distinguishing genuine complex POs from arbitrary combinations (Green, 2018, p. 2); or determining how a collection of parts must be arranged to constitute an object (Green, 2018; Koslicki, 2008).

On the other hand, psychologists and cognitive scientists examine perceiving objects in behavioural, psychophysical studies or by neuroimaging. In the first case, they observe participants' responses to presented stimuli (e.g., Hochberg and Peterson, 1987; Wolfe *et al.*, 2002). Neuroimaging allows researchers to assess information processing within a particular perceptual system—for example, in audition (see Bizley and Cohen, 2013; Griffiths and Warren, 2004), or in vision (see Cichy *et al.*, 2014; Fleming, 2014). By investigating percepts in various sense modalities, such as vision (Peterson and Rhodes, 2003), audition (Lotto and Holt, 2011), or olfaction (Gottfried, 2010), empirical studies on human perception reveal what stimuli can be its objects. Additionally, the multimodality of POs has been examined in the last decades (e.g., Auvray and Spence, 2008; Kubovy and Schutz, 2010).

Still, there are also two main approaches towards the study of perception itself. The first is related to the (un)conscious perceptual processing of sensory information, which leads to a perceptually guided action and a formation of perceptual beliefs. The second concerns perceptual phenomenology and focuses on conscious perceptual experience and its features (Nanay, 2017). The difference between these approaches lies in their purpose and method. Perceptual processing merges information from anatomical and functional levels of description in neurosciences and cognitive sciences. Perceptual phenomenology appeals more to philosophical methods involving conceptual analysis, introspection, and subjective reports.

In this chapter, we examine the topic of POs from an interdisciplinary perspective, including methods of philosophy of cognitive science and empirically informed philosophy of perception. By considering the conscious perception of objects (perceptual phenomenology) across different sense modalities, we aim at avoiding visuocentrism. Just as vision dominated the focus of research on perception (Stokes and Biggs, 2014), both in philosophy and science, the same has happened in the study of objects of perception (e.g., Alvarez and Franconeri, 2007; Brewer, 2007; Lycan, 2014; Scholl *et al.*, 2001; Siegel, 2006; Wolfe *et al.*, 2002). However, generalizing from visual objects to objects from other modalities fails to adequately formulate a comprehensive theory of POs. Different modalities target different POs. Vision picks out visual objects, audition sounds, olfaction odours, touch tactual objects, and so on. This phenomenon has been labelled the 'puzzle of diversity' (O'Callaghan, 2016, p. 1270). Our perceptual systems deal with diverse inputs and some stimuli are distinguished only by one modality, while others may be processed by more than one modality. The specificity of objects in other modalities has been confirmed by numerous investigations (see Batty, 2014; Lycan, 2018; Matthen, 2010; Mroczko-Wąsowicz and Grush, 2023; O'Callaghan, 2008, p. 804; Stevenson, 2014).

2.2 Different senses of 'object'

Many terms related to objects occur in philosophical works on object perception, such as 'perceptual objects', 'ordinary objects', 'proper objects', etc. Since they do not mean the same, the term 'object' is used in many senses. The list selected in this section is not a complete typology of objects. However, it sets our point of departure in providing a glossary of the most frequently used terms concerning objects in the literature on object perception and shows their relation to POs, which is a necessary step if one aims at defining the latter.

Ordinary objects and material objects are sometimes used interchangeably, even though every ordinary object is a material object, but not vice versa (O'Callaghan, 2008, pp. 815–16; 2016, p. 1285). Material objects have been described as everyday, physical, middle-sized, dry goods (Casati, 2015, p. 393; O'Callaghan, 2008, p. 803; 2016, p. 1271). Material objects that are not ordinary objects (e.g., a ping-pong ball glued to a fork) (O'Callaghan, 2008, p. 815) are sometimes called 'unnatural objects' (Scholl, 2007, p. 565).

Unnatural objects are surely POs. However, POs cannot be equated with material objects due to the existence of POs such as rainbows, clouds, holes, beams of light, holograms, or shadows (O'Callaghan, 2008, pp. 816, 825; 2016, p. 1274), namely, ephemeral visibilia (non-material, non-ordinary, visual objects).[1] Visible spatio-temporal features, in particular the persistence of an entity, allow such non-material objects to be treated as POs (O'Callaghan, 2008, p. 825). Nevertheless, the spatiality of objects such as rainbows is debatable and depends on how one defines what material spatial objects are (Schwartz, 2001, p. 4).

Some associate the concept of POs with the notion of sensory individuals (see O'Callaghan, 2016, p. 1276). The concept of sensory individuals marks two distinctions. It differentiates properties attributed perceptually (e.g., shape) from properties that may be attributed otherwise (e.g., the authorship of a given piece; Nanay, 2017, pp. 9, 13), and 'features from that to which they are attributed' (O'Callaghan, 2008, p. 807). The accounts which identify POs with sensory individuals, however, may differ. Cohen (2004) talks about ordinary objects, while Matthen (2004) about material objects.[2] Since POs (either ordinary or material) bear perceptible features, this understanding of sensory individuals stands in contradiction to the definition of sensory individuals as featureless bearers of features.

Intentional objects are more extensive in scope than POs. They comprise whatever a perceptual state represents or is directed at (O'Callaghan, 2008, pp. 817–18). They are POs and other entities such as (perceptual) properties or relations.[3] Such merely intentional objects

[1] In contrast to visual objects, tactile objects exclusively comprise material objects; any tactile material object is at the same time a visual material object (O'Callaghan, 2008, pp. 804, 816).

[2] Perceptual objects, understood as ordinary or material objects, are not the only candidates for sensory individuals. Some suggest spatio-temporal regions (Clark, 2000, 2004), while others mention events (Nanay, 2017, pp. 12–13; O'Callaghan, 2008, p. 824).

[3] Properties can be understood as attributes, qualities, or features (such as shape, motion, pitch, and tigerhood), while relations among objects can be temporal, spatial, or causal (such as the relation of sitting upon) (O'Callaghan, 2008, p. 818; 2016, p. 1273).

'exist only if they are actual objects that are represented in experiences' (Chediak, 2016, p. 89). Therefore, objects of hallucinations would count as intentional objects, even though there is no extramental object behind these phenomenal states.

Aristotle's 'special objects of sense' are contrasted with 'common sensibles' (Aristotle, 1984, p. 672) and correspond to the distinction between qualities of modality-specific and modality non-specific objects of perception. The first have been later called 'proper objects'—that is, intentional objects accessible only to a single sensory modality (O'Callaghan, 2008, p. 818). For instance, tones and sounds are proper objects of audition. The latter, also known as common or multimodal objects, can be perceived by more than one modality, for example, audiovisual rhythm (O'Callaghan, 2014, p. 155). Another example proving that not all objects of perception are sense-specific is the phenomenon of synaesthesia (Mroczko-Wąsowicz and Werning, 2012).

2.3 Differentiating three levels of description of perceptual objects

The way in which science describes the physical world and the way in which humans perceive physical objects are different (Rock, 1975, p. 4; Strawson, 2002, pp. 98–9). The scope of a study should be explicitly delineated to prevent making (meta)physical statements while speaking of the perceptual appearance of objects. This has not always been implemented within the study of object perception, thus risking taking ontological presuppositions for granted (Russell, 2009, p. 81).

Close inspection of the existing literature reveals three levels of description when speaking of POs.

The first level is concerned with the *ontology* of POs. It articulates the characteristics of perceptual objecthood presented in ontological debates—that is, being a subject of properties, having a mereological structure, or persisting through time (Vernazzani *et al.*, 2021, p. 1823). At this level, one can distinguish physical particulars, including features of the external world: (localized) properties; property bundles; substrata; regions of space; parts of objects; and collections of objects (Casati, 2015, p. 397).

The second level is concerned with *perceptual phenomenology* in the sense of *phenomenality*, namely the *experience* of POs. This covers the phenomenal character of conscious perceptual experience of such extramental items as concrete material objects/bodies and events.

The third level is concerned with *perceptual capacities* and mechanisms for object processing relevant to a particular sense modality. This comprises abilities of perceptual systems like achieving perceptual constancy; the object perception suite supporting object-based attention, object recognition, and perception-based demonstrative thought; and the object-directed capacities (ODCs) targeting POs (Section 3.2).

What is the relationship between those levels? Ontology investigates objects by using different parameters than studies of perception. The difference can be exposed by their views on object individuation. Ontology considers a variety of possible criteria for specifying what makes things separate entities instead of one. Ontological solutions for object individuation sometimes offer entities that are not perceivable, for example, bare particulars (see, for example, Moreland, 1998, p. 252). In perception studies, however, the criterion is one: POs are those individuals which are individuated perceptually. This is why claims about POs need not be committed to any ontological view because it is rather said of how objects are perceived, and not how they really are.

Ideally, individuating POs is consistent with the capacities of our perceptual systems. Therefore, even though different, the second and third levels are interrelated. Perceptual capacities target POs as such. For instance, sensorimotor capacities to perceive objects' action-affording properties seem to contribute to the spatial structure of POs and the spatial content of our perceptual experience of objects (Mroczko-Wąsowicz, 2015, p. 638; Vernazzani *et al.*, 2021, p. 1823). Perceptual capacities of particular perceptual systems may have been designed to follow certain principles of perceptual objecthood and thus to pick out some entities, and not the others, but the capacities do not necessarily represent these principles (Green, 2018, p. 13).

In sum, the autonomy of the first level and the interrelation between the second and third levels have often been overlooked. However, our guideline to distinguish the three levels of description of POs does not deny possible advantages from interdisciplinary approaches. As Scholl (2007, p. 564) remarks, studies on object persistence in psychology and philosophy should join their forces, because revealing the nature of underlying psychological mechanisms may support some metaphysical arguments. Yet, merely switching from one level of description to another does not entail an actual interdisciplinary approach. The lack of distinction in methodological assumptions may prevent from introducing a coherent account of POs.

There are some attempts at providing a list of criteria for POs (e.g., Stevenson, 2014, p. 1361). However, it has not been clarified what exactly such a list would regard: 'criteria for objecthood' (Stevenson, pp. 1361, 1375); 'features characterizing objecthood' (Stevenson, pp. 1362, 1365); or rather 'capacities for object processing' (Stevenson, p. 1375). Mixing the levels of description of POs disperses the focus on either the phenomenology of POs or the processing of perceptual systems. Moreover, some apply the principles of object perception as if one already possessed the knowledge of how to single out an object from a stimulus array (see the 'alien argument' in Section 2.4).

2.4 Object perception and object concept

This section is devoted to the relationship between object perception and object concept.[4] Reliance of a theory on object perception alone generates some problems, for example, the question about the role of principles of object perception (see Section 3.1). Consider the following scenario: we introduce the principles of uniformity and connectedness (POs should be uniform in features and bounded in form) (see Section 3.1.2) to an alien unfamiliar with planet Earth's environment. We present it with a half yellow and half black plastic pen on a yellow sheet of paper. On the one hand, by none of these principles, the alien would be able to judge how many POs are before it. By the principle of uniformity in the colour feature, our alien would regard the yellow part of the pen and the paper as a single object and distinguish it from the black part of the pen as another object; by uniformity in the texture feature, the alien would distinguish the plastic pen from the sheet of paper. By the principle of connectedness, it would treat the pen–paper collection as a single object. On the other hand, the two

[4] By object concept, we mean the concept of POs.

principles seem to stand in a reciprocal relation—to determine the uniformity of an object, one must know its boundaries, and to determine the connectedness of an object, one must be able to perceive at which point its uniformity in features ends; we have a vicious circle as a result.

Therefore, an additional assumption would be crucial to figure out which of the principles should be dominant against another at a particular moment, or how to apply the principles in the first place (as in the case of the pen–paper example, one must decide which uniform feature should be picked). Such an assumption cannot be missing and should be made explicit. It would need to be grounded in the concept of an object. Specifying the conditions for object perception is useless for the alien if it lacks the object concept. The human perceptual system has developed in a way to individuate objects. This enables humans to treat something as a PO by implementing an implicit concept of an object, which in turn allows, for example, to mark the object's boundaries (Schwartz, 2001, p. 8). That is also how human cognition operates when a single entity breaks apart (e.g., a torn sheet of paper). Even though it would violate some principles of object perception, such as cohesion or boundedness, we would treat the broken item as an originally single PO.

The above suggests that a concept of POs seems to be commonsensical. But even though humans may apply the object concept correctly, they would be unable to explicate it. Below we present leading frameworks articulating what a concept of POs is.

2.4.1 Conceptualizing perceptual objects

In developmental sciences, two main accounts have dominated the debate on the origins of object concepts—that is, nativist and empiricist (perceptual).

On the nativist account of object knowledge, humans are born with initial core knowledge—that is, a set of skills that allow to explore the world successively as they develop other abilities (Baillargeon, 2008; Spelke, 1994, 2000). One of the systems of this innate knowledge is responsible for forming object representations and reasoning about observed physical events (Carey, 2011; Kinzler and Spelke, 2007, p. 257; Leslie, 1995). On Spelke's (1994) core knowledge of objecthood, an object is an entity that obeys the principles of cohesion, continuity, and contact. For Baillargeon (2008, p. 3), the core principle shaping the innate idea of object is object persistence over space and time. This account advances object concepts as innate, experience-independent principles, through which events in the world are interpreted.

On the traditional empiricist account, the object concept is developed via active exploration and sensorimotor experience (Piaget, 1954; see Johnson *et al.*, 2023). Infants encounter objects as continuous, fleeting sensations, interact with them via these sensations, and gradually organize them into more structured entities. The object concept as coherent and enduring is constructed while infants gain sensorimotor experience by manually interacting with physical objects and by observing the outcomes of their object-oriented actions.

A more recent, perceptual account (Johnson, 2004; Johnson *et al.*, 2003, 2023) seems to supplement the classical interactionist/empiricist account. According to the perceptual account, infants' perception of object unity and that of object persistence are fundamental for mature object concepts. By a combination of cues over the first 6 months, infants become capable of increasingly complex acts of perceptual completion—that is, the object concept guarantees object representation in case of lack of direct perceptual accessibility to the object as in occlusion (Johnson, 2004, pp. 175, 199; 2018, pp. 2, 12; Johnson *et al.*, 2003, p. 10573). In other words, visual learning, and perceptual development in general, supports further development of cognitive principles, including object concepts. Accordingly, humans perceive objects by interaction of bottom-up and top-down processes. This enables development of the essential cognitive principles of cohesion and continuity which shape the concept of POs.

Acknowledging the importance of object conception is not to diminish the role of object perception. Supporting the role of object perception in object knowledge, Pylyshyn (2001, p. 154; original emphasis) indicated that 'while it is clear that you cannot individuate objects in the full-blooded sense without a conceptual apparatus, it is also clear that you cannot individuate them with *only* a conceptual apparatus' because concepts do not pick out tokens of individual things (Pylyshyn, pp. 129–30).[5] Object perception is what stands behind object knowledge and the formation of object representations. Without it, humans would be unable to experience objects as permanent entities with properties, and would not know that these entities can be manipulated (Santos and Hood, 2009), which contributes to the development of object knowledge. Moreover, Scholl and Leslie (1999, p. 45) claim that 'the origins of the object concept lie at the attentional interface between perception and cognition', which emphasizes the need of object perception's exploration.

Schwartz (2001, pp. 3, 16) expresses his concern on an implicit assumption in developmental studies that the object concept is innate, which results in taking for granted what a PO is. This worry may apply specifically to nativist approaches. Besides, even if one already were to have made a distinction between object conception and object perception, the way to determine their relationship in various sense modalities is unclear. For instance, there is no agreement on whether a single concept of POs should be applied to different modalities (Vernazzani *et al.*, 2021, p. 1823).

2.4.2 Operationalizing the concept of perceptual objects

This section presents influential operationalizations of the concept of POs. Examining how different frameworks apply their research methods to approach the external entities that we perceive advances our understanding of their stance on the object concept. This enables a better assessment of the role of object conception in theories of POs.

One of the proposals for how to operationalize the concept of POs is based on the theory of visual indices (Pylyshyn, 2001). The evidence for visual indices has come from the multiple object tracking (MOT) paradigm pioneered by Pylyshyn and Storm (1988) and conducted in multiple variations (Scholl *et al.*, 2001, p. 161). According to this theory, at the early stages of sensory processing, our perceptual system can successfully track up to five objects simultaneously by attaching visual indices to them. Indexing is fundamental for developing the object concept since it provides object individuation, identification, and enumeration (Leslie *et al.*, 1998, p. 11). In the case of infants, perceptual access to an object is necessary for object representation, and that representation is supplied by a visual index, which ceases to exist shortly after the object represented by this index is removed (Leslie *et al.*, 1998, p. 11).

With their theory of the object concept in infancy, Leslie *et al.* (1998, p. 12) build a bridge between perceptual approaches in the study of adult visual attention and conceptual approaches in developmental studies. They juxtapose perceptual objecthood with conceptual

[5] Contrarily, sortal theorists think that object individuation is conceptual (Casati, 2004, pp. 385–6; see Xu, 1997, pp. 365–7).

objecthood and suggest that these are intertwined by referring to the same indexing mechanism. Accordingly, an essential component of object cognition is an internal representation working as a 'sticky index' to a physical object. In this way, the interdependent necessity of both object conception and object perception has been settled.

A different operationalization is related to the previous one. Kahneman *et al.* (1992) hypothesized object-files, viz. temporary representations of perceived objects, which include an index and a feature store (see Green and Quilty-Dunn, 2021). These representations were proposed to explain object reidentification despite changes of properties. Once external entities are perceptually individuated and attended, respective perceptual systems activate object-files, which store and update information about the entities and their properties.

Additional operationalization comes from studies on cognitive development, especially experiments on infants' responses to visual stimuli behaving either like ordinary objects or not (surprise condition) (Spelke, 1990). The findings show that infants expect objects to be: connected, bounded, impenetrable, and moving on contact on a continuous route (see Section 3.1.2). Thus, object parts and animals are not experienced by infants in this way.

3 What constitutes a perceptual object?

Some categories of things are naturally excluded from considerations as POs. And even though objects of perception are already taken to be wide in scope, the assumption invariably adopted by the various studies on POs is that things that are not material or spatial are excluded from their considerations. Abstract entities (e.g., numbers), properties, mental items (e.g., dreams, qualia), and so forth are thus left out of this discussion (Casati, 2004, p. 383; Schwartz, 2001, p. 4). The topic of objects of perception, equated with POs by some (Casati, 2015), but not all (O'Callaghan, 2016; and ourselves),[6] concerns the question of how we perceive extramental entities such as 3D entities and 2D stimuli, but also events, properties, property bundles, spatial locations, and boundaries (Casati, 2015, p. 393; Clark, 2004, p. 463; Schwartz, 2001, p. 5). The scope of object perception is narrower—it includes extramental, concrete objects only—that is, stable 3D entities and 2D stimuli (Casati, 2015, p. 393).

Although the visual modality has had a privileged status in studying POs (Section 2.1), some proposals have attempted to form a unified theory of POs applicable across the modalities (see Section 3.2), or at least to consider a 'perceptual object' as a gradable notion, so that POs would seem to be objects in a weaker or stronger sense, depending on the modality (Vernazzani *et al.*, 2021).

3.1 Principles of object perception

One of the main problems of object perception is to determine what counts as a PO—that is, a unit of perception. In the following section, we present an overview of different principles of how environmental objects are individuated by our perceptual systems.

[6] O'Callaghan argues against equating objects of perception with POs. The former are intentional objects (see Section 2.2). The latter, POs, or O'Callaghan's 'objects for perception', are 'items that bear perceptible features and have perceptible parts arranged to form a unified whole' (O'Callaghan, 2016, pp. 1271–2), and so can be targets of perceptual capacities deployed in object perception (2016, p. 1269).

Solving this problem faces issues of various kinds. There is a question of how to properly characterize perceptually individuated objects. Numerous proposals lead to terms such as 'bodies'[7] (Burge, 2010; Quine, 1973, p. 54ff; Schwartz, 2001) or 'individuals' (Cohen, 2004; Green, 2018, pp. 2, 10; Mroczko-Wąsowicz and Grush, 2023; Nanay, 2017; O'Callaghan, 2016), which bear a connotation of a singular entity. At the same time, there is a question of whether the notion of POs can be applied to all sense modalities or rather whether it is modality-dependent—for instance, how to construe a conceptual scheme for POs regarding spatiality if arguably some sense modalities, such as audition, do not involve experiences of spatial character (Snowdon, 2006, p. 48).

Other issues are methodological. The principles we selected have been gathered from various traditions where different theorists provide different characterizations for POs. For instance, the same concept seems to come with different labels and some principles seem to follow others. By providing a comprehensive arrangement of principles of object perception, we question whether the claimed principles are disjunctive.

3.1.1 Principles of perceptual organization—Gestalt principles

The Gestalt approach to object perception offers explanations of the organization of sensory inputs. Nonetheless, there is an ongoing debate on what its principles concern (Todorovic, 2008, p. 5345); in particular, do they regard basic features and capacities of our perceptual system as gestaltists suggest, or do the principles concern features of the external world (Rock, 1975; Stevenson, 2014, pp. 1362, 1364) (see Section 2.3)?

Gestalt principles were introduced by Wertheimer (1923/38) to describe grouping elements of the perceptual scene. They include: proximity, symmetry, good continuation, good form, common fate, similarity, and other, more recent derivative principles (Palmer, 1992; Palmer and Rock, 1994). All of them facilitate processing of sensory stimuli. Gestalt principles have been applied typically to the visual modality (but see Bregman, 1990; Chakrabarty and Elhilali, 2019 for audition; and Chang *et al.*, 2007a, b for haptic and visual modalities).

The principle of proximity claims that 'the strength of grouping between two elements increases as these elements are brought nearer to each other' (Wagemans *et al.*, 2012, p. 1189). The principle of symmetry refers to the good symmetrical shape of an object (Wagemans *et al.*, 2012, p. 1190). The principle of good continuation says how we tend to group items that form a smooth pattern or contour (Wagemans *et al.*, 2012, p. 1189; Wertheimer, 1923/38) (see Overvliet *et al.*, 2012, p. 817 for tactile modality; Denham and Winkler, 2014, p. 602 for audition). The principle of good form (law of Prägnanz or simplicity) directs to some redundancy in form to extract simplicity and order in complex shapes (Spelke, 1990, p. 31). The principle of common fate is about uniformity in motion (Sekuler and Bennett, 2001; Luria and Vogel, 2014), temporal coherence of sensory features, and correlated changes thereof (see Shamma *et al.*, 2011 for audition). The principle of similarity with respect to some features results in a tendency to group items sharing given features (Wagemans *et al.*, 2012, p. 1180; Yu *et al.*, 2019). In audition, similarity between features of 'successive events such as

[7] There seems to be an implicit assumption that the best candidate for a PO is an object. And while this sounds tautological, this assumption has been explicitly investigated in the field of metaphysics—when searching for the best candidate for the most basic kind of object for linguistic reference. Strawson (1959, p. 55), for example, pointed to material bodies to be the best basic candidates for identification of particulars, rather than states, processes, or events. Therefore, the point of departure for the study of perceptual objects is rarely considering states, processes, or events.

pitch, timbre, loudness and location provides a basis for linking them' and so similarity can be associated with good continuation (Denham and Winkler, 2014, p. 602). The two principles can be found in the haptic modality as well (Overvliet *et al.*, 2012, p. 820). The principle of similarity could also be compared with the concept of uniformity (Casati, 2015, p. 395; Palmer and Rock, 1994, p. 30).

Another type of perceptual grouping is figure–ground organization, considered by some (Kubovy and van Valkenburg, 2001; Xu and Chun, 2009) as one of the most significant perceptual capacities involved in perceiving objects. A figure has a borderline (Wagemans *et al.*, 2012, p. 1194) or a psychological boundary (Stevenson, 2014, p. 1363) allowing to distinguish the figure from its ground. The process of figure–ground segregation uses perceptual cues such as convexity and concavity, symmetry, or narrow top–wide bottom polarity (Wagemans *et al.*, 2012, pp. 1195–6). Although the figure–ground organization has been mostly investigated within vision (Craft *et al.*, 2007), the literature includes illustrations of figure–ground segregation in other sense modalities such as audition (Matthen, 2010) and olfaction (Millar, 2019). Also, a multimodal figure–ground differentiation is possible, when an object is segregated from the surrounding by more than one modality (Vernazzani *et al.*, 2021).

3.1.2 Principles of object individuation—body principles

An approach towards object individuation, which appeals to body principles, has been offered by Green (2018). Body principles are principles of object perception which pertain to bodies (Green, p. 6), a concept defined by appealing to Quine (1973, p. 55) and Burge (2010, p. 437ff)—that is, 3D spatially extended, cohesive, and bounded entities. This proposal has implications in the metaphysics of POs, which are not necessarily accepted by all theorists. Hence, we shall talk here about 'principles of object perception', specifically 'principles of object individuation'.

Green (2018, p. 6) enlisted three body principles: cohesion, boundedness, and three-dimensionality. While the first two are drawn from Spelke (1990), the third one seems to be an overarching assumption in object perception, since it presupposes the volumetric nature of the external world. Nevertheless, the three-dimensionality principle should not be taken as a universal and necessary condition for object perception across the senses because it applies at most to visual and tactile objects.

Note that bodies alone do not exhaust the scope of POs in visual (Green, 2018, p. 6) and auditory experiences (O'Callaghan, 2016, p. 1275). Also, in a more liberal approach (O'Callaghan, 2016, p. 1274), three-dimensionality is not necessary for something to be a body. Among bodies, there are non-material ones, among which some are 2D ephemeral visibilia (see Section 2.2). Moreover, not all bounded objects are bodies. Some would view auditory POs as coherent collections of temporally bounded tones and sounds, and consider them as event-like individuals rather than bodies (O'Callaghan, 2008, p. 825; 2016, p. 1275).

As for the principles of cohesion and boundedness, these came into juxtaposition with two other principles: that of rigidity and that of no action at a distance (Spelke, 1990, p. 50). The cohesion principle states that the surface points on a single object exist continuously when they are connected (Spelke, 1990, p. 49). Now there are various interrelated understandings on what cohesion is. The first is related to spatio-temporal continuity, or topological continuity of objects (Green, 2018, p. 6). The second is related to connectedness: 'cohesion is spatial connection of all points in, or parts of, a shape or object' (Burge, 2010, p. 446). This view on cohesion was also applied in Spelke's (1994, p. 433; Kinzler and Spelke, 2007, p. 257) definition of cohesion, in which the aspects of spatial connection (and boundedness) are emphasized.

The original boundedness principle states that two points lie on distinct objects only if they are not linked by a path of connected surface points (Spelke, 1990, p. 49). The principle of boundedness is on a par with the notion of connectedness (Casati, 2015, p. 395; Palmer and Rock, 1994, p. 31), according to which as long as the surface of an object is connected, one should be able to move it in one piece; otherwise it cannot be considered as a single unit. The concept of boundedness resembles the notion of maximality formulated by Casati (2015, p. 396) as a complement to Palmer and Rock's (1994) condition for individuating POs—that is, uniform connectedness. The principle of maximality states that the area of uniform quality must be maximal (Casati, 2015, p. 396). Maximality excludes the possibility of picking out individuals within another individual.

The principle of rigidity states that surface points undergoing different motions tend to be perceived as distinct objects (Spelke, 1990, p. 50). The principle of no action at a distance states that if two surfaces are moving together, then they are interpreted as being connected, unless there is a perceptible gap between them (Spelke, 1990, p. 50). In more recent work, the principle of no action at a distance was labelled 'the principle of contact' (see Kinzler and Spelke, 2007, p. 257).

Spelke's (1990, p. 50) principles of object perception for concrete visual objects (bodies) are strongly motivated by the motion cue. The motion cue is primary due to boundedness of objects. Being restricted to vision prevents these principles from applying to objects of other senses. It remains an open question of whether these principles can or cannot be considered as universal for a unified theory of object perception, irrespective of the sense modality, or how far we may generalize them. Nonetheless, Spelke's proposal did not aim to develop a unified approach across the senses and to investigate POs other than bodies.

3.2 Theories of perceptual objects

The Gestalt approach is insufficient for a proper theory of POs. Even though it segregates a stimulus array, it does not segment it into individuated units of perception. Moreover, it does not serve as a systematic framework, but rather consists of loosely related principles. Body principles also cannot fulfil the requirements for a proper theory of POs, mainly because of their visuocentricity. Lastly, both have a shortcoming stemming from being studied in particular modalities. This section introduces two comprehensive theories of POs which embrace object perception in a modality-independent manner while specifying conditions for object individuation.

3.2.1 O'Callaghan's structural principles—mereologically complex individuals

O'Callaghan (2008, pp. 825–6; 2016, p. 1276) proposes to take POs as structured, mereologically complex individuals. O'Callaghan postulates that the entities in our environment must have some fundamental common characteristics that allow them to be identified as POs across the senses.

What has been rejected by O'Callaghan (2016, pp. 1278, 1280, 1286) as a candidate for the characteristics shared by all POs is their metaphysical status, for auditory objects are event-like individuals, while visual objects are most often object-like individuals. This has two metaphysical implications. Firstly, time plays a crucial role in recognizing the identity of auditory objects, while in the case of visual objects, it is the spatial relations between their parts (O'Callaghan, 2008, pp. 823, 825; 2016, p. 1279). Secondly, persisting visual objects endure—they are wholly present at each time, whereas persisting auditory objects perdure—they are not fully present at any one given time, but they unfold in time (O'Callaghan, 2008, p. 824).

Overall, the problem of diverse metaphysical status of POs stems from the diverse typology of objects represented by POs. This means that tactual objects and visual objects, apart from ephemeral visibilia, correspond uncontroversially to material objects/bodies. Auditory and olfactory objects correspond instead to events, and material objects generating sounds and odours are perceived only indirectly (O'Callaghan, 2008, p. 825; 2016, p. 1270). This is related to the lack of immediate awareness of perceiving ordinary material objects in audition (O'Callaghan, 2008, p. 805), olfaction, and gustation (O'Callaghan, 2016, p. 1270). Sounds, odours, and tastes are POs, despite not sharing 'the spatial or temporal characteristics of ordinary or material objects' (O'Callaghan, 2008, p. 824). If that is the case, then the overriding similarities enabling these different objects of various modalities to be POs, according to O'Callaghan, must be found elsewhere—in their shared structure as mereologically complex individuals.

For an individual to be considered a mereologically complex individual, it must meet the following criteria. Being an individual means it must be a 'singular, first-order feature bearer' (O'Callaghan, 2016, p. 1277). For an object to be mereologically complex means that we can distinguish its qualitative parts, but only collectively do they form a cohesive whole. This means that scattered objects, like flocks, can be recognized as POs as well (O'Callaghan, 2016, p. 1287). As for the structurality of an object, it implies that each PO has an internal and an external structure determined by certain intrinsic and extrinsic relations that can be discerned by a given sense modality. The internal structure, which plays an important role in recognizing POs, can be based on spatial boundaries between parts as it is for visual objects (coherent collections of spatially bounded surfaces), or it can be based on temporal boundaries between parts as it is for auditory objects (coherent collections of temporally bounded sounds) (O'Callaghan, 2008, p. 825; 2016, p. 1277). The external structure, determined by spatial location or the location in pitch space, will allow the individuation of visual or auditory objects, respectively. The relationship between spatially or temporally bounded parts of an object is important because POs are individuated and recognized only if their parts remain in a specific configuration—that is, parts stand in a constitutive, idiosyncratic relation (O'Callaghan, 2016, p. 1277). The structurality of POs seems to constitute a notable condition, especially that the structure of the stimulus array is detected by the brain (Stevenson, 2014, p. 1364).

3.2.2 Green's structural principles—structural unity schema

Green's (2018, p. 5) unified theory of POs postulates that perceptual organization principles, together with ODCs, are the foundation for object individuation. By 'perceptual organization principles', Green means three Gestalt principles (similarity, common fate, and good continuation; see Section 3.1.1) and the hierarchical organization principle (the capability to contrast wholes with their parts—that is, to group and parse; see Green, 2018, pp. 4–6). The perceptual organization principles are less restrictive than the body principles in identifying POs. However, because the perceptual organization principles do not serve as a holistic theoretical system (see Section 3.1.1), and the body principles are too restrictive in identifying POs, none of the two suffice for an account of POs. Therefore, additional criteria for what makes a PO were introduced, namely ODCs (Green, 2018, p. 5).

ODCs are supposed to suffice in singling out an object in perception (Green, 2018, p. 3). A list of three ODCs has been offered, namely, circumscription, reidentification (tracking), and ascription (Green, 2018, p. 3). Circumscription is a type of segmentation, when individuals are perceptually discriminated from their surroundings due to their boundaries. Reidentification is related to object tracking and acknowledging object persistence; thus it is about identifying an object throughout its distinct stages over time. Ascription is a perceptual process of attributing features, not necessarily of a single modality, to the represented object (Green, 2018, p. 3; O'Callaghan, 2016, pp. 1286–7). While the three capacities do not have to be exercised jointly to function properly, they are said to at least target uncontroversial cases of object perception—that is, if an entity complies with the three capacities, there is a good reason to treat it as a PO (Green, 2018, pp. 3–4).

Green (2018, p. 5) does not wish to treat ODCs as capacities picking out individuals different from those singled out by other principles. This means that ODCs are supposed to target the same types of objects as perceptual organization principles. If that is the case, one may ask in what relation they stand with respect to body principles. At first glance, it seems that ODCs do single out bodies, especially evident in vision, and hence they could be identified with the body principles (Section 3.1.2). Nonetheless, our perceptual capacities, in both vision and audition, can circumscribe, track, or ascribe features to non-bodies as well (Green, 2018, pp. 7–9). It follows that ODCs are less restrictive than the body principles, because the latter rule out individuals like flocks of birds and limbs, which after all are POs (Green, 2018, p. 13). In general, while body principles govern individuals which are bodies (body-like individuals), perceptual ODCs '*function to pick out* entities whose parts are causally related' (Green, 2018, p. 13; original emphasis).

Now, what does it mean for parts of an entity to be 'causally related'? An answer to this question has been advanced in Green's theory of POs and it appeals to the so-called 'structural unity schema'. What underlies Green's theory is that POs which are targeted by our perceptual systems are individuals (Green, 2018, p. 1). Individuals include something more than just bodies, and at the same time exclude arbitrary parts or combinations. Within the proposed structural unity schema, our perceptual system targets an object if it meets certain conditions. Firstly, the parts of a perceptual object 'participate in relatively direct, sustained causal interactions with one another' (Green, 2018, p. 13). It means that POs must be characterized by some internal causal relatedness. But because, in the end, these are the very same individuals which are targeted by ODCs, a second condition is needed—parts of a PO must be characterized by causally sustained regularities (Green, 2018, p. 18). Such a regularity may manifest itself in a causally sustained first-order resemblance—that is, involving repetition of a feature across space or time. In the third condition, the causally supported regularity should regard the '*largest possible* portion of the environment that conforms to the regularity' (Green, 2018, p. 20; original emphasis); otherwise arbitrary parts of a PO could be taken as distinct objects. This condition resembles Casati's idea of maximality (Section 3.1.2); however, Green's condition does not require topological maximality (maximizing the extent of connected surface points). It is about the maximal extent of regularity, which may be instantiated by a cluster of disconnected elements. The final condition eventually brings us to the maximal regularity-supporting causation (MRSC) schema for object perception.

Although O'Callaghan's and Green's structural theories are less restrictive than the body theory in specifying what counts as POs, they still do the job of individuating POs. They are also capable of addressing atypical POs, like flocks or sounds. Since the theories are not limited to any modality, one can apply them to POs in various senses. Therefore, these comprehensive approaches grasp the nature of POs across modalities better than Gestalt or body principles.

4 Conclusions

Defining POs faces several difficulties. The first stems from the differences in the notion of the object itself, because different disciplines make different ontological assumptions (Section 2.1). Moreover, the term 'object' is not used consistently throughout the literature (Section 2.2). Another difficulty concerns generalizing from visual objects to all POs (Section 2.1). Furthermore, it has been argued that an account of POs cannot ignore object conception (Section 2.4). Clarifying the relation between object perception and conception may require further investigation, especially since provided frameworks for the object concept are not compatible with one another (Section 2.4.1).

We presented O'Callaghan's (2008, 2016) and Green's (2018) unified theories of POs (Section 3.2) which note the fact that various sense modalities operate differently and reveal diverse aspects of object perception. What these theories have in common is that they embrace both Gestalt (Section 3.1.1) and body principles (Section 3.1.2). Nevertheless, in defining POs, the Gestalt research tradition is insufficient. It provides general principles of perceptual organization, which must be complemented with principles of object individuation (e.g., body principles), which individuate 3D spatially extended, cohesive, and bounded entities. Yet, it has also been shown that complementing Gestalt with body principles is not enough for establishing a unified theory of POs. The visuocentric approach of those principles does not solve the puzzle of diversity, and bodies do not exhaust the scope of POs. Both unified theories of POs address these two issues. O'Callaghan's (2008, 2016) structured, mereologically complex individuals meet the necessity of a modality non-specific theory by accounting for body-like individuals and event-like individuals (Section 3.2.1). Green's (2018) structural unity schema postulates that parts of POs need to exhibit causally sustained regularities to form POs (Section 3.2.2). By providing a solution for the majority of problems exposed in Section 2, the two theories deliver the most complete accounts of POs for the time being.

Both unified theories acknowledge the difference between levels of description of POs. Gestalt and body approaches seem to not distinguish between the first and third level, thus mixing features of the physical world and those imposed on the perceptual world by human perceptual systems. In contrast, O'Callaghan (2016, p. 1286) is explicit about separating the ontological status of POs, stating that his theory is independent from any metaphysical assumptions. And while Green (2018) does not explicitly relate to the ontological level of describing POs, he takes a clearer standpoint on the nature of perceptual capacities. Green's ODCs are compatible with O'Callaghan's suite of perceptual capacities targeting POs (O'Callaghan, 2016, pp. 1285–7). The latter consists of differentiation from the surrounding

(Green's circumscription), feature attribution (Green's ascription), and persistence or tracking (Green's reidentification). O'Callaghan, however, does not specify whether the suite concerns perceptual systems or POs (see Section 2.3). On the one hand, O'Callaghan (2016, pp. 1285–6) writes that 'the suite's targets are differentiated from their surroundings, have features and parts, and persist through time' (level 2). On the other hand, he writes that 'such capacities include [...] distinguishing items from their surroundings, attributing features [...], and tracking items over time through change. [...] perceptual objects are the targets on which such capacities converge' (2016, pp. 1286–7; level 3). Green then makes a more decisive standpoint regarding this issue and takes O'Callaghan's perceptual suite as pertaining to perceptual systems (level 3).

What seems to be missing in both proposals is an explicit discussion of the relation between their theories of POs and object conception. The empirical adequacy of Gestalt and body principles has been evaluated by the object perception suite and ODCs. However, by accepting body principles, both proposals appear to implicitly follow the nativist approach where object perception seems to be equated with innate object conception. This arbitrary assumption results in taking the concept of POs for granted, what might make a theory of POs incomplete. In general, it seems that in the field of object perception, the relation between POs and object concepts needs to be acknowledged.

Acknowledgements

The work was supported by the National Science Centre, Poland (Grant 2019/35/B/HS1/04386). The authors thank the editors and reviewers for their helpful comments on earlier versions of this chapter.

References

Alvarez, G. A. and Franconeri, S. L. (2007). How many objects can you attentively track?: evidence for a resource-limited tracking mechanism. *Journal of Vision*, 7(13), 1–10.

Aristotle (1984). *De Anima*, trans. J. A. Smith. In *The Complete Works of Aristotle: The Revised Oxford Translation*, edited by J. Barnes, Vol. 1, 641–92. Princeton, NJ: Princeton University Press.

Auvray, M. and Spence, C. (2008). The multisensory perception of flavor. *Consciousness and Cognition*, 17, 1016–31.

Baillargeon, R. (2008). Innate ideas revisited: for a principle of persistence in infants' physical reasoning. *Perspectives on Psychological Science*, 3, 2–13.

Batty, C. (2014). Olfactory objects. In *Perception and Its Modalities*, edited by D. Stokes, M. Matthen, and S. Biggs, 222–46. Oxford: Oxford University Press.

Bregman, A. (1990). *Auditory Scene Analysis: The Perceptual Organization of Sound*. Boston, MA: MIT Press.

Bizley, J. K. and Cohen, Y. E. (2013). The what, where and how of auditory-object perception. *Nature Reviews Neuroscience*, 14(10), 693–707.

Brewer, B. (2007). Perception and its objects. *Philosophical Studies*, 132, 87–97.

Burge, T. (2010). *Origins of Objectivity*. Oxford: Clarendon Press.

Carey, S. (2011). Précis of the origin of concepts. *Behavioral and Brain Sciences*, 34(3), 113–67.

Casati, R. (2004). Is the object concept formal? *Dialectica*, 58(3), 383–94.

Casati, R. (2015). Object perception. In *Oxford Handbook of Philosophy of Perception*, edited by M. Matthen, 393–404. Oxford: Oxford University Press.

Chakrabarty, D. and Elhilali, M. (2019). A Gestalt inference model for auditory scene segregation. *PLoS Computational Biology*, 15(1), e1006711.

Chang, D., Nesbitt, K. V., and Wilkins, K. (2007a). The Gestalt principle of continuation applies to both the haptic and visual grouping of elements. *WHC'07: Proceedings of the* Second Joint EuroHaptics Conference and Symposium on Haptic Interfaces for Virtual Environment and Teleoperator Systems, 15–20. Washington, DC: IEEE Computer Society.

Chang, D., Nesbitt, K. V., and Wilkins, K. (2007b). The Gestalt principles of similarity and proximity apply to both the haptic and visual grouping of elements. *AUIC '07: Proceedings of the Eight Australasian Conference on User Interface*, 64, 79–86. Darlinghurst, NSW: Australian Computer Society, Inc.

Chediak, K. (2016). Intentionalism and the problem of the object of perception. *Trans/Form/Ação, Marília*, 39(2), 87–100.

Cichy, R. M., Pantazis, D., and Oliva, A. (2014). Resolving human object recognition in space and time. *Nature Neuroscience*, 17(3), 455–64.

Clark, A. (2000). *A Theory of Sentience*. Oxford: Oxford University Press.

Clark, A. (2004). Feature-placing and proto-objects. *Philosophical Psychology*, 17(4), 443–69.

Cohen, J. (2004). Objects, places, and perception. *Philosophical Psychology*, 17(4), 471–95.

Craft, E., Schütze, H., Niebur, E., and von der Heydt, R. (2007). A neural model of figure-ground organization. *Journal of Neurophysiology*, 97(6), 4310–26.

Denham, S. L. and Winkler, I. (2014). Auditory perceptual organization. In *The Oxford Handbook of Perceptual Organization*, edited by J. Wagemans, 601–20. Oxford: Oxford University Press.

Fleming, R. W. (2014). Visual perception of materials and their properties. *Vision Research*, 94, 62–75.

Green, E. J. (2018). A theory of perceptual objects. *Philosophy and Phenomenological Research*, 99(3), 1–31.

Green, E. J. and Quilty-Dunn, J. (2021). What is an object file? *British Journal for the Philosophy of Science*, 72(3), 665–99.

Griffiths, T. D. and Warren, J. D. (2004). What is an auditory object? *Nature Reviews Neuroscience*, 5, 887–92.

Gottfried, J. A. (2010). Central mechanisms of odour object perception. *Nature Reviews Neuroscience*, 11, 628–41.

Hochberg, J. and Peterson, M. A. (1987). Piecemeal organization and cognitive components in object perception: perceptually coupled responses to moving objects. *Journal of Experimental Psychology: General*, 116(4), 370–80.

Johnson, S. P. (2004). Theories of development of the object concept. In *Theories of Infant Development*, edited by G. N. Bremner and A. Slater, 174–203. Hoboken, NJ: Blackwell Publishing.

Johnson, S. P. (2018). Object perception. In *Oxford Research Encyclopedia of Psychology*, edited by O. Braddick (Series Ed.) and C. von Hofsten (Vol. Ed.), 1–17. New York: Oxford University Press. doi: 10.1093/acrefore/9780190236557.013.62

Johnson, S. P., Amso, D., and Slemmer, J. A. (2003). Development of object concepts in infancy: evidence for early learning in an eye-tracking paradigm. *Proceedings of the National Academy of Sciences of the United States of America*, 100(18), 10568–73.

Johnson, S. P., Bremner, J. G., and Slater, A. M. (2023). Origins of object concepts in infancy: a perceptual account. In *Sensory Individuals: Unimodal and Multimodal Perspectives*, edited by A. Mroczko-Wąsowicz and R. Grush (Chapter 1). Oxford: Oxford University Press.

Kahneman, D., Treisman, A., and Gibbs, B. J. (1992). The reviewing of object files: object-specific integration of information. *Cognitive Psychology*, 24, 175–219.

Kinzler, K. D. and Spelke, E. S. (2007). Core systems in human cognition. *Progress in Brain Research*, 164, 257–64.

Koslicki, K. (2008). *The Structure of Objects*. Oxford: Oxford University Press.

Kubovy, M. and Schutz, M. (2010). Audio-visual objects. *Review of Philosophy and Psychology*, 1, 41–61.

Kubovy, M. and Van Valkenburg, D. (2001). Auditory and visual objects. *Cognition*, 80, 97–126.

Leslie, A. M. (1995). A theory of agency. In *Causal Cognition: A Multidisciplinary Debate*, edited by D. Sperber, D. Premack, and A. J. Premack, 121–49. Oxford: Oxford University Press.

Leslie, A. M., Xu, F., Tremoulet, P. D., and Scholl, B. J. (1998). Indexing and the object concept: developing 'what' and 'where' systems. Trends in Cognitive Sciences, 2(1), 10–18.

Lotto, A. and Holt, L. (2011). Psychology of auditory perception. *WIREs Cognitive Science*, 2(5), 479–89.

Luria, R. and Vogel, E. K. (2014). Come together, right now: dynamic overwriting of an object's history through common fate'. *Journal of Cognitive Neuroscience*, 26(8), 1819–28.

Lycan, W. G. (2014). What does vision represent? In *Does Perception Have Content?*, edited by B. Brogaard, 311–28. Oxford: Oxford University Press.

Lycan, W. G. (2018). What does taste represent? *Australasian Journal of Philosophy*, 96(1), 28–37.

Matthen, M. 2(004). Features, places, and things: reflections on Austen Clark's theory of sentience. *Philosophical Psychology*, 17(4), 497–518.

Matthen, M. (2010). On the diversity of auditory objects. *Review of Philosophy and Psychology*, 1, 63–89.

Millar, B. 2(019). Smelling objects. *Synthese*, 196, 4279–303.

Moreland, J. P. (1998). Theories of individuation: a reconsideration of bare particulars. *Pacific Philosophical Quarterly*, 79, 251–63.

Mroczko-Wąsowicz, A. (2015). What can sensorimotor enactivism learn from studies on phenomenal adaptation in atypical perceptual conditions? In *Open MIND: Philosophy and the Mind Sciences in the 21st Century*, edited by T. Metzinger and J. M. Windt, 633–49. Cambridge, MA: MIT Press.

Mroczko-Wąsowicz, A. and Grush, R. (Eds). (2023). *Sensory Individuals: Unimodal and Multimodal Perspectives.* Oxford: Oxford University Press.

Mroczko-Wąsowicz, A. and Werning, M. (2012). Synesthesia, sensory-motor contingency and semantic emulation: how swimming style-color synesthesia challenges the traditional view of synesthesia. *Frontiers in Psychology*, 3, 279.

Nanay, B. (2017). Philosophy of perception: a road map with lots of bypass roads. In *Current Controversies in Philosophy of Perception*, edited by B. Nanay, 1–20. New York, NY: Routledge.

O'Callaghan, C. (2008). Object perception: vision and audition. *Philosophy Compass*, 3(4), 803–29.

O'Callaghan, C. (2014). Not all perceptual experience is modality specific. In *Perception and Its Modalities*, edited by D. Stokes, M. Matthen, and S. Biggs, 133–63. Oxford: Oxford University Press.

O'Callaghan, C. (2016). Objects for multisensory perception. *Philosophical Studies*, 173, 1269–89.

Overvliet, K. E., Krampe, R. T., and Wagemans, J. (2012). Perceptual grouping in haptic search: the influence of proximity, similarity, and good continuation. *Journal of Experimental Psychology: Human Perception and Performance*, 38(4), 817–21.

Palmer, S. (1992). Common region: a new principle of perceptual grouping. *Cognitive Psychology*, 24, 436–47.

Palmer, S. and Rock, I. (1994). Rethinking perceptual organization: the role of uniform connectedness. *Psychonomic Bulletin and Review*, 1(1), 29–55.

Peterson, M. A. and Rhodes, G. (Eds). (2003). *Perception of Faces, Objects, and Scenes: Analytic and Holistic Processes.* Oxford: Oxford University Press.

Piaget, J. (1954). *The Construction of Reality in the Child.* New York, NY: Basic Books.

Pylyshyn, Z. W. (2001). Visual indexes, preconceptual objects, and situated vision. *Cognition*, 80, 127–58.

Pylyshyn, Z. W. and Storm, R. (1988). Tracking multiple independent targets: evidence for a parallel tracking mechanism. *Spatial Vision*, 3(3), 179–97.

Quine, W. V. (1973). *The Roots of Reference.* La Salle: Open Court.

Rock, I. (1975). *An Introduction to Perception (Book on Psychophysiological Mechanisms for Vision).* New York, NY: Macmillan Publishing Co., Inc.

Russell, B. (2009). *Our Knowledge of the External World as a Field for Scientific Method in Philosophy.* London and New York, NY: Routledge.

Santos, L. R. and B. M. Hood. (2009). Object representation as a central issue in cognitive science. In *The Origins of Object Knowledge*, edited by B. M. Hood and L. R. Santos, 1–23. Oxford: Oxford University Press.

Scholl, B. J. (2007). Object persistence in philosophy and psychology. *Mind and Language*, 22(5), 563–91.

Scholl, B. J. and Leslie, A. M. (1999). Explaining the infant's object concept: beyond the perception/cognition dichotomy. In *What is cognitive science?*, edited by E. Lepore and Z. Pylyshyn, 26–73, Oxford: Blackwell.

Scholl, B. J., Pylyshyn, Z. W., and Feldman, J. (2001). What is a visual object? Evidence from target merging in multiple object tracking. *Cognition*, 80, 159–77.

Schwartz, R. (2001). The concept of 'an object' in perception and cognition. In *From Fragments to Objects: Segmentation and Grouping in Vision*, edited by T. F. Shipley and P. J. Kellman, 3–17, Amsterdam: Elsevier.

Sekuler, A. B. and Bennett, P. J. (2001). Generalized common fate: grouping by common luminance changes. *Psychological Science*, 12(6), 437–44.

Shamma, S. A., Elhilali, M., and Micheyl, C. (2011). Temporal coherence and attention in auditory scene analysis. *Trends in Neurosciences*, 34(3), 114–23.

Siegel, S. (2006). How does visual phenomenology constrain object-seeing? *Australasian Journal of Philosophy*, 84(3), 429–41.

Skrzypulec, B. (2018). *Empiryczna Ontologia Percepcji. Struktura i Tożsamość Przedmiotów Wzrokowych w Perspektywie Nauk Kognitywnych*. Warszawa: IFIS PAN.

Snowdon, P. (2006). P. F. Strawson: *Individuals*. In *Central Works Philosophy Volume 5. The Twentieth Century: Quine and After*, edited by J. Shand, 40–63. Chesham: Acumen.

Spelke, E. S. (1990). Principles of object perception. *Cognitive Science*, 14, 29–56.

Spelke, E. (1994). Initial knowledge: six suggestions. *Cognition*, 50(13), 431–45.

Spelke, E. S. (2000). Core knowledge. *American Psychologist*, 55(11), 1230–43.

Stevenson, R. J. (2014). Object concepts in the chemical senses. *Cognitive Science*, 38, 1360–13.

Stokes, D. and Biggs, S. (2014). The dominance of the visual. In *Perception and Its Modalities*, edited by D. Stokes, M. Matthen, and S. Biggs, 350–78. Oxford: Oxford University Press.

Strawson, P. F. (1959). *Individuals*. London and New York, NY: Routledge.

Strawson, P. F. (2002). Perception and its objects. In *Vision and Mind: Selected Readings in the Philosophy of Perception*, edited by A. Noë and E. Thompson, 91–110. Cambridge, MA: MIT Press.

Todorovic, D. (2008). Gestalt. *Scholarpedia*, 3(12), 5345.

Vernazzani, A., Skrzypulec, B., and Schlicht, T. (2021). Structure of perceptual objects: introduction to the synthese topical collection. *Synthese*, 199 (1), 1819–30.

Wagemans, J., Elder, J. H., Kubovy, M., *et al.* (2012). A century of Gestalt psychology in visual perception: I. Perceptual grouping and figure-ground organization. *Psychological Bulletin*, 138(6), 1172–217.

Wertheimer, M. (1923/38). Untersuchungen zur Lehre von der Gestalt II. *Psychologische Forschung*, 4, 301–50 (Translation published in Willis Ellis. 1938. *A Source Book of Gestalt Psychology*, pp. 71–88. London: Routledge and Kegan Paul).

Wolfe, J. M., Oliva, A., Horowitz, T. S., Butcher, S. J., and Bompas, A. (2002). Segmentation of objects from backgrounds in visual search tasks. *Vision Research*, 42(28), 2985–3004.

Xu, F. (1997). From Lot's wife to a pillar of salt: evidence that *physical object* is a sortal concept. *Mind and Language*, 12(3/4), 365–92.

Xu, Y. and Chun, M. M. (2009). Selecting and perceiving multiple visual objects. *Trends in Cognitive Sciences*, 13(4), 167–74.

Yu, D., Xiao, X., Bemis, D. K., and Franconeri, S. L. (2019). Similarity grouping as feature-based selection. *Psychological Science*, 30(3), 376–85.

3

Visual completion and intermediate representations in object formation

Philip J. Kellman and Viyehni Fuchser

1 Introduction

Seeing is easy to take for granted. The apparent simplicity of obtaining knowledge of objects, surfaces, and events from light they reflect conceals deep complexities. Because light moves in straight lines and most materials are opaque, we typically receive light from only fragmentary parts of objects' surfaces. From any vantage point, most viewed objects are partly occluded by others and every object occludes much of itself. When an observer or a partly occluded object moves, complex and continuously changing patterns of occlusion are projected to the eyes.

Figure 3.1 illustrates this *fragmentation problem* in vision. Without special effort, we see a car with a well-defined shape, two trees, a roadway, sidewalks, and a leafy plant. This ecologically coherent and meaningful inventory of the scene arises from a very large number of patches of colour in the image. As a first pass at the problem of object perception, if we were to suppose that the objects in the world correspond to areas of the retinal projection that are relatively homogeneous in colour and texture, we would find that the scene in Figure 3.1 contains literally hundreds of such 'objects'. Stimulation of many non-contiguous retinal areas may arise from a single object or a continuous surface, and parts of a single object in the world may vary considerably in lightness, colour, and texture.

In this chapter, we discuss visual processes that overcome fragmentation and produce perception of complete objects. Foundations of this work have been developed over many years, but our treatment introduces new results and theory that represent, in our view, considerable advances in resolving apparent anomalies and controversies and in unifying diverse research areas. An important theme in this work is the importance of intermediate representations in object formation.

2 Contour and surface interpolation in object formation

Two separable and complementary processes connect visible regions across gaps to produce perception of coherent objects. *Contour interpolation* connects visible fragments of edges across gaps, based primarily on their geometric relationships (Grossberg and Mingolla, 1985; Heitger, von der Heydt *et al.*, 1998; Kalar *et al.*, 2010; Kellman, Garrigan, and Shipley, 2005; Kellman and Shipley, 1991). *Surface interpolation* connects separate visible areas on the basis

Philip J. Kellman and Viyehni Fuchser, *Visual completion and intermediate representations in object formation* In: *Sensory Individuals.* Edited by: Aleksandra Mroczko-Wąsowicz and Rick Grush, Oxford University Press. © Oxford University Press 2023.
DOI: 10.1093/oso/9780198866305.003.0004

Figure 3.1 Illustration of the fragmentation problem in visual perception.

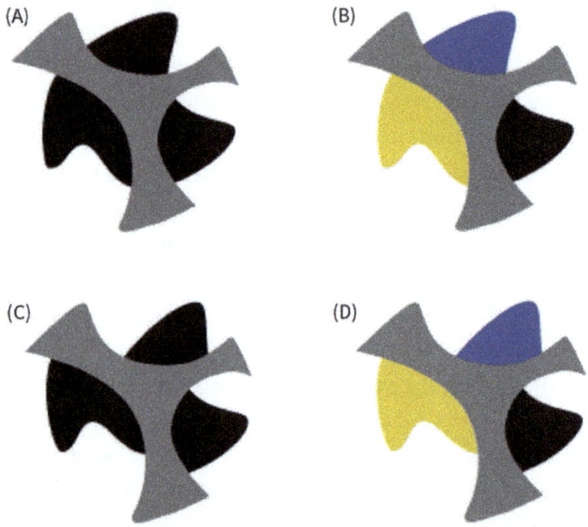

Figure 3.2 Contour and surface interpolation in object formation. (A) The three black regions are perceived to be visible parts of a single object, as their relations satisfy the requirements for both contour and surface interpolation. (B) Contour interpolation alone. Mismatched surface colours of visible regions block surface interpolation. The three visible regions have the same edge relations as in (A), such that contour interpolation still produces perception of a complete object with precise boundaries and shape. (C) Surface interpolation alone. By disrupting contour relatability, contour interpolation is blocked. Surface interpolation, caused by the matching surface colour of the visible regions, produces an impression that the three fragments connect behind the occluder, but with indeterminate shape. (D) Neither contour nor surface interpolation. As in (C), edge relations block contour interpolation, and mismatching surface colours also block surface interpolation, leading to perception of three disconnected coloured regions, with uncertain boundaries behind the occluder.

of surface properties (Yin, Kellman, and Shipley, 2000). Both processes are illustrated in Figure 3.2.

Contour interpolation plays a primary role in establishing object boundaries and shape, and it will be our focus in this chapter. When an object is perceived from fragmentary input, its shape is defined by both its physically specified edges and edges created by interpolation. Surface interpolation can complement contour interpolation. Surface areas separated by occlusion, if similar in lightness, colour, and/or texture, 'spread' behind the occluder up to boundaries defined by real and interpolated edges (Kellman and Shipley, 1991; Yin, Kellman, and Shipley, 2000). Surface interpolation, however, also works in the absence of constraining contours, as when we see the continuous blue sky extending behind branches and leaves of a tree.

2.1 Contour interpolation

Contour interpolation is governed by well-known geometric constraints. In general, contour interpolation can operate only on visible edge fragments that terminate in a contour junction or tangent discontinuity (Rubin, 2001; Shipley and Kellman, 1990). An important invariant in visual perception is that such contour junctions will always be present in the optical projection whenever one object occludes another (Kellman and Shipley, 1991).

Contour interpolation requires rather precise geometric relations between pairs of edge fragments. These were formally described by Kellman and Shipley (1991) and labelled *contour relatability*. Figure 3.3 illustrates the relevant position and orientation relations between contour fragments that support interpolation. (More formal geometric (Kellman and Shipley, 1991) and algebraic (Kalar *et al.*, 2010; Kellman, Garrigan, and Shipley, 2005; Singh and Hoffman, 1999) accounts can be found elsewhere.) Other research shows that essentially the same geometric relations, generalized to 3D space, govern 3D contour completion (Carman and Welch, 1992; Kellman *et al.*, 2005) and spatio-temporal interpolation, in which the two members of an edge pair are not visible simultaneously (Palmer, Kellman, and Shipley, 2006).

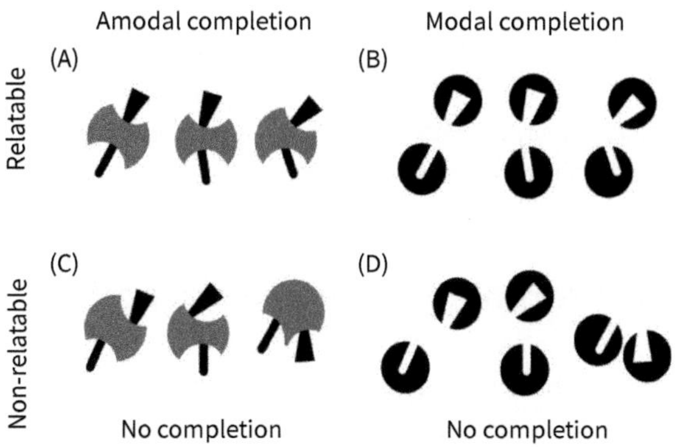

Figure 3.3 Contour relatability in amodal and modal completion. (A) and (B) Relatable edges produce perception of complete objects despite partial occlusion (amodal completion) and in illusory objects (modal completion). (C) and (D) Violation of the geometry of contour relatability produces perception of disconnected black objects (C) or lack of illusory contours (D).

The two columns in Figure 3.3 distinguish two appearances of visible fragments that are completed across gaps—*amodal* and *modal* completion (Michotte, Thines, and Crabbe, 1964). *Amodal completion* refers to the ecologically pervasive situation where objects are perceived as complete despite partial occlusion by nearer objects. Modal completion refers most often to completion of objects across backgrounds that are similar to, or match, the surface qualities of object parts (camouflage). These are more commonly referred to as *illusory contours* or *illusory objects*.

The similar geometry governing modal and amodal completion is consistent with the idea that they share a common underlying process (Kellman and Loukides, 1987). Considerable psychophysical evidence supports this *identity hypothesis* (Albert, 2007; Gold *et al.*, 2000; Kellman, Yin, and Shipley, 1998; Palmer, Kellman, and Shipley, 2006; Ringach and Shapley, 1996). Neurophysiological data, especially from electroencephalographic (EEG) methods permitting relatively precise temporal resolution, indicate common initial brain responses in the lateral occipital complex (LOC) and parietal cortical areas to both types of completion emerging at about 140 ms after stimulus presentation, consistent with psychophysical estimates of the time course of interpolation (Guttman and Kellman, 2004; Keane, Lu, and Kellman, 2007), followed by differential responses for modal and amodal completion displays evident after 240 ms (Murray, Foxe, Javitt, and Foxe, 2004). Other behavioural research has pointed to differences in outcomes in modal and amodal completion (Anderson, Singh, and Fleming, 2002; Singh, 2004; Spehar and Halim, 2016; Zhou, Tjan, Zhou, and Liu, 2008). We will see that these divergent views can be reconciled by the notion that contour interpolation is not a one-shot process from stimulus to percept, but instead depends on an intermediate representation of linked contours followed by implementation of scene constraints that affect perceptual appearance.

3 Intermediate representations in object formation

The dependence of contour interpolation on fixed geometric constraints is consistent with an automated perceptual mechanism. Keane (2018) has argued that it provides a salient example of a modular perceptual process in the sense described by Fodor (1983) in *The Modularity of Mind*. Neurally plausible models of contour completion based on contour relatability or similar geometry have been developed that show substantial agreement with human perception (Heitger *et al.*, 1998; Kalar *et al.*, 2010). Part of object formation, then, includes a relatively automatic mechanism that links visible contour segments based on their geometric relationships. However, this is not the whole story.

3.1 Variations in interpolated contour strength

A number of phenomena show that the strength, or even the existence, of perceived contour connections varies based on certain factors, even for relatable edges. In Figure 3.4, both displays have edges that are geometrically relatable, yet the outline-inducing elements produce weak or absent illusory contours.

The weakness or absence of visually completed contours in outline displays such as shown in Figure 3.4B has been often noted, with explanations centering around the differences of

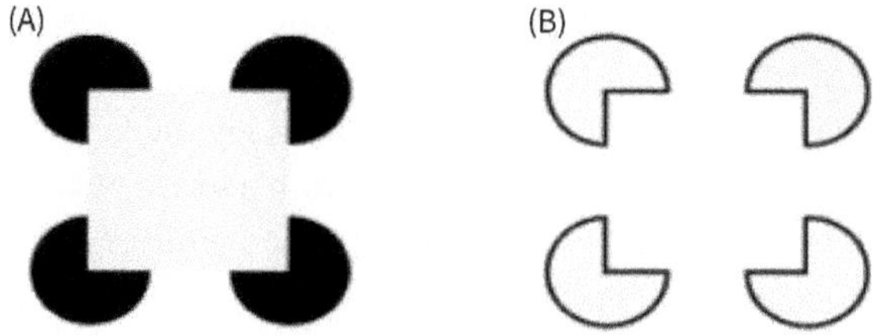

Figure 3.4 Comparison of filled (A) and outline (B) inducing elements for illusory contours illustrate differences in perception of completed contours despite unchanging geometric relations of inducing edges. (See text.)

outline edges from ordinary surface edges and a likely perceptual constraint that closed outlines perceptually determine a bounded object in their interior, with consequences for border ownership (Kanizsa, 1976; Kellman and Shipley, 1991; Koffka, 1935).

Another case in which relatable edges alone do not determine perceptual outcomes is shown in Figure 3.5, which presents a pair of displays from Su, He, and Ooi (2010). Note the difference in appearance between Figure 3.5A and Figure 3.5B. In Figure 3.5A, interpolated edges are perceived to form an illusory grey ring behind which amodally completed rectangles pass. Although some interpolated edges can be seen in Figure 3.5B around the inner and outer circles, these do not become organized into a coherent grey ring as in Figure 3.5A. Both patterns have relatable edges in the same positions. The difference suggests that other constraints, especially the differences in surface lightness of the two rectangles in each radial arm in Figure 3.5B, play a role in determining the final organization of the display.

3.2 Separating contour linking from scene description: a model of object formation

The phenomena illustrated above represent examples where visual completion is modulated by factors beyond the geometric relations of edge fragments; others have also been documented (e.g., Gerbino, 2020; Gillam, 1987; Kellman and Shipley, 1991). Geometric relations

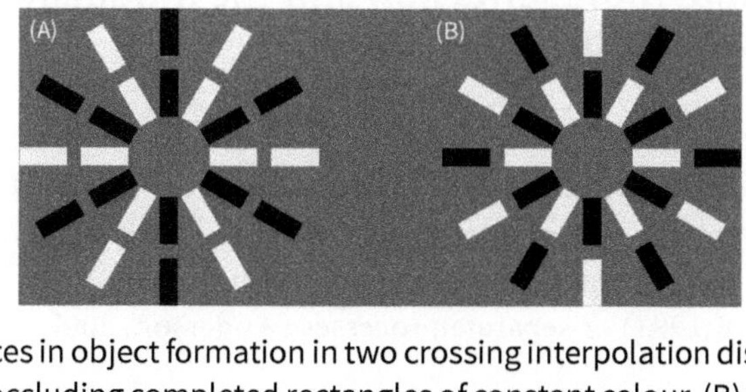

Figure 3.5 Differences in object formation in two crossing interpolation displays. (A) A grey ring is clearly seen, occluding completed rectangles of constant colour. (B) Opposite contrast polarity of the rectangles along each radial arm diminishes perception of a clear, bounded grey ring.
Adapted from Su, He & Ooi, 2010.

define which edges participate in contour interpolation, but perceptual outcomes also depend on other factors.

These findings can be explained by a theory positing separable stages in object formation. The idea that perceptual phenomena may emerge from a series of processes and representations was articulated by Marr (1982); in object formation, multiple representations and their interactions were hypothesized by Grossberg and Mingolla (1985), whose *boundary contour* process connected edges across gaps but did not lead to a *perceived* edge, unless the *feature contour* process generated a featural difference across the edge.

Kellman, Guttman, and Wickens (2001) proposed that contour interpolation involves two distinct and sequential processes, an idea elaborated in subsequent work (e.g., Carrigan, 2019; Carrigan and Kellman, 2020; Kalar *et al.*, 2010; Kellman, Garrigan, and Shipley, 2005; Kellman *et al.*, 2007a, b; Kellman, Erlikhman, and Carrigan, 2016). An initial operation is the linking of visible contour fragments based on their geometric relations. This 'promiscuous interpolation' stage produces a map of both physically specified and linked potential edges. This is an intermediate representation, as it is not what is perceived. Interpolated contours that do appear in perception will be in this representation, but not all connections given here will appear in the final scene description. Which connections survive depends on a subsequent stage in which a variety of scene constraints operate; these constraints compete and combine to reconcile, among other things, issues of depth order of crossing interpolations, closure of object boundaries, and border ownership.

Figure 3.6 depicts our model. The graphics on the right show the application to a simple display, with the information extracted at each stage. In the last section, we described some of the evidence and arguments for such a model, with a specific focus on the existence of two distinct processing stages in contour interpolation, their sequential dependency, and the existence of an intermediate representation produced by the first and operated upon by the second. The focus on these two processes is not meant to imply the absence of others. For example, processes that detect edges and junctions are shown in the model diagram in Figure 3.6. We have also mentioned the surface spreading process that often complements contour interpolation. Proposals that additional processes may be involved in completion of volumetric representations (Tse, 1999), as well as generalizations of interpolation accounts to spatio-temporal interpolation (Palmer, Kellman, and Shipley, 2006), may well interact with processes described here, but we do not take up those questions here.

3.3 Intermediate representation and the modal or amodal appearance of interpolated contours

Although we describe new evidence here for these ideas, a number of phenomena and logical considerations from earlier work support this view. We illustrate the explanatory value of an intermediate representation in resolving conflicting views about whether modal completion and amodal completion are related (Albert, 2007; Kellman *et al.*, 2005; Kellman and Shipley, 1991) or separate processes (Anderson, 2007; Anderson, Singh, and Fleming, 2002).

The idea of an intermediate representation explains both the obvious similarities in the geometric constraints on modal and amodal completion, as well as differences in their appearance and sensitivity to scene constraints other than contour relatability. On this account,

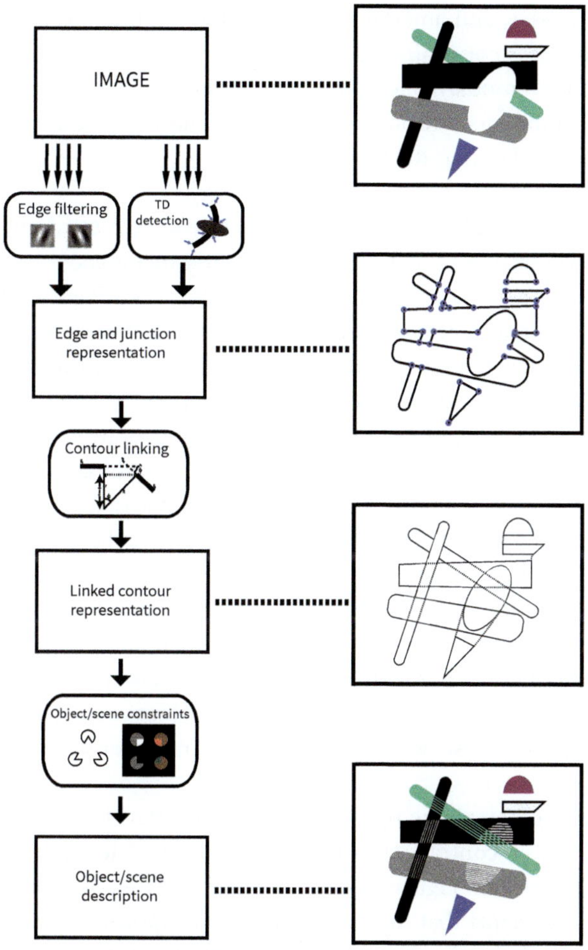

Figure 3.6 A model of contour interpolation in object formation. Left column: representations (rectangles with sharp corners) and processes (rectangles with round corners) that lead to perception of complete objects despite fragmentary information. Right column: an example image, with the information made explicit for each representation. *Edge and junctions*: dots show locations of tangent discontinuities (TDs or key points). *Linked contours*: all of the contour linkages resulting from relatable edges, including some (e.g., between the ellipse and the triangle) that will not survive into the final scene description. *Scene description*: perceptual representation of interpolated contours is determined by the interaction of a number of scene constraints (e.g., resolution of border ownership for crossing interpolations, closure, surface continuity). Occluded regions are depicted with a parallel line texture.

'modal completion' and 'amodal completion' name two visual appearances (depending on whether completed regions are situated behind other visible surfaces or not) rather than two distinct processes. When initial linkages of contours occur, those linkages are neither modal nor amodal; their ultimate appearance in perceptual representations and perceptual experience depends on a second stage. In that stage, scene constraints involving closure, assignment of border ownership, and resolution of relative depth where interpolations cross are jointly resolved, leading to modal appearance for completed contours and surfaces that appear in front of others, and amodal appearance for completions behind a nearer object or surface. Some of these constraints may differ for modal appearance vs description of an object passing behind (amodal); one example might be that the strength of the former may be enhanced when luminance relations conform to the ecological optics of transparency (Anderson, Singh, and Fleming, 2002).

3.3.1 Crossing interpolations, Petter's effect, and self-splitting objects

Figure 3.7 shows two examples of a phenomenon first investigated by Petter (1956), later termed *self-splitting objects* (Kellman and Shipley, 1991). Both Figure 3.7A and Figure 3.7B show a single continuous area of homogeneous colour and lightness, yet each display appears to split into two complete, bounded objects. Understanding why the two separate objects in each display exist in perception, despite not being given in the stimulus, can help illustrate what we mean by scene description constraints that must operate in a second stage. Contour linking, along with the physically specified edges given by luminance boundaries, produces two closed, overlapping regions in the intermediate representation. The visual system appears to obey constraints that: (1) closed contours define an object; and (2) the two objects formed by closed contours, which share overlapping areas in a 2D map, cannot in the world occupy the same space at the same time. As a result, two objects are perceived, and where their projections overlap, one appears in front of the other. The need for a depth order, from a stimulus that did not contain any depth differences, comes from these scene constraints operating on the results of initial contour linking. The object that appears nearest is given a modal appearance (it has illusory contours), whereas the other object appears behind (it is amodally completed).

In Figure 3.7A, you may notice that the 'rod' or the quadrilateral appear to switch their depth order over time. In contrast, the display in Figure 3.7B has a more stable, if complex, appearance. Neither object in Figure 3.7B is completely in front of the other. For the two 'golf clubs', most observers report that the head of each club appears in front of the shaft of the other. Petter (1956) discovered a simple rule that accounts for the appearance of displays like these. The rule is: where two interpolated edges cross, the interpolated edge that is shorter in length appears to lie in front of the other. In Figure 3.7B, where interpolations cross, the thicker part of each object appears in front of the thinner part.

If the perceptual outcome from Petter's rule is determined by a relation of the extents of two crossing interpolations (Singh, Hoffman, and Albert, 1999), it implies that the information about the locations and extents of interpolation precede and determine the depth

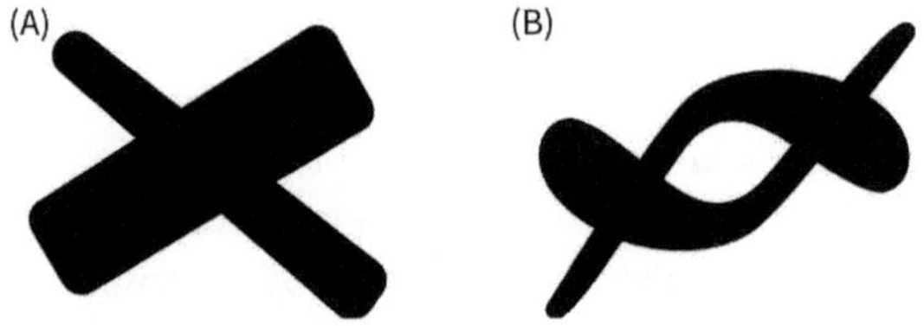

Figure 3.7 Self-splitting objects. The displays in (A) and (B) each consist of a single continuous, homogeneous black surface. Contour interpolation produces two perceived objects in each case. A scene constraint discovered by Petter (1956) determines the predominant appearance of depth order of the two objects in each display. (See text.)

order, and hence the amodal or modal appearance. In our theory, the extents and locations of interpolation are given in the contour-linking representation. Petter's rule operates in the subsequent scene description process to determine the final appearance of crossing interpolated contours. In other cases, scene description constraints may have the consequence that a single interpolated contour may appear as modal along part of its length and amodal along other parts, a fact that coheres naturally with our account but is incompatible with any view of modal and amodal completion as distinct perceptual interpolation processes (see Kellman, Yin, and Shipley, 1998; Kellman, Garrigan, Shipley, and Keane, 2007a, b).

3.3.2 Depth spreading and crossing interpolations in object formation

The importance of an intermediate representation can be seen in another fascinating phenomenon, shown in Figure 3.8, involving depth spreading. The two views comprise a stereoscopic display that can be free-fused by either crossing or diverging the eyes. In the discussion that follows, we assume the reader is using crossed fusion. (If diverging, the depth relationships observed will be the opposite of what is described.)

When the two images in Figure 3.8 are fused by crossing the eyes, the left edge of the central rectangular object appears as the nearest edge to the observer, and its right edge appears furthest away. Note that the views given to the two eyes differ only in the positions of the leftmost and rightmost edges of the central rectangular area. The perceived slant of the rectangle is a result of *depth spreading*; depth at the left and right edges, specified by binocular disparity, spreads along the horizontal edges (where there are no disparity signals) to produce perceptions of intermediate depth. The spreading of depth and the relationship of depth values to those of the vertical edges in the display determine whether completed parts of the rectangle's edges take on a modal or an amodal appearance. Depth spreading proceeds only along continuous edges (e.g., Häkkinen, Liinasuo, Kojo, and Nyman, 1998), but in this display, these edges are not continuous in the stimulus; across the gaps, they must be provided by interpolation. The perceived objects and spatial relations are a consequence of depth positions given by depth spreading, which, in turn, depends on contour linking. The appearance of interpolated edges and surfaces as either modal or amodal in the gaps emerges simply from the relative depths of crossing interpolations.

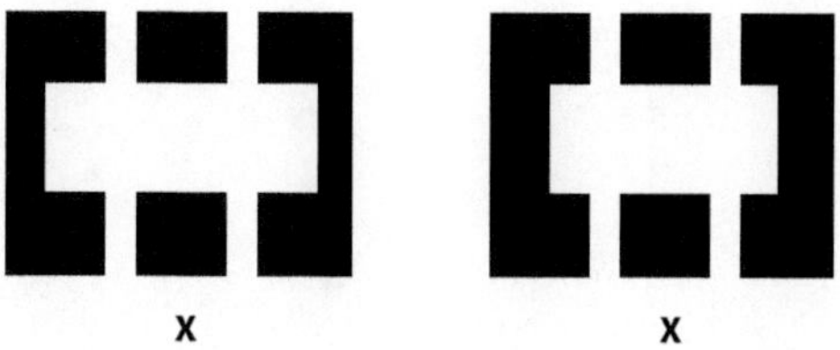

Figure 3.8 Object formation case in which interpolation necessarily precedes determination of a modal or an amodal appearance of contours. The display is a stereo pair; free fuse by crossing the eyes. (Cross eyes until the fixation 'X' becomes a single character.) At all of the crossing interpolations in this display, the modal or amodal appearance of completed contours depends on stereoscopic depth spreading of the slanted rectangle. Depth spreading occurs within boundaries established by interpolation. (See text.)
Adapted from Kellman (2003).

4 Recent research on a two-process theory of contour interpolation in object formation

The model we have presented fits with a great deal of evidence but would be strengthened by experimental investigations that further distinguish the two stages and show their properties. This has been a goal of our recent research, which we review briefly in this section.

4.1 New evidence for an intermediate representation: path detection

One consequence of the model is that it offers a coherent account of the many intriguing phenomena of *path detection*. Path detection has been the focus of many studies since it was introduced by Field, Hayes, and Hess (1993). Figure 3.9 illustrates a sample path detection stimulus. Figure 3.9A is a sequence of Gabor elements whose positions and orientations define a target 'path'. The observer's goal is to detect such paths embedded in noise, as in Figure 3.9B, typically in a two-interval forced choice paradigm.

Up to now, the relation to other phenomena has been vague. Path detection is sometimes described as *contour integration*, but the meaning of that phrase and its relation to other phenomena, such as contour interpolation or object formation, have been unclear.

Two interesting clues suggest both a deep relationship and also some crucial difference between path detection and contour interpolation. Regarding the former, research indicates that the geometric relations that determine path detection correspond to the geometry of relatability that governs contour interpolation (Field, Hayes, and Hess, 1993; Hess and Field, 1999; Kellman, Erlikhman. and Carrigan, 2016). A crucial difference is that no interpolated contours are perceived in path images. The elements of a path seem to form a perceptual group somehow, but observers do not perceive continuous connections between

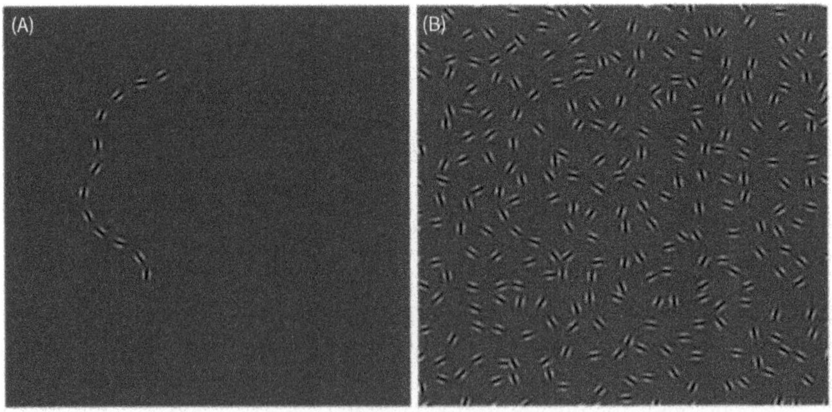

Figure 3.9 Path detection stimuli. (A) A 'path' composed of a sequence of Gabor elements. (B) The same path embedded in a noisy background of randomly arranged Gabor elements. Reprinted with permission from Field, Hayes & Hess, 1993, *Vision Research*, 33(2), 173–93.

the elements. This makes the perception of a 'path' different from the perception of illusory or occluded contours.

Kellman, Erlikhman, and Carrigan (2016) hypothesized that this close similarity and obvious perceptual difference could be explained by the two separable processes in contour interpolation. Detection of paths could be a direct consequence of linkages in the intermediate representation in contour interpolation. Scene constraints in the second stage, however, block the appearance of these contour connections in the final scene representation and perceptual experience.

What might these constraints be? Gabor elements appear as black and white against a grey background. If contours are linked in the first stage, it is not clear what bounded surfaces they enclose. The surface mismatch between the white/black contour fragments and grey gaps would not support perception of a bounded figure; moreover, the Gabor elements have multiple parallel-oriented areas that should interpolate, leading to potential conflicts of border ownership. These issues of scene description render the perceptual outcome indeterminate in terms of connected objects. Figure 3.10 suggests the validity of this analysis, using examples that remove the issues of conflicting surface information, in two different ways. The display in Figure 3.10A shows a path made of Gabor elements; no connecting edges are seen between elements, but both of the manipulations in Figure 3.10B and Figure 3.10C that remove conflicting surface information lead to perception of complete objects with continuous (amodally or modally completed) boundaries.

If this hypothesis is correct, then the 'pop-out' in path detection arises because the elements are linked in the intermediate representation.

To test this hypothesis, we studied path detection and illusory contour perception with Gabor elements and modified elements. We manipulated systematically the geometry of element relations and observed the effects on both path detection and illusory contour perception. Specifically, we used varying degrees of misalignment of collinear edges. Earlier, we

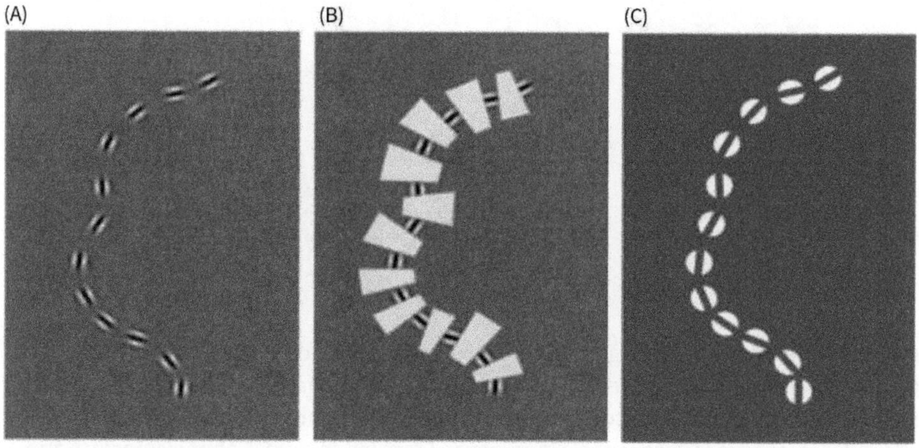

Figure 3.10 (A) Standard Gabor element path. (B) The path in (A) with occluders placed over all of the gaps between successive elements. A single partly occluded object (amodal completion) is seen. (C) Changing from Gabors to elements whose centres match the surround colour leads to perception of clear illusory contours (modal completion) connecting the elements.
From Carrigan, Erlikhman & Kellman, in preparation.

mentioned that the geometry of relatability in contour interpolation also defines the element relations that produce path detection. One factor we did not mention is that in contour interpolation, there is a small tolerance for misalignment of collinear edges within which interpolation still occurs. Strength of interpolation is a decreasing function of misalignment up to around 15–20 arcmin of visual angle, at which point interpolation is completely disrupted (Hilger and Kellman, 2005; Kellman and Shipley, 1991; Palmer, Kellman, and Shipley, 2006; Shipley and Kellman, 1992). To our knowledge, this constraint on contour relatability had not been tested in path detection. Figure 3.11A illustrates the effect of misalignment on contour interpolation.

Kellman, Erlikhman, and Carrigan (2016; Carrigan, Erlikhman, and Kellman, in preparation) used Gabor elements and three modifications, shown in Figure 3.11B. The set of four types of elements was created by using displays with and without step edges (abrupt dark to light transitions in the element centres) and by retaining the central regions of elements as white or replacing them with the surrounding grey colour.

These choices enabled us to test two ideas about paths made from Gabor elements: (a) the perceptual salience of a path occurs because the elements connect in the linked contour representation; and (b) no continuous contours are perceived in Gabor paths because these linkages fail at the scene description stage, specifically because of surface colour differences that inhibit perception of a continuous object. If these ideas are correct, we expected the following:

1. In the path detection paradigm, paths made from any of the four element types would show approximately the same performance, specifically the same decline in detection performance as a function of misalignment.

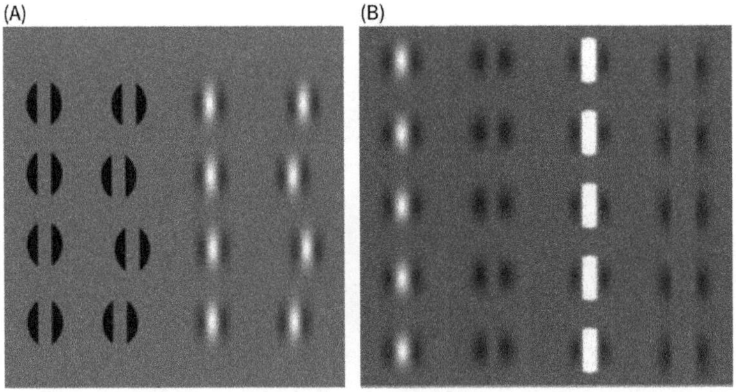

Figure 3.11 (A) Effects of misalignment of collinear elements for step edges and Gabor elements. If the figure size is at least 5 cm per side and viewed from any normal reading distance, the misalignment for columns 2 and 4 will substantially exceed 15 arcmin of visual angle. (B) Element types used in the experiments. From left to right: Gabor elements; Gabor elements modified by changing pixel values greater than the grey surround to match the surround; elements transformed so that pixel values greater than the grey surround have been replaced by a uniform white region; and Gabors with step edges and interiors matching the surround.
From Carrigan, Erlikhman & Kellman, in preparation.

2. Illusory contour perception would never occur for the elements with white centres against the grey surround.
3. For elements with grey centres, illusory contour perception would occur for collinear contours and for small amounts of misalignment. The function relating subjects' ratings of illusory contour magnitude to misalignment should mirror the decline of path detection with misalignment.

The last prediction derives from the model's prediction that contour linking is a *prerequisite* for both path detection and illusory contour perception. For elements that also satisfy scene constraints favoring object formation (i.e., a consistent surface colour within linked contours), illusory contours should be seen for aligned elements but should be disrupted at the same levels of misalignment that disrupt path detection. To test these predictions, we carried out two experiments. One was a path detection experiment (two-interval forced choice) that used the four kinds of elements and a range of misalignments. The other was a task that assessed participants' rated strength of illusory contours connecting elements, using paths made from the same element types and misalignments (but without a noisy background).

Figure 3.12A shows path detection performance as a function of misalignment for all four element types. All element types showed approximately the same overall pattern, falling to near chance performance (0.5) by about 15 arcmin of misalignment. Crucially, the relation of the centres of the elements to the surface made little or no difference in path detection. One particular comparison is highlighted by inserted pictures of the standard Gabor elements and step-edge elements with grey centres; these showed essentially the same function relating misalignment to path detection in the objective task.

Figure 3.12B plots mean observer ratings of subjective contour clarity in a magnitude estimation task (c.f., Dumais and Bradley, 1976; Shipley and Kellman, 1992a, b). Here, there is a clear bifurcation in the results. Illusory contours were robustly perceived at zero or low levels of misalignment, but only for the elements with grey centres. Illusory contour ratings for these element types declined along a function of the same shape as in the path detection results, asymptoting at about 15 arcmin of misalignment. In contrast, the two element types with white centres (Gabors and step-edge Gabors) showed low illusory contour ratings across *all* misalignments.

These results comprise a signature of the separable stages in object formation that we have theorized. First, none of the properties of the path elements, other than their geometric relations, substantially affected path detection performance in this experiment, indicating that path detection depends on geometric relations of oriented fragments, but not on the other properties varied. Second, path element types *did* make a crucial difference in illusory contour perception: only paths consistent with a possible continuous surface connection between elements produced clear illusory contours. This result shows that scene characteristics that do not matter for initial contour linking are crucial for perceptual outcomes. Finally, illusory contour perception, for the elements capable of supporting it, drops off with misalignment along the same function that relates path detection to misalignment. This shows that contour perception in the final scene description occurs *only* for contours that are connected in the prior linked contour representation. The results are consistent with the sequential dependency we have described. Contour relatability, but not coherent scene properties, allows path detection. Successful path formation is necessary, but not sufficient, for perceived

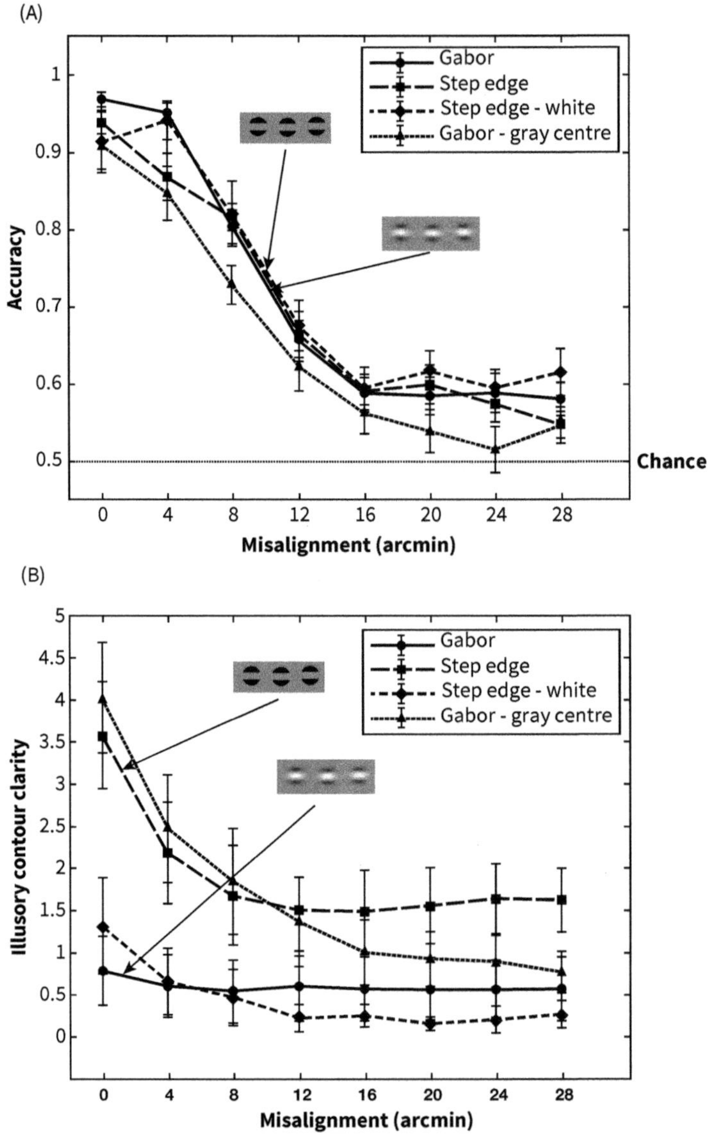

Figure 3.12 Path detection and illusory contour perception as a function of element type and misalignment. (A) Accuracy of path detection performance. (B) Ratings of illusory contour clarity. Ratings were given on a 0–5 scale, with 0 indicating no contours perceived between elements and 5 indicating the strongest possible contours. Inserts showing paths are shown for Gabor elements and modified Gabors whose centres matched the surround colour. These two element types showed nearly identical path detection performance, but very different illusory contour ratings.
Results from Kellman, Erlikhman, & Carrigan 2016; Carrigan, Erlikhman & Kellman, in preparation.

contours. Only those contours that are linked in the initial stage *and* that satisfy certain scene constraints produce visible interpolated contours.

4.2 Understanding scene description after contour linking

A limitation of previous accounts of separable stages in object formation has been a lack of work indicating how multiple scene constraints interact in the scene description stage. In recent work, we have begun to investigate these issues. A variety of cues have been investigated that make perceptual impressions of interpolated contours stronger or weaker. Their interactions are not well understood but seem to differ from the fast, automatic, and modular operation of contour linking based on relatability. Scene description processes may be probabilistic, subject to mutual constraint satisfaction, and may involve simplicity, probability, and maximization of coherence, all ideas that have figured strongly in classic and recent theoretical views of perception.

4.3 Probing scene description constraints: the character recognition paradigm

How might we probe scene description processes separately from contour linking? The problem is difficult. As we have recently found with path detection, other paradigms developed to measure contour interpolation are sensitive to relatable edges, even when the appearance of contour connections in the final scene description is weak or absent (e.g., Gold, Murray, Bennett, and Sekuler, 2000; Guttman and Kellman, 2001). We believe this occurs because contour linkages in the intermediate representation, despite not being the final outputs of perception, influence performance on various tasks.

It would be valuable to have an assay, using an objective performance method, that measures the strength of contours in the final scene description. We recently developed and tested such a method (Carrigan, 2019).

The logic of these experiments is based on the idea that *object recognition* depends most heavily on the final scene representation. Object formation processes serve to identify and describe the bounded objects in scenes. These should be those entities that not only participate in contour linking, but also survive through a subsequent stage of description of coherent, bounded objects and their positions. Consider the triangle and ellipse in the lower right panel of the model schematic in Figure 3.6. Although there are actually tangent discontinuities and relatable edges connecting these objects, these connections are deleted in the final representation. Thus, if seeing a perceptual unit connecting these two objects were important for object recognition, we would be poor at it.

These considerations led us to revisit a classic demonstration by Bregman (1990) and subsequent experimental work by Brown and Koch (1993). Figure 3.13 shows a version of the classic 'Bregman Bs'. Both Figure 3.13A and Figure 3.13B have similar fragments of the letter 'B' in the same arrangements; however, the black occluding object in Figure 3.13B clearly facilitates recognition of the letters, relative to A.

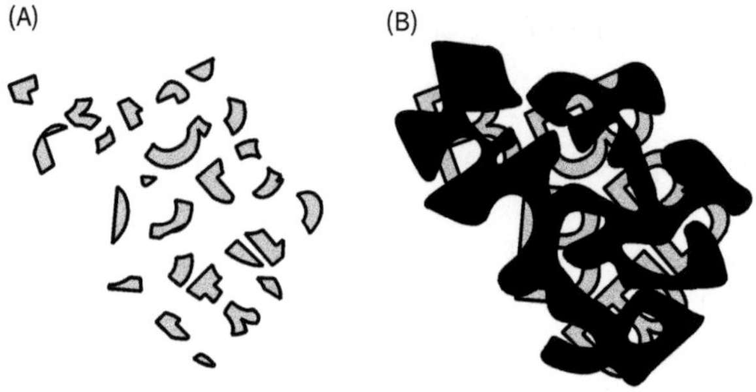

Figure 3.13 Effect of an occluding surface vs outlined parts in completion. The 'Bs' are readily recognized in (B), but not in (A). (See text.)
From Fuchser & Kellman, in preparation. After Bregman, 1990.

Brown and Koch (1993) investigated these effects experimentally. Their results suggested that the addition of completed outlines around the fragments, rather than the presence or absence of an occluder, led to slower detection times for alphanumeric characters.

We designed a related task that required recognition of fragmented alphanumeric characters. We manipulated several scene constraints relevant to the perception of an interpolated contour's strength. We used displays in which only one complete character could be formed by interpolation; the rest of the field contained 'noise' fragments. This design prevented observers from giving a correct response based only on some recognizable piece of a character.

Figure 3.14A illustrates a type of display used as a reference stimulus. The display contained filled fragments identical in surface colour, with step edges, and the space between relatable fragments occluded. These displays were assumed to be optimal in terms of the scene constraints we studied. Note how the capital 'B' hidden in the upper right of Figure 3.14A is easily recognized. This type of display was compared to the other conditions. In Figure 3.14B, the same kinds of images were used, but the occluding surfaces in Figure 3.14A were rotated and set behind the fragments. This disruption of ordinary occlusion relationships would be expected to weaken perceived connections between relatable fragments.

In Figure 3.14C, edges of half of the fragments were blurred. Assignment as a connected object would require connecting blurred edges with step edges (abrupt luminance changes). Blurred edges are detectable primarily in relatively low spatial frequency channels of the visual system. There are some indications that interpolation may proceed separately in different channels. If so, it can occur here in low spatial frequency channels (shared by blurred and step edges), but not in high spatial frequency channels.

In Figure 3.14D, fragments' luminance contrast polarity relationships with the background are reversed for pairs of inducing edges. Figure 3.14E and Figure 3.14F present filled and outline fragments that were used to measure object formation in the absence of occlusion.

A summary of some of our results is presented in Figure 3.15. Functions labelled A through to F correspond to the display types illustrated in Figure 3.14A–F. Recognition performance was best for the reference display and worse for the other conditions, which disrupted one or more scene constraints. All of the manipulations reduced recognition accuracy, with some manipulations having larger effects than others. Though not presented here, subsequent

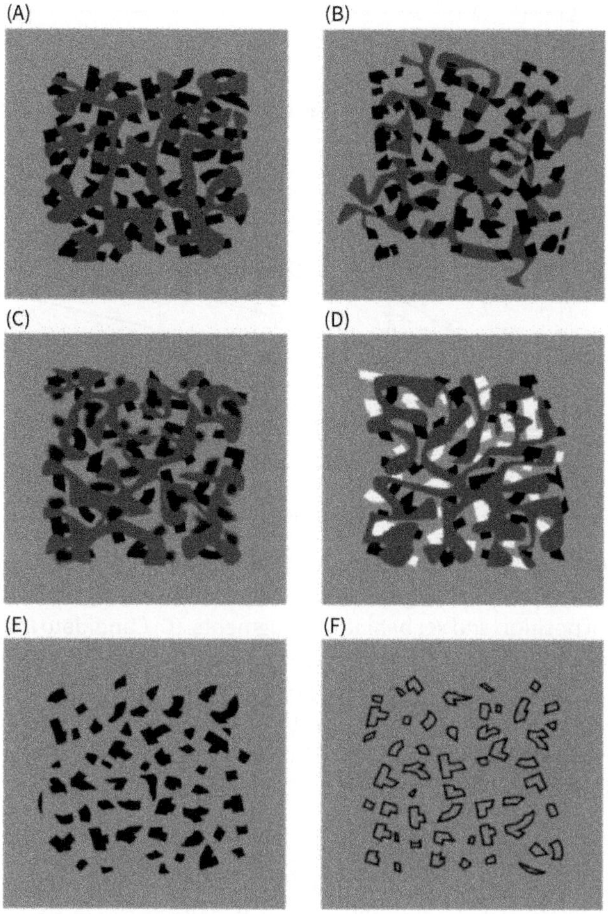

Figure 3.14 Sample of stimuli from Carrigan, 2019. (See text.) For panels A through F, the target characters are B, 5, K, P, B, and R, respectively.

experiments that tested cue combinations showed that information provided by these cues is combined in a simple additive fashion at this stage. This means that the decrease in perceptual strength when two cues were combined, in most cases, matched the sum of the decreases produced by the individual cues when presented alone (Carrigan, 2019). These results indicate relative independence of the scene description constraints tested and also that they contribute to the same perceptual outcome—the strength of object formation, here assessed by an objective measure of performance.

5 Conclusion

We have focused in this chapter on contour interpolation in the formation of objects from fragmentary information. The arguments and evidence presented here support a clear theoretical framework: contour interpolation is not a one-step process from stimulus input to perceived objects, but proceeds in multiple stages and produces intermediate outputs prior to the output of a final scene representation. In an early stage, pairs of edges terminating in

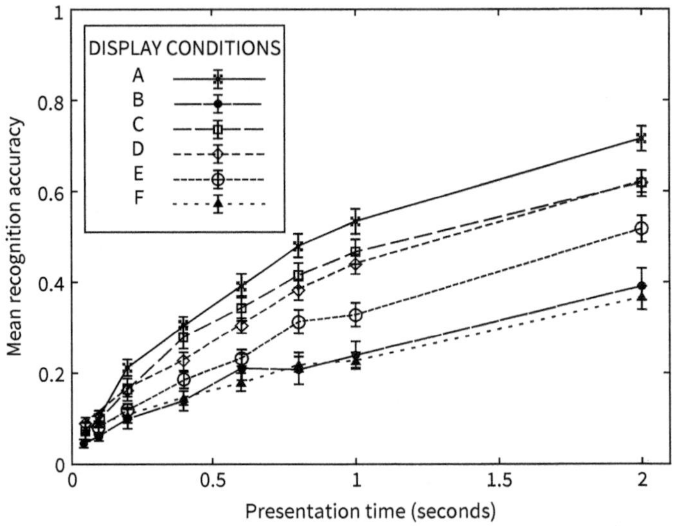

Figure 3.15 Mean recognition accuracy as a function of presentation time in character recognition experiments. (A) The fragments were filled and normally occluded. (B) Occluders were rotated out of position and set behind the fragments. (C) Candidate fragments for interpolation were always mismatched in spatial frequency composition. (D) Candidate fragments for interpolation were always of opposite luminance contrast polarity, relative to the surround. (E) Identical to (A), except that no occluder was present. (F) Identical to (E), except that fragments were outlined rather than filled. (See text.)
From Carrigan, 2019.

junctions are identified, and the geometric relations described as contour relatability determine whether an interpolated edge will connect the edge pairs in an intermediate output. At this stage, connections are produced *promiscuously* across all pairs of relatable edge fragments.

In a subsequent scene description stage or stages, information from a variety of scene constraints jointly determines, in largely additive fashion, the appearance and strength of objects formed. The sequential dependency of stages is shown in that only linked contours from the earlier processing are candidates for perceived contour connections in the final output (percept); variables that influence scene description have little or no effect on contour interpolation unless contours were linked earlier. As in other research (Guttman and Kellman, 2001), the work on path detection and illusory contour perception described here showed that for relatable edges, element characteristics relating to possible surface connections were important for illusory contour perception, but these had no ability to evoke illusory contour perception in the absence of relatable edges. Similarly, the scene constraint variables described in the character recognition work would have little effect on character recognition in the absence of relatable edges.

Why does object formation work in this way? This arrangement appears to make sense ecologically and computationally. Whereas contour linking is over-inclusive in some ways ('promiscuous'), it vastly narrows the field of which visible areas can be connected, based on a geometry that will tend to link areas that project from connected objects in the world, as suggested by studies of scene statistics (Geisler and Perry, 2009). That initial contour linking occurs without incorporating much other information about the likely objects in the scene makes it efficient and automatic. This constrains the domain considerably for application of various scene constraints. If contour interpolation were unconstrained geometrically, interpolation would require consideration of a great many more potential connections among spatially separated areas projected to the eyes.

A major advance in recent work is to characterize more substantively the scene description process. The information taken into consideration includes, at least, cues to border ownership, luminance contrast polarity relative to the background, and spatial frequency differences among inducing edges. These cues determine the perceptual strength of interpolated connections, resolve border ownership and certain issues of depth ordering, as in crossing interpolations, and determine modal (illusory) or amodal (occluded) appearance.

In closing, we mention several 'big picture' themes of the work described here. Visual perception of objects from fragmentary information involves intermediate representations, specifically a contour-linking representation based primarily on geometric relations that is formed prior to resolution of various scene constraints. In our view, this theoretical approach has the potential to resolve most controversies in the literature on contour interpolation. Further, as our recent work has shown, these ideas reveal a previously unsuspected connection to an important body of work on path detection: path detection is accessing the contour-linking representation, even when path elements are not seen as being connected by continuous illusory or occluded contours. Finally, our approach here involves aspects and strengths of what have been long-standing, competing ideas in perceptual theories. The first stage (contour-linking process) embodies virtually all of the characteristics of a modular process (Fodor, 1983; Keane, 2018)—it is fast, automatic, and informationally encapsulated, and it matures in an innately paced manner (Keane, 2018; Kellman and Arterberry, 1998). At the second stage of our model, a wider variety of information sources exert their effects jointly. Some of these might be cues described as 'top-down' or may reflect probabilistic information gained from experience with objects and natural scenes (see also the interesting suggestion of Nanay (2010) about the relation of amodal completion and imagery). Our updated model of object formation suggests that classic bottom-up and top-down theories of perception could 'meet in the middle', with each playing an important explanatory role.

This multi-stage model also coheres with findings from neuropsychological research. Work with visual agnosics reveals a double dissociation between visual agnosias in which basic contour perception is impaired (visual form agnosia) and cases in which more global segmentation and integration are selectively impaired (integrative agnosia) (Giersch, Humphreys, Boucart, and Kovacs, 2000; Riddoch, Humphreys, Akhtar, and Allen, 2008), and other work suggests that similar integration deficits can be induced by administration of lorazepam (Giersch, Humphreys, Barthaud, and Landmann, 2006). Finally, work with schizophrenic patients supports a distinction between basic contour linking and a later shape integration stage (Keane, Joseph, and Silverstein, 2014).

Acknowledgements

We gratefully acknowledge expert assistance from Rachel Older and support from NIH Grant RO1 CA236791 to P. J. K. Portions of this work have been presented at annual meetings of the Vision Sciences Society (VSS) and the Configural Processing Consortium (CPC).

References

Albert, M. K. (2007). Mechanisms of modal and amodal interpolation. *Psychological Review*, 114, 455–69.

Anderson, B. L. (2007). Filling-in models of completion: rejoinder to Kellman, Garrigan, Shipley, and Keane (2007) and Albert (2007). *Psychological Review*, 114(2), 509–25.

Anderson, B. L., Singh, M., and Fleming, R. W. (2002). The interpolation of object and surface structure. *Cognitive Psychology*, 44(2), 148–90.

Bregman, A. S. (1990). *Auditory Scene Analysis*. Cambridge, MA: MIT Press.

Brown, J. M. and Koch, C. (1993). Influences of closure, occlusion, and size on the perception of fragmented pictures. *Perception and Psychophysics*, 53(4), 436–42.

Carman, G. J. and Welch, L. (1992). Three-dimensional illusory contours and surfaces. *Nature*, 360, 585–7.

Carrigan, S. B. (2019). *Promiscuity and Pruning: Investigations of a Two-Stage Theory of Contour Interpolation*. (Doctoral dissertation, UCLA). https://escholarship.org/uc/item/41j2x52p

Carrigan, S., Erlikhman, G., and Kellman, P. J. (in preparation). Path integration and illusory contours: Evidence for an intermediate representation in visual contour interpolation.

Carrigan, S. B. and Kellman, P. (2020). Combined effects of multiple scene cues on the perceptual strengths of promiscuously interpolated contours [Abstract]. *Journal of Vision*, 20(11), 1276.

Dumais, S. T. and Bradley, D. R. (1976). The effects of illumination level and retinal size on the apparent strength of subjective contours. *Perception and Psychophysics*, 19(4), 339–45.

Field, D. J., Hayes, A., and Hess, R. F. (1993). Contour integration by the human visual system: evidence for a local 'association field'. *Vision Research*, 33(2), 173–93.

Fodor, J. A. (1983). *The Modularity of Mind*. Boston, MA: MIT Press.Geisler, W. S. and Perry, J. S. (2009). Contour statistics in natural images: grouping across occlusions. *Visual Neuroscience*, 26(1): 109–21.

Gerbino, W. (2020). Amodal completion revisited. *i-Perception*, 11(4), 1–26.

Giersch, A., Humphreys, G. W., Barthaud, J. C., and Landmann, C. (2006). A two-stage account of computing and binding occluded and visible contours: evidence from visual agnosia and effects of lorazepam. *Cognitive neuropsychology*, 23(2), 261–77.

Giersch, A., Humphreys, G. W., Boucart, M., and Kovács, I. (2000). The computation of occluded contours in visual agnosia: evidence for early computation prior to shape binding and figure-ground coding. *Cognitive neuropsychology*, 17(8), 731–59.

Gillam, B. (1987). Perceptual grouping and subjective contours. In *The Perception of Illusory Contours*, edited by S. Petry and G. E. Meyer, 268–73. Springer-Verlag.

Gold, J. M., Murray, R. F., Bennett, P. J., and Sekuler, A. B. (2000). Deriving behavioural receptive fields for visually completed contours. *Current Biology*, 10(11), 663–6.

Grossberg, S. and Mingolla, E. (1985). Neural dynamics of form perception: Boundary completion, illusory figures, and neon color spreading. *Psychological Review*, 92(2), 173–211.

Guttman, S. E. and Kellman, P. J. (2001). Contour interpolation: necessary but not sufficient for the perception of interpolated contours [Abstract]. *Journal of Vision*, 1(3), 384.

Guttman, S. E. and Kellman, P. J. (2004). Contour interpolation revealed by a dot localization paradigm. *Vision Research*, 44(15), 1799–815.

Häkkinen, J., Liinasuo, M., Kojo, I., and Nyman, G. (1998). Three-dimensionally slanted illusory contours capture stereopsis. *Vision Research*, 38(20), 3109–15.

Heitge, F., von der Heydt, R., Peterhans, E., Rosenthaler, L., and Kübler, O. (1998). Simulation of neural contour mechanisms: representing anomalous contours. *Image and Vision Computing*, 16(67), 407–21.

Hess, R. and Field, D. (1999). Integration of contours: new insights. *Trends in Cognitive Sciences*, 3(12), 480–6.

Hilger, J. D. and Kellman, P. J. (2005). Tolerance for misalignment in contour interpolation: retinal or relational? [Abstract]. *Journal of Vision*, 5(8), 571.

Kalar, D. J., Garrigan, P., Wickens, T. D., Hilger, J. D., and Kellman, P. J. (2010). A unified model of illusory and occluded contour interpolation. *Vision Research*, 50, 284–99.

Kanizsa, G. (1976). Subjective contours. *Scientific American*, 234(4), 48–52.

Keane, B. P. (2018). Contour interpolation: a case study in modularity of mind. *Cognition*, 174, 1–18.

Keane, B. P., Joseph, J., and Silverstein, S. M. (2014). Late, not early, stages of Kanizsa shape perception are compromised in schizophrenia. *Neuropsychologia*, 56, 302–11.

Keane, B. P., Lu, H., and Kellman, P. J. (2007). Classification images reveal spatiotemporal contour interpolation. *Vision Research*, 47(28), 3460–75.

Kellman, P. J. (2003). Perceptual processes that create objects from fragments. *Proceedings of the 2003 IEEE International Joint Conference on Neural Networks*, 1, 2–7.

Kellman, P., Erlikhman, G., and Carrigan, S. (2016). Is there a common mechanism for path integration and illusory contour formation? [Abstract]. *Journal of Vision*, 16(12), 311.

Kellman, P. J., Garrigan, P., and Shipley, T. F. (2005). Object interpolation in three dimensions. *Psychological Review*, 112, 586–609.

Kellman, P. J., Garrigan, P., Shipley, T. F., and Keane, B. P. (2007a). Interpolation processes in object perception: reply to Anderson (2007). *Psychological Review*, 114(2), 488–502.

Kellman, P. J., Garrigan, P., Shipley, T. F., and Keane, B. P. (2007b). Postscript: identity and constraints in models of object formation. *Psychological Review*, 114(2), 502–8.

Kellman, P. J., Garrigan, P., Shipley, T. F., Yin, C., and Machado, L. (2005). 3-d interpolation in object perception: evidence from an objective performance paradigm. *JEP: Human Perception and Performance*, 31(3), 558–83.

Kellman, P. J., Guttman, S. E., and Wickens, T. D. (2001). Geometric and neural models of object perception. In *From Fragments to Objects: Segmentation and Grouping in Vision*, edited by T. F. Shipley and P. J. Kellman, 183–245. NY: Elsevier Science.

Kellman, P. J. and Loukides, M. G. (1987). An object perception approach to static and kinetic subjective contours. In *The Perception of Illusory Contours*, edited by S. Petry and G. E. Meyer, 151–64. NY: Springer-Verlag.

Kellman, P. J. and Shipley, T. F. (1991). A theory of visual interpolation in object perception. *Cognitive Psychology*, 23, 141–221.

Kellman, P. J., Yin, C., and Shipley, T. F. (1998). A common mechanism for illusory and occluded object completion. *Journal of Experimental Psychology: Human Perception and Performance*, 24(3), 859–69.

Koffka, K. (1935). *Principles of Gestalt Psychology*. NY: Harcourt Brace.

Marr, D. (1982). *Vision: A Computational Investigation into the Human Representation and Processing of Visual Information*. San Francisco: W. H. Freeman and Company.

Michotte, A., Thines, G., and Crabbe, G. (1964). *Les complements amodaux des structures perceptives. Studia Psychologica*. Louvain: Publications Universitaires de Louvain.

Murray, M. M., Foxe, D. M., Javitt, D. C., and Foxe, J. J. (2004). Setting boundaries: brain dynamics of modal and amodal illusory shape completion in humans. *Journal of Neuroscience*, 31, 6898–903.

Nanay, B. (2010). Perception and imagination: amodal perception as mental imagery. *Philosophical Studies*, 150(2), 239–54.

Palmer, E. M., Kellman, P. J., and Shipley, T. F. (2006). A theory of dynamic occluded and illusory object perception. *Journal of Experimental Psychology: General*, 135(4), 513–41.

Petter, G. (1956). Nuove richerche sperimentali sulla totalizzazione percettiva. *Revista di Psicologia*, 50, 213–27.

Riddoch, M. J., Humphreys, G. W., Akhtar, N., Allen, H., Bracewell, R. M., and Schofield, A. J. (2008). A tale of two agnosias: distinctions between form and integrative agnosia. *Cognitive Neuropsychology*, 25(1), 56–92.

Ringach, D. L. and Shapley, R. (1996). Spatial and temporal properties of illusory contours and amodal boundary completion. *Vision Research*, 36(19), 3037–50.

Rubin, N. (2001). The role of junctions in surface completion and contour matching. *Perception*, 30(3), 339–66.

Shipley, T. F. and Kellman, P. J. (1990). The role of discontinuities in the perception of subjective contours. *Perception and Psychophysics*, 48(3), 259–70.

Shipley, T. F. and Kellman, P. J. (1992). Perception of partly occluded objects and illusory figures: evidence for an identity hypothesis. *Journal of Experimental Psychology: Human Perception and Performance*, 18(1), 106–20.

Singh, M. (2004). Modal and amodal completion generate different shapes. *Psychological Science*, 15(7), 454–9.

Singh, M. and Hoffman, D. D. (1999). Completing visual contours: the relationship between relatability and minimizing inflections. *Perception and Psychophysics*, 61(5), 943–51.

Singh, M., Hoffman, D. D., and Albert, M. K. (1999). Contour completion and relative depth: Petter's Rule and support ratio. *Psychological Science*, 10(5), 423–8.

Spehar, B. and Halim, V. A. (2016). Created unequal: temporal dynamics of modal and amodal boundary interpolation. *Vision Research*, 126, 97–108.

Su, Y. R., He, Z. J., and Ooi, T. L. (2010). Boundary contour-based surface integration affected by color. *Vision Research*, 50, 1833–44.

Tse, P. U. (1999). Volume completion. *Cognitive Psychology*, 39, 37–68.

Yin, C., Kellman, P. J., and Shipley, T. F. (2000). Surface integration influences depth discrimination, *Vision Research*, 40(15), 1969–78.

Zhou, J., Tjan, B. S., Zhou, Y., and Liu, Z. (2008). Better discrimination for illusory than for occluded perceptual completions. *Journal of Vision*, 8(7), 2.

4

Sensory binding without sensory individuals

Jake Quilty-Dunn

> *What was worked out at Ford was the practice of moving the work from one worker to another until it became a complete unit, then arranging the flow of these units at the right time and the right place to a moving final assembly line from which came a finished product.*
>
> Charles E. Sorensen (1956), *My Forty Years with Ford*, p. 116

1 Introduction

One might think of visual processing as operating like an assembly line: subsystems work out particular subcomponents that are combined to generate a final product, which is then shipped out to consumers outside the visual system such as cognition, action, and memory. All the information that is properly *perceived* is encoded in a single ultimate percept. One point in favour of the assembly line idea is that, when we attend to our perceptual experience, we do not seem to find distinct, redundant perceptual representations. Instead, many would argue, we find one unified perceptual experience (Bayne, 2010), perhaps even transcending perceptual modalities (O'Callaghan, 2014; cf. Bayne and Spence, 2015).

Tempting though the assembly line idea may be, when construed as a claim about the architecture of perceptual processing, it is very likely false. Instead, various stages of visual processing output various products that are individually available to be consumed by downstream mental processes. For example, detailed early visual representations are delivered to iconic memory, while representations of a limited number of objects are delivered to visual working memory (cf. Gross and Flombaum, 2017; Luck and Vogel, 1997; Quilty-Dunn, 2020b; Sperling, 1960). Likewise, while we can only explicitly report on a handful of objects at a time with any precision, the visual system is able to compute statistical summaries that range over dozens of objects (Dakin, 2001; Whitney and Yamanashi Leib, 2018). We frequently categorize objects based on the visual perception of their contours, but we may plan to interact with them physically, based on the visual perception of their medial axes (Firestone and Scholl, 2014). Instead of a single assembly line, therefore, we might invoke the metaphor of an entire factory generating a multiplicity of products that may or may not share features with one another and are shipped out to consumers separately (cp. Burnston and Cohen, 2013; Henke, 2021).

Jake Quilty-Dunn, *Sensory binding without sensory individuals* In: *Sensory Individuals.* Edited by: Aleksandra Mroczko-Wąsowicz and Rick Grush, Oxford University Press. © Oxford University Press 2023. DOI: 10.1093/oso/9780198866305.003.0005

The heterogeneity of vision is important to keep in mind when considering general questions like 'What is the format of perceptual representation?' or 'How does the visual system bind features together?'. The visual system may represent the world in multiple formats—some language-like and shared with paradigmatic instances of cognition, and others iconic and proprietarily visual (Quilty-Dunn, 2020c). Likewise, it may have multiple distinct ways of binding features together and understanding one sort of feature binding may fail to shed light on others (Lande, 2021). Even the same kinds of features—for example, shape and colour—might be bound in different ways by different visual subsystems. We might thus adopt a default pluralist attitude towards strategies for feature binding, which will be the topic of the speculative ideas developed in this chapter.

The basic problem of feature binding is the problem of how combinations of features are represented, such that combinations like *red square* and *blue circle* are distinguished from combinations consisting of the same primitive features, like *red circle* and *blue square*. Call the distinguishability of distinct feature combinations containing the same primitives the *core phenomenon* of feature binding.

When we dive deeper into the meaning of terms like 'feature' and 'combine', we find many distinct questions that fall under the heading of 'The binding problem' (Treisman, 1996). For example, the *computational* binding problem concerns what algorithms are implemented to construct representations of feature combinations. The *neural* binding problem concerns how feature binding is neurally instantiated. We can also ask, concerning a particular representation, what structural properties enable it to encode a feature combination.[1] This *representational* binding problem itself breaks down into two subproblems. First, the *semantic* binding problem concerns how the contents of complex representations unify simpler elements.[2] Second, the *syntactic* binding problem concerns how a feature combination is represented in one representational vehicle.

My topic here is the syntactic binding problem. The question I'll focus on is 'What kinds of structures underlie visual representations of feature combinations?'. The aim of this brief chapter is modest and programmatic. Rather than arguing that existing approaches to feature binding are completely wrong, my goal will be merely to sketch and provide preliminary motivation for a pluralist approach.

There is very good reason to think an important class of feature binding involves attributing features to objects (Cohen, 2004; Kahneman *et al.*, 1992; Matthen, 2004; Pylyshyn, 2003). There may also be reason to think of some episodes of feature binding as accomplished through attribution to locations (Clark, 2000, 2004; Treisman, 1996). I will suggest that there may be a distinctly *iconic* form of feature binding that need not be characterized in terms of attribution at all, at least syntactically (i.e., the way in which binding is reflected in the structure of the representational vehicle). I will argue that the holistic character of iconic representations is best understood not in terms of attribution, but rather in terms of

[1] O'Callaghan (2014, p. 75) distinguishes 'feature binding awareness', or the conscious experience of an object as having multiple features, from 'feature binding processes', which are processes that take representations of individual features and combine them, yielding a bound representational output. Thus, the computational binding problem may be solved by a 'feature binding process' in O'Callaghan's sense, but characterizing the structure of a bound representation does not presuppose this.

[2] The semantic binding problem only arises if representational contents are structured, as on Fregean and Russellian views of content. The problem of the 'unity of the proposition' might be seen as a form of the semantic binding problem (Gaskin, 2008; Ostertag, 2019). One might argue that the problem dissolves if contents are unstructured (Stalnaker, 1984).

coordination—that is, icons plot vectors in multidimensional property spaces rather than selecting individuals and attributing properties to them. Binding via coordination is not intended to replace binding via attribution. Instead, I suggest that both forms of binding are generated by different elements of the diverse collection of processes that constitutes the human visual system.

2 Varieties of binding

2.1 Binding as attribution

One common way of understanding binding is by appeal to property attribution. In binding through attribution, one picks out an individual, either *de re* or through quantification, and attributes multiple properties to it. A demonstrative thought exhibiting predicate conjunction like THAT OBJECT IS RED AND SQUARE is a paradigmatic example. The properties *red* and *square* are bound because they are attributed to the same individual: the referent of THAT OBJECT. Distinct constituents of the representation correspond to the individual and to its various features, and the way in which those constituents are organized instantiates property attribution, thereby binding the represented features together. Call this idea the *attribution model* of binding.

The attribution model provides an intuitive, satisfying answer to the syntactic binding problem. At its core is the idea that distinguishing feature combinations requires figuring out what the *things* are that have the relevant properties—that is, for sensory binding, there must be a class of sensory individuals that function as the bearers of combinations of properties. There is controversy about what sensory individuals are: are they something like locations (Clark, 2000, 2004), are they material objects (Cohen, 2004; Matthen, 2004), can they vary across modalities (Green, 2019; O'Callaghan, 2008), do they fail to correspond to a particular ontological class (Green, 2018), are they only represented after the deployment of attention (Treisman and Gelade, 1980), etc.?

Attribution-based approaches to feature binding are so prominent that it is common for theorists to explicate the very idea of binding in terms of attribution. Treisman (1996, p. 171) describes feature binding as an operation in which 'different properties (e.g., shape, colour, and motion) must be bound to the objects that they characterize'. Clark explains that the binding problem crops up under many different names, including the 'many-property problem' (Jackson, 1977), and understands the underlying problem as a matter of 'how to represent one thing as having more than one feature' (Clark, 2004, p. 447; cp. Clark, 2000, p. 26ff).

Despite the ubiquity of the attribution model, we ought not to characterize binding in general as simply an instance of property attribution. An immediate problem for the generality of the attribution model is that binding seems closely tied to *compositionality*, which outstrips property attribution. For example, a complex concept like RED SQUARE represents *red* and *square* together, without attributing them to an individual. In this case, something like a Merge operation is deployed to create a complex representation out of simpler constituents through set formation (Chomsky, 1995, p. 223; cp. Camp, 2018). Complex concepts can plausibly be stored in, and retrieved from, memory without figuring in a propositional structure and can therefore bind features without property attribution.

Binding through Merge thus need not involve picking out an individual and attributing properties to it.

I suggest we interpret the attribution model as providing an *explanation* of binding rather than an initial characterization of the phenomenon (cp. Cohen, 2004, p. 472). The syntactic binding problem is how visual representations package information in such a way that feature combinations are distinguished from non-coextensive combinations that incorporate the same primitive features. Attribution of multiple features to the same individual is one way that this might be accomplished. Merge is another.

The fact that the syntactic binding problem is characterizable without appeal to attribution, together with the fact that Merge provides an example of binding without attribution, suggests that we ought to be open to the possibility of non-attributive binding in the visual system. I do not believe that, as a matter of fact, Merge provides the best model for thinking about non-attributive binding in vision. But it encourages exploration of other ways of thinking about binding beyond the attribution model.

2.2 Holistic icons

Merge is an operation that takes two separate syntactic items, A and B, and transforms them into a new, complex syntactic object with A and B as constituents (Chomsky, 1995, p. 223). The concepts RED and SQUARE are both constituents of RED SQUARE. There are two key components to Merge as I'll understand it: (1) the constituents are bound together into one representation, but (2) they are *discrete*—that is, even once bound into a complex, they remain separate vehicles.

The ability to form complexes out of constituents that remain discrete even once composed is characteristic of *discursive*, or language-like, representational formats. For example, the propositional thought THAT OBJECT IS A RED SQUARE binds RED and SQUARE through predication, but these constituents remain discrete.[3] Thus, a purely formal logical rule could allow one to deduce THAT OBJECT IS A SQUARE through predicate deletion (Quilty-Dunn and Mandelbaum, 2020).

This aspect of discursive formats furnishes concrete empirical predictions. For example, a complex like RED SQUARE that is held in working memory should be able to lose each feature independently, such that performance might decay for one feature while remaining robust for the other. Green and Quilty-Dunn (2021) cite evidence that visual object representations exhibit just this kind of 'independent forgetting' of distinct feature dimensions like colour and orientation (Fougnie and Alvarez, 2011; see also Markov *et al.*, 2019). Thus, Green and Quilty-Dunn argue that visual object representations, or 'object files', have a discursive representational format.[4]

[3] One might think of attribution, at least of the kind that is instantiated in propositional thoughts with a predicate–argument structure, as one kind of Merge. Chomsky allows for distinct types formed through Merge in order to accommodate syntactic phenomena like headedness, and a predicate–argument structure may be one such type. I do not think anything in the text hangs on this issue. What matters is that not all cases of Merge are cases of predication or attribution, such that Merge can accomplish binding without attribution. For more on the predication/attribution distinction, see Burge, 2010 and Quilty-Dunn, 2020a.

[4] There are independent reasons to think object files are discursive. Space limitations and fear of self-plagiarism prevent me from rehearsing them here, but see Quilty-Dunn, 2020a, c.

Figure 4.1 Green marble cylinder.

In Treisman's Feature Integration Theory (FIT), values along distinct feature dimensions, like colour, orientation, motion, texture, etc., are initially detected independently of one another (Treisman and Gelade, 1980). This initial independence creates the need for a later binding operation or 'feature integration' process. Object files are posited to be the outputs of that binding operation: representations of individual objects that track them through time and space and attribute multiple features to them, binding and storing those features in visual working memory (Kahneman *et al.*, 1992).

The role object files play in FIT is the most widely accepted instance of binding via attribution in perception. I take the fact that we sometimes bind feature representations by integrating them in object files, and thereby attributing the features to the same object, to be as near to established fact as anything in this neck of the woods. Thus, Cohen's (2004) and Matthen's (2004) point that feature binding in vision involves attributing properties to objects is well taken, as is Clark's (2000, 2004) and Matthen's (2004) appeal to a predicate–argument structure.

If we assume both that binding in vision is *only* accomplished by attribution via object files and that object files have a discursive format, then it follows that binding in vision is only accomplished by representations in a discursive format. This conclusion is intuitively suspicious. If vision is really couched in multiple formats, including proprietary (e.g., iconic) perceptual formats, it would be surprising if binding could only occur in one of them. We should therefore consider what binding looks like outside the context of a discursive predicate–argument structure and Merge.

Perhaps the most salient alternative to discursive representational formats are *iconic* representational formats. Consider an image of a green marble cylinder (Figure 4.1).

The first thing to notice about Figure 4.1 for present purposes is that it binds features together. It does not merely represent *green*, *marble*, and *cylinder*, but rather encodes the combination of those features.[5] An image of a marble cylinder and a wooden cube, for example, differs from an image of a wooden cylinder and a marble cube. Ordinary icons like Figure 4.1

[5] I use 'marble' and 'wooden' to denote diagnostic visual textures of the relevant materials, not the materials themselves.

thus exhibit what I called above the 'core phenomenon' of binding: they distinguish different combinations of the same primitive features.

Another thing to notice is that icons like Figure 4.1 are *holistic*—that is, the multiple features instantiated by some part of the depicted scene are represented in one and the same corresponding part of the icon. In the phrase 'That is a green marble cylinder', each feature is represented by a separate constituent, but in Figure 4.1, the same part of the icon that represents the visible colour also represents visible texture and shape properties. Bundles of properties are not laid out in an orderly, discrete fashion in an icon, as they are in a sentence. Instead, features are somehow packed into the same part of the representational vehicle. This aspect of iconic representation is at once obvious and puzzling, especially if we think of representation (mental and otherwise) on the model of language and predication.

The claim that icons are holistic does not entail that a visual icon mandatorily encodes every visible feature. The idea of mandatory encoding of many features is perhaps most influentially embodied in Dretske's (1981) notion of 'analogue' representation. (His notion of analogicity is sufficiently unusual that I will use 'Dretske-analogue' when discussing it.) For Dretske, all representations carry *nested* information—that is, a representation that says *the cup is scarlet* also thereby carries the information that *the cup is red* and that *the cup is coloured*. But what distinguishes Dretske-analogue from Dretske-digital representations is that Dretske-analogue representations invariably carry *non-nested* information.[6] The sentence 'the cup is maroon' is silent on properties of the represented state of affairs beyond what can be logically inferred from the fact that there is a cup and it is maroon. But a photograph of the same scene must tell you more about it: for example, depending on the viewpoint, it will also represent the shape of the cup, whether it has liquid in it, what surfaces and objects surround it, etc.

But the holistic character of icons does not require that they encode many primitive features. Icons are holistic because, *when* two features are represented at a particular part of the represented scene, the same part of the icon that corresponds to that part of the scene and represents one feature will do double-duty and represent the other as well. An icon may not represent a given feature, as when a greyscale photograph fails to represent chromatic colour. But if a particular part of the scene is represented in a colour photograph, the same part of the photograph that represents a value along the brightness dimension will also represent a value along the hue dimension.

The holisticity of icons is underemphasized in discussions of iconicity. Theories of icons (at least *qua* mental representations) tend to focus on some version of the 'parts principle' (Kulvicki, 2015): parts of icons correspond to parts of the represented scene (Carey, 2009; Fodor, 2007; Kosslyn, 1980; Kulvicki, 2015; Sober, 1976; Toribio, 2011). This principle does not entail holisticity. But if we add an additional constraint to the parts principle—namely, that icons are isomorphic to what they represent, such that each represented part of the scene corresponds to exactly one part of the icon, and vice versa—then holisticity follows (Quilty-Dunn, 2020b). For any part of the scene, exactly one part of the icon corresponds to it. Therefore, if two properties are instantiated at that part of the scene and are represented by the icon, the same part of the icon must represent both properties.

[6] More precisely, for Dretske, representations *carry bits of information* in an analogue/digital way; thus, one representation may encode some facts analogically, and others digitally.

The holistic format of icons precludes certain solutions to the syntactic binding problem. For example, icons lack separate vehicles for objects and their features. The attribution model thus seems not to apply, at least as far as the representational *vehicle* is concerned. Whether attribution might still be the best model of the representational *content* of the icon is unclear (see below). Moreover, since a complex representation formed through Merge like RED SQUARE maintains discrete constituents, Merge also seems ill-suited to characterize icons. We have therefore run into a puzzle: how should we think about feature binding for iconic representations?

One might be tempted to dismiss the relevance of holistic icons to vision, and instead argue that we can avoid this puzzle entirely by sticking to the attribution model. But even if one is sceptical of the idea that mental representations are iconic in the sense I have outlined, the puzzle about feature binding remains, because *some* holistic icons exist. Figure 4.1 is an example; ample photographs and drawings are examples as well. The fact that icons exist and bind features together without explicit demonstrative-like representations of individuals, quantifiers, or Merge is a datum that needs to be explained.

Thus far, I have explicated the basic idea of holistic icons and argued that: (1) they exist; (2) they bind features; and (3) they do not seem well suited to an attribution- or Merge-based account. I will now argue that philosophers of perception have special reason to be interested in how icons bind features, because there is evidence that the visual system constructs and computes over icons.

3 Holistic icons in vision

As discussed above, there is evidence that object files—representations that pick out individuals, track them across space and time, and attribute properties to them—are discursive rather than iconic. According to one natural interpretation of FIT (Treisman and Gelade, 1980), the visual system first detects features in parallel, then binds them only once attention is deployed and an object file is created, at which time the features are entered into the same object file and thereby attributed to the same object. If we combine the ideas contained in the previous two sentences, we arrive at the conclusion that there are no holistic icons in vision after all (Clarke, 2022b).

I suspect something like the interpretation of FIT just sketched is common among philosophers of perception.[7] But it provides an impoverished picture of the kinds of representations constructed in vision. And it yields a clear falsifiable prediction: there is no feature binding independently of attention.

FIT enjoys many diverse forms of empirical support, far too many to survey here.[8] One of the most striking forms of evidence for FIT is the phenomenon of *illusory conjunctions*. In

[7] Treisman herself continually modified FIT (e.g., Treisman and Sato, 1990), and the characterization I describe here is a caricature of the nuanced views she developed over decades. This caricature is widespread, however (as Wolfe (2021) notes), and is thus worth discussing.

[8] One crucial kind of evidence I regrettably cannot discuss here is visual search, which was a focus of Treisman and Gelade's (1980) original presentation of FIT and Wolfe's influential FIT-based approach to visual attention, Guided Search (Wolfe, 2021; Wolfe *et al.*, 1989). The standard findings are that search latency increases as the number of items increase and, more importantly, this item-based increase is markedly higher for conjunctions than individual features. According to Prinzmetal (2012, p. 214), this conjunction-based decline in search efficiency is 'the evidence most often cited in favor of' FIT. But as Prinzmetal also points out, the visual search literature provides, at best, ambiguous support for FIT (see, for example, Dosher *et al.*, 2004; Duncan and Humphreys, 1989; and Hulleman and Olivers, 2017).

one classic experiment, subjects reported briefly presented objects, which varied in features like shape and colour (e.g., a green 'N' and a brown 'T') (Treisman and Schmidt, 1982). They found that, while subjects would occasionally misreport features that were not present (e.g., saying 'red' when nothing in the display was red), they were even more likely to incorrectly integrate present features together (e.g., reporting a green 'T'). And as FIT predicts, these so-called illusory conjunctions are more likely when attention is taxed and are less likely for objects that fall within the focus of spatial attention (Prinzmetal *et al.*, 1986). Treisman and Schmidt (1982, p. 108) noted that the existence of illusory conjunctions in the absence of attention is 'the central claim' of FIT, and Prinzmetal (2012, p. 215) calls it the 'clearest manifestation' of FIT's notion of feature binding.

While illusory conjunctions provide powerful evidence that *some form* of accurate binding requires attention, this does not entail that *all* cases of binding require attention. It might be that the specific tasks used, such as verbal report and other overt responses, require the use of a certain kind of representation. As Mordkoff and Halterman (2008, p. 385) put it, attention may not always be needed for binding, but rather 'is needed only to select and produce the task-related, overt response'. If icons are truly couched in a different format from the discursive format ordinarily used for cognition and purposeful action, then they may only surface indirectly in behaviour and be masked by other representations more directly available to decision and report.

Object files are known to operate across mental systems. For example, they operate crossmodally and play a key role in crossmodal folk physics, particularly as studied in the minds of infants (Carey, 2009; Spelke, 1988). These findings make sense. The discursive format of object files makes them well suited to play an interfacing role between perception and other cognitive systems, including language. Another reason why this kind of interfacing role makes sense for object files is that they are normally available to visual working memory (Gao *et al.*, 2011; Quilty-Dunn, 2020b), allowing them to function in a shared cognitive workspace.

Perhaps attention is normally required for correct binding of features *into object files*, thereby attributing them to a particular object. And perhaps this representational format is required for task performance, due to the role object files play at the interface of perception and cognition. But it may still be that other experimental evidence, involving different task demands, can provide evidence for binding outside the realm of attention and object files. The hypothesis that there are holistic icons in vision predicts that we should find such evidence.

I have argued elsewhere that evidence of this sort exists (Quilty-Dunn, 2020b). Iconic memory, a high-capacity visual short-term memory store, seems to process feature conjunctions in a holistic fashion (Bronfman *et al.*, 2014; Burns, 1987; Pinto *et al.*, 2013),[9] and in greater number than we can attend to, or bind into, object files (Landman *et al.*, 2003). Other evidence comes from ensemble perception, the rapid extraction of statistical summaries of large numbers of objects. The representations used to compute these summaries represent many more items than can be attended to, or represented via, object files (Utochkin and Tiurina, 2014)—perhaps by at least one order of magnitude (Dakin, 2001)—and proceed independently of visual working-memory load (Epstein and Emmanouil, 2017). They also

[9] I am equating iconic memory and so-called 'fragile visual short-term memory'. The putative distinction between the two should not matter here.

seem to encode many perceptible features at once (see Whitney and Yamanashi Leib, 2018), including feature conjunctions (Boduroglu and Yildirim, 2020).

For the purposes of this chapter, however, I want to focus on one probative piece of evidence from Mordkoff and Halterman (2008). Their goal was to develop a task that tested for binding outside attention but did not require encoding the feature conjunction in a format usable for decision or report. The solution was to use a version of the *flankers task* (Eriksen and Eriksen, 1974). In the standard flankers task, a target appears in the middle of the screen and is surrounded by 'flankers' (e.g., two on each side). The task typically involves discriminating the target by hitting a button, and typically requires different responses for different targets. So for example, the task might be to hit a left-hand button if the target is a '1' or '2', and to hit a right-hand button if it is a '3' or '4'. Like the famous Stroop effect, the standard effect is that performance is better on trials where the flankers are 'congruent' with the target (e.g., target = 1, flankers = 2) and worse on 'incongruent' trials (e.g., target = 1, flankers = 3).

Flankers are useful in this context because they must be perceived to affect performance, even though the subject is attending only to the target. Thus, the efficacy of flanking feature conjunctions could provide evidence for feature conjunctions outside attention. In the standard flankers task, however, both target and flankers have a particular 'response code' (i.e., right- or left-hand button press). Using the standard flankers task to probe the effects of feature conjunctions would thus raise the worry that attention is necessary for representing a feature conjunction in a format that can trigger the response code, rather than for binding per se.

As a result, Mordkoff and Halterman also used a *correlated* flankers task (Miller, 1987). The targets and flankers were coloured shapes (i.e., feature conjunctions). One crucial feature of their task is that the conjunction-defined flankers were never used as targets (though flankers and targets shared primitive features). Conjunction-defined flankers therefore had no assigned behavioural response. Trials were balanced so that conjunctions were *correlated* with responses; for example, trials on which flankers were yellow diamonds were also trials on which the correct response to the target was a left-hand key press. This correlation allowed them to look for an effect of feature conjunctions on performance when the conjunctions are both unattended and not assigned an overt response.

FIT predicts that, in a *standard* flankers task, conjunction-defined flankers should fail to affect performance since they fall outside the focus of attention and thus cannot be processed as conjunctions. Mordkoff and Halterman found this null effect as predicted, replicating earlier findings (Cohen and Shoup, 2000). This result is also predicted, however, by the more restricted hypothesis that attention is necessary for binding features into object files that are available for decision and report, which is compatible with holistic icons representing feature conjunctions outside attention.

In the *correlated* flankers task, however, this hypothesis predicts that the feature conjunctions *should* have an effect, whereas FIT should again predict that conjunctions cannot be processed outside of attention, and thus conjunction-defined flankers should have no impact. Mordkoff and Halterman did find an impact, however: flankers that were correlated with targets grouped by a certain behavioural response improved discrimination of those targets (and impaired discrimination of others), *even when the correlation held only for feature conjunctions.*

One might object that participants attended to the flankers, perhaps because they noticed their correlations with response codes. But Mordkoff and Halterman replicated the effect in blocks of trials that interspersed the standard and correlated flankers test, eliminating the usefulness of adopting a general strategy of attending to flankers. They also replicated the effect while moving the flankers further away from targets and placing other flankers in between the targets and correlated flankers, making it yet less likely that the correlated flankers were attended to. Furthermore, the idea that subjects attend to flankers contradicts FIT's explanation of why conjunction-defined flankers fail to affect performance in the standard flankers task.

These results provide striking evidence that unattended feature combinations are bound together in vision. Equally importantly, they also show that this fact is masked in tasks that require explicit behavioural response to the relevant combinations. There seem to be multiple distinct forms of binding at work in the visual system, which differ in their relationship to cognition and report.

The attribution model is plausible in the case of object files, which bind features in a discursive format and thereby make feature combinations available to be exploited by cognition and purposeful action, including conceptualized response codes like *red square → left button*. Tasks that require this sort of availability require attention to process feature combinations. But other visual processes construct holistic icons, which encode feature combinations in a fundamentally different format. We cannot expect feature combinations represented iconically to exhibit the same sort of global availability distinctive of object files. Feature combinations in visual working memory ('VWM') are not holistic (Fougnie and Alvarez, 2011), but object-based (Markov *et al.*, 2019), suggesting that discursive object files are the vehicles of feature binding in VWM. It is thus possible that icons are not available to VWM.[10] In that case, the various cognitive and motor processes that are mediated by VWM may not be able to exploit iconically represented feature conjunctions. Thus, the fact that we can only find evidence of holistic binding in iconic memory, or indirectly through influences on other processes (e.g., ensemble perception and discriminating correlated targets), is to be expected.

Two objections before moving on.

First objection: perhaps holistic icons are *neurally* implausible, because we know that early areas in visual cortex represent features independently. The idea that early vision consists solely of 'specialized populations of receptors that respond selectively' (Treisman and Gelade, 1980, p. 97) is widespread (Livingstone and Hubel, 1988; Zeki, 1993). Such independent areas are thought to correspond to distinct 'feature maps' in primary and secondary visual cortices, a claim which has frequently been a starting point for philosophical discussions of feature binding (e.g., Campbell, 2002, p. 30ff; Clark, 2000, p. 27ff; Clarke, forthcoming; Matthen, 2005, p. 67ff).

However, despite having attained the status of received wisdom, the hypothesis that distinct feature dimensions like colour and orientation are only represented independently in early visual cortex is false. Leventhal *et al.* (1995) found that, contrary to Livingstone and Hubel's (1988) widely cited claim that features are segregated in the primate primary visual cortex (also known as striate cortex, or V1), 'it appears that most cells in layers 2 and 3 are

[10] This generates another puzzle: how could icons not be available to VWM, given that visual imagery uses icons and apparently requires VWM resources (Hyun and Luck, 2007)? Visual imagery, however, is not simply a matter of tokening icons. Even on iconic models, visual imagery requires fluid interaction with discursive representations of imaged objects, scenes, part-whole structures, and various other contents (Kosslyn, 1980). Furthermore, icons outside of VWM might be 'pointed' to by object files in VWM (Balaban *et al.*, 2019), establishing functional interaction without making icons themselves available to VWM.

selective concomitantly for aspects of form, motion, and color' and thus 'cells exhibiting the sort of stimulus selectivity required by the Livingstone and Hubel model do not exist in significant numbers in monkey striate cortex' (Leventhal *et al.*, 1995, 1817).

More recently, Garg *et al.* (2019) used two-photon calcium imaging to examine whether cells in V1 show dimension-specific selectivity.[11] They found that, among cells that were strongly selective for colour, the *majority* were also strongly selective for orientation; among cells that were selective for orientation, nearly half were also selective for colour. Garg *et al.* (2019, p. 1278) conclude that 'shape and color are mutually and unambiguously extracted and represented in a substantial population of V1 neurons'. The widespread view that shape and colour are primarily neurally segregated in early vision appears to be wrong.[12]

Second objection: perhaps what looks like representation of feature combinations here is not really representation at all—that is, perhaps these states are merely registrations of proximal stimulus at the earliest stages of perceptual processing, before the visual system purports to represent the distal environment. Treisman (1988, p. 204) argues that there may be early feature binding, but that such conjunctions represent only 'properties of the retinal stimuli', not 'real-world properties, after constancy mechanisms have operated'. Likewise, Matthen (2004, p. 507ff) argues that vision prior to attention and object individuation corresponds to spatial properties of the retina, not the distal environment.

Fortunately, there is evidence that bears directly on this issue. As mentioned above, ensemble perception seems to run statistical summaries over icons that represent more items than can be attended to. Im and Chong (2009) tested ensemble perception of average size of circles using stimuli that instantiated the Ebbinghaus illusion, in which a circle looks larger or smaller depending on whether it is surrounded by small or large circles, respectively. They found that ensemble-coding processes averaged over the illusory *perceived* size, not retinal size. Likewise, several recent experiments have found that average size is computed for perceived size, integrating depth cues prior to ensemble coding (Haberman and Suresh, 2020; Markov and Tiurina, 2021; Tiurina and Utochkin, 2019). Cells in V1 also show size constancy (Sperandio *et al.*, 2012) and orientation constancy (Sauvan and Peterhans, 1999).

4 The coordination model

I have been arguing that there is reason to posit iconic representations in vision, and that neither the attribution model nor Merge capture how these representations solve the syntactic binding problem. I will now propose an alternative model: icons plot coordinates in multidimensional spaces, binding values in those spaces without exemplifying the predicate–argument structure characteristic of the attribution model.

The notion of iconic representation is often discussed together with the notion of *analogue* representation (e.g., Carey, 2009, p. 135; Dretske, 1981, p. 137; Kosslyn and Pomerantz, 1977). However, many paradigmatically analogue representations are not iconic, such as mercury thermometers. I lack space to discuss previous philosophical work on this topic in

[11] Two-photon calcium imaging is a technique that involves using a two-photon microscope to observe changes in fluorescence in individual cells caused by fluctuations in calcium concentration.

[12] The fact that there is widespread integration of colour and form in V1 is compatible with there *also* being segregated colour-specific and form-specific processing (Liu *et al.*, 2020). In keeping with the rejection of the 'assembly line' model, it would make sense for there to be both parallel feature detection and holistic icons in early vision.

anything like adequate detail (e.g., Beck, 2019; Clarke, 2020; Goodman, 1968; Haugeland, 1981; Kulvicki, 2015; Lewis, 1971; Maley, 2011, 2020a; Peacocke, 1986, 2019). In Goodman's (1968) seminal discussion of the analogue/digital distinction, he argued that analogue representations are *dense*—that is, between any two values along some semantically significant dimension of variation in the vehicle, there lies a third value. But analogue representations arguably need not be dense (Lewis, 1971).

Maley (2011) argues that density is unimportant, and that what is distinctive of analogue representations is that variation along some dimension of the vehicle reliably corresponds to variation along the represented feature dimension. For example, whether a clock is analogue does not seem like it should hang on whether the second hand glides from second to second (dense) or instead ticks abruptly (non-dense). The relevant joint in nature seems to be between both kinds of clocks and truly digital clocks, which represent changes in time by updating numerals (Quilty-Dunn, 2017). What makes the abruptly ticking clock analogue is that variation along a dimension of the vehicle (how far the second hand has moved) functions to represent variation along the represented dimension (number of seconds elapsed), and an increase/decrease in movement of the hand corresponds to an increase/decrease in the number of seconds elapsed. The same is not true of digital clocks where numerals vary along dimensions (e.g., on an old digital clock, the number of vertical or horizontal bars contained in the numeral), but such variation fails reliably to correspond to variation in the temporal dimension.

Other ways of characterizing analogue representation exist, such as Peacocke's (2019, p. 52) notion of 'representation of magnitudes, by magnitudes' and Lewis' (1971) claim that analogue representations employ primitive physical magnitudes. But a common thread, exemplified in Maley's work, is that analogue vehicles vary along dimensions in ways that mirror variation along represented feature dimensions (Beck, 2019; Clarke, 2020). It is this common thread that I want to use to think about binding in holistic icons.

Recent work on analogue format, almost without exception, treats analogue features individually. This focus may be partly due to background assumptions consistent with the assembly line model of vision, particularly in the context of FIT. One might assume that analogue feature representations are deployed separately, and then an object file is deployed to attentively bind features together. But as we have seen, this model is ill-equipped to account for binding outside of attention and object files. More generally, it is odd to think that analogue representations of features do not compose, or that their composition is somehow dependent on object file-like representations. An image like Figure 4.1 seems to bind features together *in a distinctly analogue way*—the texture, colour, and position of a part of the surface of the cylinder are represented by means of dimensions in the vehicle that correspond to feature dimensions and they are bound together. An icon is a certain sort of *complex analogue representation*. We need some meaningful notion of analogue feature binding. Moreover, even the purported parallel feature detection posited by FIT requires binding outside object files for complex features that are treated as primitive: colour values contain values along multiple dimensions at once (e.g., hue and saturation), and this binding is accomplished prior to attention and attribution to objects.

So how are features bound in analogue icons? One might imagine many elaborate ways that analogue feature values could be bound, but I think there is one extremely simple structure that suffices. Key to the notion of individual analogue representations is the idea of variation along *dimensions*. Coordinate systems (e.g., Cartesian coordinates) allow a point to be plotted along multiple dimensions simultaneously. For example, the coordinates (3, 4) provide a representation of a point that has *both* a value of 3 along the x-axis and a value of 4 along the y-axis. Thus, coordination provides a framework for conceptualizing how a single point can be plotted along multiple dimensions at once.

Now consider part of an icon. For simplicity's sake, consider a primitive part, like a pixel of Figure 4.1.[13] That part has values along multiple dimensions at once, including hue, brightness, saturation, and the spatial x- and y-axes of the image. We can therefore model that part of the icon as a point in a five-dimensional space, defined by a set of coordinates of the form $(a_i, b_j, c_k, d_l, e_m)$, where each subscript denotes a particular feature dimension (e.g., hue) and each letter from a to e denotes a particular value along that dimension. Call this model *the coordination model* of binding in holistic icons.[14]

The primitive icon part need not be understood as having a constituent structure consisting of five vehicles just because the set of coordinates we use to model it does. Instead, it merely exemplifies values along independently varying dimensions. We can distinguish *constituents* of a representation from its semantically exploitable *properties*, just as we more generally distinguish parts of objects from their properties. Consider a brown chair with four legs; while one could use 'part' to refer to both its back and its brown colour, there is clearly some relevant distinction between the chair's relation to these two things. The colour is not a part of the chair in the same sense that the back is (Olson, 2017).[15] Likewise, the saturation of a pixel is not a part of it in the same sense that a pixel is part of the whole icon (cf. Lande, 2020).

The core ideas behind the coordination model are familiar. The idea that aspects of perception can be understood by appeal to multidimensional feature spaces has long been discussed in the literature on perceptual *quality spaces* or *similarity spaces* (Berger, 2018, 2021; Clark, 1993; Gauker, 2012; Rosenthal, 2010). This literature focuses primarily on mental properties that surface in the qualitative character of conscious experience, and much of it assumes something like FIT (Clark, 2000; Rosenthal, 2005, p. 200). I have (deliberately) said nothing at all about conscious experience, and holistic icons require binding outside the constructs of FIT. But the basic idea of thinking of visual representations as plotting points in multidimensional feature spaces has long been discussed in this literature (cp. vectors and tensor products in connectionist models; Eliasmith, 2013, p. 387ff; Smolensky, 1990). It also echoes Haugeland's (1998, p. 192) notion of iconic contents as 'variations of values along certain dimensions with respect to locations in certain other dimensions'.

The coordination model itself is very simple, and my word limit looms. I will move directly to considering some objections, explicating the model further as I go.

One might object that the coordination model is not meaningfully distinct from the attribution model. Clark's (2000) version of the attribution model allows features to be attributed not only to objects (Clark, 2004), but also to locations. For Clark, visual representations engage in *feature placing*, placing features at locations and thereby attributing features to those

[13] This really is just for simplicity's sake. Icons can have non-pixel-like primitives (Davies, 2021), and non-primitive parts of icons can bind properties; the latter is arguably required for binding of some properties like global shape.

[14] 'Coordination' is also used, with a different meaning, in discussions of crossmodal binding (see Cohen's chapter in this volume is Chapter 17; Fulkerson, 2014; O'Callaghan, 2017).

[15] Some metaphysicians do hold that properties can be parts of objects in some sense, for example, as 'logical parts' (Paul, 2002). My claim is only that there are *some* meaningful senses of 'part' in which properties are not parts.

locations. Why not then construe the non-locational features of an icon as attributed to locations as represented along spatial axes?

However, the apparatus of coordination provides a model of bound elements in icons that does not require taking any elements to be attributed to any others. A key feature of attribution is *asymmetry*: in the thought BANANAS ARE YELLOW, the property of being yellow is attributed to bananas while the property of being bananas is not attributed to yellow. But coordination does not have this asymmetry built in. A part of an icon varies along multiple dimensions at once and thereby plots a position in a multidimensional space. Nothing in the coordination model requires, or even suggests, that values along some feature dimensions are attributed to values along other coordinated dimensions. This is clear in the case of Cartesian coordinates: the x-axis and y-axis values are bound together, but neither is attributed to the other. Likewise, there is nothing in the syntactic features of an icon that compel us to construe a value along the hue dimension as attributed to a value along spatial dimensions anymore than to construe spatial values as attributed to colour values.

One might press further that there is something special about spatial properties that prevents them from being represented by means of values along dimensions that can be coordinated with values along, for example, colour and orientation dimensions without attribution. Clark (2000) rejects the idea of vehicle properties that correspond to location, instead holding that features are simply placed at *external* locations, with no quasi-spatial relations holding between representations of features. Much of Clark's (2009) discussion relies on the idea that binding must be accomplished through attribution to external locations, and on his rejection of visual icons.

However, phenomena like mental rotation (Shepherd and Metzler, 1971), instructed scanning (Kosslyn *et al.*, 1978), and spontaneous scanning (Finke and Pinker, 1982) suggest that there *are* visual icons that encode spatial relations via functional relations between representations of features that covary with objective spatial relations—in other words, space is encoded by yet another analogue dimension. It can therefore be coordinated with other analogue dimensions without attribution.

One virtue of the attribution model is that it provides a straightforward account of *accuracy conditions*: the representation is accurate if the selected individual has the properties attributed to it. I have not said how coordination fixes accuracy conditions. But my topic here has been the *syntactic* binding problem: how is the vehicle structured such that it represents feature combinations? An answer to this question need not come prepackaged with an answer to questions about how accuracy conditions are determined. Perhaps causal relations to environmental individuals on occasions of use provide 'targets' for iconically encoded feature combinations without elements of the vehicle corresponding to those individuals (Cummins, 1996). But in the absence of such an answer, we are left with a familiar question: how do perceptual representations come to be accurate or inaccurate with respect to the perceiver's environment?

References

Balaban, H., Drew, T., and Luria, R. (2019). Neural evidence for an object-based pointer system underlying working memory. *Cortex*, 119, 362–72.

Bayne, T. (2010). *The Unity of Consciousness*. New York, NY: Oxford University Press.

Beck, J. (2019). Perception is analog: the argument from Weber's Law. *Journal of Philosophy*, 116(6), 319–49.
Berger, J. (2018). A defense of holistic representationalism. *Mind and Language*, 33(2), 161–76.
Berger, J. (2021). Quality-space functionalism about color. *Journal of Philosophy*, 18(3), 138–64.
Boduroglu, A. and Yildirim, I. (2020). Statistical summary representations of bound features. *Attention, Perception, and Psychophysics*, 82, 840–51.
Bronfman, Z. Z., Brezis, N., Jacobson, H., and Usher, M. (2014). We see more than we can report: 'Cost free' color phenomenality outside focal attention. *Psychological Science*, 25(7), 1394–403.
Burge, T. (2010). Steps toward origins of propositional thought. *Disputatio*, 4(29), 39–67.
Burns, B. (1987). Is stimulus structure in the mind's eye? An examination of dimensional structure in iconic memory. *The Quarterly Journal of Experimental Psychology*, 39(A), 385–408.
Burnston, D. and Cohen, J. (2013). Perception of features and perception of objects. *Croatian Journal of Philosophy*, 12(3), 283–314.
Camp, E. (2018). Why maps are not propositional. In *Non-propositional Intentionality*, edited by A. Grzankowski and M. Montague, 19–45. Oxford: OUP.
Campbell, J. (2002). *Reference and Consciousness*. Oxford: Oxford University Press.
Carey, S. (2009). *The Origin of Concepts*. New York, NY: Oxford University Press.
Chomsky, N. (1995). *The Minimalist Program*. Cambridge, MA: MIT Press.
Clark, A. (1993). *Sensory Qualities*. Oxford: Oxford University Press.
Clark, A. (2000). *A Theory of Sentience*. Oxford: Oxford University Press.
Clark, A. (2004). Feature-placing and proto-objects. *Philosophical Psychology*, 17(4), 443–69.
Clark, A. (2009). Location, location, location. In *Computation, Cognition, and Pylyshyn*, edited by D. Dedrick and L. Trick, 281–302. Cambridge, MA: MIT Press.
Clarke, S. (2020a). Beyond the icon: core cognition and the bounds of perception. *Mind and Language*, 37(1), 94–113.
Clarke, S. (2022b). Mapping the visual icon. *Philosophical Quarterly*, 72(3), 552–77.
Cohen, J. (2004). Objects, places, and perception. *Philosophical Psychology*, 17(4), 471–95.
Cohen, A. and Shoup, R. (2000). Response selection processes for conjunctive targets. *Journal of Experimental Psychology: Human Perception and Performance*, 26(1), 391–411.
Cummins, R. (1996). *Representations, Targets, and Attitudes*. Cambridge, MA: MIT Press.
Dakin, S. C. (2001). Information limit on the spatial integration of local orientation signals. *Journal of the Optical Society of America A*, 18(5), 1016–26.
Davies, W. (2021). Colour relations in form. *Philosophy and Phenomenological Research*, 102(3), 574–94.
Dosher, B. A., Han, S., and Lu, Z.-L. (2004). Parallel processing in visual search asymmetry. *Journal of Experimental Psychology: Human Perception and Performance*, 30(1), 3–27.
Dretske, F. (1981). *Knowledge and the Flow of Information*. Cambridge, MA: MIT Press.
Duncan, J. and Humphreys, G. W. (1989). Visual search and stimulus similarity. *Psychological Review*, 96(3), 433–58.
Eliasmith, C. (2013). *How To Build a Brain*. New York: Oxford University Press.
Epstein, M. L. and Emmanouil, T. A. (2017). Ensemble coding remains accurate under object and spatial visual working memory load. *Attention, Perception, and Psychophysics*, 79, 2088–97.
Eriksen, B. A. and Eriksen, C. W. (1974). Effects of noise letters upon the identification of a target letter in a nonsearch task. *Perception and Psychophysics*, 16(1), 143–9.
Finke, R. A. and Pinker, S. (1982). Spontaneous imagery scanning in mental extrapolation. *Journal of Experimental Psychology: Learning, Memory, and Cognition*, 8(2), 142–7.
Firestone, C. and Scholl, B. (2014). 'Please tap the shape, anywhere you like': shape skeletons in human vision revealed by an exceedingly simple measure. *Psychological Science*, 25(2), 377–86.
Fodor, J. (2007). The revenge of the given. In *Contemporary Debates in Philosophy of Mind*, edited by B. McLaughlin and J. Cohen, 105–16. Oxford: Blackwell.
Fougnie, D. and Alvarez, G. A. (2011). Object features fail independently in visual working memory: evidence for a probabilistic feature–store model. *Journal of Vision*, 11(12), 1–12.
Fulkerson, M. (2014). *The First Sense: A Philosophical Study of Human Touch*. Cambridge, MA: MIT Press.

Gao, T., Gao, Z., Li, J., Sun, Z., and Shen, M. (2011). The perceptual root of object-based storage: an interactive model of perception and visual working memory. *Journal of Experimental Psychology: Human Perception and Performance*, 37(6), 1803–23.

Garg, A. K., Li, P., Rashid, M. S., and Callaway, E. M. (2019). Color and orientation are jointly coded and spatially organized in primate primary visual cortex. *Science*, 364, 1275–9.

Gaskin, R. (2008). *The Unity of the Proposition*. Oxford: Oxford University Press.

Gauker, C. (2012). Perception without propositions. *Philosophical Perspectives*, 26, 19–50.

Goodman, N. (1968). *Languages of Art: An Approach to a Theory of Symbols*. Indianapolis, IN: Bobbs-Merrill.

Green, E. J. (2018). What do object files pick out? *Philosophy of Science*, 85(2), 177–200.

Green, E. J. (2019). A theory of perceptual objects. *Philosophy and Phenomenological Research*, 99(3), 663–93.

Green, E. J. and Quilty-Dunn, J. (2021). What is an object file? British Journal for the Philosophy of Science, 72(3), 665–99.

Gross, S. and Flombaum, J. (2017). Does perceptual consciousness overflow cognitive access? The challenge from probabilistic, hierarchical processes. *Mind and Language*, 32(3), 358–91.

Haberman, J. and Suresh, S. (2020). Ensemble size judgments account for size constancy. *Attention, Perception, and Psychophysics*, 83(3), 925–33.

Haugeland, J. (1981). Analog and analog. *Philosophical Topics*, 12(1), 213–25.

Haugeland, J. (1998). *Having Thought*. Cambridge, MA: Harvard University Press.

Henke, B. (2021). A fresh look at the two visual streams. *Journal of Consciousness Studies*, 28(5-6), 198–207.

Hulleman, J. and Olivers, C. (2017). The impending demise of the item in visual search. *Behavioral and Brain Sciences*, e132, 1–20.

Hyun, J.-S. and Luck, S. J. (2007). Visual working memory as the substrate for mental rotation. *Psychonomic Bulletin and Review*, 14, 154–8.

Im, H. Y., and Chong, S. C. (2009). Computation of mean size is based on perceived size. *Attention, Perception, and Psychophysics*, 71(2), 375–84.

Jackson, F. (1977). *Perception: A Representative Theory*. Cambridge: Cambridge University Press.

Kahneman, D., Treisman, A., and Gibbs, B. J. (1992). The reviewing of object files: object-specific integration of information. *Cognitive Psychology*, 24, 175–219.

Kosslyn, S. M. (1980). *Image and Mind*. Cambridge, MA: Harvard University Press.

Kosslyn, S. M., Ball, T. M., and Reiser, B. J. (1978). Visual images preserve metric spatial information: evidence from studies of imagery scanning. *Journal of Experimental Psychology: Human Perception and Performance*, 4, 47–60.

Kosslyn, S. M. and Pomerantz, J. R. (1977). Imagery, propositions, and the form of internal representations. *Cognitive Psychology*, 9, 52–76.

Kulvicki, J. (2015). Analog representation and the parts principle. *Review of Philosophy and Psychology*, 6(1), 165–80.

Lande, K. (2021). Mental structures. *Noûs*, 55(3), 649–77.

Landman, R., Spekreijse, H., and Lamme, V. A. F. (2003). Large storage capacity of integrated objects before change blindness. *Vision Research*, 43, 149–64.

Leventhal, A. G., Thompson, K. G., Liu, D., Zhou, Y., and Ault, S. J. (1995). Concomitant sensitivity to orientation, direction, and color of cells in layers 2, 3, and 4 of monkey striate cortex. *The Journal of Neuroscience*, 15(3), 1808–18.

Lewis, D. (1971). Analog and digital. *Noûs*, 5, 321–7.

Liu, Y., Li, M., Zhang, X., *et al.* (2020). Hierarchical representation for chromatic processing across macaque V1, V2, and V4. *Neuron*, 108, 538–50.

Livingstone, M. and Hubel, D. (1988). Segregation of form, color, movement, and depth: anatomy, physiology, and perception. *Science*, 240, 740–9.

Luck, S. J. and Vogel, E. K. (1997). The capacity of visual working memory for features and conjunctions. *Nature*, 309, 279–81.

Maley, C. J. (2011). Analog and digital, continuous and discrete. *Philosophical Studies*, 155, 117–31.

Markov, Y. A. and Tiurina, N. A. (2021). Size-distance rescaling in the ensemble representation of range: study with binocular and monocular cues. *Acta Psychologica*, 213, 103238.
Markov, Y. A., Tiurina, N. A., and Utochkin, I. S. (2019). Different features are stored independently in visual working memory but mediated by object-based representations. *Acta Psychologica*, 197, 52–63.
Matthen, M. (2004). Features, places, and things: reflections on Austen Clark's theory of sentience. *Philosophical Psychology*, 17(4), 497–518.
Matthen, M. (2005). *Seeing, Doing, and Knowing: A Philosophical Theory of Sense Perception*. Oxford: Oxford University Press.
Miller, J. (1987). Priming is not necessary for selective-attention failures: semantic effects of unattended, unprimed letters. *Perception and Psychophysics*, 41, 419–34.
Mordkoff, J. T. and Halterman, R. (2008). Feature integration without visual attention: evidence from the correlated flankers task. *Psychonomic Bulletin and Review*, 15(2), 385–9.
O'Callaghan, C. (2008). Object perception: vision and audition. *Philosophy Compass*, 3(4), 803–29.
O'Callaghan, C. (2014). Intermodal binding awareness. In *Sensory Integration and the Unity of Consciousness*, edited by D. J. Bennett and C. S. Hill, 73–103. Cambridge, MA: MIT Press.
O'Callaghan, C. (2017). Enhancement through coordination. In *Current Controversies in Philosophy of Perception*, edited by B. Nanay, 109–20. New York, NY: Routledge.
Olson, E. T. (2017). Properties as parts of ordinary objects. In *Being, Freedom, and Method: Themes from the Philosophy of Peter van Inwagen*, edited by J. A. Keller, 62–79. Oxford: Oxford University Press.
Ostertag, G. (2019). Structured propositions and the logical form of predication. *Synthese*, 196, 1475–99.
Paul, L. A. (2002). Logical parts. *Noûs*, 36(4), 578–96.
Peacocke, C. (1986). Analogue content. *Proceedings of the Aristotelian Society*, Supplementary Volume 60, 1–17.
Peacocke, C. (2019). *The Primacy of Metaphysics*. Oxford: Oxford University Press.
Pinto, Y., Sligte, I. G., Shapiro, K. L., Lamme, V. A. F. (2013). Fragile visual short-term memory is an object-based and location-specific store. *Psychonomic Bulletin and Review*, 20, 732–9.
Prinzmetal, W. (2012). At the core of feature integration theory: on Treisman and Schmidt (1982). In *From Perception to Consciousness: Searching with Anne Treisman*, edited by J. Wolfe and L. Robertson, 211–16. New York, NY: Oxford University Press.
Prinzmetal, W., Presti, D. E., and Posner, M. I. (1986). Does attention affect visual feature integration? *Journal of Experimental Psychology: Human Perception and Performance*, 12, 361–9.
Pylyshyn, Z. W. (2003). *Seeing and Visualizing: It's Not What You Think*. Cambridge, MA: MIT Press.
Quilty-Dunn, J. (2017). *Syntax and Semantics of Perceptual Representation*. PhD thesis, The Graduate Center. NY: City University of New York.
Quilty-Dunn, J. (2020a). Concepts and predication from perception to cognition. *Philosophical Issues*, 30(1), 273–92.
Quilty-Dunn, J. (2020b). Is iconic memory iconic. *Philosophy and Phenomenological Research*, 101(3), 660–82.
Quilty-Dunn, J. (2020c). Perceptual pluralism. *Noûs*, 54(4), 807–38.
Quilty-Dunn, J. and Mandelbaum, E. (2020). Noninferential transitions: imagery and association. In *Inference and Consciousness*, A. Nes and T. Chan, 151–71. London: Routledge.
Rosenthal, D. (2005). *Consciousness and Mind*. Oxford: Oxford University Press.
Rosenthal, D. (2010). How to think about mental qualities. *Philosophical Issues*, 20(1): 368–93.
Sauvan, X. M. and Peterhans, E. (1999). Orientation constancy in neurons of monkey visual cortex. *Visual Cognition*, 6(1), 43–54.
Shepard, R. N. and Metzler, J. (1971). Mental rotation of three-dimensional objects. *Science*, 171, 701–3.
Smolensky, P. (1990). Tensor product variable binding and the representation of symbolic structures in connectionist systems. *Artificial Intelligence*, 46, 159–216.
Sober, E. (1976). Mental representations. *Synthese*, 33(1), 101–48.
Sorensen, C. E. (1956). *My Forty Years with Ford*. With S. T. Williamson. New York, NY: Norton.

Spelke, E. (1988). Where perceiving ends and thinking begins: The apprehension of objects in infancy. In *Perceptual Development in Infancy: Minnesota Symposium on Child Psychology, Vol. 20*, edited by A. Yonas, 197–233. Hillsdale, NJ: Lawrence Erlbaum.

Spence, C. and Bayne, T. (2015). Is consciousness multisensory? In *Perception and Its Modalities*, edited by D. Stokes, M. Matthen, and S. Biggs, 95–132. New York, NY: Oxford University Press.

Sperandio, I., Chouinard, P. A., and Goodale, M. A. (2012). Retinotopic activity in V1 reflects the perceived and not the retinal size of an afterimage. *Nature Neuroscience*, 15(4), 540–2.

Sperling, G. (1960). The information available in brief visual presentations. *Psychological Monographs: General and Applied*, 74(11), 1–29.

Stalnaker, R. (1984). *Inquiry*. Cambridge, MA: MIT Press.

Tiurina, N. A., and Utochkin, I. S. (2019). Ensemble perception in depth: correct size-distance rescaling of multiple objects before averaging. *Journal of Experimental Psychology: General*, 148(4), 728–38.

Toribio, J. (2011). Compositionality, iconicity, and perceptual nonconceptualism. *Philosophical Psychology*, 24(2), 177–93.

Treisman, A. (1988). Features and objects. *The Quarterly Journal of Experimental Psychology, Section A: Human Experimental Psychology*, 40(2), 201–37.

Treisman, A. (1996). The binding problem. *Current Opinion in Neurobiology*, 6(2), 171–8.

Treisman, A. and Gelade, G. (1980). A feature-integration theory of attention. *Cognitive Psychology*, 12, 97–136.

Treisman, A. and Sato, S. (1990). Conjunction search revisited. *Journal of Experimental Psychology: Human Perception and Performance*, 16(3), 459–78.

Treisman, A. and Schmidt, H. (1982). Illusory conjunctions and the perception of objects. *Cognitive Psychology*, 14, 107–41.

Utochkin, I. S. and Tiurina, N. A. (2014). Parallel averaging of size is possible but range-limited: a reply to Marchant, Simons, and De Fockert. *Acta Psychologica*, 146, 7–18.

Whitney, D. and Yamanashi Leib, A. (2018). Ensemble perception. *Annual Review of Psychology*, 69, 105–29.

Wolfe, J. M. (2021). Guided Search 6.0: an updated model of visual search. *Psychonomic Bulletin and Review*, 28, 1060–92.

Wolfe, J. M., Cave, K. R., and Franzel, S. L. (1989). Guided search: an alternative to the feature integration model of visual search. *Journal of Experimental Psychology: Human Perception and Performance*, 15(3), 419–33.

Zeki, S. (1993). *A Vision of the Brain*. Oxford: Blackwell.

5
Figments of imagination: 'Scaffolded attention' creates non-sensory object and event representations

Joan Danielle K. Ongchoco and Brian J. Scholl

1 Introduction: individuals

What do we see? The answer seems obvious: we see the objects and events that populate our local environments. And just what constitutes an object or event? This question leads directly down a well-known metaphysical rabbit hole, but an intuitive answer might begin by noting that the world behaves according to certain highly regular statistical patterns—or recurring relationships between incoming stimuli. For example, if two stones are sitting next to each other, and you push the left side of the rightmost stone, the right side of the rightmost stone moves along with it—whereas the right side of the leftmost stone stays put. It is tremendously useful to be able to predict such patterns when interacting with our environments, and so we directly see them in the form of discrete objects. (As Steven Pinker (1997, p. 317–18) once noted, if you want to know what an object is, just grab some and pull; the stuff that comes with your hand is usually the object, and the stuff that does not is not.) Similarly, temporal events intuitively exist insofar as the local statistics of the environment tend to change only gradually at most times, but then relatively suddenly at other times. Consider, for example, a student attending various classes throughout the schoolday: what was going on at the end of one class might have been rather similar to what was going on at the beginning of that same class, yet perhaps radically different from what was going on at the beginning of their next class. And this same point can be phrased in terms of predictability: the statistics that describe our local environments will be far more reliable predictors of what is about to happen at some times (while in an ongoing event, e.g., in the same class) compared to others (across an event boundary, e.g., when switching classes).

A great deal of work in cognitive science has focused on the nature and construction of mental object and event representations, simply because they *matter*. In particular, these discrete representations serve as the underlying units over which many other perceptual and cognitive processes operate. Visual attention, for example, is often irresistibly object-based, in the sense that it is far easier to select multiple features of a single visual object than to select those same features (perhaps even equally spaced) from different visual objects—the so-called 'same-object advantage' (e.g., Duncan, 1984; for a review, see Scholl, 2001). In the idealized display depicted in Figure 5.1, for example (which borrows a famous stimulus configuration from Egly *et al.* (1994)), observers are better at discriminating whether two probes

Joan Danielle K. Ongchoco and Brian J. Scholl, *Figments of imagination* In: *Sensory Individuals*. Edited by: Aleksandra Mroczko-Wąsowicz and Rick Grush, Oxford University Press. © Oxford University Press 2023. DOI: 10.1093/oso/9780198866305.003.0006

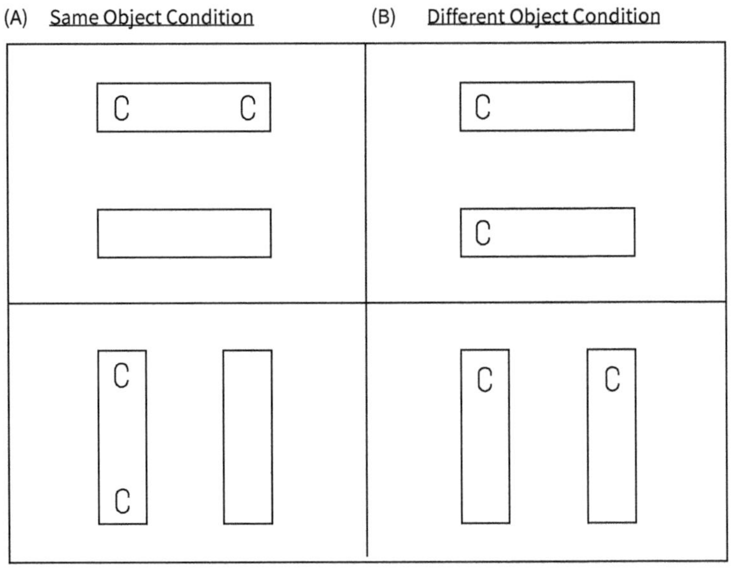

Figure 5.1 The traditional two-rectangle stimuli from 'same-object advantage' studies. (A) Potential probe locations on different ends of the same rectangle. (B) Potential equidistant probe locations spanning multiple rectangles.
Based on Egly *et al.*, 1994.

are identical or not when the probes appear on opposite ends of the same rectangle (as in Figure 5.1A) compared to when they appear (just as far from each other) on the same ends of different rectangles (as in Figure 5.1B). (Such effects are sometimes characterized in terms of the facilitated spreading of attention within objects, and sometimes by the cost of switching from one object representation to another.) Or similarly (moving from space to time), it is easier to compare two probes presented at different times when they occur within the same temporal event (e.g., a single musical phrase) than when they occur with the same delay but across an event boundary (e.g., in different musical phrases (De Freitas *et al.*, 2014)).

Or consider the nature of visual working memory: such memories have strict capacity limits (as in the classic 7 ± 2 formulation, or the more recent 'magic number 4' (Cowan, 2001; Miller, 1956)), but what are the units over which those limits are determined? Here too, the answer may involve discrete object representations—rather than, say, the brute amount of to-be-remembered information (for a review of this controversial topic, see Suchow *et al.*, 2014). For example, in some situations, it is more difficult to store the colours of six objects than to store the colours of only three objects—but it may be no more difficult to store four features from a single object (e.g., its colour, shape, size, and texture) than to store only two of these features (e.g., Luck and Vogel, 1997; Vogel *et al.*, 2001; for a review, see Fukuda *et al.*, 2010). Or similarly (moving from space to time), your ability to recall information (say, a list of words) from working memory is a function not just of the (continuous amount of) time elapsed since you studied them, but also of the number of discrete intervening events. For example, performance on such simple tasks is impaired when one walks through a doorway—compared to walking an equivalent distance within a single room (e.g., Pettijohn and Radvansky, 2016; Radvansky and Copeland, 2006; for a review, see Radvansky, 2012). (And far beyond mental processes such as attention and memory, this move from the continuous to the discrete has

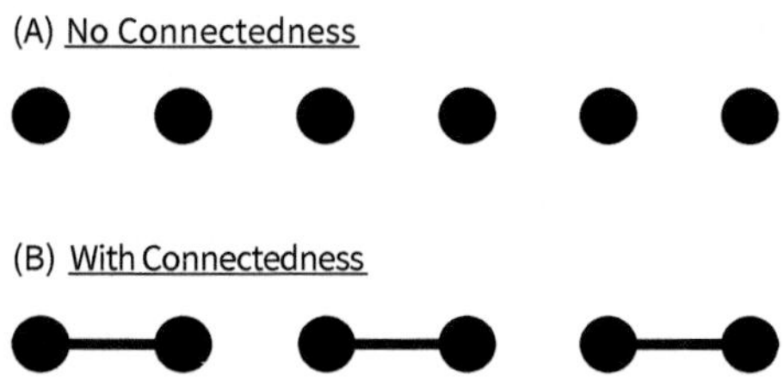

Figure 5.2 An example of grouping by connectedness. (A) Six independent discs. (B) Three dumb-bells.
After Palmer & Rock, 1994.

also played a prominent role in other well-known scientific advances—perhaps most famously in the discovery that apparently continuous change over evolutionary time is mediated by an entirely discrete genetic code.)

1.1 Sensory individuals

So far we have determined that objects and events exist in the world (at least in an intuitive statistical sense), and that discrete object and event representations are central to many cognitive processes (such as attention and memory). But how does the visual system identify these discrete individuals in the first place? In nearly all cases, objects are formed via segmentation and grouping mechanisms that operate over specific sensory cues—'sensory' in the sense that they derive (however indirectly) from signals present in the incoming light or sound itself.

One such cue, for example, is connectedness (e.g., Palmer and Rock, 1994): when two elements are connected by a contour, they tend to be grouped into a single object representation—as in the contrast between the six discs you see in Figure 5.2A vs the three dumb-bells you see in Figure 5.2B. Another example is closure: contours are more likely to be seen and represented as parts of single discrete objects (and same-object advantages for such displays are stronger) when they have closed boundaries, as in the contrast between the groups of parallel lines in Figure 5.3A vs the rectangles in Figure 5.3B (Marino and Scholl, 2005). Still other cues are 'sensory' but may only be present in statistical patterns over time. For example, if two shapes repeatedly co-occur in the same arrangement together, then they may come to exhibit object-based effects even when they are not explicitly grouped together in any single display (Zhao and Yu, 2016).

1.2 The current focus: *non-sensory* individuals?

The specific grouping and segmentation cues just reviewed may be of some interest, but it is surely no surprise that some such cues exist. After all, if there were no sensory cues, it might not seem possible for there to be substantive object or event representations at all. (We see objects and events all day long, but in an important sense, they seem to disappear when we close our eyes!) As a result, it might seem initially surprising, or even paradoxical, that our

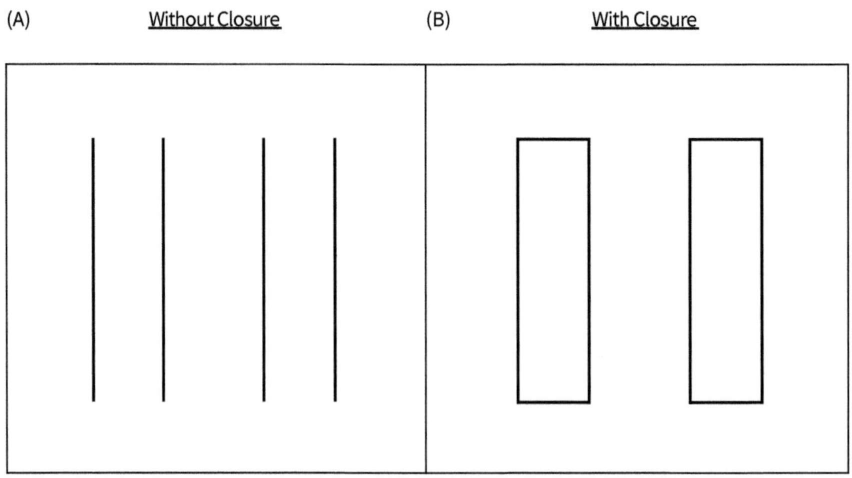

Figure 5.3 Lines grouped by proximity (A) or via both proximity and closure (B), as explored by Marino and Scholl (2005).

focus here will be on the possibility of *non*-sensory individuals in everyday perception—on the vivid perception of structured objects and events from displays that have no explicit sensory structure whatsoever.

That such phenomena are even possible can be appreciated by staring for a minute at the blank grid in Figure 5.4. (Take a break from reading and really do this: just try to clear your

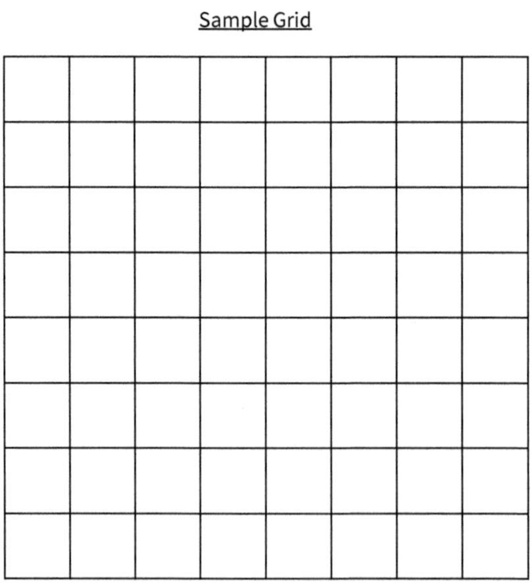

Figure 5.4 An example of visuospatial scaffolded attention. Clear your mind and simply stare at this regular grid for a minute. When doing so, many people find that they start seeing a shifting array of structured shapes and patterns (such as a block-letter H).

mind, and simply stare at this grid without trying to see anything in particular—and keep it up for at least a minute.) This is surely among the least exciting figures that you will encounter while reading today (or in this book): this figure consists of an array of identical (horizontal and vertical) lines arranged in an entirely regular pattern. These lines form squares, but the resulting array of squares has no *structure*—with each square on the grid identical to every other square.

Nevertheless, a curious thing can happen when viewing such a grid. Despite the lack of structure, many people find that they *see* structure anyway, in a spontaneous and stochastic manner: certain squares seem to be grouped with each other (despite the lack of classical grouping cues), such that these groups of squares seem to appear as a sort of 'figure' against the 'ground' of the rest of the grid. For example, you might suddenly see a simple geometric shape (such as a '+' sign) or a pattern (such as a pair of vertical lines), or even a recognizable form (such as a block letter 'H'). These patterns seem unambiguous—in the sense that you are never uncertain about whether you are seeing an 'H' or a '+', but the experience is inherently fleeting: different percepts (corresponding to different groups of squares being brought into the 'figure') seem to come and go spontaneously, and even uncontrollably. (When we stare at Figure 5.4 while actively trying to see, say, a block letter 'E', we sometimes fail for a moment, and unintentionally see a block letter 'S' instead.) Thus, while the grid itself of course is sensory, the more complex (and clear and concrete) structured patterns that people experience seeing have no sensory trace at all.

We have dubbed this phenomenon scaffolded attention (Ongchoco and Scholl, 2019), insofar as the unstructured squares seem to form the 'scaffold' from which these curious shifting percepts are built. (You might find, for example, that you readily see such shapes when viewing Figure 5.4, but not when staring at an empty page—when this scaffold is not available.) Scaffolded attention seems rather different from the broader category of hallucinations, in that it arises in everyday non-clinical observers. And the resulting 'imagined objects' also depart from more familiar notions of mental imagery, in that the shapes and patterns people experience seeing seem especially clear and concrete (sometimes even hard to distinguish from actual visual structure)—and they may also arise without (and even despite!) any conscious intention.

In this chapter, we explore the nature of scaffolded attention, with a special focus on the possibility that the patterns people experience correspond to the creation of bona fide object (and event) representations in perception. First, we review the *power* of visuospatial scaffolded attention, as in the 'grid' example, by showing how the perceived shapes and patterns are represented as bona fide objects in visual processing. Second, we explore the *scope* of this phenomenon, reviewing how it also operates in time (rather than in space) and in audition (rather than in vision). And finally, we characterize the *prevalence* of scaffolded attention, exploring how it varies across individuals and may be related to imagery and to other forms of attention. Collectively, this work demonstrates how not all individuals are 'sensory' and how we can effectively create objects and events with our own minds.

2 The power of scaffolded attention: object-based effects

Let us first characterize the phenomenon of (spatial) scaffolded attention in more detail, focusing in particular on its power to influence other aspects of perception and attention.

2.1 Phenomenology

We tried the exercise from Figure 5.4 on ourselves, colleagues, family, friends, and other members of our university community as an informal experiment. We presented them with an 8×8 grid of squares on a web browser, and asked them to clear their mind and stare at the grid for a minute. Afterwards, we asked them to indicate the particular squares that formed shapes and patterns that they saw on the grid, by using their mouse cursor (if on a desktop computer) or finger (if on a tablet). Upon interaction, the squares of the grid turned grey to give the participant feedback. When they were done drawing a pattern, participants could submit this pattern, after which the grid would refresh (so that they could submit multiple patterns).

The responses to this simple exercise were incredibly rich and varied. Some people saw horizontal and vertical lines (as in Figure 5.5A and B). Others saw symbols such as a 'plus' or 'equal' sign (as in Figure 5.5C and D). There were shapes, letters, and digits (Figure 5.5E–G), and even letter-like (or digit-like) configurations that do not exist as English characters (Figure 5.5H and I). And some people even saw more complex and extensive patterns (Figure 5.5J–L). In addition, some people saw only a single figure (such as a block letter H) while staring at the grid (though they might sometimes see that figure on different parts of the grid). Other people, in contrast, reported seeing many different (and distinct) shapes and patterns during their minute of staring at the grid. And anecdotally, these percepts seemed especially stochastic: people often mentioned seeing these patterns appear, shift, and disappear in quick succession, and often without any corresponding intention.

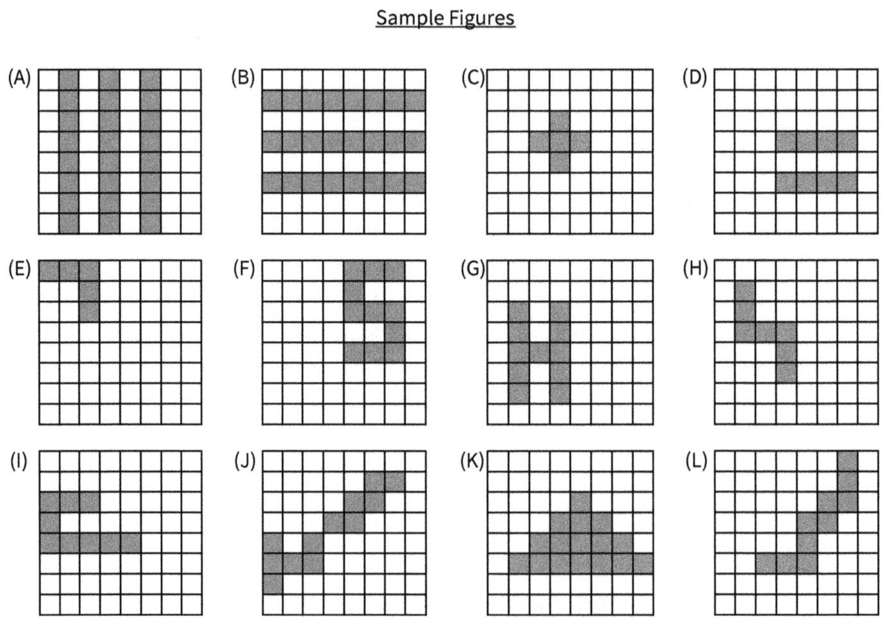

Figure 5.5 Sample participants' drawings of the patterns and shapes they saw while staring at the grid in Figure 5.4: horizontal and vertical lines (A–B), various symbols and shapes (C–E), block-letters/digits (F–G), and more abstract patterns (H–I).

Beyond this diversity of perceived shapes and patterns, we can also characterize the nature of those percepts in more detail. Although there was some variance in how observers described their experiences, they all seemed to refer in one way or another to a form of perceptual selection or 'highlighting'. Some phrased this in terms of figure/ground segregation, noting that they were 'bringing the relevant squares forward'—or even that they 'stared at the grid until the rest of the squares went away' (as if irrelevant squares were being suppressed or inhibited as the figure came to the foreground). Others described an even more vivid phenomenology, in which the squares that were parts of the perceived patterns appeared brighter or differently shaded. And yet others described their experiences more actively, with descriptions reminiscent of 'visual routines' (Ullman, 1984), as they talked about the squares being 'outlined' or 'traced' while they imagined.

2.2 Previous research

We know of only a handful of previous studies that directly explored this sort of phenomenon, all from 25 to 45 years ago (Crassini, 1986; Farah, 1989; Heil *et al.*, 1993; Kosslyn *et al.*, 1988; Podgorny and Shepard, 1978, 1983). (In addition, several other papers have since correlated intentional imagery performance using such stimuli with individual traits or characteristics such as age (e.g., Kosslyn *et al.*, 1990), dyslexia (e.g., Koenig *et al.*, 1991), and a variety of clinical symptoms such as social anxiety (e.g., Morrison *et al.*, 2011) or depression (e.g., Zarrinpar *et al.*, 2006).) Critically, these early studies all differed from the current projects in several key ways. First, and perhaps most importantly, these studies all focused not on the global patterns that observers perceived per se, but rather on the individual squares themselves. Second, these studies were always constrained to vision and to space (whereas below we will introduce this same phenomenon in audition and in time). Finally, in these studies, participants were always explicitly asked to imagine a particular pattern, whereas we will also describe and probe the spontaneity of such phenomena.

In the earliest previous studies (Podgorny and Shepard, 1978, 1983), observers saw 5 × 5 grids, and were asked to imagine a particular shape, such as a block letter I or E, superimposed on the squares. As they imagined, a target probe dot could appear at the centre of a square, and observers simply had to indicate whether the probe was 'on' or 'off' the imagined shape. The result was just what you would expect in this situation if observers had been attending to actual (sensory) figures: probe detection was better (and response times were faster) for probes on squares that were part of the imagined shapes, compared to squares that were not—and in addition, probe detection was modulated by the compactness and size of the imagined shapes, such that responses were faster for imagined shapes with fewer squares (e.g., a vertical line with only five squares), compared to those with more squares (e.g., a block letter E with 11 squares).

These results seemed fully consistent with the leading attentional 'spotlight' metaphor of the day (for a retrospective review, see Cave and Bichot, 1999): just as a spotlight illuminates a given spatial region, so too attention may be selecting the particular squares of the relevant pattern. And just as a more diffuse spotlight would be correspondingly dimmer, the attentional benefit (as measured by the probe detection performance) was inversely related to the number of squares involved.

2.3 Bona fide object representations?

While this previous work focused on the individual squares, we were interested in the perceived shapes and patterns themselves. In particular, we sought to determine whether this phenomenology corresponded to the creation of actual discrete object representations in visual processing. If so, we reasoned that such experiences should give rise not only to spatial benefits for the relevant squares (as in Podgorny and Shepard's work), but also object-based effects (as in studies of object-based attention).

We tested this in the most direct way possible (Ongchoco and Scholl, 2019), by looking for the existence of traditional same-object advantages (as described above) with the imagined shapes. Observers saw 3 × 3 grids (see Figure 5.6A) and were asked to imagine specific patterns. Sometimes those patterns consisted of multiple discrete objects—such as two vertical (or horizontal) lines (as in the top row of Figure 5.6B). Other times, those patterns were connected by additional squares, such that observers now imagined single objects (as in the bottom row of Figure 5.6B). We asked them to attend to the squares until they could 'see

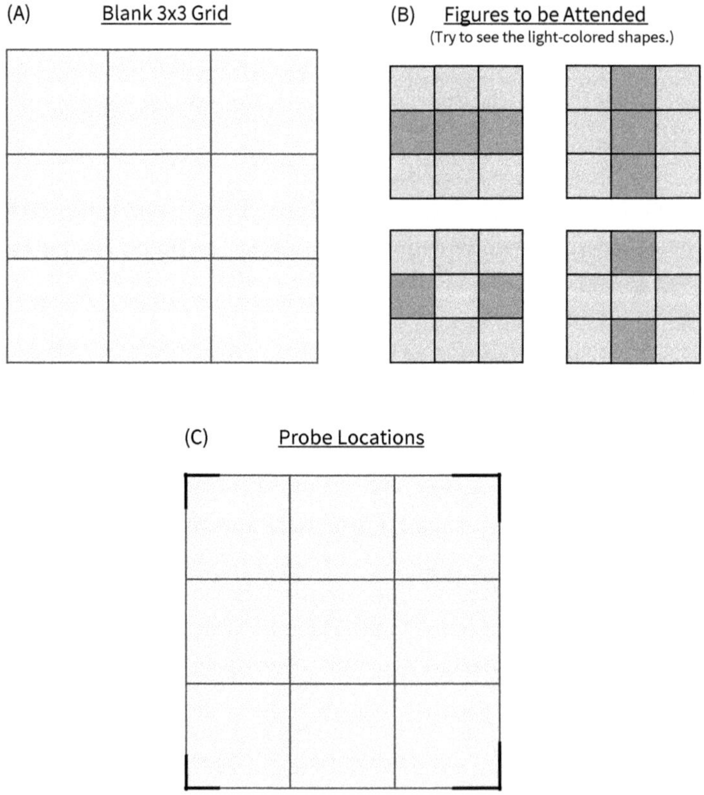

Figure 5.6 Sample stimuli from the studies of visuospatial scaffolded attention explored in Section 2 (Ongchoco and Scholl, 2019). (A) A blank 3 × 3 grid. (B) Sample patterns that observers were asked to imagine while viewing the grid. (C) Sample 'corner' probe locations (with 'short' probes on the left, and 'long' probes on the right). (See the text for details.)

a pattern, almost as if it were superimposed on the grid'. While they imagined on each trial, two probes appeared, each on a different corner of the grid (see Figure 5.6C). These 'corner' probes varied in the lengths of their line segments—sometimes with a relatively 'short' corner (i.e., the corner probes on the left side of the grid in Figure 5.6C), and sometimes with a relatively 'long' corner (i.e., the corner probes on the right side of the grid in Figure 5.6C)—and observers simply had to report whether the two corner probes on each trial had matching segment lengths or not.

Of course, in the actual displays themselves, there were no objects—only a structureless array of squares. Might we nevertheless observe robust same-object advantages for the purely imagined patterns? In fact, we did (Ongchoco and Scholl, 2019): observers were better (and faster) at discriminating between probes that were on different ends of the same object (e.g., a probe on each end of one of the rectangles from the top row of Figure 5.6B), compared to equidistant probes that spanned multiple objects (e.g., with one probe on each of the two rectangles from the top row of Figure 5.6B). And critically, these robust differences disappeared when these same patterns were connected into single objects via a central square (as in the block letter I and H shapes from the bottom row of Figure 5.6B)—further emphasizing the object-based nature of these effects. And these differences could not be explained in terms of the brute number of squares involved in the imagined patterns (i.e., six squares for the lines vs seven squares for the block letters), since in high-powered replications, we also observed robust same-object advantages with different numbers of squares involved when observers viewed 4 × 4 grids (and when the connected patterns did not involve central squares anymore, but rather U-like patterns).

2.4 From attention to grouping

The experiments reviewed above collectively demonstrate that scaffolded attention involves the creation of bona fide object representations, beyond the squares themselves. This seems especially striking, not only because there were no actual structured objects in the visual stimuli, but also because this seems to invert the usual explanatory relationship between objects and attention. Even the label *object-based attention* seems to imply that the objects come first, and then attention selects these pre-segmented units. But in scaffolded attention, it seems to be attention that is creating the object representations in the first place. (For this reason, we subtitled the original experimental report of such effects 'From object-based attention to attention-based objects' (Ongchoco and Scholl, 2019).)

What might explain this mental creation of 'non-sensory' objects? We speculate that the missing link is perceptual grouping. Attention and grouping have sometimes been characterized as two sides of the same coin, such that attentional selection of two elements automatically groups them together. Indeed, it has been noted that 'you can modulate subjective grouping at will to some extent in many grouping displays, especially when the stimulus factors are relatively subtle' (Driver *et al.*, 2001, p. 90). Here, the structureless scaffold of squares may serve as the subtlest possible 'stimulus factors': observers intentionally attend to the (otherwise identical) squares in the pattern, and as a result they are grouped together, forming perceptual objects that have both vivid phenomenological consequences and performance benefits.

3 The scope of scaffolded attention: events (in time) vs objects (in space)

So far we have characterized scaffolded attention as a visuospatial phenomenon, and all of the previous work in this vein has employed visually presented scaffolds in the form of spatial grids. But the connections between grouping and attention that were discussed in the previous paragraph do not seem specific in this way. Accordingly, to determine the nature and scope of this phenomenon, we have also explored such connections in the 'audiotemporal' domain (Ongchoco and Scholl, 2022a)—looking to see if *temporally* regular auditory patterns may also serve as a scaffold, within which people perceive temporal structure (perhaps due to a similar attention-to-grouping mechanism).

3.1 Phenomenology

Our studies of visuospatial scaffolded attention (as reviewed in the previous section) were initially motivated less by the early papers on this topic (which we discovered later), and more by our own 'everyday hallucinations' of seeing structure when looking at grids (as on regularly tiled floors). Similarly, the work we turn to now was motivated by our everyday experiences of *hearing* structure in uniform beats (such as a metronome or regular dripping water). Although each individual 'beat' in such patterns is identical to each of the others in every respect (duration, pitch, intensity, etc.), we nevertheless find ourselves hearing rhythms in such stimuli—for example, hearing the beats as coming in groups, which may even be highlighted by phantom 'downbeats' every three or four notes.

We also note in passing that such structure can be phenomenologically salient when *producing* such patterns oneself, as well. In a previous life, for example, the second author of this chapter was a jazz drummer, and in this context, one may sometimes have to produce many regular beats (e.g., on a drum or cymbal) without changing their intensities—but one nevertheless often experiences producing these beats in groups of various types and sizes. Meanwhile, in another previous life, the first author of this chapter was a competitive ballroom dancer, and when choreographing a piece, one may sometimes tune in to the regular beats underlying the melody of a waltz or a quickstep—that are each identical and simply provide the tempo of the dance—but one nevertheless can hear these beats in various rhythmic patterns that go on to inspire the actual movements that are produced.

When we recreated such stimuli in the laboratory, we discovered that these experiences were shared by many other observers. In one sense, this should have been an especially boring exercise, since listening to fully regular beats normally is not so engaging. But our observers were fascinated by how they did indeed experience higher-order structure. Some, for example, reported hearing phantom temporal gaps between perceived rhythmic phrases, while others reported that individual notes sounded somehow different based on whether they were parts of the perceived rhythmic phrases (a type of 'figure') or not (the respective 'ground'). How people perceived these two types of beats also varied: some mentioned that the 'notes they didn't imagine faded in volume' relative to those beats that were part of the experienced rhythmic patterns, while others mentioned that these two types of beats were heard differentially as 'thumps' vs 'booms', and that one 'could almost hear a ticking where the rhythm was'.

3.2 Bona fide event representations

Is this shared experience of hearing rhythmic patterns in regular beats a merely phenomenological curiosity—perhaps a type of epiphenomenon? Or might this experience go on to structure the nature of perceptual representations in a way that has other downstream consequences for attention and memory? To find out, we conducted a dynamic 'event-based' analogue of the (static) 'object-based' experiments reviewed in the previous section. Previous work revealed that same-'object' advantages do indeed operate as well in time and audition: comparing probes on two beats presented at different times is facilitated when both beats are part of the same perceived rhythmic phrase, controlling for factors such as absolute inter-beat latency (De Freitas *et al.*, 2014). Here we explored whether a similar 'same-event advantage' would result from patterns wherein the rhythmic structure was entirely non-sensory, due to scaffolded attention in time (Ongchoco and Scholl, 2022a).

Just as observers in the visuospatial experiments were explicitly asked to 'imagine' certain particular patterns of squares on a grid, here observers were first introduced to explicit (sensory) rhythms that they were told to memorize—after which they were asked to imagine those rhythms as repeating during subsequent regular tone sequences (Figure 5.7A). And just as observers in the visuospatial experiments had to respond to the 'corner' probes (Figure 5.6C), here (in a detection task) observers simply had to press a key as soon as they heard the second of two higher-pitched tones (Figure 5.7B). In this design, the first higher-pitched tone effectively served as a 'cue' to the imminent arrival of the second 'probe' tone (to which observers had to respond). And indeed, the results of this study were accordingly parallel. Akin to the 'same-object advantage' from visuospatial scaffolded attention, these experiments yielded 'same-*event* advantages': observers responded faster when the cue and probe tones occurred in the same (purely imagined!) rhythmic phrase, compared to when they spanned a phrase boundary (controlling for their brute temporal separation; see Figure 5.7C). (And critically, this effect disappeared in control experiments in which the probe tones were played at the same moments, but the cue tones were never played—demonstrating that such effects are specific to the relationship within vs across event representations.) This demonstrates that the experience of scaffolded attention in time (as in space) is not just an epiphenomenon, but rather changes the resulting perceptual representations in ways that have downstream consequences for performance.

3.3 Attention and grouping redux

Despite the drastic changes in these experiments in both modality (from vision to audition) and dimension (from space to time), we suggest that the underlying explanation remains the same. Although the beats themselves were all identical, the task effectively requires observers to attend to some beats more than to others. This attention then groups these beats together, giving rise to segmentation, boundaries, and figure/ground relationships. And the result is another form of 'everyday hallucination': we experience non-sensory events that once again have consequences for both phenomenology and performance.

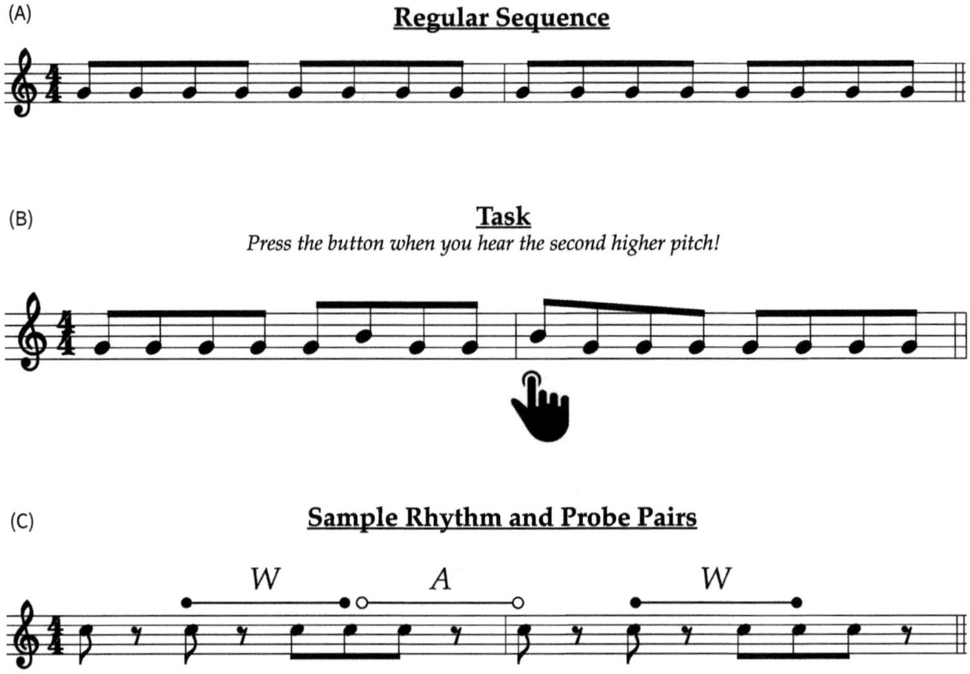

Figure 5.7 Musical notation of the stimuli from the studies of 'audiotemporal' scaffolded attention explored in Section 3 (Ongchoco and Scholl, 2022a). (A) A regular sequence of tones that observers actually heard. (B) A regular sequence of tones, two of which have higher pitches. Observers simply had to press a key as quickly as possible when they heard the second higher-pitched tone. (C) A sample rhythm that observers were asked to imagine. Higher-pitched tone pairs could occur within the same rhythmic phrase (here labelled 'W') or could span multiple rhythmic phrases (labelled 'A'), equating for brute temporal latencies and the number of intervening tones. (See the text for details.)

4 The prevalence of scaffolded attention: individual differences and their correlates

Throughout this chapter, we have been writing about these curious non-sensory individuals that people perceive when looking at (or listening to) regular 'scaffold' patterns, as if everyone experienced this phenomenon—and to the same degree. But is this true? And more generally, do such phenomena occur only in the context of experiments in which observers are asked to (try to) perceive such patterns? The answer to both of these questions is clearly 'no'.

4.1 Anecdotal observations

In the first place, we know that many people experience scaffolded attention even outside of the laboratory context, because they have been eager to tell us so! Indeed, while conducting the experiments described above, many of our observers have mentioned how they experience such percepts during their everyday lives—for example, while staring at the regular tiles on a bathroom floor (or on an office ceiling) or when staring at a grid of regular objects as in the game of *Go*. And these are often the same observers who note that they often cannot help spontaneously seeing such patterns (beyond the individual squares) as soon as they are shown a grid.

Other observers, however, report not perceiving any such patterns at first—when just asked to view the grid (or listen to the beats) with no specific instructions—but then as soon as we tell them to try to actively imagine a structured pattern or two, they eagerly start describing such new-found experiences on their own. And still other observers have noted that they have not had such experiences before, and have noted that actually *perceiving* the patterns is difficult (or even impossible) for them, despite any instructions. (One laboratory member who has aphantasia—the lack of visual mental imagery (Keogh and Pearson, 2018)—even suffered through multiple laboratory meetings filled with freewheeling discussions of the richness and variety of spontaneous scaffolded attention in different contexts before noting that they had no idea what everyone else was talking about. To them, the empty grid was perceived as just that, no matter how long they stared or how hard they tried to see structured patterns.)

4.2 Experimental measurements

Inspired by such anecdotes, we conducted an individual-differences study with hundreds of observers, in order to empirically determine the prevalence of scaffolded attention as both an ability and a spontaneous experience (Ongchoco and Scholl, 2022b). This study had three key parts.

First, as a measure of spontaneity, we retreated from specific 'leading' instructions: instead, we showed observers a grid and just asked them to 'clear your mind and pay close attention to the grid' for 30 seconds (Figure 5.8A). And then afterwards, we simply asked them (in a series of explicit questions) whether they had experienced seeing any of the squares as being grouped into larger structured shapes or patterns—and those who reported 'yes' were then asked to describe their percepts and to indicate how many distinct patterns they saw. Second, beyond spontaneity, we also asked (in a separate experiment) whether observers *could* experience scaffolded attention when explicitly asked to try. Finally, to explore the factors that may mediate any individual differences in scaffolded attention, we also had observers complete a battery of off-the-shelf measures, the results of which we could then correlate with both the ability and spontaneity responses. We focused on tests of attention (given our attention-based explanations for this phenomenon) and mental imagery (which is perhaps the most common variety of intentional visual 'hallucination'). In particular, we tested these four measures:

1. *VVIQ*. The 'Vividness of Visual Imagery Questionnaire' (Marks, 1973; Figure 5.8B) is perhaps the most common measure of mental imagery. It asks people (for example) to 'think of a loved one whom they do not frequently see' and to report how vividly they could imagine the contours of their head, shoulders, or body.
2. *SUIS*. The 'Spontaneous Use of Imagery Survey' (Reisberg *et al.*, 2003; Figure 5.8B) measures how spontaneously people use imagery in their everyday lives. It asks people

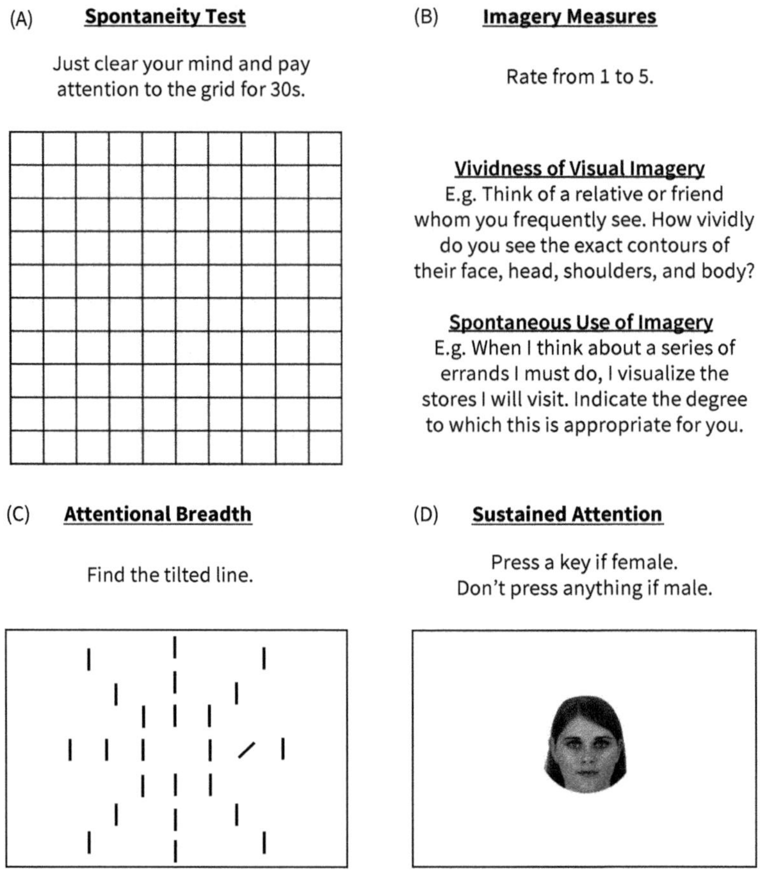

Figure 5.8 Various tasks discussed in the study of individual differences reviewed in Section 4 (Ongchoco and Scholl, 2022b). (A) A test of spontaneous scaffolded attention. (B) Sample prompts from measures of mental imagery (the VVIQ and SUIS), where observers responded using scales from 1 (no image at all/never appropriate) to 5 (clearly and vividly/always appropriate). (C) A caricatured display from the FFOV task (a measure of attentional breadth), in which observers had to report (in a subsequent selection display) the location of a single tilted target line. (D) A caricatured display from the GradCPT task (a measure of sustained attention and vigilance), in which observers simply pressed a key every time they saw a female face, but withheld responses to male faces.

(for example) whether 'they visualize two-digit numbers (e.g., 24 and 31) in order to add them' or whether 'they visualize the stores they will visit when thinking about a series of errands to run'.

3. *FFOV*. The 'Functional Field of View' task (Pringle *et al.*, 2001; Figure 5.8C) is a standard measure of attentional breadth. Observers see an array of vertical lines flash on screen on each trial and must simply report the location of a single tilted target line—which can vary in its distance from the array's centre. The function linking performance to the target's eccentricity then serves as a measure of the spatial breadth of each observer's attentional window.

4. GradCPT. The 'Gradual Continuous Performance Task' (Rosenberg *et al.*, 2013, 2016; Figure 5.8D) is a measure of vigilance and sustained attention. Observers see a series of faces (Ma *et al.*, 2015) appear, one after another, and must press a key whenever they see a female face (while withholding responses to male faces). The male faces appear only 10% of the time, rendering the task (purposefully and maximally) boring. The better an observer is at sustaining attention, the better they should be at withholding responses despite the task's repetitive nature.

This broad survey of individual differences revealed that 40% of observers spontaneously experienced scaffolded attention (seeing structured shapes and patterns beyond the squares themselves), and that 78% of observers could experience it when they were explicitly asked to try. The spontaneity of scaffolded attention was correlated with higher scores on the VVIQ, SUIS, and FFOV tasks, but not with GradCPT performance. But only attentional breadth significantly predicted spontaneity over and above the variance explained by the other measures. The ability to experience scaffolded attention when trying to do so, meanwhile, was uniquely predicted by both attention measures (FFOV and GradCPT), but not by either of the imagery measures (VVIQ and SUIS).

These results surprised us, at least in part—especially the correlations (or lack thereof) involving mental imagery. To be sure, the experience of scaffolded attention feels different from mental imagery in many ways (e.g., as mental images can often seem less distinct, and more closely driven by conscious intentions). Nevertheless, mental imagery is surely the most salient and well-studied example of a sort of putative 'non-sensory' perception—and so it would have been natural to expect that people with especially vivid or active imagery would also be more likely to experience scaffolded attention (at all or to a greater degree). (Indeed, some early research simply assumed that such tasks were primarily tapping mental imagery (e.g., Kosslyn *et al.*, 1988).) Accordingly, it was striking to discover that familiar measures of mental imagery predicted neither the proclivity nor the ability to experience scaffolded attention. And this may also prove problematic for some philosophical treatments which have claimed that hallucination is primarily (or even always) a matter of imagery (e.g., Nanay, 2016).

Instead, these data supported the notion that scaffolded attention is indeed an attentional phenomenon (which we admit was something of a relief, since we had already used the term 'scaffolded attention' in print before collecting these data!). And at least in retrospect, we think that these particular connections (and attentional predictors) make some sense. The ability to see such 'everyday hallucinations' may depend on the corresponding ability to focus intently on the empty 'scaffold' grid in the first place, which might explain the link to measures of (especially sustained) attention. And then the spontaneity of such percepts might naturally also depend on just how many squares one is attending to: people with relatively narrow attentional breadth (as reflected by low FFOV scores) may be more likely to attend only to individual squares when looking at the grid, while people with relatively wide attentional breadth (as reflected by high FFOV scores) may be more typically or naturally attending to many squares at once. If scaffolded attention intrinsically involves different regular units of the scaffold (the identical squares in a grid, or the identical beats in a sequence) being grouped together by attention (as we suggested above), then this might naturally require that people be selecting multiple squares/beats at a time. And our repeated references to scaffolded attention in terms of 'everyday hallucinations' might also find some synergy with such attentional explanations, given that explanations for more extreme forms of hallucination (of the sort observed in clinical settings, e.g., with schizophrenic patients) have also sometimes alluded to a key role for attention (e.g., Collerton *et al.*, 2005).

5 Conclusions

The phenomenon of scaffolded attention that we have reviewed in this chapter suggests a new way of thinking about object and event representations—and where they come from. We typically think of objects and events as being (obviously, naturally) *sensory* in nature, with the exceptions (dreaming, imagery, hallucinations) all being 'fringe' cases of one form or another. (And surely the other chapters in this volume—not to mention the book's very title—are a testament to this, as they all focus on the nature of 'sensory individuals'.) But the experiences and experiments explored here suggest at least one instance (across multiple modalities and dimensions) wherein the everyday perception of many (or even most) people involves clear and concrete non-sensory individuals. They are *non-sensory* in the sense that they are 'figments of the imagination' (not deriving from any structure in the incoming sensory signals, beyond the existing uniform 'scaffold'), and they are *individuals* in the sense that these experiences correspond to bona fide object and event representations.

References

Cave, K. R. and Bichot, N. P. (1999). Visuospatial attention: beyond a spotlight model. *Psychonomic Bulletin and Review*, 6, 204–23.

Collerton, D., Perry, E., and McKeith, I. (2005). Why people see things that are not there: a novel perception and attention deficit model for recurrent complex visual hallucinations. *Behavioral and Brain Sciences*, 28, 737–57.

Cowan, N. (2001). The magical number 4 in short-term memory: a reconsideration of mental storage capacity. *Behavioral and Brain Sciences*, 24, 87–114.

Crassini, B. (1986). On the spatial distribution of visual attention. *Journal of Experimental Psychology: Human Perception and Performance*, 12, 380–2.

De Freitas, J., Liverence, B. M., and Scholl, B. J. (2014) Attentional rhythm: a temporal analogue of object-based attention. *Journal of Experimental Psychology: General*, 143, 71–6.

Driver, J., Davis, G., Russell, C., Turatto, M., and Freeman, E. (2001). Segmentation, attention and phenomenal visual objects. *Cognition*, 80, 61–95.

Duncan, J. (1984). Selective attention and the organization of visual information. *Journal of Experimental Psychology: General*, 113, 501–17.

Egly, R., Driver, J., and Rafal, R. D. (1994). Shifting visual attention between objects and locations: evidence from normal and parietal lesion subjects. *Journal of Experimental Psychology: General*, 123, 161–77.

Farah, M. J. (1989). Mechanisms of imagery-perception interaction. *Journal of Experimental Psychology: Human Perception and Performance*, 15, 203–11.

Fukuda, K., Awh, E., and Vogel, E. K. (2010). Discrete capacity limits in visual working memory. *Current Opinion in Neurobiology*, 20, 177–82.

Heil, M., Rösler, F., and Hennighausen, E. (1993). Imagery-perception interaction depends on the shape of the image: a reply to Farah (1989). *Journal of Experimental Psychology: Human Perception and Performance*, 19, 1313–20.

Keogh, R. and Pearson, J. (2018). The blind mind: no sensory visual imagery in aphantasia. *Cortex*, 105, 53–60.

Koenig, O., Kosslyn, S. M., and Wolff, P. (1991). Mental imagery and dyslexia: a deficit in processing multipart visual objects? *Brain* and *Language*, 41, 381–94.

Kosslyn, S. M., Cave, C. B., Provost, D. A., and von Gierke, S. M. (1988). Sequential processes in image generation. *Cognitive Psychology*, 20, 319–43.

Kosslyn, S. M., Margolis, J. A., Barrett, A. M., Goldknopf, E. J., and Daly, P. F. (1990). Age differences in imagery abilities. *Child Development*, 61, 995–1010.

Luck, S. J. and Vogel, E. K. (1997). The capacity of visual working memory for features and conjunctions. *Nature*, 390, 279–81.

Ma, D. S., Correll, J., and Wittenbrink, B. (2015). The Chicago face database: a free stimulus set of faces and norming data. *Behavior Research Methods*, 47, 1122–35.

Marino, A. C. and Scholl, B. J. (2005). The role of closure in defining the 'objects' of object-based attention. *Perception and Psychophysics*, 67, 1140–9.

Marks, D. F. (1973). Visual imagery differences in the recall of pictures. *British Journal of Psychology*, 64, 17–24.

Miller, G. A. (1956). The magical number seven, plus or minus two: some limits on our capacity for processing information. *Psychological Review*, 63, 81–97.

Morrison, A. S., Amir, N., and Taylor, C. T. (2011). A behavioral index of imagery ability in social anxiety. *Cognitive Therapy* and *Research*, 35, 326–32.

Nanay, B. (2016). Hallucination as mental imagery. *Journal of Consciousness Studies*, 23, 65–81.

Ongchoco, J. D. K. and Scholl, B. J. (2019). How to create objects with your mind: from object-based attention to attention-based objects. *Psychological Science*, 30, 1648–55.

Ongchoco, J. D. K. and Scholl, B. J. (2022a). Scaffolded attention in time: 'everyday hallucinations' of rhythmic patterns from regular auditory beats. *Attention, Perception, and Psychophysics*, 84, 322–40.

Ongchoco, J. D. K. and Scholl, B. J. (2022b). Hallucinating visual structure: individual differences in 'scaffolded attention'. *Cognition*, 225, 105129.

Palmer, S. and Rock, I. (1994). Rethinking perceptual organization: the role of uniform connectedness. *Psychonomic Bulletin and Review*, 1, 29–55.

Pettijohn, K. A. and Radvansky, G. A. (2016). Walking through doorways causes forgetting: event structure or updating disruption? *The Quarterly Journal of Experimental Psychology*, 69, 2119–29.

Pinker, S. (1997). *How the Mind Works*. Penguin Books. New York, USA.

Podgorny, P. and Shepard, R. N. (1978). Functional representations common to visual perception and imagination. *Journal of Experimental Psychology: Human Perception and Performance*, 4, 21–35.

Podgorny, P. and Shepard, R. N. (1983). Distribution of visual attention over space. *Journal of Experimental Psychology: Human Perception and Performance*, 9, 380–93.

Pringle, H. L., Irwin, D. E., Kramer, A. F., and Atchley, P. (2001). The role of attentional breadth in perceptual change detection. *Psychonomic Bulletin and Review*, 8, 89–95.

Radvansky, G. A. (2012). Across the event horizon. Current Directions in Psychological Science, 21, 269–72.

Radvansky, G. A. and Copeland, D. E. (2006). Walking through doorways causes forgetting: situation models and experienced space. *Memory and Cognition*, 34, 1150–6.

Reisberg, D., Pearson, D. G., and Kosslyn, S. M. (2003). Intuitions and introspections about imagery: the role of imagery experience in shaping an investigator's theoretical views. *Applied Cognitive Psychology*, 17, 147–60.

Rosenberg, M., Noonan, S., DeGutis, J., and Esterman, M. (2013). Sustaining visual attention in the face of distraction: a novel gradual-onset continuous performance task. *Attention, Perception, and Psychophysics*, 75, 426–39.

Rosenberg, M. D., Finn, E. S., Scheinost, D., *et al.* (2016). A neuromarker of sustained attention from whole-brain functional connectivity. *Nature Neuroscience*, 19, 165–71.

Scholl, B. J. (2001). Objects and attention: the state of the art. *Cognition*, 80, 1–46.

Suchow, J. W., Fougnie, D., Brady, T. F., and Alvarez, G. A. (2014). Terms of the debate on the format and structure of visual memory. *Attention, Perception, and Psychophysics* 76, 2071–9.

Ullman, S. (1984). Visual routines. *Cognition*, 18, 97–159.

Vogel, E. K., Woodman, G. F., and Luck, S. J. (2001). Storage of features, conjunctions, and objects in visual working memory. *Journal of Experimental Psychology: Human Perception and Performance*, 27, 92–114.

Zarrinpar, A., Deldin, P., and Kosslyn, S. M. (2006). Effects of depression on sensory/motor vs. central processing in visual mental imagery. *Cognition and Emotion*, 20, 737–58.

Zhao, J. and Yu, R. Q. (2016). Statistical regularities reduce perceived numerosity. *Cognition*, 146, 217–22.

6
Perceptual variation in object perception: A defence of perceptual pluralism

Berit Brogaard and Thomas Alrik Sørensen

1 Introduction

Before reading this chapter, pause for a minute and have a look at your surroundings. Notice the details of the room and the various objects scattered around it. Perhaps you have just brewed a cup of coffee, which now sits next to you on your desk. As you direct your attention towards it, you notice the smell of the coffee and the vapour that slowly ascends from it, conveying to you that it is hot and freshly brewed. Taking a sip, you register the coffee's warmth, liquidity, and smoothness on your tongue and the bitterness of the mocha flavours.

This brief exercise bears witness to our remarkable abilities to rapidly process, categorize, and act on the sensory information we extract from our surroundings. In this chapter, we offer new evidence from philosophy, psychology, and neuroscience that suggests that each of us harbours unique sensory strategies which constitute the basis of our perceptual abilities to categorize sensory input. We devise these strategies to solve the different tasks we encounter in our environment, and our perceptual system structurally adjusts to the strategies that are most successful (Mannix and Sørensen, 2019). Although many of us encounter similar environmental challenges, and therefore devise similar strategies to solve them, strategies can vary substantially across different perceivers. So while you and your colleague may agree that the liquid in the mug is coffee, you may harbour radically different colour, smell, and taste categories for coffee.

On a common characterization, a sensory individual is the output of perceptual categories that represent basic or causally unified features (e.g., Green, 2019). Going by this characterization, our view entails that when different perceivers view the same distal object in the same viewing conditions, their perceptual experiences may represent radically different sensory individuals (i.e., perceptual objects or features).

The thesis we set forth is what we might call 'pluralism about sensory individuals', or 'perceptual pluralism' for short.[1] Perceptual pluralism encompasses the following three

[1] As sensory individuals are the products of perceptual categorization, we could also refer to pluralism about perceptual sensory individuals as 'pluralism about perceptual categorization'. We hasten to say that pluralism about sensory individuals should not be confused with pluralism about the format of perceptual content (e.g., iconic vs symbolic) (Quilty-Dunn, 2020).

Berit Brogaard and Thomas Alrik Sørensen, *Perceptual variation in object perception* In: *Sensory Individuals*.
Edited by: Aleksandra Mroczko-Wąsowicz and Rick Grush, Oxford University Press. © Oxford University Press 2023.
DOI: 10.1093/oso/9780198866305.003.0007

propositions: (1) how sensory input is categorized is highly variable across different perceivers as a result of them using different perceptual strategies to solve the perceptual problems they encounter; (2) this variability in perceptual categorization need not be the result of misperception; and (3) contingent on minor constraints (e.g., not being the product of deviant brain processes), the resulting perceiver-specific sensory individuals represent actual features and objects in the external, mind-independent world.

Perceptual pluralism contrasts with what we will call 'perceptual monism'. To a first approximation, this is the view that for any sensory input, there is a single correct way to categorize it. We take this view—in its most extreme iteration—to imply that correctly processed and categorized sensory individuals represent external, mind-independent features or objects in a 1:1 fashion.

Before proceeding, we should clarify what we mean by 'perceptual strategies'. Perceptual strategies should be kept apart from the cognitive tasks they subserve. For example, an important cognitive task in everyday life is that of recognizing other people. To solve this task, one may apply a range of different perceptual strategies, for example, the recognition of facial features, gait, clothing, voice, context, etc. (Sørensen and Overgaard, 2018). With increased expertise, some strategies may be more successful than others (e.g., using facial features for identification). In a Hebbian fashion (Hebb, 1949), the brain may increasingly rely on these strategies. The array of strategies of a particular perceiver thus depends on her specific expertise.

Acquiring expertise involves a learning process, typically in the form of repetition and practice, rather than mere instruction. Some forms of expertise require enhanced abilities to remember, imagine, or control the body. For example, excelling at memory-matching smartphone games requires developing enhanced memory skills; excelling at mental rotation games requires developing enhanced imaginative skills, and excelling at line dancing requires enhanced bodily coordination and balancing skills. A different kind of expertise is perceptual expertise. Learning processes that result in perceptual expertise—that is, perceptual learning—enhance perceptual capacities by altering the structure or function of the perceptual system. Three distinct perceptual learning mechanisms have been identified (Chudnoff, 2018; Goldstone, 1998), which we will call 'differentiation', 'unitization', and 'anchoring'.[2]

2 Differentiation as a perceptual learning strategy

Differentiation alters the attentional weight assigned to parts or features of a stimulus (Cecchi, 2014; Ransom, 2020). On the level of the brain, the receptive fields of the neurons that initially encode the stimulus adapt to become more sensitive to important features and structural relations (Gilbert and Sigman, 2007).

This kind of learning was first examined in Gibson and Gibson's (1955) classic pattern-matching experiment. The participants (adults, children aged 8.5–11, and children aged 6–8) were presented with a target card depicting a nonsense scribble for 5 seconds and then a deck of 34 cards, one at a time, for 3 seconds, to determine for each whether it was a match.

[2] An older categorization divides perceptual learning into: 'imprinting', 'attentional weighting' (differentiation), and 'chunking' (unitization) (see, for example, Goldstone (1998)). According to Goldstone (1998), imprinting is the generation of an internal model, or memory representation, based on repeated exposure to a stimulus.

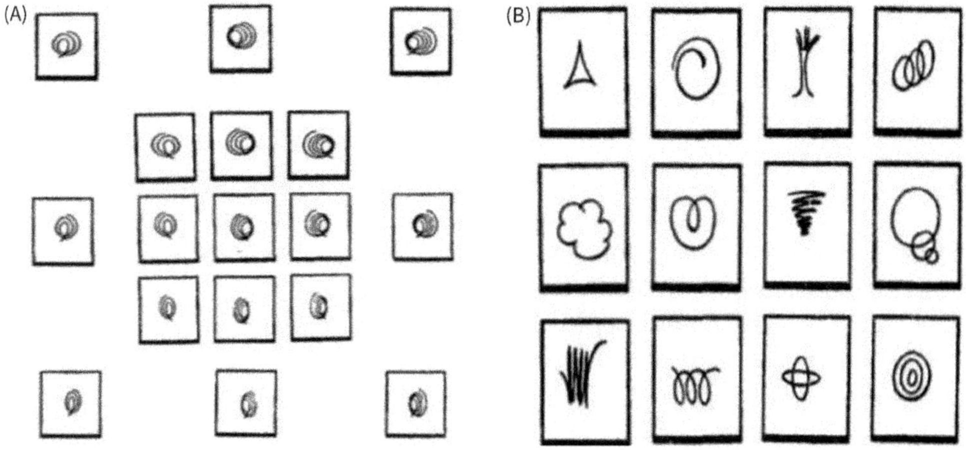

Figure 6.1 (A) Nonsense scribbles differing in three dimensions of variation (i.e., number of coils, compression, and orientation). (B) Nonsense scribbles differing in many dimensions of variation.
From Gibson and Gibson, 1955.

Seventeen of the 34 cards were highly similar to the target (Figure 6.1A); 12 were easily distinguishable from the target (Figure 6.1B), and four matched the target exactly. This process was repeated (without feedback on accuracy) until the participants made the four correct identifications in a single trial.

As the results show, each age group had a different learning curve (Table 6.1). For the adult group, the number of undifferentiated items at the outset was small (mean = 3.0), and they only needed a few trials to identify the target (mean = 3.1). For the youngest group, the number of trials needed to identify the target could not be fully determined. The older

Table 6.1 Performance across three age groups

Variable	Adults ($N = 12$)	Older children ($N = 10$)	Younger children ($N = 10$)
Mean number of undifferentiated items on first trial	3.0	7.9	13.4
Mean number of trials required for completely specific response	3.1	4.7	6.7*
Percentage of erroneous recognitions for items differing in *one* quality	17	27	53
Percentage of erroneous recognitions for items differing in *two* qualities	2	7	35
Percentage of erroneous recognitions for items differing in *three* qualities	0.7	2	28

* Only two of the younger children achieved a completely specific identification. The mean number of undifferentiated items on the last trial ($n = 6.7$) was still 3.9.

From Gibson and Gibson, 1955.

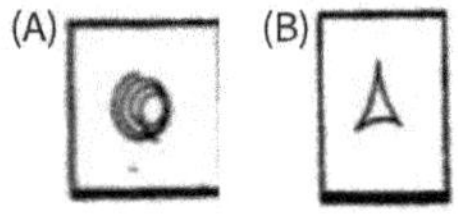

Figure 6.2 A target scribble (A) and random scribble (A). (Adapted from Figure 6.1A.)

children performed somewhere in between, with a mean of 7.9 undifferentiated items on the first trial and a mean of 4.7 trials required to identify a target.

Prior to the learning process, participants were able to distinguish the centre scribbles (Figure 6.2A) from a clear outlier (Figure 6.2B). But after the learning process, the older children and adults had acquired the ability to distinguish between pairs of scribbles, which only differed in the number of coils, compression, and orientation (Figure 6.2A).

The experiment thus demonstrates that detected (or detectable) features of a stimulus (e.g., number of coils, compression, orientation) can come to serve as a diagnostic for recognition after a perceptual learning process.

3 Unitization as a perceptual learning strategy

Unitization, the second variety of perceptual learning, can be viewed as the opposite of differentiation. Whereas differentiation divides wholes into more differentiated features, unitization integrates parts into units. An example of expertise that depends on unitization can be found in chess experts. Studies comparing experts and novices indicate that while novices can only encode the position of individual chess pieces in short-term memory, experts are able to encode whole chess configurations (e.g., Gobet and Simon, 1996). The basic unit encoded is the 'chunk' (Miller, 1956; Wenger and Rhoten, 2020), which consists of a configuration of pieces that are frequently encountered together and that are related by type, colour, role, and position (Chase and Simon, 1973a, b).

There seem to be no patent differences in thought process statistics (e.g., number of moves considered, search heuristics, etc.) between expert and novice chess players (de Groot, 1966). However, studies have found differences in perceptual processing between experts and novices: experts are able to recall a chess position almost perfectly after viewing it for 5 seconds. What seems to account for this difference is not the experts' superior memory, but their ability to perceive the structure of certain positions and encode them as units. Similar expertise-related enhancements of short-term memory have also been reported with more general stimuli (e.g., Sørensen and Kyllingsbæk, 2012).

Chase and Simon (1973b) tested the hypothesis that expert chess players extract more information from a chess position than less skilled players, using two tasks. In the perceptual task, players were asked to reconstruct a position in plain view as quickly and as accurately as possible. The players were assumed to encode one chunk per glance. The number and duration of glances at the board were used as an indicator of the size of chunks encoded in memory. In the memory task, players were asked to reconstruct a position from memory following a 5-second exposure. If they did not reconstruct the position correctly on the first try, their board was cleared and the task repeated with the same position until they got it right. The time between the placement of successive pieces was used as an indicator of the size of the chunks. The findings indicated that the experts performed superiorly because of their ability to encode positions into larger perceptual chunks.

In a separate experiment, Chase and Simon (1973a) asked beginners, Class A players, and Masters to first memorize the steps in a game with 25 positions and then to reconstruct a specific game position from memory. The initial hypothesis was that novice players would rely more on the function of individual pieces (e.g., an attack) than spatial patterns of positions. This was confirmed by the results, which showed that beginners had limited access to larger patterns in long-term memory: 80% of novices' retrieved patterns consisted of a single piece, indicating that they had to reconstruct positions one piece at a time. Masters and Class A players recalled patterns consisting of only a single piece, only 26% and 37% of times, respectively, suggesting access to larger patterns in long-term memory. The more skilled players were furthermore able to retrieve encoded configurations up to four times faster than beginners.

A study by Reingold *et al.* (2001) points to the enhanced skill set of expert chess players being perceptual rather than conceptual. Here, a minimized 5 × 5 chessboard was displayed to novice, intermediate, and expert chess players. In session one, configurations fell into two types: (1) configurations with 2–3 pieces in a checking ('yes') condition (e.g., the bishop in one corner, the king in the diagonal corner); (2) configurations with 2–3 pieces in a non-checking ('no') condition (e.g., the rook in one corner, the king in the diagonal corner) (Figure 6.3).

The players were told to determine as quickly and as accurately as possible whether the king was in check. The results showed that non-experts responded more slowly when there

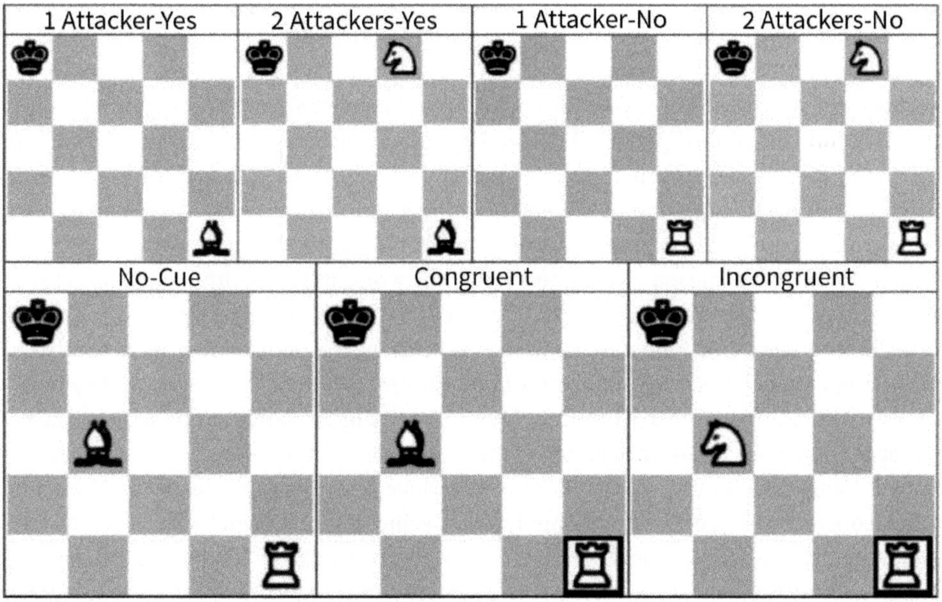

Figure 6.3 Examples of the check configurations. The top row demonstrates 'yes' (check) vs 'no' (non-check) conditions. The bottom row illustrates the no-cue condition ('no' trials) and conditions in which a cued non-checking attacker appeared together with an attacker that was either congruent (i.e., non-checking) or incongruent (i.e., checking).
From Reingold *et al.*, 2001.

were two attackers than one. The extra piece did not affect experts, which indicates holistic processing for experts, but non-holistic processing for non-experts. The finding points to the expert chess players possessing enhanced perceptual skills, enabling them to process chess configurations perceptually as units.

In the second session, only the two-attacker positions (e.g., bishop/rook plus the king) from the first part were used and double-check positions were added to create four possible combinations of checking for both attackers. The non-checking configuration was the congruent condition, and the checking configuration the incongruent condition. On half of the trials, one of the attackers was cued (in red). Participants were instructed to proceed as before if there was no cue. If a cue was present, they should ignore the non-coloured attacker. If the configurations were processed as individual piece, cueing should improve performance because the player would not need to examine the non-cued checking configuration. But if the configurations were processed holistically, cueing should not improve performance. The results showed that cueing only helped experts, suggesting that they process chess configurations holistically. Surprisingly, when a cue was present, experts were faster in the congruent (non-checking) than the incongruent (checking) condition. This may be due to a Stroop-like interference (Stroop, 1935), owing to the incongruent (check) relation that was to be ignored grabbing the experts' attention. In expert chess players, the Stroop-like interferences may occur because experts' enhanced skills, which enabled them to perceive chess configurations holistically, interfered with tasks requiring non-holistic processing of individual chess pieces. Thus, training to become an expert alters how we perceptually process stimuli.

4 Anchoring as a perceptual learning strategy

Anchoring, the third type of perceptual learning, improves performance on perceptual categorization (or recognition) tasks by linking a target category and an unrelated, but biologically more fundamental or personally more significant, category (Hubel and Wiesel, 1962). Anchoring can serve as a mnemonic for learning meaningless information (e.g., random strings of numbers) (e.g., Elmera *et al.*, 2017; Sutherland, 2017). As we will see, it also lies at the core of the perceptual abnormality known as 'synaesthesia'.

Synaesthesia is an unusual way of perceiving the world in the healthy population that involves automatic connections between seemingly unrelated experiences, images, or concepts. For example, the number 3 printed in black ink may look like it is covered with a thin film of copper green, the word 'street' may flood the mouth with the flavour of lentil soup, and the key of C# minor may elicit a bright pink spiral radiating from the centre of the visual field. In grapheme–colour synaesthesia, one of the most common forms, perceiving or thinking about an achromatic grapheme (known as the 'inducer') gives rise to the experience (known as the 'concurrent') that the grapheme has a specific colour (Armel and Ramachandran, 1999; Ásgierson, Nordfang, and Sørensen, 2015; Brogaard, 2014; 2020; Grossenbacher and Lovelace, 2001).

Most cases of synaesthesia are developmental, starting in early childhood. Whether synaesthesia is a hereditary trait has been the subject of fierce debate, partly because the question of its genetic basis has been thought to be intertwined with that of whether the condition can be learned. Evidence cited in favour of the heredity of synaesthesia includes family clustering data (Baron-Cohen *et al.*, 1996) and family linkage studies (e.g., Tilot *et al.*, 2018).

As family members often are exposed to similar environmental factors, family clustering data fail to demonstrate heredity (Brogaard, Marlow, and Rice, 2014; 2016; Mannix and Sørensen, 2019; Sørensen, 2019). Family linkage studies provide more compelling evidence for a genetic basis in some forms of synaesthesia (e.g., Tomson, Avidan, Lee, *et al.*, 2011). Genetic markers from family linkage studies have also been implicated in developmental pruning errors (Carmichael and Simner, 2013), suggesting that synaesthesia may sometimes occur when insufficient pruning results in atypical information processing.

But even if all forms of synaesthesia should turn out to involve a hereditary component, this would not preclude that learning plays a central role in its genesis. This is because even if a condition is hereditary, it is not inevitable that a genetically predisposed person will develop it. For example, a variety of personal and environmental factors have been found to contribute to the development of breast cancer, including, gender, age, ethnicity, diet, lifestyle, and hormonal and reproductive factors. Likewise, several personal and environmental factors have been found to contribute to the development of synaesthesia, including normal tendencies to associate certain items, for instance, high pitch and lighter colours (Mondloch and Maurer, 2004), and pitch and taste (Holt-Hansen, 1968), as well as synaesthesia-specific effects like the inducer's visual or spectral shape (Brang and Ramachandran, 2011), the inducer's contextual frequency (Beeli *et al.*, 2007; Simner *et al.*, 2005), and reliance on inducer–concurrent connections during early learning (e.g., Witthoft, Winawer, and Eagleman, 2015).

Witthoft and Winawer (2013) matched data from 11 grapheme–colour synaesthetes who had completed a synaesthesia battery. Ten of the 11 subjects reported having owned one of three sets of refrigerator magnets with the same colour scheme sold by FisherPrice between 1972 and 1989. The researchers found a remarkable similarity of letter–colour associations among the synaesthetes. The fewest matches among the synaesthetes were 14 letters. The probability of finding 14 or more letter matches in 26 chances is less than one in a billion, which strongly suggests that the subjects acquired their letter–colours associations from the Fisher-Price toys. The toys may thus have played a role in the development of grapheme–colour synaesthesia because the formation of synaesthetic connections between the letters and colours turned out to be a successful strategy for learning to categorize letters (Mannix and Sørensen, 2021). Being a biologically more fundamental category than letters, the colours may have served as anchors for the categorization of letters.

This raises the question of whether neurotypical adults can be trained to become synaesthetes. Studies have shown that when non-synaesthetes are repeatedly exposed to reading materials with letters printed in different colours, and subsequently are asked to name the grapheme ink colours that are incongruent with their synthetic colours, this results in Stroop interference, a mark of grapheme–colour synaesthesia (e.g., Bor *et al.*, 2014).

However, these studies did not induce genuine grapheme–colour synaesthesia. First, the incongruency effects were typically smaller than those found in grapheme–colour synaesthesia (Elias *et al.*, 2003). Second, neuroimaging studies have shown that trainees do not have the same brain activation patterns as synaesthetes (Nunn *et al.*, 2002). The limited success of training studies could be due to the lack of a genetic basis for synaesthesia in participants. Alternatively, it may have been that learning to associate graphemes with colours failed to serve as a strategy for mastering a useful or personally significant task, such as acquiring the alphabet, which is only a useful task for school-aged children (Brogaard, 2017; 2020; Mannix and Sørensen, 2021). Finally, although several studies provide extensive training, this is still not on a par with the amount of training required for learning new skills (e.g., reading).

Sørensen and Kyllingsbæk (2012) raised a similar concern about studies claiming that training does not affect short-term memory capacity. These studies apparently demonstrated that extensive training did not affect short-term memory capacity (e.g., Chen *et al.*, 2006). However, in Sørensen and Kyllingsbæk (2012), we used a change–detection paradigm (Luck and Vogel, 1997) to examine short-term memory capacity for letters and line drawings across age groups—the logic being that the different age groups had increasing familiarity and expertise with letters, whereas none of the participants had any overt training with the line drawings. Across the different groups, there was no change in memory capacity for line drawings. The youngest age group was only able to retain two letters, but with increasing age, we saw a systematic capacity increase up to adulthood where participants could retain approximately four letters. This pattern was replicated in a slightly different design by Dall *et al.* (2016) in adults, who demonstrated stable short-term memory capacity for both line drawings and letters. Nevertheless, in a critical condition using Japanese hiragana, there was a systematic short-term memory increase across subjects who did not know Japanese, studied Japanese, or were experts, corroborating the link between expertise and short-term memory capacity.

As we saw, synaesthesia can be learnt from environmental cues, which may reflect prioritization of a particular perceptual strategy that can be applied during learning. The notion that we develop different perceptual strategies to enhance learning is supported by the fact that grapheme–colour synaesthesia is one of the most common types (Ward, 2013). It is not far-fetched to imagine that when faced with the task of learning new abstract categories (e.g., letters), linking the new categories to a previously established category (i.e., a colour) would aid learning (Mannix and Sørensen, 2021).

Evidence for this theory also comes from the observation that synaesthetic colours sometimes change (e.g., Sørensen, Nordfang, and Ásgeirsson, 2016), emerge over time (Simner and Bain, 2013), or—anecdotally at least—completely disappear during the teen years or in early adulthood (Keane, 2017). As synaesthetes would have learnt to read, write, and manipulate numbers at this age, the synaesthetic associations would have lost their usefulness as learning strategies. Relatedly, the number '0' often lacks a colour or has an unusual colour (e.g., gold) in young synaesthetes until they reach school age (Brogaard, Marlow, and Rice, 2014), reflecting that most children do not develop a good understanding of the meaning of '0' until they face maths problems in school. At this point, strategically associating '0' with a basic hue may facilitate grasp.

5 Perceptual categorization and expertise

Perceptual processing goes through various stages during which sensory information is interpreted and encoded into memory. Goldstone (1998) regards learning based on the imprinting of sensory information in memory and the repurposing of this information in categorization as a kind of perceptual learning. Traditionally, perceptual processing has been described as an initial stage where attention prioritizes sensory information for encoding in short-term memory, followed by a stage of extended processing which increases the probability of encoding into more durable long-term memory representations (e.g., Atkinson and Shiffrin, 1968). Although this model was criticized by clinical studies in the early 1970s, it aligns with a folk psychological interpretation of how information flows from short-term memory to long-term memory.

An alternative to the traditional model holds that sensory information initially is matched with object and feature representations in long-term memory (Brogaard and Sørensen, in press), and that the most relevant representations then are selected for representation in short-term memory. The theory of visual attention (TVA; Bundesen, 1990) proposes this type of relationship. According to the TVA, the sensory input is modulated by two attentional top-down parameters prioritizing selection for features (filtering) or objects (pigeonholing). This model allows for flexible parallel attentional prioritization based on filtering and pigeonholing, as well as the match between the object/features and the perceiver's memory templates (for a formal mathematical description, see Bundesen, 1990). In fact, the mechanisms for perceptual learning we described earlier bear resemblances to the TVA's three core principles of selection: feature selection (*differentiation*), object selection (*unitization*), and template matching between sensory input and long-term memory (*anchoring*).

If models like the TVA that assume a more central role for long-term memory in perceptual processing are correct, then we should expect the consolidation of new perceptual categories to facilitate the matching of sensory signals to these categories. Differences in perceptual processing as new categories are acquired in a variety of different domains seem to support this notion, for example, face processing (Curby and Gauthier, 2007), car experts (Curby et al., 2009), and cartoons (e.g., Xie and Zhang, 2017).

Recently, we investigated two aspects of complexity: one varying number of feature elements and another varying stimulus familiarity (Dall et al., 2021). Jackson and Raymond (2005) described this distinction as one between physical (feature elements) and perceived complexity (familiarity). Using the TVA to isolate the specific components of attention, we showed that Chinese expert readers[3] are not affected by early parameters like threshold for visual perception. We nevertheless saw a systematic influence of familiarity on processing speech and short-term memory capacity. Interestingly, this modulation of processing speed and memory depended only on familiarity and not at all on the increase in feature elements. So this latter study points to a more holistic processing of Chinese characters by experts, akin to the holistic processing of other stimulus types (e.g., faces or chess configurations). For novices, the number of feature elements seemed to be the driving factor.

In light of the previously discussed findings supporting holistic processing of chess configurations in experts, this evidence indicates that memory representations influence how sensory information is perceptually categorized. These memory representations, in turn, depend on prior perceptual encounters with the relevant categorization tasks.

Like prior experience, the languages we speak can also shape perceptual categorization. According to the category adjustment model of colour memory, colour representations in short-term memory are the result of combining information in different proportions from two sources (Cibelli et al., 2016). One is the fine-grained bottom-up representation of the colour seen, and the other is a coarse-grained representation of colour as depicted by a language-specific category (e.g., blue). The combination of the two sources of information results in a colour representation in short-term memory that is modulated by a language-specific colour category.

[3] Combining a whole report design (Sperling, 1960) with varied exposures, Dall *et al.* (2021) were able to fit the TVA model to the exponential performance of Chinese experts and extract a range of attentional parameters, thus investigating Jackson and Raymond's (2005) processing dimensions. Stroke count in Chinese characters were used to capture physical complexity and character frequency as a measure of perceived complexity over four variables: low vs high stroke count and low vs high familiarity.

The TVA offers a plausible explanation of the category adjustment model. If, as the TVA suggests, incoming colour signals are initially matched with colour categories in long-term memory, and the most relevant colour categories are selected for representation in short-term memory, then we should expect the fine-grained bottom-up colour representation to be modulated by language-specific colour categories.

This, in turn, points to a form of linguistic relativity. The Sapir–Whorf hypothesis, a strong form of linguistic relativity, states that the languages we speak can modulate our perceptual, affective, and cognitive states. While we may not fully align with the strong interpretation of this hypothesis, the TVA and the category adjustment model suggest that prior knowledge modulates perceptual categorization. In the Section 6, we explore how colour language can influence early perceptual processing.

6 Linguistic effects on colour categorization

Originally set forth by Sapir (1921) and Whorf (1956), the hypothesis that our native language shapes what we think, remember, and perceive is also known as 'linguistic relativity'. Linguistic relativity about colour categorization came under harsh criticism in the late 1960s when Berlin and Kay (1969) popularized the view that despite documented cultural variation in colour language (e.g., Bornstein, 2007), there is a set number of basic colour categories. Writing during an era when academics increasingly believed that language and cognition were universal, innate phenomena,[4] Berlin and Kay maintained that their basic colour categories were innate, and hence culturally ubiquitous. Cultural variation in colour language, they argue, reflect that cultural processes influence which colour terms people use to depict the basic colour categories.

Although few thinkers are in the grip of Berlin and Kay's universality thesis today, the Sapir–Whorf hypothesis remains highly controversial (e.g., Regier and Xu, 2017). This is so despite relatively sparse evidence against it. Historically, the main challenge to the Sapir–Whorf hypothesis has been the finding that speakers of languages with very few colour words sometimes perform as well as English speakers on colour categorization tasks, which would suggest that non-English speakers' perceptual colour categories align with those of English speakers (Rosch, 1978).

However, studies in psycholinguistics have since then provided compelling evidence for thinking that the linguistic concepts people possess in their native language can modulate early perceptual processing (Athanasopoulos et al., 2015; Thierry et al., 2009). Thierry et al. (2009) demonstrated this effect in native English and Greek speakers. Greek contains two unique hue concepts for blue: '*ghalazio*' and '*ble*', corresponding to what native English speakers regard as two shades of the unique hue blue (Figure 6.4).

Using electroencephalography, the team recorded brain waves from the skull of the participants while they were performing a basic oddball shape discrimination task in which they had to press a button when seeing a square shape (the target) within a regularly paced stream of circles (Figure 6.5).

[4] Universality theses that gained enormous influence during the 1960s include most famously Chomsky's generative grammar, but also Ekman's six basic emotional categories.

Figure 6.4 Greek contains two colour concepts for blue: darker blue *ble* (left) and lighter blue *ghalazio* (middle).

This discrimination task helped keep the participants' attention focused on something other than what was measured, namely the strength of the brain's unconscious response to a colour change during a trial. In one block of trials, circles were *ghalazio* by default, and 'oddballs' were *ble*. In the second block, this pattern was reversed, and in the final two 'control' blocks, the experiments from the first two blocks were repeated, but with light and dark green figures.

The findings showed a significantly greater brain wave response to a colour shift from *ghalazio* to *ble*, or vice versa, in Greek vs English speakers. A stronger response to a colour shift signifies a greater perceptual difference between the colours.

Importantly, responses to a change from light green to dark green, or vice versa, were found to be the same in English and Greek speakers. Moreover, the Greek participants' responses to a shift in a shade of green were significantly smaller than their responses to a shift in a shade of blue. These findings provide strong support for the claim that not only does possessing the concepts *ghalazio* and *ble* shape how Greek speakers memorize and think about the two colour shades, but it also modulates colour categorization at the lowest level of perceptual processing.

These findings suggest that very early perceptual processing is shaped by linguistic concepts we acquired as children. This, in turn, provides renewed credence to the Sapir–Whorf hypothesis. Taken together with the previous sections, this points to a highly individualized

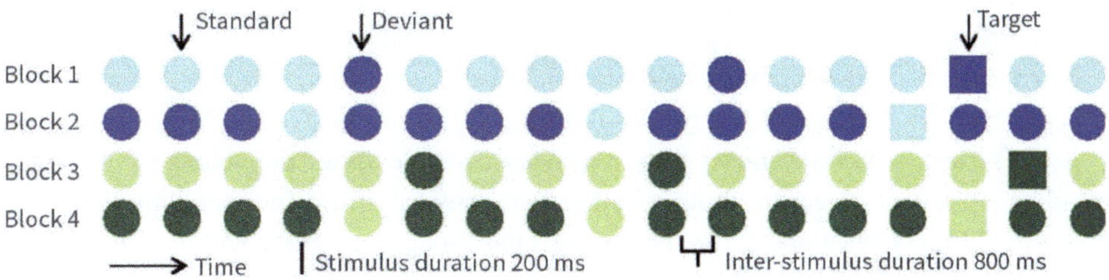

Figure 6.5 Participants were asked to press a button when they saw a square within a regularly paced stream of circles. Meanwhile, the researchers measured the visual mismatch negativity effect in response to a light colour/dark colour transition.
From Thierry *et al.*, 2009.

variability in perceptual categorization in different domains, including, but not limited to, language.

7 Perceptual pluralism defended

The evidence reviewed above jointly indicates that how sensory information is processed and categorized can vary greatly across perceivers relying on different perceptual strategies to solve the perceptual tasks they encounter. This is the first conjunct of perceptual pluralism, which—as noted in the introduction—we take to encompass the following three claims: (1) how sensory input is categorized is highly variable across different perceivers as a result of them using different perceptual strategies to solve the perceptual problems they encounter; (2) this variability in perceptual categorization need not be the result of misperception; and (3) contingent on minor constraints (e.g., not being the product of deviant brain processes), the resulting perceiver-specific sensory individuals represent actual features and objects in the external, mind-independent world.

As perceptual pluralism encompasses all three claims, it should come as no surprise that the first claim (1) is perfectly compatible with perceptual monism, the view that there is only one correct way of categorizing perceptual information. If indeed there were a single correct way to categorize sensory input, then the empirical evidence reviewed above would merely be documenting that we are accurately categorizing sensory input in some circumstances, but not in others, which would be old hat.

Several philosophers have made a case for some restricted version of perceptual monism (although not under that name). For example, Byrne (2006) offers a defence of what we term perceptual monism which is restricted to colour perception. According to Byrne, there are objective facts about the colour of objects, but there is irresolvable disagreement about what those facts are. Byrne (2006, p. 337) expresses the view as follows (in response to Cohen, 2006):

> Suppose that [two] normal human observers Mya and Bo are viewing a chip C . . . C looks unique green to Mya, and bluish green to Bo. The problem, as Cohen has it, is to explain 'what would (metaphysically) make it the case' that Mya, say and not Bo, is perceiving C correctly. [Cohen] purports to find the explanation 'extremely hard to imagine', and so concludes that both Mya and Bo are perceiving C correctly. . . . [I maintain that] what 'makes it the case' that Mya, not Bo, is perceiving C correctly, is that Mya is representing C as being unique green, Bo is representing C as being bluish green (no problem so far), and C is unique green, not bluish green (likewise no problem).[5]

Byrne's view is this: whenever two perceivers disagree about the colour of an external, mind-independent object or about whether two external, mind-independent objects have the same colour, at least one of them is wrong. One apparent problem with this view is that it entails that there are unknowable colour facts. For any coloured object, there are bound to be individuals with no apparent visual system defects who would disagree about its colour, were they to view it in the same viewing conditions. Yet we would have no basis for deciphering

[5] We have replaced S_1 with Mya, and S_2 with Bo.

who is right and who is wrong. So for any coloured object, the truth about its colour would be unknowable. This form of radical colour epistemicism seems highly implausible.

Our objection to Byrne generalizes to similar defences of other types of perceptual monism: generally speaking, it is implausible to think that, for any sensory input, people fall neatly into groups of normal and abnormal perceivers, and it is also improbable that perception gives us no reliable insight into the structure of the external, mind-independent world.

Similarly, our objections lend credence to the second claim of perceptual pluralism, viz. that perceptual variability in a population of speakers regarding the same sensory input need not be the result of misperception, which is to say that overlapping and disjoint categorizations of a given perceptual input may all represent real categories in the external, mind-independent world (see also Cohen, 2009).

Thus, the perceptual categorization of a certain sensory input as the unique hue blue in native English speakers represents an actual category of blue objects. Likewise, the perceptual categorization of the very same sensory input as the two unique hues *ghalazio* and *ble* in native Greek speakers represents two real categories of coloured objects.

The same can be said for other variations in perceptual categorization. In Gibson and Gibson's (1955) pattern-matching experiment, the untrained participants categorized the scribbles as belonging to a single category of 'scribbles' (Figure 6.1A). But after training, the same participants made more fine-grained categorizations of the scribbles. Likewise, in the studies of chess expertise, experts and novices were found to categorize sensory input from chess pieces on a chessboard in different ways (e.g., Chase and Simon, 1973a). Novices classified the sensory input in terms of individual chess pieces, whereas experts classified the same input in terms of holistic chess configurations. But while we can say that the chess experts are better at chess, we cannot conclude that only the experts' perceptual categorizations are the correct ones—they are merely different due to the modulation of expertise.

Finally, we argued that anchoring is the predominant perceptual learning mechanism in synaesthesia. Synaesthetes improve on categorization tasks by creating perceptual connections or semantic memory associations between a target category and an unrelated category that is biologically more fundamental or of greater personal significance. Consider, for example, a grapheme–colour synaesthete who connects single digits with unique colours. For her, the number '3', say, does not just belong to a mathematical category, but also to the category of green objects.

8 Conclusion

Here we have defended pluralism about perceptual categorization based on studies suggesting that variability in perceptual categorization and the resulting sensory individuals reflects that we use different strategies for solving the different tasks we encounter in our environment, and that the perceptual system functionally and structurally adjusts to the strategies that are most successful.

Perceptual pluralism entails that prior sensory input encoded in memory modulates our perceptual objects (or sensory individuals). This may raise a concern about whether perceptual pluralism entails that low-level perception is cognitively penetrable (e.g., Firestone and Scholl, 2016; Pylyshyn, 1999). Controversies about this thesis concern its methodological basis (Brogaard and Gatzia, 2017; Firestone and Scholl, 2016), its implications for a popular view about perceptual justification (Chudnoff, 2020; Siegel, 2017), and its implications in perceptual bias (Brogaard, 2021; Siegel, 2017).

Rather than providing a definitive answer to the question of whether perceptual pluralism entails the cognitive penetrability thesis, however, our argument shows that not much hinges on how we answer it, because whatever disquieting conclusions can be derived from it can also be deduced, given more innocuous top-down influences.

Acknowledgements

We are grateful to two reviewers for invaluable comments on a previous version of this paper. Sørensen was supported by the Independent Research Fund, Denmark. Brogaard was supported by a Cooper Fellowship from the University of Miami.

References

Armel, K. C. and Ramachandran, V. S. (1999). Acquired synesthesia in Retinitis Pigmentosa. *Neurocase*, 5, 293–96.

Ásgeirsson, Á. G., Nordfang, M., and Sørensen, T. A. (2015). Components of attention in grapheme-color synesthesia: a modeling approach. *PLoS One*, 10(8), e0134456.

Athanasopoulos, P., Bylund, E., Montero-Melis, G., *et al.* (2015). Two languages, two minds: flexible cognitive processing driven by language of operation. *Psychological Science*, 26(4), 518–26.

Atkinson, R. C. and Shiffrin, R. M. (1968). Human memory: a proposed system and its control processes. In *The Psychology of Learning and Motivation: Advances in Research and Theory*, edited by K. W. Spence and J. T. Spence, 89–195. New York: Academic Press.

Baron-Cohen, S., Burt, L., Smith-Laittan, F., Harrison, J., and Bolton, P. (1996). Synaesthesia: prevalence and familiality. *Perception*, 25(9), 1073–9.

Beeli, G., Esslen, M., and Jäncke, L. (2007). Frequency correlates in grapheme–color synaesthesia. *Psychological Science*, 18(9), 788–92.

Berlin, B. and Kay, P. (1969). *Basic Color Terms: Their Universality and Evolution*. Berkeley, CA: University of California Press.

Bor, D., Rothen, N., Schwartzman, D. J., Clayton, S., and Seth, A. K. (2014). Adults can be trained to acquire synesthetic experiences. *Scientific Reports*, 4, 7089.

Bornstein, M. H. (2007). Hue categorization and color naming: cognition to language to culture. In *Anthropology of Color: Interdisciplinary Multilevel Modeling*, edited by R. E. MacLaury, G. V. Paramei, and D. Dedrick, 3–27. Amsterdam/Philadelphia: John Benjamins Publishing Company.

Brang, D. and Ramachandran, V. S. (2011). Survival of the synesthesia gene: why do people hear colors and taste words? *PLoS Biology*, 9(11), e1001205.

Brogaard, B. (2014). Varieties of synesthetic experience. In *Phenomenology and the Neurophilosophy of Consciousness*, edited by R. Brown, 409–12. Springer.

Brogaard, B. (2017). Synesthetic binding and the reactivation model of memory. In *Sensory Blending: On Synaesthesia and Related Phenomena*, edited by O. Deroy, 126–50. Oxford: Oxford University Press.

Brogaard, B. (2020). Color synaesthesia and its philosophical implications. In *The Routledge Handbook of Philosophy of Colour*, edited by D. Brown and F. Macpherson. Francis & Taylor.

Brogaard, B. (2021). Implicit biases in visually guided action. *Synthese*, 198, 3943–67.

Brogaard, B. and Gatzia, D. E. (2017). Is color experience cognitively penetrable? *Topics in Cognitive Science*, 9(1), 193–214.

Brogaard, B., Marlow, K., and Rice, K. (2014). The long-term potentiation model for grapheme–color binding in synesthesia. In *Sensory Integration and the Unity of Consciousness*, edited by D. Bennett and C. Hill, 37–72. Cambridge, MA: MIT Press.

Brogaard, B., Marlow, K. and Rice, K. (2016). Do synesthetic colors grab attention in visual search? *Review of Philosophy and Psychology*, 7(4), 701–14.
Brogaard, B. and Sørensen, T. A. (in press). Predictive processing and object recognition. In *Expected Experiences: The Predictive Mind in an Uncertain World*, edited by T. Cheng and J. Hohwy. New York, NY: Routledge.
Bundesen, C. (1990). A theory of visual attention. *Psychological Review*, 97(4), 523–47.
Byrne, A. (2006). Comments on Cohen, Mizrahi, Maund, Levine. *Dialectica*, 60, 337–40.
Carmichael, D. A. and Simner, J. (2013). The immune hypothesis of synesthesia. *Frontiers in Human Neuroscience*, 7, 563.
Cecchi, A. S. (2014). Cognitive penetration, perceptual learning and neural plasticity. *Dialectica*, 68(1), 63–95.
Chase, W. G. and Simon, H. A. (1973a). The mind's eye in chess. In *Visual Information Processing*, edited by W. G. Chase, 215–81. New York, NY: Academic Press.
Chase, W. G. and Simon, H. A. (1973b). Perception in chess. *Cognitive Psychology*, 4, 55–81.
Chen, D., Yee Eng, H., and Jiang, Y. (2006). Visual working memory for trained and novel polygons. *Visual Cognition*, 14(1), 37–54.
Chudnoff, E. (2018). The epistemic significance of perceptual learning. *Inquiry*, 61(5–6), 520–42.
Chudnoff, E. (2020). Experience and epistemic structure: can cognitive penetration result in epistemic downgrade? In *Inference and Consciousness*, edited by A. Nes with T. Chan, 255–75. New York, NY: Routledge.
Cibelli, E., Xu, Y., Austerweil, J. L., Griffiths. T. L., and Regier, T. (2016). The Sapir–Whorf hypothesis and probabilistic inference: evidence from the domain of color. *PLoS One*, 11(7), e0158725.
Cohen, J. (2006). Color and perceptual variation revisited: unknown facts, alien modalities, and perfect psychosemantics. *Dialectica*, 60(3), 307–19.
Cohen, J. (2009). *The Red and the Real*. Oxford: Oxford University Press.
Curby, K. M. and Gauthier, I. (2007). A visual short-term memory advantage for faces. *Psychonomic Bulletin and Review*, 14(4), 620–8.
Curby, K. M., Glazek, K., and Gauthier, I. (2009). A visual short-term memory advantage for objects of expertise. *Journal of Experimental Psychology: Human Perception and Performance*, 35(1), 94–107.
Dall, J. O., Wang, Y., Cai, X., Chan, R. C., and Sørensen, T. A. (2021). Visual short-term memory and attention: an investigation of familiarity and stroke count in Chinese characters. *Journal of Experimental Psychology: Learning, Memory, and Cognition*, 47(2), 282–94.
Dall, J. O., Watanabe, K., and Sørensen, T. A. (2016). Category specific knowledge modulates capacity limitations of visual short-term memory. In *2016 8th International Conference on Knowledge and Smart Technology (KST)*. IEEE, 275–80.
de Groot, A. D. and Gobet, F. (1996). *Perception and Memory in Chess*. Assen: Van Gorcum.
Elias, L. J., Saucier, D. M., Hardie, C., and Sarty, G. E. (2003). Dissociating semantic and perceptual components of synaesthesia: behavioural and functional neuroanatomical investigations. *Cognitive Brain Research*, 16(2), 232–7.
Elmera, S., Greber, M., Pushparaj, A., Kühnis, J., and Jäncke, L. (2017). Faster native vowel discrimination learning in musicians is mediated by an optimization of mnemonic functions. *Neuropsychologia*, 104, 64–75.
Firestone, C. and Scholl, B. J. (2016). Cognition does not affect perception: evaluating the evidence for 'top-down' effects. *Behavioral and Brain Sciences*, 39, e229.
Gibson, J. J. and Gibson, E. J. (1955). Perceptual learning: differentiation or enrichment? *Psychological Review*, 62(1), 32–41.
Gilbert, C. D. and Sigman, M. (2007). Brain states: top-down influences in sensory processing. *Neuron*, 54, 677–96.
Gobet, F. and Simon, H. A. (1996). Templates in chess memory: a mechanism for recalling several boards. *Cognitive Psychology*, 31, 1–40.
Goldstone, R. L. (1998). Perceptual learning. *Annual Review of Psychology*, 49, 585–612.
Green, E. J. (2019). A theory of perceptual objects. *Philosophy and Phenomenological Research*, 99(3), 663–93.

Grossenbacher, P. G. and Lovelace, C. T. (2001). Mechanisms of synesthesia: cognitive and physiological constraints. *Trends in Cognitive Sciences*, 5(1), 36–41.

Hebb, D. O. (1949). *The Organization of Behavior: A Psychological Theory*. New York: Wiley.

Holt-Hansen, K. (1968). Taste and pitch. *Perceptual and Motor Skills*, 27(1), 59–68.

Hubel, D. H. and Wiesel, T. N. (1962). Receptive fields, binocular interaction and functional architecture in the cat's visual cortex. *Journal of Physiology*, 160, 106–54.

Jackson, M. C. and Raymond, J. E. (2005). Visual working memory: capacity is dependent on perceived, not physical, stimulus complexity. *Journal of Vision*, 5, 621.

Keane, D. (2017). Philosopher David Chalmers on consciousness, the hard problem and the nature of reality. ABC News, 6 July 2017. Available from: https://www.abc.net.au/news/2017-07-07/david-chalmers-and-the-puzzle-of-consciousness/8679884

Luck, S. J. and Vogel, E. K. (1997). The capacity of visual working memory for features and conjunctions. *Nature*, 390(6657), 279–81.

Mannix, T. K. and Sørensen, T. A. (2019). Do we experience the same world? *Psykologi Information*, 47(1), 22–7.

Mannix, T. K. and Sørensen, T. A. (2021). Colours and category learning: implications for grapheme–colour synaesthesia. In *Sálubót: Afmælisrit til heiðurs Jörgen L. Pind*, edited by Á. Kristiánsson, H. M. Sigurdardottir, and K. Arnason, 221–30. Reykjavik: University of Iceland Press.

Miller, G. A. (1956). The magical number seven, plus or minus two: some limits on our capacity for processing information. *Psychological Review*, 63(2), 81–97.

Mondloch, C. J. and Maurer, D. (2004). Do small white balls squeak? Pitchobject correspondences in young children. *Cognitive, Affective, and Behavioral Neuroscience*, 4, 133–6.

Nunn, J. A., Gregory, L. J., Brammer, M., *et al.* (2002). Functional magnetic resonance imaging of synesthesia: activation of V4/V8 by spoken words. *Nature Neuroscience*, 5, 371–5.

Pylyshyn, Z. W. (1999). Is vision continuous with cognition? The case for cognitive impenetrability of visual perception. *Behavioral and Brain Sciences*, 22, 341–423.

Quilty-Dunn, J. (2020). Perceptual pluralism. *Noûs*, 54(4), 807–38.

Ransom, M. (2020). Attentional weighting in perceptual learning. *Journal of Consciousness Studies*, 27(7–8), 236–48.

Regier, T. and Xu, Y. (2017). The Sapir–Whorf hypothesis and inference under uncertainty. *WIREs Cognitive Science*, 8, e1440.

Reingold, E. M., Charness, N., Schultetus, R. S., and Stampe, D. M. (2001). Perceptual automaticity in expert chess players: parallel encoding of chess relations. *Psychonomic Bulletin and Review*, 8(3), 504–10.

Rosch, E. (1978). Principles of categorization. In *Cognition and Categorization*, edited by E. Rosch and B. B. Lloyd, 27–48. Hillsdale, NJ: Lawrence Erlbaum.

Sapir, E. (1921). *Language: An Introduction to the Study of Speech*. New York, NY: Harcourt Brace.

Siegel, S. (2017). *The Rationality of Perception*. Oxford: Oxford University Press.

Simner, J. and Bain, A. E. (2013). A longitudinal study of grapheme–color synesthesia in childhood: 6/7 years to 10/11 years. *Frontiers in Human Neuroscience*, 7, 603.

Simner, J., Ward, J., Lanz, M., *et al.* (2005). Non-random associations of graphemes to colours in synaesthetic and non-synaesthetic populations. *Cognitive Neuropsychology*, 22(8):1069–85.

Sørensen, T. A. (2019). Synaesthesia: atypical sensory experiences. In *Sense Me*, 142–51. Kolding: Trapholt.

Sørensen, T. A. and Kyllingsbæk, S. (2012). Short-term storage capacity for visual objects depends on expertise. *Acta Psychologica*, 140(2), 158–63.

Sørensen, T. A., Nordfang, M., and Ásgeirsson, Á. (2016). Gradual consolidation of synesthesia during adolescence: a case study. *Journal of Vision*, 16(12), 460.

Sørensen, T. A. and Overgaard, M. (2018). Prosopagnosia or prosopdysgnosia: facing up to a change of concepts. 10th International Conference on Knowledge and Smart Technology: Cybernetics in the Next Decades, KST 2018, IEEE Press (August 2018), 260–3.

Sperling, G. (1960). The information available in brief visual presentations. *Psychological Monographs: General and Applied*, 74(11), 1–29.

Stroop, J. R. (1935). Studies of interference in serial verbal reactions. *Journal of Experimental Psychology*, 18, 643–62.

Sutherland, N. (2017). *Cross-Modal Perceptual Learning: A Novel Shape Tasting Method for Sensory Discrimination of Wine*. [Thesis] Reed College.

Thierry, G., Athanasopoulos, P., Wiggett, A., Dering, B., and Kuipers, J. R. (2009). Unconscious effects of language-specific terminology on preattentive color perception. *Proceedings of the National Academy of Sciences of the United States of America*, 106(11), 4567–70.

Tilot, A. K., Kucera, K. S., Vino, A., Asher, J. E., Baron-Cohen, S., and Fisher, S. E. (2018). Rare variants in axonogenesis genes connect three families with sound–color synesthesia. *Proceedings of the National Academy of Sciences of the United States of America*, 115(12), 3168–73.

Tomson, S. N., Avidan, N., Lee, K., *et al.* (2011). The genetics of colored sequence synesthesia: suggestive evidence of linkage to 16q and genetic heterogeneity for the condition. *Behavioral and Brain Research*, 223, 48–52.

Ward, J. (2013). Synesthesia. *Annual Review of Psychology*, 64(1), 49–75.

Wenger, M. J. and Rhoten, S. E. (2020). Perceptual learning produces perceptual objects. *Journal of Experimental Psychology: Learning, Memory, and Cognition*, 46(3), 455–75.

Whorf, B. L. (1956). Language, thought, and reality. In edited by J. B. Carroll. Cambridge, MA: MIT Press.

Witthoft, N. and Winawer, J. (2013). Learning, memory, and synesthesia. *Psychological Science*, 24(3), 258–65.

Witthoft, N., Winawer, J., and Eagleman, D. M. (2015). Prevalence of learned grapheme–color pairings in a large online sample of synesthetes. *PLoS One*, 10(3), e0118996.

Xie, W. and Zhang, W. (2017). Familiarity increases the number of remembered Pokémon in visual short-term memory. *Memory and Cognition*, 45(4), 677–89.

PART 2

SENSORY INDIVIDUALS IN UNIMODAL PERCEPTION

7
Looming perception: Seeing in a dynamic world

Frédérique de Vignemont

1 Introduction

The world is presented to us as being dynamic, the spatial relationship between us and the objects that we perceive constantly changing. The task for the perceptual system is to distinguish between changes that come from the subject moving (self-motion) and changes that come from the objects moving (object motion). Once it has been determined that the object is in motion, the perceptual system then differentiates biological motion from non-biological motion. It also needs to determine whether the object is looming towards the subject or not, and if so, when it may come in contact with her, what is known in the literature as the *time-to-collision* (TTC). These last 40 years or so, there has been a large experimental literature on how one reacts to looming stimuli on a collision course, either testing in the field with sportsmen, car drivers, and pilots or in the laboratory using psychophysical tasks (for reviews, see Regan and Gray, 2000; Tresilian, 1999; Yan *et al.*, 2011). Under optimal conditions (involving binocular perception), one is extremely accurate, overestimating by no more than a few milliseconds (Gray and Regan, 1998). TTC information can then be directly exploited by the motor system to programme either avoidance or interceptive responses. In addition, seeing looming objects can elicit a relatively primitive *sense of impending collision*, typically just a few seconds before the impact. Many times in movies (especially in 3D), video games, and virtual reality, we can have the strong impression—though mistaken—that something is going to hit us, though we know perfectly well that nothing can reach us, seated in our safe living room. We then experience something like 'This is going to crash into me!'. One may say that it expresses more than the mere impression that an object is currently moving towards us. It informs us about what might happen to us in the *close future*.

Here I shall propose that to best understand the sense of impending collision, we need to switch from an object-oriented and space-oriented conception of perception to a more dynamic event-oriented and time-oriented conception. There have been many discussions on apparent motion. Participants report an illusory impression of movement when seeing two brief stimuli presented successively at two distinct locations (e.g., Dainton, 2008; Dennett and Kinsbourne, 1992; Grush, 2006; Hoerl, 2012). Though it bears some similarities with apparent motion, we shall see that the sense of impending collision raises its own set of questions. Indeed, unlike standard cases of motion illusion, it involves the end of the motion, instead of intermediary stages. Secondly, the temporal scale is much larger. Apparent motion works for a time lag of 100 ms between successive stimuli,

Frédérique de Vignemont, *Looming perception* In: *Sensory Individuals*. Edited by: Aleksandra Mroczko-Wąsowicz and Rick Grush, Oxford University Press. © Oxford University Press 2023. DOI: 10.1093/oso/9780198866305.003.0008

whereas the TTC is of a few seconds in most experiments. Thirdly, the sense of impending collision does not consist in an illusory motion experience. As I shall propose, it results from *amodal* completion, which enables one to be visually aware of more than what one can actually see. Here I shall argue that amodal completion applies not only for objects, but also for dynamic events.

I shall then explore the mechanisms involved in amodal completion. Do they qualify as perceptual or should we rather conceive them as being imaginative? Perception is factive: it is about what happens. One may thus immediately reject the proposal that one can perceive something that has not happened yet and that may never happen. By contrast, imagination is often characterized as the ability that allows representing what needs not be actual. It thus seems particularly appropriate for representing the future. One may then suggest that anticipating the collision is just running offline the rest of the ongoing looming motion till it ends 'in one's head'. There are many ways, however, this view can be spelled out and we shall see that none of them can fully account for the sense of impending collision. Instead, I will argue that there is evidence that points towards a perceptual account. If this is right, then it means that there is a sense in which one can be said to have a perceptual sense of the future.

2 A future-directed awareness

Compare the following two scenarios:

1. You see a ball coming towards you from the other side of the field.
2. You see a ball 1 metre away coming towards you.

In both cases, the direction of the ball motion is exactly the same: you experience that the ball is moving and you also experience that it is moving towards you. Thus, in both cases, you have a dynamic egocentric experience: you locate the object in motion in its spatial relation to you. Still, there seems to be a phenomenological difference between (1) and (2). It cannot be only because of a sense of danger. Arguably, cricket players do not feel afraid each time they play and they see a ball approaching them. Instead, I suggest that it is because you experience a sense of impending collision only in (2). Roughly speaking, you experience that the ball will crash into you any second now. This goes beyond the basic awareness of the ongoing looming motion because it has a future orientation. True, the collision will happen very soon, but still, it is not now. If it were not future-oriented, the sense of impending collision would be relatively useless because one needs to anticipate the event before it might occur to have the time to be ready. Future orientation is crucial for action planning. Behavioural responses to a risk of collision are not mere reflexes. Though often automatic, they are still flexible, depending on a variety of parameters that determine the type of response (avoidance or interception) and the way to execute it (with the hand or a tool like a racket). We have now known for a long time that action is essentially predictive (Wolpert and Flanagan, 2001), but it is even more so for looming stimuli. When the object that is the target of an action is simply in motion, as in the game of duck hunter in attraction parks, one needs to visually predict where it will be in the soon future in order to plan the right motor command, such as shooting it. Visual prediction then follows from one's intention to act. When the object is looming towards the subject, visual prediction must come even before the intention is generated. The anticipation of the impact explains that one starts moving, instead of staying still.

Future orientation, however, is not only a sensorimotor problem of logistics. One needs to ascribe a future location to the event to be motivated to act on it.[1] If one were aware of the future collision as happening now, then it would deprive its motivational force from the sense of impending collision. By representing the collision as located in the future, one is aware that it is still open, whether there will be a collision or not. This may not be the case from a metaphysical perspective, but from our epistemological perspective, we do not know whether the event will happen or not, and thus we feel that we can still do something about it. Roughly speaking, the die has not been cast yet and the future still seems up to us. As noted by Debus (2014, 2016) about future-directed mental time travel, it exhibits 'agential openness' (Debus, 2016, p. 232):

> Agential openness claim: when a subject S-imagines a future event, the subject is aware that she herself, or others, might act in ways that could bring about, or prevent, the actual onset of the very event that she S-imagines now.

Thanks to the sense of impending collision, we are aware that we can act in ways that could prevent the occurrence of the very event that we anticipate. The role of the sense of impending collision is not for the control of fine-grained visuo-motor action. TTC information can be directly exploited by the motor system without one being aware of it. The role of the sense of impending collision is rather at the executive level. It can guide the allocation of attention and the selection of actions. In brief, the sense of impending collision can justify why we abruptly drop everything we were doing in order to duck. Most philosophical interest has focused on the motivational role of pain for bodily protection, but pain strikes when it is too late: something wrong has already happened. The sense of impending collision aims at preventing the pain to even occur, and for that, it needs to have a future orientation. I shall now propose that it is thanks to the mechanism of *amodal completion* that we can be aware of what has not (yet) happened.

3 A temporal version of amodal completion

Perceptual completion is typically defined as the filling-in of the missing features of a stimulus that is physically present. One can distinguish two types of completion, which obey different laws (Singh, 2004).

In *modal* completion, one experiences a sensory impression in the absence of corresponding sensory signals. For example, you have an illusory visual experience of a bright white square, defined by a sharp illusory contour, occluding four black Pac-Man shapes (Figure 7.1A). Unlike modal completion, *amodal* completion does not involve phenomenal filling-in. One perceives the completeness of the object, but without having the impression of actually seeing the missing parts. In the shown example, it looks to you that there is a full black square partly occluded by a grey circle (Figure 7.1B). Thanks to amodal completion,

[1] This does not entail that there is a feeling of futureness associated with the awareness of the collision (for discussion, see Debus, 2016; Fernandez, 2020).

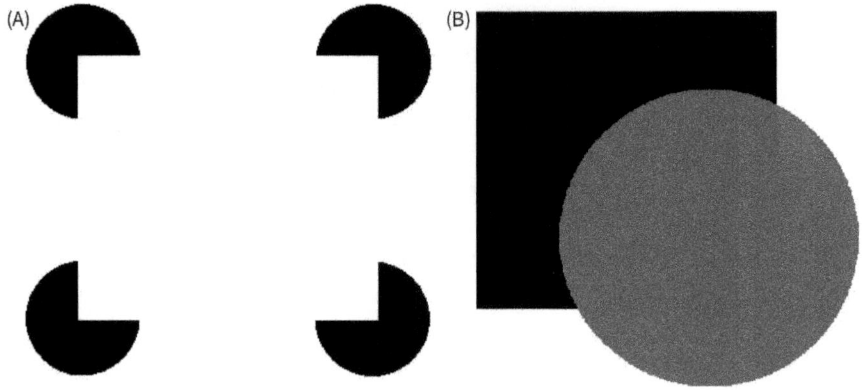

Figure 7.1 (A) Modal completion. (B) Amodal completion.

you are aware of the square as a whole, and not of some parts of it only. Nonetheless, what it is like to see the occluded parts of the square is not the same as what it is like to see its visible parts. It is in this sense that it qualifies as being amodal.

As most of the literature on perception, the phenomenon of perceptual completion has been studied mainly in static scenes. It has been described for shape and object perception, but a similar phenomenon can occur for dynamic events such as motion (Burge, 2010; Viera and Nanay, 2020). For example, if you see two photographs depicting a simple movement, with the two photographs differing only in the position of one limb segment relative to a joint, their rapid alternation gives rise to an apparent motion percept, which represents the shortest possible path for the limb. Execution of a limb movement along this seen trajectory is, however, not anatomically possible, as it would violate natural joint constraints. When the presentation rate slows down, participants perceive paths of apparent limb movement that follow natural human limb trajectories (Shiffrar and Freyd, 1990). One way to interpret this result is that when participants have the time, they can represent the intermediary stages of the biological motion that are not visible, thanks to kinaesthetic knowledge about biologically possible movements. Put it another way, the biological motion is completed. Another example of motion completion can be found in the cutaneous rabbit illusion (Geldard and Sherrick, 1972). Repeated rapid tactile stimulations at the wrist, then near the elbow, can create the illusion of touches at intervening locations along the arm, as if a rabbit were hopping along it. There is a tactile filling-in mechanism for the intermediary motion stages, though they have never happened. One may then say that both examples involve modal completion: one has an illusory visual or tactile experience of motion.

The sense of impending collision is another example of motion completion, though as mentioned earlier, it differs in several respects from standard cases of motion illusion. Most importantly, as I shall now argue, it results from *amodal* completion. Consider what happens in most TTC studies. The experimenter artificially generates the impression that a sphere is moving towards the participants at constant speed. Then the sphere becomes no longer visible. Participants are then asked whether it will arrive before or after another looming sphere presented in the same manner (Field and Wann, 2005; Yan *et al.*, 2011) or an auditory stimulus (Gray and Regan, 1998). For them to do so, they need to 'complete' the end of the motion. This does not entail that they have a visual experience of the forthcoming collision. In classic cases of amodal completion, one does not visually experience the occluded part of the object in the same way as when the part is not occluded. Likewise, what it is like to experience the collision when one anticipates it is not the same as when the collision really happens. Consider Chemero's (2009, pp. 123–4) detailed description of a video game simulation of a car collision:

1. As your car approaches the wall, the image of the wall on your monitor expands.
2. When you get close enough, individual bricks will become visible.
3. As you continue toward your virtual crash, the image of the wall will cover the entire monitor, and images of individual bricks will expand.
4. As you get closer to the wall, the images of the bricks will expand so that only a few of them are actually able to fit on the monitor, and they will appear textured.
5. Moving closer still, the images of the texture elements on the bricks will expand as well;
6. Then there is the loud crash noise and the cracked virtual windshield.

When we simply anticipate the collision, we clearly do not experience something like that. We do not experience the looming object filling more and more our visual field the closer the object gets, until it occupies most of our visual field and appears all blurred (if it is in a direction path towards our eyes). There is only an amodal filling-in of the motion when anticipating contact. What, then, do we experience when we have a sense of impending collision?

When one sees an object partially hidden from sight, the object still looks complete. One may then say that one is visually aware of the *whole object*, even if one does not see some of its *spatial parts*. Now, when one sees a looming object on a collision course, I argue, one can be visually aware of the *whole event*, even if one does not see some of its *temporal parts*. Thanks to amodal completion, one is aware of more than the temporal part of the event that one is currently seeing; one is aware of the unfolding of the whole event until its natural conclusion (meaning its collision).[2] The sense of impending collision consists in this awareness. It is not just the current motion; it is not just its future ending; it is the whole of it—how what is happening now will continue happening until it eventually stops.

Now one may worry that, unlike most examples of perceptual completion (including apparent motion), the part to complete is the end of the event, and not some intermediary stage. However, there is evidence that the perceptual system can extrapolate ongoing motion, as illustrated by the phenomenon of representational momentum. It has been found that when one has to indicate the remembered final location of a moving target, it is usually displaced from the actual final location in the direction of anticipated motion (i.e., the target is judged as travelling slightly farther than it actually travelled) (Freyd and Finke, 1984, Hubbard, 2014). This indicates that one does not simply represent the motion with a certain direction; one can represent the next step, which has not happened yet and which may never happen. Motion extrapolation can be used to compensate neural processing times: the stimulus keeps moving after the sensation was initiated, but before that sensation reaches perceptual awareness, and thus without extrapolation, the visual experience would always lag behind the actual position of that target (Nijhawan, 1994). Motion extrapolation, I propose, can also be used for computing TTC. It is important to understand though the completion is about the future, it is not about a new event. It is only the hidden end of an occurring event. Debus

[2] By natural conclusion, I mean the conclusion that would normally be reached if nothing happens that prevents it.

(2014) claims that one cannot be experientially aware of a future particular event because we lack the right causal connection with it. However, though one has no direct informational link with the collision itself, one has a direct link with the motion of the object, to which the collision is only a potential endpoint. We do see the looming motion; we simply do not see all of it. One is only predicting the climax of a particular event already happening, an event that one is currently perceiving. Thanks to motion completion, one does not simply represent 'a collision'. One represents 'this collision', which is spatially and temporally anchored in virtue of its direct spatio-temporal continuity with the current event of motion. This is what makes its timing predicable with such an accuracy.

4 Imagining the future

To propose that the sense of impending collision is made possible thanks to amodal completion, however, does not answer all the questions. One still needs to understand how amodal completion operates. Nanay (2010) considers four possible accounts: in terms of sensory inputs, in terms of perceptual belief, in terms of sensorimotor access, and in terms of mental imagery. He disqualifies the first three and argues in favour of an imaginative conception of amodal perception (see also Brown, 2018 for a similar conception). On this view, one perceives the full square despite it being partially occluded in Figure 7.1B, because one's visual imagery represents its occluded parts and the visual image is projected in egocentric space—that is, in the space in which one sees the visible parts of the square. Visual imagery then fills in the gap, so to speak. Likewise, Viera and Nanay (2020) claim that motion completion involves temporal mental imagery. Mental imagery seems particularly appropriate for predicting a forthcoming event. According to Moulton and Kosslyn (2009, p. 1274), this anticipatory role is even the primary function of mental imagery:

> Imagery allows us to answer 'what if' questions by making explicit and accessible the likely consequences of being in a specific situation or performing a specific action. For example, imagery can allow us to predict the path of a projectile.

But what exactly does one imagine when one predicts the path of a projectile coming towards one's body? We shall now see that the imaginative account is open to several interpretations and that no matter how one tries to spell it out, it faces a series of difficulties.

Let us first simply assume that the sense of impending collision involves running offline the simulation of the motion of the stimulus approaching until impact. This is what was actually asked from participants in a series of studies on TTC (Vagnoni *et al.*, 2012, 2015, 2017, 2020). They were asked to imagine the object continuing to approach at the same rate and to press a button when they judged that it would collide with them. However, there is an obvious basic problem with this interpretation. In the studies, the participants started imagining when the looming stimulus stopped being visible. But in real life, one generally keeps on seeing the motion that is unfolding under one's eyes. Why mentally running in parallel the simulation of the motion that one is actually seeing? There is no information to gain, especially since it cannot afford anticipation of what is forthcoming. Indeed, running such simulation takes time, too much time. More precisely, the time that it takes for the participants in the TTC studies to finish running their mental simulation of the motion has been found to be a reliable proxy for the time it would take for the motion to be actually completed (Vagnoni *et al.*, 2012). Imagination is then explicit, but the same dynamics should be found at the implicit level (Jeannerod, 1994). Hence, on the basis of motion simulation, one can know about the collision only at the end, when it is supposed to occur. This cannot give rise to the sense of *impending* collision, which is anticipatory by nature.

One may then suggest that the sense of impending collision involves only visually imagining the collision itself, and not the motion that leads to it. But is visual imagery the best modality for that? Let us go back to our previous example of the approaching ball. When the ball is really close to you, you experience that it is going to crash right on your face. But what is it like to visually imagine the crash? If you were really hit by the ball, your visual experience would be quite limited, or even non-existent. The problem is not only that for objects to be properly seen, they must be not too close to the eyes and collision necessarily means the absence of distance between the perceiver and the object. The problem is also that one automatically closes one's eyes, what is known as the blink reflex, which is a defensive response to threats. You thus have little past experience of seeing the ball hitting you, at least from the inside. It is only if you adopt a third-person perspective (from the outside) that you can visually represent the collision, by visualizing the ball next to your face, from the outside. Allocentric imagination, however, does not appear as the right candidate for the sense of impending collision, which can be conceived as an anticipatory form of egocentric awareness.

There is, however, an alternative imaginative account, which appeals to *touch*, instead of vision, and which can preserve the first-personal character of the sense of impending collision. On this interpretation, when one sees a looming object, one spontaneously engages in a task of tactile imagination, simulating the impact from the inside. The specificity of touch is to give access to the relational property of pressure from a first-person perspective (Vignemont and Massin, 2015)—and what is collision if not a form of pressure exerted by an object against one's own body? In favour of this view, it has been shown that merely seeing objects can trigger tactile expectations. For instance, one is slower and makes more mistakes to judge the location of tactile stimuli if, at the same time, one sees a visual stimulus close to the body, but at a location that is incongruent with the tactile stimulation (Spence *et al.*, 2004). This kind of crossmodal effect can be explained by the fact that the perceptual system anticipates the potential impact of what is seen, thus generating tactile expectations, which then compete with the experience of the actual tactile stimulus. However, though this new imaginative account shows promise, it cannot be the full story for the sense of impending collision. It is only when visual stimuli enter the immediate surrounding of one's body (up to 50 cm), what is known as peripersonal space, that tactile anticipation occurs.[3] When the approaching stimulus has not reached this stage yet, there is no visuo-tactile effect. Imaging studies confirm that the perception of looming stimuli partly engages the network involved in the processing of peripersonal space but goes beyond it (Clery *et al.*, 2020). Since functional magnetic resonance imaging (fMRI) studies have a bad temporal resolution, one can only speculate that there is first a distinct network activated by looming stimuli before the activation of the peripersonal network, which occurs at the very end of the looming.[4] Put it another way, the tactile system waits until the last minute to kick in or, more accurately, until the last milliseconds when the looming object is located at less than 50 cm from the body. By

[3] For a comprehensive multidisciplinary investigation of peripersonal space, see Vignemont *et al.* (2021).

[4] Thanks to Justine Clery for her help on the relation between looming perception and peripersonal space.

contrast, one starts experiencing that the ball is going to hit one's body before. Hence, though a better candidate than visual imagining, tactile imagining cannot account for the sense of impending collision, unless one wants to restrict it to peripersonal space.

One may then propose that one is aware of the forthcoming collision not in virtue of visually or tactually imagining the object in contact with one's body, but in virtue of *motorically* imagining one's response to it. In line with this motor interpretation, activity was found in sensorimotor areas in the dorsal pathway normally involved in reaching and catching, though the TTC task was purely perceptual, with subjects remaining perfectly still in a magnetic resonance imaging (MRI) machine and the approaching sphere being only shown on a screen (Field and Wann, 2005). Another study showed that participants underestimated more TTC if they had reduced freedom of movements (Vagnoni *et al.*, 2017). It may then seem that motor imagery is the most appropriate form of imagination for the sense of impending collision. This would be in line with other findings on the role of motor imagery for perception. For instance, it has been found that to judge the laterality of a hand visually presented with different orientations, participants mentally rotate their own hand to match the visible hand, as shown by the fact that the larger the angle of the rotation to perform, the more time it takes to answer (Parsons, 2003).

There is a major difference, however, between the laterality task and the TTC task. Motor imagery can be used to determine whether one sees a left or a right hand, but it cannot be used to determine whether there will be a collision or when the collision will occur.[5] It is actually the other way around: one first needs to anticipate the collision before running offline one's motor system. Otherwise, there would be no reason to plan one's motor response. Nonetheless, though motor imagery is not the source of our anticipatory awareness, it can still accompany it. As said earlier, the sense of impending collision plays a motivational role and it should not come as a surprise that it can trigger the mental rehearsal of avoiding or intercepting behaviours.

At first sight, imagination seemed to be the best candidate to account for the anticipatory awareness of collision, but it is actually not clear how it can account for what one experiences. As argued, motion imagination takes too much time, whereas collision imagination is best imagined tactually and motorically than visually. However, tactile and motor imagination cannot ground the sense of impending collision; they can only accompany it. No matter the modality, one may further object that imagination cannot play the motivational role that is played by the sense of impending collision. The relation between imagination and action is highly debated, but it seems that even if it can play some role in guidance, its relation to the motor system is not direct (Currie and Ravenscroft, 2002; Kind, 2011; O'Brien, 2005). If so, simply imagining a collision does not suffice to get one to react. Finally, the imaginative account of amodal completion is not without controversies. For Briscoe (2011, 2018), a key feature of visual imagery is its distinctive phenomenal character, which is similar to visual experiences, so that one can fail to distinguish between them. Visually imagining the square involves having a visual impression of it with a specific brightness. But amodal completion precisely qualifies as amodal because there is no such visual impression. Furthermore, Briscoe (2011) claims that imagery generally fades rapidly, whereas amodal completion is

[5] The effect of one's bodily incapacities on TTC estimation may just be the result of a general bias of the type 'allow you more time'. But the bias does not have to be the result of an offline motor simulation. The fact is that the timing of action preparation is slightly different from TTC.

relatively stable, even if you look at the partly hidden object for a long time. He also notes that though imagery does not have to be voluntary, it still normally remains under control and one can modify its content. By contrast, amodal completion is obligatory. One cannot help but being visually aware of the full object, or of the full event in our case.

There is, however, a further interpretation of the imaginative account—one defended by Nanay (2010, 2016, 2020) and Brown (2018)—which may not run into all these difficulties. On this view, visual imagery should be clearly distinguished from visualizing. Instead, it should be conceived as a specific type of visual representation that supplements and enriches perceptual processes. Brown (2018, p. 149) concludes that there is 'a seamless infusion of perception with imagination'. As a consequence, it may be difficult to take apart the respective contribution of each. Nanay (2016, p. 129), for instance, suggests the following criterion: content comes 'in a bottom-up way in perception whereas it comes in a top-down way in mental imagery'. This use of the notion of mental imagery is relatively liberal and many may question the definition not only of imagination but also of perception. In a sense, what is really at stake in this debate is the boundary of perception, where it ends and where something else (call it mental imagery if you want) starts. Depending on the answer, the sense of impending collision may be conceived as perceptual or as imaginative. Clearly, the debate on the boundary of perception goes beyond the scope of this chapter, and here I will simply review the evidence that points in the direction of a perceptual interpretation of the sense of impending collision. I shall now argue that though there is no corresponding sensory stimulation for the collision, the sense of impending collision still results from a bottom-up process of motion completion that is stimulus-driven. This conception is in line with the way TTC estimates are described in the experimental literature. For many indeed, one 'directly perceives' TTC (e.g., Gray and Regan, 1999; Tresilian, 1999; Vagnoni *et al.*, 2012; Yan *et al.*, 2011).

5 Perceiving the future?

Amodal completion has been shown to occur very early on in the visual system, in a pre-attentive manner. It is thus generally conceived as a purely perceptual bottom-up mechanism (Briscoe, 2011; Burge, 2010; Kalpokas, 2017; Phillips, 2017). Most discussions, however, have focused on object completion, and not on motion completion. Here also it involves early visual mechanisms, which have been found in many animal species and in human infants as early as 12 months old (van der Meer *et al.*, 2012). In the primary visual cortex, there is a specific neural mechanism that fixes the direction of the object motion in 3D space and that is preferentially sensitive to increasing or decreasing object size, what has been called a looming detector (Regan and Beverley, 1978). It has been shown, for instance, that participants' attention was automatically drawn towards a looming stimulus in visual search tasks, though it was not the target, but only a distractor (Lin *et al.*, 2008). The estimation of TTC is then generated on the basis of early visual mechanisms, which extrapolate the rest of the motion on the basis of low-level visual cues (Nijhawa *et al.*, 1994). More specifically, TTC can be computed using only the optical variable tau (τ), defined as the inverse of the relative rate of expansion of the incoming object's image on the retina (Lee, 1976):

$$\tau = \theta / (\Delta\theta / \theta t)$$

Where θ represents the projected angular size of the approaching object, and $\Delta\theta/\Delta t$ represents the rate of expansion of the image.

The optical variable τ decreases approximately linearly with time and it is at its minimum shortly before TTC. On only the basis of τ, one can predict TTC with remarkable accuracy. The optical variable, however, works only for a constant suprathreshold speed rate, neglecting possible acceleration, and it does not work for very small objects or asymmetrically shaped objects (Tresilian, 1999; Yan *et al.*, 2011). Other parameters can then contribute to TTC estimates, but they remain relatively low-level visual features such as size and distance.

Still, it is worth noting that although the estimation of TTC does not require additional knowledge to be computed, it can still be modulated by cognitive factors. A series of studies showed that participants underestimated more TTC for threatening or large objects relative to non-threatening or small ones (Vagnoni *et al.*, 2012, 2015, 2020). However, the fact that TTC judgements can be influenced by semantic information does not show that motion completion is generated by top-down mechanisms. Firstly, it is unclear whether this modulatory effect occurs at the perceptual level or at the post-perceptual level only, as acknowledged by the authors of the studies themselves (Vagnoni *et al.*, 2017). Secondly, TTC can be computed independently of any recognition of the looming object. In this sense, TTC involves only what Briscoe (2011, 2018) calls stimulus-driven completion, by contrast with cognitive completion. It is equivalent to the example of the occluded square previously described (Figure 7.1A). Its completion is based on sensory cues only and does not rely on stored knowledge. It is different from cognitive completion, which depends on categorial recognition. For instance, if you see just the front of a horse, with the rest appearing to be occluded, you can perceive the horse as a whole, but only if you know what a horse normally looks like. Without one identifying what one sees, the object cannot look complete. To compute TTC does not require such top-down processing.

To recapitulate, when seeing a looming stimulus, the visual stimulation on the retina and its corresponding optical expansion reliably trigger computation of the optical variable τ, and there is no need to recognize the stimulus. Some may even conclude that we can directly pick up the information about the forthcoming collision, in Gibson's (1979) sense:

> Note that τ need not be computed by the gannet. It is available at the retina. τ, in other words, can be perceived directly [...] sensitivity to the ratio of optical angle to the expansion of optical angle is sensitivity to the timing of approaching collision.
>
> (Chemero, 2009, p. 124)

More precisely, one still needs to *extrapolate* the TTC on the basis of the available information. The sensory stimulation needs to be enriched by knowledge about the physical principles governing motion. For some, this shows that motion completion is not perceptual. According to Brown (2018), for instance, there can be mental imagery even if operations are performed on incoming stimuli (a bottom-up process, thus), as long as the computation involves the use of 'stored perceptual contents' that go beyond stimulus information and that enrich it. In Brown's view, however, the contribution of mental imagery does not disqualify the sense of impending collision to be perceptual. Still, one may question this interpretation. It is true that the sense of impending collision requires knowledge about the physical principles governing motion, but it is not clear how this knowledge is implemented. Are the physical principles governing motion the content of a dedicated representation, or are they simply incorporated into the functional properties of the cognitive architecture? One may further challenge the assumption that all internal models that enrich perception qualify as mental imagery. For instance, Grush (2004) proposes that we have what he calls amodal environment emulators, which anticipate changes in our surrounding on the basis of our own movements and of the dynamics of the objects themselves. So far, this is compatible with the imaginative conception. In Grush's view, however, emulators should be conceived as imagery only when they operate offline, and not when they are exploited to interpret sensory inputs.

I shall not settle this fundamental debate on perception and imagery here. All we can conclude is that on some conceptions of perception at least, the sense of impending collision can be conceived as perceptual. It is a bottom-up process, causally controlled by visual stimulation and based on perceptual invariants, and it represents a particular event as completed, even though it is not yet finished. If this suffices for the sense of impending collision to qualify as perceptual, then this seems to lead us to an interesting, but possibly challenging, conclusion—that is, that one can be perceptually aware of the close future. More precisely, I propose that in the sense of impending collision, one perceives the whole event, including not only its present parts, but also its future parts. There is therefore a sense in which one can be said to perceive the future: in virtue of perceiving a whole event with a future part, one can be said to have visual awareness of the close future. This future that we anticipate is close in time—it corresponds to the first instants after what we experience as being now, but it is still beyond now.

Acknowledgements

I would like to thank Ned Block, Geraldine Carranante, Justine Clery, Mohan Matthen, and Wayne Wu for their feedback and support. This work was accomplished thanks to the ANR-17-EURE-0017 FrontCog and the ANR-10-IDEX-0001-02 PSL.

References

Briscoe, R. E. (2011). Mental imagery and the varieties of amodal perception. *Pacific Philosophical Quarterly*, 92(2), 153–73.

Briscoe, R. E. (2018). *Superimposed mental imagery: on the uses of make-perceive*. In *Perceptual Imagination and Perceptual Memory*, edited by F. Macpherson and F. Dorsch, 161–85. Oxford: Oxford University Press.

Brown, D. H. (2018). Infusing perception with imagination. In *Perceptual Imagination and Perceptual Memory*, edited by F. Macpherson and F. Dorsch, 133–60. Oxford: Oxford University Press.

Burge, T. (2010). *Origins of Objectivity*. Oxford: Oxford University Press.

Chemero, A. (2009). *Radical Embodied Cognitive Science*. Cambridge, MA: MIT Press.

Cléry, J. C., Schaeffer, D. J., Hori, Y., Gilbert, K. M., Hayrynen, L. K., Gati, J. S., ... Everling, S. (2020). Looming and receding visual networks in awake marmosets investigated with fMRI. *Neuroimage*, 215, 116815.

Currie, G. and Ravenscroft, I. (2002). *Recreative Minds: Imagination in Philosophy and Psychology*. New York, NY: Oxford University Press.

Dainton, B. (2008). Sensing change. *Philosophical Issues*, 18, 362–84.

Debus, D. (2014). 'Mental time travel': remembering the past, imagining the future, and the particularity of events. *Review of Philosophy and Psychology*, 5(3), 333–50.

Debus, D. (2016). Temporal perspectives in imagination: on the nature and value of imagining the future. In *Seeing the Future: Theoretical Perspectives on Future-Oriented Mental Time Travel*, edited by K. Michaelian, S. Klein, and K. Szpunar, 217–40. Oxford: Oxford University Press.

Dennett, D. C. and Kinsbourne, M. (1992). Time and the observer: the where and when of consciousness in the brain. *Behavioral and Brain Sciences*, 15, 183–247.

de Vignemont, F. and Massin, O. (2015). Touch. In *The Oxford Handbook of Philosophy of Perception*, edited by M. Matthen, 294–313. Oxford: Oxford University Press.

de Vignemont, F., Serino, A., Wong, H. Y., and Farnè, A. (Eds.) (2021). *The World at Our Fingertips: A Multidisciplinary Investigation of Peripersonal Space*. Oxford: Oxford University Press.

Fernandez, J. (2020). Self-referential memory and mental time travel. *Review of Philosophy and Psychology*, 11(2), 283–300.

Field, D. T. and Wann, J. P. (2005). Perceiving time to collision activates the sensorimotor cortex. *Current Biology*, 15(5), 453–8.

Freyd, J. J. and Finke, R. A. (1984). Representational momentum. *Journal of Experimental Psychology: Learning, Memory, and Cognition*, 10(1), 126.

Geldard, F. A. and Sherrick, C. E. (1972). The cutaneous rabbit: a perceptual illusion. *Science*, 178(4057), 178–9.

Gibson, J. J. (1979). *The Ecological Approach to Visual Perception*. Boston: Houghton Mifflin.

Gray, R. and Regan, D. (1998). Accuracy of estimating time to collision using binocular and monocular information. *Vision Research*, 38(4), 499–512.

Gray, R. and Regan, D. (1999). Do monocular time-to-collision estimates necessarily involve perceived distance? *Perception*, 28(10), 1257–64.

Grush, R. (2004). The emulation theory of representation: motor control, imagery, and perception. *Behavioral and Brain Sciences*, 27(3), 377–96.

Grush, R. (2006). How to, and how not to, bridge computational cognitive neuroscience and Husserlian phenomenology of time consciousness. *Synthese*, 153, 417–50.

Hoerl, C. (2012). Seeing motion and apparent motion. *European Journal of Philosophy*, 23(3), 676–702.

Hubbard, T. L. (2014). Forms of momentum across space: representational, operational, and attentional. *Psychonomic Bulletin and Review*, 21(6), 1371–403.

Jeannerod, M. (1994). The representating brain: neural correlates of motor intention and imagery. Behavioural and *Brain sciences*, 17, 1–34.

Kalpokas, D. (2017). Experience and justification: revisiting McDowell's empiricism. *Erkenntnis*, 82(4), 715–38.

Kind, A. (2011). The puzzle of imaginative desire. *Australasian Journal of Philosophy*, 89(3), 421–39.

Lee, D. N. (1976). A theory of visual control of braking based on information about time-to-collision. *Perception*, 5, 437–59.

Lin, J. Y., Franconeri, S., and Enns, J. T. (2008). Objects on a collision path with the observer demand attention. *Psychological Science*, 19(7), 686–92.

Moulton, S. T. and Kosslyn, S. M. (2009). Imagining predictions: mental imagery as mental emulation. *Philosophical Transactions of the Royal Society B: Biological Sciences*, 364(1521), 1273–80.

Nanay, B. (2010). Perception and imagination: amodal perception as mental imagery. *Philosophical Studies*, 150(2), 239–54.

Nanay, B. (2016). Imagination and perception. In *The Routledge Handbook of Philosophy of Imagination*, edited by A. Kind, 124–34. Routledge.

Nijhawan, R. (1994). Motion extrapolation in catching. *Nature*, 370, 256–7.

O'Brien, L. (2005). Imagination and the motivational role of belief. *Analysis*, 65(1): 55–62.

Parsons, L. M. (2003). Superior parietal cortices and varieties of mental rotation. *Trends in Cognitive Sciences*, 7(12), 515–17.

Phillips, B. (2017). Inscrutability and visual objects. *Synthese*, 194, 8, 2949–71.

Regan, D. and Beverley, K. I. (1978). Looming detectors in the human visual pathway. *Vision Research*, 18(4), 415–21.

Regan, D. and Gray, R. (2000). Visually guided collision avoidance and collision achievement. *Trends in Cognitive Sciences*, 4(3), 99–107.

Shiffrar, M. and Freyd, J. J. (1990). Apparent motion of the human body. *Psychological Science*, 1(4), 257–64.

Singh, M. (2004). Modal and amodal completion generate different shapes. *Psychological Science*, 15(7), 454–9.

Spence, C., Pavani, F., and Driver, J. (2004). Spatial constraints on visual-tactile cross-modal distractor congruency effects. *Cognitive, Affective, and Behavioral Neuroscience*, 4(2), 148–69.

Tresilian, J. R. (1999). Visually timed action: time-out for 'tau'? Trends in Cognitive *Sciences*, 3(8), 301–10.

Vagnoni, E., Andreanidou, V., Lourenco, S. F., and Longo, M. R. (2017). Action ability modulates time-to-collision judgments. *Experimental Brain Research*, 235(9), 2729–39.

Vagnoni, E., Lingard, L., Munro, S., and Longo, M. R. (2020). Semantic modulation of time-to-collision judgments. *Neuropsychologia*, 147, 107588.

Vagnoni, E., Lourenco, S. F., and Longo, M. R. (2012). Threat modulates perception of looming visual stimuli. *Current Biology*, 22(19), R826–7.

Vagnoni, E., Lourenco, S. F., and Longo, M. R. (2015). Threat modulates neural responses to looming visual stimuli. *European Journal of Neuroscience*, 42(5), 2190–202.

Van der Meer, A. L. H., Svantesson, M., and van der Weel F. R. R. (2012). Longitudinal study of looming in infants with high-density EEG. *Developmental Neuroscience*, 34, 488–501.

Viera, G. and Nanay, B. (2020). Temporal mental imagery. In Cambridge Handbook of the Imagination, edited by Anna Abraham, 227–40. Cambridge: Cambridge University Press.

Wolpert, D. M. and Flanagan, J. R. (2001). Motor prediction. *Current Biology*, 11(18), R729–32.

Yan, J. J., Lorv, B., Li, H., and Sun, H. J. (2011). Visual processing of the impending collision of a looming object: time to collision revisited. *Journal of Vision*, 11(12), 7.

8
Shine of bronze and sound of brass: The relational perceptual constructs of timbre and gloss

Alessandra Buccella and Mazviita Chirimuuta

1 Introduction

Objects around us have many perceptible qualities, which we experience through our various sense modalities, such as vision, hearing, touch, olfaction, etc. These perceptible qualities, though intuitively familiar, are, however, not yet fully understood. Two cases in point are gloss (i.e., the visual quality that makes material objects look 'shiny') and timbre (i.e., the auditory quality that generally would distinguish two musical instruments playing the same note at the same loudness). However, these intuitive definitions do not answer the question: 'What are gloss and timbre, *really*?'. Ontological questions about these perceptible qualities of auditory and material objects have not generated as much vexation for philosophers as more commonly discussed qualities such as colour. This may be because it has been assumed that the sensations associated with the former qualities track clearly identifiable physical properties of the external world, namely the form of a sound wave or the *specularity* of a surface (i.e., how 'mirror-like' it is). As we will show, the course of perceptual psychology since the early twentieth century must undermine this complacency.

A central goal of this chapter is to show that the psychology of timbre and gloss perception gives additional grounds for rejecting a *detection framework* for understanding sensory systems, one which takes these systems in humans and other animals to be appropriately modelled by devices built to detect, as accurately as possible, specific physical properties.

With gloss perception, the problem for the detection framework is the context dependency of the effect that each physical feature is shown to have on the more global perception of a surface or object. With the perception of timbre, the same issues emerge. We will show that attempts to reduce timbre to a cluster of physical properties, independent of context, have so far failed, and that, like in the case of gloss, timbre perception depends on more global features of the scene in which the primary physical cause of the experience is embedded.

Another central claim is that without the detection framework, relationist ontologies of the perceptual qualities of timbre and gloss are much more plausible than physicalist ones which identify sensory qualities with perceiver-independent physical properties. By 'relationist' ontology, we mean the thesis that the very existence of timbre and gloss *constitutively* depends not only on the context, but also on the perceiver (and, in particular, the features of her sensory systems). Relationist colour ontologies come in various stripes (Cohen, 2009). Most

Alessandra Buccella and Mazviita Chirimuuta, *Shine of bronze and sound of brass* In: *Sensory Individuals*.
Edited by: Aleksandra Mroczko-Wąsowicz and Rick Grush, Oxford University Press.
DOI: 10.1093/oso/9780198866305.003.0009

well-known is the classic Lockean theory that colours are dispositional properties, though more relevant here is the Gibsonian view that colours are properties which can only be specified with reference to an animal's ecology (Thompson, 1995), or the *adverbialist* theory that colours are properties of perceptual interactions (Chirimuuta, 2015). The shared commitment of relationists is that colours are real properties genuinely instantiated in the world, but they are perceiver-dependent ones. Likewise, we will here make the case that timbre and gloss are real, perceiver-dependent properties. We will call them *interaction properties*.[1]

2 Gloss

The psychologists Chadwick and Kentridge (2015) recount the history of the science of gloss perception, showing how the early assumption that there is one relevant physical stimulus for gloss came to be untenable. They themselves endorse the view that 'gloss is a complex interaction of illumination, surface, environment and observer' (Chadwick and Kentridge, 2015, p. 224). This shift in the conception of what gloss fundamentally is—from a physicalist to an 'interactionist' one—tracks the fall of the detection framework for understanding gloss perception. The relationship between these two conceptual shifts—the first away from a physicalist ontology, the second away from a detection framework—is philosophically significant.

2.1 Gloss physicalism

It is telling that research on the perception of gloss begins with the invention of devices for measuring the sheen of manufactured products, such as paper, around the time of the First World War.[2] Such instruments aid product standardization, which is, of course, an essential requirement of commerce and industry. As stated in a publication from the National Bureau of Standards (Hunter, 1937, p. 20):

> Gloss is one of the major factors in appearance, and a number of methods for determining it have been developed. The purpose of these methods is usually to enable the producers and vendors of articles in commerce to rate them according to glossiness.

Ingersoll's (1921) *glarimeter* worked by measuring the ratio of light reflected in a mirror-like (*specular*) way, as in the highlights or reflection visible on a glossy surface, to light reflected diffusely in all directions. This was achieved simply by measuring the proportion of polarized light, since specularly reflected light is almost completely polarized.

Such measuring devices for gloss are useful in industrial settings where it can be ensured that the kinds of objects measured (i.e., paper) and the setting (e.g., illumination conditions)

[1] In what follows, we will often talk about context dependence and perceiver dependence as if they were, if not interchangeable, at least very tightly related. While, in principle, it is true that a quality can be context-dependent without being perceiver-dependent, and vice versa, relationist ontologies like the one we are defending all rely on the Gibsonian idea that perceiver and environment form a unified system (Gibson, 1979), and that, for the purposes of explaining perception, the boundary between 'internal' factors relating to a perceiver and 'external' features of the environment cannot be clearly marked.

[2] Ingersoll (1921, fn1) references his own publication of 1914 and 1919 of a similar device developed independently by Karl Kieser.

do not vary from one occasion to the next. As we will see, the task of perceptual science differs crucially in that psychologists are attempting to explain how vision works in the range of conditions encountered in everyday life. Once conditions are not so strictly standardized, the importance of viewing context on the perception of gloss becomes salient.

Gloss physicalism requires there be a clear-cut distinction between gloss as an intrinsic physical property of a surface and the appearance of that property to a human perceiver. Thus, Hunter (1937, p. 20), reviewing methods for the determination of gloss, follows Jones (1922) in defining *gloss* as the 'actual physical properties of surfaces responsible for their glossy appearance', whereas *glossiness* is 'the appearance that results because the surface possesses that property'. While Ingersoll (1921) and Pfund (1930) presume that gloss is one kind of physical property, for example, 'per cent of polarization' of reflected light (Ingersoll, 1921, p. 215), Hunter (1937) goes on to propose that there are six different kinds of gloss, each of which generates a different kind of glossiness appearance. This is still a physicalist notion of gloss, but with multiple kinds of physical properties and multiple kinds of instruments for detecting them.

This move to a pluralized version of physicalism was necessitated by the attempt to compare the glossiness of different sorts of materials. The definition developed for the measurement of glossiness of paper was not found to be suitable, or to yield consistent results, when its application was extended to other materials. It is apt to make a comparison here with the way that theories of colour physicalism had to abandon the assumption that each shade of colour can be identified with one particular microstructural or spectral property of a surface, and to concede that each shade is to be identified with a disjunction of microstructural properties or class of spectral surface reflectances (SSRs).[3] The disunity of colour, when defined in this way, is the impetus for many of the important objections to physicalism (e.g., Yablo, 1995).

An alternative to the splitting of gloss, according to the compendium of different physical stimuli, is to emphasize the phenomenological unity of perceived gloss and argue that these appearances are the result of the combination and interaction between various physical cues. This is the proposal of Harrison and Poulter (1951, p. 97), who write:

> Gloss is judged by an appraisal of a physical situation *as a whole*. That is, gloss is a *gestalt*. It is likely that specular reflexion and contrast are not the only components of this *gestalt*.

This 'gestalt' approach[4] to gloss may be contrasted with the atomistic assumption embedded in the detection framework, which is the assumption that a physical stimulus (e.g., the ratio of specular to diffuse reflection) will give rise to the same percept of gloss, independently of context (i.e., the surrounding objects in the scene). This implies that the percept of the object is just the rendered aggregate of these independently determined qualities: shape + colour +

[3] Some relevant references are: Byrne and Hilbert, 2003; Hilbert, 1987; and Smart, 1975.

[4] We follow the authors themselves in referring to this as the 'gestalt' approach, while not assuming any identity with the classic Gestalt psychology of Wertheimer, Koffka, and Köhler, among others. Chadwick and Kentridge (2015, p. 233) characterize the gestalt approach as a multi-cue, rather than single-cue, explanation of gloss perception (cfr. Wendt *et al.* (2010), and see Fleming (2014) for more on the multi-cue explanation). But in their review, Chadwick and Kentridge also emphasize that the human observer is an ineliminable factor in the explanation of what gloss is. The question of whether the gestalt approach needs to be any more than a multi-cue one is somewhat reminiscent of Merleau-Ponty's challenge to classic Gestalt psychology—that the move from 'single-cue' to 'multi-cue' is not enough to break free of the 'constancy hypothesis' (the assumption of a one-to-one relationship between stimulus and sensation), and so out of the physicalist and atomistic mould.

lightness + gloss + texture. By contrast, the gestalt approach gives a primacy to the perception of the scene as a whole, such that each isolatable perceptual quality, like gloss, is taken to be influenced not by one localized physical property, but by various attributes of the context.

One recent expression of the gestalt approach comes from Leloup *et al.* (2014, p. 560), who write that gloss perception has a 'multidimensional nature' and 'perception of surface gloss implies much more than an estimation of the physical reflectance properties of the surface alone'. However, the goal of their research is still to measure gloss, achieving a 'better correspondence between gloss measurement and gloss perception' (Leloup *et al.*, 2014, p. 560). Indeed, one technological driver for the achievement of better models of human gloss perception comes from computer graphics—the goal being to make computer-rendered scenes in video games as realistic as possible. The difference is that the response of the human visual system becomes a target of modelling, broadening the investigation beyond, merely, the effect of an optical stimulus on a mechanical device. As Leloup *et al.* (2014, p. 561) write, '[s]ince gloss is stressed to be a perceptual attribute, the measurement of gloss must necessary [sic] take into account the response of the visual system'.

2.2 Towards an interactionist account of gloss

We have contrasted the gestalt approach with the one associated with the detection framework for understanding gloss perception. The gestalt approach sees a need to acknowledge features of the whole scene, and their organization, on the various perceptual qualities of any one object. Visual features cannot be analysed just as independent stimuli that generate the same perceptual response whatever the context, nor can the response of the human visual system be duplicated by a simple mechanical device like the glarimeter.

In conflict with the atomistic assumptions held by early developers of glossmeters, what psychologists later found is that the visual system does not determine ('measure') visual qualities independently of one another. For instance, lightness (Harrison and Poulter, 1951; Pfund, 1930), illumination (Olkkonen and Brainard, 2011), texture—depth of relief (Ho *et al.*, 2008), colour (Wendt *et al.*, 2010), motion (Hartung and Kersten, 2002), and shape (Chadwick and Kentridge, 2015; Marlow *et al.*, 2012; Olkkonen and Brainard, 2011) have all been found to influence perceived glossiness (see Figure 8.1, for example). Even tactile perception has been reported to affect judgements of gloss and visual texture (Bergmann Tiest and Kappers, 2007).

Also worth noting is the mutuality of some of the effects. For example, the darker of two surfaces with matched spectral properties will appear glossier, but also glossier surfaces appear darker than they would otherwise (Harrison and Poulter, 1951). Another important point is that the effect of manipulation of one stimulus variable does not easily predict the effect of the manipulation in combination with a change in an additional variable (Olkkonen and Brainard, 2011, p. 1031). This is how Leloup *et al.* (2014, p. 566) summarize the problem:

> The presented overview of object-intrinsic and object-extrinsic parameters that influence the way in which the human visual system arrives at a particular gloss sensation, illustrates the complexity of gloss perception. The prophecy of the joint effect of changes in these influencing parameters seems not straightforward. While simple conjoint models may

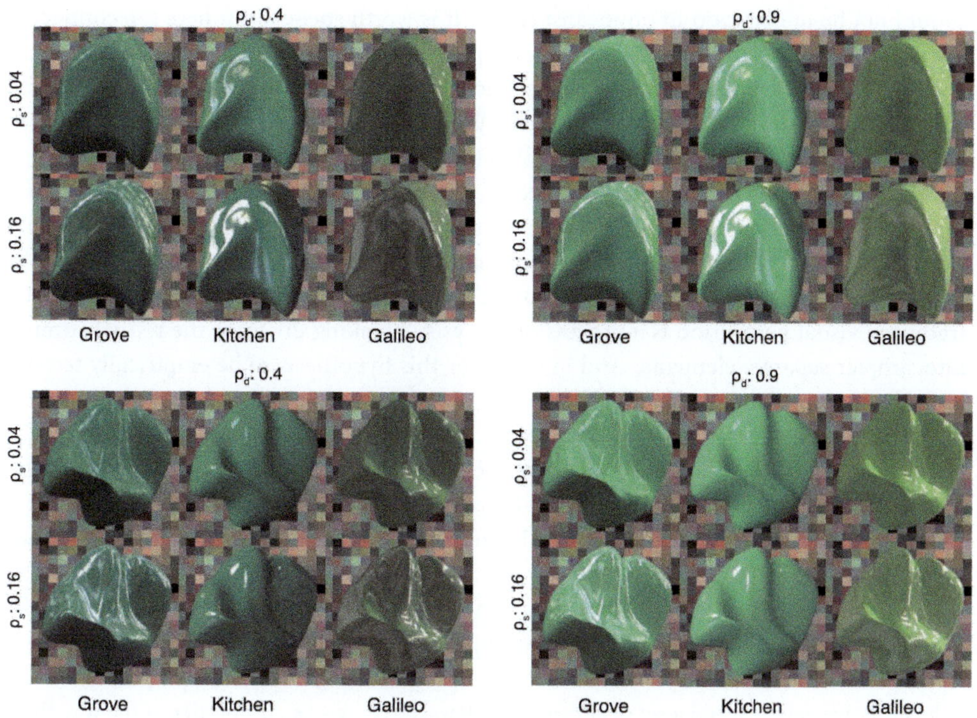

Figure 8.1 Illustration of the interaction between gloss, lightness, shape, and ambient illumination (stimuli used by Olkkonen and Brainard (2011)). ρ_s is the specular component of reflectance (the higher the value, the glossier a surface will tend to appear), and ρ_d is the diffuse component of reflectance (the higher the value, the lighter the surface will tend to appear) in the *bidirectional reflectance distribution function* (BRDF)* of the computer-rendered object. Grove, Kitchen, and Galileo are three different 'light fields'—renderings of the ambient illumination captured at three different locations.
*BDRF 'describes the amount of light reflected as a function of the directions of the incident and reflected light' (Olkonnen and Brainard, 2011, p. 1014).

> predict particular influences to a certain extent [. . .] the effect of a combined change in multiple object-extrinsic factors cannot be easily predicted from their individual effects.

Citing the results of Marlow *et al.* (2012) on the interactions between surface properties and illumination in gloss perception, Fleming (2014) argues that the visual system is not performing *inverse optics*, or some other kind of unconscious inference, in order to estimate the BRDF (*bidirectional reflectance distribution function*) of the surface. Rather, Fleming (2014, p. 69) writes:

> the goal of material perception is to identify and measure statistically informative appearance attributes—like the size, contrast and distinctness of highlights—that capture how variations in material properties manifest themselves in the image.

Far from being something that must be 'inferred away', contextual and perceiver-dependent features of objects of perception are, in fact, fundamental to vision, which, accordingly, should not be understood in physicalist terms. It is worth spelling out here the connection between physicalism and the inverse optics hypothesis, associated originally with Helmholtz and latterly with vision researchers such as Richard Gregory, Irvin Rock, and David Marr. It was hypothesized that the goal of the visual system was to estimate intrinsic physical properties of objects in the environment, and that this task was executed by estimating the magnitude and spectrum of ambient illumination, and employing this estimate to disambiguate the retinal inputs that confound intrinsic physical surface properties and properties of illumination. Inverse optics, therefore, relied on an underlying physicalist assumption regarding the nature of the representational targets of the visual system. According to the inverse optics tradition, visual perception is a process of analysis, 'breaking down' of the retinal stimulus into simpler separate elements. And in order for this hypothesis to be empirically testable, such elements must be intrinsic features of the physical world (i.e., the kind of things that ordinary measuring instruments detect).

It is because of this intimate connection between physicalism, the detection framework of sensory systems, and inverse inference-based approaches to perception (of which inverse optics is only one of many examples) that Fleming (2014, p. 73) can conclude that perceptual qualities are *not* physical properties:

> we have suggested that when we look at an object and experience a vivid subjective impression of its material properties, we are not actually perceiving its physical properties at all. Instead, we have learned a set of appearance characteristics—i.e., properties of the way the material tends to appear in the image—that capture its distinctive 'look'. (p. 73)

But this leaves the conception of perceptual qualities under-specified. If not a physical property, *what is gloss*? According to Chadwick and Kentridge (2015, p. 225), the current consensus among psychologists is that percepts of gloss are determined by 'pseudocues', by which they mean 'cues that the visual system extracts from the scene' (Chadwick and Kentridge, 2015, fn1). So it seems that the scientific consensus has shifted towards treating gloss as a perceiver-dependent property. It might seem, furthermore, that the science recommends an anti-realist conception of gloss—one in which gloss is a construct of the visual system, projected onto real physical stimuli—a view like anti-realism about colour (Hardin, 1988). Indeed, if one takes as criteria for reality (1) absolute constancy (illumination independence of the percept) and (2) correlation with specific physical cues, one will have to deny that gloss is a real property. Yet gloss is, without doubt, part of our manifest image, and a useful one since gloss aids the reidentification of objects and materials (cfr. Fleming, 2014), which counts against an anti-realist conception of gloss. Thus, it is preferable to give up those two criteria for the reality of a property, in favour of one where a property can be both real and perceiver-dependent (i.e., determined by cues extracted somewhat idiosyncratically by the perceptual system). The parallel here would be with colour relationism, in general—more specifically, an *interactionist* theory in which gloss is a property of the process consisting in a viewer perceptually interacting with an object embedded in a scene and bathed in ambient illumination, like the adverbialism defended by Chirimuuta (2015). Later, following our discussion of timbre, we will say why this interactionist ontology fits both of these cases.

3 Timbre

Most of us have at least intuitive, implicit knowledge of what timbre is, and this knowledge seems to be rooted in the fact that we *use* timbre perception to recognize sounds and sound sources (cfr. Isaac, 2018, pp. 506–7). We are indeed capable of recognizing the sound of a violin, regardless of whether a high- or low-pitched, loud or soft note is being played, whether it is being played by an amateur or a professional, whether it is located near or far from us (within a certain range, of course), and so on, thus suggesting that there is some acoustic quality, different from pitch and loudness, stable enough to ground source recognition. This quality is what we generally refer to as timbre. Moreover, these similarity and dissimilarity judgements about sounds tend to be reliable and consistent not only across time and spatial context, but also across perceivers. However, establishing that there are recognitional tasks, the success of which is best explained by the existence of a certain property being stably represented, still falls short of specifying the nature of this property. So: *what is timbre?*

The oldest proposal regarding the nature of timbre identifies it as the experiential correlate of spectral properties of sound waves. However, this view cannot explain the fact that we are able to experience sounds as very similar in timbre despite the sound waves having different spectral compositions, like in the case of a violin played in the same room as the listener, in a different room of the house, or in a large open space outdoors.

Some have suggested a negative definition, according to which timbre is 'the property that distinguishes two sounds of identical pitch, intensity, duration and location' (Griffiths and Warren, 2004, p. 888). A purely negative definition of this kind could work if only we did not have evidence that pitch, loudness, location, etc. affect timbre perception and timbre similarity judgements in quite substantial ways (e.g., Bizley and Cohen, 2013).

The empirical study of timbre perception is divided into at least three categories: speech timbre, musical timbre, and environmental timbre (Town and Bizley, 2013). However, it seems that in all three domains, the perceptual cues underlying timbre similarity judgements are more or less the same. In particular, perceived timbre similarity seems to correlate quite strongly with spectral and temporal features of sounds, which in turn depend on physical features of the sound source (usually a material object) and of the medium through which sound waves travel (usually air).

Multidimensional scaling (MDS) is used to obtain a low-dimensional representation of the similarity space from subjects' reports of similarity or dissimilarity between pairs of sounds of same loudness, pitch, and duration (Krumhansl, 1989; McAdams and Cunible, 1992; Samson *et al.*, 2002).[5] MDS is a data analysis technique that transforms relations of similarity and dissimilarity—the basis of which is still unknown at the data collection stage—into distances among points in a shared, lower-dimensional space. For instance, Grey (1977),

[5] While in other fields, MDS has been mostly replaced by principal component analysis (PCA) as the main data analysis technique, perceptual psychology still relies heavily on MDS for the following reason. The initial data perceptual psychologists collect are usually subjects' reports such as, in the case of timbre research, judgements of similarity and dissimilarity between pairs of tones. Because similarity and dissimilarity judgements do not themselves contain information about the dimensions along which similarity and dissimilarity are being assessed, scientists need to effectively *create* a similarity space before such space can be analysed further. While MDS allows researchers to make subjects' reports comparable with one another in the first place, in order to subsequently derive a low-dimensional feature representation (of timbre, for instance) from them, a technique like PCA *starts* with a feature representation. Perceptual psychologists still use MDS because they do not have a feature representation to begin with—only a measure of similarity.

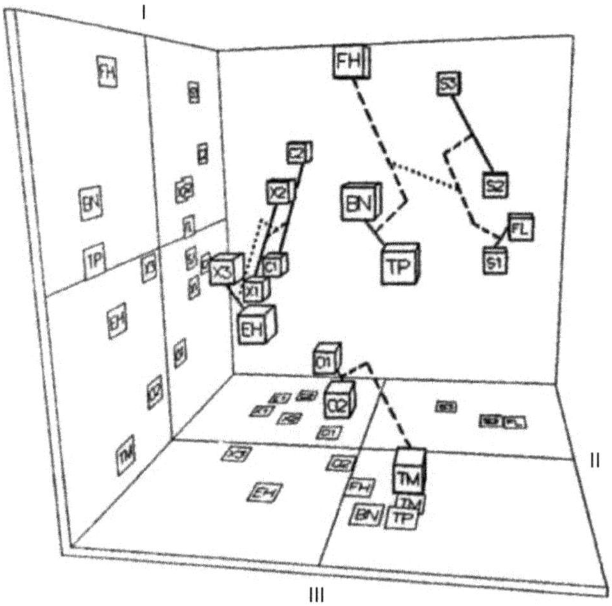

Figure 8.2 Timbre space derived through multidimensional scaling combining subjects' similarity judgements (Grey, 1977). The blocks represent different perceived timbre qualities, which were then labelled using the names of specific musical instruments: BN (bassoon), FH (French horn), TM (trombone), TP (trumpet), FL (flute), EH (English horn), C (clarinet), O (oboe), S (string instruments), and X (saxophone).

the 'pioneer' of MDS, managed to represent different timbres as locations in a 3D similarity space, based on his subjects assigning degrees of similarity (from 'very dissimilar' to 'very similar') to various sound pairs. As a consequence of this procedure, the target feature (in our case, timbre) is associated with a low-dimensional and more tractable (i.e., easier to manipulate experimentally) physical correlate, represented by a location in the quality space,[6] and consisting in a set of spectral and temporal properties of the stimuli. For example, Grey (1977) used the following three properties as the dimensions of his similarity space (Figure 8.2): the shape of the spectral envelope—that is, the overall 'look' of the sound's spectrogram (Axis 1); the combination of spectral fluctuation—that is, how much a sound's spectral shape changes throughout its duration, and synchronous high frequency transients—that is, higher harmonics entering and exiting together (Axis 2); and the frequency and amplitude of the sound wave's beginning and end—that is, how the sound starts and ends (Axis 3).

Creating low-dimensional timbre similarity spaces through MDS requires a lot of abstraction and idealization, especially with respect to the heterogeneity of perceivers' responses. For example, it is generally assumed that all subjects in their similarity assessments assign the same 'weight' to the different perceptual dimensions, while we know that substantial individual

[6] Isaac (2018, pp. 507–8) suggests we understand Grey's similarity space in analogy with the colour solid, the three dimensions of which correspond to hue, saturation, and lightness. However, differently from the colour case, there are no already established perceptual qualities corresponding to the dimensions along which timbre experience can vary.

differences in sensory processing, attention, listening strategies, etc. among subjects affect perceived timbre similarity (e.g., McAdams, 2013; McAdams *et al.*, 1995).

While applying simplifying strategies such as MDS has allowed psychologists to identify the most relevant physical properties with which perceived timbre correlates, timbre as a perceptual quality remains very hard to define (Siedenburg and McAdams, 2017, p. 3). Timbre has been indeed treated as 'the terminological wastebasket of music psychology and psychoacoustics' (Siedenburg and McAdams, 2017, p. 3). Even though scientists can find reasonably adequate physical pseudo-correlates that work for specific experiments and for the purpose of answering specific questions, too many aspects of perceived timbre escape a strictly physical explanation. MDS algorithms and new sound-synthetizing technology can allow researchers to perform sophisticated experiments, to collect enormous amounts of data, and perhaps even to discover physical structures that our auditory system is at least partially sensitive to,[7] but they are not of much help if the goal is to understand how humans track and recognize sounds on the basis of their timbre when neither the stimuli nor the context are controlled by an experimenter. As a consequence, some scientists now relegate timbre 'in the mind of the listener' (Siedenburg and McAdams, 2017, p. 3), especially in virtue of the fact that the auditory system itself has a way of picking up and combining cues that actually makes experienced sound very weakly correlated with perceiver-independent physical properties. Here is where we turn to the second research route, the conceptual and ontological one.

There are different ways in which the latter claim could be interpreted, which in turn correspond to different ontological views of timbre. These range from radically anti-realist positions, such as eliminativism (we should stop talking about timbre altogether and turn to more empirically tractable notions) or fictionalism (timbres are fictional properties needed for some explanatory purposes, but not instantiated out there in the world), to a variety of realist ones with different kinds of physicalism attached to them.[8] In particular, two families of views stand out: timbre as a property of those material objects that we ordinarily identify as sound sources (e.g., musical instruments) and timbre as a property of a more 'global' system, like a process or an event involving sound sources in addition to other elements, such as the medium and other material objects in the scene. The two families of views can be seen as departing in different directions from this claim by Griffiths and Warren (2004, p. 888): 'Timbre is integral to the identity of acoustic sources (for example, musical instruments), combinations of sources […] or acoustic events (for example, vowels). Therefore, the terms "timbre" and "object" can both be applied to sound sources or events.'

3.1 Source-property views

By 'source-property view', we mean the view according to which timbres are properties of sound sources—that is, the material objects whose vibration in a medium causes the formation of sound waves (e.g., bells, guitar strings). This view can be traced back to Helmholtz (1954), even though he initially suggested that timbres depended on geometrical properties

[7] This is Isaac's (2018) structuralist solution.

[8] It is worth noting that the different hypotheses regarding what timbre is run parallel to the hypotheses regarding what sounds more generally are. As Isaac (2018) also points out, a lot of the debate around timbre is going to depend on what we think sounds themselves are, since everybody generally agrees that timbre should be seen as a property of a sound, or a sound type.

of sound waves. However, it is clear that what Helmholtz (1954, p. 10) was trying to capture is a quality that is very closely tied to sound sources: 'By the quality of a tone [*Klangfarbe*, also translated as "timbre"] we mean that peculiarity which distinguishes the musical tone of a violin from that of a flute or that of a clarinet or that of the human voice, when all these instruments produce the same note at the same pitch.' This view is currently the most popular among psychologists studying auditory perception and relying on psychophysics. The main reason for its popularity is that it is consistent with the already mentioned experimental methodology commonly used to isolate the physical correlates of auditory percepts: MDS. Indeed, the locations in low-dimensional quality space derived from MDS are usually labelled in terms of the material objects (e.g., musical instruments) producing the sound, as families of instruments with similar material composition tend to produce sounds that appear similar to the listener along the dimensions used in MDS models. But what *kind* of property of sound sources can timbre be? Two options that we consider are intrinsic and dispositional. We will evaluate each in turn.

Firstly, it may be argued that timbre cannot be an intrinsic property of sound sources. Intrinsic properties do not change, depending on the context in which their bearer is; however, the same object in different contexts can produce sounds with different timbres. The same violin played in a concert hall 'sounds' very differently from when it is played in a subway train: how can timbre be an intrinsic property of the violin, then? Here, the intrinsic-property theorists could insist that timbre is for material sound sources what surface spectral reflectance is for a material surface. Just as a colour realist (Byrne and Hilbert, 2003; Hilbert, 1987) maintains that colour is an intrinsic property of surfaces, regardless of the fact that surface is experienced differently with respect to colour under different illuminations, an analogous timbre realist can claim that the timbre of an object is identical to some physical feature of that object, and that the fact that timbre is experienced differently by perceivers in different contexts is consistent with this idea.

But while the colour realist has a relatively easy time connecting a surface's veridical colour with its SSR, there is no measurable physical property of sound sources that clearly correlates with timbre experiences in the same way SSRs correlate with experienced colours. Indeed, for timbre experiences to be generated, it is not enough that the object is immersed in an appropriate medium (e.g., air), but something *else* must happen to the object for it to produce a sound: a violin's strings need to be bowed, a drum needs to be hit, etc. On the other hand, to continue the analogy with colour, colour experiences only depend on the object being hit by light; no additional change in the object or its relationship with a perceiver is required. Thus, if the intrinsic-property theorist wants to maintain the analogy with colour realism, it should be between SSRs and the physical features of the sound waves generated by the source when the appropriate kind of thing 'happens' to it (i.e., the kind of thing that would cause a sound to be generated), not between a surface's SSR and the sound source itself.

But if this is the appropriate analogy, then we must ask: do sound sources have the propensity to emit only one fixed set of sound waves with measurable physical properties in the same way as a material surface has one fixed measurable SSR? For more obvious sound sources such as musical instruments, one could appeal to the appropriate way in which the instrument is supposed to be played. But musical instruments are not the only material objects that produce sounds, and that therefore have timbres.

How could we decide what the actual sound (analogous to a surface's SSR) of, say, a glass is, given that the glass can produce very different sounds, depending on whether, for instance, someone gently hits it with a fork or drags her wet finger around its upper border? Both seem to be plausible candidates, as we might argue that both are equally likely to come to mind if one asked us what the standard way to get a glass to emit a sound is. And what about the actual sound of a table, a pot, or a refrigerator? Every material object is, at least potentially, a sound source and, as such, according to the intrinsic-property theorist, possesses a specific intrinsic timbre which, following the analogy with colour realism, should be experienced whenever the source emits its actual sound. However, in most cases, determining which one among the many possibilities is the actual sound, which in turn determines the timbre of a material object, seems arbitrary.

However, in response, one might argue that timbres are not intrinsic properties of sound sources, but rather intrinsic properties of the event in which the sound source is involved and that results in it emitting a sound, in close analogy with what Pasnau (2009) argued about colours. Even though this view seems prima facie more plausible, minimally because it is less 'odd' to say that timbres are properties of sound events (e.g., a violin's strings being bowed) than it is to say that they are properties of tables and refrigerators, it still does not fully account for the way in which our timbre experiences relate to the physical world. In particular, this is still an intrinsic-property view which, just like the one discussed above, assumes that timbre is a perceiver—and a context-independent property. Once again, thus, timbre-as-experienced-by-a-perceiver-in-a-context is separated from what timbre really is.

Let us move on to why timbre does not seem to be a dispositional property of sound sources to affect perceivers, either.[9] Consider the case in analogy with colour dispositionalism—that is, the view according to which colours are dispositional properties of material surfaces to cause colour sensations (e.g., Johnston, 1992; Levin, 2000; Peacocke, 1984). Dispositions are properties possessed by objects in virtue of which the object will behave in certain specific ways when placed in certain specific circumstances. In the case of colour, a dispositional view would claim that colours are dispositions of material surfaces to appear in a certain way to a standard observer in standard conditions, where what counts as 'standard' is usually stipulated by the theorist based on certain criteria. This view faces a version of the same problem pointed out for the intrinsic-property view: it is hard to establish in a non-arbitrary way what the 'standards' are, especially considering the variety of conditions that, if asked, one would judge to be 'normal'.

While a dispositionalist account is certainly a step in the right direction, compared to the intrinsic-property view, it is not fully satisfying yet. Timbre is usually identified as the 'quality' of a sound, and sounds themselves, especially when considered as objects of auditory perception, are not entities whose ontological status seems to depend exclusively on their source—though some philosophers, for example, Nudds (2010) and Kulvicki (2014), hold such a view. But once we accept that sounds (and timbres) ontologically depend on more than just the material sources and the physical changes occurring to them, we have effectively opened the door to what we call 'global-property' views.

[9] See also Isaac (2018) for other objections against dispositional views, and in particular Kulvicki (2008, 2014).

3.2 Global-property views

By 'global-property view', we mean the view according to which timbres are properties of more global and complex entities. These entities are usually events, or processes: the sound source undergoes some sort of time-extended change that generates sound waves, which travel and interact with other elements in the scene, such as the surrounding medium and other material objects.

One version of this view was broadly inspired by Gibson and is adopted within the field of ecological psychoacoustics (Neuhoff, 2004). On this view, timbres are properties we attribute to complex, higher-order event/situation types that are salient for us as creatures living in a certain environment (e.g., a predator's walking towards us). These events are not constitutively auditory: timbre is just one way to gather information about them using one of many available sense modalities. However, there is also a less Gibsonian version of the global-property view, according to which timbres are properties of sounds, in turn seen as events that are distinct from the event that caused the sound (Casati *et al.*, 2013; O'Callaghan, 2007; Roden, 2010) and, according to some, constitutively auditory (Scruton, 2009). On this view, timbre categories (or kinds) consist of 'recurrent constellations of features of sound generation processes', without entailing any 'essential limit on what kinds of relationships between more basic physical features can be picked out through identification of timbres' (Roden, 2010, p. 145).

Despite being quite influential, global-property views are still a minority position among experimental psychologists, mostly because they do not easily fit the MDS experimental paradigm. Indeed, the similarity clusters captured by MDS algorithms seem to align well with those spectral and temporal properties of sounds that most directly depend on the material composition of their sources (e.g., woodwinds vs string instruments). On the other hand, however, the global-property view seems more plausible than the source-property view from a phenomenological standpoint: when we experience timbre, we tend to associate it not with the sound source itself, but with what the source is 'doing' in a particular context (i.e., with the sound event in which the source is participating).

This view is admittedly very close to what we favour, but something is still missing: in particular, global-property views (both the Gibsonian and the more contemporary one) do not seem to consider the perceiver—and, in particular, her perceptual system—as playing a decisive role in determining the nature of timbre. Indeed, Roden identifies timbres with features of sound *generation* processes, which, being essentially mechanical interactions between material objects and a surrounding medium, do not include the perceiver.

However, recall what we learned from the case of gloss (Harrison and Poulter, 1951): the fact that gloss affords a unified phenomenological appearance to human subjects, regardless of the material constitution of the surface in question, seems to be fundamental to what gloss *is*.[10] Even though there is not the same amount of empirical work on timbre perception as there is for gloss perception, the fact that we auditorily reidentify objects on the basis of timbre, combined with the objections to source-property views, provides at least prima facie evidence in favour of timbre being a global property, but of a perceiver-dependent kind: an

[10] As opposed to, for instance, what gloss is 'used for', or the role it plays in determinate recognitional tasks that, potentially, a robotic device could be trained to accomplish successfully.

interaction property. In analogy with gloss, timbre would be a property of the process that occurs when a listener perceptually interacts with a sound event embedded in a scene and surrounded by a medium (e.g., air).

Furthermore, the 'gestalt' approach to gloss perception maintained that gloss is not a visual quality that is recovered or recoverable independently of the perception of a whole scene. This is analogous to the 'global' account of timbre. Just as gloss is not simply a property of surfaces, independent of context and perceiver, timbre is not simply a property of sound sources, sound waves, or vibration events taken individually, independent of context (the medium) and perceiver. Both gloss and timbre are relational constructs, or interaction properties: they are properties of processes of perceptual interaction between a perceiver and an object, embedded in a scene and in a certain context.

4 Conclusion

The perceivable qualities of gloss and timbre have much more in common than one might initially think. In this chapter, we argued that both fit an interactionist ontology—that is, they are best understood as properties of perceptual interactions between an object and a perceiver, embedded in a certain context. Because gloss and timbre form such an 'odd' couple, we believe that what we have argued about them can potentially be extended to many other secondary qualities.

Although more work is needed to establish the latter point, it should be noted that the reasons why we defended an interactionist account of gloss and timbre are basically the same reasons why Chirimuuta (2015) has proposed an interactionist (or 'adverbialist') account of colour, and why, more generally, the majority of perceptual scientists tend to reject the detection framework for understanding sensory systems, thus steering away from reductive physicalism with respect to sensory qualities. Together with the detection framework, the cases of gloss and timbre invite us to abandon the atomistic conception of the relationship between physical stimuli and percepts. According to the atomistic conception, indeed, physical stimuli had precise psychological correlates. This, in turn, entails that mental contents (and, in particular, conscious sensations) could be manipulated, measured, and predicted by intervening on physical stimuli.

Timbre and gloss are clear examples of sensory qualities which do not conform to the atomistic conception. What goes on in our minds stands in a very complicated, multifaceted, and flexible relationship with what goes on in the world: our sensory systems do not detect, scan, measure, and reproduce the physical features of the world as they are; they elaborate, combine, select, and sometimes 'invent' ways in which those physical features can be useful for the creature perceiving them.

Lastly, the commonalities between gloss and timbre suggest that sense modalities like vision and hearing, so different both physiologically and anatomically, might nonetheless be quite similar in the way in which our internal sensory states connect us to certain physical features of the external world (Handel, 2006). That the philosophy of perception is very visuocentric is no news, though we acknowledge that, in recent years, philosophers' interest for other sense modalities has increased substantially. The next step, now, would be to reflect on similarities and differences among sense modalities from a more holistic perspective (Cavedon-Taylor, 2021; O'Callaghan, 2014, 2019). Our discussion of timbre and gloss is meant to support this endeavour by showing the way to a realm where the properties investigated by perceptual science can be both real *and* perceiver-dependent, empirically tractable *and* infused with philosophical depth.

References

Bergmann Tiest, W. M. and Kappers, A. M. L. (2007). Haptic and visual perception of roughness. *Acta Psychologica*, 124(2), 177–89. Available from: https://doi.org/https://doi.org/10.1016/j.actpsy.2006.03.002.

Bizley, J. K. and Cohen, Y. E. (2013). The what, where and how of auditory-object perception. Nature Rev*iews* Neurosci*ence*, 14(10), 693–707. Available from: https://doi.org/10.1038/nrn3565.

Byrne, A. and Hilbert, D. R. (2003). Color realism and color science. *Behavioral and Brain Sciences*, 26(1), 3–21.

Casati, R., Di Bona, E., and Dokic, J. (2013). The Ockhamization of the event sources of sound. *Analysis*, 73(3), 462–6. Available from: https://doi.org/10.1093/analys/ant035.

Cavedon-Taylor, D. (2021). High-level perception and multimodal perception. In *Purpose and Procedure in Philosophy of Perception*, edited by Heather Logue and Louise Richardson, 147–73. Oxford, UK: Oxford University Press. Available from: https://doi.org/10.1093/oso/9780198853534.003.0008

Chadwick, A. C. and Kentridge, R. W. (2015). The perception of gloss: a review. *Vision Research*, 109(PB), 221–35. Available from: https://doi.org/10.1016/j.visres.2014.10.026

Chirimuuta, M. (2015). *Outside Color: Perceptual Science and the Puzzle of Color in Philosophy*. Cambridge, MA: MIT Press. Available from: https://doi.org/10.7551/mitpress/9780262029087.001.0001

Cohen, J. (2009). *The Red and the Real: An Essay on Color Ontology*. Oxford, UK: Oxford University Press. Available from: https://doi.org/10.1093/acprof:oso/9780199556168.001.0001

Fleming, R. W. (2014). Visual perception of materials and their properties. *Vision Research*, 94, 62–75. Available from: https://doi.org/https://doi.org/10.1016/j.visres.2013.11.004

Gibson, J. J. (1979). *The Ecological Approach to Visual Perception*. Hillsdale, New Jersey: Lawrence Erlbaum Associates, Inc.

Grey, J. M. (1977). Multidimensional perceptual scaling of musical timbres. *Journal of the Acoustical Society of America*, 61(5), 1270–7. Available from: https://doi.org/10.1121/1.381428

Griffiths, T. D. and Warren, J. D. (2004). What is an auditory object? *Nature Reviews Neuroscience*, 5(11), 887–92.

Handel, S. (2006). *Perceptual Coherence: Hearing and Seeing*. Oxford, UK: Oxford University Press. Available from: https://books.google.com/books?id=Y_WekgEACAAJ

Hardin, C. L. (1988). *Color for Philosophers: Unweaving the Rainbow*. Indianapolis, Indiana: Hackett Publishing Company.

Harrison, V. G. W. and Poulter, S. R. C. (1951). Gloss measurement of papers: the effect of luminance factor. *British Journal of Applied Physics*, 2(4), 92–7. Available from: https://doi.org/10.1088/0508-3443/2/4/302

Hartung, B. and Kersten, D. (2002). Distinguishing shiny from matte. *Journal of Vision*, 2(7), 551.

Helmholtz, H. L. F. (1954). *On the Sensations of Tone as a Physiological Basis for the Theory of Music*, 2nd English edition. New York: Dover Publications.

Hilbert, D. R. (1987). *Color and Color Perception: A Study in Anthropocentric Realism*. Stanford, California: Csli Press.

Ho, Y.-X., Landy, M. S., and Maloney, L. T. (2008). Conjoint measurement of gloss and surface texture. *Psychological Science*, 19(2), 196–204. Available from: https://doi.org/10.1111/j.1467-9280.2008.02067.x

Hunter, R. S. (1937). Methods of determining gloss. *NBS Research Paper RP*, 958, 19–39.

Ingersoll, L. R. (1921). The glarimeteran instrument for measuring the gloss of paper. *Journal of the Optical Society of America*, 5(3), 213–17. Available from: https://doi.org/10.1364/JOSA.5.000213

Isaac, A. M. C. (2018). Prospects for timbre physicalism. *Philosophical Studies: An International Journal for Philosophy in the Analytic Tradition*, 175(2), 503–29. Available from: https://doi.org/10.1007/s11098-017-0880-y

Johnston, M. (1992). How to speak of the colors. *Philosophical Studies*, 68(3), 221–63. Available from: http://www.jstor.org/stable/4320356

Jones, L. A. (1922). The gloss characteristics of photographic papers. *Journal of the Optical Society of America*, 6(2), 140–61. Available from: https://doi.org/10.1364/JOSA.6.000140

Krumhansl, C. L. (1989). Why is musical timbre so hard to understand. *Structure and Perception of Electroacoustic Sound and Music*, 9, 43–53.

Kulvicki, J. (2008). *The Nature of Noise*. Ann Arbor, MI: Michigan Publishing.

Kulvicki, J. (2014). Sound stimulants: defending the stable disposition view. In *Perception and Its Modalities*, edited by D. Stokes, 205–21. Oxford, Oxford University Press. Available from: https://doi.org/10.1093/acprof:oso/9780199832798.003.0009

Leloup, F., Obein, G., Pointer, M., and Hanselaer, P. (2014). Toward the soft metrology of surface gloss: a review. Color Research and Application, 39, 559–70. Available from: https://doi.org/10.1002/col.21846

Levin, J. (2000). Dispositional theories of color and the claims of common sense. *Philosophical Studies*, 100(2), 151–74. Available from: http://www.jstor.org/stable/4321071

Marlow, P. J., Kim, J., and Anderson, B. L. (2012). The perception and misperception of specular surface reflectance. *Current Biology*, 22(20), 1909–13. Available from: https://doi.org/10.1016/j.cub.2012.08.009

McAdams, S. (2013). 2 – Musical timbre perception. In *The Psychology of Music*, third edition, edited by D. Deutsch, 35–67. Cambridge, MA: Elsevier Academic Press. Available from: https://doi.org/https://doi.org/10.1016/B978-0-12-381460-9.00002-X

McAdams, S. and Cunible, J.-C. (1992). Perception of timbral analogies. *Philosophical Transactions of the Royal Society of London. Series B: Biological Sciences*, 336(1278), 383–9.

McAdams, S., Winsberg, S., Donnadieu, S., De Soete, G., and Krimphoff, J. (1995). Perceptual scaling of synthesized musical timbres: common dimensions, specificities, and latent subject classes. *Psychological Research*, 58(3), 177–92. Available from: https://doi.org/10.1007/BF00419633

Neuhoff, J. G. (2004). *Ecological Psychoacoustics*. Cambridge, MA: Elsevier Academic Press.

Nudds, M. (2010). What are auditory objects? *Review of Philosophy and Psychology*, 1, 105–22. Available from: https://doi.org/10.1007/s13164-009-0003-6

O'Callaghan, C. (2007). *Sounds: A Philosophical Theory*. Oxford, UK: Oxford University Press. Available from: https://doi.org/10.1093/acprof:oso/9780199215928.001.0001

O'Callaghan, C. (2014). Not all perceptual experience is modality specific. In *Perception and Its Modalities*, edited by D. Stokes, pp. 133–65. Oxford, UK: Oxford University Press. Available from: https://doi.org/10.1093/acprof:oso/9780199832798.003.0006

O'Callaghan, C. (2019). *A Multisensory Philosophy of Perception*. Oxford, UK: Oxford University Press. Available from: https://doi.org/10.1093/oso/9780198833703.001.0001

Olkkonen, M. and Brainard, D. H. (2011). Joint effects of illumination geometry and object shape in the perception of surface reflectance. *I-Perception*, 2(9), 1014–34. Available from: https://doi.org/10.1068/i0480

Pasnau, R. (2009). The event of color. *Philosophical Studies*, 142(3), 353–69.Available from: https://doi.org/10.1007/s11098-007-9191-z

Peacocke, C. (1984). Colour concepts and colour experience. *Synthese*, 58, 365–81.

Pfund, A. H. (1930). The measurement of gloss. *Journal of the Optical Society of America*, 20(1), 23–5. Available from: https://doi.org/10.1364/JOSA.20.000023

Roden, D. (2010). Sonic art and the nature of sonic events. *Review of Philosophy and Psychology*, 1, 141–56. Available from: https://doi.org/10.1007/s13164-009-0002-7

Samson, S., Zatorre, R. J., and Ramsay, J. O. (2002). Deficits of musical timbre perception after unilateral temporal-lobe lesion revealed with multidimensional scaling. *Brain*, 125(3), 511–23. Available from: https://doi.org/10.1093/brain/awf051

Scruton, R. (2009). *Understanding Music: Philosophy and Interpretation*. London, UK: Continuum.

Siedenburg, K. and McAdams, S. (2017). Four distinctions for the auditory 'wastebasket' of timbre. Frontiers in Psychology, 8. Available from: https://doi.org/10.3389/fpsyg.2017.01747

Smart, J. J. C. (1975). On some criticisms of a physicalist theory of colors. In *Philosophical Aspects of the Mind–Body Problem*, edited by C. L. Y. Cheng, 54–63. Honolulu, Hawaii: University Press of Hawaii.

Thompson, E. (1995). *Colour Vision: A Study in Cognitive Science and the Philosophy of Perception*. New York: Routledge.

Town, S. M. and Bizley, J. K. (2013). Neural and behavioral investigations into timbre perception. *Frontiers in Systems Neuroscience*, 7, 88. Available from: https://doi.org/10.3389/fnsys.2013.00088

Wendt, G., Faul, F., Ekroll, V., and Mausfeld, R. (2010). Disparity, motion, and color information improve gloss constancy performance. *Journal of Vision*, 10(9), 7.

Yablo, S. (1995). Singling out properties. *Philosophical Perspectives*, 9, 477–502. Available from: https://doi.org/10.2307/2214231

9

Mechanisms underlying auditory-object perception

Ronald W. DiTullio and Yale E. Cohen

1 Introduction

A great joy in life for many of us is hearing our favourite band perform a song live. We do this with seemingly little to no mental energy despite the fact that, upon closer inspection, our brain has to engage in a variety of very complicated processes for our listening experience to be successful. For example, to hear the song clearly, we have to separate the band's music from the background noise of the audience (and those rowdy folks yelling 'Free Bird!'). During a jamming section, we may choose to guide our attention to the mandolin or ignore the mandolin so that we can focus on the fiddle. If the song is a new song, we may also engage memory processes to compare the new song's structure with others sung by the band or by other bands.

We can conceptualize some of these processes as computations, acting in parallel and serial, in the brainstem and cortex. These processes include, but are not limited to: (1) perceptual grouping; (2) categorization; (3) attention; (4) and decision-making (Bizley and Cohen, 2013; Griffiths and Warren, 2004). Perceptual grouping is a form of feature-based stimulus segmentation that determines whether acoustic events are grouped into a single auditory object or are segregated into distinct auditory objects. Categorization is a process in which stimuli and sounds are classified based not only on their acoustic features, but also on other types of available information. Although attention is not always necessary, our awareness of a sound can be influenced by attention. Auditory decision-making involves the brain's interpretation of information within and across discrete stimuli to detect, discriminate, or identify their source or content (Gold and Shadlen, 2007). Consequently, it is a fundamental part of the neural computations underlying auditory-object perception, as it is arguably the end goal of auditory-object perception (Nudds, 2010). Additionally, because decision-making depends on the temporal structure of an auditory stimulus, it is a deliberative process in which listeners often accumulate and interpret auditory information over time to form categorical judgements (Brunton *et al.*, 2013; Green *et al.*, 2010; Mulder *et al.*, 2013).

In this chapter, we highlight a subset of these computations and discuss their neural mechanisms. We focus on the rhesus macaque as a model system to study cortical processing, with a particular emphasis on computations and mechanisms that subserve auditory perception. We pose challenges to the field as to how to best study auditory perception and define some key open questions that remain unanswered.

Ronald W. DiTullio and Yale E. Cohen, *Mechanisms underlying auditory-object perception* In: *Sensory Individuals*. Edited by: Aleksandra Mroczko-Wąsowicz and Rick Grush, Oxford University Press. © Oxford University Press 2023.
DOI: 10.1093/oso/9780198866305.003.0010

2 What is an auditory object?

Before going any further, we should pause and take some time to define an 'auditory object'. At its most basic level, an auditory object is the perceptual result of the auditory system's interpretation of acoustic stimuli (Bizley and Cohen, 2013; Griffiths and Warren, 2004). That is, what we 'hear' is an auditory object: the honk of a car, the speech sounds of a politician, or the bleating of a goat. We note that this definition then raises questions about the ability of the auditory system to interpret acoustic events, often referred to as auditory scene analysis; we will discuss auditory scene analysis more fully later in this chapter. Because of this definition of an auditory object, it implies that auditory objects are emergent rather than existing in and of themselves. In other words, although acoustic stimuli physically exist, auditory objects only emerge because these stimuli interact with the auditory system. So, to provide a glib answer to a well-worn question: when a tree falls in the woods and no one is around to hear it, there is an acoustic stimulus, but no auditory object. Relatedly, another attribute of an auditory object is that it must be assigned to a source (Nudds, 2010). For example, if you hear mooing while walking through a pasture, this object needs to be assigned either to the cow in the pasture or to our nearby friend who was doing an impression, or to a very oddly hidden speaker playing recordings of mooing. This is unlike visual objects: the visual object and the source are nominally identical.

Another interesting aspect of auditory objects is that, by definition, they unfold over time (Bregman, 1990), and thus, time will play a critical role in the perception of auditory object. Indeed, commonly encountered auditory objects, such as speech sounds and other natural stimuli, are well characterized by the specific structure of their spectrotemporal dynamics (Cohen *et al.*, 2007; Singh and Theunissen, 2003).

Other objects can be formed by temporal regularities—that is, how smoothly and consistently the acoustic events unfold over time. For example, if a person randomly hits a desk with their hand, we perceive each 'thunk' as an independent auditory object and do not link one thunk with the next. However, if the thunks come quickly enough and the time between thunks is consistent, we might link the thunks together and hear the auditory object as a beat or as a rhythm. This grouping is often called an 'auditory stream'. The individual thunks are not the unit of perception, but instead, the temporal sequence of the thunks is the unit, the auditory object (Griffiths and Warren, 2004; Schnupp *et al.*, 2012).

Thus, depending on the manner in which auditory stimuli unfold over time, the temporal sequence of an auditory stimulus may be considered the auditory object, or each stimulus may be considered as an object. However, at least for human listeners, the temporal manipulations that can affect this perceptual grouping (and a listener's perception of the auditory object) can be independent of the accumulation process that underlies a particular auditory decision. For example, a human listener's report of the frequency content of a sequence of tone bursts is independent of whether this series of tone bursts is heard as 'one auditory object' or 'multiple auditory objects' (Liu *et al.*, 2015).

3 Parsing the auditory scene

As we have alluded to, a single auditory stimulus is defined by its spectrotemporal properties. But in real-world acoustic environments, such as in a bustling restaurant or our band analogy from above, there are multiple sources, each of which are generating their own auditory stimuli. To add further complexity, as these stimuli propagate from their sources (e.g., the banjo and the mandolin) to our ears, they are mixed together and enter the ear and auditory system as an unlabelled stimulus mixture. How then does the brain unmix those stimuli that comprise the auditory scene, so that we perceive our auditory world as stable, distinct auditory objects (perceptual units)?

In general, this process is called 'auditory scene analysis' (Bregman, 1990). Auditory scene analysis is the means by which the auditory system identifies the features of an acoustic stimulus that have common relationships with one another (either at a particular time or over time) and then 'grouping' together the neural representations of these features to form an auditory object. Acoustic features that are distinct are segregated from one another and form different auditory objects. These ideas follow closely from those discussed by the Gestalt psychologists of the early 1900s. In general, we can think of two types of grouping cues: simultaneous (which occur at a particular time, such as pitch, harmonicity, and spatial location) and sequential (which occur over time).

Because of the way the brain interprets grouping cues, any auditory stimuli whose frequencies change coherently over time are grouped together into a single sound. This is true regardless of whether the frequencies are harmonically related or not. A harmonic stimulus is made up of frequencies that are integer multiples of the lowest (or *fundamental*) frequency. One classic example of an auditory stimulus that has simultaneous grouping cues is human speech, which is a harmonic stimulus. Harmonicity is such a potent grouping cue that even if one frequency component is mistuned, the mistuned frequency band is segregated from the rest of the harmonically related frequencies and is heard as a separate auditory object (Bregman, 1990). Pitch is another important simultaneous grouping cue and is the primary means by which we can separate the voices of simultaneously talking speakers (Bregman, 1990; Griffiths and Warren, 2004).

There is an interesting relationship between harmonicity and pitch. The pitch of a harmonic stimulus is based on its fundamental frequency. We can, though, construct an auditory stimulus that is missing its fundamental frequency. Despite having this unique acoustic structure, the pitch of this stimulus is the same as if it did have the fundamental frequency. So, although all harmonic stimuli have pitches, harmonicity is not a requirement for pitch.

Pitch-selective areas have been identified in the human brain, in a region just anterior to Heschl's gyrus (Patterson *et al.*, 2002; Penagos *et al.*, 2004; Warren and Griffiths, 2003; Norman-Haignere *et al.*, 2016). However, there is some debate as to whether pitch is encoded by specialized brain areas or whether pitch perception is, instead, subserved by a broad network of brain areas (Bizley and Cohen, 2013). Indeed, both sets of findings may be viable. A specialized area might subserve pitch perception, whereas a broad brain network might enable pitch to be used as a cue for organizing the auditory scene. Another possibility is that the specialized areas and the broad network might represent a hierarchy of pitch processing (Griffiths and Hall, 2012), culminating in a representation that is independent of the listening context (i.e., an 'invariant' representation of pitch).

Sequential grouping cues are those cues by which the brain can track information that changes smoothly and continuously across time. Indeed, speech sounds and vocalizations tend to be defined by such smooth slow changes (Cohen *et al.*, 2007; Woolley *et al.*, 2005). Sequential grouping cues enable temporal sequences of auditory stimuli to be assigned to the same sound source. For example, two sequences of tone bursts may be assigned to different sources when their frequency content is very different. However, the means by which the brain learns to extract such slow features and to use them for perception remains an open question.

A consequence of sequential grouping cues is a concept called *amodal completion*. Imagine the following experiment. An experimentalist presents a tone burst that is followed by a silent period and then another presentation of the tone burst. Most listeners would report hearing a first sound, followed by a silent gap, and then a second sound. If the silent period is filled with a noise burst, listeners often report hearing a single continuous tone burst along with the noise burst. It is as if the brain has 'filled in' the missing stimulus. The brain assumes (guesses) that there is actually one long tone burst (one sound source), and a second sound source (that generates the noise burst) happens to co-occur with the first source. In other words, we perceive one auditory object (the long tone burst), even though, in reality, only a part of it is actually physically present in the environment. In everyday life, we experience this phenomenon when we are talking with someone and another person sneezes, which blocks (or 'masks') the speaker's voice. Even though the voice is blocked, we have the perception that we can still hear their words.

4 Cortical processing of auditory information

The fundamental organization of the auditory cortex across all mammals is remarkably similar (Hackett, 2011; Kaas and Hackett, 2000). Here, we focus on the organization in humans and non-human primates; in particular, we focus on rhesus macaques, due to the homology between macaques and humans (Hackett, 2011). The auditory cortex is defined primarily by the thalamocortical projection from the ventral division of the auditory thalamus to the cortex. In humans, this projection generates at least two primary auditory fields in Heschl's gyrus that together are referred to as the 'core' (Baumann *et al.*, 2013). These fields may be analogous to the macaque primary auditory cortex (A1) and the rostral field (R). The macaque cortex includes a third core area, the rostral–temporal (RT) field.

In both human and non-human primates, surrounding the core auditory cortex is a second series of auditory fields that is called the 'belt', which, in humans, overlaps with Wernicke's area. We consider these to be secondary auditory fields because their major source of neural input is from the core auditory cortex, and not from the auditory thalamus. There is another series of auditory fields that surrounds the belt. These fields are called the 'parabelt' and are tertiary auditory fields because their main source of auditory input arises from the belt (secondary) fields.

So, from a simplified anatomical point of view, we can see that information flows from the thalamus to the core fields, to the belt, and ultimately to the parabelt regions. This picture is highly suggestive of a hierarchy of information flow in which representations of auditory stimuli become increasingly more sophisticated from the thalamus to the parabelt. However, a more realistic picture of the core, belt, and parabelt would include multiple serial, parallel, convergent, and divergent patterns of connectivity. There are also feedback connections between different regions of the auditory cortex (e.g., from belt regions of the cortex to the core auditory cortex), from higher-order cortical areas (e.g., the prefrontal cortex) back to the auditory cortex, and from the auditory cortex back to the subcortical structures, like the thalamus and inferior colliculus (Winer and Lee, 2007). Our understanding of how these patterns of connectivity relate to auditory perception—especially the role of feedback projections—remains elusive, despite several important sets of studies (Blackwell *et al.*, 2020; Kusmierek *et al.*, 2012; Kusmierek and Rauschecker, 2009; Recanzone and Sutter, 2008; Tabas *et al.*, 2020).

In cortical regions downstream from the core auditory cortex, we can identify a pattern of connectivity that is called the 'ventral auditory pathway' (Rauschecker, 2012, 2018; Rauschecker and Scott, 2009; Romanski *et al.*, 1999; Romanski and Averbeck, 2009). In both humans and non-human primates, the ventral pathway begins in the core auditory cortex and terminates in the frontal lobe—the inferior frontal cortex in humans and the ventrolateral prefrontal cortex (vlPFC) in monkeys. In monkeys, the pathway is known in more detail and includes the medial lateral (ML) and anterolateral (AL) belt regions of the auditory cortex, as well as the parabelt. A second pathway, in both humans and non-human primates, can also be identified, which is referred to as the 'dorsal auditory pathway'.

5 Function of the ventral auditory pathway in auditory-object perception

The ventral auditory pathway is thought to mediate auditory-object perception and auditory scene analysis. By contrast, the dorsal auditory pathway mediates spatial processing, sensorimotor processing, and temporal sequencing, and in humans, it may have a fundamental role in language (Alain *et al.*, 2001; Rauschecker, 2012, 2018; Rauschecker and Scott, 2009; Romanski *et al.*, 1999; Romanski and Averbeck, 2009).

A variety of lines of evidence support this differentiation of information processing in the human brain; we highlight some of the more important ones. For example, a functional magnetic resonance imaging (fMRI) study noted more activation in the auditory cortex and inferior frontal gyrus when subjects participated in a pitch task than when they participated in a location task (Alain *et al.*, 2001). By contrast, during the location task, there was increased activation in regions of the dorsal auditory pathway (e.g., the parietal cortex). Other studies highlight the development of categorical representations, and invariant representations of auditory objects occur along the ventral auditory pathway, but not along the dorsal pathway (DeWitt and Rauschecker, 2012; Klein and Zatorre, 2015; Leaver and Rauschecker, 2010). By contrast, anticipation of the next stimulus in a temporal sequence of auditory stimuli generates more activation in the dorsal auditory pathway, consistent with its role in temporal sequencing (Leaver *et al.*, 2009; Rauschecker, 2012).

These findings from the human literature are augmented by neural findings from non-human animals. Indeed, Rauschecker and colleagues found that between A1 and AL, neural sensitivity to vocalizations and other kinds of complex auditory stimuli increased, whereas between A1 and CL/CM (which are belt regions of the auditory cortex in the dorsal pathway), there is increased sensitivity to the spatial location of an auditory stimulus (Rauschecker *et al.*, 1995; Rauschecker and Tian, 2000; Tian *et al.*, 2001). In support of this set of findings,

Miller and Recanzone (2009) found that the sensitivity of pooled auditory cortex activity in the dorsal pathway—but not in the ventral pathway—was sufficient to explain the psychophysical sensitivity of a listener's ability to localize a sound source.

6 Tests of the causal role of the ventral auditory pathway in auditory-object perception

Is there causal evidence for the contribution of the ventral auditory pathway in auditory-object perception? Tests of causality have a rich history in the motor and visual systems but are rarely applied to tests of the auditory perception and behaviour. Further, causal tests are even rarer in human subjects. Nonetheless, patient lesion studies do suggest a causal role for the ventral stream in auditory-object processing (Poliva *et al.*, 2015).

Causal tests are better facilitated in non-human animals due to the broader range of causal manipulations that are available to experimenters. For example, in a classic study by Lomber and Malhotra (2008), they found that, in the cat, the dorsal and ventral pathways were independent and causally contributed to different auditory behaviours: the dorsal pathway (without the ventral pathway) contributes to sound localization, but the ventral pathway (without the dorsal pathway) contributes to sound identification. Whether this finding would translate to primates is unclear because the organization of the auditory cortex in cats is somewhat different than in primates (Hackett, 2011).

Other studies have directly identified causal roles for the ventral auditory pathway in auditory perception. For example, in one study, Lizabeth Romanski and colleagues identified a causal contribution of the vlPFC to working memory in the context of monkey communication signals, with either their vocalizations or the distinct facial expression associated with a vocalization (Plakke *et al.*, 2015). In this study, they found that vlPFC inactivation (through reversible cooling) did not inhibit the monkeys' capacity to conduct a visual working memory task. However, the monkeys had significant problems when they were asked to do an auditory–visual or auditory working memory task.

A second set of studies from our group examined the specific causal contributions of different regions of the ventral auditory pathway to auditory-object decisions (Tsunada *et al.*, 2016, 2019). In one study, we applied microstimulation to clusters of neurons in the AL and ML belt regions of the auditory cortex while monkeys reported whether they heard a temporal sequence of tone bursts as 'low frequency' or 'high frequency'. Although we found that microstimulation did not, on average, affect performance in the ML belt region, we identified a significant effect in the AL belt region. That is, when we microstimulated an AL neuron whose frequency tuning was associated with the high-frequency content of the tone burst sequence, the monkey's reports were biased towards 'high-frequency' reports. In contrast, when we microstimulated an AL neuron whose frequency tuning was associated with the low-frequency content of the tone burst sequence, the monkey's reports were biased towards 'low-frequency' reports. Because the primary source of auditory input to the AL belt region is the ML belt region (Hackett, 2011), this finding indicates that stimulus representations, such as those found in the ML bel region, can be transformed into representations of sensory evidence, like those seen in the AL belt region, in a single processing stage (Tsunada *et al.*, 2016).

More recently, in a second study, we identified a causal role for vlPFC in decision-making when monkeys participated in a similar categorization task. As discussed in detail below, vlPFC activity predicts monkeys' choice biases on the next trial. Consistent with this finding, when we microstimulated vlPFC sites, microstimulation did not affect the current decision but instead causally affected monkeys' choices on the next trial.

7 Neural correlates of auditory decision-making relating to auditory objects

A critical manipulation during decision-making studies is to probe a listener's choices when a stimulus is 'ambiguous'. For example, if a listener is asked to report whether a temporal sequence of tone bursts contains more 'low-frequency' or 'high-frequency' tone bursts, an ambiguous or 'noisy' stimulus would be one that randomly contains 50% of low-frequency tone bursts and 50% of high-frequency tone bursts. Because this stimulus is noisy, a listener's reports vary randomly, on a trial-by-trial basis. The experimenter can then separate changes in stimulus features from those related to changes in the behavioural report (i.e., what type or kind of auditory object did you hear?) and tease apart the information represented by particular neurons and/or neural populations. If a neuron primarily represents a stimulus feature, its response should be relatively invariant to the behavioural choice. By contrast, if a neuron reflects a behavioural choice, then neural activity should differ as a function of the choice. In other words, neural activity that is modulated by a behavioural choice is more closely aligned with auditory-object perceptual processing than with stimulus processing.

Several studies have noted such A1 choice-related activity (Bizley *et al.*, 2013; Ceballo *et al.*, 2019; Francis *et al.*, 2018; Kilian-Hütten *et al.*, 2011; Niwa *et al.*, 2012, 2013; Riecke *et al.*, 2009). For example, during a pitch discrimination task, both local-field potentials and spiking activity appear to be modulated by a listener's pitch choices. Other work has identified choice activity during a task in which the listeners had to detect amplitude-modulated noise. Finally, choice activity has been found in human studies when the subjects were asked to discriminate between ambiguous speech sounds.

However, another set of literature provides a different set of findings. This literature has failed to identify choice-related activity in A1 and in several belt regions of the auditory cortex (Christison-Lagay *et al.*, 2017; Erlich *et al.*, 2015; Hanks *et al.*, 2015; Lemus *et al.*, 2009a, b; Tsunada *et al.*, 2011, 2016). For example, Lemus and colleagues failed to identify A1 choice activity while monkeys were participating in an acoustic-flutter task, but did find choice-related activity in the ventral premotor cortex. Across a variety of tasks and stimuli, we also have not identified choice activity in single neurons in the early auditory cortex, with the exception of one streaming study (Christison-Lagay and Cohen, 2018).

How can we reconcile these apparently contradictory findings? We do not have an authoritative answer but posit one idea for further consideration. Studying auditory decisions is not simply cataloguing that region X has choice activity, whereas region Y does not. Instead, it is critical to attempt to identify whether choice activity reflects a feedforward computation that contributes causally to the eventual auditory decision (Nienborg and Cumming, 2009) or a feedback computation that is correlated with, but not (causally) related to, the decision. Indeed, if choice-related activity in a particular brain region occurs after the actual choice, one's interpretation of its functional utility would be substantially different than if it occurred at the time of the choice or just prior to the choice. Thus, experimental paradigms need to address not only if choice activity is present, but also *when* it is present.

A recent approach from our group has championed this idea for studying auditory decisions by combining neurophysiological recordings, behaviour, and computational modelling of variants of sequential sampling models, like the drift diffusion model (Gold and Shadlen, 2007). These models conceptualize perceptual decisions as a process of converting incoming auditory sensory evidence, which we hypothesize is represented as the noisy spiking activity of neural populations of auditory neurons, into a decision variable that guides behaviour. Such models are advantageous because they allow us to make predictions about a listener's choices and reaction times as a function of the manipulated stimulus parameter (e.g., the percentage of high-frequency tone bursts in a temporal tone burst sequence). Accordingly, model fits can be directly compared with listener accuracy and reaction time to yield insights into the decision process and how various manipulations affect behaviour. Most critical to our earlier point, these fits parcel reaction time into a 'decision' time (i.e., the time needed for the listener to make a perceptual decision) and a 'non-decision' time (i.e., the time period reflecting post-decisional processes, motor preparation, etc.). This parcellation allows for an analysis of the causal relationship between neural activity and perceptual choice.

Using this approach, we have made several insights into the causal temporal flow of auditory information along the ventral auditory pathway (Tsunada *et al.*, 2016, 2019). When we compared the responses of AL neurons relative to the decision time, we found a correlation between stimulus sensitivity (i.e., nominally frequency tuning) and choice-related activity. Importantly, this correlation became significant just before the decision time. In contrast, we failed to find such a correlation in the ML data. This suggests that AL neurons—but not ML neurons—may transmit information about the current decision.

Neural activity in vlPFC had a different dynamic profile than that seen in the AL and ML belt regions. In this brain region, we found that while the decision was being formed, vlPFC activity was not modulated by any of the task and stimulus variables. However, after the decision had been formed, vlPFC activity was modulated by the sensory evidence, the monkeys' choices, and the trial outcome (i.e., was the trial correct or incorrect?). These findings are consistent with the hypothesis that vlPFC post-decision activity may play a role in evaluating the just-completed decision and biasing the next decision. Indeed, vlPFC activity predicts the monkeys' subject-specific choice biases on the next trial.

The ability to identify the temporal dynamics of a decision-making process and then relate these dynamics to neural activity is a powerful way to identify a flow of information. We encourage others to use a similar approach in order to distinguish between feedforward and feedback hierarchical contributions to perception.

8 Auditory categorization

Auditory categorization is a cognitive process in which auditory objects are classified and grouped based on their acoustic features and other types of information (e.g., semantic knowledge about the sounds) (Freedman *et al.*, 2001, 2002; Miller *et al.*, 2002, 2003; Russ *et al.*, 2007; Spence, 1937, 1987). For example, when we hear the word 'hello' from different speakers, we can categorize the speaker's country of origin (e.g., the United States or Britain) based on the accent of their voice. On the other hand, if we solely want to analyse the linguistic content transmitted by the speech stimulus, we do not consider the unique accent, pitch, etc. of each speaker and categorize the sound into the distinct word category 'hello'.

Thus, auditory categorization enables humans, as well as non-human animals, to extract, manipulate, and efficiently respond to sounds (and other stimuli). Categories can also be conceptualized as forming a hierarchy, with the most general category at the top of the hierarchy. For example, the Beatles' '*I want to hold your hand*' can be categorized generally as 'music', more specifically as 'popular music', and even more specifically as 'rock and roll'.

A specific type of categorization is called *categorical perception* (Doupe and Kuhl, 1999). What is fascinating about categorical perception is that our perception of a sound does not linearly track with changes in its acoustic features. That is, small changes in the acoustic features of a stimulus can lead to large changes in perception, while in other situations, large changes in the acoustic features of a stimulus do not change perception at all. Our perception can be deemed 'categorical' in that despite potentially large changes in the acoustic features of a stimulus, we only hear one sound—that is, one category.

We can also form more complicated types of categories, so-called *abstract* categories (Eckardt and Züberbuhler, 2004; Fabre-Thorpe *et al.*, 1998; Nieder *et al.*, 2002; Orlov *et al.*, 2000; Russ *et al.*, 2007; Vogels, 1999; Züberbuhler, 2000a, b; Züberbuhler and Seyfarth, 1997). These categories are not formed wholly on the basis of shared acoustic similarities. Instead, an abstract category is one in which a group of auditory stimuli are linked together based on shared features, as well as on semantic information or acquired knowledge. For example, a standard office chair and a bean bag chair look very different. However, because we can sit on both, we can place them in an abstract category of 'things we can sit on'. In another example, a combination of physical characteristics and knowledge about their reproductive processes put dogs, cats, and killer whales into one category ('mammals'), whereas canaries are placed into a separate category. However, if we use different criteria to form a category of 'pets', dogs, cats, and canaries would be members of this 'pet' category, but perhaps not killer whales.

This type of categorization is not limited to human listeners. Indeed, non-human primates form abstract categories based on their vocalizations (Gifford *et al.*, 2003; Hauser, 1998). For example, rhesus monkeys produce a vocalization called a 'harmonic arch' and another one called a 'warble'. These vocalizations are acoustically distinct (they sound different). Yet, to our monkey friends, they transmit the same type of information: the discovery of high-quality rare food. Another pair of vocalizations, 'coos' and 'grunts' are also acoustically distinct and transmit a different type of information: the discovery of low-quality common food. In another example, Diana monkeys can form abstract categories based on communication calls from different species: they respond similarly to information regarding the presence of predators, regardless of whether the alarm call originated from one of their kind, from a different monkey species, or even from a guinea fowl, a type of bird that lives in the same region as these monkeys (Cheney and Seyfarth, 1990; Züberbuhler and Seyfarth, 1997).

In both human and non-human listeners, as one progresses from the core auditory cortex to downstream brain regions, neurons become more sensitive to more complex categorical representations (Gauthier *et al.*, 1999; Leaver and Rauschecker, 2010; Marschark *et al.*, 2004; Miller and Cohen, 2010; Rosch, 1978; Tsunada and Cohen, 2014). In the core auditory cortex, we find category representations for simple auditory stimuli. For example, when monkeys are asked to categorize a type of frequency sweep as either 'increasing' or 'decreasing', neurons in the core auditory cortex have categorical responses. That is, they respond similarly, for example, to all 'decreasing' stimuli, even though the specific frequency content of the stimuli is different (Selezneva *et al.*, 2006).

More complex categories are found outside of the core auditory cortex; these include category representations for certain speech stimulus (e.g., /ba/ vs /da/), bird song and other types of vocalizations, musical instruments, and even tool sounds. There are even representations of abstract categories: neurons in the vlPFC are modulated by the valence of the vocalization (high-quality food vs low-quality food) (Gifford *et al.*, 2005). That is, these neurons coded the 'meaning' of the vocalizations, as opposed to their acoustic properties. Similarly, other studies in humans and non-human primates have found abstract category representations of the number of objects, emotional valence of a speaker's voice, and semantic information (e.g., gender) (Joassin *et al.*, 2011; Nieder, 2012). Thus, along the ventral auditory pathway, cortical areas encode categories for more complex stimuli and more abstract information (Leaver and Rauschecker, 2010).

9 Conclusion and open questions

Many fundamental questions remain about the neural processing supporting auditory objects. For example, recent auditory studies have analysed the differential contribution of feedforward and feedback information flow (Blackwell *et al.*, 2020; Clayton *et al.*, 2021; Williamson and Polley, 2019; Yin *et al.*, 2020). However, it is critical to conduct simultaneous recordings in different cortical regions to more rigorously understand how brain regions interact and contribute to information flow, especially with respect to auditory perception. Relatedly, the degree to which subcortical processes interact with decision-making and how these feedforward and feedback loops contribute to auditory-object processing is an open question (Cope *et al.*, 2015; King *et al.*, 2018). Secondly, at least in mice, the primary auditory cortex contributes to simple and more complex choices in different ways (Ceballo *et al.*, 2019). However, the degree to which different types of auditory objects differentially engage the core auditory cortex and brain regions downstream of the core auditory cortex has yet to be fully articulated. Thirdly, is the ventral auditory pathway the only pathway that contributes to perceptual decisions? Do other pathways (e.g., the dorsal auditory pathway) interact with the ventral pathway to form a consistent and coherent representation of the auditory scene? These interactions seem especially important for language processing in humans (Rauschecker, 2012). Finally, future work, in both humans and non-human listeners, should investigate the degree to which, and by what mechanisms, circuits of neurons (within a cortical area or across cortical areas) contribute to auditory-object processing (Downer *et al.*, 2015, 2017; Fitzgerald *et al.*, 2013; Francis *et al.*, 2018; Ganguli *et al.*, 2008; Gao *et al.*, 2017; Ni *et al.*, 2018; Norman-Haignere *et al.*, 2019; Norman-Haignere and McDermott, 2018).

References

Alain, C., Arnott, S. R., Hevenor, S., Graham, S., and Grady, C. L. (2001). 'What' and 'where' in the human auditory system. *Proceedings of the National Academy of Sciences of the United States of America*, 98, 12301–6.

Baumann, S., Petkov, C., and Griffiths, T. (2013). A unified framework for the organization of the primate auditory cortex. *Frontiers in Systems Neuroscience*, 7, 11.

Bizley, J. K. and Cohen, Y. E. (2013). The what, where, and how of auditory-object perception. *Nat Reviews Neuroscience*, 14, 693–707.

Bizley, J. K., Walker, K. M., Nodal, F. R., King, A. J., and Schnupp, J. W. (2013) Auditory cortex represents both pitch judgments and the corresponding acoustic cues. *Current Biology*, 23, 620–5.

Blackwell, J. M., Lesicko, A. M., Rao, W., De Biasi, M., and Geffen, M. N. (2020). Auditory cortex shapes sound responses in the inferior colliculus. *eLife*, 9, e51890.

Bregman, A. S. (1990). *Auditory Scene Analysis*. Boston, MA, MIT Press.

Brunton, B. W., Botvinick, M. M., and Brody, C. D. (2013). Rats and humans can optimally accumulate evidence for decision-making. *Science*, 340, 95–8.

Ceballo, S., Piwkowska, Z., Bourg, J., Daret, A., and Bathellier, B. (2019). Targeted cortical manipulation of auditory perception. *Neuron*, 104, 1168–79.e5.

Cheney, D. L. and Seyfarth, R. M. (1990). *How Monkeys See the World*. Chicago, IL, University of Chicago Press.

Christison-Lagay, K. L., Bennur, S., and Cohen, Y. E. (2017). Contribution of spiking activity in the primary auditory cortex to detection in noise. *Journal of Neurophysiology*, 118, 3118–31.

Christison-Lagay, K. L. and Cohen, Y. E. (2018). The contribution of primary auditory cortex to auditory categorization in behaving monkeys. *Frontiers in Neuroscience*, 21, 601.

Clayton, K. K., Williamson, R. S., Hancock, K. E., *et al.* (2021). Auditory corticothalamic neurons are recruited by motor preparatory inputs. *Current Biology*, 31(2), 310–321.e5.

Cohen, Y. E., Theunissen, F., Russ, B. E., and Gill, P. (2007). Acoustic features of rhesus vocalizations and their representation in the ventrolateral prefrontal cortex. *Journal of Neurophysiology*, 97, 1470–84.

Cope, T. E., Baguley, D. M., and Griffiths, T. D. (2015). The functional anatomy of central auditory processing. *Practical Neurology*, 15, 302–8.

Dewitt, I. and Rauschecker, J. P. (2012). Phoneme and word recognition in the auditory ventral stream. *Proceedings of the National Academy of Sciences of the United States of America*, 109, E505–14.

Doupe, A. J. and Kuhl, P. K. (1999). Birdsong and human speech: common themes and mechanisms. *Annual Review Neuroscience*, 22, 567–631.

Downer, J. D., Niwa, M., and Sutter, M. L. (2015). Task engagement selectively modulates neural correlations in primary auditory cortex. *Journal of Neuroscience*, 35, 7565–74.

Downer, J. D., Rapone, B., Verhein, J., O'Connor, K. N., and Sutter, M. L. (2017). Feature-selective attention adaptively shifts noise correlations in primary auditory cortex. *Journal of Neuroscience*, 37, 5378–92.

Eckardt, W. and Züberbuhler, K. (2004). Cooperation and competition in two forest monkeys. *Behavioral Ecology*, 15, 400–11.

Erlich, J. C., Brunton, B. W., Duan, C. A., Hanks, T. D., and Brody, C. D. (2015). Distinct effects of prefrontal and parietal cortex inactivations on an accumulation of evidence task in the rat. *eLife*, 4, e05457.

Fabre-Thorpe, M., Richard, G., and Thorpe, S. J. (1998). Rapid categorization of natural images by rhesus monkeys. *Neuroreport*, 9, 303–8.

Fitzgerald, J. K., Freedman, D. J., Fanini, A., Bennur, S., Gold, J. I., and Assad, J. A. (2013). Biased associative representations in parietal cortex. *Neuron*, 77, 180–91.

Francis, N. A., Winkowski, D. E., Sheikhattar, A., Armengol, K., Babadi, B., and Kanold, P. O. (2018). Small networks encode decision-making in primary auditory cortex. *Neuron*, 97, 885–97.e6.

Freedman, D. J., Riesenhuber, M., Poggio, T., and Miller, E. K. (2001). Categorical representation of visual stimuli in the primate prefrontal cortex. *Science*, 291, 312–16.

Freedman, D. J., Riesenhuber, M., Poggio, T., and Miller, E. K. (2002). Visual categorization and the primate prefrontal cortex: neurophysiology and behavior. *Journal of Neurophysiolology*, 88, 929–41.

Ganguli, S., Bisley, J. W., Roitman, J. D., Shadlen, M. N., Goldberg, M. E., and Miller, K. D. (2008). One-dimensional dynamics of attention and decision making in LIP. *Neuron*, 58, 15–25.

Gao, P., Trautmann, E., Yu, B., *et al.* (2017). A theory of multineuronal dimensionality, dynamics and measurement. *bioRxiv*, 214262.

Gauthier, I., Tarr, M. J., Anderson, A. W., Skudlarski, P., and Gore, J. C. (1999). Activation of the middle fusiform 'face area' increases with expertise in recognizing novel objects. *Nature Neuroscience*, 2, 568–73.

Gifford, G. W., 3rd, Hauser, M. D., and Cohen, Y. E. (2003). Discrimination of functionally referential calls by laboratory-housed rhesus macaques: implications for neuroethological studies. *Brain, Behavior, and Evolution*, 61, 213–24.

Gifford, G. W., 3rd, Maclean, K. A., Hauser, M. D., and Cohen, Y. E. (2005). The neurophysiology of functionally meaningful categories: macaque ventrolateral prefrontal cortex plays a critical role in spontaneous categorization of species-specific vocalizations. *Journal of Cognitive Neuroscience*, 17, 1471–82.

Gold, J. I. and Shadlen, M. N. (2007). The neural basis of decision making. *Annual Review Neuroscience*, 30, 535–74.

Green, C. S., Pouget, A., and Bavelier, D. (2010). Improved probabilistic inference as a general learning mechanism with action video games. *Current Biology*, 20, 1573–9.

Griffiths, T. D. and Hall, D. A. (2012). Mapping pitch representation in neural ensembles with fMRI. *Journal of Neuroscience*, 32, 13343–7.

Griffiths, T. D. and Warren, J. D. (2004). What is an auditory object? *Nature Reviews Neuroscience*, 5, 887–92.

Hackett, T. A. (2011). Information flow in the auditory cortical network. *Hearing Research*, 271, 133–46.

Hanks, T. D., Kopec, C. D., Brunton, B. W., Duan, C. A., Erlich, J. C., and Brody, C. D. (2015). Distinct relationships of parietal and prefrontal cortices to evidence accumulation. *Nature*, 520, 220–3.

Hauser, M. D. (1998). Functional referents and acoustic similarity: field playback experiments with rhesus monkeys. *Animal Behavior*, 55, 1647–58.

Joassin, F., Maurage, P., and Campanella, S. (2011). The neural network sustaining the crossmodal processing of human gender from faces and voices: an fMRI study. *Neuroimage*, 54, 1654–61.

Kaas, J. H. and Hackett, T. A. (2000). Subdivisions of auditory cortex and processing streams in primates. *Proceedings of the National Academy of Sciences of the United States of America*, 97, 11793–9.

Killian-Hutten, N., Valente, G., Vroomen, J., and Formisano, E. (2011). Auditory cortex encodes the perceptual interpretation of ambiguous sound. *Journal of Neuroscience*, 31, 1715–20.

King, A. J., Teki, S., and Willmore, B. D. B. (2018). Recent advances in understanding the auditory cortex. *F1000Res*, 7.

Klein, M. E. and Zatorre, R. J. (2015). Representations of invariant musical categories are decodable by pattern analysis of locally distributed BOLD responses in superior temporal and intraparietal sulci. *Cerebral Cortex*, 25, 1947–57.

Kusmierek, P., Ortiz, M., and Rauschecker, J. P. (2012). Sound-identity processing in early areas of the auditory ventral stream in the macaque. *Journal of Neurophysiology*, 107, 1123–41.

Kusmierek, P. and Rauschecker, J. P. (2009). Functional specialization of medial auditory belt cortex in the alert rhesus monkey. *Journal of Neurophysiology*, 102, 1606–22.

Leaver, A. M. and Rauschecker, J. P. (2010). Cortical representation of natural complex sounds: effects of acoustic features and auditory object category. *Journal of Neuroscience*, 30, 7604–12.

Leaver, A. M., van Lare, J., Zielinski, B., Halpern, A. R., and Rauschecker, J. P. (2009). Brain activation during anticipation of sound sequences. *Journal of Neuroscience*, 29, 2477–85.

Lemus, L., Hernandez, A., and Romo, R. (2009a). Neural codes for perceptual discrimination of acoustic flutter in the primate auditory cortex. *Proceedings of the National Academy of Sciences of the United States of America*, 106, 9471–6.

Lemus, L., Hernandez, A., and Romo, R. (2009b). Neural encoding of auditory discrimination in ventral premotor cortex. *Proceedings of the National Academy of Sciences of the United States of America*, 106, 14640–5.

Liu, A. S., Tsunada, J., Gold, J. I., and Cohen, Y. E. (2015). Temporal integration of auditory information is invariant to temporal grouping cues. *eNeuro*, 2.

Lomber, S. G. and Malhotra, S. (2008). Double dissociation of 'what' and 'where' processing in auditory cortex. *Nature Neuroscience*, 11, 609–16.

Marschark, M., Convertino, C., McEvoy, C., and Mastellier, A. (2004). Organization and use of the mental lexicon by deaf and hearing individuals. *American Annals of the Deaf*, 149, 51–61.

Miller, C. T. and Cohen, Y. E. (2010). Vocalization processing. In *Primate Neuroethology*, edited by A. Ghazanfar and M. L. Platt. Oxford: Oxford University Press.

Miller, E. K., Freedman, D. J., and Wallis, J. D. (2002). The prefrontal cortex: categories, concepts, and cognition. *Philosophical Transactions of the Royal Society of London Series B, Biological Sciences*, 29, 1123–36.

Miller, E. K., Nieder, A., Freedman, D. J., and Wallis, J. D. (2003). Neural correlates of categories and concepts. *Current Opinion in Neurobiology*, 13, 198–203.

Miller, L. M. and Recanzone, G. H. (2009). Populations of auditory cortical neurons can accurately encode acoustic space across stimulus intensity. *Proceedings of the National Academy of Sciences of the United States of America*, 106, 5931–5.

Mulder, M. J., Keuken, M. C., van Maanen, L., Boekel, W., Forstmann, B. U., and Wagenmakers, E. J. (2013). The speed and accuracy of perceptual decisions in a random-tone pitch task. *Attention, Perception, and Psychophysics*, 75, 1048–58.

Ni, A. M., Ruff, D. A., Alberts, J. J., Symmonds, J., and Cohen, M. R. (2018). Learning and attention reveal a general relationship between population activity and behavior. *Science*, 359, 463–5.

Nieder, A. (2012). Supramodal numerosity selectivity of neurons in primate prefrontal and posterior parietal cortices. *Proceedings of the National Academy of Sciences of the United States of America*, 109, 11860–5.

Nieder, A., Freedman, D. J., and Miller, E. K. (2002). Representation of the quantity of visual items in the primate prefrontal cortex. *Science*, 297, 1708–11.

Nienborg, H. and Cumming, B. G. (2009). Decision-related activity in sensory neurons reflects more than a neuron's causal effect. *Nature*, 459, 89–92.

Niwa, M., Johnson, J. S., O'Connor, K. N. and Sutter, M. L. (2012). Activity related to perceptual judgment and action in primary auditory cortex. *Journal Neuroscience*, 32, 3193–210.

Niwa, M., Johnson, J. S., O'Connor, K. N. and Sutter, M. L. (2013). Differences between primary auditory cortex and auditory belt related to encoding and choice for AM sounds. *Journal Neuroscience*, 33, 8378–95.

Norman-Haignere, S. V., Albouy, P., Caclin, A., McDermott, J. H., Kanwisher, N. G., and Tillmann, B. (2016). Pitch-responsive cortical regions in congenital amusia. *Journal of Neuroscience*, 36, 2986–94.

Norman-Haignere, S. V., Kanwisher, N., McDermott, J. H., and Conway, B. R. (2019). Divergence in the functional organization of human and macaque auditory cortex revealed by fMRI responses to harmonic tones. *Nature Neuroscience*, 22, 1057–60.

Norman-Haignere, S. V. and McDermott, J. H. (2018). Neural responses to natural and model-matched stimuli reveal distinct computations in primary and nonprimary auditory cortex. *PLoS Biology*, 16, e2005127.

Nudds, M. (2010). What are auditory objects? *Review of Philosophy and Psychology*, 1, 105–22.

Orlov, T., Yakovlev, V., Hochstein, S., and Zohary, E. (2000). Macaque monkeys categorize images by their ordinal number. *Nature*, 404, 77–80.

Patterson, R. D., Uppenkamp, S., Johnsrude, I. S., and Griffiths, T. D. (2002). The processing of temporal pitch and melody information in auditory cortex. *Neuron*, 36, 767–76.

Penagos, H., Melcher, J. R., and Oxenham, A. J. (2004). A neural representation of pitch salience in nonprimary human auditory cortex revealed with functional magnetic resonance imaging. *Journal of Neuroscience*, 24, 6810–15.

Plakke, B., Hwang, J., and Romanski, L. M. (2015). Inactivation of primate prefrontal cortex impairs auditory and audiovisual working memory. *Journal of Neuroscience*, 35, 9666–75.

Poliva, O., Bestelmeyer, P. E., Hall, M., Bultitude, J. H., Koller, K., and Rafal, R. D. (2015). Functional mapping of the human auditory cortex: fMRI investigation of a patient with auditory agnosia from trauma to the inferior colliculus. *Cognitive and Behavioral Neurology*, 28, 160–80.

Rauschecker, J. P. (2012). Ventral and dorsal streams in the evolution of speech and language. *Frontiers in Evolutionary Neuroscience*, 4, 7.

Rauschecker, J. P. (2018). Where, when, and how: are they all sensorimotor? Towards a unified view of the dorsal pathway in vision and audition. *Cortex*, 98, 262–8.

Rauschecker, J. P. and Scott, S. K. (2009). Maps and streams in the auditory cortex: nonhuman primates illuminate human speech processing. *Nature Neuroscience*, 12, 718–24.

Rauschecker, J. P. and Tian, B. (2000). Mechanisms and streams for processing of 'what' and 'where' in auditory cortex. *Proceedings of the National Academy of Sciences of the United States of America*, 97, 11800–6.

Rauschecker, J. P., Tian, B., and Hauser, M. (1995). Processing of complex sounds in the macaque nonprimary auditory cortex. *Science*, 268, 111–14.

Recanzone, G. H. and Sutter, M. L. (2008). The biological basis of audition. *Annual Review Psychology*, 59, 119–42.

Riecke, L., Mendelsohn, D., Schreiner, C., and Formisano, E. (2009). The continuity illusion adapts to the auditory scene. *Hearing Research*, 247, 71–7.

Romanski, L. M. and Averbeck, B. B. (2009). The primate cortical auditory system and neural representation of conspecific vocalizations. *Annual Reviews Neuroscience*, 32, 315–46.

Romanski, L. M., Tian, B., Fritz, J., Mishkin, M., Goldman-Rakic, P. S., and Rauschecker, J. P. (1999). Dual streams of auditory afferents target multiple domains in the primate prefrontal cortex. *Nature Neuroscience*, 2, 1131–6.

Rosch, E. (1978). Principles of categorization. In *Cognition and Categorization*, edited by E. Rosch and B. B. Lloyd. Hillsdale, NH: Erlbaum.

Russ, B. E., Lee, Y.-S., and Cohen, Y. E. (2007). Neural and behavioral correlates of auditory categorization. *Hearing Research*, 229, 204–12.

Schnupp, J. W., Nelken, I., and King, A. J. (2012). *Auditory Neuroscience: Making Sense of Sound*. Boston, MA, MIT Press.

Selezneva, E., Scheich, H., and Brosch, M. (2006). Dual time scales for categorical decision making in auditory cortex. *Current Biology*, 16, 2428–33.

Shepard, R. N. (1987). Toward a universal law of generalization for psychological science. *Science*, 237, 1317–23.

Singh, N. C. and Theunissen, F. E. (2003). Modulation spectra of natural sounds and ethological theories of auditory processing. *Journal of the Acoustical Society of America*, 114, 3394–411.

Spence, K. W. (1937). The differential response in animals to stimuli varying within a single dimension. *Psychological Review*, 44, 430–44.

Tabas, A., Mihai, G., Kiebel, S., Trampel, R., and von Kriegstein, K. (2020). Abstract rules drive adaptation in the subcortical sensory pathway. *eLife*, 9, e64501.

Tian, B., Reser, D., Durham, A., Kustov, A., and Rauschecker, J. P. (2001). Functional specialization in rhesus monkey auditory cortex. *Science*, 292, 290–3.

Tsunada, J., Cohen, Y. and Gold, J. I. (2019). Post-decision processing in primate prefrontal cortex influences subsequent choices on an auditory decision-making task. *eLife*, 8, e46770.

Tsunada, J. and Cohen, Y. E. (2014). Neural mechanisms of auditory categorization: from across brain areas to within local microcircuits. *Front Neuroscience*, 8, 161.

Tsunada, J., Lee, J. H., and Cohen, Y. E. (2011). Representation of speech categories in the primate auditory cortex. *Journal of Neurophysiology*, 105, 2634–46.

Tsunada, J., Liu, A. S., Gold, J. I., and Cohen, Y. E. (2016). Causal contribution of primate auditory cortex to auditory perceptual decision-making. *Nature Neuroscience*, 19, 135–42.

Vogels, R. (1999). Categorization of complex visual images by rhesus monkeys. Part 1: behavioural study. *European Journal of Neuroscience*, 11, 1223–38.

Warren, J. D. and Griffiths, T. D. (2003). Distinct mechanisms for processing spatial sequences and pitch sequences in the human auditory brain. *Journal of Neuroscience*, 23, 5799–804.

Williamson, R. S. and Polley, D. B. (2019). Parallel pathways for sound processing and functional connectivity among layer 5 and 6 auditory corticofugal neurons. *eLife*, 8, e42974.

Winer, J. A. and Lee, C. C. (2007). The distributed auditory cortex. *Hearing Research*, 229, 3–13.

Woolley, S. M., Fremouw, T. E., Hsu, A., and Theunissen, F. E. (2005). Tuning for spectro-temporal modulations as a mechanism for auditory discrimination of natural sounds. *Nature Neuroscience*, 8, 1371–9.

Yin, P., Strait, D. L., Radtke-Schuller, S., Fritz, J. B., and Shamma, S. A. (2020). Dynamics and hierarchical encoding of non-compact acoustic categories in auditory and frontal cortex. *Current Biology*, 30, 1649–63.e5.

Züberbuhler, K. (2000a). Causal cognition in a non-human primate: field playback experiments with Diana monkeys. *Cognition*, 76, 195–207.
Züberbuhler, K. (2000b). Interspecies semantic communication in two forest primates. *Proceedings of the Royal Society of London. Series B, Biological sciences*, 267, 713–18.
Züberbuhler, K. and Seyfarth, R. M. (1997). Diana monkey long-distance calls: messages for conspecifics and predators. *Animal Behaviour*, 53, 589–604.

10
Audition and composite sensory individuals

Nick Young and Bence Nanay

1 Introduction: the sensory individuals of audition

Perception is often thought to be, at root, a matter of (conscious or unconscious) *property attribution*, with properties like colour, shape, and texture being attributed visually, and properties like pitch, timbre, and loudness being attributed auditorily. Our focus here is not on which properties are auditorily attributed, a long-standing and high-profile debate in philosophy of perception, see, for example, Siegel, 2006 and Nanay, 2011; and in the case of audition, di Bona, 2017), but rather on what those properties are attributed to—that is, on the *sensory individuals* of auditory perception.[1]

The sensory individuals of vision are commonly thought to be ordinary material objects (for versions of this view, see Matthen, 2005, pp. 277–82; Nanay, 2013, p. 51; Cohen, 2004; for an opposing view, see Clark, 2000, Chapter 2; 2004, Nanay 2022), but when it comes to auditory individuals, there is far less consensus. We might think that this is surprising, and that it is obvious what the sensory individuals of audition are. We hear *sounds*. Such a straightforward answer, however, faces various difficulties.

For a start, there is disagreement as to what sounds are. Which type of environmental feature is picked out and represented? The most obvious, common-sense, answer to this question would be 'sound waves'. However, the wave view has been subject to various criticisms[2] (e.g., O'Callaghan, 2007) and various alternatives have been put forward, identifying sounds with vibrations (Casati and Dokic, 1994), or disturbances of the air (or some other medium) caused by these vibrations (O'Callaghan, 2007). Opinions differ as to what *type* of feature sounds are. Are they *particulars* which are attributed pitch, timbre, and loudness, or are they *properties* which are themselves attributed to particulars?[3] Rather than entering into these

[1] We use the term 'sensory individual' in the sense it was introduced by Cohen (2004), who described an idea from Clark (2000), according to which perception does not simply extract features from the environment but *attributes* them to individuals. It seems safe to say that the philosophical work that we refer to here concerning what sounds and sources are can be thought of as claims about the sensory individuals of audition, even if the authors do not use this term specifically.

[2] We should not assume that the immediate cause of a perceptual experience is necessarily what is perceptually represented. While light hitting the retinas is the immediate cause of visual experiences, this is not a good reason for thinking that we visually represent light; rather, the mainstream view in philosophy and vision science is that we perceptually represent the material objects from which light reflects. Moreover, there is an apparent mismatch between the location of sound waves and the apparent location of what we hear: sound waves produced by a bell strike will propagate through the air and surround the listener, yet the sound is not heard as 'all around', but at the location at which the striking occurred (O'Callaghan, 2007, Chapter 2; Pasnau, 1999).

[3] Leddington (2019, fn. 6), a proponent of the property view, distinguishes two ways of understanding the idea that sounds 'have' pitch, timbre, and loudness. If sounds are particulars, then 'have' can be understood in terms of *property attribution*. If sounds are properties, 'have' can be understood in terms of *constitution*; sounds have pitch, timbre, and loudness in a similar way to how colours have a certain brightness, hue, and saturation.

Nick Young and Bence Nanay, *Audition and composite sensory individuals* In: *Sensory Individuals*. Edited by: Aleksandra Mroczko-Wąsowicz and Rick Grush, Oxford University Press. © Oxford University Press 2023. DOI: 10.1093/oso/9780198866305.003.0011

debates here, we will assume in what follows that hearing a sound is a matter of representing some type of particular, and focus instead on the sensory individuals of *source hearing*. In contemporary philosophical work on audition, it is very common to say that we hear not only sounds, but also their source events: the collisions, rollings, and scrapings, which cause objects to vibrate and produce sound waves in the surrounding air. It is not obvious how such a claim can be understood in terms of sensory individuals. First, what aspect of the environment, specifically, do we represent and attribute properties to when we hear sources? Second, how does the hearing of sources connect to the hearing of sounds? That is, how does the representation of whatever aspect of the environment we identify with sounds relate to the representation of whatever aspect of the environment we identify with sources?

In what follows, we shall examine various proposals about the nature of the sensory individuals of source hearing and show that while each can account for some aspects of auditory perception, each also faces certain difficulties. We then put forward a new conception of sensory individuals according to which auditory sensory individuals are *composite individuals*. A feature shared by all existing accounts of sounds and sources is that they postulate sensory individuals that are non-composite. They identify the sensory individual of sound hearing with *one* type of entity in the environment, such as sound waves or medium-disturbing events, and the sensory individual of source hearing with another, such as vibrations or interactions. We question this assumption and argue that our perceptual systems represent two or more aspects of the environment as a *single* sensory individual. Finally, we show that taking auditory individuals to be composite sensory individuals allows for an account of audition that is less problematic than its existing alternatives.

2 Sounds and sources

2.1 Sources as interaction events or vibration events

Why think that we hear source events as well as sounds? First of all, source events stand in straightforward causal relations to auditory experiences (O'Callaghan, 2011). Second, we very often become aware, from auditory experience, of what sort of source event is taking place (e.g., a wooden window frame rattling, a glass shattering), immediately and without conscious inference (O'Callaghan, 2011). Third, it is plausible to suppose that the function of the auditory system, like all perceptual systems, is to provide us with useful information about material objects and their activities. We have an *interest* in what objects are doing what, rather than in how objects are causing the air around them to move (Nudds, 2010).[4]

While none of these reasons shows absolutely that sources are heard, they do at least show that accounts of auditory perception which allow for source hearing are desirable, and we shall assume this in what follows. We can think of the claim that we hear sources as a claim about sensory individuals. What aspect of the environment do we represent and attribute properties to when we hear sources?

[4] Indeed, it is difficult to find any contemporary philosopher working on auditory perception who does not think that sources are heard. Examples of psychological work focusing on the auditory representation of sources can be found in Gaver's (1993a,b) ecological approach (see also Carello *et al.*, 2005), as well as in what is sometimes referred to as a *psychomechanical* approach to timbre perception (e.g., McAdams *et al.*, 2004; see also Agus *et al.*, 2019).

Assuming that sounds are particulars, there are at least two approaches to answering this question. According to the first approach, source hearing is a case of representing objects interacting or moving against one another. Consider, for example, the following remarks from O'Callaghan (2011, p. 376):

> By sources I mean things like floorboards, electronics, collisions, jiggled keys and speeches. It is plausible that sound sources [. . .] are audible. Spoken language permits saying you hear floorboards, electronics, collisions, jiggled keys and speeches rather than just their sounds. Audition prompts you to form thoughts about material bodies and happenings. Certain aspects of auditory experience, such as the way things appear grouped auditorily, are attuned to features of material objects and events rather than merely to features of sounds.

There are many different possible ways in which objects can interact in such a way as to produce sound waves (and thereby elicit auditory experiences), including collisions, rollings, scrapings, and smashings. For simplicity, we will just use the label 'interaction events'. Although there are disagreements as to exactly how the details should be fleshed out,[5] thinking of source hearing as representing interaction events can be considered the mainstream view (see, for example, Leddington, 2019; Nudds, 2009; Young, 2018).

However, a slightly different conception of source hearing can be found in Casati et al.'s (2013, p. 463) 'ockhamization' of auditory perception. They say:

> The sound is not a proper part of a distinct event that is its source; it is identical with what we called the event source. The collision you hear is the sound you hear; the vibration of the stuff the tuning fork is made of is the sound. There is no difference between hearing the sound and hearing the event source, because there is no difference between the sound and the event source.

We can see that the authors are sympathetic to the idea that to hear a source is to hear an interaction event. Indeed, they say that interaction events that we hear, such as collisions, just *are* the sounds that we hear. On the other hand, they also seem to identify sounds with another type of event: 'the vibrations of the stuff tuning fork is made of'. That is, an event that occurs immediately after the interaction event. If both the collision and the vibration of the stuff are 'the sound', then Casati *et al.* could be interpreted as saying that source events, as well as being interactions, are vibration events.[6] However vibrations would seem to be quite

[5] In particular, there is disagreement as to whether hearing a source event involves hearing the objects involved (Kulvicki, 2008; Young, 2018), or hearing the events but not the objects (Leddington, 2014, p. 330; O'Callaghan, 2008).

[6] The idea that vibrations are a type of source event can also be found in O'Callaghan (2008, p. 824), who says: 'Sounds and streams furnish information about the events and happenings—the collisions, vibrations, and interactions—that commonly make or produce sounds'. Green (2019 , p. 674) also considers this way of thinking about sources, which he calls the 'minimal object approach'. A somewhat similar view has been put forward by Kulvicki (2008, 2014) on which sounds are stable dispositional properties of objects—specifically, the disposition to vibrate at certain frequencies. On this view, vibrations are not identical to sounds, but rather reveal sound properties, in the same way that reflectance theorists (e.g., Byrne and Hilbert, 2003) would say that light reflecting from an object reveals that object's colour property. Although, on this view, like that of Casati *et al.*'s, audition involves the representation of only one sensory individual (rather than one for sound hearing and one for source hearing), Kulvicki does not identify hearing a source event with hearing an object vibrate. Rather, he suggests that 'that we hear events by hearing stable properties of objects' (see Kulvicki, 2008, pp. 11–13 for details of this view; and Young, 2018 for criticism).

a different type of event from interactions. Consider a visual analogy: I strike a tuning fork behind you and you turn to see the object still quivering from the impact. While you can see this *after-effect* of the collision perfectly well, you did not see the impact itself. Regardless of whether Casati *et al.* really do think vibrations of stuff are sources instead of, or as well as, the interactions that produce them, these considerations do provide us with an alternative sensory individual for source hearing.

We might think that this alternative is less mainstream than the interaction view for good reason. Consider the motivations mentioned above for why an account of source hearing is desirable: sources stand in straightforward causal relations to auditory experiences, we become immediately aware of them through auditory experience, and we have an interest in them. While vibrations stand in straightforward causal relations to auditory experiences, we do not hear *how* objects are vibrating,[7] and it is not as if we have any interest in the vibrations of objects *per se*.

On the other hand, identifying source events with vibratory events does have one advantage over thinking of sources as interaction events. Consider a very common—perhaps the most common—type of source event: collisions. When a tuning fork is struck, the auditory experience you enjoy begins only when the fork begins to vibrate, and only lasts as long as these vibrations continue. This is because—obviously—sound waves only propagate through the air and reach the eardrums while the tuning fork is vibrating. By the time the fork begins to vibrate and the auditory experience begins, however, the interaction event is already over: the mallet has already *completed* its journey through the air to meet the fork. A reason to favour identifying vibrations with sources over the interaction events which they produce, then, is that vibration events are more *temporally aligned* with our auditory experience than interaction events.[8]

2.2 Hearing sources through sounds

Whereas Casati *et al.* argue that sound hearing and source hearing are one and the same, those who think that we hear source events, but that sounds themselves are sound waves, or medium disturbances, appear committed to the claim that audition involves the representation of *two* sensory individuals: sound and source. If this is correct, they are required to explain how the representation of one is connected to the representation of the other. As O'Callaghan (2011, p. 2) puts it:

> If you hear a source such as the jiggling of keys, then you may hear the jiggling keys to be the source of the jingling sound. When you do, you hear the sound to stand in some relation to its source. What relation?

[7] While we can hear pitch and timbre, these qualities are determined by objects vibrating at multiple frequencies simultaneously, and we are not able to hear (for example) how many of these frequency components there are nor how they change over time.

[8] Given that it takes time for sound waves to propagate, and time for our auditory systems to produce a representation once they have reached our ears, the vibrations are, of course, not *completely* in sync with our auditory experience. The point here is that they are closer to being synchronous than the collision event.

Nudds (2010, p. 18) has argued that we attribute sounds with the property of *having been caused by* a source of a particular sort. He goes on to suggest that representing this sort of property allows for the representation of sources themselves:

> When we experience a sound as having been produced by a source, our experience represents it as having that non-intrinsic property. Therefore, our auditory experience represents sounds and the sources of sounds and it represents sources as the sources of sounds by representing sounds as having a non-intrinsic property—the property of having been produced by a source of a certain kind.[9]

However, even if we grant that sounds are heard to bear these types of properties, we might still doubt that this amounts to the *sources* of sounds themselves being *represented* in auditory experience. This is not simply the objection that perceiving the product of some cause is often not sufficient to see the cause itself (seeing smoke does not allow for the seeing of the fire; seeing a footprint does not allow for the seeing of the foot) (Dretske, 1981; O'Callaghan, 2011); rather, even if the result of some cause were perceived as bearing the non-intrinsic property of having been produced by that cause, this would not amount to the perceptual representation of that cause. Imagine, for example, that apples from the south of Italy had the perceivable property of coming from the south of Italy; seeing apples to bear this property would give you good reason to think that they came from the south of Italy, but nothing here amounts to the perceptual representation of the south of Italy. In other words, regardless of the merits of Nudds' proposal, we might prefer to find an account on which sources are auditorily represented in their own right.

O'Callaghan (2011) considers at great detail, albeit without endorsing it, an alternative *mereological* account of source hearing, on which we hear sounds as parts of sources and, through hearing these parts, *mediately* perceive sources. That is, hearing a sound allows for the hearing of a source in a manner somewhat similar to how we might think that seeing the front-facing parts of an object allows for the perception of the *whole* object (e.g., Bermúdez, 2000; Clarke, 1965). Put in terms of sensory individuals, we can say that, on this approach, we represent two features of our environment, sound and source, and the representation of the latter is secondary to, or depends on, the representation of the former.

However, one worry we might have here is that seeing part of an object as a part would seem to involve seeing it as connected to other parts of the object, which could themselves potentially be brought into view: if you are looking at a table in front of you, you might see the tabletop and two of the table legs, and you see them as connected to other, occluded parts of the object, such as the underside of the tabletop and the other two legs. If you changed your viewing angle, these parts could be brought into view (Nanay, 2010, 2018). But audition seems quite different: if we hear the sound that the rattling window makes, it is not as if we can move (or do anything else) so that other parts of the source (such as the window frame or the collision event) are brought within earshot. If this is not possible, what reason do we have to think that we hear these sounds as parts connected to other—unheard—parts of sources, and therefore, what reason do we have to think that we are mediately perceiving the source after all? O'Callaghan (2011, p. 400) responds to this type of worry by appealing to

[9] Nudds (personal communication) has since moved away from this way of reconciling sound hearing with source hearing.

the possibility that auditory experience can be influenced by experiences in other perceptual modalities. He says:

> the relevant possibilities for perception are not restricted to one sense. The events you hear are perceptible other than through audition, and this shapes the perceptual experience of hearing them. Hearing as of something that is or could be seen or touched makes a difference to the phenomenology of auditory experience [...] Hearing as of something that has visible or tactile features differs phenomenologically from hearing as of something without such features. Thus you might hear as of events that have inaudible parts. The audibility of an individual does not require that each of its parts is audible.

O'Callaghan goes on to develop two ways in which experience in other modalities 'shapes' auditory experience, so that it allows for mereological source hearing. First, if we have visual or tactile access to the source event which is currently making the sound we hear, this could lead to a multimodal perceptual experience on which an individual is represented as bearing both visual and auditory features. Second, if we can only hear a source but have had 'a long history of both seeing and hearing' it, we 'perceptually experience it to be something that has unseen but nonetheless visible features' (O'Callaghan 2011, p. 401). While experiences in one modality are almost certainly influenced by present or past experiences in another, we might think that this type of response has an unappealing consequence: it limits the ability to hear sources to those who are able to have experiences in other perceptual modalities. Imagine a person who can hear perfectly well but was born blind and with complete hypothesia (no sense of touch). As their auditory experiences are not, and would never have been, shaped by visual or tactile experience, it seems as if O'Callaghan's account would preclude them from being able to hear sounds as parts of sources, and so also preclude them from ever hearing sources. We might think that this consequence is too big a bullet to bite. Would a person in this position have such radically impoverished auditory experiences, compared to ordinary perceivers? This seems especially pressing if we accept that the *function* of auditory experiences is to represent sources.

A second worry is that explaining the relation between hearing sources and hearing sounds in a way that does not appeal to perceived causation gives up too much. O'Callaghan (O'Callaghan 2011, p. 401) makes clear that he is not denying that sounds are effects of sources, but is sceptical that they are heard as such. But is this approach too revisionary? If we accept that we hear sounds and sources, and hear them as related, it is at least tempting to say that the relation we hear between the two is causal. Of course, we should be wary of putting too much weight on phenomenological seemings, but, all else considered, an account which retains the idea that we hear sounds as caused by sources seems preferable to one, such as the mereological view, which does not.

We have outlined two issues about the auditory individuals of source hearing. First, there is the question of whether hearing sources consists in auditorily representing interaction events or vibration events. Second, there is the question as to the nature of the audible relation between sounds and sources.

3 Composite individuals

We can now see the difficulties that current accounts of source hearing face. Implicit in all of these views is the idea that the sensory individuals of audition match up with one type of environmental feature: source hearing is the representation of vibrations or of interaction events; we hear sources through sounds either by representing sounds to have a 'having been caused by' property or by hearing sources via the hearing of sounds.

We suggest that if this assumption is given up, if we allow that two or more different types of environmental features can be represented as a single *composite* sensory individual, these problems can be avoided. Different theories of sounds and sources can be united in such a way as to retain their advantages and eliminate their disadvantages. In this section, we shall outline how composite sensory individuals should be understood; in the next, we shall show how they can be applied to the issues concerning sound and source hearing that have been introduced in the previous section.

Some particulars are not ontological simples—they are composed of different parts. In fact, most particulars are—maybe even all of them. A table is composed of the tabletop and the legs of the table. But the tabletop itself is composed of atoms, subatomic particles, etc. This is an ontological distinction between ontological simples and their mereological sums.

But we can ask a related question about how particulars are represented perceptually. When you see a table, what you see is a composite particular—it is composed of various parts. Is the sensory individual of this perceptual episode a composite sensory individual? It is, as long as your perceptual system represents the table as being composed of different parts. A sensory individual is a composite sensory individual if and only if the perceptual system represents it as being composed of different parts. Remember that a sensory individual is part of the perceptual content: it is what properties are perceptually attributed to. If properties are perceptually attributed to what is represented as a composite entity, this constitutes a composite sensory individual. Lots of things are composite. But only if we perceptually represent them as composite can we talk about a composite sensory individual.

More needs to be said about what it means to represent something as composite. Suppose you represent particular X and X has some proper parts A and B. If representing X entails representing A, and it also entails representing B, then X is represented as composite. Not all representations of composite particulars represent them as composite. If X, a composite particular, has some proper parts A and B, but representing X does not entail representing A or B, then while X is a composite particular, it is not represented as composite. To go back to the table example, a table is a composite particular, but we can represent a table without representing it as composite: if you tell me that you bought a table, I can form a belief that you bought a table without having any kind of representation about the parts of the table. Representing a table does not entail representing parts of the table.

The last paragraph was about representation in general: we can represent a composite particular without representing it as composite. But how about perceptual representations? Can we perceptually represent a composite particular without representing it as composite? While a case could be made that one of the important differences between perceptual and non-perceptual representations is that the former entails representing proper parts (if there are any), whereas the latter does not, we will not defend this claim here.

But while representing a table does not entail representing parts of the table, *perceptually* representing a table does seem to entail representing parts of the table. If you see a table, you represent both the tabletop and the legs. Maybe the legs are occluded, but you still represent them. If you did not represent the legs, you would not perceptually represent the table (but something else—the tabletop). So at least in many (but maybe even all) cases of perceptual representation, perceptually representing something entails representing its proper parts. A quick note about the scope of this claim: does it entail representing all of its proper parts? Presumably not, but it entails representing at least some of its proper parts.

Given that, as we have seen, most (and maybe all) particulars are composite and in most (and maybe all) cases of perceptual representations, perceptually representing a particular entails representing its proper parts, it follows that most sensory individuals are composite. Here is a way of keeping composite and non-composite sensory individuals apart. If your perceptual state represents a composite sensory individual, you can attend to its proper parts. When you see the table, you can perceptually attend to the tabletop or the legs—or some part of the tabletop. All these would amount to attending to parts of the composite sensory individual. If you could not attend to parts of a sensory individual, this would mean that either it does not have proper parts or, if it does, its parts are not represented. Either way, the sensory individual would fail to qualify as composite.

The question we now turn to is the ways in which a sensory individual can be composite. If a particular has proper parts and these parts are represented, we can ask what the relation is between these parts. We will talk about three options: spatial, temporal, and causal relations. The last of these will be what plays an important role in our account of auditory sensory individuals.

In the case of the table example, the table is a composite particular because it is spatially extended. And when you look at the table, your perceptual state represents it as composite because it is spatially extended. If an object takes up some of your visual field, it is perceptually represented as spatially extended. You can attend to its top and then to its bottom, to its left side and to its right side. Seeing it entails representing these proper parts of it. In short, this would amount to your perceptual state attributing properties to a composite sensory individual, where this composition amounts to spatial composition.

Most cases of visual perception are like this; hence, most of our visual states attribute properties to composite sensory individuals. But not all of them. Sometimes it can happen that an object is so far away from you that you only see it as a dot and, as a result, you just cannot visually represent its proper parts. In this case, your perceptual state attributes properties to a non-composite sensory individual.

But even when it comes to distant dots, we can get composite sensory individuals, as examples of ensemble perception show (see Ariely, 2001 for a summary of the ensemble perception literature). Consider first a simple display consisting of two spatially separated dots, moving entirely synchronously with each other: as one moves left, right, etc., so does the other, each accelerating and decelerating at the same rate as the other. In this case, we attribute the property of moving left, say, to not just the individual dots, but also the composite individual of the ensemble of dots. Further, if the trajectory of these dots is slightly different, the perceptual system averages the movements and attributes this average to the ensemble. The ensemble of dots is a composite sensory individual (whose proper parts are the individual dots).

Our second example is about temporal composition. We have seen that perceiving spatially extended particulars, as long as they are perceptually represented as spatially extended (i.e., if they take up more than just a dot on your visual field), counts as attributing properties to composite sensory individuals. We can make a similar claim about temporally extended particulars: perceiving temporally extended particulars, as long as they are perceptually represented as temporally extended (i.e., if they take more than a split second), counts as attributing properties to composite sensory individuals.

Given that auditory stimulus is temporally extended, the sensory individual of most auditory states is composite. Again, there are exceptions that would be the analogue of the dot case in the discussion of spatial composition above. If you hear a brief beep, while the stimulus does have a temporal extension, your auditory state might not represent this sensory individual as having temporal extension. But when you hear the word 'custard', this auditory state entails hearing the first syllable and the second one. And you can attend to the first syllable or the second. The sensory individual of most auditory states is temporally composite.

Finally, the relation between the parts of the sensory individual can also be a causal one. Here is a non-perceptual warm-up case. According to John Searle (1983), an action consists of the intention-in-action triggering the bodily movement. Action should not be identified either with the intention-in-action or with the bodily movement, but with the complex particular of the former causing the latter.[10] So an action is a complex particular, which consist of causally related parts. As we also represent it as composite, it also counts as a composite individual.

Moving on to perceptual examples, when you see the cue ball hitting the eight ball, this is a temporally extended event and you perceptually represent it as such. You perceptually represent its parts, its beginning and its end. These parts are temporally related. But it is also an event that has two causally related parts. If we accept (for the sake of the argument) the view according to which causal relation can be perceptually represented (Michotte, 1963; Scholl and Tremoulet, 2000), the event of one ball hitting another is perceptually represented and it is represented as causally composite. Composite sensory individuals can be spatially, temporally, or causally composite.[11]

Our claim is that auditory sensory individuals are temporally extended, causally composite sensory individuals in exactly this sense: we attribute properties auditorily to causally composite sensory individuals. And this is a radically new account of auditory individuals inasmuch as the existing accounts of audition seem to presuppose that the sensory individual of audition is not causally composite.

While this is a new account of auditory individuals, at least one influential account of what sounds are, at least in some of its formulations, could be thought to be a special case of it. O'Callaghan (2010, p. 36; see also 2007, Chapter 5; 2011, p. 395) considers sounds to be *medium-disturbing events*: 'the events in which a medium is disturbed or changed or set

[10] Needless to say, our argument about auditory individuals does not presuppose the truth of Searle's claims about the philosophy of action; it is merely a demonstration that positing causally composite individuals is not a completely unprecedented move.

[11] On Green's theory of perceptual objects, our perceptual systems 'function to pick out entities whose parts are causally related', although he stops short of suggesting that such relations are represented in perceptual content.

into motion in a wave-like way by the motions of bodies. Events such as collisions and vibrations of objects cause the sound events. Among the effects of sounds may be sound waves propagating through a medium and the auditory experiences of perceivers'. Although the second and third sentences of this quote suggest that we should think of sounds as monadic, non-composite individuals, which are the products of vibrations and the producers of sound waves, it is, arguably, difficult to conceive of medium-disturbing events without conceiving them as composites, consisting of a disturber and a disturbed. While vibrations can occur in a vacuum, and air, in the form of sound waves, and they can move without being immediately adjacent to an object, sounds, as O'Callaghan conceives of them, cannot occur in isolation—they require two different aspects of the environment for their existence. If this is correct, then to hear a medium disturbance is to auditorily represent a composite sensory individual.[12]

Before moving on, it is worth emphasizing that hearing a composite individual is *not* a form of mediate perception. It is not that one part of the individual is perceived *through* the hearing of another part; rather, the different parts of a composite individual are representationally on a par. We do not think that this is a particularly controversial idea. There is no reason why a single perceptual modality cannot represent multiple types of environmental feature simultaneously. Consider, for example, Matthen's work on sensory perception. In Matthen (2010a), for example, he suggests that an experience of a white wall looking both white and, somehow, pink in red light need not be explained by referring to two types of properties being attributed to the wall (its real white colour and its apparent pink colour, for example), but rather the wall and *the illumination itself* being represented as separate individuals, with different colour properties being attributed to each. Or, to take an auditory example, he argues that when we hear a group of notes playing a melody, we do not hear the melody in virtue of hearing the notes (nor the notes in virtue of the melody), but represent both the notes *and* the melody (Matthen, 2010b). Where our account departs from one like Matthen's is that separate environmental aspects are not simply represented, but represented as causally related parts of a composite.

In the Section 2, we outlined two debates concerning the nature of source perception: what are the sensory individuals of source hearing? How does the representation of sounds relate to the representation of sources? We also saw that, in both cases, the solutions that have been proposed face certain problems. Thinking of source hearing in terms of composite sensory individuals provides, we argue, novel answers to both these questions, which avoid the difficulties of the competing views.

[12] Some remarks by O'Callaghan (2007, pp. 68–9) suggest that he is sceptical that hearing a medium disturbance must involve perceiving the medium and the disturber. He says: 'one might maintain that to perceive an interaction—for that interaction to count among the things one perceptually experiences—requires perceiving its relata [...] While this might be the norm in the case of visible objects, nothing prevents interaction itself from having perceptual significance in absence of perceptual awareness of what interacts [...] one might be aware of a disturbance event through experiencing its distinctive audible qualities, while immediately experiencing neither the object nor the medium'. However, one worry we might have with this is that audible qualities *supervene* on the relata. Pitch, timbre, and loudness depend on, respectively, frequency, spectral composition, and amplitude—these are properties instantiated by the vibrating object and the disturbed air, but not the causal interaction between the two.

4 Composite individuals and the problems

4.1 Sources

Composite sensory individuals allow us to unite both candidates that have been put forward for source hearing, thereby retaining the benefits of each. We saw in the Section 2 that a problem of identifying source hearing solely with the representation of vibrations is unsatisfactory because vibrations, in and of themselves, do not seem to be an aspect of the world in which listeners have an interest. However, thinking of source hearing as just the representation of interaction events themselves runs into difficulties when we consider collision events: in these cases, the objects have ceased to interact by the time we enjoy an auditory experience, something better explained by the vibration account.

We suggest that hearing a source should be thought of as the representation of a composite individual, consisting of an interaction event part and a vibration part, with a causal relation between the two. As such, an individual includes an interaction event as a part, and it also includes the properties of those interacting objects. Hence, it is an item that listeners would have an interest in representing. Such an account therefore easily avoids the problems faced by the vibrations account of source hearing.

Representing a composite also allows for a novel way of dealing with collision events. We can think of the representation of interaction events and the vibrations they produce as a *temporal* composite, and the fact that the latter are represented as a part of composite which includes the former allows for the *representation* of the former. Another way of putting this would be to say that the representation of the vibration part of the individual allows for the *post-dictive* representation of its earlier interaction event part.[13]

4.2 Sources and sounds

Finally, the idea of composite sensory individuals suggests a novel way of linking sound hearing with source hearing. We saw in the Section 2.1 that whereas the view put forward by Casati *et al.* collapsed the distinction between sound and source, on views which identify sounds as sound waves, or medium disturbances, there is a distinction between representing a sound and representing a source event. In the Section 2.2, we saw that according to Nudds, source hearing is possible because sounds are heard as bearing the property of having been produced by a source of a certain type, but that it was not obvious why representing such a property would allow for bona fide *representation* of the source; and that according to O'Callaghan, we hear sounds as parts of sources, but that this approach is weakened if we consider some disanalogies between hearing sounds and seeing the front-facing surfaces of objects.

We suggest that asking how we hear sources *through* hearing sounds posits a false dilemma: we do not hear the former *via* the hearing of the latter; rather, hearing consists in the representation of an individual with component parts, none of which is perceived *prior* to any other. Consider what it is to hear sources as composite individuals: hearing a source

[13] Choi and Scholl (2006) present empirical work which they suggest shows that collision events can be perceived post-dictively.

is to represent a particular consisting of vibrations and an interaction event. Neither individual is heard through the other: although the interaction event is heard as the cause of the vibrations, there is no order of perceptual priority—the parts are simply represented as causally related parts of a composite. We can think of the relation between hearing sources and hearing sounds in a similar way: hearing a sound and its source consists in representing a composite with causally related parts. We can think of the relation between hearing sources and hearing sounds in a similar way: hearing a sound and its source consists in representing a composite with causally related parts. Although this proposal involves both perceived causality and parthood, the lack of perceptual priority between its causal components allows us to avoid the difficulties that Nudds' and O'Callaghan's proposals face.

If hearing a source is a matter of representing a composite with a vibration part and an interaction part, and the relation between hearing sounds and sources is a matter of hearing the two as composite parts of the same individual, there is an obvious final step that could be taken. Hearing involves the representation of a *single* composite sensory individual, consisting of a sound part, a vibration part, and an interaction part, all represented as causally related to one another: interactions events are heard as causing vibrations, which are heard as causing sounds.

5 Conclusion

The resulting account has a fairly unorthodox view of what could count as auditory individuals. Matt Soteriou (2018, p. 36) in a recent paper argues that we should endorse a more pluralistic approach to what we call auditory individuals (and what he calls 'the bearer of acoustic properties'). Our move could be considered to be the exact opposite of Soteriou's, albeit the picture we end up with is somewhat similar to Soteriou's.

For us, there is only one auditory individual: a mereologically complex composite sensory individual. In different auditory episodes, and depending on our momentary interests, different parts of this mereologically complex composite sensory individual may be more salient—you can attend to various parts of this composite sensory individual—to the sound, to the vibration event, and so on. But there is no plurality when it comes to the sensory individual itself. But our approach, like Soteriou's, opens the way to accommodating seemingly conflicting intuitions about the nature of auditory individuals—not by allowing a variety of auditory individuals, but by allowing only one, albeit a composite one.

Finally, a word about the scope of our strategy. We have been focusing on auditory individuals. But similar questions can, and should, be asked about the other sense modalities as well. Olfaction is an obvious target, for example, where structurally similar questions are asked about the nature of olfactory individuals: are they odours (the equivalent of sounds) or source events (see Batty, 2011 for an overview)? Do we smell the odour of burnt toast or of the source event of the burning of toast? Again, one novel way of answering this question would be to claim that the olfactory individual is a composite sensory individual (comprising of both the source event and the odour).[14]

[14] This work was supported by the ERC Consolidator grant [726251], the FWF-FWO grant [G0E0218N], the FWO research grant [G0C7416N] and the FWO-SNF research grant G025222N. We are grateful to Jonathan Cohen, Jason Leddington, Casey O'Callaghan and two anonymous reviewers for very helpful comments.

References

Agus, T. R., Suied, C., and Pressnitzer, D. (2019). Timbre recognition and sound source identification. In *Timbre: Acoustics, Perception, and Cognition*, edited by K. Siedenburg, C. Saitis, S. McAdams, A. N. Popper, and R. R. Fay, 59–85. Cham: Springer.

Ariely, D. (2001). Seeing sets: representation by statistical properties. *Psychological Science*, 12(2), 157–62.

Batty, C. (2011). Smelling lessons. *Philosophical Studies*, 153, 161–74.

Bermúdez, J. L. (2000). Naturalized sense data. *Philosophical and Phenomenological Research*, 61(2), 353–74.

Byrne, A. and Hilbert, D. (2003). Colors and reflectances. *Behavioral and Brain Sciences*, 26(1), 3–21.

Carello, C., Wagman, J. B., and Turvey, M. T. (2005). Acoustic specification of object properties. *In Moving image theory: Ecological considerations*, 79–104.

Casati, R., Di Bona, E., and Dokic, J. (2013). The Ockhamization of the event sources of sound. *Analysis*, 73(3), 462–6.

Casati, R. and Dokic, J. (1994). *La philosophie du son*. Nîmes: Chambon.

Choi, H. and Scholl, B. J. (2006). Perceiving causality after the fact: postdiction in the temporal dynamics of causal perception. *Perception*, 35, 385–99.

Clark, A. (2000). *A Theory of Sentience*. Oxford: Oxford University Press.

Clark, A. (2004). Feature-placing and proto-objects. *Philosophical Psychology*, 17(4), 443–69.

Clarke, T. (1965). Seeing surfaces and physical objects. In *Philosophy in America*, edited by M. Black, 98–114. Ithaca, NY: Cornell University Press.

Cohen, J. (2004). Objects, places, and perception. *Philosophical Psychology*, 17(4), 471–95.

Di Bona, E. (2017). Towards a rich view of auditory experience. *Philosophical Studies*, 174(11), 2629–43.

Dretske, F. (1981). *Knowledge and the Flow of Information*. Cambridge, MA: MIT Press.

Gaver, W. W. (1993a). How do we hear in the world? Explorations in ecological acoustics. *Ecological Psychology*, 5(4), 285–313.

Gaver, W. W. (1993b). What in the world do we hear?: an ecological approach to auditory event perception. *Ecological Psychology*, 5(1), 1–29.

Green, E. J. (2019). A theory of perceptual objects. *Philosophy and Phenomenological Research*, 99(3), 663–93.

Kulvicki, J. (2008). The nature of noise. *Philosophers' Imprint*, 8(11), 1–16.

Kulvicki, J. (2014). Sound stimulants: defending the stable disposition view. In *Perception and Its Modalities*, edited by D. Stokes, M. Matthan, and S. Biggs, 205–21. Oxford: Oxford University Press.

Leddington, J. (2014). What we hear. In *Consciousness Inside and Out*, edited by R. Brown. 321–34. Dordrecht: Springer.

Leddington, J. P. (2019). Sounds fully simplified. *Analysis*, 79(4), 621–9.

Matthen, M. (2010a). How things look (and what things look that way). In *Perceiving the World*, edited by B. Nanay, 226–53. New York, NY: Oxford University Press.

Matthen, M. (2010b). On the diversity of auditory objects. *Review of Philosophy and Psychology*, 1, 63–89.

Matthen, M. P. (2005). *Seeing, Doing, and Knowing: A Philosophical Theory of Sense Perception*. Oxford: Oxford University Press.

McAdams, S., Chaigne, A., and Roussarie, V. (2004). The psychomechanics of simulated sound sources: material properties of impacted bars. *Journal of the Acoustical Society of America*, 115(3), 1306–20. Available from: https://doi. org/10.1121/1.1645855.

Michotte, A. (1963). *The Perception of Causality*. New York, NY: Basic Books.

Nanay, B. (2010). Perception and imagination: amodal perception as mental imagery. *Philosophical Studies*, 150, 239–54.

Nanay, B. (2011). Do we see apples as edible? *Pacific Philosophical Quarterly*, 92, 305–22.

Nanay, B. (2013). *Between Perception and Action*. Oxford: Oxford University Press.

Nanay, B. (2018). The importance of amodal completion in everyday perception. *i-Perception*, 9(4), 1–16.

Nanay, B. (2022). The sensory individuals of picture perception. *Philosophical Studies* 179: 3729-3746.
Nudds, M. (2009). Sounds and space. In *Sounds and Perception: New Philosophical Essays*, edited by M. Nudds and C. O'Callaghan, 69–96. Oxford: Oxford University Press.
Nudds, M. (2010). What are auditory objects? *Review of Philosophy and Psychology*, 1(1), 105–22.
O'Callaghan, C. (2007). *Sounds: A Philosophical Theory*. Oxford: Oxford University Press.
O'Callaghan, C. (2008). Object perception: vision and audition. *Philosophy Compass*, 3(4), 803–29.
O'Callaghan, C. (2010). Sounds and events. In *Sounds and Perception: New Philosophical Essays*, edited by M. Nudds and C. O'Callaghan, 26–49. Oxford: Oxford University Press.
O'Callaghan, C. (2011). Hearing properties, effects, or parts? *Proceedings of the Aristotelian Society*, 111(3), 375–405.
Pasnau, R. (1999). What is sound? *The Philosophical Quarterly*, 49(196), 309–24.
Scholl, B. J. and Tremoulet, P. D. (2000). Perceptual causality and animacy. *Trends in Cognitive Sciences*, 4, 299–309.
Searle, J. R. and Willis, S. (1983). *Intentionality: An Essay in the Philosophy of Mind*. Cambridge: Cambridge University Press.
Siegel, S. (2006). Which properties are represented in perception. In *Perceptual Experience*, edited by T. S. Gendler and J. Hawthorne, 481–503. Oxford University Press.
Soteriou, M. (2018). Sound and illusion. In *Perceptual Ephemera*, edited by T. Crowther and C. MacCumhaill, 31–49. Oxford: Oxford University Press.
Young, N. (2018). Hearing objects and events. *Philosophical Studies*, 175(11), 2931–50.

11
What is it we touch?

William G. Lycan

1 What is it we touch?

> She stood awhile, breathing. She followed the counter all the way to the wall, sweeping its surfaces. Her fingers found objects, small things she could identify like a stapler and a nameplate and larger things she dropped upon ascertaining they were not what she wanted and did not think about . . .
>
> She circled the counter twice. She made her way back to where the bareword had been and dropped to her hands and knees and began to feel around on the floor. Almost immediately she found cloth and hair and her hum became a shriek and she couldn't do it any more.
>
> Max Barry, *Lexicon*

Among the traditional 'five senses', touch is an outlier, and in more than one respect.

2 Ways in which touch differs

Aristotle himself noted that, while each of vision, hearing, smell, and taste has its single 'proper sensible', touch seems to have 'several kinds of what is tangible', raising the question of whether it 'is a single sense or a group of senses'.[1] More recently, in the preface to his masterly survey of the senses, Moreland Perkins (1983, pp. 3–4) writes, 'Touch presents rather a special case . . . [It is] the sense which offers the most formidable obstacles to [Perkins' 348-page case against 'direct realism'] . . . [I]t was with dismay that I discovered that the argument for touch had to be carried to the quantum mechanical level'.[2]

There are further distinctive features:

— Touch has no proper gross sense organ, no analogue of eyes, ears, nose, and tongue. (The skin—what Perkins calls the 'whole skin'—is a bodily organ, but not itself a sense organ.)

[1] *De Anima*, Book II, Chapter 11, p. 672 (*Complete Works of Aristotle: The Revised Oxford Translation, Vol. 1*, ed. Jonathan Barnes, 1984; Princeton, NJ: Princeton University Press).

[2] No, sorry, in this paper, I will not divulge my own rich views on the quantum mechanics of touch.

William G. Lycan, *What is it we touch?* In: *Sensory Individuals*. Edited by: Aleksandra Mroczko-Wąsowicz and Rick Grush, Oxford University Press. © Oxford University Press 2023. DOI: 10.1093/oso/9780198866305.003.0012

- In one way, touch is well known to be 'a group of senses' rather than a single sense: there are distinct sensory subsystems, each of whose receptors are in the skin, and 'touch' at best generalizes over those. Touch 'seems to possess a greater functional complexity than the other senses, a complexity that codes a large range of distinct phenomenal qualities through easily distinguished sensory channels' (Fulkerson, 2015, p. 194).
- Accordingly, touch presents trickier issues of individuation than does any of the other senses. (Smell is *slightly* tricky, due to the difference between orthonasal and retronasal pathways, but at least each of those paths leads directly to the olfactory epithelium.)
- The verbs 'see', 'hear', 'smell', and 'taste' are explicitly perceptual. Non-perceptual uses of them are metaphorical or otherwise derivative. 'Touch' is not a perceptual verb: 'It was crowded and I sat touching the wall'; 'Don't touch the items on display'; 'She touched Charles on the shoulder'; 'Did you touch any poison ivy?'. ('Feel' in its exploratory sense would perhaps be explicitly perceptual; try substituting it for 'touch' in the examples just given.) 'S can no longer see/hear/ . . .' announces sensory impairment, but 'S can no longer touch' is just odd. The other verbs have non-success uses and can take non-existents as their objects—'She saw green lights', 'I heard a voice just like my father's'—but you cannot touch something that does not exist.
- For each of the other four senses, we can ask what (if anything) that sense represents; for example, 'What does smell represent?' The question is not always straightforwardly answerable,[3] but 'What does touch represent?' encounters special complications (see Section 5).
- One of the complications is that in order for touch to represent much of anything, the subject's body must be moving with respect to the relevant chunk of the environment.
- Fulkerson (2014a, pp. 100–4) sees a 'duality' in the knowledge afforded by touch: touch tells us not only about objects in the environment, but equally things about our own bodies.[4]
- It is widely held that touch has a kind of epistemic privilege, epistemic superiority to the other four senses. It is felt to be more reliable; for example, if we are in doubt as to a visual appearance, we check it by way of touch, as when Macbeth says, 'Is this a dagger which I see before me/The handle toward my hand? Come, let me clutch thee' (*Macbeth*, II, I, 44–5). (For discussion, see Massin and de Vignemont, 2020.)
- A technical item: following Gibson (1962), Mohan Matthen (2014, p. 275) contends that touch uniquely deploys two different spatial coordinate systems.[5]

[3] I do not mean just that there is no undisputed answer. Lycan (2014) argues that we have no way of finding a complete answer for the case of vision; Lycan (2018) contends that it is doubtful whether taste (strictly so-called) does represent at the level of the whole person; if it does, the person usually cannot detect that introspectively. In this paper, I simply assume that perception does represent, though in '"What touch represents": preliminaries' and 'What touch represents', I shall mention a couple of needed qualifications.

[4] That idea goes back, we are told by Krueger (1970), at least to Katz (1925). (There are some philosophers who see such a duality in other sense modalities, for example, O'Dea, 2011.)

[5] He goes on to note an interesting multimodal effect of exploring food texture with one's tongue: *flavour* (as opposed to taste, narrowly speaking) is partly tactile.

3 Individuating the senses

Grice (1962) kicked off what has become a huge literature on distinguishing and individuating the senses (see Matthen, 2015 for an overview). I do not mean to swell it. Metaphysically, in general, I think there is only rarely a fact of the matter regarding individuation or counting, the only exceptions being when someone has provided an unusually precise sortal predicate like 'molecule' or 'six-pack'. ('Sortal' is sometimes defined as a predicate that carries determinate individuating conditions as part of its meaning, but not all predicates that have been called sortals qualify: for example, 'building' does not completely individuate (Lycan, 2021).) Rather, I am a pragmatist about individuating: we can count things as one or as two or more, depending on our purposes, explanatory or financial or whichever.

Accordingly, in the matter of the senses in particular, I am what Fulkerson (2014b) sympathetically calls a 'sensory pluralist'. To take a currently popular example, for some purposes, orthonasal and retronasal smells together constitute one sense, while for others, they are two. So I do not think there is a determinate way of dividing touch into a number of sensory modalities or systems.

Lycan (2000, p. 277) offered a different sort of picture for categorizing senses: a large map defined by 'general respects of similarity and differences between senses', on which a given sense modality (however individuated) could be located by its particular pattern of similarities and differences; I offered a list of 16 such 'vectors'. (In that, I anticipated the far more sophisticated and detailed method accomplished by Fiona Macpherson (2011, 2015), who posits a big multidimensional space which 'might reveal unconsidered features of the actual and possible senses, help us to define what a sense is, and provide grounds that we might use to decide what is one token sense rather than multiple token senses' (Macpherson, 2015, p. 432); despite that last clause, she denies that the senses 'come in a limited number of discrete kinds' (p. 435). And she adds a 'principal component analysis' with which to perform formal operations on her space.)

I will not list all my vectors, but merely mention a few: (1) informational richness; (2) degree of objectification; (7) characteristic features of the quality space; (8) variety and kind of besetting illusions; (10) credentials as presenting a phenomenal individual; (13) phenomenal distinctness from other senses; and (14) cognitive penetrability. My main purpose in this section is to add three more, each occasioned by thinking about touch in particular:

(17) Importance of diachronicity, especially of exploration (Fulkerson, 2014a, Chapter 3).[6] In particular, the relation between synchronic representation (that is, representation that is accomplished at a single instant) and diachronic representation that takes a stretch of time to be built (see Section 5.1).

(18) Feature-binding power (Fulkerson, 2014a, Chapter 2).

(19) Importance of representational layering (which I shall define in Section 5.5 and go on to discuss at some length).

6 Famously, Alva Noë (2004, p. 1) argues that touching, not seeing, should be the paradigm of perceiving; we investigate our surroundings 'not all at once, but through time, by skillful probing and movement'. He does not succeed in showing that that is as thoroughly true for vision as for touch (Block, 2005; Lycan, 2006; Prinz, 2006), but no one should deny that vision does have an exploration importance rating. Schwenkler (2014) calls Fulkerson on that.

4 The present volume's title issue

Sensory individuals, as the editors of this volume use the term, are the events or entities that we perceive; they are those individuals to which *perceptual* mental states attribute properties. This is meant commonsensically, as opposed to technical and controversial details regarding what a given sense modality represents, strictly speaking. For example, we see ordinary middle-sized objects such as rocks and chairs and newspapers and houses and billboards and lakes and other people, as well as 'ephemera' such as rainbows and shadows; we hear sound events such as thuds, extended crashes, clocks striking, people talking, and songs being sung; we smell substances like coffee and skunks and other people, in addition to odours in the air. Those are uncontroversial facts, given the everyday uses of the perception verbs.

In the same way, what we touch and feel are primarily middle-sized physical objects or just their surfaces. Not ephemera; we cannot feel a rainbow, and it is at best odd to talk of feeling a shadow. But we can also feel liquids and gaseous vapours, in the sense of exposing our body parts to them as a way of discerning things like their temperatures, their densities, and such textures as they may have.[7]

When we restrict our discussion to perceiving properties, as opposed to the substance whose properties they are, 'feels' seems the appropriate analogue of the appearance verbs 'looks', 'sounds', 'smells', and 'tastes'. (Obviously that sense of the word differs from the use of 'feel' as an action verb applying to persons.)

In many or most of the examples I have listed in this section, the use of the perceptual verb is appropriate and correct, but it does not follow, and would not be true to say, that the relevant *sensory system* represents the objects and properties as such. The visual system represents colours, shapes, and volumetric distances, and *possibly* it applies Rosch's (1975) 'basic categories' such as 'chair' and 'dog'; but it does not have the concepts 'lake' or 'star', much less 'billboard', 'five-dollar bill', or 'post office', even though we see billboards, five-dollar bills, post offices, and arguably the President of the United States. The auditory system represents sounds of such-and-such volume, timbre, and relative location, but not D♭ minor triads or missile launches. In saying that, I am assuming a weak modularity thesis (weaker than that of Fodor (1983)), roughly that there is a useful distinction between what is represented by the sensory system alone and what may be represented only when the system's output has been operated on by inferential or other cognitive processes that draw on background information not possessed by the system itself. (The boundaries of the system may be structured in grades or stages (Lycan (2014)). And shortly I shall introduce a more liberal notion of sensory representation.)

Similarly, not every thing or property you can detect by tactual-exploratory means is represented by touch—that is, by any of the multifarious sensory systems that fall under the heading of 'touch'. Let us move on to the question of what touch does represent in that stricter way.

[7] I believe we can also feel *absences* as such, like a missing tooth with our tongue or (in the dark) a missing picket in a picket fence or an empty space on a shelf where a valuable ornament should be (cf. Sorensen, 2009; Farennikova, 2013).

5 'What touch represents': preliminaries

A guiding, but in some ways misleading, question is that of what we can tell/learn about a physical object by touching it and feeling it. (Here I focus on properties rather than on the 'sensory individuals' whose properties they are. And think of touch *unaided by vision*, as in the epigraph to this paper.) I shall not attempt a complete catalogue, but at least: the object's hardness or solidity; the texture of its surface; its shape and size; its comparative temperature; whether it is wet or dry, slippery, sticky, and such; its heft or weight; its location with respect to us; whether it is pulsating or vibrating; whether it is moving in space.

If we are to mine that list to say 'what touch represents', even when it is clear we mean touch 'per se' (unaided by any of the other four senses), we need to observe some distinctions.

5.1 Synchronic vs diachronic

Touching a physical object at a single moment (assuming there were such a specious present) tells us very little about the object, probably not even about its hardness or texture. It can be argued that without some passage of time, touch cannot represent anything at all, though I will not make that argument here. Even when touch is not exploratory, it normally requires contrasts that can be made as between one instant and a succeeding instant. (Thermal representation may be the exception.)

5.2 Passive vs active

Synchronically or not, touch can represent properties of objects that are in contact with your skin, without your having to move or do anything at all. The feel of your clothing, such as gloves or a hat, represents properties of the clothing. But unless you move and do something at least a bit exploratory, your tactual representational contents will not be very interesting. (Lederman and Klatzky (1987) document and classify a small set of 'exploratory procedures', stereotyped patterns of movement with respect to an object. An intermediate case is one in which an object moves against your skin while you remain stationary.) So far as I am aware, the term 'haptic' has no single authoritative definition, but it roughly designates touch that is at least minimally exploratory, and most of what is written about tactual representation is about haptic touch.[8]

[8] Klatzky and Lederman (1995, p. 1111) argue that a surprising amount of object-identificatory information can be gleaned in a single 'haptic glance', a period of 200 ms incorporating 'little or no movement of the fingers'. (Thanks to each of two anonymous referees for steering me to that article.) But I do not think the authors would maintain that strictly synchronic touch can reveal more than I have allowed; they speak of 'very fast search rates' (p. 1112).

5.3 Using skin sensors alone vs using them in close conjunction with associated internal sensing

(Such as proprioception, kinaesthesis, and registering of muscular and/or tendon tension.) This may or may not coincide perfectly with passive vs active; that is an empirical question.

5.4 Skin contact vs not

In one way, it may seem, touch by definition implies contact between the skin and something else (bar cases in which you are touching a part of your own body and what is touched is more skin). But going both by the ordinary use of the words and by functional considerations, one can touch and feel objects in the environment while wearing thin surgical gloves, or even heavy gloves, or socks or shoes if one is feeling around with one's foot, etc. (More on that below.)

5.5 Basic vs derivative representation

This takes a bit of explaining.

We cannot assume that a single mental state that is a perceptual representation has just one intentional object or content. Peacocke (1992), Lycan (1996), Noë (2004), and Schellenberg (2008) have each argued that a representation can have more than one content, with the contents systematically related to each other by some priority or dependence relation. More specifically, the perceptual state may represent one object or property *by* representing a more primitive or less ambitious one. I have called views of this kind 'layering' views.[9] For example (I borrow these from Lycan (2020) on audition): by hearing a sound of such-and-such volume, timbre, and duration, you hear a bang, and by hearing the bang, you hear a collision; you hear a major triad by hearing each of three tones sounded simultaneously or in close succession, and *perhaps* you hear a D♭ minor triad by hearing the triad; by hearing noises as of a human voice, you hear words, and, some boldly contend, by hearing those, you literally hear meanings—NB I am not reverting to the broad commonsensical use of 'hear'. I am still talking about what the auditory system represents, and in offering these examples, I have made highly substantive empirical claims. The last one, about hearing meanings, is defended by a few philosophers, but by no psycholinguists to my knowledge. Some psychologists of music have ambitiously attributed significant harmonic quasi-concepts to the auditory system itself.

Now, such layering plays each of two special roles in tactual representation, or so I will contend. First, I shall argue that nearly all tactual representation is derivative from representation of conditions of or in the subject's own skin. Second, I shall argue that even among

[9] My own model for layering of this sort is that of deferred linguistic referring: uncontroversially, we sometimes refer to a thing by referring to a distinct, but in some way more accessible, thing that is saliently associated with it. The layering can be multiple: in the same gesture, we can refer to a chalk mark, thereby referring to a numeral, thereby to a number, thereby to an office in the building, thereby to its occupant. (I point to the chalk mark and say 'not gonna like this'.)

the skin conditions, some are represented only by representing more primitive ones. In the heading of this subsection, I used 'basic' to mean a representing that is done by a psychological state, but not *by* that state's also representing something more fundamental.

6 What touch represents

A further guide to help in narrowing down what a sense modality represents in our strict, weak-modular sense is to ask what properties would have qualified as belonging to a Russellian sense-datum of the relevant type, that is, a non-physical entity with which we are directly acquainted, that actually has a property that subjectively appears to us in perception, such as a colour, timbre, or smell. (It is entirely fair to consider such properties on their own, whether or not one believes in sense-data. Just assume the first-person perspective, put yourself back in Russell's movie theatre wondering how to 'get to' the alleged external world, observe the passing show as such, and *pretend* that it is composed of actual phenomenal individuals.) Obviously visual sense-data would have colours and shapes; auditory ones have volume and pitch; olfactory ones have particular smells.

Matthen (2014) and others speak of 'tactile qualities'; Matthen (p. 275) cites 'hardness, coldness, flatness, etc.'. Fulkerson (2014a, Chapter 5) identifies a class of 'basic tangible features', which he specifies as 'intensive' features in a carefully defined and motivated sense (pp. 114ff): thermal profile, smoothness, hardness, weight, and perhaps more (p. 119). They are 'basic' in that more complex tactile qualities are composed of them, which is also characteristic of a Russellian sense-datum, but there is no canonical list of such qualities, and I do not know how Russell, or we, might pick out such a set as containing all and only the distinctive qualities of a supposed tactual sense-datum considered as an individual thing.

Instead, I shall appeal to my layering thesis. Where it is plausible to say that touch represents a quality q_1, but only by representing q_2, and not vice versa, I shall not count q_1 as a basic tactual representatum. But even if we restrict our attention to the basic, I will not attempt to give a complete, much less definitive, answer to the question of what touch basically represents. The literature is still in ferment and the phenomena are too complicated. I shall defend the one general idea I have mentioned.

Perkins (1983, p. 242) argues convincingly that what we 'most directly feel' by touch is 'pressure or stress'.[10] (He has set aside thermal perception in a previous chapter; I shall return to thermal below.) But more specifically (p. 243):

> [T]he sensory receptors that mediate tactual experience are located within our skin. What most directly stimulates these tactual receptors within the skin is not the pressure at the surface of the skin but the resultant stress within the skin, with its concomitant deformation of the skin. The hat presses upon the skin, a stress is imparted to the skin, the skin is deformed, and the tactual sensory receptors located within the skin respond to this stress within and deformation of the skin itself.

[10] He does not couch his arguments in terms of representation; I will be adapting them into representational language.

What we most directly feel is 'this physical distortion of our flesh'. In particular, 'we feel the skin's stress or strain more directly than we feel the contact or pressure of the alien object . . .' (p. 244).

But Perkins acknowledges that there is an intimate relation between the 'stress or strain' within the flesh and 'the pressure that we feel to be exerted at the surface of our flesh': '[I]t is reasonable to construe pressure upon our skin as the most direct object of tactual awareness located before our *whole* skin' (pp. 246–7, italics original). I suggest that this is a very small instance of the 'by' relation that organizes representational layering. We represent pressure *on* the skin by representing compressive stress and/or deformation within the skin.

Perkins goes on to point out that various things we learn about external objects by touching and feeling them manifest themselves as patterns of pressure. We detect shape by perceiving such patterns over time. Felt vibration is obviously a pattern of pressure.

> [R]oughness and smoothness are merely fine-grained differences in the shapes of the surfaces of the objects in contact with our skin. A hard object exerts more pressure against us than a soft one, induces less compression in our flesh, distorts our flesh less, and yields less ground to *our* pressure than a soft object (p. 247, italics mine).

And once we add perception of 'things' resistance', which is 'partially mediated by our feeling the tension in our tendons' (p. 247), still other tangible qualities can be accommodated, including wetness, weight, and relative motion. All these representata too are clear candidates for non-basicness; we represent hardness, vibration, wetness, and motion *by* representing patterns of pressure on the skin (which, in turn, is done by representing stress or distortion within the skin).

Thus my thesis: first and foremost, touch represents conditions of one's own skin. Call this the Skin view.

That idea is, or sounds, a bit paradoxical. Touch in the perceptual sense is, well, perception, not mere bodily sensation or even proprioception. As I magnanimously granted in Sections 4 and 5 above, when we use touch to find out things, the things we find out are things about physical objects in the environment, not (usually) things about our own skin. Moreover, tactual *experience* is that of how external objects feel to us (though, obviously, touching the objects sometimes causes bodily sensations also). But that is layering, or so I have argued following Perkins. If I am right, touch differs from the other senses in involving a great deal more and finer-grained representational layering.

(But is touch 'exteroceptive' in Louise Richardson's (2013) sense? Unequivocally, yes. See Section 8.)

A further argument for the Skin view is based on the fact noted in Section 5.4, that we touch and feel physical objects without our skin making contact with them.[11] Layering again: we already know that the boundary between skin and what is immediately external to it is per se of little epistemological significance, because we tactually represent external things nearly as readily through thin gloves or socks or other items of clothing as we do using naked skin. The difference between the clothing/skin relation and the skin/receptors relation is not dramatic; from the representational point of view, the skin is much more like clothing than it

[11] The point is very well made by Fulkerson (2012).

is like a sense organ.[12] The skin/external boundary is between two non-basic representational layers, not between the basic one and the first derivative one.[13]

On the second point heralded at the end of Section 5.5: in my work on other sense modalities, representational layering has always corresponded to levels of distality (i.e., the degree of causal remoteness from the direct stimulation of the sense receptors) and/or grades of conceptual sophistication; typically the 'by' relation relates a representatum at one level to the one at the next level inward and/or conceptually more primitive. As noted in Section 5.1 (see especially fn. 9), that is because I have always thought of perceptual representation synchronically. With touch, a great deal of layering is merely diachronic. The obvious examples are the representation of vibration, and that of an object's overall shape, which I represent by representing its component geometric contours over time.

7 Heat and cold

Despite the obvious fact that feeling a substance immediately discloses the substance's relative temperature, thermal perception has usually been separated from the more central forms of tactual perception. Perkins makes it an entirely separate chapter of his book (Four, 'Feeling Heat and Cold'), simply excluded from 'The Sense of Touch' (Seven). But the most dramatic separation is urged by Kathleen Akins (1996), who contends that it is no function of thermoreception to represent the world at all.[14]

Akins (1996. p. 342) argues in great detail and convincingly that our sensations of warmth and cold are not 'the products of skin receptors that are finely tuned to surface temperature'; the thermal receptors are not thermometers, not detectors of external conditions. Rather,

[12] Fulkerson (2012, p. 3, fn 3): '[T]actual information is reliably transmitted through inert layers of skin, nails, and teeth ...'. An anonymous referee has pointed out that some of our tactual detections, notably that of heft or weight, depend on signals from receptors in the muscles and joints, and so our representations of those external properties do not derive solely from the representing of skin deformation. That is an important point, but for simplicity, I shall continue to speak of the 'Skin' view.

[13] Fulkerson (2014a, Chapter 4) defends a position closely related to the Skin view, but the relation is a little hard to detail, due to some terminological differences and to the fact that he says little about representation. The obviously similar tenets are these: (1) 'Perceptual touch experience [always] depends on bodily awareness' (p. 90). But (2) the bodily awareness can be 'implicit', 'not consciously attended', but 'primed for attention' (p. 90). And (3) 'perceptual touch involves the representation of distal information that ... [depends upon] bodily information made available through ... bodily engagement with the world' (p. 91). Fulkerson does not mention particular conditions within the skin. He certainly does not hold the layering doctrine. There is no suggestion of a single representation having multiple contents. He emphasizes that the dependence of touch experience on bodily awareness is causal only; in fact, he says he 'believe[s] a causal dependency provides the best chance of capturing what we mean in saying that perceptual touch depends on bodily awareness' (p. 78). A further difference is that the status of the bodily 'information' is downplayed: the 'potential [bodily] experience ... is an informational node that connects later processing systems involved in perceptual touch to information transmitted from a distal source.... The (changing and variable) perception of [the body] provides the input to an informational system in which (stable) information about [an external object] is extracted over time' (p. 95). (Though just at one point, Fulkerson says that the 'background' informational states 'are essentially representations in the appropriate perceptual format and capable of performing the functional roles of attended experiences'; no disagreement there, but the difference is still that, for me, the full-scale tactual representations are of skin conditions and merely *double as* representations of external objects in a psychosemantically derivative way.) Finally, Fulkerson (2015, pp. 195–6) has to put in a bit of work to say how tactile sensors differ from pain receptors, while for me, that question does not arise, since no one has ever suggested that nociceptive representations themselves also represent conditions in the external world. Campbell (2020) nicely refutes the suggestion that bodily *sensations* are needed for tactual experience of external objects (assuming that 'sensation' means something more than Fulkerson's 'implicit' awareness).

[14] It is not quite accurate to cast Akins' view as *separating* thermal perception from touch proper, because her project is more radical: to question the whole idea that what sensory systems are for is to represent the environment. But her arguments apply noticeably better to thermoreception than to other senses.

the thermoreceptive system is 'narcissistic': its function is to warn its owner 'when thermal *damage [to the body]* is occurring or before it is likely to occur, when temperature *changes* are likely to have specific consequences [for the owner's health]' (p. 350, italics mine). Thermal sensations are more like pains than like visual or auditory sensations. That function does not require accurate reflection, much less measurement of the environmental temperature, and the prevalence of Protagorean thermal illusions (see below) shows that we do not get any such reflection.

Akins has made a very strong case, and I accept each of the claims recorded in the preceding paragraph. But it does not follow either that thermoreception is not really part of touch, that it does not represent the external world, or even that it does not have the function of representing external temperature. On the first point, recall Fulkerson's (2014a) argument that feature-binding is a strong individuating vector for sensory modalities, and that among the features bound by touch into the representation of an individual physical object—bound so closely that it takes introspective work to distinguish them—is the object's comparative temperature. (And see Ratcliffe, 2012 on oiliness and wetness.)

On the second point: layering again. Akins does not explicitly address the position that touch represents external temperature *by* representing thermal conditions within the skin. No doubt she would argue that if we try to understand touch as doing that, we would have to admit that the alleged external representations are sadly inaccurate. To which I reply that 'sadly' is too strong. When you step outdoors in the morning, you know whether to put on a jacket or take off the one you have already put on; in fact, you can probably give a fair estimate of the air temperature, say within 10°F. Much the same goes for feeling bathwater with your foot or a radiator with your hand.

More specifically, the tactual feel of an external object can be incorrect. If, for whatever reason, the bathwater feels cold when it is warm, or the radiator feels warm when it is stone cold, you are misperceiving and, I infer, *mis*representing. For that matter, in standard cases of thermal illusion, external substances are not as they feel to be. In Protagoras' example of hands in water, the water feels to be both warm (with the one hand) and cool (with the other). And consider 'paradoxical' heat and cold (e.g., Susser, Sprecher, and Yarnitsky (1999)).[15] Notice further that if we think in terms of possible worlds, there are worlds that count as what Hintikka called 'perceptual alternatives': ones which *are* as our world sensorily appears to be; thus, there is a corresponding function from worlds to truth values, and any such function corresponds to a proposition expressed.[16] And so a thermal feeling actually has world-directed semantical properties: reference and a truth condition.

Finally, on the third point, I hope Akins would not infer from an organ's having one function that it does not also have another. Even if the first function is the organ's main or principal function, an organ often has one or more subfunctions.[17]

[15] Actual thermal hallucinations are rare, but Mauguière (1999) reports that they have occurred in epileptics.

[16] This is a standard account of the propositional meanings of natural-language sentences. For example, 'There is an elk in the second-floor men's rest room in Manchester Hall' is false in our actual world but is true in some other possible worlds; the function from worlds to truth-values represents the range of circumstances under which that sentence would be true.

[17] Richard Gray (2013) too appreciates Akins' case, and joins her in rejecting the thermometer model and the commonsensical claim that touch represents thermal properties of external substances. But he does not join in her entirely anti-representational stance; he argues ingeniously and in detail that heat and cold sensations represent processes of thermal energy being transmitted to and from the body. Which may well be true, but like Akins, Gray does not consider either layering or the view that thermoreception has the representing of external properties as a

8 Tactual consciousness and experience

I have always maintained that in one of the 'c-'word's 18 or 19 senses, consciousness is statistically rare. The sense I mean is as in 'a conscious memory' or 'a conscious decision': a mental state is 'conscious' in that its subject is internally aware of it as such. In that 'state' sense, as it is sometimes called, some mental states are conscious states; some are not but could have been, had the subject attended to them; some (such as language-processing states) are structurally inaccessible to consciousness; and just possibly some are barred from consciousness by a Freudian Censor or the like.

The word 'experience' is sometimes used liberally, to include all cases of perceiving or sensing whether state-conscious or not. I prefer to reserve the term, as I think most writers do, for sensing consciously—that is, for sensings of which their subjects are aware.

There are several competing theories of what makes a mental state a conscious one in the present 'state' sense, for example, the 'higher-order perception' account defended by Armstrong (1981) and Lycan (1996). None is very interesting, because unlike 'consciousness' of some more exciting kinds that provoke mind–body dualism and such, the phenomenon is of little ontological importance. The difference between a type of mental state's occurring non-consciously and that same type of state's occurring consciously is superficial; in particular, it would not normally affect the state's own representational content. And as I have said, awareness of our own mental states is statistically atypical: though, granted, quite often we are aware of one or more of our own mental states, we are much more often not aware of them, because our attention is focused on the external goings-on with which we must constantly deal.[18]

Here endeth the mini-lecture. The only reason I have brought up 'consciousness' and 'experience' in their internal-awareness senses is that they bear directly on a few specific issues regarding touch.

First, there is an obvious objection to my view that touch first and foremost represents conditions of our own skin: the Skin view is sharply at odds with the phenomenology of touch. When we touch and feel a physical object, it is normally just that object we are aware of, not either the (I say represented) skin condition that mediates that awareness *or* for that matter the awareness itself. To see this, consider that throughout nearly every millisecond of your entire life you are touching something or other, and you hardly ever notice. You *certainly* hardly ever notice deformations of your skin as such.

Reply: the operative word in the objection is 'phenomenology', meaning what mental items we are *aware of* during a tactual exploration. And the premise is true: normally we do not notice the mediating skin condition, and even less often are we aware of any representation

subsidiary function. I would also remind him of the thermal illusions and repeat my semantical argument. Perkins' chapter on heat and cold runs 65 pages. Since he does not favour representational language—though he does speak of '[thermoperception's] mechanism of reference to an instance of temperature before the skin' (Gray 2013, p. 138), and on p. 161 he calls his account 'representational'—I am not quite sure what position he would take on the present issue, but he definitely holds that thermal perceptions attribute comparative temperatures to external substances. I think he would say, further, that (1) the attribution is partly projection ('[T]he sensuous quality instantiated within thermal consciousness . . . functions as the sensuous content of our sensuous attribution of intensity of heat to the object before our skin' (p. 139); but (2) the attributed property is also characterized by its causal powers (p. 140).

[18] As Armstrong (1981, p. 55) says '. . . in the ["state"] . . . sense of the word "consciousness" consciousness is the cream on the cake of mentality, a special and sophisticated development of mentality. It is not the cake itself.'

of it. That is because the relevant representations are not 'state'-conscious. But they are always there; our skin sensors are quietly filing their reports. If they were not, you would feel nothing tactually. What interest us and, for obvious reasons, get our attention are our derivative worldly representata, the properties of the object perceived. So my view is not *at odds with* phenomenology, for there is no reason to expect that we would be introspectively aware of contents at the more proximal and primitive representational layers.[19]

Now, to return to Richardson (2013, p. 405) and exteroceptivity: a modality is exteroceptive when the sensible qualities inhering in a sensation 'seem ... to be qualities of objects distinct from our bodies' and the seeming is not just a matter of cognitive association. Also, exteroceptivity is connected to *finding out*: 'In exteroceptive experience we find out about the [relevant] qualities of objects ... by their seeming to have the qualities in question' (Richardson, 2013, p. 406), though, n.b., exteroceptivity does not entail representation but is phenomenological only—all absolutely true of touch, and perfectly consistent with the Skin view.

Finally, Fulkerson (2012, p. 14) points out that 'we can and do experience objects through touch, even when those objects are not in contact with [even] the apparent limits of our bodies'. He gives two good examples: driving a car, you notice through tactual experience alone that the road surface has changed from smooth asphalt to gravel; and, lying in bed with eyes still closed, you feel a cat leap onto the foot of the bed and make its inexorable march towards your face. And I have been told that congenitally blind people often experience the sensations produced by ground and wall surfaces as in the tips of their canes. This seems paradoxical: how is it that we can have vivid touch experiences, when the relevant objects are neither in contact with our skin nor even within inches, feet, or yards of the '*apparent* limits' of our bodies which include gloves and socks?

Surprise: layering vs phenomenology again. Our awareness, our state-conscious feeling, is determined by our attention, which obviously is driven by our interests. We do not notice the mediating representata. (We could and would notice them were we motivated to attend to those less remotely distal layers.)

9 Tactual aspect-perception

Sense modalities differ in the varieties of aspect-perception they afford. Of course, nearly everything ever written by a philosopher about aspect-perception has been on vision, on '*seeing* as' (though Lycan (2020) argues that auditory aspect-perception plays several central roles in human life). Philosophers' introduction to the subject was primarily by Wittgenstein (1953, sec. xi) and through popular culture's fun with ambiguous figures such as the duck–rabbit, the Necker cube, the old/young lady, and such. Space does not permit background discussion here (Lycan (1971, 2020) discussed these phenomena at length), but after a few reminders, I shall comment briefly on tactual aspect-perception and its lessons for our approach to the general phenomenon.

19 Tactual layering was greatly extended by the invention of Braille. Blind people can read using their fingers alone, feeling letters and perhaps whole words as such. I shall return briefly to Braille in the Section 9. J. D. Trout (2019) has discussed the novel idea that there is purely tactual perception of *oral* speech.

The reminders: (1) perceiving-as is part of sensory phenomenology, not just a matter of thinking or cognitive interpreting. (Hence Wittgenstein's (1953, p. 196) famous puzzle: 'The expression of a change of aspect is the expression of a *new* perception and at the same time of the perception's being unchanged . . . .' He means roughly that while the configuration represented in our sensory experience remains the same, the experience itself changes somehow perspectivally, though the relevant 'perspective' is still sensory in nature.) (2) Alternate aspects compete; we cannot see both the duck and the rabbit simultaneously, but must flip between them. (3) Aspects can be consciously and deliberately flipped, at will.

The closest appropriate analogue for the verbs 'see as' and 'hear as' would be 'feel as': 'This fabric is a synthetic but I'm feeling it as silk.' When considering examples, it will again help to close your eyes.

A surface can be felt as being of a particular kind of stuff, or as of a different kind. A rock could be felt as a doorknob or a knife, or more commonly as a rock. A larger and oddly shaped rock could be imaginatively felt as a human head with a face; even more imaginatively, a flat rock with some scoring could be felt as a souvenir copy of a tablet brought down from Mt Sinai. Feeling-as is greatly enriched by the existence of Braille: a trained blind person feels a pattern of small bumps on a page as a letter or as a whole word or more. (That fact does not decide the question of whether the person tactually represents words; not all aspect-perception is *purely* sensory representation. Think of seeing a cloud as a Tibetan antelope having stomach trouble.)

In the dark—see again the epigraph to this paper—you can be genuinely uncertain what sort of object you have encountered, and aspect-flip. Even once you have made certain, you could go on aspect-flipping just for fun, much as you do with the old/young lady.

I shall close this paper by considering the implications for each of two substantive empirical theses that have been defended regarding aspect-perception. (Lycan (2020) discusses each of these for the case of hearing-as.)

EXPECTATION. Perceiving X under a particular aspect is, at least in part, a matter of expectation, of how one would tentatively expect X to behave or its parts to be perceived should circumstances alter. Gombrich (1960) speaks of how we 'project' possibilities from what we perceive.

ATTENTION. Aspect-perception is an attentional phenomenon. (Empirical studies reveal that the neural correlates of ordinary attention are activated during their subjects' aspect-flipping activities, and in the visual case, it has been shown that the aspect seen in an ambiguous figure depends on which part of the figure the subject focuses on (Chastain and Burnham, 1975; Kleinschmidt *et al.*, 1998; Ricci and Blundo, 1990). Orlandi (2011, 2012) specifically uses ATTENTION to argue that seeing-as requires neither concepts nor interpretation.

What we know of touch very strongly favours EXPECTATION, because touch is so centrally exploratory or at least diachronic. Feeling-as surely is a matter of expectation and projection, at least in part. When we feel with our fingers or another body part, we are literally expecting encounters that have not yet occurred.

But feeling-as gives an equivocal verdict on ATTENTION. As just noted, we do explore a surface; and we can revisit parts or aspects of a surface at will, again and again, scanning it in this order or that. (That distinguishes feeling a surface from hearing a musical performance.) In that sense, we can attend by *re-perceiving*. But what the advocates of ATTENTION were thinking of were our ways of attending to pictures: focusing on one or more parts of a picture that we had continued to see during the whole interval. That is a matter of shifting attention *without* changing what we are perceiving, and it was that feature that offered a partial solution to Wittgenstein's puzzle: since attention is considered a matter of 'heightening' or 'strengthening' or 'enhancing' an existing representation, as opposed to creating a new representation, what has not changed when an aspect dawns is the content of the existing representation. '[T]he expression of a *new* perception and at the same time of the perception's being unchanged.' Re-perceiving does not deliver that feature.

10 Summary of main contentions

- 'Touch' per se is not a sense, though of course there is a category of tactual perception.
- Representational layering plays a much bigger role in perceiving by touch than it does in perceiving of any other kind.
- First and foremost, or non-derivatively, what touch represents is stress and/or deformation within one's own skin.
- External things and their properties are represented derivatively, though by the very perceptual states that represent conditions of the skin.
- Contra Akins, touch can represent comparative temperatures in external objects.
- A large proportion of tactual representation is non-conscious; and, normally, what we are *aware of* feeling tactually is represented only derivatively.
- Tactual 'feeling-as' strongly confirms EXPECTATION, a widely held thesis regarding aspect-perception, but lends little or no support to ATTENTION, the view that aspect-perception is to be explained as an attentional phenomenon.

References

Akins, K. (1996). Of sensory systems and the 'aboutness' of mental states. *Journal of Philosophy* 93, 7, 337–72.
Armstrong, D. M. (1981). What is consciousness? In *The Nature of Mind and Other Essays*, edited by D. M. Armstrong, 55–67. Ithaca, NY: Cornell University Press.
Block, N. J. (2005). Review of Noë (2004). *Journal of Philosophy*, 102(5), 259–72.
Campbell, J. (2020). Is tactual knowledge of space grounded in tactual sensation? In *The Epistemology of Non-visual Perception*, edited by B. Brogaard and D. E. Gatzia, 146–64. Oxford: Oxford University Press.
Chastain, G. and Burnham, C. A. (1975). The first glimpse determines the perception of an ambiguous figure. *Perception and Psychophysics*, 17(3), 221–4.
Farennikova, A. (2013). Seeing absence. *Philosophical Studies*, 166(3), 429–54.
Fodor, J. A. (1983). *The Modularity of Mind*. Cambridge, MA: Bradford Books, MIT Press.
Fulkerson, M. (2012). Touch without touching. *Philosophers' Imprint*, 12(5), 1–15.
Fulkerson, M. (2014a). *The First Sense*. Cambridge, MA: MIT Press.
Fulkerson, M. (2014b). Rethinking the senses and their interactions: the case for sensory pluralism. *Frontiers in Psychology*, 5(457), 1426.
Fulkerson, M. (2015). What counts as touch? In *Perception and Its Modalities*, edited by D. Stokes, M. Matthen, and S. Biggs, 191–204. Oxford: Oxford University Press.
Gibson, J. J. (1962). Observations on active touch. *Psychological Review*, 69(6), 477–91.
Gombrich, E. H. (1960). *Art and Illusion*. London: Phaidon.
Gray, R. (2013). What do our experiences of heat and cold represent? *Philosophical Studies*, 166(1 Suppl), 131–51.

Grice, H. P. (1962). Some remarks about the senses. In *Analytical Philosophy, First Series*, edited by R. J. Butler, 248–68. Oxford: Oxford University Press.

Katz, D. (1925). *Der Aufbau der Tastwelt*. Leipzig: Verlag von Johann Ambrosius Barth.

Klatzky, R. L. and Lederman, S. J. (1995). Identifying objects from a haptic glance. *Perception and Psychophysics*, 57(8), 1111–23.

Kleinschmidt, A., Buchel, C., Zeki, S., and Frackowiak, R. S. (1998). Human brain activity during spontaneously reversing perception of ambiguous figures. *Proceedings of the Royal Society of London, Series B: Biological Sciences*, 265(1413), 2427–33.

Kreuger, L. E. (1970). David Katz's *Der Aufbau der Tastwelt* (The world of touch): a synopsis. *Perception and Psychophysics*, 7(6), 337–41.

Lederman, S. J. and Klatzky, R. L. (1987). Hand movements: a window into haptic object recognition. *Cognitive Psychology*, 19(3), 342–68.

Lycan, W. G. (1971). Gombrich, Wittgenstein and the Duck-Rabbit. *Journal of Aesthetics and Art Criticism*, 30(2), 229–37. Reprinted in: J. V. Canfield (ed.), *The Philosophy of Wittgenstein: Aesthetics, Ethics and Religion* (New York, NY: Garland Publishing, 1985).

Lycan, W. G. (1996). *Consciousness and Experience*. Cambridge, MA: Bradford Books/MIT Press.

Lycan, W. G. (2000). 'The Slighting of Smell (with a brief note on the slighting of chemistry)'. In *Of Minds and Molecules*, edited by N. Bhushan and S. Rosenfeld, 273–89. Oxford: Oxford University Press. Originally delivered in 1989, under the title 'Philosophy and Smell', as the Presidential Address at the Fifteenth Annual Meeting of the Society for Philosophy and Psychology, Tucson, AZ.

Lycan, W. G. (2006). Enactive intentionality. *Psyche*, 12, 3. Available from: http://journalpsyche.org/files/0xaaf8.pdf

Lycan, W. G. (2014). What does vision represent? In *Does Perception Have Content?*, edited by B. Brogaard, 311–28. Oxford: Oxford University Press.

Lycan, W. G. (2018). What does taste represent? *Australasian Journal of Philosophy*, 96(1), 28–37.

Lycan, W. G. (2020). Hearing as. In *The Epistemology of Non-visual Perception*, edited by B. Brogaard and D. E. Gatzia, 118–45. Oxford: Oxford University Press.

Lycan, W. G. (2021). How far is there a fact of the matter (as regards split brains, mind, agency, and personhood)? Presented (remotely) in an 'Author Meets Critics' session on Elizabeth Schechter's *Self-Consciousness and 'Split' Brains*, at the APA (Pacific Division) meetings, April, 2021.

Macpherson, F. (2011). Taxonomizing the senses. *Philosophical Studies*, 153(1), 123–42.

Macpherson, F. (2015). The space of sensory modalities. In *Perception and Its Modalities*, edited by D. Stokes, M. Matthen, and S. Biggs, 432–61. Oxford: Oxford University Press.

Massin, O. and de Vignemont, F. (2020). 'Unless I put my hand into his side, I will not believe': the epistemic privilege of touch. In *The Epistemology of Non-visual Perception*, edited by B. Brogaard and D. E. Gatzia, 165–87. Oxford: Oxford University Press.

Matthen, M. (2014). Image content. In *Does Perception Have Content?*, edited by B. Brogaard, 265–90. Oxford: Oxford University Press.

Matthen, M. (2015). The individuation of the senses. In *The Oxford Handbook of Philosophy of Perception*, edited by M. Matthen, 567–86. Oxford: Oxford University Press.

Mauguière, F. (1999). Scope and presumed mechanisms of hallucinations in partial epileptic seizures. *Epileptic Disorders*, 1(2), 81–91.

Noë, A. (2004). *Action in Perception*. Cambridge, MA: MIT Press.

O'Dea, J. (2011). A proprioceptive account of the senses. In *The Senses: Classical and Contemporary Philosophical Perspectives*, edited by F. Macpherson, 297–310. New York, NY: Oxford University Press.

Orlandi, N. (2011). The innocent eye: seeing-as without concepts. *American Philosophical Quarterly*, 48(1), 17–32.

Orlandi, N. (2012). Embedded seeing-as: multi-stable visual perception without interpretation. *Philosophical Psychology*, 25(4), 1–19.

Peacocke, C. (1992). *A Study of Concepts*. Cambridge, MA: MIT Press.

Perkins, M. (1983). *Sensing the World*. Indianapolis, IN: Hackett Publishing.

Prinz, J. (2006). Putting the brakes on enactive perception. *Psyche*, 12(1), 1–19. Microsoft Word - Prinz.doc (journalpsyche.org)

Ratcliffe, M. (2012). What is touch? *Australasian Journal of Philosophy*, 90(3), 413–32.

Ricci, C. and Blundo, C. (1990). Perception of ambiguous figures after focal brain lesions. *Neuropsychologia*, 28(11), 1163–73.

Richardson, L. (2013). Sniffing and smelling. *Philosophical Studies*, 162(2), 401–19.

Rosch, E. (1975). Principles of categorization. In *Cognition and Categorization*, edited by E. Rosch and B. B. Lloyd, 27–48. Hillsdale, NJ: Lawrence Erlbaum.

Schellenberg, S. (2008). The situation-dependency of perception. *Journal of Philosophy*, 105(2), 55–84.

Schwenkler, J. (2014). Review of Fulkerson (2014a). Notre Dame. *Philosophical Reviews, 7*.

Sorensen, R. (2009). Hearing silence: the perception and introspection of absences. In *Sounds and Perception: New Philosophical Essays*, edited by M. Nudds and C. O'Callaghan, 126–45. Oxford: Oxford University Press.

Susser, E., Sprecher, E., and Yarnitsky, D. (1999). Paradoxical heat sensation in healthy subjects: peripherally conducted by Aδ or C fibres? *Brain*, 122(2), 239–46.

Trout, J. D. (2019). Speech and its relations to touch. Talk delivered at Australian National University, 11 June 2019.

Wittgenstein, L. (1953). *Philosophical Investigations*. Oxford: Basil Blackwell.

12
Human olfaction: A view from the top

Stephen Pierzchajlo and Jonas K. Olofsson

1 Introduction

Whenever you enter someone's home for the first time, you are for a brief moment struck by how it smells—the unique molecular composition of lingering food odours, furniture and building materials, perfumes, and body odours. Detecting abrupt changes constitutes a distinctive feature of olfaction in daily life. In the following, we will discuss how olfactory perception is generated at the interface between sensory stimulation and cognition—the mental processes of acquiring knowledge and understanding through thought, experience, and the senses. The research literature on olfaction has often been concerned with the limitations of cognition and the dominance of bottom-up sensory processes (e.g., Jönsson and Olsson, 2003; Yeshurun and Sobel, 2010). And while olfaction is, in some ways, cognitively limited relative to visual or auditory senses (Cain *et al.*, 1998; Olofsson and Gottfried, 2015), here we will argue that people are highly capable of using environmental cues to predict incoming odours (Zelano *et al.*, 2011), and that odour perception is, to a significant extent, shaped by the process of matching sensory inputs to these predicted inputs. We will focus on evaluating a 'novelty detection' account (Köster *et al.*, 2014) that explicitly posits no major role for explicit cognitive representation of odours. We will evaluate how novelty detection and predictive coding frameworks may apply to olfaction, and provide suggestions for future research.

2 Novelty processing

Distinct properties of stimuli and events influence attentional allocation; large, fast-moving, or emotionally salient stimuli are prioritized (Kuraoka and Nakamura, 2007; Öhman *et al.*, 2001; Sander and Scheich, 2001; Treisman and Gormican, 1988; Zald and Pardo, 2002). Unexpected stimuli represent another form of attentional capture (Ranganath and Rainer, 2003). Brain responses typically change as a stimulus is repeated (Figure 12.1A; Kafkas and Montaldi, 2018).

Early neuroimaging research (Tulving *et al.*, 1994) used positron emission tomography (PET) to estimate brain activation while participants viewed previously encoded (familiar) or novel images. Instead of just subtracting the PET activation of 'new' images from 'old' images to understand the increased brain activation associated with memory, Tulving and colleagues explored the opposite contrast. Seemingly at odds with fundamental assumptions of memory research at the time, the hippocampus (an area vitally

Stephen Pierzchajlo and Jonas K. Olofsson, *Human olfaction* In: *Sensory Individuals.* Edited by: Aleksandra Mroczko-Wąsowicz and Rick Grush, Oxford University Press. © Oxford University Press 2023. DOI: 10.1093/oso/9780198866305.003.0013

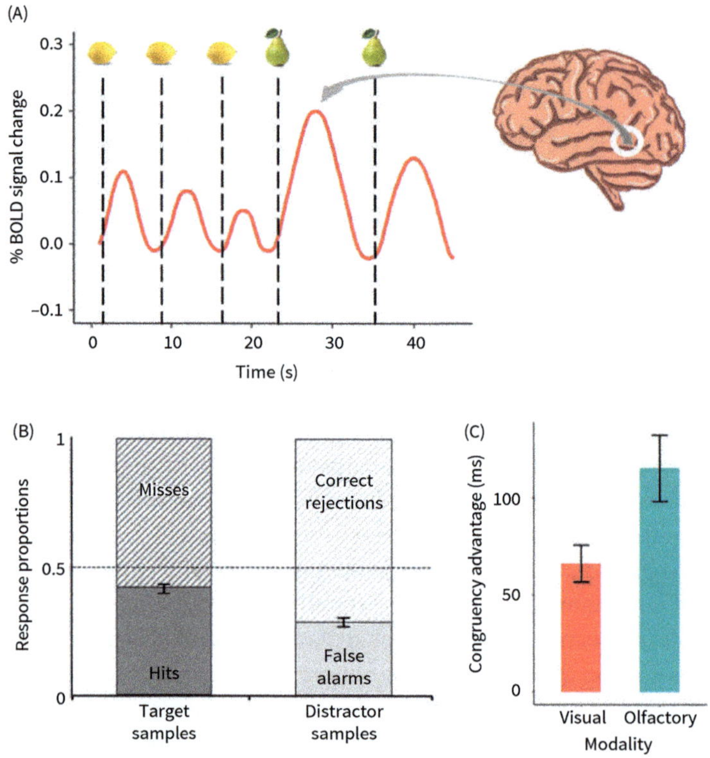

Figure 12.1 (A) An illustration of novelty processing. The red line represents the haemodynamic response function where visual objects are processed; a reduced cortical activation to a successively presented visual stimulus (lemon), followed by a greater activation to a novel stimulus (pear). (B) Combined data from the meta-analysis by Morin-Audebrand and colleagues (2012; adapted with permission). On average, participants are better at determining when an odour was not previously present (correct rejections), compared to their ability to determine when an odour was previously present (hits). (C) Reaction times (in milliseconds) are plotted for visual and olfactory targets (differences between cue-incongruent and cue-congruent trials (Pierzchajlo *et al.*, 2021).

important for memory) was more activated for new images (Tulving *et al.*, 1994). Around the same time, several single-cell recording primate studies reported similar results (Fahy *et al.*, 1993; Li *et al.*, 1993). Tulving *et al.* (1994) thus proposed that the hippocampus was activated by novel stimuli; this later became an established feature of memory function (Tulving, 2010).

The 'novelty/encoding hypothesis' (Markowitsch *et al.*, 1996; Tulving and Kroll, 1995) proposed that novel visual stimuli are detected by the hippocampus/limbic system, which triggers complex higher-level encoding operations, while familiar visual stimuli are stored without additional processing. Kumaran and Maguire (2007) instead hypothesized that 'the hippocampus may generate predictions about how future events will unfold, and critically detect when these expectancies are violated, even when task demands do not require it'. In response, Tulving (2010) argued that repeated stimulation leads to reduced efficacy in cortical processing, and that novelty-related increases result from automatic processes.

Novelty is often studied in repetition suppression experiments where stimulus repetitions result in the decreasing amplitude of cortical responses (Ringo, 1996), a reflection of the decrease in stimulus novelty. Repetition suppression effects are thought to be caused by neuronal adaptation; the specific location of a repetition suppression effect depends on the feature being adapted to (Sanchez-Vives *et al.*, 2000; Sohal and Hasselmo, 2000).

Research on novelty has highlighted medial temporal lobe (MTL) regions (including the hippocampus, perirhinal cortex, entorhinal cortex, parahippocampal cortices, and amygdala) in processing of novel words, objects, scenes, and faces (Kafkas *et al.*, 2017; Staresina *et al.*, 2012). Since the 1960s, electroencephalographic (EEG) research has also established that event-related potentials such as the P300 are affected by novelty, including stimuli that are unfamiliar, deviate from expectation, or are task-irrelevant (Barry *et al.*, 2016; Sutton *et al.*, 1965); novelty does not just refer to one type of event. Similar effects of novelty are found in human and non-human mammals (Paller, 1994). Novelty responses do not necessarily confer a behavioural advantage; in one study, responses to novel targets (previously unseen pictures) were associated with delayed P300 brain responses and slower behavioural response times, compared to when the same targets were repeated subsequently (Olofsson and Polich, 2007). Different senses have similar brain responses to novel stimuli (Knight, 1996) and cortical responses are similarly diminished when attention is directed away from the stimuli—this includes olfaction (Andersson *et al.*, 2018). We conclude that novelty and its basis in adaptation processes are fundamental features of the central nervous system, that there are similarities across senses, and that novelty may confer a behavioural disadvantage.

2.1 Olfactory novelty detection

How does novelty detection apply to the olfactory system? The most developed account of olfactory novelty detection is, to our knowledge, that of Köster and colleagues (2014). The foundation of this account is the prominent sensory adaptation and habituation observed in olfactory perception. Odour exposure leads to a rapidly decreased sensitivity for the stimulated odour, and to some extent other odours. While olfactory adaptation varies across odours, high odour concentrations and long exposure times will increase adaptation effects (Dalton, 2000). Full recovery might take up to a minute or more. Early research also showed that stimulating one nostril with an odour dampens sensitivity in the other, even though their receptors and cortical projections are anatomically separated (Elsberg, 1935). This led researchers to realize that central mechanisms were partly responsible for the decreased sensitivity and called this central effect 'habituation', which was distinguished from sensory-level 'adaptation' (Köster and de Wijk, 1991). The novelty detection account that we will focus on here may be viewed as a generalization of findings from olfactory adaptation and habituation research, to account also for other odour memory research findings. According to the novelty detection account, we typically experience odours implicitly and without much conscious attention. The sense of smell functions mainly as a warning system that alerts us of airborne pathogens or stops us from ingesting contaminated food. Familiar or expected odours are thus ignored, and only novel or unexpected odours are noticed. Another indication that olfaction does not seem to prioritize known odours is the finding that olfaction is a sense with 'complete adaptation' following sustained stimulation (Köster and de Wijk, 1991). Conversely, full visual adaptation does not occur because, the authors argue, visual object representation is always important for spatial navigation and object identification (Köster *et al.*, 2014). The authors thus formulated the 'misfit theory of spontaneous conscious odour perception' (henceforth called the 'misfit' account) to outline what they perceive to be olfaction's major (or potentially only) contribution to human perception.

To evaluate the misfit account, we first need to define 'odour identification'—a cognitive process that involves explicitly knowing the odour source. While such knowledge can be achieved, in principle, with or without contextual cues (Schab, 1991), experimental researchers distinguish between 'odour naming', when an odour source is generated verbally without any supportive cues, and 'odour identification', when an odour is correctly associated to a non-olfactory (e.g., written, spoken, or depicted) response alternative (see, for example, Olofsson and Pierzchajlo, 2021). These tasks are considered declarative, explicit, knowledge-based assessments of odour memory, and odour identification is incorporated in standardized olfactory assessments (Oleszkiewicz *et al.*, 2019).

According to the misfit account, odour identification and naming are not a crucial function of human olfaction. Köster and colleagues (2014) state that 'Odors are not meant to be objectified and identified and therefore we are so bad at it.' Instead, they argue that odours are typically processed implicitly—as unconscious associations between the odour and the context where it occurred—rather than involving explicit knowledge of odours themselves. Odours that are congruent with these associations are not consciously evaluated. 'Misfits' only occur when there is a mismatch between the implicit association between the odour and the context. These contextually deviant odours demand cognitive processing, whereas expected odours remain implicit and unnoticed (Köster *et al.*, 2014). The misfit account seemingly denies the existence or relevance of proactive, explicit cognitive processes on odour perception. Instead, olfactory perception is viewed as reactive, driven by bottom-up change detection processes. Köster and colleagues question the need for an explicit odour identification/representation system, stating that 'characteristics of the olfactory engram do not come into play in recognition at all'.

The strongest evidence in support of the misfit framework (Köster *et al.*, 2014) comes from a specific food memory paradigm. In this paradigm, participants incidentally encode odour memories during breakfast without knowing their memory of the odours will later be tested. Participants are later presented with odours or tastes of food items that either were (target) or were not (distractor) present at the breakfast and must try to correctly remember as many targets as possible. The results from such experiments are interpreted as demonstrating that even though people have weak abilities to recognize odours that were previously present, we are simultaneously proficient at determining when an odour is novel. The authors argue, based on the results, that representations of odour identities are not important for olfactory recognition. As Köster *et al.* (2014) put it, 'what is not there ... cannot be interfered with or forgotten, but novelty is always functional'. Because the authors argue that memory traces of previously encountered olfactory stimuli are not present, we take this view as the authors suggesting that odour identities are not represented in the mind.

We argue that although the misfit account of olfaction captures some important characteristics of odour perception, the account can be questioned on empirical and theoretical grounds. Our concerns are outlined in the following. A meta-analysis was published by Morin-Audebrand and colleagues (2012), who synthesized results from six research articles using the food memory paradigm—a key source of evidence for the misfit framework (Köster *et al.*, 2014). A consistent result in all studies is that participants, when asked, had difficulty determining which odours and tastes of food were present during the previous meal, but they were more accurate at determining the odours and tastes that were not present (Figure 12.1B). Morin-Audebrand *et al.* (2012) argue this demonstrates that olfactory memory is not concerned with storing representations of odours, and instead evolved only to detect odours that are novel.

2.2 A critique of the misfit account

We argue that the evidence provided (Morin-Audebrand *et al.*, 2012), even if accepted at face value, is insufficient to downplay the importance of explicit cognitive olfactory representations. Converging evidence is required from a range of experimental contexts to substantiate such general claims (Yarkoni, 2020). Here, the claim that olfactory cognition uniquely prioritizes novel over familiar stimuli relies primarily on evidence from a specific experimental design. The key finding of better performance for 'correct rejections' compared to 'hits' (Figure 12.1B) is, in fact, not found in other types of odour memory experiments. Traditional odour recognition memory experiments show instead that participants often have a high 'hit rate' (recognizing that an odour was 'old'—that is, presented during a prior encoding phase), but that novel odours are often misclassified as 'old'; some experiments even show that olfaction is even more prone to such outcomes than vision (e.g., Cornell Kärnekull *et al.*, 2015). Odour recognition memory is often facilitated when the odour is familiar—presumably, relying on a well-established declarative memory trace (Larsson, 1997; Murphy *et al.*, 1991; Rabin and Cain, 1984). These results suggest that the misfit account might not even generalize within the episodic odour memory domain.

Novelty detection is unquestionably an important feature of olfaction—novel odours attract attention, while sustained odours lead to adaptation. However, we argue that the misfit account erroneously invokes contextual novelty (i.e., where an odour does not fit expectations about the environment) as a similar bottom-up process (Köster *et al.*, 2014), when this should, in fact, be conceptualized as a top-down mechanism—a (explicit or implicit) cognitive expectation that may be confirmed or contradicted by an incoming odour stimulus. Thus, while the novelty detection framework denies explicit cognitive odour representations, we will instead argue that such 'top-down' processes are present and potentially important in odour perception.

3 Predictive coding

Hermann von Helmholtz (1860) reasoned that perception is a probabilistic process where sensory systems attempt to infer the causes of bodily effects. His work influenced later scholars (Gregory, 1980; Neisser, 1967; Yuille and Kersten, 2006) who argued that the brain does not build models of the world by piecing together bottom-up information, but instead predicts current sensory inputs from a list of possible causes (i.e., creating models that can be compared). Predictive coding describes a process whereby higher-level brain areas predict inputs occurring at lower levels (Clark, 2013). These predictions are compared to the actual activation in lower-level sensory circuits. This comparison typically results in some level of discrepancy—the prediction error, which is summed and transmitted back up to the level in the hierarchy where the prediction originated (Rao and Ballard, 1999). This leads to a reformulation of the current model by updating the prior probability of the sensory data occurring at the level where the prediction error was made. This process is repeated iteratively and continuously until the discrepancy between prediction and sensory input has been sufficiently minimized. Such computational error correction mechanisms are, according to this framework, central to mammalian brain function.

How do prior expectations about stimuli shape perception? They are known to bias perception; for instance, in a visual illusion based on the expectation that light comes from above, circles appear concave or convex, depending on where a shadow is cast (Oliva and Torralba, 2007). Organisms form prior probability models about frequent events/stimuli to aid their survival (de Lange *et al.*, 2018). Prior probability models are likely encoded in the neural tuning properties of the sensory cortices themselves (Cloherty *et al.*, 2016). For example, cardinal (horizontal and vertical) orientations are more prevalent than other orientations in natural images and, unsurprisingly, visual neurons that are tuned for recognizing cardinal orientations are most represented in the early visual cortex (Albright, 1989). Additionally, prior experience within a specific context can give an ambiguous stimulus a more concrete identity. For instance, an otherwise amorphous blob can appear to be the Statue of Liberty when appropriately placed in an image of Manhattan and its surrounding rivers (Lavenex and Amaral, 2000).

3.1 Cross-sensory predictions in olfaction

There is strong evidence that olfactory perception and cognitive processes make use of predictive coding mechanisms. Some olfactory capabilities, such as odour detection and recognition, are typically much slower than their visual analogues (Laing and MacLeod, 1992; Robinson, 1968). However, support from visual information is greatly beneficial for enhancing performance in olfactory tasks. That human olfaction might rely on other systems to make up for its limitations in identification is not a new idea. The comparative neurologist Charles Judson Herrick (1933) noted that mammalian olfaction strongly relies on non-olfactory cues to identify the source or location of an olfactory stimulus. In mammalian evolutionary development, the olfactory cortex declined relative to the total brain size (Moulton, 1967). The size of the olfactory cortex in primates thus became overshadowed by the neocortex, a consequence of evolution favouring higher cognitive functions and multisensory associations over sensory-specific analysis (Ribeiro *et al.*, 2014). Herrick (1933) hypothesized that the interconnectivity of olfactory with non-olfactory regions might point to a perceptual strategy unique to olfaction—that mammals, to a large extent, use information from other sensory systems to enhance olfactory processes.

A controversial issue in olfactory research is whether we perceive odours as 'objects'. In everyday language, an 'object' is typically something we see, and perception research has focused on visual objects and their features. Objects are similar to the notion of 'sensory individuals' discussed in other parts of this volume. Adherents of odour object perception points to the fact that odour perception usually involves complex blends which are processed configurally, such that neural responses to individual molecules are associated and integrated through a learning process that eventually results in a holistic 'object' (Gottfried, 2010; Stevenson and Wilson, 2007). Individual odour components are typically not perceptually separable (Thomas-Danguin *et al.*, 2014). Odour objects are considered a form of perceptual memory representation, involving the piriform cortex (Wilson and Stevenson, 2003; Wilson and Sullivan, 2011).

Olfactory coding differs from vision in a number of ways, leading some philosophers to dismiss any notion of perceptual 'odour objects' (Barwich, 2019; Keller, 2017). It is obvious that many of the criteria for 'objecthood' in visual perception will not be met by olfactory perception. The visual cortex is, for example, topologically organized, and stimuli are processed in an extensive hierarchy of visual associative areas which retains their edges, colours, and other individual features (Singer and Gray, 1995; Tanaka, 1996). However, we emphasize that objects are things in the world, and our perceptual systems—visual, auditory, olfactory, etc.—only generate representations thereof. Different senses may provide different types of representations, but they may nevertheless represent the same object. A more inclusive definition of objecthood might instead encompass perceptual content sufficient to establish a representation of the external object and support its identification. As Gottfried (2010) notes, an object does not always have to be seen in order to be identified; a lion, for example, can also be identified from its distinctive smell or roar; these senses provide *information about the object*. Although a more liberal definition of objecthood might not, in itself, be strong enough to distinguish odour *objects* from mere *qualities*, the existence of configural odour processing, as discussed above, does add explanatory power, and thus relevance, to the notion of odour objects. Furthermore, odours are often extremely difficult to identify in the absence of any other sensory information (Cain, 1979; Engen and Ross, 1973; Laing and Francis, 1989; Livermore and Laing, 1996). We suggest that olfaction might be uniquely dependent on top-down processes and multisensory integration, in order to compensate for its perceptual limitations. In our view, the notion of odour objects captures these characteristic features of olfaction.

How might novelty detection accounts be empirically differentiated from predictive coding accounts? We believe that response time studies with predictive cues provide useful evidence. In support of the misfit account, Möller and colleagues (2007) reported that when two odours are presented in succession and participants are asked to decide whether they were the same or different, the different odour pairs were detected more rapidly than when the odours were the same. The results were interpreted as supporting the preferential processing of novel odours (Moller *et al.*, 2007). In our view, these results are instead likely influenced by sensory adaptation when the same odour is repeatedly presented. The first odour, desensitizing the olfactory receptors to the following stimulus, will cause a delayed response when the same odour is presented a second time. Sensory adaptation explains parsimoniously why different odour pairs are more rapidly detected. Thus, this result is not informative regarding the existence or roles of mental representations of odours or their use in perceptual predictions.

If novel odours were indeed prioritized behaviourally, as suggested in the misfit account, odours that are unexpected, given the experimental context, would be rapidly processed when sensory adaptation was eliminated as a potential confound. However, the opposite result is typically observed. Crossmodal cue-target experiments have repeatedly shown that when participants are primed with a word (e.g., 'lemon'), they are faster in responding 'yes' to a subsequent matching target odour (lemon) relative to responding 'no' to a non-matching (and hence contextually novel) target odour (e.g., rose; Olofsson *et al.*, 2012, 2013). In these experiments, participants do not have to retrieve the identity of the odour upon perceiving it, but just to indicate whether the percept matches or not with the expected odour percept, which was indicated by the identity cue. Inconsistent with a novelty detection advantage, there is a sizable delay (>100 ms) for incongruent (contextually novel) odours. In this experimental setup, the risk of sensory adaptation is avoided because the cue is presented in another sensory modality than the target. The same non-olfactory cues and olfactory targets are presented, with only their known semantic relationships varying. The advantageous behavioural processing for expected odours is proof that participants translate the meaning of the word cue (e.g., 'lemon') into a perceptual template of lemon odour, even before the odour is perceived, and this prediction facilitates the processing of the odour. This result would not be observed if the olfactory system had a general tendency to prioritize novel odours.

Cross-sensory cue-target experiments provide a method of comparing olfactory processing efficiency in terms of response times. They can be modified in a flexible way to study a variety of features, such as effects of sensory modality, stimulus set size, cue-target delay, stimulus probability, and/or which stimulus features are prompted by the cues. Research within this experimental paradigm has shown that odour source objects (e.g., 'lemon') are particularly effective in facilitating rapid and accurate responses, whereas cuing based on odour valence or edibility are less effective, suggesting that the source object may constitute a fundamental unit of odour processing (Olofsson, 2014). When odours are instead used as cues for a subsequent verbal target (e.g., a lemon odour cue primes the participant to a matching or non-matching target word), the unexpected word targets reactivate olfactory-associated brain regions, seemingly providing prediction error feedback to the olfactory system (Olofsson *et al.*, 2014). When odour cues are non-matching with the word targets, there is a behavioural delay, suggesting a need for more extensive processing of non-matching combinations (Olofsson *et al.*, 2014, 2018). Overall, congruent trials are faster than non-matching trials. Under some circumstances, but not always, this rapid processing might come at the expense of decreased accuracy (Olofsson *et al.*, 2014, 2018). But all our studies show a speed advantage for congruent crossmodal odour stimuli. In a recent study, Pierzchajlo *et al.* (2023) let participants determine whether or not a verbal cue matched a subsequent visual or olfactory target stimulus. The greatest congruency advantage occurred when the word cue was object-based (e.g., 'lemon') and the target was olfactory, rather than visual (Figure 12.1C). The olfactory system may be even more dependent on predictive cues (e.g., a word that cues perceptual objects) than the visual system, at least in object identification tasks, in line with the proposal that the human olfactory system effectively uses information from other senses in order to identify an odour (Herrick, 1933). This influence, we argue, comes in the form of crossmodal predictions that are imposed on the olfactory cortices.

3.2 Olfaction, viewed from the top

Recent theoretical work suggests that the combination of representational coarseness in olfaction, coupled with known top-down feedback connections from other cortices to the piriform cortex, indicates a route for odour-predictive visual information to enhance olfactory identification (Olofsson and Gottfried, 2015). The piriform cortex is considered critical for establishing configural odour percepts (Gottfried *et al.*, 2006; Howard *et al.*, 2009). In the absence of any other information, olfactory stimuli presumably activate a set of potential odour identity representations because the relatively coarse, configural nature of olfactory responses does not allow precise and multifaceted odour representations. Conversely, because visual stimuli are processed in an elaborate hierarchy of feature-selective cortical regions (Distler *et al.*, 1993; Saygin *et al.*, 2012; Van Essen *et al.*, 1992), the identity of a visual stimulus is rarely in doubt. However, because the ventral stream of the visual cortex contains feedback projections to various olfactory regions, Olofsson and Gottfried (2015) hypothesized that visual information effectively communicates predictions to the piriform cortex. Thus, an anticipatory template in the piriform cortex for a potential odour could be generated via congruent visual or auditory cues that allows the piriform cortex to increase the probability of successful odour identification by simply generating a pattern of activation that represents an analogous olfactory representation (see Zelano *et al.*, 2011). Importantly, these patterns of activation may be sustained for a few seconds during an anticipatory period, but decay quite rapidly when the odour does not match the neural template (Zelano *et al.*, 2011). Recent functional magnetic resonance imaging (fMRI) evidence suggests the posterior part of the piriform cortex functions as a 'multisensory binding site' where auditory and visual cues are integrated with concurrent odours (Porada *et al.*, 2019). In a recent study by Zhou *et al.* (2019), intracranial electrophysiological recordings were monitored in awake patients who matched target odours (rose and mint odours) to a preceding verbal cue ('rose' vs 'mint'). Results suggest that such multimodal integration is mediated by the phase-locking of low-frequency waves across auditory and olfactory brain regions when participants utilize a verbal cue to predict an olfactory target. The effect was observed in the period when the odour was anticipated, and its strength was positively associated with the behavioural accuracy on the matching task (Zhou *et al.*, 2019). These results strongly support our notion of predictions in the olfactory system, and suggests a cortical mechanism underlying top-down olfactory predictions.

3.3 Novelty detection or predictive coding?

How do we reconcile our account with the empirical finding that people are better at recognizing 'new' compared to 'old' (i.e., previously presented) food items in a flavour recognition task (Morin-Audebrand *et al.*, 2012)? We note that the flavour recognition task design does not involve any mental stimulus predictions. The 'misfit' account only recognizes odours from a bottom-up, change detection perspective. We argue that olfaction can also rely on explicit predictions regarding incoming odours; these predictions may engage odour templates to aid successful identification. Under circumstances tested within the 'misfit' account, in the absence of robust contextual predictions, it is possible that novel odours/flavours attract more processing resources than familiar odours/flavours. But such features of novel stimuli might also perhaps be observed for visual and auditory events, and it does not point to novelty being uniquely prioritized in olfaction. Additionally, our results from cue-target experiments do not fit with the general claims of novelty as a dominant feature in olfaction; instead, they support the notion that olfactory processing is highly dependent on cognitive top-down representations.

We conclude that the misfit account is, in our view, overgeneralized to the cognitive domain where it does not fit with available evidence. Cognitive olfactory processing is instead highly dependent on prior experiences and contextual predictions, making active use of crossmodal cues and memory traces to facilitate performance. However, we think that one cognitive aspect of the misfit account is accurate; we agree that novel, unexpected, or contextually incongruent odours demand more elaborate and effortful processing, sometimes at the expense of non-olfactory stimuli in a multisensory environment (Hörberg *et al.*, 2020). But importantly, olfactory novelty processes do not typically confer a behavioural advantage—instead it is characterized by effortful and slow processing.

4 Future directions

Olfactory researchers typically use 'novelty' as an umbrella term. It seems that researchers studying olfactory-related novelty detection often tie novelty to contextual parameters (as in Morin-Audebrand *et al.*, 2012) rather than to familiarity. In fact, authors of several papers even try to use targets and foils that are as similar as possible in order to remove the familiarity of stimuli as a confound (i.e., Mojet and Köster, 2005). Thus, these paradigms capture 'novelty' only in the sense that the novel stimuli are novel to the participants *in that specific context* and the familiar stimuli are familiar *in that specific context*. These paradigms do not necessarily capture expectation violations, however, as there is no reason to assume participants only expect to encounter food they previously ate. Complementary approaches are needed to study different aspects of novelty.

Predictive coding effects are similarly observed for familiar and unfamiliar stimuli, and they appear to often be generated automatically. In many predictive coding paradigms, regardless of the behavioural task, expectation violation is typically what drives the effects. The expectation, based on prior sensory cues or probabilities, is violated (i.e., Egner *et al.*, 2010; Kok *et al.*, 2012). We recommend that research on olfactory novelty detection involves systematic comparisons across sensory systems and uses insights from research outside of the olfactory domain—otherwise it will remain unclear if olfaction is really different from other senses. As we described, cross-sensory cue-target experiments might provide a generic platform to flexibly study several of these effects.

If olfaction is, in fact, primarily an alarm system used to detect potential threats in the environment, as is often claimed, it is useful to reflect on what data would support this conclusion. Outdoors, or in our homes, a sudden, unexpected odour would alert us to the presence of something unusual and potentially dangerous. It is difficult to experimentally assess human olfactory threat detection with high ecological validity. But if the ecological demands have shaped our core olfactory abilities, we should see evidence of that even in well-controlled experiments. The argument that all olfactory research needs to be conducted in situations closely resembling ecological situations (e.g., Köster *et al.*, 2014) is, in our view, mistaken. Ecological scenarios are informative, but often insufficient to understand the nature of the olfactory system, and carefully controlled experiments remain indispensable (Olofsson and Pierzchajlo, 2021). For example, an 'ecological approach' risks confusing statements about the *properties of olfaction* with statements about the *properties of our typical environments*. For example, we do not typically track a scented trail to its source, yet we are able to do so with some limited training (Porter *et al.*, 2007). In fact, for individuals living in dense rainforest environments where the visual environment is obscured, scent tracking might be a critical olfactory function (Majid and Kruspe, 2018). It would be mistaken to deny this olfactory capability as absent from the human repertoire. Denying explicit odour cognition is similarly mistaken; there are many cases in our everyday lives where explicit odour predictions likely come into play, for example in meal preparation and cooking, in monitoring clothes and body parts for malodour, and in attaching emotionally to our family members and pets.

In future research, different types of olfactory novelty and related concepts should be carefully operationalized and contrasted. We propose that familiarity (based on prior experience) and expectancy (based on experimental probability) should be assessed independently and compared—this will help us understand how novelty aspects influence olfactory processing. Furthermore, we should only accept theoretical claims of broad generalizability if they are broadly supported by broadly converging experimental results (Yarkoni, 2020); in our opinion, the olfactory misfit account does not meet this criterion.

In sum, we propose that the olfactory misfit account is fraught with shortcomings related to the small set of conditions under which novelty appears to be prioritized, issues regarding definitions of the novelty concept, and lack of consideration of suitable experiments to test the generalizability of results. Instead, we argue that the human mind is highly capable of integrating cues from the audiovisual senses to form explicit predictions that guide olfaction. We propose that top-down factors play profound roles in olfactory perception and cognition, and that adopting this view would benefit the understanding of human olfaction.

References

Albright, T. D. (1989). Centrifugal directional bias in the middle temporal visual area (MT) of the macaque. *Visual Neuroscience*, 2(2), 177–88. Available from: https://doi.org/10.1017/s0952523800012037

Andersson, L., Sandberg, P., Olofsson, J. K., and Nordin, S. (2018). Effects of task demands on olfactory, auditory, and visual event-related potentials suggest similar top-down modulation across senses. *Chemical Senses*, 43(2), 129–34. Available from: https://doi.org/10.1093/chemse/bjx082

Barry, R., Steiner, G., and De Blasio, F. (2016). Reinstating the novelty P3. *Scientific Reports*, 6, 31200. Available from: https://doi.org/10.1038/srep31200

Barwich, A.-S. (2019). A critique of olfactory objects. *Frontiers in Psychology*, 10, 1337. Available from: https://doi.org/10.3389/fpsyg.2019.01337

Cain, W. S. (1979). To know with the nose: keys to odor identification. *Science*, 203(4379), 467–70. Available from: https://doi.org/10.1126/science.760202

Cain, W. S., de Wijk, R., Lulejian, C., Schiet, F., and See, L.-C. (1998). Odor identification: perceptual and semantic dimensions. *Chemical Senses*, 23(3), 309–26. Available from: https://doi.org/10.1093/chemse/23.3.309

Clark, A. (2013). Whatever next? Predictive brains, situated agents, and the future of cognitive science. *Behavioral and Brain Sciences*, 36(3), 181–204. Available from: https://doi.org/10.1017/S0140525X12000477

Cloherty, S. L., Hughes, N. J., Hietanen, M. A., Bhagavatula, P. S., Goodhill, G. J., and Ibbotson, M. R. (2016). Sensory experience modifies feature map relationships in visual cortex. *eLife*, 5, e13911. Available from: https://doi.org/10.7554/eLife.13911

Cornell Kärnekull, S., Jönsson, F. U., Willander, J., Sikström, S., and Larsson, M. (2015). Long-term memory for odors: influences of familiarity and identification across 64 days. *Chemical Senses*, 40(4), 259–67. Available from: https://doi.org/10.1093/chemse/bjv003

Dalton, P. (2000). Psychophysical and behavioral characteristics of olfactory adaptation. *Chemical Senses*, 25(4), 487–92. Available from: https://doi.org/10.1093/chemse/25.4.487

de Lange, F. P., Heilbron, M., and Kok, P. (2018). How do expectations shape perception? *Trends in Cognitive Sciences*, 22(9), 764–79. Available from: https://doi.org/10.1016/j.tics.2018.06.002

Distler, C., Boussaoud, D., Desimone, R., and Ungerleider, L. G. (1993). Cortical connections of inferior temporal area TEO in macaque monkeys. *Journal of Comparative Neurology*, 334(1), 125–50. Available from: https://doi.org/10.1002/cne.903340111

Egner, T., Monti, J. M., and Summerfield, C. (2010). Expectation and surprise determine neural population responses in the ventral visual stream. *Journal of Neuroscience*, 30(49), 16601–8. Available from: https://doi.org/10.1523/JNEUROSCI.2770-10.2010

Elsberg, C. (1935). Olfactory fatigue. *Bulletin of the Neurological Institute of New York*, 4, 479–95.

Engen, T. and Ross, B. M. (1973). Long-term memory of odors with and without verbal descriptions. *Journal of Experimental Psychology*, 100(2), 221–7. Available from: https://doi.org/10.1037/h0035492

Fahy, F. L., Riches, I. P., and Brown, M. W. (1993). Neuronal activity related to visual recognition memory: long-term memory and the encoding of recency and familiarity information in the primate anterior and medial inferior temporal and rhinal cortex. *Experimental Brain Research*, 96(3), 457–72. Available from: https://doi.org/10.1007/BF00234113

Gottfried, J. A. (2010). Central mechanisms of odour object perception. *Nature Reviews Neuroscience*, 11(9), 628–41. Available from: https://doi.org/10.1038/nrn2883

Gottfried, J. A., Winston, J. S., and Dolan, R. J. (2006). Dissociable codes of odor quality and odorant structure in human piriform cortex. *Neuron*, 49(3), 467–79. Available from: https://doi.org/10.1016/j.neuron.2006.01.007

Gregory, R. L. (1980). Perceptions as hypotheses. *Philosophical Transactions of the Royal Society of London. Series B, Biological Sciences*, 290(1038), 181–97.

Herrick, C. J. (1933). The functions of the olfactory parts of the cerebral cortex. *Proceedings of the National Academy of Sciences of the United States of America*, 19(1), 7–14. Available from: https://doi.org/10.1073/pnas.19.1.7

Hörberg, T., Larsson, M., Ekström, I., Sandöy, C., Lundén, P., and Olofsson, J. K. (2020). Olfactory influences on visual categorization: behavioral and ERP evidence. *Cerebral Cortex*, 30(7), 4220–37. Available from: https://doi.org/10.1093/cercor/bhaa050

Howard, J. D., Plailly, J., Grueschow, M., Haynes, J.-D., and Gottfried, J. A. (2009). Odor quality coding and categorization in human posterior piriform cortex. *Nature Neuroscience*, 12(7), 932–8. Available from: https://doi.org/10.1038/nn.2324

Jönsson, F. U. and Olsson, M. J. (2003). Olfactory metacognition. *Chemical Senses*, 28(7), 651–8. Available from: https://doi.org/10.1093/chemse/bjg058

Kafkas, A., Migo, E. M., Morris, R. G., Kopelman, M. D., Montaldi, D., and Mayes, A. R. (2017). Material specificity drives medial temporal lobe familiarity but not hippocampal recollection: familiarity-based recognition in the MTL. *Hippocampus*, 27(2), 194–209. Available from: https://doi.org/10.1002/hipo.22683

Kafkas, A. and Montaldi, D. (2018). How do memory systems detect and respond to novelty? *Neuroscience Letters*, 680, 60–8. Available from: https://doi.org/10.1016/j.neulet.2018.01.053

Keller, A. (2017). Olfaction as the Paradigm for Perceptual Philosophy. [Doctoral dissertation, City University of New York, NY]. Available from:https://academicworks.cuny.edu/gc_etds/1004/

Knight, R. T. (1996). Contribution of human hippocampal region to novelty detection. *Nature*, 383(6597), 256–9. Available from: https://doi.org/10.1038/383256a0

Kok, P., Jehee, J. F. M., and de Lange, F. P. (2012). Less is more: expectation sharpens representations in the primary visual cortex. *Neuron*, 75(2), 265–70. Available from: https://doi.org/10.1016/j.neuron.2012.04.034

Köster, E. P., and de Wijk, R. A. (1991). Olfactory Adaptation. In *The Human Sense of Smell*, edited by D. G. Laing, R. L. Doty, and W. Breipohl, 199–215. Berlin: Springer. Available from: https://doi.org/10.1007/978-3-642-76223-9_10

Köster, E. P., Møller, P., and Mojet, J. (2014). A 'Misfit' Theory of Spontaneous Conscious Odor Perception (MITSCOP): reflections on the role and function of odor memory in everyday life. *Frontiers in Psychology*, 5, 1–12. Available from: https://doi.org/10.3389/fpsyg.2014.00064

Kumaran, D. and Maguire, E. A. (2007). Match mismatch processes underlie human hippocampal responses to associative novelty. *Journal of Neuroscience*, 27(32), 8517–24. Available from: https://doi.org/10.1523/JNEUROSCI.1677-07.2007

Kuraoka, K. and Nakamura, K. (2007). Responses of single neurons in monkey amygdala to facial and vocal emotions. *Journal of Neurophysiology*, 97(2), 1379–87. Available from: https://doi.org/10.1152/jn.00464.2006

Laing, D. G. and Francis, G. W. (1989). The capacity of humans to identify odors in mixtures. *Physiology and Behavior*, 46(5), 809–14. Available from: https://doi.org/10.1016/0031-9384(89)90041-3

Laing, D. G. and MacLeod, P. (1992). Reaction time for the recognition of odor quality. *Chemical Senses*, 17(3), 337–46. Available from: https://doi.org/10.1093/chemse/17.3.337

Larsson M. (1997). Semantic factors in episodic recognition of common odors in early and late adulthood: a review. *Chemical Senses*, 22(6), 623–33. Available from: https://doi.org/10.1093/chemse/22.6.623

Lavenex, P. and Amaral, D. G. (2000). Hippocampal–neocortical interaction: a hierarchy of associativity. *Hippocampus*, 10(4), 420–30. Available from: https://doi.org/10.1002/1098-1063(2000)10:4<420::AID-HIPO8>3.0.CO;2-5

Li, L., Miller, E. K., and Desimone, R. (1993). The representation of stimulus familiarity in anterior inferior temporal cortex. *Journal of Neurophysiology*, 69(6), 1918–29.

Livermore, A. and Laing, D. G. (1996). Influence of training and experience on the perception of multicomponent odor mixtures. *Journal of Experimental Psychology: Human Perception and Performance*, 22(2), 267–77.

Majid, A. and Kruspe, N. (2018). Hunter–gatherer olfaction is special. Current *B*iology, 28(3), 409–13.e2. Available from: https://doi.org/10.1016/j.cub.2017.12.014

Markowitsch, H. J., Craik, F. E., Habib, R., and Houle, S. (1996). Novelty and familiarity activations in PET studies of memory encoding and retrieval. *Cerebral Cortex*, 6(1), 71–9. Available from: https://doi.org/10.1093/cercor/6.1.71

Mojet, J. and Köster, E. P. (2005). Sensory memory and food texture. *Food Quality and Preference*, 16, 251–66. Available from: https://doi:10.1016/j.foodqual.2004.04.017

Moller, P., Mojet, J., and Koster, E. P. (2007). Incidental and intentional flavor memory in young and older subjects. *Chemical Senses*, 32(6), 557–67. Available from: https://doi.org/10.1093/chemse/bjm026

Morin-Audebrand, L., Mojet, J., Chabanet, C., *et al.* (2012). The role of novelty detection in food memory. *Acta Psychologica*, 139(1), 233–8. Available from: https://doi.org/10.1016/j.actpsy.2011.10.003

Moulton, D. G. (1967). Olfaction in mammals. *American Zoologist*, 7(3), 421–9. Available from: https://doi.org/10.1093/icb/7.3.421

Murphy, C., Cain, W. S., Gilmore, M. M., and Skinner, R. B. (1991). Sensory and semantic factors in recognition memory for odors and graphic stimuli: elderly versus young persons. *American Journal of Psychology*, 104(2), 161–92.

Neisser, U. (1967). *Cognitive Psychology*. New York, NY: Appleton-Century-Crofts.

Öhman, A., Lundqvist, D., and Esteves, F. (2001). The face in the crowd revisited: a threat advantage with schematic stimuli. *Journal of Personality and Social Psychology*, 80(3), 381–96. Available from: https://doi.org/10.1037/0022-3514.80.3.381

Oleszkiewicz, A., Schriever, V. A., Croy, I., Hähner, A., and Hummel, T. (2019). Updated Sniffin' Sticks normative data based on an extended sample of 9139 subjects. *European Archives of Oto-Rhino-Laryngology*, 276(3), 719–28. Available from: https://doi.org/10.1007/s00405-018-5248-1

Oliva, A. and Torralba, A. (2007). The role of context in object recognition. *Trends in Cognitive Sciences*, 11(12), 520–7. Available from: https://doi.org/10.1016/j.tics.2007.09.009

Olofsson, J. K. (2014). Time to smell: a cascade model of human olfactory perception based on response-time (RT) measurement. *Frontiers in Psychology*, 5, 1–8. Available from: https://doi.org/10.3389/fpsyg.2014.00033

Olofsson, J. K., Bowman, N. E., and Gottfried, J. A. (2013). High and low roads to odor valence? A choice response-time study. *Journal of Experimental Psychology: Human Perception and Performance*, 39(5), 1205–11. Available from: https://doi.org/10.1037/a0033682

Olofsson, J. K., Bowman, N. E., Khatibi, K., and Gottfried, J. A. (2012). A time-based account of the perception of odor objects and valences. *Psychological Science*, 23(10), 1224–32. Available from: https://doi.org/10.1177/0956797612441951

Olofsson, J. K. and Gottfried, J. A. (2015). The muted sense: neurocognitive limitations of olfactory language. *Trends in Cognitive Sciences*, 19(6), 314–21. Available from: https://doi.org/10.1016/j.tics.2015.04.007

Olofsson, J. K., Hurley, R. S., Bowman, N. E., Bao, X., Mesulam, M.-M., and Gottfried, J. A. (2014). A designated odor–language integration system in the human brain. *Journal of Neuroscience*, 34(45), 14864–73. Available from: https://doi.org/10.1523/JNEUROSCI.2247-14.2014

Olofsson, J. K. and Pierzchajlo, S. (2021). Olfactory language: context is everything. Trends in Cognitive Sciences, 25(6), 419–20. Available from: https://doi.org/10.1016/j.tics.2021.02.004

Olofsson, J. K. and Polich, J. (2007). Affective visual event-related potentials: arousal, repetition, and time-on-task. *Biological Psychology*, 75(1), 101–8. Available from: https://doi.org/10.1016/j.biopsycho.2006.12.006

Olofsson, J. K., Syrjänen, E., Ekström, I., Larsson, M., and Wiens, S. (2018). 'Fast' versus 'slow' word integration of visual and olfactory objects: EEG biomarkers of decision speed variability. *Behavioral Neuroscience*, 132(6), 587–94. Available from: https://doi.org/10.1037/bne0000266

Paller, K. (1994). The neural substrates of cognitive event-related potentials: a review of animal models of P3. In *Cognitive Electrophysiology*, edited by H.-J. Heinze, T. F. Münte, and G. R. Mangun, 300–3. Boston, MA: Birkhauser.

Pierzchajlo, S., Jernsäther, T., Fontana, L., Zampini, M., Almeida, R., and Olofsson, J. (2023, January 11). A predictive nose: Olfactory identification is more reliant than visual identification on predictive object cues. Retrieved from psyarxiv.com/tjzyp

Porada, D. K., Regenbogen, C., Seubert, J., Freiherr, J., and Lundström, J. N. (2019). Multisensory enhancement of odor object processing in primary olfactory cortex. *Neuroscience*, 418, 254–65. Available from: https://doi.org/10.1016/j.neuroscience.2019.08.040

Porter, J., Craven, B., Khan, R. M., *et al.* (2007). Mechanisms of scent-tracking in humans. *Nature Neuroscience*, 10(1), 27–9. Available from: https://doi.org/10.1038/nn1819

Rabin, M. D. and Cain, W. S. (1984). Odor recognition: familiarity, identifiability, and encoding consistency. *Journal of Experimental Psychology. Learning, Memory, and Cognition*, 10(2), 316–25. Available from: https://doi.org/10.1037//0278-7393.10.2.316

Ranganath, C. and Rainer, G. (2003). Neural mechanisms for detecting and remembering novel events. *Nature Reviews Neuroscience*, 4(3), 193–202. Available from: https://doi.org/10.1038/nrn1052

Rao, R. P. N. and Ballard, D. H. (1999). Predictive coding in the visual cortex: a functional interpretation of some extra-classical receptive-field effects. *Nature Neuroscience*, 2(1), 79–87. Available from: https://doi.org/10.1038/4580

Ribeiro, P. F. M., Manger, P. R., Catania, K. C., Kaas, J. H., and Herculano-Houzel, S. (2014). Greater addition of neurons to the olfactory bulb than to the cerebral cortex of eulipotyphlans but not rodents, afrotherians or primates. *Frontiers in Neuroanatomy*, 8, 1–12. Available from: https://doi.org/10.3389/fnana.2014.00023

Ringo, J. L. (1996). Stimulus specific adaptation in inferior temporal and medial temporal cortex of the monkey. *Behavioural Brain Research*, 76(1), 191–7. Available from: https://doi.org/10.1016/0166-4328(95)00197-2

Robinson, D. A. (1968). Eye movement control in primates. *Science*, 161(3847), 1219–24. Available from: https://doi.org/10.1126/science.161.3847.1219

Sanchez-Vives, M. V., Nowak, L. G., and McCormick, D. A. (2000). Cellular mechanisms of long-lasting adaptation in visual cortical neurons in vitro. *Journal of Neuroscience*, 20(11), 4286–99. Available from: https://doi.org/10.1523/JNEUROSCI.20-11-04286.2000

Sander, K. and Scheich, H. (2001). Auditory perception of laughing and crying activates human amygdala regardless of attentional state. *Cognitive Brain Research*, 12(2), 181–98. Available from: https://doi.org/10.1016/S0926-6410(01)00045-3

Saygin, Z. M., Osher, D. E., Koldewyn, K., Reynolds, G., Gabrieli, J. D. E., and Saxe, R. R. (2012). Anatomical connectivity patterns predict face selectivity in the fusiform gyrus. *Nature Neuroscience*, 15(2), 321–7. Available from: https://doi.org/10.1038/nn.3001

Schab, F. R. (1991). Odor memory: taking stock. *Psychological Bulletin*, 109(2), 242–51. Available from: https://doi.org/10.1037/0033-2909.109.2.242

Singer, W. and Gray, C. M. (1995). Visual feature integration and the temporal correlation hypothesis. *Annual Review of Neuroscience*, 18(1), 555–86. Available from: https://doi.org/10.1146/annurev.ne.18.030195.003011

Sohal, V. S. and Hasselmo, M. E. (2000). A model for experience-dependent changes in the responses of inferotemporal neurons. *Network: Computation in Neural Systems*, 11(3), 169–90. Available from: https://doi.org/10.1088/0954-898X_11_3_301

Staresina, B. P., Fell, J., Do Lam, A. T. A., Axmacher, N., and Henson, R. N. (2012). Memory signals are temporally dissociated in and across human hippocampus and perirhinal cortex. *Nature Neuroscience*, 15(8), 1167–73. Available from: https://doi.org/10.1038/nn.3154

Stevenson, R. J. and Wilson, D. A. (2007). Odour perception: an object-recognition approach. *Perception*, 36(12), 1821–33. Available from: https://doi.org/10.1068/p5563

Sutton, S., Braren, M., Zubin, J., and John, E. R. (1965). Evoked-potential correlates of stimulus uncertainty. *Science*, 150(3700), 1187–8. Available from: https://doi.org/10.1126/science.150.3700.1187

Tanaka, K. (1996). Inferotemporal cortex and object vision. *Annual Review of Neuroscience*, 19(1), 109–39. Available from: https://doi.org/10.1146/annurev.ne.19.030196.000545

Thomas-Danguin, T., Sinding, C., Romagny, S., *et al.* (2014). The perception of odor objects in everyday life: a review on the processing of odor mixtures. *Frontiers in Psychology*, 5, 1–18. Available from: https://doi.org/10.3389/fpsyg.2014.00504

Treisman, A. and Gormican, S. (1988). Feature analysis in early vision: evidence from search asymmetries. *Psychological Review*, 95(1), 15–48.

Tulving, E. (2010). How do brains detect novelty? In *Memory, Aging and the Brain: A Festschrift in Honour of Lars-Göran Nilsson*, edited by L. Bäckman and L. Nyberg, 92–112. New York, NY: Psychology Press.

Tulving, E. and Kroll, N. (1995). Novelty assessment in the brain and long-term memory encoding. *Psychonomic Bulletin and Review*, 2(3), 387–90. Available from: https://doi.org/10.3758/BF03210977

Tulving, E., Markowitsch, H. J., Kapur, S., Habib, R., and Houle, S. (1994). Novelty encoding networks in the human brain: positron emission tomography data. *Neuroreport*, 5(18), 2525–8. Available from: https://doi.org/10.1097/00001756-199412000-00030

Van Essen, D., Anderson, C., and Felleman, D. (1992). Information processing in the primate visual system: an integrated systems perspective. *Science*, 255(5043), 419–23. Available from: https://doi.org/10.1126/science.1734518

Von Helmholtz, H. (1860). Treatise on physiological optics, Bd. 3: the perceptions of vision. Available from: https://echo.mpiwg-berlin.mpg.de/ECHOdocuView?url=/permanent/library/HS7FH69N/pageimgandpn=5andmode=imagepath

Wilson, D. A. and Stevenson, R. J. (2003). Olfactory perceptual learning: the critical role of memory in odor discrimination. *Neuroscience and Biobehavioral Reviews*, 27(4), 307–28. Available from: https://doi.org/10.1016/S0149-7634(03)00050-2

Wilson, D. A. and Sullivan, R. M. (2011). Cortical processing of odor objects. *Neuron*, 72(4), 506–19. Available from: https://doi.org/10.1016/j.neuron.2011.10.027

Yarkoni, T. (2020). The generalizability crisis. *Behavioral and Brain Sciences*, 45, e1. Available from: https://doi.org/10.1017/S0140525X20001685

Yeshurun, Y. and Sobel, N. (2010). An odor is not worth a thousand words: from multidimensional odors to unidimensional odor objects. *Annual Review of Psychology*, 61(1), 219–41. Available from: https://doi.org/10.1146/annurev.psych.60.110707.163639

Yuille, A. and Kersten, D. (2006). Vision as Bayesian inference: analysis by synthesis? *Trends in Cognitive Sciences*, 10(7), 301–8. Available from: https://doi.org/10.1016/j.tics.2006.05.002

Zald, D. H. and Pardo, J. V. (2002). The neural correlates of aversive auditory stimulation. *NeuroImage*, 16(3), 746–53. Available from: https://doi.org/10.1006/nimg.2002.1115

Zelano, C., Mohanty, A., and Gottfried, J. A. (2011). Olfactory predictive codes and stimulus templates in piriform cortex. *Neuron*, 72(1), 178–87. Available from: https://doi.org/10.1016/j.neuron.2011.08.010

Zhou, G., Lane, G., Noto, T., *et al.* (2019). Human olfactory-auditory integration requires phase synchrony between sensory cortices. *Nature Communications*, 10(1), 1168. Available from: https://doi.org/10.1038/s41467-019-09091-3

13
A world of odours

Clare Batty and Barry C. Smith

Can we conceive of an objective world of reidentifiable particulars that are not material bodies? A world that consisted entirely of sounds or odours? The former is considered by P. F. Strawson in Chapter Two of *Individuals* and discussed at length by Gareth Evans (1985) in "Things Without the Mind". A world that consists entirely of odours is considered both theoretically and empirically by Jay Gottfried and colleagues in Bao *et al.* (2019) who ask whether humans can navigate a sensory space informed by odour cues and, on that basis alone, form a cognitive map of a 2D space of odours. To investigate this idea, they 'posit a world populated exclusively with odor stimuli to determine whether and how a navigator—with only the luxury of the sense of smell—can traverse an olfactory landscape' (Bao *et al.*, 2019, p. 1066).

In considering the question of whether a subject, endowed with hearing alone, could conceive of an objective world of sounds, Strawson and Evans rely on thought experiments: an 'armchair' method used to test the possibility of things being a certain way under certain imagined conditions. Thought experiments enable us to test the commitments of different theories—what they allow for, or rule out—and help us to determine what is and is not possible. Non-philosophers are often suspicious of thought experiments, questioning their validity and their 'point'. However, one could consider empirical experiments to be thought experiments carried out by researchers who get out of the armchair to determine not just what is possible, but also what is actual. After all, such studies begin with an experimental design. We can see these parallels when considering the striking similarities between Strawson's discussion in *Individuals*, of whether a being whose experience is entirely auditory can conceive of an objective world of sounds, and the intuition that frames the discussion in Bao et al's article, of whether an agent equipped only with olfaction could navigate a world exclusively populated with odours.

What could be the philosophical upshot of comparing Strawson's thought experiment with this actual experiment? Contemporary philosophical discussions of olfaction have been dominated by two issues: (1) whether olfactory experience is spatial; and (2) whether it represents objects. Although it might seem that these two questions are related, discussion has centred on whether the representation of space is necessary for object representation in olfaction. In this paper, we introduce an olfactory analogue of Strawson's Sound World into this discussion. The comparison between Strawson's thought experiment and a real empirical experiment is instructive here. By drawing the parallel between Strawson's discussion of an aspatial world in *Individuals* and the results of the experiments of Bao *et al.*'s study, we are able to consider the role that olfactory experience itself plays, or could play, in a conception of an objective world of odours. We conclude that, in order to form a conception of

Clare Batty and Barry C. Smith, *A world of odours* In: *Sensory Individuals*. Edited by: Aleksandra Mroczko-Wąsowicz and Rick Grush, Oxford University Press. © Oxford University Press 2023. DOI: 10.1093/oso/9780198866305.003.0014

persisting and reidentifiable odour-particulars, we need, like Strawson, to take seriously the idea of exercising many senses while inhabiting a 3D space in which material objects (i.e., odour sources) are basic.

1 From sound world to odour world

In Chapter Two of *Individuals*, Strawson asks whether there could be a conceptual scheme for a world consisting of objective and reidentifiable particulars that are not material bodies. This question comes hot on the heels of his establishing, in Chapter One, that material bodies are the basic particulars of our 'conceptual scheme as it is' (Strawson, 2003 [1959], p. 11). His question in Chapter Two is whether it is *necessary* that such a system of objective and reidentifiable particulars is one in which material bodies are basic.

According to Strawson, an objective particular is a particular than can exist independently of anyone's experience of it. In setting out what it means to have experience of such a particular, Strawson distinguishes between experiences (i.e., states of one's own consciousness) and what those experiences are *of* (i.e., what we are conscious *of*). As Strawson claims, an experience of a certain type is of an objective particular if there is a 'use for the distinction between [the subject] and his states on the one hand, and something not himself or a state of himself of which he has experience on the other' (Strawson 2003 [1959], p. 69). In this way, Strawson ensures that experience or awareness of a mental event, such as a pain, cannot constitute experience of an *objective* particular. Presumably, his reasons for ruling out mental particulars as objective is the idea that pains and other mental events (controversial as this now may be) are not the kinds of particulars that can exist unperceived. However, it does not follow that experience of objective particulars need be experience of *material bodies*. What is meant by 'objective particular' is simply an entity that does not depend for its existence on our experience of it. As we will see, there are other candidates for experienced particulars of this kind than material bodies. In addition to sounds, which we will come to soon, Strawson also mentions shadows, flashes, and bangs.[1]

Before considering his discussion of sounds, it is important to also say something about what he means by 'reidentifiable'. Setting out 'criteria of reidentification' is a task of Chapter One, where Strawson canvasses the various ways in which we individuate particulars. For our purposes, it is not necessary to delve deeply into the discussion of Chapter One. What is important to note is what Strawson claims any criteria for the reidentification of particulars must capture. He claims: 'Our methods, or criteria, of reidentification must allow for such facts as these: that the field of our observation is limited; that we go to sleep; that we move. That is to say, they must allow for the facts that we cannot at any moment observe the whole of... [our environment], that there is no part of it that we can observe continuously, and that we ourselves do not occupy a fixed position within it' (p. 32). If these facts obtain (as they surely seem to do), and if there are objective particulars that are *reidentifiable* by us, then we can identify those particulars at different times despite gaps in our experience of them. For example, leaving the office brings about a gap in your visual experience of your mug sitting on the desk. If you identify the mug when you next see it, as you certainly would do

[1] See, for example, Strawson, 2003 [1959], p. 46.

upon re-entering the room, then there must be something that counts as your identifying, or failing to identify, it as the very same mug (and not simply a qualitatively identical one). According to Strawson, the fact that we are able to reidentify such objects through gaps in our experiences of them is due to their continuing to exist in the gaps and, to make sense of this fact, our conceptual scheme presupposes an objective space which contains them. What is more, given the presupposition of such a space in which particulars continue to exist in the gaps between our experience of them, Strawson argues that the basic particulars of our conceptual scheme *must* be material bodies.

The details of Strawson's argument need not concern us here since they would take us too far afield from our real focus: Strawson's discussion of the conditions of particular individuation and what this can tell us about those achievements in the olfactory case. We will address only those aspects of Strawson's discussion in Chapter One that are necessary for this aim. The majority of Strawson's discussion in Chapter Two involves investigating whether there could be other conceptual schemes positing objective and reidentifiable particulars that are not material bodies. Given that aim, Strawson considers the possibility of a 'No-Space World'—a world in which our experiences are structured temporally, but not spatially. Given the connection between material bodies and space, Strawson claims that a conception of such a world could not include material bodies. Material bodies, for Strawson, are occupiers of space; as perceived, they stand in spatial relations to one another. Strawson's question then becomes: could we conceive of a No-Space World that still contained objective and reidentifiable particulars? This question gives rise to his renowned 'Sound World' thought experiment. The Sound World is one in which the subject's experiences are purely auditory; it is a world *as conceived by the subject who has exclusively auditory experiences* and its supposition makes no assumption about the kinds of particulars that are *in fact* in the world.[2] According to Strawson, the Sound World would be a No-Space World. Although it might first appear that auditory experience presents sounds as coming from a certain direction and/or distance (e.g., 'sounds seem to come from the right or left, from above or below, to come near and recede' (p. 65)), Strawson urges that experiences of those kinds are not just auditory, but also tactile and kinaesthetic. If we are to consider the possibility of a Sound World, we must consider experience stripped of those elements, leaving simply the auditory.[3] When we do, according to Strawson, we see that sounds 'have no intrinsic spatial characteristics: such expressions as "to the left of", "spatially above", "nearer", "farther" have no intrinsically auditory significance' (p. 65).[4]

The final formulation of Strawson's question is this: would a being whose experiences are purely auditory have the materials to conceive, or make sense, of a world of reidentifiable,

[2] This leaves it open that an error theory of auditory experience is true—that is, that, while auditory experience might tell us that there are certain particulars (and properties) in the world, the world contains no such particulars (or properties). Whether such a theory is true is not Strawson's question. His concern is with the deliverances of auditory experience and is prior to any consideration of its accuracy.

[3] It is controversial whether we can rule out the contribution of any other modality in this way. But such a method is certainly Strawson's aim and has been the aim of many working in the philosophy of olfaction.

[4] We recognize that this claim is controversial and open to dispute. For example, O'Callaghan (2010) argues that, except in certain highly impoverished cases, Strawson's claim that auditory experience is aspatial is false. In arguing for this claim, he draws on empirical work such as that by Bregman (1990), Hartmann and Wittenberg (1996), Mills (1972), and Shinn-Cunningham (2001a, b). For the purposes of this paper, however, we set aside assessing Strawson's claim in the face of arguments to the contrary. In constructing an analogy with olfaction and connecting it to discussions of sensory individuation in the olfaction literature, we can grant Strawson his claim. See also fn. 8, p. XXX.

objective particulars? In the Sound World—again, the world as conceived by the subject of exclusively auditory experiences—candidates for such particulars are sounds. One might think that they could just as easily have been odours in an Odour World—that is, a world as conceived by the subject who has exclusively olfactory experiences. However, Strawson rejects that example, claiming that 'taste and smell are strikingly more trivial than the other [senses]'. As he puts it, 'to suppose our experience free of gustatory and olfactory elements would [not], by itself, be to invite a significant conceptual revolution' (p. 64). Strawson adds, as a parenthetical remark: 'One does not *see* the world differently if one has a cold' (p. 64). It is a curious remark since Strawson appeals to the impact on other sense modalities (i.e., vision) in ruling out olfactory experience as a suitable candidate for his thought experiment. He will, however, go on to tell us, in his response to the initial worry about auditory experience, that we must not do this. In any 'Modality-Exclusive World', we must only consider the type of experience at hand, stripped of any other type of sensory element. We take it that the best way to interpret Strawson is as claiming that, unlike sounds in the case of audition, there are no eligible particulars encountered in olfactory experience with which to explore his question. His evidence for this is the alleged fact that, when one has a cold, the world does not seem any different. Why doesn't it? Because no objects (widely conceived as reidentifiable particulars, as set out above) appear to be absent from it.[5]

But *that* is just the question. And there *are* candidate particulars in the olfactory case—namely, odours. By 'odour' we mean a collection of airborne volatile molecules, or a gaseous emanation.[6] Although not material bodies in any usual sense, we certainly think of ourselves as encountering odour-particulars in olfactory experience. And as we see from what Strawson himself claims about sounds, there is no reason to rule out odours as something to ask after in the Odour World. According to Strawson, a conceptual scheme which posits objective, reidentifiable particulars is one for which the universal/particular distinction makes sense. This is why, he claims, the auditory case provides, at least on the face of it, an eligible particular. As Strawson (p. 67) tells us, for the case of audition:

> We may imagine, for example, that the same piece of music is being simultaneously played in two different concert-halls. We may imagine a moment at which a certain chord is played. Then two different normal hearers, each in a different hall, have, roughly simultaneously, roughly similar auditory experiences. But though in one sense the sound they hear is the same—it is the same chord for each—in another sense, the sense we are concerned with, the sounds they hear are distinct. They hear different sound-particulars.... Two listeners in the same concert-hall, however, hear the same sound particulars as each other, as well as hearing the same chord, i.e. the same sound-universal...

In the case in which two listeners hear the same sound-particular, Strawson (p. 67) claims that 'the conditions of particular-identity of setting and source are both fulfilled'. The listeners, after all, are in the same concert hall and listening to the same orchestra. Whether in

[5] In fact, there is reason to question Strawson's assertion. Those who lose their sense of smell suddenly through post-viral infection or head trauma often say they perceive the world differently, as though a dimension is missing or as if they are *seeing* the world from behind glass. Even those with congenital anosmia may have very different perceptual experiences of the spaces in which they are (see Tafalla, 2014).

[6] 'Odour' is also sometimes used in property-talk, to denote the distinctive properties presented to one in olfactory experience. We will only use 'odour' in particular-talk and reserve 'smell', in its nominal form, to denote such a property. We will never use 'odour' or 'smell' to refer to olfactory experience, or any aspect of it.

the Sound World, conceived of as a No-Space World, we could arrive at conditions of identity that distinguish particulars from one another is precisely the question Strawson is asking. And the fact that we can, albeit at the outset, understand the notion of an apparent sound-particular in examples like this provides us with a particular to ask after in the purely auditory Sound World.

But, if that is true, then it is hard to see why the same is not true for the olfactory case. We can easily adapt Strawson's example of the concert hall for it. Like the music played in Strawson's two concert halls, a case for which Strawson's condition of 'identity of setting' is fulfilled, we can imagine two people smelling the same perfume at the same time in different stores. Similarly, we could find ourselves wondering whether the odour of skunk in the garage is the same odour that we smelled before (even though we have scrubbed or attempted to mask the odour with another), or whether a new skunk has made its way into the garage and become the source of a second odour-particular of the same kind—despite the success of our deodorizing efforts the first time around. In the first skunk case, we would likely ask 'is it still here?'; in the second, we would exclaim 'not again!'. What the perfume stores and skunk case show is that the particular/universal distinction makes sense for the phenomena that we can discriminate in olfactory experience.[7] Take away the skunk odour and the world *would* seem different! Like Strawson's Sound World, then, the odour cases provide us with something to ask after in a pure Odour World.

Moreover, as the case of failed deodorizing shows, we are perfectly happy to conceive of odours as existing *unperceived*. An odour may be imperceptible while it is being overpowered by another odour, though when the overpowering odour fades, the original is still there. Similarly, odours can be present when we fall asleep and still there upon waking (e.g., the smell of damp in the room). We can, and would, conceive that the odour has continued to exist unperceived while we slept.

Similarly, like the Sound World, there is at least prima facie reason to think that the Odour World would be a No-Space World. As we know, in the Odour World, the world as it is for the subject of experience, we are meant to consider olfactory experience, on its own, stripped of any other modality's contribution to the subject's conscious state. Just like sounds, however, we think of odours as coming from certain directions and as covering more or less space as we estimate the distance of their sources. Arguably like sounds (and perhaps more plausibly than the case of sound), we do so on the basis of olfactory experience, together with the inputs of other sensory modalities, as when we track odours to their sources. This idea is echoed in some of the recent philosophical literature on olfaction. Lycan (2000, p. 278) claims: 'Phenomenologically, an odor is just with us, happening right in the center of our minds. If we bracket the external source of the smell and waive our contingent knowledge of that source, smell has no directional or other spatial aspect.' Similarly, Batty (2010, p. 524) claims: 'If we bracket information gained from movement and any other sensory modality, and consider olfactory experience at-a-time, then we see that any locatedness of these properties —other than simply 'here'—goes as well.' Although the Batty's claim suggests that olfactory experience is *minimally* spatial, helping itself to the location of 'here', comments like

[7] Martin (2014) and Mizrahi (2014) deny that such a distinction can be made for the case of olfactory experience. As their claims trade on more substantive philosophical commitments (about the meaning of perceptual language in the former case and about the nature of olfactory objects in the latter), we set them aside. Here we want only to make the prima facie case that such a distinction for olfactory experience makes sense.

Lycan's and Batty's still suggest that there is arguably little difference between the Odour World and Strawson's Sound World. Recall what Strawson (2003[1959], p. 65) claims about experienced sounds: they 'have no intrinsic spatial characteristics: such expressions as "to the left of", "spatially above", "nearer", "farther" have no intrinsically auditory significance'. Arguably an analogous claim could be made about experienced odours, and both Lycan's and Batty's claims capture that insight. Moreover, experiences of odours, stripped of the contribution of other sensory inputs, are arguably ones for which 'expressions [such] as "to the left of", "spatially above", "nearer", "farther" have no intrinsically [olfactory] significance.'(ibid, p. 65) If the Sound World thought experiment gets going as a result of Strawson's observations about sounds, then so ought an Odour World thought experiment with relevantly similar observations about odours.[8] Similarly, if the Sound World is one in which perceived sounds fail to count as perceived material bodies because of their lack of spatial representation, then perceived odours fail to do so as well.

Our question is, then: could a being whose experience was purely olfactory conceive of a world of objective, reidentifiable particulars—such as odours—existing independently of that being's experience of them? It is surprising that philosophers of olfaction have failed to consider an olfactory analogue of Strawson's Sound World. There are significant connections to draw between Strawson's aims in Chapter Two of *Individuals* and the central questions that have marked the philosophy of olfaction since it began in earnest in the early 2000s. 'The' debate in the philosophy of olfaction has concerned the question of whether, and how, objects are represented in olfactory experience. Of particular concern has been whether objects are individuated by, or in, olfactory experience. Like Strawson's aims for the Sound World thought experiment, this discussion has largely been undertaken in modality-exclusive terms, considering olfactory experience in and of itself, absent the contribution of any other sensory modality.[9] There is an obvious affinity, here, between a Strawsonian examination of an Odour World and discussions of object representation in the philosophical literature thus far.

In particular, discussion has centred on whether olfactory experience ever achieves figure-ground segregation and, if it does, whether that accomplishment must necessarily be spatial. Can a view according to which olfactory experience achieves aspatial figure-ground segregation support an object-based view or, to put it in Strawson's terms, a view of odours as particulars? No matter where one lands on this question, philosophers have begun to recognize that focusing merely on synchronic olfactory experience does not provide us with a robust notion of olfactory perception—certainly not one that reflects our actual behaviour in smelling. Looking to Strawson, then, can provide us with a further means of addressing issues of object individuation for the olfactory case. His consideration of the Sound World is an exercise in examining the deliverances of diachronic auditory experience; it is also a way of considering whether space is integral to individuation of particulars. The Odour World presents a way to pursue these issues for olfactory experience. To be sure, some philosophers have considered the role that diachronic experience has to

[8] In the lead-up to the quoted passage above, Lycan (2000, p. 278) claims: 'Hearing . . . comes to us spatially organized, even phenomenally speaking; normally, to hear a sound at all is to hear it as coming from this direction or that and from such-and-such a distance. By contrast, smell is aspatial.' Here, not only do we see resistance to Strawson's claim that auditory experience, stripped of other elements, would be non-spatial, but we also have the suggestion that olfactory experience is even *better suited* to the Strawsonian task.

[9] See fn. 3.

play in the representation of objects in olfactory experience.[10] Millar (2019), for example, considers the role that diachronic experience plays in achieving *spatial* figure-ground segregation. To this end, however, she argues that, in order to see this, 'we need to look to empirical evidence showing that tactile stimulation and bodily movements play a crucial role in olfactory phenomenology' (p. 4280). What is distinctive about the introduction of a Strawsonian Odour World into the contemporary philosophical discussion is that, in asking what is uniquely olfactory and perhaps *aspatial*, it focuses our attention on whether olfaction plays, or could play, a more fundamental role in our conception of an objective world than has hitherto been appreciated. As we shall see, this is precisely what Bao *et al.* are asking in their research.

2 Objectivity in sound and odour worlds

In order to draw the relevant connections between the Strawsonian Odour World and Bao *et al.*'s empirical research, it is helpful to say some more about the outcomes of Strawson's Sound World thought experiment. Strawson rejects the Sound World as a world of objective and reidentifiable particulars. In order to have a conception of such reidentifiable particulars, an inhabitant of the Sound World must be able to distinguish between later presentations of what is *numerically* the same sound and those of distinct, yet *qualitatively* identical, sounds (i.e., distinct sounds with the very same pitch, timbre, and volume).[11] According to Strawson, we cannot achieve this in the Sound World as he sets it out initially; there is nothing in auditory experience itself that allows us to distinguish between a case in which we hear the same sound again and one in which we hear another sound with the same properties. As Strawson argues, it is only if we appeal to some analogue of space that we are able to make use of any distinction between qualitative and numerical identity for the Sound World. The analogue of space that he introduces is the master sound—a continuous background sound of the same timbre and volume, but varying in pitch over time. The master sound accompanies all other sounds, and its shifting pitch provides the Sound World's quasi-spatial dimension, going up or down in frequency like the sounds we hear when tuning a radio to each station. Particular sounds have as their 'locations' the pitch of the master sound that accompanies them, as if one could tune into them on a radio by 'moving' to a certain pitch of the master sound. According to Strawson, this would allow us to distinguish between a case in which we hear the same sound again and another in which we hear a different sound with the same properties. If a sound we hear at a later time is accompanied by the same pitch level of the master sound, its 'location' at that pitch of the master sound would allow us to conceive it as numerically the same sound as the one heard previously and one always to be found at that pitch. If we hear a sound later accompanied by a different pitch level of the master sound, that later sound may

[10] This is not to say that there is no other discussion of diachronic olfactory experience in the philosophical literature. See, for example, Skrzypulec (2019) and Young (2020). We cite Millar (2019) in the body of the paper, as her concerns closely align with those of Strawson and Bao *et al.*

[11] As we have seen, Strawson states this claim in terms of the universal/particular distinction—that is, that we must be able to make sense of re-encountering one and the same particular and encountering a different particular of the same universal. We choose here to state the claim in terms of the distinction between qualitative and numerical identity. This aligns more directly with Strawson's own reliance on *reidentifiability*, as well as Evans' (1985) discussion of Strawson, to which we turn in this section.

be experienced as qualitatively identical to the one heard earlier; however, its different 'location' in the master sound would have it as numerically distinct from that earlier sound.

What about the Odour World? Like the Sound World, it would seem that, left only with the deliverances of olfactory experience, we would not, as in the case of the Sound World, be able to employ the distinction between numerical and qualitative identity. Or, it is at least not clear why the case of the purely olfactory world should differ from that of the purely auditory one. But could there be an olfactory analogue of the master sound—a 'master odour'? What, if anything, could provide an analogue of space in the olfactory case? We know that odours can fade as we move away from them, diminishing to almost nothing, but increase again in perceived intensity if we retrace our steps to where we were before. Here, with the help of a spatial dimension, we have continuity of an odour through changes in its intensity—like a sound that gets louder as we approach and fainter as we retreat. This leaves open the possibility of establishing an analogue to Strawson's master sound—a continuous background odour of abiding olfactory quality, but one that varies in terms of its intensity over time. To be sure, sometimes a change in the intensity of an odour stimulus makes for a wholesale change in perceived quality. For example, Gross-Isseroff and Lancet (1988) show that, at low levels of concentration, methyl heptanoate smells like violets, while at high concentration, it smells fecal. Similarly, they show that diphenylmethane smells like geranium at low concentrations, and like oranges at high ones. In other cases, however, there is no stark difference in odour quality with changes in intensity. These are the cases we help ourselves to here, as do Bao *et al.* in the empirical work we consider in Section 3.

With the addition of a master odour to the Odour World, any other perceived odour would have as its 'location' a particular intensity of the master odour that accompanies it. This would allow us to distinguish between a case in which we smell the same odour again and another in which we smell another odour with the same properties accompanying a different intensity of the 'master odour'. To see that this is so, consider the familiar case in which one odour is overpowered by a second, only to reappear after the second odour recedes. If the odour we smell at this later time is accompanied by the same intensity level of the master odour (as was experienced earlier), its 'location' at that intensity of the master odour would allow us to conceive of it as numerically the same odour as the one smelled previously. If this later odour is accompanied by a different intensity level of the master odour, that later odour may be experienced as qualitatively identical to the one smelled earlier; however, its different 'location' in the master odour would have it as a numerically distinct odour. In sum, other odours could appear, linger, fade, and be masked by others—all the while leaving the perceived intensity of the master odour as a means of establishing a distinction between numerical and qualitative identity for the Odour World. So there could be an analogue for space for the Odour World.

For Strawson, the master sound might allow us to employ the distinction between numerical and qualitative identity, but the necessity of its introduction amounts to an undesirable imposition of space, or something akin to space, on a No-Space World—the Sound World, for Strawson. Our aim is not to adjudicate whether this is a legitimate worry for Strawson, whether the introduction of the master sounds threatens a consideration of a No-Space World.[12] What *is* important to note here is that, setting aside his rejection of a purely olfactory

[12] For example, one might think that Strawson is too cautious here, given that the master sound plays a role *analogous* to space but is not itself spatial.

world (a rejection which we have argued is illegitimate), Strawson's considerations would suggest that nothing other than an analogue of space would suffice to establish a distinction between numerical and qualitative identity in the Odour World.

In "Things Without the Mind" (1985), Gareth Evans famously examines Strawson's Sound World thought experiment. In doing so, he contemplates ways, other than those on which Strawson relies, to conceive of sounds as existing independently of a perceiver's experience of them—that is, other than their being 'located' somewhere else when not experienced by us, as with the master sound. Evans rejects Strawson's master sound as a legitimate analogue for space for (at least) two reasons; he believes that its addition has the effect of 'wrecking the argument' (p. 255) for the claim that an analogue of space is necessary for objectivity in the Sound World. Firstly, he argues that the addition of the master sound into the thought experiment does not provide the necessary distinction between numerical and qualitative identity. As any sound in the Sound World will appear alongside a certain pitch of the master sound, Evans argues that no pair of alleged qualitatively identical, yet numerically distinct, sounds will ever be truly qualitatively identical. Secondly, Evans argues that the master sound is not, in fact, an analogue of space, as it does not possess *dimensionality*—an ordering of its various pitches. As he notes, the same effect could be obtained if the master sound was comprised of an unordered series of pitches. As a result, it fails to provide an analogue for space.

What this means is that it remains an open question whether the Sound World, with the addition of Strawson's master sound, would provide an instance of a No-Space World in which subjects are able to form a conception of objective and reidentifiable particulars. For the sake of drawing connections to the relevant empirical literature, we set aside this question. What is important for drawing those connections, however, is what Evans proposes as an alternative criterion of objectivity. For Evans, it is extended *continuity*, rather than *identity* (and reidentifiability), that is the clue to a sound's objective status. He claims (p. 255)

> For it may be held that a one-dimensional space which is genuinely parallel to our own may be constructed out of an auditory experience (or, indeed, out of any experience) provided that experience exhibits such order and regularity that its course can be seen as simultaneously due to the way the world is laid out and to the subject's continuous motion through it. Provided the subject's experience is sufficiently regular to enable him to establish short-term generalizations of the form:
>
> An experience of kind k will intervene between any experience of kind k' and any experience of kind k".
>
> it might appear that he can distil from its changing course a more or less detailed map of his world, with an object of kind K between (in a 'travel -based' sense of this word) objects of kind K' and K".

Evans suggests that this proposal works for any kind of sensory experience (any 'Sensory World'), provided that the relevant continuity obtains. And, as it turns out, the case of odour is particularly well suited to Evans' framing of the question. We are all familiar with experiencing the long decay of an odour (e.g., the gradual after-effect of a fireworks display or an extinguished cigarette). In such a case, we experience an odour to tail off, changing intensity and fading over time, but nonetheless remaining continuously (numerically) the same odour. That is, it is a fact about our olfactory experience that we know how an odour is likely to be experienced as it recedes in time (or moves away); this is knowledge of how that odour behaves through time, during a period of decay (or at a sufficient distance from its source, or from us). And equally, it would seem, this model holds for cases in which we have gaps in our experiences of the odour. When we re-enter the room after having left it for an hour, we certainly seem to recognize the odour of the lavender spray that was present in greater concentration before we left, or that the skunk odour has faded when we return to the garage to check on it. Can cases like this help us in framing Strawson's question for the Odour World? That is, could a being whose experience was purely olfactory, and exhibited such regularities, conceive of odours as existing independently of them?

Taking this case at face value, and following Evans' lead, perhaps what we need to sustain the idea of an objective world of odours for the Odour World is for there to be enough order and regularity in experience to establish a generalization of the following form:

> between any olfactory experience *x* and an olfactory experience *z*, where there is an affinity between the experience of *x* and *z*, there is an olfactory experience *y* that intervenes along a continuum linking *x* and *z*.

Following Evans, this would be to conceive of *y* as lying *between* *x* and *z*, in a 'travel-based sense' of *between*, indicating a conception of what is experienced as the joint upshot of a world of odours and where one is in it. This would allow subjects of experience to construct an odour 'landscape' for odours existing unperceived. However, unlike Evans' proposal, it is not different odours appearing on a continuum from *x* to *z*, but different properties of the same odour appearing on a continuum between stages *x* and *z*.[13]

In this case, the intensity an odour has at a given time could be recognized as continuous with an intensity it had, or will have, at other times, so that one would be able to imagine intermediate stages of intensity on a continuum of properties of that odour that were not perceived but *perceivable* (e.g., how intense that odour would be at a later, or earlier, stage of sampling). If this is right, we can ask whether it would be possible to construct a one-dimensional space on the basis of olfactory experience alone where patterns in experience could be thought of as due to the structure of that odour space and where one was in it. Could such a space be conceived as made up of an independently existing odour whose intensity profile changed relative to one's position in the space?

3 Generating an odour map from experience

On the present conception of the Odour World, then, our experiences could be sufficiently regular so that we can expect that an olfactory experience of kind k' with intensity I' will intervene between any experience k with I and any experience k" with I". This continuous space of possible experience looks a lot like the odour landscape navigated solely on the basis of purely olfactory stimuli that the Gottfried Lab hypothesize and set out to test, as reported on in Bao *et al.*, 2019:

[13] We are grateful to an anonymous reviewer for urging this clarification.

> [It] is unknown whether humans can navigate a sensory space informed only by odor cues and how the brain might internalize a representation of two-dimensional olfactory space.... Here, we posit a world populated exclusively with odor stimuli to determine whether and how a navigator—with only the luxury of the sense of smell—can traverse an olfactory landscape.

The *whether* and the *how* here are: whether humans can navigate a sensory space informed only by odour cues *and* how the brain might represent such a navigable space. At this stage, it is moot whether a navigable 'sensory space' is *spatial*, and whether *navigating* (or mapping) that space should count as *spatial* navigation (or *spatial* mapping). What is important at this point is that they take their 'Odour World' to be a 'navigable space'—a space of continuous variations in the intensities of a combined pine and banana odour mixture. It thus incorporates the required dimensionality that Evans argues is absent from Strawson's conception of such a world. What's more, the only things that the participants have access to in the experimental set-up are changing intensities of the odour stimuli (pairs of banana and pine odours at varying concentration ratios). Bao *et al.* ask *whether* participants exposed to these pairs are able to extract enough information from the stimuli to represent, or map, a 2D olfactory 'landscape', in which each stimulus exists at some point in that space. If participants can predict their experiences of odour mixture pairs of varying intensities, the hypothesis about *how* they do it is that grid cells in the entorhinal cortex (ERC) known to underlie navigation, path finding, and abstract conceptual structure are recruited and configured as a map as participants learn to perform the task.

The thought guiding Bao *et al.* is that the central function of the olfactory system is spatial navigation and their hope is to demonstrate this role of the olfactory system by proving that humans have the means to navigate a 2D odour landscape *solely* on the basis of olfactory inputs. But if humans can acquire this navigation skill when performing the task, does this also enable them to conceive of a continuous layout of odour points inferred from the predictable changes in odour intensity pairs experienced as they 'navigate' that space? This issue mimics Strawson's, and even more so Evans', concerns.

The analogy between a Strawsonian thought experiment and Bao *et al.*'s real experiment can be brought into focus as follows. Strawson begins with auditory (it might as well be olfactory) experience and, from that standpoint, asks questions about what auditory (olfactory) experience gives us experience *of*. The neurobiology of olfaction has a similar starting point, beginning with the brain and the inputs it receives, and from that standpoint, asks what the causal structure of the world would have to be like to give rise to the predicted patterns found in those inputs.[14] In both cases, we have a focus on how it is with the participant—*from the point of view of experience* in the case of philosophy, and *from representations in the brain* in the case of neurobiology. We can view both the thought experiment and the real experiment as probing the content of those experiences or representations—that is, the question of what, 'from the inside', we could legitimately assume about the world *on the basis of olfactory stimuli alone*. The analogy between these two tasks means the result of one can inform the other, providing a novel way of interpreting Bao *et al.*'s results *and* of addressing the objectivity of olfactory experience.

[14] This is precisely what the predictive processing approach sets out to address.

As we indicated above, driving Bao *et al.*'s hypothesis is the idea that the central function of the olfactory system is navigation. So what entitles Bao *et al.* to see their experiment as casting light on how creatures like us (with other sensory modalities) can nonetheless map out a space of odours when in receipt of olfactory inputs alone? Notice that Bao *et al.* are not merely asking whether humans can navigate using olfaction. Previous studies have already established that humans are able to follow odour trails, and can locate odour sources in their environment without the aid of vision. Porter et al. (2005) demonstrated that blindfolded subjects are able to follow a chocolate trail on grass, and Jacobs *et al.*, 2015 demonstrated that blindfolded and disoriented subjects are able to navigate back to the location of the source of a previously sniffed odour. All of this shows that we can use olfaction, *together with bodily movement*, to navigate a physical odour space. However, these tasks already build in the spatial awareness of an odour-sensitive agent as they move through a physical landscape. Participants rely on tactile and kinaesthetic inputs to determine whether experience presents odours as coming from a certain direction and/or distance, from the left or the right, from straight ahead, and so on. These are also the sorts of inputs that Strawson explicitly rules out any appeal to in setting up the possibility of a pure Sound World or, in our case, a pure Odour World. By ruling out cues based on other modalities, Bao *et al.* try something analogous to our version of Strawson's task. They ask whether an agent can, when relying on olfactory experience alone, construct a stimulus space conceived of as a world of independently existing odours (in this case, banana and pine) of varying intensities, in the absence of vision, movement, or other sensory cues. Indeed, their absence is a feature of the experimental design, as it is a functional magnetic resonance imaging (fMRI) study with participants lying immobile on their backs in the scanner.

The task is as follows. Participants are exposed to banana odour and pine odour at different intensities, then exposed to odour mixtures combining different ratios of banana and pine odours. Participants are trained to discriminate between odour pairs that differ in terms of an increase or a decrease in intensity of one or the other of the odours in the mixture (i.e., banana or pine). Participants are then given an odour pair to sniff while looking at a screen indicating the intensity levels of banana and pine odours in the mixture. The screen then indicates a change in the intensity of the banana and/or the pine odours (up or down) and participants are asked to imagine the target odour mixture. They are then presented with a stimulus odour mixture and asked to judge whether it corresponds to the target odour they were imagining (i.e., whether it is on or off target). Participants receive feedback on whether their judgements are correct or incorrect. After being trained in this task for four days, on the fifth day, participants perform this task in the magnetic resonance imaging (MRI) scanner as part of a functional neuroimagery study of *how* they perform it. By this time, participants show the effects of learning and are increasingly accurate in recognizing whether the presented odour mixture corresponds to the imagined odour mixture (target odour). In each trial, there were easy (far from target), hard (close to target), and medium (intermediate distance from target) odour mixtures to judge. The set-up is represented in Figure 13.1. There was significant effect of training particularly on the easy trials as shown in Figure 13.2.

While participants perform the task, fMRI results show activation in the ERC, indicating the recruitment of grid-like place cells enabling participants to complete the task successfully. How, then, should we interpret these results? The authors are keen to stress that the representations amount to a cognitive map, as opposed to showing mere odour discriminatory capacities, with respect to changes in concentration experienced across sequential sniffs of the

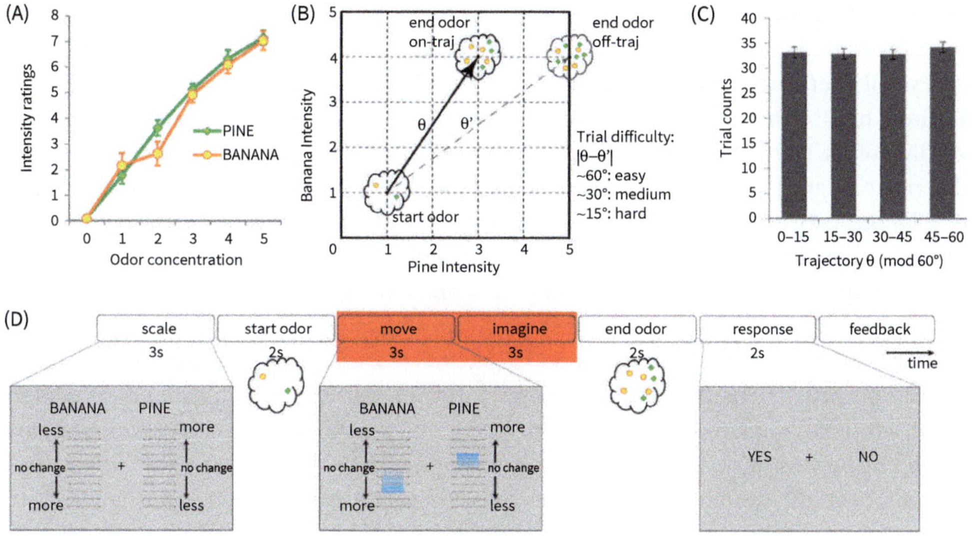

Figure 13.1 Experimental design and navigation performance: (A) intensity ratings, (B) coordinates on the odour map, (C) trajectories, (D) example trials of the odour prediction task.With permission from Elsevier (Bao *et al.*, *Neuron*, 2019).

odour pairs. This matters, of course, because as Evans (1985, p. 255) reminds us for Strawson's auditory case, a candidate notion of objectivity can be constructed 'provided that experience exhibits such order and regularity that its course can be seen as simultaneously due to the way the world is laid out and to the subject's continuous motion through it'. Building up a map of odour pairs when asked to navigate from one to another could, perhaps, make it possible to conceive of trajectories through an odour landscape for which generalizations can be generated along the following lines, as adapted from Evans above:

> An olfactory experience of kind k' with intensity I' will intervene between any experience k with I and any experience k" with I".

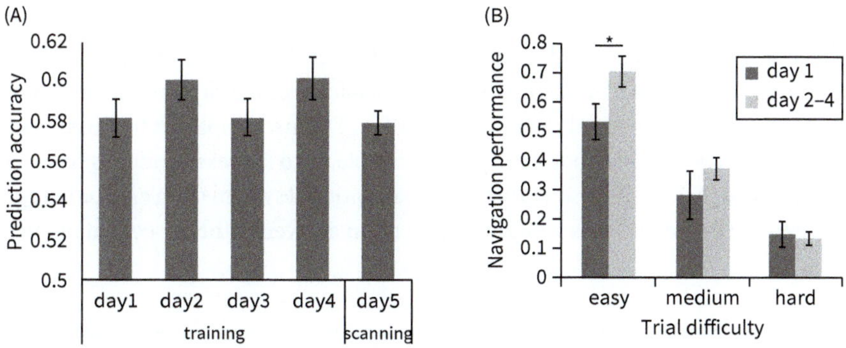

Figure 13.2 Olfactory navigation performance: (A) navigation performance accuracy over 5 days, (B) navigation performance on easy to hard trials.
With permission from Elsevier (Bao *et al.*, *Neuron*, 2019).

But this is to move too quickly. What entitles us to say that participants are doing any more than predicting correctly what an odour pair of any particular ratio of pine and banana will smell like? Perhaps the training induces something like an internalized 'look-up table' for how to discriminate any odour pair from any other. However, that is too minimal a reading of the learned discriminatory skill. The participants are not only able olfactorily to discriminate any pair of pine and banana mixtures; they are also able to impose a structure of relations between them. It is *this* ability that warrants talk of mapping for the representations in the ERC. The claim for a map is motivated by the way in which participants learn to orientate themselves in, and predict points on, a 2D array where the X-axis is banana intensities and the Y-axis is pine intensities. According to Bao *et al.*, participants map out a structure of the stimulus landscape and follow trajectories in that abstract space based on sampling odour intensity gradients along the two dimensions of banana and pine intensities, which is compatible with them conceiving the space they navigate as continuous and extending beyond any point they currently encounter.[15]

But this orientation in a sensory space does not yet amount to 'navigating' an objective space, however tempting the objective interpretation may be. Bao *et al.* admit that the neural mechanism that plots positions in this 2D landscape is not how humans navigate their olfactory environment. What is more, we know from other experimental findings that the ERC is involved in encoding non-spatial conceptual maps in humans (Garvert *et al.*, 2017) and in navigating two dimensions of economic choices, such as the magnitude of reward and the probability of getting it, in the monkey brain (Bongioanni *et al.*, 2021). Maps in the ERC can enable participants to compute conceptual relations or expected value by navigating that 2D space, but there is no reason to think of these as spatial maps. In the case of participants in Bao *et al.*'s experiment, an 'odour landscape' or sensory space represented on the basis of olfactory stimuli alone could be a much more abstract space like these.[16]

So why think that this mechanism serves to show how the 'olfactory system tracks and locates odor sources in an odiferous environment' (Bao *et al.*, 2019, p. 1071)? As we have seen, interpretation of the data is compatible with Evans' candidate for objectivity based on the idea of continuity--that is, a continuous space of possible experiences that the subject traverses in the mind and whose place in that space is seen as responsible for those experiences. But equally data from participants' performance of the task, and the configuration in the ERC, are compatible with their navigating a 2D sensory or odour space that carries no such assumptions about an objective odour landscape. In fact, the results produced from the data that Bao *et al.* give us leave it underdetermined whether the order and regularity in experience support a conception of a continuous world of *odours* or a more abstract mapping to the relations between sensory input data points in that 2D sensory space. Of course, the mechanism that supports the latter possibility could enable us to represent odours in space when embedded in a larger and multisensory spatial navigational system of an embodied agent that is able to explore space. But this would not give us an answer to the Strawsonian question of

[15] It might also be thought that the visual display that cues participants to imagine new ratios of the banana-pine mixture is essential to their mapping an odour space. Bao *et al.* worry about this point too; however they note that different visual presentations in different up-down orientations elicit the same responses. Instructions could also have been given auditorily, so it is more likely that these cues are causally enabling, rather than constitutive of, the resulting mapping abilities.

[16] The authors are very grateful to Daniel Yon for bringing these points to their attention.

whether a subject whose only experience was purely olfactory could conceive of a world of objective particulars.

If reasons can be found to interpret participants in Bao *et al.*'s experiment as thinking of odours as reidentifiable particulars, what would they be? As mentioned above, if the qualities an odour has at a given time, such as intensity, could be recognized as continuous with qualities it could have at other times, and one could imagine intermediate stages of qualities (intensities) on that continuum, these would be qualities that were not currently perceived but were perceptable. If that is right, the brain's task is to determine what the causal structure of the environment must be to give rise to regularities in the changes of perceived odour pair intensities that can be experienced, predicted, and confirmed (or disconfirmed) on the basis of olfactory inputs and expected inputs of pine and banana mixtures. An answer to how the brain performs this task can be provided by a predictive processing model in which the brain is constantly generating and revising an internal model of the environment. According to this framework, the brain uses a cascade of top-down processes to generate fine-grained predictions about the flow of sensory data, given its expectations about the causes of those sensory signals. Predictions are compared to actual sensory inputs, and mismatches produce prediction errors that are passed upwards in the processing hierarchy to refine these expectations (Friston, 2009; Hohwy, 2012). In this way, generated predictions that minimize mismatch with sensory input (i.e., prediction errors) increase the likelihood of the model (of the environmental causes of those inputs), while prediction errors lead to fine-tuning of the model that predicts the next sensory signals and ultimately the source of those signals. The important part of this story is the nature and origin of the fine-tuned model that generates predictions of sensory inputs (like those predicted by the participants in Bao *et al.*'s task). The assumptions made by the model about the environmental causes of the olfactory signals would have to be—as Bao *et al.* acknowledge—that the odour intensity gradients, by which participants learn to navigate, can be construed in terms of changes of odour concentration (and hence perceived odour intensity) due to the odour's distance from the participant. As Bao *et al.* (2019, p. 1066) say:

> We took advantage of the fact that odor concentration decreases with distance from its source and that perceived odor intensity monotonically scales with concentration.

This additional assumption is remarkably similar to something suggested by Strawson (2003 [1959], p. 75):

> we want an analogy of distance-of nearer to and farther away from-for only, at least, under this condition would we have anything like the idea of a dimension other than the temporal in which unperceived particulars could be thought, of as simultaneously existing in some kind of systematic relation to each other and to perceived particulars.

But notice that Bao *et al.*'s added assumption question-beggingly builds in not just an analogue of space, as Strawson suggests would be required, but also a straightforward spatial presupposition of a structured stimulus landscape as a *physical space*, with distance of odours to and from the subject. It is an assumption about the causes in the world of the pattern of sensory inputs we receive and crucially it is a spatial assumption. If the odour intensity gradients do reflect changes in odour concentrations—due to distance from an odour source or odour cloud approaching or receding—then a mechanism with the ability to predict these intensity gradients along the banana and pine axis, when presented with odour pairs, will serve as a cognitive map of a continuous world of banana and pine odours. But this is only when embedded within a larger system, fed by other multisensory inputs, that provides appropriate clues about our physical surroundings.

As a result of their imaging of the ERC, Bao *et al.* (2019, p. 1071) conclude that 'the human brain has access to internalized representations of odor mixture arrays to guide spatial orientation and route planning [in a 2D layout]'. However, that conclusion does not show that the activation of ERC grid cells by themselves encode *spatial* information about the layout of odour intensities in a landscape, but only that the structure those cells induces on the stimuli would allow that mechanism to encode odour gradients in a physical landscape under the right embedding conditions.

With the assumption about perceived intensity/concentration, distance, and direction in place, a neural mechanism with the ability to predict positions along the intensity gradients of the banana and pine axis could represent a continuous olfactory landscape that we could navigate by smell alone. However, the systematically varying odour mixture inputs can only serve as a cognitive map of such an odour landscape when embedded within a larger system that gives us further sensory information about our surroundings, including a spatial layout for nearer and further odour sources. To conceive of an objective, but purely temporal, world of odours, we would need an analogue of nearer and further to make sense of continuously existing odours with systematically varying intensities. And it would have to do that without presupposing that the changes in intensity were due to varying *distance* from the odour source. As it stands, Bao *et al.* have not provided Strawson with a demonstration that it is possible to conceive of an objective world consisting entirely of odours known on the basis of olfactory experience *alone*. What's more, it may be an over-interpretation of their interesting empirical finding to claim that 'a navigator—with only the luxury of a sense of smell—can traverse an olfactory landscape' (Bao *et al.*, p. 1066). To navigate by olfaction, we need, like Strawson, to take seriously the idea of inhabiting a 3D space in which material objects (i.e., odour sources) are basic—a world in which, as embodied creatures, we jointly exercise many senses to navigate our environment.

Acknowledgements

This paper was written during the coronavirus pandemic, in which we had to say goodbye to some but greeted others anew. A particular blessing was the emergence of the University of London Institute of Philosophy's olfaction group, a weekly online meeting of olfactory researchers from many disciplines. We would particularly like to thank Ann-Sophie Barwich, John Behan, Mika Hyman and Harry Sherwood for their help and insightful contributions to this paper. We would also like to thank audiences at the What the Nose Tells the Brain workshop in Bielefeld and the Institute of Philosophy's Lab Meeting for discussions of earlier presentations of this paper.

References

Bao, X., Gjorgieva, E., Shanahan, L. K., Howard, J. D., Kahnt, T., and Gottfried, J. A. (2019). Grid-like neural representations support olfactory navigation of a two-dimensional odor space. *Neuron*, 102(5), 1066–75.e1065.

Batty, C. (2010). A representational account of olfactory experience. *Canadian Journal of Philosophy*, 40(4), 511–38.

Bongioanni, A., Folloni, D., Verhagen, L., Sallet, J., Klein-Flügge, M. C., and Rushworth, M. F. S. (2021). Activation and disruption of a neural mechanism for novel choice in monkeys. *Nature*, 59, 270–4.

Bregman, A. S. (1990). *Auditory Scene Analysis: The Perceptual Organization of Sound*. Cambridge, MA: MIT Press.

Evans, G. (1985). Things without the mind: a commentary upon Chapter Two of Strawson's Individuals. In *Collected Papers*, 249–90. Oxford: Oxford University Press. Available from: https://global.oup.com/academic/product/collected-papers-9780198236276?cc=gb&lang=en&#

Friston, K. (2009). The free-energy principle: a rough guide to the brain? *Trends in Cognitive Sciences*, 13, 293–301.

Garvert, M. M., Dolan, R. J., Behrens, T. E. (2017). A map of abstract relational knowledge in the human hippocampal-entorhinal cortex. *eLife*, 6, e17086.

Gross-Isseroff, R. and Lancet, D. (1988). Concentration-dependent changes of perceived odor quality. *Chemical Senses*, 13(2), 191–204.

Hartmann, W. A. and Wittenberg, A. (1996). On the externalization of sound images. *Journal of the Acoustical Society of America*, 99(6), 3678–88.

Hohwy, J. (2012). Attention and conscious perception in the hypothesis testing brain. *Frontiers in Psychology*, 3(96), 1–14. Available from: https://internal-journal.frontiersin.org/articles/10.3389/fpsyg.2012.00096/full

Jacobs, L. F., Arter, J., Cook, A., and Sulloway, F. J. (2015). Olfactory orientation and navigation in humans. *PLoS One*, 10, e0129387.

Lycan, W. (2000). The slighting of smell. In *Of Minds and Molecules: New Philosophical Perspectives on Chemistry*, edited by N. Bhushan and S. Rosenfeld, 273–90. Oxford: Oxford University Press.

Martin, M. G. F. (2014) What's in a look? In *Perceiving the World*, edited by B. Nanay, 190–225. New York, NY: Oxford University Press.

Millar, B. (2019). Smelling objects. *Synthese*, 196, 4279–303.

Mills, A. W. (1972). Auditory localization. In *Foundations of Modern Auditory Theory*, Vol. II, edited by J. V. Tobias, 303–48. New York, NY: Academic Press.

Mizrahi, V. (2014). Sniff, smell and stuff. *Philosophical Studies*, 171, 233–50.

O'Callaghan, C. (2010). Perceiving the locations of sounds. *Review of Philosophy and Psychology*, 1, 123–40.

Porter, J., Anand, T., Johnson, B., Khan, R. M., and Sobel, N. (2005). Brain mechanisms for extracting spatial information from smell. *Neuron*, 47, 581–92.

Shinn-Cunningham, B. (2001a). Creating three dimensions in virtual auditory displays. In *Usability Evaluation and Interface Design: Cognitive Engineering, Intelligent Agents and Virtual Reality*, edited by M. J. Smith, G. Salvendy, D. Harris, and R. J. Koubek, 604–8. Mahwah, NJ: Lawrence Erlbaum.

Shinn-Cunningham, B. (2001b). Localizing sound in rooms. In *Acoustic Rendering for Virtual Environments*. Proceedings of the ACM SIGGRAPH and EUROGRAPHICS Campfire: Acoustic Rendering for Virtual Environments, 17–22. Snowbird, Utah.

Skrzypulec, B. (2019). Olfactory objecthood. *Philosophia*, 47, 881–900.

Strawson, P. F. [1959] (2003). *Individuals*. New York, NY: Routledge.

Tafalla, M. (2014). Smell and anosmia in the aesthetic: appreciation of gardens. *Contemporary Aesthetics (Journal Archive)*, 12, Article 19. Available from: https://digitalcommons.risd.edu/liberalarts_contempaesthetics/vol12/iss1/19

Young, B. (2020). Perceiving smellscapes. *Pacific Philosophical Quarterly*, 101, 203–23.

14
Smelling odours and tasting flavours: Distinguishing orthonasal smell from retronasal olfaction

Benjamin D. Young

1 Introduction

The olfactory system contains two separate senses (Smith, 2015): orthonasal and retronasal olfaction. The simplest example of a primary difference between these senses is the different anatomical pathways by which stimuli reach the olfactory receptors. Orthonasal olfactory perception occurs when we smell chemical stimuli (odorants) that are inhaled from the front of the nose. Retronasal olfaction occurs when odorants within the mouth are exhaled through the back of the throat and out the front of our nostrils. The difference in anatomical routes of stimulus delivery makes it natural to think that despite each system transducing the same type of distal stimulus (i.e., an odorant plume), they nonetheless differ in their kinds of perceptible objects,[1] such that we exteroceptively smell odours within a smellscape via orthonasal transduction (Young, 2020) and retronasal olfaction plays an essential part in the multisensory perception of flavourful objects contained within our headspace (Auvray and Spence, 2008; Prescott, 2015; Small, 2012). It is this pretheoretic intuition that the paper seeks to explore—orthonasal olfaction counts as the perceptual system responsible for our perception of smells within our environment, while retronasal olfaction is one of the many sensory systems that combine to yield our perception of flavour.

What is fascinating about the olfactory system is that it contains two separate pathways of stimulus presentation that share the same type of distal stimuli, the same sensory transduction mechanism, the same sensory processing centres, and some of the same cortical processing areas, yet they differ in terms of what types of perceptible objects they transduce and the kinds of sensory qualities they generate. Elsewhere it has been argued that accounting for the question of 'what are smells?' requires handling three nested issues. What

[1] Throughout this chapter, the term perceptible object should be understood as being the same as O'Callaghan's (2008, 2016) locution of 'object for perception' where this is to be understood as a wider, neutral, and more inclusive conception of perceptual objecthood that requires a suite of capacities that form the basis of object perception relative to a modality. O'Callaghan's framework allows for a broader notion for demarcating the objects of perception relative to a modality—his conception is broader than the notion of ordinary material object, yet narrower than that of intentional object, while being neutral enough to encompass relationalist theories of perception. However, Young (2020) argued, contrary to O'Callaghan (2016), that temporal structure must also be considered when conceiving of the object of olfactory distal perception in the same way that temporal mereological structure is responsible for determining the objective nature of auditory perceptual objects. Additionally, the conclusion of this chapter speculatively argues against his claim that flavour perception does not have an object for perception (the conclusion can similarly be used against the same claim argued for by Stevenson (2014)).

Benjamin D. Young, *Smelling odours and tasting flavours* In: *Sensory Individuals*. Edited by: Aleksandra Mroczko-Wąsowicz and Rick Grush, Oxford University Press. © Oxford University Press 2023. DOI: 10.1093/oso/9780198866305.003.0015

are the perceptible distal objects transduced by the system, what is it about these stimuli that generate the olfactory quality of smells, and what is the intentional object of our experience of smells (Young, 2016)? A comprehensive account of our perception of smell requires accounting for the olfactory quality of the object of perception, its distal nature as an odour plume, and how we intentionally represent smell experiences. It has been argued that the Molecular Structure Theory (MST) of smells can viably provide an answer to each of the aforementioned nested issues based on its determination of smells as the molecular structure of chemical compounds within odour plumes composing a smellscape (Young, 2019a, 2020). The molecular structure of the chemical compound and the odour plume account for the smell's olfactory quality and distal nature (Young, 2016, 2019a), which can then explain the perceptible[2] object of olfactory experiences as odours within an overlapping gaseous sea conceived of as a smellscape (Young, 2019b, 2020).

By considering the perceptible object of olfactory perception to be the molecular structure of chemical compounds composing odour plumes, MST provides the means for explaining what determines a smell's olfactory quality, how smells are identified and individuated, and how we can perceive smell as odour plumes against a background smellscape (Young, 2016, 2019a, b, 2020; Young *et al.*, 2020). The bonus of accepting this account is the further explanatory purchase of providing a means of individuating the perceptual modality of smell. What sets olfaction apart from the other modalities is not merely its proprietary sensory system, but the perceptible object of smell with its unique olfactory qualities that cannot be sensed using other sensory systems (Young, 2019a, 2020). We employ smell to gain access to the intrinsic olfactory qualities of a certain range of chemical stimuli in the environment. We smell odours, which is shorthand for we perceive the olfactory quality of groupings of types of chemicals within the gaseous sea of chemical currents in which we are immersed. Given this background theoretical framework and the overlap in shared transduction mechanisms, stimuli, and sensory processing centres, the question becomes whether we employ retronasal olfaction to perceive smells.

What is known about retronasal olfaction will be surveyed to establish the claim that, despite their underlying similarities, orthonasal olfaction and retronasal olfaction do not share the perceptible objects and olfactory qualities. Surveying the differences between orthonasal and retronasal olfaction in terms of their separate anatomical routes of odour delivery, stimulus transduction, sensory encoding, cortical processing, and perception provides the basis for the central argument that our sense of smell is relative only to orthonasal olfaction. The paper progresses in the following sections to reach this conclusion. Section 2 introduces retronasal olfaction. Section 3 surveys how the different anatomical pathways generate further differences in stimuli transduction, with implications for demarcating the perceptible objects and their sensory qualities. Section 4 continues the focus on how the differences noted in the previous section generate further distinctions in odorant perception as it is encoded at the receptor level. Section 5 progresses the argument further by looking at how the two systems project into separate and non-convergent cortical processing areas. Section 6 wraps up the discussion of the differences between orthonasal and retronasal olfaction by focusing on perceptual states and how these differ in the perception

[2] The terminology of 'intentional object' used in earlier versions of MST has been updated to 'perceptible object' throughout this chapter, in accordance with the assumption of O'Callaghan's framework and conception of objects for perception (see fn. 1).

of sensory qualities. The paper concludes by arguing that orthonasal and retronasal olfaction do not share the same perceptible object and that the differences in stimulus transduction and encoding yield perceptually different types of olfactory qualities. Despite both transducing the same type of distal stimuli (odorant plumes), they do not share the same type of perceptible object, as shown by differences in perceived olfactory qualities, which is the primary means of olfactory object identification according to MST.[3] Thereby, a disparity is established between retronasal and orthonasal olfaction on at least two of the three central questions required for a comprehensive account of 'what are smells?'

2 What is retronasal olfaction?

It comes as a surprise to most people that what we ordinarily report as taste experiences are partially generated by the olfactory system in our nose. Yet not everything that goes on inside the nose should be considered part of our sense of smell. Somatosensory and trigeminal stimulation within the nasal cavity might play a role in our everyday experience of smelling, yet (without associative learning) they do not yield perceptual olfactory qualities, which has been argued is an essential aspect of smell (Young, 2016, 2017). It has been argued elsewhere that the olfactory system, beginning with olfactory receptor neurons (ORNs) and projecting to cortical areas of encoding, is the primary determinate of smell experiences of olfactory qualities, and thereby the basis of our perception of smell (Young, 2017, 2019a). However, there are two distinct pathways for the odour plume to reach the olfactory epithelium, and this frames the conundrum of this paper. Do these distinct pathways yield two senses of smell (Smith, 2015), one perceptual modality of smell despite the duality (Shepherd, 2012), or perhaps two olfactory sensory systems, only one of which yielding the perceptual modality of smell?

One pathway for odours to reach the olfactory system is via the front of the nose. Orthonasal olfaction, as it is so labelled, is what we primarily refer to as our sense of smell. Our olfactory experiences are usually of external odours within the environment that have been brought into our nose through normal inhalation or actively sampling the surrounding air by sniffing. In addition to the orthonasal pathway, there is a second pathway for airflow from our throat via the nasopharynx upwards to the back of the olfactory epithelium. A great deal of the sensory qualities that we assign to eating and drinking are transduced by the retronasal olfactory system. The activity of chewing and swallowing causes an odorant laced plume to flow from the mouth into the nose, thereby traversing the olfactory epithelium and out of the nostrils. Often what we call the flavour of food and drink is not just its gustatory qualities, but a complex multisensory experience including retronasal olfaction, gustation, somatosensation (including thermal and nociception), and chemothesis (Prescott, 2015; Small, 2012; Stevenson, 2009). Each of these separate sensory channels provides access to the multiphasic object for flavour perception, which compose a unified perceptual experience of flavour. So how many senses of smell do we have? Despite sharing the same range of chemical stimuli and receptor transduction, what will be argued throughout this paper is

[3] Smells, according to MST, are primarily identified and individuated based on their olfactory quality, with their sensory qualities of intensity and hedonics considered to be further features of the smell (for a further discussion of concentration, intensity, hedonics, and valence of smells as properties of the odour object, see Young, 2011, 2019b).

that we only have one sense of smell—orthonasal olfaction. Under laboratory conditions, we can elicit pure retronasal olfactory perception, yet normal everyday retronasal experiences are of multisensory flavour objects. Aside from exceptional cases, it is arguably the case that retronasal sensations do not generate perceptions of smells.

3 Background effects: airflow, mucus, and placement of stimuli contact with epithelium

The separate anatomical pathways of odorant delivery do not exhaust the full range of differences between orthonasal and retronasal olfaction. The direction of airflow, how the stimuli traverse the mucosa layer covering the olfactory epithelium, and where the stimuli make contact with ORNs along the epithelium all play a role in generating distinct perceptual states, despite sharing an identical set of chemical stimuli. This section explores the background stimulus transduction effects that are operant in generating differences based on the separate anatomical pathways.

The direction of air flowing through the nostrils, and respiratory patterns determine whether the chemical stimulus will be perceived orthonasally or retronasally. Whether the stimulus is being brought in through the front of the nostrils or being pushed out from the back of the nasopharynx will determine how it is perceived. Chemical stimuli that are inhaled from the front of the nostrils are perceived as orthonasally, whereas exhalation of odorants from the back of the throat through the front of the nostril generates retronasal olfaction (Shepard, 2009; reviewed in Goldberg, 2018). Exhalation has been shown to be essential for retronasal olfaction, such that cortical areas responsible for encoding of retronasal perception are phase-locked with exhalation breathing patterns (Masaoka *et al.*, 2010).

If we consider the types of perceptible objects by which each system is ordinarily stimulated (odour plumes within smellscapes vs flavourful objects placed within the oral cavity), it makes sense that each system is sensitive to a different direction of airflow. Assuming that the objects of smell perception are odorous gaseous plumes generated by environmental turbulence, then it would make sense that orthonasal olfaction as an exteroceptive sense would be sensitive to airflow and odorants based on inhalation and sniffing originating from stimuli in front of our nostrils, whereas retronasal olfaction is sensitive to the flow of odorants originating within the oral cavity via the back of the throat. The distinct pathways begin to suggest that the types of object each system is designed to sense might be used to distinguish orthonasal olfaction from retronasal olfaction. It might not only be that the systems are sensitive to different types of objects, but also how the same set of chemical stimuli interact with the receptors based on the difference in pathways can shift the perceived olfactory quality, such that the same type of odorant might be perceived as having different olfactory qualities based on the path of delivery.

How the odour stimulus is transported to the olfactory epithelium impacts the nature of olfactory perception. Difference in how the odorant is transported to, and interacts with, the olfactory receptors partially explains the perceptual differences between the systems despite being presented with identical stimuli. The olfactory epithelium is segmented into areas of different types of ORNs, with slight shifts in sensitivity to different types of molecules (Schoenfeld and Cleland, 2005a, b). Moreover, the olfactory epithelium is coated in a layer of mucosa that increases in depth, progressing away from the nostrils. Segmentation of receptor sensitivity to different chemical properties of the stimulus, combined with shifts in the depth of the mucosa, provides further reasons for considering the olfactory system as composed of two systems for stimulus encoding and transduction (Scott *et al.*, 2014). Smaller molecules traverse the mucous layer faster at the front of the olfactory epithelium and more slowly towards the back of the epithelium. Larger molecules take longer to traverse the back of the mucous layer but have a greater chance of making it through the mucus intact and being transduced at the olfactory epithelium. Even if the stimulus is held constant, odorants will have different perceptual profiles based upon where they traverse the mucosa and what area of the olfactory epithelium they interact with. How a stimulus interacts with the anatomical structure of the olfactory epithelium might partially explain why the concentration of an odour needs to be increased from an orthonasal presentation to a retronasal presentation to achieve the same perceived intensity.

How the odorant traverses the mucus partially explains the difference between orthonasal and retronasal perception. Mozell (1966) and Mozell and Jagodowicz (1973), using gas chemotopography, showed that different-size molecules transduced by the olfactory system are absorbed by the mucosa at different rates. Their research generated the sorption hypothesis that the rate at which a molecule is absorbed by the mucus and transported to the olfactory epithelium affects the perception of smells. While there has been some debate regarding whether or not sorption yields a determination of perceived odorant identity or concentration, as modulated by sniffing rates (Cenier *et al.*, 2013), Kent *et al.* (2003) extend Mozell's findings by showing that the sorption rate also affects where the odorant will interact with the ORNs. Where the stimulus places on the olfactory epithelium further influences how it is transduced by the olfactory system. Using a range of aldehydes of different carbon lengths, they showed differential rates of ORN activity within the olfactory epithelium brought about by something as small as a change in a single carbon molecule. The different sorption rates for stimuli are further borne out by Wilkes *et al.*'s (2009) results that the rate of sorption varies even with binary mixtures. Their research shows that stimulus presentation can be distinguished using differences in concentration rates between orthonasal and retronasal perception, which is corroborated by Scott (2007) and Scott *et al.* (2014) that sorption and intranasal airflow show differentiations for different types of molecular structures. Not only does mucus play a role in the rate at which the stimulus can interact with the epithelium, but also the sorption rates and where the odorant comes into contact with the olfactory epithelium make a difference.

In addition to direction of airflow, sorption rate, and place of transduction on the olfactory epithelium, Nagashima and Touhara (2010) provide evidence that the enzymatic reactions within the mucus affect the olfactory stimuli for *in vivo* orthonasal olfaction. Whether this is applicable to retronasal olfaction and how it might effect perceived olfactory quality requires further research, but it seems likely that the mucus itself plays a role (in terms of its enzymatic reactions) in the transduction of olfactory stimuli, which will have a differential effect upon odorant transduction based on the depth of mucus that the odorant traverses to reach the olfactory receptor's cilia—assuming the quantity of mucus interacted with by an odorant will increase the rate of enzymatic reactions.

What is of further interest is that we can identity odorant onset relative to orthonasal or retronasal presentation, yet without training (Negoias *et al.*, 2013), we are not able to localize the lateralization of the odorant to a nostril (Frasnelli *et al.*, 2009, 2010). We might be able to localize and identify whether or not we are having an orthonasal or a retronasal experience, but we cannot differentiate odorant onset between nostrils. These results might depend upon the role of the trigeminal nerve in the nostrils; nevertheless, in combination with other sensory systems, these separate pathways generate a distinct capacity to identify the origin of the perceptible object. Thus, each system is tuned to transduce a different range of perceptible objects in a manner that might yield differences in the perceived olfactory quality of the same type of odorant based on how the stimulus reaches, and is transduced by, the receptor neurons.

Given differences of sensory transport, stimulus transduction, odour encoding, and perception, it is arguably the case that olfaction has two different sensory systems despite sharing the same transduction mechanisms and range of stimuli (for a review of further evidence supporting the claim that olfaction has two sensory systems, see Bojanowski and Hummel, 2012; for a review with clinical data, see Goldberg *et al.*, 2018). Moreover, in reviewing differences between the two olfactory systems, Goldberg *et al.* (2018) survey further evidence that the olfactory quality of an odorant might depend upon the route of delivery. Aside from the differences in airflow pathways, the placement of the stimuli on the olfactory epithelium, and sorption rates, further differences distinguish orthonasal and retronasal olfaction.

4 Differences in sensory transduction

The two systems of olfaction can be distinguished in terms of their respective activation patterns on the olfactory epithelium relative to the direction airflow, which might play a role in explaining Frasnelli *et al.*'s (2009) finding that we can localize the onset of an odorant between orthonasal and retronasal presentations. However, ORN sensitivity also differs between orthonasal and retronasal transduction of the same stimulus. Small *et al.* (2005) showed that regardless of the chemical structure of the odorant, receptivity differs, as measured by the threshold for orthonasal and retronasal stimuli. Orthonasal olfaction has a lower threshold and a greater sensitivity for the same stimuli than retronasal olfaction, which requires higher concentration levels to achieve the same level of judged intensity. Further differences regarding olfactory sensitivity relative to route and placement of presentation on the epithelium can also be found in Frasnelli *et al.* (2004), such that the anterior portion of the epithelium shows increased chemosensitivity whereas the posterior area shows increase sensitivity towards mechanical simulation. Even ORN sensitivity shows a direct difference, as measured by event-related potential (ERP), with decreased amplitudes and longer latency when measuring retronasal olfactory sensory transduction. Taken together, these studies suggest that even if the same stimulus is presented via the two pathways, the sensitivity of ORNs within the olfactory epithelium yields differences in perceived sensory qualities.

There are further differences in transduction and processing at the sensory encoding stage, depending upon how the olfactory stimulus is combined with a gustatory stimulus. Orthonasal and retronasal presentations of the same odorant, combined with congruent or incongruent gustatory stimuli, yielded differences in olfactory sensitivity and processing at the initial transduction stage of olfactory processing (Welge-Lussen *et al.*, 2009). Incongruent gustatory stimuli, combined with an orthonasally presented odorant, were interpreted as possibly increasing arousal based on conflict priming, while congruent gustatory stimuli, combined with retronasal odorants, enhanced processing, as determined by ERP measures. Not only do olfactory stimuli show differential processing, when combined with gustatory stimuli, but the olfactory system also shows differential responses relative to the route of a presentation.[4]

Sensory habituation also cleaves the systems apart. Using salivary responses, as a measure of habituation, Bender *et al.* (2009) showed that salivary responses display a rebound effect after habituation to food odours that is different between orthonasal and retronasal stimulation. Additionally, habituation would not occur if the odorant was presented first to one system and then to the other, which indicates a lack of overlap in the encoding between the two sensory transduction systems in olfaction. Based on these results, the authors concluded that at least for food odours, orthonasal and retronasal olfaction should not be considered the same perceptual system. Thus, at least relative to what we consider the perceptible object of retronasal olfaction—that is, consumable entities placed within the mouth, the two systems treat the same set of stimuli differently. The research surveyed thus far suggests that the same set of chemical stimuli and their resultant odour plumes might generate different perceptual qualities, depending upon whether they are transduced and encoded by orthonasal or retronasal olfaction.

5 Differences in cortical encoding

Orthonasal and retronasal olfaction can also be contrasted based on their distinct cortical pathways and processing centres (cf. Hummel (2008) who surveyed both electroencephalographic (EEG) and ERP differences, as well as cortical processing differences). Orthonasally presented odorants are initially transduced at the olfactory epithelium, encoded within the olfactory bulb and then processed within the piriform cortex, which then projects to the primary olfactory centres within the orbitofrontal cortex (OFC) (for an overview of the unique cortical pathways of orthonasal olfaction in connection with consciousness, see Young, 2012). While there is some convergence between retronasal and orthonasal cortical processing within the OFC, retronasal olfaction is mostly processed in separate cortical areas that show a sensitivity to food-related odorants and a convergence with other sensory systems that transduce objects placed within the oral cavity.

An early study showed variance in cortical processing by identifying activation in separate cortical systems for the transduction of chocolate odour.[5] Small *et al.* (2005) showed that there is an experience-dependent difference between retronasal and orthonasal presentation of food odours. Unsurprisingly there are also differences in cortical activation relative to breathing patterns for inhalation vs exhalation. Employing EEG, Masaoka *et al.* (2010) measured differential activation patterns—different cortical areas are activated for retronasal

[4] Further anatomical differences related to the route of delivery in sensory transduction also separate orthonasal from retronasal olfaction. Using a limited model, which did not take into account the chemodynamics of the odorants, Verhagen (2015) showed that lung retention might affect retronasal olfaction. Their results indicate that lung retention might cause around a 74% reduction in the concentration of the odour profile for retronasal olfaction.

[5] It should be noted that chocolate and possibly lavender are outlier stimuli that are reported as having the same olfactory quality, regardless of presentation to retronasal or orthonasal olfaction. While seemingly counterexamples to half of the thesis of the paper that olfactory qualities differ between the two systems, thereby allowing the individuation of smell from flavour perception, the contradiction is only apparent. The claim of this paper should not be taken as a necessary truth that all stimuli will yield different qualities, but rather the weaker claim that in the vast majority of cases, the difference in olfactory quality can be taken as defeasible empirical evidence, combined with the other evidence that the systems are also poised to perceive different types of perceptible objects.

olfaction, including the anterior ventral insula (AVI) that shows sensitivity to the type and direction of airflow.

Different kinds of olfactory stimulation also show differences in cortical processing. For instance, activation of the central sulcus (CS) for retronasal olfaction possibly indicates a mechanism for oral referral, whereby the perception of a retronasal stimulus is reported as arising from activity in the oral cavity (Small *et al.*, 2005). The activation pattern in the CS relative to retronasal presentation is independent of the encoding of the direction of airflow, as this was controlled for by continuous airflow for both retronasal and orthonasal stimulus delivery. Small *et al.*'s findings also show activation of the Rolandic operculum, at the base of the CS, that corresponds to the representation of the oral cavity. Activation of the CS was predominantly measured during retronasal olfaction and was not apparent for orthonasal olfaction. Their study documents differential activation within the primary areas of cortical olfactory processing for food odours, depending upon which of the two olfactory systems is transducing the stimulus. These findings indicate that there is cortical sensitivity for the oral presentation of a stimulus relative to retronasal olfaction that does not show activation when the same stimuli are delivered orthonasally. Furthermore, their results identify key olfactory areas of the cortex that have preferential firing patterns for food odours when the stimulus is presented retronasally. Retronasal cortical processing is deeply intertwined with the encoding of sensory information from other transduction channels responsible for perceiving consumable flavourful objects; hence it seems that retronasal olfaction is primed to function as a sensory transduction system whose components work in cohort to track and perceive a different type of perceptible object from orthonasal olfaction. Taken together, these results indicate a difference in the type of perceptible object each system transduces, as well as a differential sensitivity to ranges of chemical stimuli identified in terms of their type of sensory quality.

The anterior cingulate cortex (ACC) seems to be responsible for the representation of retronasal olfaction when combined with gustatory information. In addition to activation of the AVI, there are measurable differential responses in the olfactory cortex for retronasal olfaction (Small *et al.*, 2005). The OFC and ACC show increased activation for the combination of a retronasal stimulus with a gustatory stimulus, yet decreased activation when the stimulus is presented orthonasally coupled with a gustatory stimulus. In more recent studies, Small *et al.* (2012) argue that the AVI is responsible for the representation of flavour perception, as it is the first area to show integration between gustatory, tactile, and perhaps thermal perception, together with retronasal olfaction. Their results build upon research done by Cerf-Ducastel and Murphy (2001) measuring increased AVI activation for retronasal olfactory stimulants. The studies of the cortical processing centre of retronasal olfaction as partially located in the AVI, together with the ACC, indicate that retronasal olfaction is mostly encoded and processed as different patterns of cortical activation. Further research supporting the claim that the two olfactory systems employ distinct cortical networks for their encoding and processing is reviewed in Shepherd (2012) who documents cells within the OFC that display sensitivity for the combination of tastes, odours, and touch.[6]

[6] Stevenson (2009) postulates that shared cells of this type within the OFC combine the different sensory channels of flavour perception not as the mere concatenation of kinds, but in a synthetic fashion, which he suggests might explain wine tasters' difficulties in perceiving flavour components. If the cortex encodes the sensory profile of a complex stimulus as wine as a unified percept, it is quite reasonable to think that perceptually teasing the components apart might require a specialized skill set and perceptual probing (Smith, 2009). Additionally, it shows a similarity

The differences in cortical encoding of retronasal olfactory stimuli in the AVI, inferior cingulate cortex, and OFC are echoed by research on the role of attention relative to orthonasal and retronasal olfaction, when combined with gustatory stimuli. Veldhuizen and Small (2011) showed that attentional modulation could be utilized to separate cortical systems for orthonasal and retronasal olfaction, as well as gustatory perception. Their study demarcated separate attentional areas serving orthonasal and retronasal olfaction within the cortex, as well as identifying cortical areas in the insula that are responsive to both retronasal and gustatory attentional modulation. Their results further validate the claim that the two systems employ different cortical processing pathways and centres that are sensitive to different types of perceptible objects and sensory qualities. Additionally, given the evidence surveyed above regarding cortical processing underlying retronasal olfaction, it is arguably the case that the primary function of retronasal processing is not to perceive odour plumes independent of the other features of flavourful objects that are transduced by gustatory, somatosensory, and thermal sensory system.

6 Perceptual differences

The literature surveyed thus far suggests that olfaction has two sensory pathways that are sensitive to different types of perceptible objects, which might explain differences in the perception of olfactory quality, even with the presentation of the same odorants. Until recently, it was not possible to deliver an odour to the retronasal olfactory system independent of other sensory channels. As Shepherd (2012, p. 17) puts it, 'retronasal smell is never sensed by itself'. While this might be true under naturalistic conditions, whereby a stimulus must be placed within the oral cavity, thereby stimulating somatosensory, thermal, or gustatory receptors, Heilmann and Hummel (2004) developed a technique for the direct stimulation of orthonasal and retronasal olfactory systems. Tubing is placed through the nasal cavity, ending either at the front of the nostrils or directly below the olfactory epithelium close to the nasopharynx. By placing the tubing in the nostrils in these two locations, a stream of odorant-laced air can be delivered directly to each system. Even in these extreme laboratory conditions, different olfactory qualities are reported based on the route of delivery. The same set of pure odorants delivered directly to each system yields differences in perceived olfactory qualities.[7]

Heilmann and Hummel's (2008) research lends support to previous findings of a decreased threshold for orthonasal perception, as compared to retronasal olfaction, which is also observed in congenitally blind individuals. Gagnon *et al.* (2015) documented that congenitally blind individuals have increases in orthonasal perception, as measured by odorant threshold, yet there is no change in thresholds for retronasal perception.[8] These differences in threshold

with the format of the smell percept that has been argued to be functionally compositional, yet often not synchronically decomposable (Young, 2019c).

[7] Hummel *et al.* (2006) provide a fantastic review of the differences in perception between orthonasal and retronasal perception, including context of presentation, airflow, and trigeminal stimulation, as well as ORN receptivity.

[8] Their results might be partially attributed to using orthonasal olfaction for spatial perception, while lack of increase in retronasal perception might be attributed to lack of food foraging behaviour and the range of types of food consumed by blind individuals.

help explain the difference in perception of overall odorant intensity between orthonasal and retronasal olfaction. Retronasal olfaction has a higher odorant threshold, such that the same stimulus is judged as less intense as the same concentration when presented orthonasally (Heilmann and Hummel, 2004). Additionally, Diaz (2004) reported differences in judged concentration of an odour based on the pathway of delivery. To achieve the same perceived intensity between the two systems, the concentration cannot be held constant between the route of delivery but must be modulated to control for the overall concentration of the headspace of the odorant relative to the route of delivery.

The disparities in intensity and threshold might explain variations in perceptual matching that indicate the olfactory percept depends upon the route of delivery[9] and familiarity of the odorant (Hannum *et al.*, 2018). The differences of perceived intensity and difficulties in matching are further supported by research on odour identification with the same set of stimuli for orthonasal vs retronasal presentations. Rozin (1982) was the first to indicate the difficulty in identifying the same stimulus with retronasal presentations that were originally learnt via orthonasal delivery. His findings are corroborated by more recent research that shows odour identification is better when the stimulus is learnt and matched in the same pathway of delivery, and decreases for heterogeneous pairings of perceptual presentations (Sun and Halpern, 2005).

Orthonasal olfaction and retronasal olfaction show further dissociations in clinical cases involving the loss of smell. Both Hummel (2008) and Smith (2015) have interesting discussions of clinical dissociations that provide reasons to think olfaction has two independent sensory systems. Furthermore, Landis *et al.* (2005) document 18 patients with no complaints of taste (including flavour) dysfunction, yet they display a deficit in orthonasal olfaction. Using an identification task, the average subject failed the orthonasal task at 36%, as well as having no ERPs recorded for the orthonasal presentation. By contrast, the retronasal identification test yielded 56% of correct identification with recorded ERPs of ORNs. Based on these findings, Landis *et al.* argue that the two systems might not just be functionally distinct, but possibly structural distinguishable. They hypothesize that the olfactory epithelium is structurally organized to be sensitive to orthonasal presentations vs retronasal presentations of the same set of odorants, which would both explain their clinical data, as well as substantiate previous claims (see Section 3) that the direction and placement of the odorant on the epithelium based on the route of delivery generate differences in perceived olfactory quality.

The differences in perception between orthonasal and retronasal perception include context of presentation, airflow, and trigeminal stimulation, as well as ORN receptivity (for a full review, see Hummel *et al.*, 2006). However, for the purposes of the thesis of the paper, what needs to be noted is that the perceptible object of each system is different and that the further pathways of delivery of the odorant to the olfactory system generate upstream changes in the way the molecular structure of the chemical compounds that compose the perceptible object are transduced at the receptor level, encoded by the receptors, processed within the cortex, and perceived by the individual. Given all of these differences, it seems safe to conclude that for at least two out of the three questions involved in answering the question 'what are smells?', the answers will differ with respect to orthonasal vs retronasal olfactory stimulation.

[9] There is some limited evidence suggesting that olfactory adaptation might also depend upon the route of delivery. Pierce and Simons (2018) provide experimental evidence that orthonasal adaptation occurs at a different rate than retronasal olfaction and that there is no cross-adaptation between each route of delivery.

We do not use retronasal olfaction to perceive the same type of perceptible object, nor do the same set of stimuli transduced by the two systems yield the same olfactory qualities.

7 Conclusion

The dissociation between the olfactory sensory systems, combined with the fact that pure retronasal olfactory perception occurs only under laboratory conditions, makes it reasonable to think that retronasal olfaction does not independently give rise to conscious percepts. Rather, retronasal olfaction plays a dominant role in generating our uniform perception of flavour (Stevenson, 2009). Retronasal olfaction is sensitive to food-related odorants, which are bound into a uniform conscious perception based upon experience-dependent learning of congruent stimulus presentation (Bender *et al.*, 2009).

Flavour is its own perceptual modality that synthetically binds multiple sensory channels. Small (2012) and Prescott (2015) each argue that the uniform perception of flavour is generated by the brain based upon superadditive responses in the cortex, as indicated by research identifying cortical activation within processing centres sensitive to gustatory qualities, retronasal odours, somatosensation, and thermal properties of perceptible objects that we consume. The most compelling evidence to support this line of argument is that oral referral (the phenomena whereby we locate the flavour experience in the mouth, despite predominantly being generated from retronasal sensation) is based upon the congruent presentation of gustatory tastes, together with retronasal olfactory sensations (Lim and Johnson, 2012).[10] While other perceptual modalities, such as vision might influence our experience of flavours, vision might not be considered an essential part of the flavour perception (Small, 2012). Similarly, Prescott (2015) argues that while gustatory, retronasal olfaction, and somatosensory stimulation in the oral cavity are all essential to flavour perception, vision, audition, and orthonasal olfaction are not essential and merely influence our perception of flavour.[11]

[10] Lim and Johnson (2011, 2012) across two studies showed that oral referral might not be mediated by simultaneous tactile stimulation when an odorant is presented to the retronasal system, but more predominantly by taste sensations on the tongue. Their initial study showed that it is not necessary to thread tubing through the nose to generate retronasal olfactory perceptions of odorants localized as occurring within the nose, but also that tactile stimulation on its own is not always sufficient to generate oral referral. In the later study, they further showed that the congruency of a tastant with the odorant modulated oral referral. What is of further interest for the purposes of the argument of this paper is that more than 50% of the participants reported they localized the stimulus location of a pure odorant present retronasally as occurring within the nose, which might be taken as indicating that retronasal olfaction under some experimental conditions transduces the same perceptible object as orthonasal olfaction. Yet, this conclusion might not be warranted, given their studies' focus and design. In the first study, while the stimulus was sometimes localized to the nose, this might be attributed both to the method of reporting (chart with options) and learnt association as borne out by the role of congruency of tastants. In fact, their study seems to support the claim that under normal perceptual conditions, we treat the tactile, gustatory, and retronasal olfactory sensory channels as contributing to the perception of a unified flavour object. And in the later study, about half of the time, the participants' reports could also be explained, as the authors noted in their discussion, based on learnt association, such that under normal naïve conditions, the participant might have always localized the stimulus to the oral cavity or tongue, but given that in this instance, there was a tasteless jelly on their tongue, they inferred that the perceptible qualities of the aberrant stimulus could only be transduced by the olfactory system.

[11] The argument on offer throughout this chapter is in keeping with O'Callaghan's framework with its distinction between object for perception and object of perception, but does not explicitly frame the argument as such. What follows explicates this fit, since the evidence offered suggests that it is not clear that retronasal olfaction under naturalistic conditions has an object for perception. Moreover, retronasal olfaction's suite of perceptual abilities required to track even an object of perception is not the same as those required for orthonasal object perception. Furthermore, it is speculatively the case, given the convergence of cortical processing from a host of different sensory systems that track consumable entities within the oral cavity/headspace, that flavour perception is objective in the stronger sense of having an object for perception. Thus, retronasal olfaction does not share the same perceptible

Orthonasal perception allows us to smell, whereas retronasal olfaction is a sensory channel used for perceiving flavourful entities. Only under experimental conditions can we experience pure retronasal perception. Even in situations in which we use experimental methodologies to present odorants to the retronasal system independently of sensations in the oral cavity, these odorants do not generate the same olfactory qualities. Orthonasal olfaction should be considered the modality of smell, while the retronasal system is best thought of as a sensory pathway bound together with other sensory systems that, when combined in the right sort of way, yields our unified perception of flavourful objects.

Our everyday experiences of odours as individuated in terms of the perceptible object of smell and olfactory qualities derive from orthonasal olfactory perception, while those that might be derived from retronasal olfaction not only differ in olfactory quality, but also are not attributed to smell, but to the perceptual activities in the oral cavity. We do not claim to smell retronasal olfactory qualities. Even when attention is brought to the dominant role of olfaction in generating our experience of consuming food or drink, we are more likely to refer to these as parts of our flavour experience. We do not smell flavours.

Employing orthonasal olfaction, we perceive the olfactory quality (i.e., smell) of odorous objects (conceived of as the molecular structure of chemical compounds within odour plumes), which can be corroborated based on our everyday experiences of smells (with some slight modulations to our naïve assumptions concerning ordinary objects). Retronasal olfaction allows us to perceive neither the olfactory qualities nor the objects of smell in this sense, but rather it forms a sensory system that provides a constitutive component of flavour perception. The retronasal olfactory system can be differentiated based on its difference in both olfactory qualities and perceptible object. Having established that retronasal olfaction can be distinguished from orthonasal smell based on anatomical pathways, sensory transduction, cortical encoding, and conscious perception, as well as the perceived olfactory quality of the same stimulus differing between modes of delivery, it seems reasonable to conclude that retronasal olfaction does not smell.

Acknowledgements

Portions of this paper have been presented at the Olfactus 2019 meeting held at the Stockholm University Brain Imaging Centre, the Ruhr-Universität Bochum, Philosophy Colloquium, and the work in progress seminar at the Centre for Philosophical Psychology at Antwerp University. I would like to thank all of the members of the audience on these occasions for their helpful questions and feedback. In particular, I would like to thank Asifa Majid, Barry Smith, Bence Nanay, Jonas Olofsson, Solveig Aasen, Tobias Schlict, and Keith Wilson for their comments on different aspects of this research project.

object as orthonasal olfaction, despite sharing the same type of distal stimulus. With this part of the argument in place, it then becomes apparent that flavour perception would not be multimodal, as each sensory system would not be providing its own proper parts, but rather it would be considered as having a uniform object of perception that can only be perceived as such by multisensory channels, each transducing features of the flavourful entity conceived of as multiphasic dynamic entities placed within our mouths and permeating our headspace.

References

Auvray, M. and Spence, C. (2008). The multisensory perception of flavor. *Consciousness and Cognition*, 17, 1016–31.

Bender, G., Hummel, T., Negoias, S., and Small, D. M. (2009) Separate signals for orthonasal vs. retronasal perception of food but not nonfood odors. *Behavioral Neuroscience*, 123, 481–9.

Bojanowski, V. and Hummel, T. (2012). Retronasal perception of odors. *Physiology and Behavior*, 107, 484–7.

Cenier, T., McGann, J. P., Tsuno, Y., Verhagen, J. V., and Wachowiak, M. (2013). Testing the sorption hypothesis in olfaction. *Journal of Neuroscience*, 33(1), 79–92.

Cerf-Ducastel, B. and Murphy, C. (2001). fMRI activation in response to odorants orally delivered in aqueous solutions. *Chemical Senses*, 26, 625–37.

Diaz, M. E. (2004). Comparison between orthonasal and retronasal flavour perception at different concentrations. *Flavour and Fragrance Journal*, 19, 499–504.

Frasnelli, J., Ariza, V. Charbonneau, G., Collignon, O., and Lepore, F. (2010). Localisation of unilateral nasal stimuli across sensory systems. *Neuroscience Letters*, 478, 102–6.

Frasnelli, J., Charbonneau, G., Collignon, O., and Lepore, F. (2009). Odor localization and sniffing. *Chemical Senses*, 34(2), 139–44.

Frasnelli J., Heilmann, S., and Hummel, T. (2004). Responsiveness of human nasal mucosa to trigeminal stimuli depends on the site of stimulation. *Neuroscience Letters*, 362(2004), 65–9.

Frasnelli, J., Ungermann, M., and Hummel, T. (2008). Ortho- and retronasal presentation of olfactory stimuli modulates odor percepts. *Chemosensory Perception*, 1, 9–15.

Gagnon, L., Ismaili, A. R. A., Ptito, M., and Kupers, R. (2015). Superior orthonasal but not retronasal olfactory skills in congenital blindness. *PLoS One*, 10(3), e0122567.

Goldberg, E. M., Wang, K., Goldberg, J., and Aliani, M. (2018). Factors affecting the ortho- and retronasal perception of flavors: a review. *Critical Reviews in Food Science and Nutrition*, 58(6), 913–23.

Hannum, M., Stegman, M. A., Fryer, J. A., and Simons, C. T. (2018) Different olfactory percepts evoked by orthonasal and retronasal odorant delivery. *Chemical Senses*, 43, 515–21.

Heilmann, S. and Hummel, T. (2004). A new method for comparing orthonasal and retronasal olfaction. *Behavioral Neuroscience*, 118, 412–19.

Hummel, T. (2008). Retronasal perception of odors. *Chemistry and Biodiversity*, 5, 853–861.

Hummel, T., Heilmann, S., Landis, B. N., Redden, J., Frasnelli, J., Small, D. N., *et al.* (2006). Perceptual differences between chemical stimuli presented through the ortho- or retronasal route. *Flavour and Fragrance Journal*, 21, 42–7.

Kent, P. F., Mozell, M. M., Youngentob, S. L., and Yurco, P. (2003). Mucosal activity patterns as a basis for olfactory discrimination: comparing behavior and optical recordings. *Brain Research*, 981, 1–11.

Landis, B. N., Frasnelli, J., Reden, J. Lacroix, J. S., and Hummel, T. (2005). Differences between orthonasal and retronasal olfactory functions in patients with loss of the sense of smell. *Archives of Otorhinolaryngology—Head and Neck Surgery*, 131, 977–81.

Lim, J. and Johnson, M. B. (2011). Potential mechanisms of retronasal odor referral to the mouth. *Chemical Senses*, 36, 283–9.

Lim, J. and Johnson, M. B. (2012). The role of congruency in retronasal odor referral to the mouth. *Chemical Senses*, 37, 515–21.

Masaoka, Y., Satoh, H., Akai, L., and Homma, I. (2010). Expiration/the moment we experience retronasal olfaction in flavor. *Neuroscience Letters*, 473, 92–6.

Mozell, M. M. (1966). The spatiotemporal analysis of odorants at the level of the olfactory receptor sheet. *Journal of General Physiology*, 50, 25–41.

Mozell, M. M. and Jagodowicz, M. (1973). Chromatographic separation of odorants by the nose: retention times measured across *in vivo* olfactory mucosa. *Science*, 181, 1247–9.

Nagashima, A. and Touhara, K. (2010). Enzymatic conversion of odorants in nasal mucus affects olfactory glomerular activation patterns and odor perception. *Journal of Neuroscience*, 30(48), 16391–8.

Negoias, S., Aszmann, O., Cory, I., and Hummel, T. (2013). Localization of odors can be learned. *Chemical Senses*, 38, 553–62.

O'Callaghan, C. (2008). Object perception. *Philosophy Compass*, 3(4), 803–29.

O'Callaghan, C. (2016). Objects for multisensory perception. *Philosophical Studies*, 173, 1269–89.

Pellegrino, R., Atchley, A., Ali, S., Shingleton, J., and Luckett, C. R. (2018). Retronasal habituation/characterization and impact on flavor perception using time-intensity. *Chemosensory Perception*, 13, 1–10.

Pierce, A. M. and Simons, C. T. (2018). Olfactory adaptation is dependent on route of delivery. *Chemical Senses*, 43, 197–203.

Prescott, J. (2015). Multisensory processes in flavour perception and their influence on food choice. *Current Opinion in Food Science*, 3, 47–52.

Rozin, P. (1982). Taste-smell confusions and the duality of the olfactory sense. *Perception and Psychophysics*, 31, 397–401.

Schoenfeld, T. A. and Cleland, T. A. (2005a). The anatomical logic of smell. *Trends in Neuroscience*, 28, 620–7.

Schoenfeld, T. A. and Cleland, T. A. (2005b). Anatomical contributions to odorant sampling and representation in rodents/zoning in on sniffing behavior. *Chemical Senses*, 31, 131–44.

Scott, W., Acevedo, H. P., Sherrill, L., and Phan, M. (2007). Responses of the rat olfactory epithelium to retronasal air flow. *Journal of Neurophysiology*, 97(3), 1941–50.

Scott, J. W., Sherill, L., Jiang, J., and Zhao, K. (2014). Tuning to odor solubility and sorption pattern in olfactory epithelial responses. *Journal of Neuroscience*, 34(6), 2025–36.

Shepherd, G. (2012). *Neurogastronomy*. New York: Columbia University Press.

Small, D. M. (2012). Flavor is in the brain. *Physiology and Behavior*, 107, 540–52.

Small, D. M., Gerber, J. C., Mak, Y. E., and Hummel, T. (2005). Differential neural responses evoked by orthonasal versus retronasal odorant perception in humans. *Neuron*, 47, 593–605.

Smith, B. (2009). The objectivity of tastes and tasting. In *Questions of Taste*, edited by B. Smith, 41–77. Oxford: Oxford University Press.

Smith, B. (2015). The chemical senses. In *Oxford Handbook of Philosophy of Perception*, edited by M. Matthen, 567–86. Oxford: Oxford University Press.

Stevenson, R. J. (2009). *The Psychology of Flavor*. Oxford: Oxford University Press.

Stevenson, R. J. (2014). Object concepts in the chemical senses. *Cognitive Science*, 38(7), 1360–83.

Sun, B. C. and Halpern, B. P. (2005). Identification of air phase retronasal and orthonasal odorant pairs. *Chemical Senses*, 30, 693–706.

Veldhuizen, M. G. and Small, D. M. (2011). Modality-specific neural effects of selective attention to taste and odor. *Chemical Senses*, 36, 747–60.

Verhagen, J.V. (2015). A role for lung retention. *Chemosensory Perception*, 8, 78–84.

Welge-Lüssen, A., Looser, G.-L., Westermann, B., and Hummel, T. (2014). Olfactory source localization in the open field using one or both nostrils. *Rhinology*, 52, 41–7.

Welge-Lussen, A., Husner, A., Wolfensberger, M., and Hummel, T. (2009). Influence of simultaneous gustatory stimuli on orthonasal and retronasal olfaction. *Neuroscience Letters*, 454, 124–8.

Wilkes, J. F., Laing, D. G., Hutchinson, I., Jinks, A. L., and Monteleone, E. (2009). Temporal processing of olfactory stimuli during retronasal perception. *Behavioural Brain Research*, 200, 68–75.

Young, B. D. (2011). *Olfaction: Smelling the Content of Consciousness*. [PhD thesis] City University of New York, NY, 3478817.

Young, B. D. (2012). Stinking consciousness! *Journal of Consciousness Studies*, 19, 223–43.

Young, B. D. (2016). Smelling matter, *Philosophical Psychology*, 29(4), 520–34. doi: 10.1080/09515089.2015.1126814

Young, B. D. (2017). Enactivism's last breaths. In *Contemporary Perspective in the Philosophy of Mind*, edited by M. Curado and S. Gouveia, 106–27. Cambridge: Cambridge University Press.

Young, B. D. (2019a). Smelling molecular strucute. In *Perception, Cognition, and Aesthetics*, edited by D. Shottenkirk, S. Gouveia, and J. Curado, 64–84. Routledge Press.

Young, B. D. (2019b). The many problems of distal olfactory perception. In *Spatial Senses: Philosophy of Perception in an Age of Science*, edited by T. Cheng, O. Deroy, and C. Spence, 148–69. Routledge Press.

Young, B. D. (2019c). Smell's puzzling discrepancy. *Mind and Language*, 35, 90–114. doi.org/10.1111/mila.12233

Young, B. D. (2020). Perceiving smellscapes. *Pacific Philosophical Quarterly*, 101(2), 203–23. Available from: https://link.springer.com/article/10.1007/s11098-019-01371-4

Young, B. D., Escalon, J., and Mathew, D. (2020). Odors: from chemical structures to gaseous plumes. *Neuroscience and Biobehavioral Reviews*, 111, 19–29. https://doi. org/10.1016/j.neubiorev.2020.01.009

PART 3

SENSORY INDIVIDUALS IN MULTIMODAL PERCEPTION

15
Material objects as the singular subjects of multimodal perception

Mohan Matthen

1 Introduction

Each of our external senses is dedicated to certain sensory *features*: vision is concerned with colour, shape, and size; audition with pitch and volume and certain phonetic units; touch delivers information about shape, size, temperature, texture, and weight; and so on. What I want to discuss in this paper is a ubiquitous framework that perception builds for these sensory features: we often, or even dominantly, perceive them as jointly bound to singular subjects.[1] Colour, for example, is dominantly perceived as belonging to a spatially delimited subject that could at the same time be perceived—the very same thing—as emitting a sound, being of a certain temperature, and having a certain smell. These multisensory logical subjects are moreover *privileged subjects* for motion and action. That is, they seem to move, act, and be acted upon as whole units—their parts and features seem to move because *they* do, not the other way around. These privileged multimodal subjects are *material objects*. (I will say more about what material objects are in the Section 3.)

Of course, material objects are not the only bearers of perceptual properties. Liquids and gases have sensed properties as well. However, the logical structure here is different. With certain exceptions (that I will discuss briefly in the Section 5), fluids are generally perceived as agglomerations of parts that have the feature in question, rather than as single causally coherent units that move, act, and are acted upon as wholes. The whole of a body of water has no greater claim to be the bearer of these properties than its parts, and no greater perceptual salience as such.

The feature-bearing role of material objects is a significant feature of our perceptual systems (and those of other higher animals). The classic philosophical analysis—what Imogen Dickie (2010) calls the 'Old Empiricist View'[2]—locates all perceptual features in a spatial 'field' or treats them as simple non-located events. This analysis is often correct—think of visual phenomena like beams or flashes of light and non-visual properties such as sounds and

[1] We also perceive detached sensory properties—the warm light of the early morning, the rustling of a breeze in the forest, the warmth of a fire, and so on.

[2] Dickie concerns herself with vision only, about which the 'Old Empiricist View' asserts that 'visual experience presents us with an array of *features* laid out around us and developing over time' (emphasis added). In a footnote, she mentions that Locke also includes 'sounds, and some tangible qualities'. These are not, in the traditional view, 'laid out around us'—sounds because they lack location, and tangible qualities because they are within us—false, but this is what the Old Empiricist View holds.

Mohan Matthen, *Material objects as the singular subjects of multimodal perception* In: *Sensory Individuals*. Edited by: Aleksandra Mroczko-Wąsowicz and Rick Grush, Oxford University Press.
DOI: 10.1093/oso/9780198866305.003.0016

smells. But it is superficial: it overlooks the crucial role of material objects as subjects for sensory features.[3] My aim in this paper is to explain this structural feature of perception.

2 External perception

Material object-perception is a kind of external perception.[4] So, I will start with this.

To arrive at an understanding of external perception, let us start with a characterization of sensing.

> An organism *O* senses circumstance-type *C* if a part of *O*'s body autonomously gathers and deploys information that emanates from an instance of *C*, and by doing so, enables *O* to respond in a functionally appropriate manner to that instance of *C*.

I sense *warmth*, for example, when temperature sensors in my skin are activated by ambient heat and my touch-sense translates the output of these receptors into an awareness of warmth that enables me to respond as is appropriate to warmth occurring in my vicinity—and when all of this occurs autonomously, that is, without the action of my conscious reasoning and volition.[5]

Understood this way, one might think that *external* perception is just the gathering of information that puts the subject in a position to respond to *external* circumstances. For since the correct response to external circumstances requires information about those circumstances, one might think that it requires external sensing. But this would not be correct. For external response may be mediated by *proximal* sensing. Here is how. Let *C* be some external circumstance to which an organism *O* can respond by sensing. Let *R(C)* be the information-carrying effect that *C* has on sensory receptors in *O*'s body. It could be that *O* is able to respond to *C* by monitoring *R(C)*.[6] Here, *R(C)* carries information about *C*, but *O* does not extract this information; *O* is sensitive to *C* only by being sensitive to its own receptors.

This kind of proxy-for-external-sensing is how plants respond to their environments. Consider, for example, a plant growing a root. Roots tend to put out more lateral roots on the moister than on the dryer side of their bodies; this is how they respond to moisture gradients in the soil. This growth pattern is mediated by the suppression of a cell reproduction factor in dry conditions, resulting in fewer lateral roots on the dry side of the root body (Orosa-Puente *et al.*, 2018).[7] Water contact or dryness on the root surface *carries* information about moisture in the soil beyond—distal moisture is the cause of proximal moisture. The plant, however, does not extract this information. It responds just to conditions on its root surface. This is the kind of case that demonstrates that we do not need to posit external sensing to explain

[3] See Clark, 2004a, b and Matthen, 2004 for a discussion of the issue.

[4] Our own bodies are material objects, but we perceive them in a different way—from the inside. I am excluding this.

[5] Perhaps, *I* don't sense warmth unless I am aware of it; *I* can't respond without such awareness. But my thermoregulatory system may autonomously deploy warmth information by activating my sweat glands, for example, and this can happen without my awareness of either the condition or the response. This shows that sensing need not be whole-organism sensing—thermoregulatory sensing can occur autonomously in my body without *my* being involved. This is a complication that need not detain us here.

[6] Sense-datum theorists take a position of this sort: they hold that all response, including response to external circumstances, is mediated by a proximal effect of those circumstances.

[7] See McElrone *et al.*, 2013 for an introductory overview of water uptake in plants.

externally directed functioning. Tyler Burge (2010, p. 422) sums it up correctly: 'There is no explanatory need to attribute representation of distal attributes, as distinguished from registration of proximal stimulation.'

Because of examples like this, external perception cannot be defined in terms of the target of response. It has to be defined in terms of *sensory content* (or 'representation', in Burge's terminology). External perception has *content* that is 'about' the external world. Though roots of plants respond to distal moisture gradients, their sensory organs do not represent these gradients; they only represent water in contact with their own body surfaces (if they represent that). However, when I hear a man singing on the street, I hear the sound coming from a more or less determinate place distant from me. The content of this auditory experience is external because it does not make me aware of an occurrence in my own body. Rather, it presents me with something happening a certain distance away.

Given that proxies-for-external-sensing exist, how can one be confident of this assessment? How can one rule out the possibility that when I seem to hear a distant singer, I am really aware only of a proximal event—for instance, the vibration of my cochlea—and that my mind passes by dint of repeated exposure from this cochlear event to the idea of something happening far away? Once again, the best answer to this question comes from Tyler Burge (2010). My awareness of the singer is *perceptually constant*. That is, there are changes in my cochlear condition as I listen to the singer—he sings different notes at different volumes; there are other noisy things on the street, and these pollute the sound stream from the singer; I move about my business and receive the sound stream from different distances and angles. But through all of these changes, the perceptual impression of a song sung from the same distant spot persists. There is, in short, a changing effect on my perceptual receptors, but a single continuing experience of the same singer at the same distant spot. This constant perceptual impression cannot simply be the same as awareness of the cochlear disturbance because its content does not vary, though the cochlear disturbance does. Despite variation in the impact on my auditory receptors, the impression covaries with the distant object and informs me of it.

Now, it has been argued that perceptual constancy is learnt, or in some other way acquired by repeated exposure. It is not my intention to contest this. All that I want to insist on is that the perceptual constancy is an autonomous function: it is not a matter of *my mind* moving discursively from cochlear disturbance to distant cause—perceptual constancy is not (as a referee helpfully remarks) a matter of *conceptual inference by the subject*. In fact, it could not be, since I have no awareness of cochlear disturbance. My phenomenal experience is of a song sung by a single singer at a distant location. Whether it is produced by an innate process or by one modified by perceptual learning, it is this experience that allows me to form beliefs about the singer, respond to him with pleasure, and so on. This is what makes it external perception.

3 Material objects

Now, what is the function of external perception? When is it required?

I believe that one can shed some light on these questions by considering the perceptual needs of higher animals. Most animals move about their environments under their own power—they are *motile*. By contrast with sessile organisms—organisms that are not self-propulsive—higher animals such as birds and mammals seek out resources at a distance from themselves. While plants reproduce by courtesy of winds or cooperative animals that spread their seed, higher animals have actually to *do* something to find reproductive opportunities—they have to seek out sexual mates, and to do so, they have first to identify them as such. Animals forage and hunt; plants merely follow nutritional gradients and absorb whatever such tropisms bring them into contact with. Animals recognize landmarks and navigate by them; plants obviously have no need for such things. For all of these functions, external perception is required. But mere external perception does not suffice—it is not enough to identify various perceived locations as external to the subject and to recognize some pattern among them. Additionally, certain material objects have to be identified as occupants of external space—food, mates, predators. So, the spatial field of perception must be so organized as to identify continuities and causal links among points that identify them as belonging to coherent material objects.

What is a material object? Minimally, it is a continuous body of matter that retains its form by internal causal connections. A rigid body is an example. Internal forces ensure that it retains its shape under external forces—when it is pushed with sufficient force from one side, for example, the whole of it moves. When it ingresses another material object—for example, when a fork enters an omelette—it does not seem as if a new composite material object has been created—at least, not unless other conditions are met. A rigid body acts *as a whole.* A baseball bat has properties as a whole that determines what happens when it hits a ball; when it breaks, the joint weight of its fragments and the shape of the spatial volume they occupy have no such relevance.[8]

Now, not all material objects are rigid, but they are all causally unified in such a way that they are the bearers of properties as wholes. My dog is not a rigid body, but his shape deforms only in certain ways: he retains his form in a broader sense of this term. As well, he is a privileged subject. *He* runs and eats: the movements of his limbs and his mouth are seen as his movements.

The perceptual needs of higher animals go beyond identifying points in space as external and attributing qualities to them. For knowing the occupants of these spatial points is not sufficient for knowing what will happen when one interacts with them. Knowing that a baseball bat-shaped volume of space is occupied by bits of wooden matter is not sufficient for knowing that there is something there one can pick up and swing. Knowing that there is a dog-shaped collection of animal parts in my vicinity is not sufficient for knowing whether it is capable of barking and jumping.

Now, of course, everybody knows that there is a difference between a material object and the collection of its detached parts. And everybody will allow that everyday activities demand that one knows which of the two one is confronted with. The classic empiricist line on this is that one is able to discern one from the other by discursively applying general differences that one has learnt to employ in these contexts. I contend that this empiricist line is inadequate for two reasons.

The first is that there is a perceptual phenomenology of whole objects—this is well known from the work of Gestalt psychologists and that of developmental psychologists who have tried to show how this phenomenology emerges in infants. As Jean Piaget (1954, p. 3) writes:

> A world without objects would not present the character of spatial homogeneity and of coherence in displacements that marks our universe. Inversely the absence of 'groups' in the

[8] Elizabeth Spelke (1990, pp. 48–54) discusses rigid bodies as paradigms of material objects.

> changes of position would be equivalent to endless transformations, that is, continuous changes of states in the absence of any permanent object.

In other words, external perception is not merely point-perception; it includes the perception of connected wholes, or 'groups'.

The second reason is that the phenomenology Piaget describes is functionally dictated. External perception would be of no use to higher animals without material object perception. For, as I argued earlier, motility introduces a complexity of action to animal life—animals that move under their own power use that movement to find food and mates and to avoid predators. Since the main point of perception beyond the skin is interaction with other material objects, it would be functionally useless to perceive externally—to have the capacity for perceptual constancy, for example—without perceiving whole material objects. Thus, the 'groups' of which Piaget talks above are not merely spatial groups; they are, and must be, causally connected groups. Piaget's paradigm is based on infants learning by active search, prehension, and interaction with objects. Obviously, these activities require objects to display a certain 'coherence in displacements'. But they have to be more than spatial 'groups'. For, as Piaget writes later (1954. p. 96), an object is a 'truly coherent whole':

> only that phenomenon constitutes a real object which is connected in an intelligible way with the totality of a spatio-temporal and causal system (p. 87)

and

> the object is fully constructed in correlation with causality to the extent that this coordination of schemata results in the formation of an intelligible spatio-temporal world endowed with permanence (p. 88).

4 Generalized perceptual completion

Object-perception has most thoroughly been studied in the case of vision. And this is appropriate, since vision is the modality best adapted to the identification of material objects. Because vision is essentially spatial—information coming from the eyes is spatially laid out from the very beginning of the visual process—its representational structure matches the spatial character of material objects, which occupy connected and discrete places. For this reason, the visual system's algorithms for object-identification are not replicable in the other modalities. In this section, however, I will be concerned with a strategy that does generalize. I call it *Generalized Completion*. I will use it to shed light on multimodal object perception.

Vision is atomistic in the following sense. Every visible object in 3D space projects to a collection of retinal cells. So, given the 2D retinal image, one can ask which collections of retinal places are projections from a single visible object. Classically, in Gestalt psychology and its successors, this question is investigated by inquiring what relations among a collection of retinal-image places mark that collection as the image of a single material object. For example, it is claimed that when we see segments of a smooth curve fractured by continuous line segments, we see the continuous segments as objects in front of, and occluding, a single object with the shape of the fractured curve (see Figure 15.1). Similarly, when a 'moving'

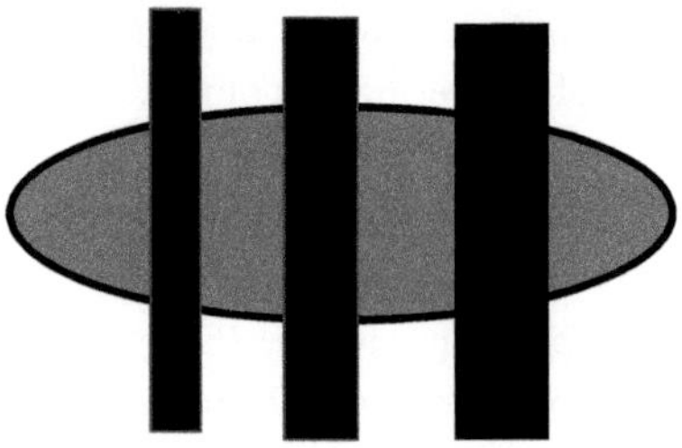

Figure 15.1 Amodal completion: occlusion inferred by continuity.

point disappears and another one later appears in a dynamically plausible continuation of the interrupted trajectory, this is seen as a single object passing behind an occluder and reappearing later. These are principles of completion and continuity among collections of visible places, and it is a commonplace of visual psychology that it is through the application of such principles that the visual system is able to identify and distinguish material objects.

Now, topological principles, such as continuity and completion, are not free-standing. Vision employs them because they enable it to work backwards from a retinal pattern to the kind of object that (likely) projected to that kind of pattern. In keeping with this reconstructive function, the phenomenology of visual object-perception corresponds to the inferred object, not the retinal pattern. In Figure 15.1 (or, rather, in real-world presentations of this pattern), the visual phenomenology is that of an oval (or possibly an ellipsoid) *behind* three rectangles. To put it in a way that sounds somewhat paradoxical, this object *looks* as if it has parts *that cannot be seen*. This is the phenomenon that psychologists call 'amodal completion', the term amodal marking the fact that the continuity of the interrupted contour is visually experienced, even though the occluded parts have no effect on visual receptors. The hidden parts are *visually* experienced in the sense that the impression essentially depends on *looking* at Figure 15.1—just as the impression of the non-occluded parts does. But the phenomenology of the hidden parts is different—their visual characteristics (such as colour) are not experienced.

This is the phenomenological pattern I want to generalize. When a perceptual system processes incident information *I*, and thereby registers the presence of an object of type *T*, perceptual *experience* will be of an object of type T, with obscured features essential to *T* registering merely as present, without their specific characteristics being presented. *Amodal* completion inserts features that would have affected sensory receptors if they had not been hidden.[9] But in what I will call *Generalized Completion*, some of the inserted features are of a kind that *cannot* have a causal effect on sensory receptors. For example, material objects look causally coherent and resistant to ingress—these are modal characteristics and, as such, they do not affect the eyes. This is the result of data processing by the visual system.[10]

[9] There is amodal completion in audition and touch. Albert Bregman (1994) showed that when audition completes a partially obscured continuous melodic line, what we seem to hear is a completed melodic line with partially determinate gaps. And Geldard and Sherrick (1972) showed that when widely separated parts of a subject's arm are stimulated with a series of taps, she experiences a connecting series of 'phantom' taps.

[10] Burge (2010, p. 444) rebukes Elizabeth Spelke for suggesting that essential properties of material objects, such as *cohesion*, *boundedness*, *solidity*, and *spatio-temporal continuity*, are imperceptible. He points out that the attribution of such features can be 'generated within perceptual systems'. I have tried to accommodate a correct premise of Spelke's thinking without taking on her mistaken inference that the representation of these properties is post-perceptual.

5 Generalized completion in non-visual modalities

Generalized Completion suggests a different approach to the perception of material objects than the one that Casey O'Callaghan (2016) develops. O'Callaghan starts with the observation that the senses have proprietary objects. Vision and touch target material objects, he says, but we 'hear sounds, smell odors, and taste flavors'—not material objects (p. 1270). But, he continues, we need objects common to the senses to accommodate how the senses work together. For example, to accommodate perception of a noisy car moving, we need to see and hear a single object. But if auditory objects are sounds, and sounds cannot be seen, what could the common object be?

So far, all of this is both correct and insightful. The problem comes when O'Callaghan (p. 1275) insists that the common object could not be a material object—we do not hear the *car*, he alleges. Why not? Because 'humans do not hear spatial edges, surfaces, occlusion, or rigidity'. He is claiming, in short, that since audition cannot spatially delimit material objects, it cannot present a material object. Accordingly, he proposes that the common objects of multisensory perception are 'mereological sums' of objects peculiar to each modality. What we see is a car participating in an 'event-like individual' that also has a sound as a constituent part (pp. 1282–3). My question is whether this really captures our phenomenology. When I see and hear a passing car, I seem to perceive one and the same individual in both modalities. I do not merely perceive the sound *fused with* a unimodal *visual* object.[11]

My proposal is that, in cases like this, our perceptual systems infer a common subject by a process of Generalized Completion. What I seem to see and hear is a single object that has unseen and unheard properties. My perceptual system autonomously inserts into perceptual awareness a multimodal, not a merely visual, subject for both the visual and auditory properties that I perceive. This common subject is a material object.

Consider a case like this. I am listening (live) to an orchestra. At a certain point, a flute enters. I cannot quite locate the player by audition alone: she is somewhere off to the left, but that is as close as my hearing lets me get. I look closely at the orchestra, however, and I see the flautist: I can tell which player she is because of how her body moves as she plays her part. Suddenly, the flute part becomes more precisely located; its location autonomously snaps to the visually identified orchestra member. Both sight and hearing now identify *her* as the source of the melody.[12]

The principle that lies behind this audiovisual coordination is simple. In order to play the melody, the flautist performs certain actions with her head, upper body, and arms. These visible bodily motions reveal her as a material object and locate her in space. The coordination of the two modalities employs detailed spatial information that audition by itself cannot access. This affects auditory phenomenology—it seems to me as if she *over there* is producing

[11] It is interesting to note that audition does seem to have access to some of what we would otherwise take to be visible properties: for example, Carello *et al.* (1998) show that subjects were able to estimate by sound the length of dropped dowels made of the same wood. Of course, this effect depends on the shape and rigidity of the dropped objects, and thus on the test objects being material objects.

[12] Spelke *et al.* (1983) suggest that in adults (but not in infants), *impact* is highly relevant to the audio-visual localization of sound: the location of a bang snaps to a visible collision. It's worth remarking that percussive sounds are the products of material objects colliding, and so Generalized Completion takes some credit for creating this appearance. A natural extrapolation would suggest that when you see a moving object interact with a stationary one—for example, a suitcase being pulled over a bumpy floor—you hear *the suitcase* making a noise, not the floor.

sound. I do not *hear* the spatial edges of the flautist, it is true. O'Callaghan is obviously right about this. And I do not *see* the melody she is playing. I also do not see the back of her head or her causal coherence. The properties that make her a material object do not have a direct effect on my visual or auditory receptors. Nevertheless, I have a somewhat indeterminate awareness of these properties. Generalized Completion presents me with a material object as the source of the sound.[13]

The same kind of thing can be said about smell. In the Section 8, I will discuss how smell helps us identify *types* of material substance. Here, let me make a remark about how we determine that it inheres in a material object. Consider the act of bringing something to your nose to smell it. Touch reveals this thing as a solid material object (perhaps containing a liquid). The smell gets stronger as you bring it closer, and weaker as you put it down. It is thereby revealed as an object that has a smell—that is, the smell is revealed as attached to a moving object, rather than free-floating in the way that the smell of flowers is after you have used a spray air freshener. Again, Generalized Completion is at work. You perceive a material object that has a smell.

Now, this account of perceiving smells in material objects is, in effect, contested by Tyler Burge (2010, p. 415). He would say that when I bring something to my nose, I am *locating* a smell by the use of 'a homing or beaconing technique that serially follows directional intensity', a procedure not unlike that followed by plants to grow roots on their moister side. (See Section 2.) I will argue in the Section 8, below that this is an oversimplification—but put this aside for now. My point for the present is that even if it *is* mere 'homing or beaconing', bringing something to my nose allows me to do more than simply *locate* the smell: it allows me to identify the object in my hand as the source of the smell. And this is a *perceptual* achievement—that is, it is an attribution of a sensory feature to a material object autonomously by a perceptual system.

Now, to be fair to Burge, he does say, a couple of paragraphs earlier, that smell is non-perceptual 'unless ... supplemented by input from other sources'. I assume that he would point to my kinaesthetic awareness of my hand moving towards my nose as an instance of 'input from other sources'. If this is his position, he subscribes to a sceptical attitude taken by Clare Batty (2014), who writes: 'everyday olfactory experiences do not possess ... robust spatial representation ... and, as a result, (do) not allow us to single out particular objects in our environment'. Batty proposes that olfactory experience has a 'weak form of abstract content', in which 'there is something or other "here," or "at" the perceiver, that has certain olfactory properties'. Burge's position is more exclusionary, but he agrees that we do not ever have olfactory experience of an object or of a substance. It is only a slight exaggeration to characterize their position this way: I never smell a mango or a glass of whiskey; I simply experience a smell in myself (Burge) or 'at' my body (Batty).

I would argue that this line of thought is unduly restrictive. Briefly, all of the senses use awareness of own-body movement as a way of enriching the spatial information available to them when the body is stationary. Vision uses motions of the eyes, head, and body to gather 3D information; audition supplements these motions with visual information (as we saw above); touch uses the movement of an object against the skin to distinguish bodily stimulation from external stimulation (see Section 7). So, if you exclude multimodal perception, all of the senses become, at the very least, less robustly perceptual.

[13] My point here is an extension of Burge's criticism of Spelke: see fn. 10.

To summarize, all of the sense-modalities that receive information from the external world are capable of representing that information as emanating from sources in the external world. They normally do this by cooperating with each other and with kinaesthesia. As a result, they do not merely *locate* these sources of information in the outside world, but also *attribute* them to discrete material objects.

6 Four kinds of perceptual objects

Firstly, material objects are, as I said at the beginning, multimodal subjects for perceptual features. They are, moreover, privileged subjects for motion and causal agency in the sense that they, not their parts, are felt in perceptual awareness to be the primary movers. That is, their parts are felt to move and to act because they are as wholes. They appear this way because they appear to be causally coherent.

Now, material objects are not the only subjects we perceive—we also perceive liquids and ephemera such as shadows and beams of light (Matthen, 2018). And in non-visual modalities, we perceive sounds, smells, and flavours. These individuals lack one of the two perceptual characteristics of material objects. Either they are not multimodal, or they are not privileged subjects.

Secondly, start with liquids: perceptual experience does not present them as privileged subjects. That is, the parts are not experienced as moving or acting because the whole is. For example, when I take a shower, the column of water does not feel as if it is acting on me as a whole—there is nothing unifying in my experience of the water comparable with how the bar of soap feels in my hand. And, except insofar as it is contained by a material object, I do not see a quantity of water as having a rearward-facing surface, as available for interaction, or as resisting ingression by other material objects—that is, not as a single whole. Finally, I do not experience it as the subject of properties in multiple modalities. I see water flowing from a faucet and hear it splashing in the sink, but I do not perceive these events as characterizing the *column* of water taken as a whole, in the way that the sound of a car and its visible movement characterize it as a whole. A waterfall makes a rushing sound all over, but again no particular bit of it, including the cascade-as-a-whole, is privileged as the primary noisemaker. It seems to make a noise because its parts do.

There is an illuminating qualification that must be made about this generalization about liquids. When they naturally clump, for example, when they fall in drops, they look more like material objects. And the more viscous they are, the more they clump. Thick paint coalesces into blobs, spreads in step-like streams, forms relatively long-lasting ripples when stirred, and forms non-vertical streams when poured with an unsteady hand. These movements are closer to those of material objects than those of less viscous liquids. They allow us to perceive viscosity (van Assen *et al.*, 2018). In short, more viscous liquids are more causally coherent and are consequently perceived to be more like material objects.

Thirdly, consider visual ephemera such as cast shadows, rainbows, and beams of light. These are by-products of boundary-drawing for material object-perception and, as such, they are privileged subjects. They look like material objects in certain respects,[14] but they

[14] Shadows should not be thought of as completely ephemeral. Cast shadows resemble, and are causally connected to, their objects and thus they serve as sources of information. See Pavani and Galfano, 2015 for a review. (Here, I am grateful to a referee.)

also look unlike them in others. Shadows, for example, are delimited, but they do not look causally coherent—it is no surprise when a moving cast shadow is momentarily stretched and distorted as it passes over something. Nor do they look as if they have a rearward-facing side. Shadows and beams of light do not look as if they resist ingression. Crucially, these objects are not subjects for multisensory feature attribution. They are not perceived as sources of sound or smell.

Fourthly, and finally, non-visual objects. These too are privileged subjects. As O'Callaghan (2016, pp. 1274–5) writes:

> ... sounds are public objects of audition, even if they are merely audible. They are audible bearers of perceptible attributes such as pitch, timbre, loudness, and duration. Multiple distinct sounds are audible at a time, sounds persist and survive change, and sounds can occlude and mask each other. You can attend to a sound in contrast to its features, its location, the material object that makes it, or its audible background. You can track a sound over time and form audition-based demonstrative thoughts about it...

However, sounds are unimodal and fall short of material objects in this other critical dimension.

7 Point-content and object-content: the case of touch

We examine things closely by holding and handling them. We visually examine them by turning them over in our hands; we listen by holding them up to our ears; we smell by bringing things up to sniff. These activities demonstrate that the things we touch are also the subjects of features in other modalities. And our manner of handling these things shows that they are privileged subjects of dynamic properties—the thing that is in our hands is the very thing we visually examine or listen to or smell. In short, they manifest both hallmarks of material objects identified in the preceding sections: they are privileged subjects for properties in different modalities.

There is, however, a special problem understanding our perceptual grasp of the object of touch. Typically, we hold an object by contact with just a small portion of its surface. How then do we know by touch that this is a complete material object? At first pass, this is just another instance of Generalized Completion. Remember how this principle was used to address O'Callaghan's worry that 'humans do not hear spatial edges'. Why not say that here too, the containing contours of material objects are inferred by the tactile system, even though they are not in contact with the skin. This is good, as far as it goes, but traditionally, another difficulty has lurked. Many philosophers of perception assume that touch is an *internal* sense-modality—that what we feel by touch is not an external object, but rather pressure on the skin. In order to accommodate both this supposed bodily focus, or interiority, of touch and also the fact that we are in contact only with fragments of external objects, some suggest that our ideas of tactile objects are constructions from sense data.[15]

[15] O'Shaughnessy (1989) and Martin (1992) start from the interiority of touch but offer nuanced accounts of *attention* that pioneer the exteriorization of this modality. Unfortunately, this is a wrong direction: the exterior targets of touch require more than a shift of attention. See Matthen, 2021 for a critique.

Now, in order to assess this thesis, we must note first that touch displays a certain duality. A stimulation applied to the skin can be felt as something happening *on the skin*; this is generally known as 'tactile' sensation. But if the subject moves her body against the intruding stimulus, or pushes back against it, the *stimulus* is (also) felt *externally*; this is known as 'haptic' perception.[16] So, think about feeling a pen in my grip. If the pen lies passively between my fingers, I may feel only the pressure at my fingertips. This is tactile sensation; it is interior. However, when I *grip* the pen—exert pressure on it in order to use it—I also perceive it as an external object. When kinaesthesia and proprioception, or awareness of the movement, muscular effort, and position of my body, combine with touch, I am aware of the external stimuli in contact with my skin.

Now, think again of what I feel in my hand when I use the pen. My fingers are exerting pressure inwards and correspondingly, they feel the resistance of this object. Moreover, each digit feels the reciprocal pressure of the others through the object of interest. Together, these coordinated feelings add up to perceiving the object as a coherent whole. This suffices for being aware of it as a privileged subject, even though the details of its shape are not known. This is Generalized Completion.

Now, what about each tactile experience taken in isolation? The experience of pressure on each digit is comparable to that if it were to press down on a hard object. Now, here is the point. Even if it makes sense for the atomist to treat of these feelings separately—that is, without taking the coordination between them into account—each still feels as if it is in contact with a material surface. Taken by itself, neither reveals a single material object that has the thickness of the pen, for you need the two sensations together for that. But each reveals a material object nonetheless, for it is only material objects that exert resistance when you press down on them. What my thumb in isolation reveals when I hold the pen is a hard something—something that is a subject for the tactual property of hardness. In short, the thumb reveals the generic material-object status of the pen, without revealing the more specific information about its size or shape.

I conclude that *haptic perception* is distributionally about material objects from individual points of contact on up. It is extremely contentious to hold that feeling the pen is a simple construct from perceptions at minimal points of contact—coordination is needed; simple concatenation does not suffice. But bracketing this point, the fact remains that I perceive the pen as a material object even at the minimal point of contact. Notice that in arriving at this conclusion, I have not assumed that every perceptible logical subject is a material object. Rather, my premise was that anything that resists pressure from my body is a material object.

8 Type-identifying material objects: smell

I argued earlier that smell can be experienced as emanating from material objects when we move relative to those objects. I want now to conclude by arguing that smell is externally directed in another way. It identifies *types* of material object.

[16] See Fulkerson, 2015 and Matthen, 2021 for further details and discussion.

Let us start with the phenomenon.

1. You turn on the coffeemaker in your kitchen. As the coffee begins to brew, you smell it. You also start the butter going to make yourself a fried egg. You smell the butter as well.
2. Later, you head off to the office. You walk by an idling delivery truck, and you smell its unpleasant exhaust fumes. At the same time, you walk by the open door of a coffee shop. You smell the coffee while still aware of the fumes.

In these incidents, you identify coffee by its smell in two different contexts. There are two points here of note.

Firstly, you experience the same smell—the smell of coffee—in two quite different contexts. This is, as I will explain in a moment, an instance of *perceptual constancy*, and thus a sign of the external significance of olfaction.

Secondly, the smell of coffee is, on both occasions, a *single* smell (though, no doubt, a bit different on the two occasions, since the coffee you brew is different from the one in the coffee shop). The generality of this kind of non-decomposability with regard to smells indicates that, as a modality, it is directed towards *types* of substance—coffee, butter, diesel fumes—not towards putative subordinate features by which types of substances are identified. In other words, the primary olfactory identificatory marker of coffee is the smell of coffee—not components of that smell.

Perceptual constancy—why do I say that (1) and (2) together constitute perceptual constancy? In any sense-modality, each sensory receptor is affected by any stimulus to which it is 'matched'. As noted earlier, vision is spatially laid out at its front end, and touch is laid out by bodily parts. This aids in the separation of visual and tactile stimuli. In the other modalities—audition, taste, and smell—the receptors are affected by everything in range.[17]

Taking olfaction in particular, the smell receptors are affected by coffee and sizzling butter in (1), and by coffee and diesel fumes in (2). That is, the activation level of each olfactory receptor is simply a sum of effects created by these two odorants (and any others in range). The total stimulatory pattern is quite different in the two incidents. Yet, olfaction is able to extract a commonality—the activation pattern characteristic of coffee. This is classic perceptual constancy. At its front end, the olfactory system's activation is a sum; yet olfactory experience reveals the type identity of one of the summands. In other words, given two multidimensional vectors $(a + b)$ and $(a + c)$, the system somehow figures out the value of a.

Tyler Burge (2010, p. 415) contradicts this, writing that 'constancy [is] not prominent in the workings of [smell]' because 'proximal registration of a chemical blend suffice[s] for most purposes'. It is a little difficult to figure out what he means because this seems so obviously false. Not only do we extract the smell of coffee from the 'blend' of coffee and diesel, but also we are able to discern the smells of coffee and hazelnut in a flavour blend, and to distinguish and match these components with those in a coffee plus orange blend. Clearly, the proximal registration of these blends does not 'suffice' for olfaction in these cases; the olfactory system breaks down the proximal stimulus into its distal components.

[17] This is also true of audition: each hair cell emanating from each part of the cochlea reflects the sum total of sound at the frequency specific to that hair cell. However, audition has some spatial discrimination, sounds have natural resonances, and voices utter phonetic streams that have natural unity conditions determined by phonetics, lexemes, and meaning. These conditions make certain decompositions of the sound stream more natural than others.

Burge seems to be influenced here by a misleading hypothesis about memory. He writes (2010, p. 416):

> Ordinary language sometimes portrays the taste of wine or the smell of banana as perception. Such cases are usually to be assimilated to belief and propositional memory derived from non-perceptual sensory states. The representation is not at the purely sensory level. The sensory system responds to certain types of proximal stimulation that in fact come from such things as wine or bananas. Discriminations can be either generic or fine-grained. But, except for the special case [of locational constancy] noted with respect to smell, I know of no perceptual constancies in the gustatory or olfactory sensory systems themselves.

Now, memory could function in two ways in smell (and flavour) identification. One way, which is irrelevant—though it may be what Burge has in mind—is 'semantic'. That is, I might put a verbal label ('coffee') on a smell because I remember that this is what it is called. This, as I say, is irrelevant because what is important is the awareness and reidentification of the smell of coffee, which precedes tagging it with a label.

The second way in which memory could be involved is sensory and non-conceptual. When I smell coffee, I am aware of a sensory profile. In the incidents recounted above, I am aware of a similar profile rising and falling in strength as I move about the kitchen or the sidewalk. When I get closer to the coffee, the olfactory components that I remember in the smell of coffee get stronger; when I move away, those components get weaker. Similarly, in the coffee–orange mixture, two olfactory profiles with which I am familiar seem to be added together. This kind of pattern recognition is of a piece with what Stevenson and Wilson (2007, p. 1824) propose.

> The brain appears to store previous patterns of glomerular output, so that when a new output arrives it may be matched to all previous encodings. If there is a good match, we consciously experience a discrete odour, distinct from its background and discriminable from other odours. It may or may not be identified (named). Where the input is unfamiliar, our experience is vague and ill-formed and the odour is hard to discriminate from other unfamiliar odours.

I know of no evidence that suggests that this kind of memory should be 'assimilated to belief and propositional memory', as Burge says. The olfactory system seems autonomously to incorporate such traces of earlier encounters.[18]

Perceptual unity—I come now to my second point. The smell of things like coffee is perceived as a single smell. As Stevenson and Wilson (2007, p. 1822) write:

> Aged Rioja red wines have over 800 volatile components, of which approximately 60 have odours detectable by humans. Parmigiano Reggiano cheese has around 160 volatile components with nearly 30 of those odorous. Despite these incredibly complex stimuli, our percept

[18] Stevenson and Wilson suggest one other mechanism. We get adapted to smells very quickly—the newly arrived guest smells the cooking right away, but the host has long become unaware of it. So, smells have a window of sole presentation.

is largely unitary (despite anecdotal evidence and folklore to the contrary), and we perceive a specific red wine or cheese.

This is a contrast with vision, where the characteristics of a type like *dog* or *cat* are discrete: legs like so, fur like this, etc. For even if the visual appearance of types of things is somewhat inexplicitly stored, there is no single non-decomposable look-of-a-cat in the way that there is a single non-decomposable smell-of-coffee. This is a powerful point. The experiences of smell generated by the olfactory system are not just the components of a classificatory system; they are actually type-markers.

What should we take away from these reflections on olfaction? Firstly, that olfaction is external perception by the criteria I borrowed from Burge: it is perceptually constant. That is, olfaction represents the presence of something external. So far, I agree with Batty (2010). Secondly, given awareness of one's own movement when one manipulates or homes in on smells, one can be aware of a smell as an attribute of a particular material object. Here, I depart from Batty, who argues, largely on grounds that overlook multimodal perception, that smell cannot be localized, except 'at' the perceiver herself. And finally, to this particularity, add the unitary nature of smells that arise from *types* of material substance. From this, I conclude that we smell instances of these types.

9 Conclusions

At the beginning of this paper, I said that the external senses characteristically provide us with awareness of material objects and their properties. I have tried to clarify and explain this claim. I showed how vision provides us awareness of properties beyond those that affect the eyes by the information carried by light. Then I proposed a process of Generalized Completion by which non-visual modalities generate experiences of material objects by recruiting sensory exploration using the perceiver's own body. Finally, I discussed the somewhat special cases of touch and smell. Touch is special because its object-awareness is bottom-up and atomistic. Smell is special because its experiences are of types of material substance. Flavour-perception might be similar in this respect.[19]

References

Batty, C. (2010). A representational account of olfactory experience. *Canadian Journal of Philosophy*, 40, 511–38.

Batty, C. (2014). The illusion confusion, *Frontiers in Psychology*, 5, 231. Available from: https://doi.org/10.3389/fpsyg.2014.00231

Bregman, A. (1994). *Auditory Scene Analysis: The Perceptual Organization of Sound*. Cambridge, MA: MIT Press.

Burge, T. (2010). *The Origins of Objectivity*. Oxford: Oxford University Press.

Carello, C., Anderson, K. L., and Kunkler-Peck, A. J. (1998). Perception of object length by sound. *Psychological Science*, 9(3), 211–14.

Clark, A. (2004a). Feature-placing and proto-objects. *Philosophical Psychology*, 17(4), 443–69.

[19] Thanks to Casey O'Callaghan for illuminating and helpful discussion.

Clark, A. (2004b). Sensing, objects, and awareness: reply to commentators. *Philosophical Psychology*, 17(4), 553–79.
Dickie, I. (2010). We are acquainted with ordinary things. In *New Essays on Singular Thought*, edited by R. Jeshion, 213–45. Oxford: Oxford University Press.
Fulkerson, M. (2015). *The First Sense: A Philosophical Study of Human Touch*. Cambridge, MA: MIT Press.
Geldard, F. A. and Sherrick, C. E. (1972). The cutaneous 'rabbit': a perceptual illusion. *Science*, 178, 178–9.
Martin, M. F. (1992). Sight and touch. In *The Contents of Experience*, edited by Tim Crane, 196–215. New York, NY: Cambridge University Press.
Matthen, M. (2004). Features, places, and things: reflections on Austen Clark's theory of sentience. *Philosophical Psychology*, 17(4), 497–518.
Matthen, M. (2018). Ephemeral vision. In *Perceptual Ephemera*, edited by T. Crowther and C. MacCumhaill, 312–36. Oxford: Oxford University Press.
Matthen, M. (2021). The dual structure of touch: the body vs peripersonal space. In *The World at Our Fingertips*, edited by F. de Vignemont, H. Y. Wong, A. Farné, and A. Serino, 197–214. Oxford: Oxford University Press.
McElrone, A. J., Choat, B., Gambetta, G. A., and Brodersen, C. R. (2013). Water uptake and transport in vascular plants. *Nature Education Knowledge*, 4(5), 6.
O'Callaghan, C. (2016). Objects for multisensory perception. *Philosophical Studies*, 173, 1269–89.
Orosa-Puente, B., Leftley, N., and von Wangenheim, D., *et al.* (2018). Root branching toward water involves posttranslational modification of transcription factor ARF7. *Science*, 362, 1407–10.
O'Shaughnessy, B. (1989). The sense of touch. *Australasian Journal of Philosophy*, 67(1), 37–58.
Pavani, F. and Galfano, G. (2015). The multisensory body revealed through its cast shadow. *Frontiers in Psychology*. 666. Available from: https://doi.org/10.3389/fpsyg.2015.00666
Piaget, J. (1954). *The Construction of Reality in the Child* (M. Cook, Trans.). New York, NY: Basic Books.
Spelke, E. (1990). Principles of object perception. *Cognitive Science*, 14, 29–56.
Spelke, E. S., Born, W. S., and Chu, F. (1983). Perception of moving, sounding objects by four-month-old infants. *Perception*, 12(6), 719–32.
Stevenson, R. J. and Wilson, D. A. (2007). Odour perception: an object-recognition approach. *Perception*, 36, 1821–33. Available from: https://doi.org/10.1068/p5563
Van Assen, J. J. R., Barla, P., and Fleming, R. W. (2018). Visual features in the perception of liquids. *Current Biology*, 28(3), 452–8.

16
Multisensory feature integration

Charles Spence

1 Introduction

The title of this review could legitimately be taken to refer to the integration of features from any combination of senses (e.g., vision, audition, touch, smell, and taste, to name but five of the most frequently mentioned). However, the multisensory literature that has been published to date has generally tended to focus either on just the integration of the higher spatial senses (namely vision, audition, and, to a lesser extent, touch; see Spence and Frings, 2020, for a recent review), or else on the integration of the chemical senses (gustatory, olfactory, and trigeminal; see Small, 2004; Viana, 2011), the latter giving rise to multisensory flavour perception (see Spence, 2015a, for a review). The fundamental question to be addressed in this review is whether the same rules of multisensory feature integration (MFI) apply in the two cases.

1.1 On the benefits of multisensory integration

At the outset, it is worth stressing that our perception of the world around us is inherently multisensory (Stein and Meredith, 1993), with all species that have multiple senses integrating those senses for the coherent control of action (Stein, London, Wilkinson, and Price, 1996). MFI typically leads to more precise and robust perception, and hence to behaviours that are more efficient (e.g., Ernst, 2012; Ernst and Bülthoff, 2004; Spence, 2018; though see also Sinnett, Soto-Faraco, and Spence, 2008). Determining which of the features from different senses should be bound together to deliver the multisensory perceptual objects, events, and experiences that fill our daily lives is a key component of, and problem for, the successful integration of multisensory information. To accomplish this goal, perceptual systems rely on both low-level features of the stimulus array (such as, for example, spatio-temporal coincidence), as well as on any multisensory associations that may have been learnt (or internalized) during the course of development (Welch and Warren, 1980, 1986), based on the statistics of the environments in which we happen to develop (Parise, Knorre, and Ernst, 2014).

1.2 Alternative accounts of multisensory feature integration

When thinking about MFI, there are at least two relatively distinct literatures to draw upon. The first, feature integration theory (FIT), was developed by Anne Treisman in order to

Charles Spence, *Multisensory feature integration* In: *Sensory Individuals*. Edited by: Aleksandra Mroczko-Wąsowicz and Rick Grush, Oxford University Press. © Oxford University Press 2023. DOI: 10.1093/oso/9780198866305.003.0017

help explain how it is that visual features are combined (e.g., Treisman, 1986; Treisman and Gelade, 1980). Over the years, FIT has been extended from the visual to the auditory modality (e.g., Takegata *et al.*, 2005), and, from there, to the case of MFI (see Spence and Frings, 2020, for a review). The other, increasingly popular, account for the (multisensory) integration of features is in terms of statistically optimal Bayesian causal inference (e.g., Körding *et al.*, 2007; Shams and Beierholm, 2010). It is, though, important to note that these two approaches to MFI tackle somewhat different computational problems.

FIT was developed in the context of the integration of non-redundant visual cues in order to provide a solution to the binding problem—that is, knowing which of the many cues (or features) that may be available at any given time belong together (e.g., Shams and Kim, 2010; Spence, 2010; Spence, Ngo, Lee, and Tan, 2010). According to Treisman, attention acts as the glue that helps bind features (coded within spatiotopically organized, yet discrete, feature maps) to create perceptual objects. Though here it is worth noting that experimental psychologists and philosophers would sometimes appear to have different things in mind when referring to objects, and objecthood beyond the visual modality (though exploring this topic unfortunately falls beyond the scope of the present chapter).

The alternative Bayesian account of MFI has been used to help explain the integration of more or less redundant cues (e.g., multiple visual cues to depth; Hillis, Ernst, Banks, and Landy, 2002), or different sensory cues providing redundant estimates of the size (Ernst and Banks, 2002), orientation (Gori, Del Viva, Sandini, and Burr, 2008; see also Ernst, Banks, and Bülthoff, 2000), or position of a stimulus (Alais and Burr, 2004; Battaglia, Jacobs, and Aslin, 2003; Meijer, Veselič, Calafiore, and Noppeney, 2019; see also Mihalik and Noppeney, 2020). In recent years, the approach has been extended to explain/predict the audiovisual integration of speech stimuli (Bejjanki, Clayards, Knill, and Aslin, 2011; Ma *et al.*, 2009; Noppeney and Lee, 2018). Maximum likelihood estimation (MLE) assumes that the binding problem had already been solved (i.e., the question addressed concerns the relative, or Bayes-optimal, weighting of each sensory cue to the final percept, rather than the question of which cues belong together in the first place; see Shams, Ma, and Beierholm, 2005; Wozny, Beierholm, and Shams, 2008). Bayesian causal inference (e.g., Körding et al., 2007; Rohe and Noppenay, 2015) incorporates a number of priors (see also Beierholm, Quartz, and Shams, 2009) that help to explain which sensory features are more likely to be bound together (see Chen and Spence, 2017, for a review).

Both FIT and Bayesian causal inference have been developed in the context of the integration of features from the spatial senses (namely vision, audition, and touch). That said, Maier and Elliott (2020) have recently provided some of the first empirical evidence suggesting that the adaptive weighting of gustatory and olfactory cues in the case of multisensory flavour perception (specifically stimulus pleasantness) may also be described by the Bayesian approach. Furthermore, the visual search task, which has been such a distinctive component of Treisman's work, has also been extended to the search for flavourless among flavoured cubes in the mouth (Delwiche, Lera, and Breslin, 2000).

1.3 What role attention in feature integration?

Many researchers consider that attention plays a key role in feature integration, no matter whether one is talking about the integration of features within a sensory modality (Treisman,

1986) or between them (e.g., Spence, 2019a; Spence and Frings, 2020). While MFI can sometimes take place in the absence of attention, in such cases the likelihood of illusory conjunctions tends to increase (Cinel, Humphreys, and Poli, 2002; Treisman and Schmidt, 1982). However, as we will see below, spatial attention appears to play a somewhat different role in the MFI of the spatial and chemical senses. According to Treisman's FIT (Treisman, 1986, 1988, 1996, 1998; see also Quinlan, 2003), spatial attention acts as the glue that helps bind the appropriate visual features (such as colour, form, etc.) together. The spatial spotlight of attention is considered central to the integration of the appropriate visual features (though see also Briand and Klein, 1987). Spatial attention also plays an important role in the case of MFI, though the latter sometimes occurs in the absence of attention (see Spence and Frings, 2020, for a review). Intriguingly, however, and contrary to the narrative that one often encounters in the literature, MFI in the spatial senses appears to occur regardless of the spatial coincidence (or otherwise) of the various inputs, at least under those conditions where only a single stimulus happens to be presented to each sensory modality, and when localization of the stimuli is task-irrelevant, as is typically the case in laboratory studies of multisensory identification (see Spence, 2013, 2019b for reviews). So, for example, it has been demonstrated that the magnitude of the McGurk effect is unaffected by whether the auditory and visual stimuli happen to be presented from the same location or not (e.g., Jones and Jarick, 2006). Meanwhile, Bayesian causal inference posits a prior that spatio-temporally coincident sensory features are likely to belong to the same object or event, hence leading to increased MFI. And, at least according to Helbig and Ernst (2008), attention only operates subsequent to MLE. That said, the rubber hand illusion, while not necessarily in any sense a 'spatial' task, might nevertheless be modulated by the distance of the rubber hand from the body.

2 Attention and multisensory flavour perception

In the case of multisensory flavour perception, gustatory, retronasal olfactory, and trigeminal cues (when present) are integrated and localized as if they originate from the oral cavity. It has been suggested that this phenomenon, known as oral referral, results from the fact that attention is always directed to the mouth whenever we eat and drink (Spence, 2016, 2019a; Stevenson, 2016; Stevenson, and Oaten, and Mahmut, 2011). The ubiquity of oral referral is surprising inasmuch as the majority of the information that contributes directly (or constitutively) to the perception of flavour is typically transduced by olfactory receptors (Spence, 2015b). As such, given that the majority of the information is transduced by/in the nose, one might have thought that it would make sense for flavours to be localized there subjectively too, rather than to the oral cavity. At the same time, however, there is also a sense in which oral referral makes perfect sense, given that the mouth is ultimately from where the chemosensory stimuli that we perceive normally originate (i.e., from the food in the mouth).[1]

While one might consider oral referral to be a kind of ventriloquism in the mouth, it is important to note that (retronasal) olfactory cues are experienced as if they were tastes following

[1] Or as Spalding (1920, p. 11) put it a century ago: 'Whether stimuli are connected or not, interactions between the human senses have long been seen "for the simple reason that they are all means of registering on our brain and imagination impressions derived from outside sources".'

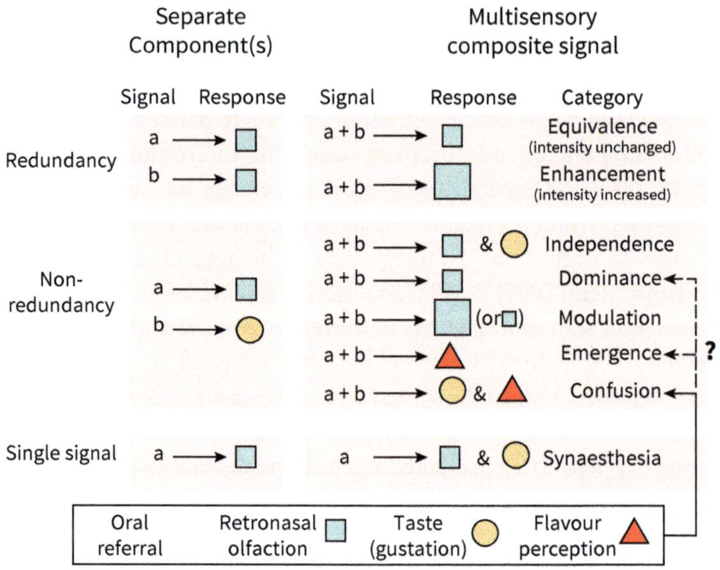

Figure 16.1 Classification of the different types of multisensory interactions. Bremner *et al.* included multisensory synaesthetic percepts, given widespread interest in the topic since the publication of Partan and Marler's original paper. I have added an additional row—'Confusion'. The argument forwarded here is that oral referral may represent the only example of such modality confusion that has been reported in the literature to date. Note also that confusion cannot be subsumed within any of the other types of multisensory interaction that have been outlined previously.
Adapted from Bremner, Lewkowicz, & Spence, 2012, in turn adapted from Partan & Marler, 1999.

their mislocalization to the oral cavity in a way that simply does not occur in the case of the audiovisual ventriloquism illusion, say (Alais and Burr, 2004; Jackson, 1953).[2] That is, we all confuse which sense the features belong to in the case of oral referral. As such, it has been suggested that oral referral may actually reflect a relatively unique form of MFI, one that is only seen in the chemical senses (Spence, 2016) (see Figure 16.1).

2.1 What role temporal synchrony in multisensory feature integration?

The simultaneity of feature onset undoubtedly plays a fundamental role in MFI (see Vatakis and Spence, 2010, for a review of the audiovisual literature). Of course, just because stimuli may be presented simultaneously does not necessarily mean that they will also be processed, or made available neurally, at the same time as well. Indeed, there are various salient differences in physical and neural transduction latencies to be aware of in the case of MFI

[2] Another reason for denying that oral referral is a form of spatial ventriloquism is cases of partial oral referral, where olfactory and gustatory cues are localized in different places to one another, but the presence of the other cue influences where each is localized does not seem to occur.

(e.g., see Schroeder and Foxe, 2002, 2004; Spence and Squire, 2003). That said, much the same problem has also been documented previously in the case of intramodal integration of visual features, such as, for example, when binding visual stimuli comprising rapidly changing colour and direction of motion cues (Moutoussis and Zeki, 1997a, b). However, in all but the most extreme cases (think thunder and lightning), neural mechanisms of temporal integration appear capable of eliminating this problem by resynchronizing desynchronized sensory inputs (Grossberg and Grunewald, 1997; see also Bergenheim, Johansson, Granlund, and Pedersen, 1996; Engel and Dougherty, 1971; Freeman *et al.*, 2013). At the same time, MFI has also been shown to reduce the apparent temporal asynchrony between congruent combinations of sensory features (Habets, Bruns, and Röder, 2017; Parise and Spence, 2009; Spence, 2011), and attention may spread more easily between those combinations of sensory features that we have learnt tend to co-occur (e.g., Fiebelkorn, Foxe, and Molholm, 2010, 2012).

As far as the chemical senses are concerned, there would appear to be a broader temporal window of multisensory integration for the combination of olfactory and gustatory stimuli than when either of these (chemical) senses are compared with visual stimuli (Gotow and Kobayakawa, 2017). However, whether this reflects a meaningful difference between the senses themselves, or rather the confound with the spatial coincidence of the chemosensory stimuli, but not of the visual stimuli, remains unclear (cf. Spence, Kettenmann, Kobal, and McGlone, 2001 on this potential confound).

3 When correlation implies causation

Potentially more important as far as facilitating multisensory integration is concerned than mere temporal synchrony is the presence of correlated inputs in different senses (e.g., Denison, Driver, and Ruff, 2013; Parise, Harrar, Ernst, and Spence, 2013; Parise, Spence, and Ernst, 2012; Radeau and Bertelson, 1987). On occasion, correlated (in terms of cadence or rhythm) sensory inputs give rise to some pretty surprising examples of MFI that make little sense conceptually—such as when synchronized stroking/tapping of an inanimate surface and an unseen part of one's body gives rise to the feeling that the inanimate surface is actually part of one's own body (Armel and Ramachandran, 2003). Correlated, but temporally desynchronized, visuo-tactile inputs can also give rise to the rubber hand illusion (e.g., see Ehrsson, Spence, and Passingham, 2004).

4 Interim summary

Spatio-temporal coincidence and temporally correlated sensory inputs facilitate MFI. Having taken a closer look at the role played by lower-level stimulus-related factors in modulating MFI in the spatial and chemical senses, it is time now to turn our attention to the increasingly important role played by an observer's knowledge (i.e., the associative learning, and subsequent internalization, of the multisensory statistics of the environment) in modulating MFI (see Barlow, 2001; Connolly, 2014; Fateminia, Ghotbabadi, and Azad, 2018; Welch and Warren, 1980, 1986).

4.1 Crossmodal correspondences and coupling priors

According to Bayesian causal inference, crossmodal correspondences can be considered as but one kind of coupling prior (Ernst, 2007; Spence, 2011). The research that has been published to date demonstrating that corresponding stimuli (or features), such as, for example, the correspondence between the pitch of a sound and the size of an object (with lower pitch and larger size matching; e.g., Evans and Treisman, 2010; Gallace and Spence, 2006; Parise and Spence, 2009; see also Bien, ten Oever, Goebel, and Sack, 2012) gives rise to enhanced MFI, as compared to the integration of those stimuli (or features) that are crossmodally incongruent. What is more, the enhanced temporal and spatial integration of associatively learnt pairings of multisensory stimuli reduces spatio-temporal incongruency, thus presumably leading to more integrated (or unified) multisensory percepts (Habets et al., 2017; Parise and Spence, 2009; Spence, 2011).

Until recently, the majority of research assessed MFI in those crossmodal correspondences that had already been established by the time that they were studied in the laboratory. There is, though, an emerging body of research investigating how multisensory perception/neural processing changes as a function of associative learning of a range of novel correspondences between the sensory features presented to different sensory modalities (e.g., Brunel, Carvalho, and Goldstone, 2015; Einhäuser, Methfessel, and Bendixen, 2017; Ernst, 2007; Kim and Kim, 2019; Lauzon, Abraham, Curcin, and Stevenson, 2020; Tanabe, Honda, and Sadato, 2005; Tong, Li, Bruns, and Röder, 2020). Once again, though, the majority of this research comparing MFI before and after associative learning has tended to be focused on the spatial senses.

When it comes to the role of coupling priors in multisensory flavour perception, enhanced oral referral (meaning that oral referral is more likely to occur) has been documented in the case of congruent, as compared to incongruent, combinations of olfactory and gustatory stimuli (e.g., Fondberg *et al.*, 2018; Lim, Fujimaru, and Linscott, 2014; Lim and Johnson, 2012; see also Frank, Shaffer, and Smith, 1991), where what is deemed congruent depends on prior co-exposure of specific olfactory and gustatory qualities (Breslin, Belanger, Tharp, and Dalton, 2003; Schifferstein and Verlegh, 1996), that is, on associative learning (see Stevenson and Boakes, 2004). Intriguingly, the simple co-occurrence of novel olfactory and gustatory stimuli has been shown to result in the former taking on some of the perceptual qualities of the latter (Blank and Mattes, 1990; Stevenson and Boakes, 2004; Stevenson, Boakes, and Prescott, 1998). So, for example, vanilla smells sweet to most people despite the fact that vanilla pods are themselves very bitter. The reason for the association with sweetness comes via the sweet food products, like vanilla ice cream and cola, that contain the combination of vanilla aroma and intense sweetness (Sakai *et al.*, 2001; Spence, 2022). There is also some emerging evidence to suggest that the nutritive status of basic tastants may determine the kinds of multisensory integration that are observed with retronasal odours (Linscott and Lim, 2016). These various phenomena, then, help to highlight some of the key differences that have been identified between MFI in the case of the spatial, as compared to the chemical, senses.

As the interest of the multisensory research community has increasingly shifted in recent years from the role of low-level stimulus features (such as spatio-temporal coincidence) in MFI to the role of observer–internal factors (namely, various kinds of learnt associations), the question of which factors determine the speed and strength of such multisensory associative learning (across the lifespan) has also started to come to the fore (e.g., Altieri, Stevenson, Wallace, and Wenger, 2015; Baier, Kleinschmidt, and Müller, 2006; Fifer, Barutchu, Shivdasani, and Crewther, 2013; Gonzalo, Shallice, and Dolan, 2000; Guellaï *et al.*, 2019). While it might be assumed that the strength of the coupling prior for co-occurring combinations of multisensory features would primarily be determined by the regularity of co-occurrence of the stimuli at various different timescales (e.g., Ernst, 2007; Knoeferle, Knoeferle, Velasco, and Spence, 2016; Murray and Sperdin, 2010; Stevenson and Boakes, 2004), there is evidence to suggest that people find it somewhat easier to internalize novel associations when the unisensory cues happen to correspond crossmodally (e.g., Barenholtz, Lewkowicz, Davidson, and Kogelschatz, 2014; cf. Imai *et al.*, 2015). For example, Barenholtz and colleagues reported on a couple of experiments showing that people exhibit enhanced learning of multisensory associations between gender-congruent face–voice pairs, compared to incongruent pairs of audiovisual stimuli. These researchers were also able to demonstrate that this 'categorical congruency' advantage extended to non-human stimuli, such as pairs of class-congruent animal pictures and vocalizations (e.g., dogs and barks) vs class-incongruent pairs (e.g., dogs and bird chirps). However, whether this represents an innate difference between corresponding and non-corresponding stimulus combinations, or rather can be explained simply in terms of the broader impact on perception of the internalization of the statistics of the environment, is currently unclear (cf. Bahrick and Lickliter, 2000; Walker-Andrews, 1994). Research over the last couple of decades or so has revealed that associative learning between olfactory and gustatory stimuli (specifically, when the former stimuli come to take on the properties of the latter) is acquired more easily in the case of novel, as compared to familiar, olfactory stimuli (see Stevenson and Boakes, 2004; though see also Fondberg, Lundström, and Seubert, 2021). In the coming years, it would seem likely that the literatures on associative learning and MFI may converge.

There is, though, an important question here as to how the learning of the statistics of the environment differs in the case of multisensory conceptual learning (e.g., the learning of a concept such as 'dog', where features of the concept/object are encountered through different senses), as compared to the case of the acquisition (or internalization) of crossmodal correspondences (Ernst, 2007), over and above the creation of a conceptual representation in the one case, but not in the other. Indeed, much research that needs to be conducted in order to determine the task dependence of such semantically congruent modulation of multisensory integration (see Barutchu and Spence, 2020; Spence, 2019c).

4.2 Crossmodal interactions in the absence of multisensory integration

There are many examples of the crossmodal influence of one stimulus (or feature) on another that do not necessarily give rise to what researchers would conventionally call MFI. There is, for example, extensive evidence of the crossmodal priming of masked visual stimuli (e.g., the outline drawing of a dog) by the presentation of the appropriate (or, rather, congruent) auditory stimulus, be it the spoken word describing an object (e.g., the word 'DOG') or the sound that is typically associated with that object (e.g., a barking sound; see Chen and Spence, 2010, 2011, 2013, 2018a, b; Vallet, Brunel, and Versace, 2010; though see also Li *et al.*, 2020). While the majority of studies have tended to focus on the auditory priming of visual perception, there is an emerging literature showing that similar crossmodal priming effects can also be observed in the opposite direction (Thomas, Nardini, and Mareschal, 2017). Taken together, therefore, the key point to note is that simply demonstrating a crossmodal interaction between different sensory modalities does not in-and-of-itself necessarily imply that one is looking at an example of MFI. Parise and Spence (2012) provide another example of the crossmodal interaction of a range of corresponding auditory and visual sensory features in the absence of MFI, as demonstrated in the Implicit Association Test (IAT). Note that in the IAT, only a single unisensory stimulus is presented in each trial.

Meanwhile, in the case of multisensory flavour perception, visual cues often set expectations regarding the likely taste and flavour of the food and drink that we are about to consume (Piqueras-Fiszman and Spence, 2015). Such visually induced flavour expectations have been shown to bias the perception of flavour (Spence, 2015c; Wang and Spence, 2019) without necessarily being accepted (by all commentators) as part of the multisensory flavour experience itself (Auvray and Spence, 2008; ISO, 2008; Spence, Smith, and Auvray, 2015). Wang and Spence, for example, were able to show that adding a pink odourless, tasteless food dye biased people's judgements of the aroma/flavour of a white wine. Perhaps even more striking are those examples where simply changing the colour of the ambient lighting have been shown to bias what people say about the taste of a wine served in a black tasting glass (Spence, Velasco, and Knoeferle, 2014). While such crossmodal priming effects may well have perceptual consequences (that are not too dissimilar from those observed in the case of MFI), the time course of the latter tends to be qualitatively different. That is, crossmodal priming effects tend to be maximal when the priming stimulus is presented some hundreds of milliseconds prior to the to-be-primed stimulus (e.g., see Chen and Spence, 2013; Shankar *et al.*, 2010; Wang, Spence, and Knoeferle, 2020). By contrast, MFI tends to be maximal when the features are presented, or rather processed, simultaneously (within what is known as the multisensory temporal binding window; e.g., Wallace and Stevenson, 2014). Here, it may also be interesting to note, in passing, that crossmodal priming (rather than integration) tends to be demonstrated for semantically meaningful stimuli, whereas MFI is perhaps more often observed in the case of simple (i.e., semantically meaningless) sensory features.

4.3 Integrating the chemical and spatial senses

In the case of the multisensory integration of chemosensory cues that gives rise to multisensory flavour perception, there is typically not much of a binding problem, given that all of the chemosensory cues are normally relevant to whatever flavourful stimuli happen to be in the oral cavity.[3] That said, there is an important distinction to be drawn here between orthonasal and retronasal olfaction (e.g., Rozin, 1982; Wilson, 2021), depending on whether the volatile chemosensory signal is experienced as having a source that is external to the body, as when we sniff (orthonasal), vs a source within

[3] This is not quite true, of course, as sometimes we are able to associate different flavours with different oral somatosensory textures (see Spence, Youssef, and Wang, 2017). However, oral somatosensory cues are not, strictly speaking, considered to be part of multisensory flavour perception (see ISO, 2008).

the oral cavity when volatile rich air is pulsed out from the back of the nose when we chew and swallow (retronasal; see also Ni *et al.*, 2015; Wilson, 2021). As such, one might imagine that retronasal (but not orthonasal) olfactory stimuli would be integrated into multisensory flavour percepts. However, this is simply (and somewhat surprisingly) not the case. Instead, psychophysical studies reveal that (sub-threshold) tastants are integrated with olfactory stimuli in essentially the same manner, regardless of whether the latter happen to be delivered via the orthonasal or retronasal routes (e.g., Dalton, Doolittle, Nagata, and Breslin, 2000; Pfeiffer, Hollowood, Hort, and Taylor, 2005; see Spence, 2012 for a review). Perhaps the olfactory stimuli were subject to oral referral by the sub-threshold tastants in Dalton *et al.*'s study. If so, this might be considered as providing an example of 'implicit' integration.

Such results would seem to run counter to the suggestion that interoceptive sensory cues might be processed (and possibly combined) in a manner that is qualitatively different than the combination of interoceptive and exteroceptive senses (Rozin and Kalat, 1972). For instance, the latter researchers talk of the principle of 'stimulus belongingness', according to which an organism can more readily associate a pair of stimuli if they both happen to be exteroceptive (i.e., impinge on the organism from external sources) or both interoceptive (i.e., if both stimuli are associated with an internal source). According to Christensen (1983, p. 790), colour and aroma (i.e., orthonasal olfaction) are exteroceptive, whereas flavour and hunger are both interoceptive. I would, however, question whether multisensory flavour perception is really interoceptive, or rather whether it is an exteroceptive signal with a source that just so happens to be located within the confines of the body. Note here that flavour perception would not seem to fit the definition of interoception as the sensing of the internal state of the body (Craig, 2002), nor is flavour included in Craig's exhaustive list of interoceptive senses.[4] (Intriguingly, it has been reported that applying a cool stimulus to the skin vs to the oral mucosa leads to different thermal compensation, perhaps suggesting that stimuli presented external to the oral cavity vs within really might be treated in a qualitatively different manner; Eccles, Du-Plessis, Dommels, and Wilkinson, 2013). Perhaps, though, a more meaningful distinction in this case might be between the integration of those features belonging to the spatial vs to the chemical senses (cf. Gotow and Kobayakawa, 2017).

However, even the latter distinction is not without its problems, given that orthonasal olfaction is both a chemical sense, but also, to a certain extent at least, a spatial sense (e.g., in the sense that people are remarkably good at tracking a scent spatially—that is, as a result of active exploration (see Porter *et al.*, 2007). That said, vision and audition (and possibly also touch) allow us to perceive objects at a distance, and arguably also allow for the perception of space itself (Martin, 1992; Young, 2017). Meanwhile, in another example of the interaction between the spatial and chemical senses, it has been reported that the spatial source of gustatory (i.e., chemosensory) stimuli are typically bound to the location of an oral somatosensory (and thus spatial) stimulus—such as a tasteless Q-tip that is drawn across the tongue (Todrank and Bartoshuk, 1991).

[4] Though perhaps the burning mouthfeel associated with chilli pepper might be considered an interoceptive signal.

5 Conclusions

MFI in the spatial senses (of vision, audition, and touch) and in the chemical senses (of gustation, retronasal olfaction, and trigeminal inputs) takes place in different parts of the brain (e.g., Driver and Noesselt, 2008; Small, 2012; see also Seubert *et al.*, 2015). What is more, somewhat different rules of MFI appear to apply in the two cases. In part, this may well be related to the fact that widespread MFI appears to take place earlier (neurally) in the case of the chemical senses of retronasal olfaction and gustation (in the secondary cortex; e.g., Blankenship, Grigorova, Katz, and Maier, 2019; Fortis-Santiago *et al.*, 2010; Veldhuizen *et al.*, 2010; see also Vincis and Fontanini, 2016) than is commonly observed when it comes to MFI in the spatial senses (Stevenson, 2014, 2016). It is interesting to consider here the relevance of the dual route of olfactory information to the cortex (both direct and via the thalamus), which may have implications for the role of attention in MFI in the chemical senses (Plailly, Howard, Gitelman, and Gottfried, 2008), and the fact that the primary olfactory cortex is considered the paleocortex (i.e., rather than the neocortex), which may add its own constraints and limitations to what olfaction can and cannot do (see also Stevenson and Attuquayefilo, 2013). It has long been argued that spatial coincidence (and hence spatial attention) plays an important role in the binding of auditory, visual, and tactile cues (Spence and Driver, 2004). Similarly, the fact that attention is always directed to the oral cavity when eating and drinking may help to explain the ubiquity of the phenomenon of oral referral (Spence, 2019a). In passing, it would certainly be interesting to know more about whether the same pool of attentional resources are used in the MFI that takes place for the spatial and for the chemical senses (see Forster and Spence, 2018; Van der Wal and van Dillen, 2013).

The crossmodal congruency of sensory features has been shown to play a modulatory role in MFI, regardless of the particular combination of senses under investigation. Coupling priors, internalizing congruency relationships based on the multisensory statistics of the environment (Chen and Spence, 2017), lead to enhanced binding, regardless of the pairing of senses under consideration. What is more, the multisensory integration of congruent combinations of sensory features typically gives rise to stimulus representations and multisensory experiences that are processed more fluently (Winkielman, Ziembowicz, and Nowak, 2015; cf. Barenholtz et al., 2014). In the case of the multisensory perception of flavour, the integration of congruent olfactory, visual cues has been shown to lead to superadditive interactions (e.g., de Araujo *et al.*, 2003; Seubert et al., 2015), in terms of the hedonic, if not necessarily the sensory discriminative, evaluation of stimuli.

Ultimately, MFI takes place among the spatial senses, among the chemical senses, and, on occasion, also between the two. So, returning to the question that was raised at the start of this piece, while many of the factors that modulate MFI have been shown to be similar in the chemical and spatial cases (such as the role of attention and crossmodal congruency), there are also a number of important differences between the kinds of integration taking place in the two cases (e.g., as evidenced by the phenomena of oral referral and smelled tastes in the chemical senses), hence meaning that one cannot necessarily use the same account to explain both kinds of MFI. Nevertheless, it would seem like a fruitful avenue to continue pursuing the similarities and differences as far as the MFI in the case of the spatial and chemical senses is concerned.

References

Alais, D. and Burr, D. (2004). The ventriloquist effect results from near-optimal bimodal integration. *Current Biology*, 14, 257–62.

Altieri, N., Stevenson, R. A., Wallace, M. T., and Wenger, M. J. (2015). Learning to associate auditory and visual stimuli: behavioral and neural mechanisms. *Brain Topography*, 28(3), 479–93.

Armel, K. C. and Ramachandran, V. S. (2003). Projecting sensations to external objects: evidence from skin conductance response. *Proceedings of the Royal Society B*, 270, 1499–506.

Auvray, M. and Spence, C. (2008). The multisensory perception of flavor. *Consciousness and Cognition*, 17, 1016–31.

Bahrick, L. E. and Lickliter, R. (2000). Intersensory redundancy guides attentional selectivity and perceptual learning in infancy. *Developmental Psychology*, 36, 190–201.

Baier, B., Kleinschmidt, A., and Müller, N. (2006). Cross-modal processing in early visual and auditory cortices depends on the statistical relation of multisensory information. *Journal of Neuroscience*, 26, 12260–5.

Barenholtz, E., Lewkowicz, D. J., Davidson, M., and Kogelschatz, L. (2014). Categorical congruence facilitates multisensory associative learning. *Psychonomic Bulletin and Review*, 21, 1346–52.

Barlow, H. (2001). The exploitation of regularities in the environment by the brain. *Behavioral and Brain Research*, 24, 602–7.

Barutchu, A. and Spence, C. (2020). An experimenter's influences on motor enhancements: the effects of letter congruence and sensory switch-costs on multisensory integration. *Frontiers in Psychology*, 11, 588343.

Battaglia, P. W., Jacobs, R. A., and Aslin, R. N. (2003). Bayesian integration of visual and auditory signals for spatial localization. *Journal of the Optical Society of America A*, 20, 1391–7.

Beierholm, U. R., Quartz, S. R., and Shams, L. (2009). Bayesian priors are encoded independently from likelihoods in human multisensory perception. *Journal of Vision*, 9(5), 23, 1–9.

Bejjanki, V. R., Clayards, M., Knill, D. C., and Aslin, R. N. (2011). Cue integration in categorical tasks: insights from audio-visual speech perception. *PLoS One*, 6(5), e19812.

Bergenheim, M., Johansson, H., Granlund, B., and Pedersen, J. (1996). Experimental evidence for a sensory synchronization of sensory information to conscious experience. In *Toward a Science of Consciousness: The First Tuscon Discussions and Debates*, edited by S. R. Hameroff, A. W. Kaszniak, and A. C. Scott, 303–10. Cambridge, MA: MIT Press.

Bien, N., ten Oever, S., Goebel, R., and Sack, A. T. (2012). The sound of size: crossmodal binding in pitch-size synesthesia: a combined TMS, EEG, and psychophysics study. *NeuroImage*, 59, 663–72.

Blank, D. M. and Mattes, R. D. (1990). Sugar and spice: similarities and sensory attributes. *Nursing Research*, 39, 290–3.

Blankenship, M. L., Grigorova, M., Katz, D. B., and Maier, J. X. (2019). Retronasal odor perception requires taste cortex, but orthonasal does not. *Current Biology*, 29, 1–8.

Bremner, A., Lewkowicz, D., and Spence, C. (Eds.). (2012). *Multisensory Development*. Oxford: Oxford University Press.

Breslin, P. A. S., Belanger, M. A., Tharp, C. D., and Dalton, P. (2003). Learned flavor congruency: the importance of experience in the sub-threshold integration of tastant/odorant pairings. Paper presented at the 5th Pangborn Symposium, Boston, MA.

Briand, K. A. and Klein, R. M. (1987). Is Posner's 'beam' the same as Treisman's 'glue'?: on the relation between visual orienting and feature integration theory. *Journal of Experimental Psychology: Human Perception and Performance*, 13, 228–41.

Brunel, L., Carvalho, P. F., and Goldstone, R. L. (2015). It does belong together: cross-modal correspondences influence cross-modal integration during perceptual learning. *Frontiers in Psychology*, 6, 358.

Chen, Y.-C. and Spence, C. (2010). When hearing the bark helps to identify the dog: semantically-congruent sounds modulate the identification of masked pictures. *Cognition*, 114, 389–404.

Chen, Y.-C. and Spence, C. (2011). Crossmodal semantic priming by naturalistic sounds and spoken words enhances visual sensitivity. *Journal of Experimental Psychology: Human Perception and Performance*, 37, 1554–68.

Chen, Y.-C. and Spence, C. (2013). The time-course of the cross-modal semantic modulation of visual picture processing by naturalistic sounds and spoken words. *Multisensory Research*, 26, 371–86.

Chen, Y.-C. and Spence, C. (2017). Assessing the role of the 'unity assumption' on multisensory integration: a review. *Frontiers in Psychology*, 8, 445.

Chen, Y.-C. and Spence, C. (2018a). Dissociating the timecourses of the crossmodal semantic priming effects elicited by naturalistic sounds and spoken words. *Psychonomic Bulletin and Review*, 25, 1138–46.

Chen, Y.-C. and Spence, C. (2018b). Audiovisual semantic interactions between linguistic and non-linguistic stimuli: the time-courses and categorical specificity. *Journal of Experimental Psychology: Human Perception and Performance*, 44, 1488–507.

Christensen, C. (1983). Effects of color on aroma, flavor and texture judgments of foods. *Journal of Food Science*, 48, 787–90.

Cinel, C., Humphreys, G. W., and Poli, R. (2002). Cross-modal illusory conjunctions between vision and touch. *Journal of Experimental Psychology: Human Perception and Performance*, 28, 1243–66.

Connolly, K. (2014). Multisensory perception as an associative learning process. *Frontiers in Psychology*, 5, 1095.

Craig, A. D. (2002). How do you feel? Interoception: the sense of the physiological condition of the body. *Nature Reviews Neuroscience*, 3, 655–66.

Dalton, P., Doolittle, N., Nagata, H., and Breslin, P. A. S. (2000). The merging of the senses: integration of subthreshold taste and smell. *Nature Neuroscience*, 3, 431–2.

De Araujo, I. E. T., Rolls, E. T., Kringelbach, M. L., McGlone, F., and Phillips, N. (2003). Taste-olfactory convergence, and the representation of the pleasantness of flavour, in the human brain. *European Journal of Neuroscience*, 18, 2059–68.

Delwiche, J. F., Lera, M. F., and Breslin, P. A. S. (2000). Selective removal of a target stimulus localized by taste in humans. *Chemical Senses*, 25, 181–7.

Denison, R. N., Driver, J., and Ruff, C. C. (2013). Temporal structure and complexity affect audio-visual correspondence detection. *Frontiers in Perception Science*, 3, 619.

Driver, J. and Noesselt, T. (2008). Multisensory interplay reveals crossmodal influences on 'sensory-specific' brain regions, neural responses, and judgments. *Neuron*, 57, 11–23.

Eccles, R., Du-Plessis, L., Dommels, Y., and Wilkinson, J. E. (2013). Cold pleasure. Why we like ice drinks, ice-lollies and ice cream. *Appetite*, 71, 357–60.

Ehrsson, H. H., Spence, C., and Passingham, R. E. (2004). That's my hand! Activity in premotor cortex reflects feeling of ownership of a limb. *Science*, 305, 875–7.

Einhäuser, W., Methfessel, P., and Bendixen, A. (2017). Newly acquired audio-visual associations bias perception in binocular rivalry. *Vision Research*, 133, 121–9.

Engel, G. R. and Dougherty, W. G. (1971). Visual-auditory distance constancy. *Nature*, 234, 308.

Ernst, M. O. (2007). Learning to integrate arbitrary signals from vision and touch. *Journal of Vision*, 7(5), 7.1–14.

Ernst, M. O. (2012). Optimal multisensory integration: assumptions and limits. In *The New Handbook of Multisensory Perception*, edited by B. E. Stein, 527–43. Cambridge, MA: MIT Press.

Ernst, M. O. and Banks, M. S. (2002). Humans integrate visual and haptic information in a statistically optimal fashion. *Nature*, 415, 429–33.

Ernst, M. O., Banks, M. S., and Bülthoff, H. H. (2000). Touch can change visual slant perception. *Nature Neuroscience*, 3, 69–73.

Ernst, M. O. and Bülthoff, H. H. (2004). Merging the senses into a robust percept. *Trends in Cognitive Sciences*, 8, 162–9.

Evans, K. K. and Treisman, A. (2010). Natural cross-modal mappings between visual and auditory features. *Journal of Vision*, 10(1), 6.1–12.

Fateminia, M., Ghotbabadi, T. D., and Azad, K. M. (2018). Taste laws in color wheel and its impact on food packaging. *Journal of Food Processing and Technology*, 9, 6.

Fiebelkorn, I. C., Foxe, J. J., and Molholm, S. (2010). Dual mechanisms for the cross-sensory spread of attention: how much do learned associations matter? *Cerebral Cortex*, 20, 109–20.

Fiebelkorn, I. C., Foxe, J. J., and Molholm, S. (2012). Attention and multisensory feature integration. In *The New Handbook of Multisensory Processing*, edited by B. E. Stein, 383–94. Cambridge, MA: MIT Press.

Fifer, J. M., Barutchu, A., Shivdasani, M. N., and Crewther, S. G. (2013). Verbal and novel multisensory associative learning in adults. F1000Res, 2, 34.

Fondberg, R., Lundström, J. N., Blöchl, M., Olsson, M. J., and Seubert, J. (2018). Multisensory flavor perception: the relationship between congruency, pleasantness, and odor referral to the mouth. *Appetite*, 125, 244–52.

Fondberg, R., Lundström, J. N., and Seubert, J. (2021). Odor-taste interactions in food perception: exposure protocol shows no effects of associative learning. *Chemical Senses*, 46, bjab003. doi.org/10.1093/chemse/bjab003.

Forster, S. and Spence, C. (2018). 'What smell?' Temporarily loading visual attention induces a prolonged loss of olfactory awareness. *Psychological Science*, 29(10), 1642–52.

Fortis-Santiago, Y., Rodwin, B. A., Neseliler, S., Piette, C. E., and Katz, D. B. (2010). State dependence of olfactory perception as a function of taste cortical inactivation. *Nature Neuroscience*, 13, 158–9.

Frank, R. A., Shaffer, G., and Smith, D. V. (1991). Taste-odor similarities predict taste enhancement and suppression in taste-odor mixtures. *Chemical Senses*, 16, 523.

Freeman, E. D., Ipser, A., Palmbaha, A., *et al.* (2013). Sight and sound out of synch: fractionation and renormalisation of audiovisual integration and subjective timing. *Cortex*, 49, 2875–87.

Gallace, A. and Spence, C. (2006). Multisensory synesthetic interactions in the speeded classification of visual size. *Perception and Psychophysics*, 68, 1191–203.

Gonzalo, D., Shallice, T., and Dolan, R. (2000). Time-dependent changes in learning audiovisual associations: a single-trial fMRI study. *Neuroimage*, 11, 243–55.

Gori, M., Del Viva, M., Sandini, G., and Burr, D. C. (2008). Young children do not integrate visual and haptic information. *Current Biology*, 18, 694–8.

Gotow, N. and Kobayakawa, T. (2017). Simultaneity judgment using olfactory–visual, visual–gustatory, and olfactory–gustatory combinations. *PLoS One*, 12(4), e0174958.

Grossberg, S. and Grunewald, A. (1997). Cortical synchronization and perceptual framing. *Journal of Cognitive Neuroscience*, 9, 117–32.

Guellaï, B., Callin, A., Bevilacqua, F., *et al.* (2019). Sensus communis: some perspectives on the origins of non-synchronous cross-sensory associations. *Frontiers in Psychology*, 10, 523.

Habets, B., Bruns, P., and Röder, B. (2017). Experience with crossmodal statistics reduces the sensitivity for audio-visual temporal asynchrony. *Scientific Reports*, 7(1486), 1–7.

Helbig, H. B. and Ernst, M. O. (2008). Visual-haptic cue weighting is independent of modality-specific attention. *Journal of Vision*, 8(10), 21, 1–16.

Hillis, J. M., Ernst, M. O., Banks, M. S., and Landy, M. S. (2002). Combining sensory information: mandatory fusion within, but not between, senses. *Science*, 298, 1627–30.

Imai, M., Miyazaki, M., Yeung, H. H., *et al.* (2015). Sound symbolism facilitates word learning in 14-month-olds. *PLoS One*, 10, e0116494.

ISO (2008). *Standard 5492: Terms Relating to Sensory Analysis. International Organization for Standardization*. Vienna: Austrian Standards Institute.

Jackson, C. V. (1953). Visual factors in auditory localization. *Quarterly Journal of Experimental Psychology*, 5, 52–65.

Jones, J. A. and Jarick, M. (2006). Multisensory integration of speech signals: the relationship between space and time. *Experimental Brain Research*, 174, 588–94.

Kim, S. and Kim, J. (2019). Effects of multimodal association on ambiguous perception in binocular rivalry. *Perception*, 48(9), 796–819.

Knoeferle, K., Knoeferle, P., Velasco, C., and Spence, C. (2016). Multisensory brand search: how the meaning of sounds guides consumers' visual attention. *Journal of Experimental Psychology: Applied*, 22, 196–210.

Körding, K. P., Beierholm, U., Ma, W. J., Tenenbaum, J. B., Quartz, S., and Shams, L. (2007). Causal inference in multisensory perception. *PLoS One*, 2(9), e943.

Lauzon, S. Á., Abraham, A. E., Curcin, K., and Stevenson, R. A. (2020). The relationship between multisensory associative learning and multisensory integration. bioRxiv preprint. doi: https://doi.org/10.1101/2020.08.28.272633

Li, Q., Wu, Q., Yu, Y., *et al.* (2020). Semantic congruency modulates the effect of attentional load on the audiovisual integration of animate images and sounds. *i-Perception*, 11(6), 1–24.

Lim, J., Fujimaru, T., and Linscott, T. D. (2014). The role of congruency in taste–odor interactions. *Food Quality and Preference*, 34, 5–13.

Lim, J. and Johnson, M. (2012). The role of congruency in retronasal odor referral to the mouth. *Chemical Senses*, 37, 515–21.

Linscott, T. D. and Lim, J. (2016). Retronasal odor enhancement by salty and umami tastes. *Food Quality and Preference*, 48, 1–10.

Ma, W. J., Zhou, X., Ross, L. A., Foxe, J. J., and Parra, L. C. (2009). Lip-reading aids word recognition most in moderate noise: a Bayesian explanation using high-dimensional feature space. *PLoS One*, 4, e4638.

Maier, J. X. and Elliott, V. E. (2020). Adaptive weighting of taste and odor cues during flavor choice. *Journal of Neurophysiology*, 124(6), 1942–7.

Martin, M. G. F. (1992). Sight and touch. In *The Contents of Experience*, edited by T. Crane, 196–215. New York, NY: Cambridge University Press.

Meijer, D., Veselič, S., Calafiore, C., and Noppeney, U. (2019). Integration of audiovisual spatial signals is not consistent with maximum likelihood estimation. *Cortex*, 119, 74–88.

Mihalik, A. and Noppeney, U. (2020). Causal inference in audiovisual perception. *Journal of Neuroscience*, 40(34), 6600–12.

Moutoussis, K. and Zeki, S. (1997a). A direct demonstration of perceptual asynchrony in vision. *Proceedings of the Royal Society London (B)*, 264, 393–9.

Moutoussis, K. and Zeki, S. (1997b). Functional segregation and temporal hierarchy of the visual perceptive systems. *Proceedings of the Royal Society London (B)*, 264, 1407–14.

Murray, M. M. and Sperdin, H. F. (2010). Single-trial learning of multisensory object associations impacts visual memory retrieval. In *Multisensory Object Perception in the Primate Brain*, edited by M. J. Naumer and J. Kaiser, 191–208. New York, NY: Springer.

Ni, R., Michalski, M. H., Brown, E., *et al.* (2015). Optimal directional volatile transport in retronasal olfaction. *Proceedings of the National Academy of Sciences of the United States of America*, 112, 14700–4.

Noppeney, U. and Lee, H. L. (2018). Causal inference and temporal predictions in audiovisual perception of speech and music. *Annals of the New York Academy of Sciences*, 1423(1), 102–16.

Parise, C. V., Harrar, V., Ernst, M. O., and Spence, C. (2013). Cross-correlation between auditory and visual signals promotes multisensory integration. *Multisensory Research*, 26, 307–16.

Parise, C. V., Knorre, K., and Ernst, M. O. (2014). Natural auditory scene statistics shapes human spatial hearing. *Proceedings of the National Academy of Sciences of the United States of America*, 111, 6104–8.

Parise, C. and Spence, C. (2009). 'When birds of a feather flock together': synesthetic correspondences modulate audiovisual integration in non-synesthetes. *PLoS One*, 4(5), e5664.

Parise, C. V. and Spence, C. (2012). Audiovisual crossmodal correspondences and sound symbolism: an IAT study. *Experimental Brain Research*, 220, 319–33.

Parise, C. V., Spence, C., and Ernst, M. (2012). When correlation implies causation in multisensory integration. *Current Biology*, 22, 46–9.

Partan, S. and Marler, P. (1999). Communication goes multimodal. *Science*, 283, 1272–3.

Pfeiffer, J. C., Hollowood, T. A., Hort, J., and Taylor, A. J. (2005). Temporal synchrony and integration of sub-threshold taste and smell signals. *Chemical Senses*, 30, 539–45.

Piqueras-Fiszman, B. and Spence, C. (2015). Sensory expectations based on product-extrinsic food cues: an interdisciplinary review of the empirical evidence and theoretical accounts. *Food Quality and Preference*, 40, 165–79.

Plailly, J., Howard, J. D., Gitelman, D. R., and Gottfried, J. A. (2008). Attention to odor modulates thalamocortical connectivity in the human brain. *Journal of Neuroscience*, 28, 5257–67.

Porter, J., Craven, B., Khan, R. M., *et al.* (2007). Mechanisms of scent-tracking in humans. *Nature Neuroscience*, 10, 27–9.

Quinlan, P. T. (2003). Visual feature integration theory: past, present, and future. *Psychological Bulletin*, 129, 643–73.
Radeau, M. and Bertelson, P. (1987). Auditory-visual interaction and the timing of inputs. Thomas (1941) revisited. *Psychological Research*, 49, 17–22.
Rohe, T. and Noppeney, U. (2015). Cortical hierarchies perform Bayesian causal inference in multisensory perception. *PLoS Biology*, 13(2), e1002073.
Rozin, P. (1982). 'Taste-smell confusions' and the duality of the olfactory sense. *Perception and Psychophysics*, 31, 397–401.
Rozin, P. and Kalat, J. W. (1972). Learning as a situation-specific adaptation. In *Biological Boundaries of Learning*, edited by M. E. P. Seligman and J. L. Hager, 66–97. New York, NY: Appleton-Century-Crofts.
Sakai, N., Kobayakawa, T., Gotow, N., Saito, S., and Imada, S. (2001). Enhancement of sweetness ratings of aspartame by a vanilla odor presented either by orthonasal or retronasal routes. *Perceptual and Motor Skills*, 92, 1002–8.
Schifferstein, H. N. J. and Verlegh, P. W. J. (1996). The role of congruency and pleasantness in odor-induced taste enhancement. *Acta Psychologica*, 94, 87–105.
Schroeder, C. E. and Foxe, J. J. (2002). The timing and laminar profile of converging inputs to multisensory areas of the macaque neocortex. *Brain Research: Cognitive Brain Research*, 14, 187–98.
Schroeder, C. E. and Foxe, J. J. (2004). Multisensory convergence in early cortical processing. In *The Handbook of Multisensory Processes*, edited by G. A. Calvert, C. Spence, and B. E. Stein, 295–309. Cambridge, MA: MIT Press.
Seubert, J., Ohla, K., Yokomukai, Y., Kellermann, T., and Lundström, J. N. (2015). Superadditive opercular activation to food flavor is mediated by enhanced temporal and limbic coupling. *Human Brain Mapping*, 36, 1662–76.
Shams, L. and Beierholm, U. R. (2010). Causal inference in perception. *Trends in Cognitive Sciences*, 14, 425–32.
Shams, L. and Kim, R. (2010). Crossmodal influences on visual perception. *Physics of Life Reviews*, 7(3), 269–84.
Shams, L., Ma, W. J., and Beierholm, U. (2005). Sound-induced flash illusion as an optimal percept. *Neuroreport*, 16, 1923–7.
Shankar, M., Simons, C., Levitan, C., Shiv, B., McClure, S., and Spence, C. (2010). An expectations-based approach to explaining the crossmodal influence of color on odor identification: the influence of temporal and spatial factors. *Journal of Sensory Studies*, 25, 791–803.
Sinnett, S., Soto-Faraco, S., and Spence, C. (2008). The co-occurrence of multisensory competition and facilitation. *Acta Psychologica*, 128, 153–61.
Small, D. M. (2004). Crossmodal integration—insights from the chemical senses. *Trends in Neurosciences*, 27, 120–3.
Small, D. M. (2012). Flavor is in the brain. *Physiology and Behavior*, 107, 540–52.
Spalding, W. R. (1920). The inter-relationship of the ear and the eye. *Etude*, 38, 11–12.
Spence, C. (2010). Multisensory integration: solving the crossmodal binding problem. Comment on 'Crossmodal influences on visual perception' by Shams and Kim. *Physics of Life Reviews*, 7, 285–6.
Spence, C. (2011). Crossmodal correspondences: a tutorial review. *Attention, Perception, and Psychophysics*, 73, 971–95.
Spence, C. (2012). Multi-sensory integration and the psychophysics of flavour perception. In *Food Oral Processing: Fundamentals of Eating and Sensory Perception*, edited by J. Chen and L. Engelen, 203–19. Oxford: Blackwell.
Spence, C. (2013). Just how important is spatial coincidence to multisensory integration? Evaluating the spatial rule. *Annals of the New York Academy of Sciences*, 1296, 31–49.
Spence, C. (2015a). Multisensory flavour perception. *Cell*, 161, 24–35.
Spence, C. (2015b). Just how much of what we taste derives from the sense of smell? *Flavour*, 4, 30.
Spence, C. (2015c). On the psychological impact of food colour. *Flavour*, 4, 21.
Spence, C. (2016). Oral referral: on the mislocalization of odours to the mouth. *Food Quality and Preference*, 50, 117–28.

Spence, C. (2018). Multisensory perception. In *The Stevens' Handbook of Experimental Psychology and Cognitive Neuroscience*, 4th edition, Vol. 2, edited by J. Wixted and J. Serences, 1–56. Hoboken, NJ: John Wiley and Sons.

Spence, C. (2019a). Attending to the chemical senses. *Multisensory Research*, 32, 635–64.

Spence, C. (2019b). Evaluating the spatial rule of multisensory integration: when exactly does spatial coincidence matter? In *Spatial Senses: Philosophy of Perception in an Age of Science*, edited by T. Cheng, O. Deroy, and C. Spence, 284–306. New York, NY: Routledge.

Spence, C. (2019c). On the relative nature of (pitch-based) crossmodal correspondences. *Multisensory Research*, 32, 235–65.

Spence, C. (2022). Odour hedonics and the ubiquitous appeal of vanilla. *Nature Food*, 3, 837–46. Available from: https://doi.org/10.1038/s43016-022-00611-x. https://rdcu.be/cXOTy.

Spence, C. and Driver, J. (Eds.). (2004). *Crossmodal Space and Crossmodal Attention*. Oxford: Oxford University Press.

Spence, C. and Frings, C. (2020). Multisensory feature integration in (and out) of the focus of spatial attention. *Attention, Perception, and Psychophysics*, 82(1), 363–76.

Spence, C., Kettenmann, B., Kobal, G., and McGlone, F. P. (2001b). Shared attentional resources for processing vision and chemosensation. *Quarterly Journal of Experimental Psychology*, 54A, 775–83.

Spence, C., Ngo, M., Lee, J.-H., and Tan, H. (2010). Solving the correspondence problem in haptic/multisensory interface design. In *Advances in Haptics*, edited by M. H. Zadeh, 47–74. Vukovar: In-Teh Publishers. Available from: http://www.sciyo.com/articles/show/title/solving-the-correspondence-problem-in-haptic-multisensory-interface-design

Spence, C., Smith, B., and Auvray, M. (2015). Confusing tastes and flavours. In *Perception and Its Modalities*, edited by D. Stokes, M. Matthen, and S. Biggs, 247–74. Oxford: Oxford University Press.

Spence, C. and Squire, S. B. (2003). Multisensory integration: maintaining the perception of synchrony. *Current Biology*, 13, R519–21.

Spence, C., Velasco, C., and Knoeferle, K. (2014). A large sample study on the influence of the multisensory environment on the wine drinking experience. *Flavour*, 3, 8.

Spence, C., Wang, Q. J., and Youssef, J. (2017). Pairing flavours and the temporal order of tasting. *Flavour*, 6, 4. Available from: http://rdcu.be/pStQ

Stein, B. E., London, N., Wilkinson, L. K., and Price, D. P. (1996). Enhancement of perceived visual intensity by auditory stimuli: a psychophysical analysis. *Journal of Cognitive Neuroscience*, 8, 497–506.

Stein, B. E. and Meredith, M. A. (1993). *The Merging of the Senses*. Cambridge, MA: MIT Press.

Stevenson, R. J. (2014). Flavor binding: its nature and cause. *Psychological Bulletin*, 140, 487–510.

Stevenson, R. J. (2016). Attention and flavour binding. In *Multisensory Flavor Perception: From Fundamental Neuroscience through to the Marketplace*, edited by B. Piqueras-Fiszman and C. Spence, 15–36. Duxford: Elsevier.

Stevenson, R. J. and Attuquayefilo, T. (2013). Human olfactory consciousness and cognition: its unusual features may not result from unusual functions but from limited neocortical processing resources. *Frontiers in Psychology*, 4, 819.

Stevenson, R. J. and Boakes, R. A. (2004). Sweet and sour smells: learned synaesthesia between the senses of taste and smell. In *The Handbook of Multisensory Processing*, edited by G. A. Calvert, C. Spence, and B. E. Stein, 69–83. Cambridge, MA: MIT Press.

Stevenson, R. J., Boakes, R. A., and Prescott, J. (1998). Changes in odor sweetness resulting from implicit learning of a simultaneous odor-sweetness association: an example of learned synaesthesia. *Learning and Motivation*, 29, 113–32.

Stevenson, R. J., Oaten, M. J., and Mahmut, M. K. (2011). The role of attention in the localization of odors to the mouth. *Attention, Perception, and Psychophysics*, 73, 247–58.

Takegata, R., Brattico, E., Tervaniemi, M., Varyagina, O., Näätänen, R., and Winkler, I. (2005). Preattentive representation of feature conjunctions for concurrent spatially distributed auditory objects. *Cognitive Brain Research*, 25, 169–79.

Tanabe, H. C., Honda, M., and Sadato, N. (2005). Functionally segregated neural substrates for arbitrary audiovisual paired-association learning. *Journal of Neuroscience*, 25, 6409–18.

Thomas, R. L., Nardini, M., and Mareschal, D. (2017). The impact of semantically congruent and incongruent visual information on auditory object recognition across development. *Journal of Experimental Child Psychology*, 162, 72–88.

Todrank, J. and Bartoshuk, L. M. (1991). A taste illusion: taste sensation localized by touch. *Physiology and Behavior*, 50, 1027–31.

Tong, J., Li, L., Bruns, P., and Röder, B. (2020). Crossmodal associations modulate multisensory spatial integration. *Attention, Perception, and Psychophysics*, 82, 3490–506.

Treisman, A. (1986). Features and objects in visual processing. *Scientific American*, 255, 106–11.

Treisman, A. (1988). Features and objects: the fourteenth Bartlett memorial lecture. *Quarterly Journal of Experimental Psychology*, 40A, 201–37.

Treisman, A. (1996). The binding problem. *Current Opinion in Neurobiology*, 6, 171–8.

Treisman, A. (1998). Feature binding, attention and object perception. *Philosophical Transactions of the Royal Society London B*, 353, 1295–306.

Treisman, A. and Schmidt, H. (1982). Illusory conjunctions in the perception of objects. *Cognitive Psychology*, 14, 107–41.

Treisman, A. M. and Gelade, G. (1980). A feature-integration theory of attention. *Cognitive Psychology*, 12, 97–136.

Vallet, G., Brunel, L., and Versace, R. (2010). The perceptual nature of the cross-modal priming effect. *Experimental Psychology*, 57, 376–82.

Van der Wal, R. C. and van Dillen, L. F. (2013). Leaving a flat taste in your mouth: task load reduces taste perception. *Psychological Science*, 24, 1277–84.

Vatakis, A. and Spence, C. (2010). Audiovisual temporal integration for complex speech, object-action, animal call, and musical stimuli. In *Multisensory Object Perception in the Primate Brain*, edited by M. J. Naumer and J. Kaiser, 95–121. New York, NY: Springer.

Veldhuizen, M. G., Nachtigal, D., Teulings, L., Gitelman, D. R., and Small, D. M. (2010). The insular taste cortex contributes to odor quality coding. *Frontiers in Human Neuroscience*, 4, 58.

Viana, F. (2011). Chemosensory properties of the trigeminal system. *ACS Chemical Neuroscience*, 2, 38–50.

Vincis, R. and Fontanini, A. (2016). Associative learning changes cross-modal representations in the gustatory cortex. *eLife*, 5, 1e16420.

Walker-Andrews, A. (1994). Taxonomy for intermodal relations. In *The Development of Intersensory Perception: Comparative Perspectives*, edited by D. J. Lewkowicz and R. Lickliter, 39–56. Hillsdale, NJ: Lawrence Erlbaum.

Wallace, M. T. and Stevenson, R. A. (2014). The construct of the multisensory temporal binding window and its dysregulation in developmental disabilities. *Neuropsychologia*, 64, 105–23.

Wang, Q. J. and Spence, C. (2019). Drinking through rosé-coloured glasses: influence of wine colour on the perception of aroma and flavour in wine experts and novices. *Food Research International*, 126, 108678. Available from: http://dx.doi.org/10.1016/j.foodres.2019.108678

Wang, Q. J., Spence, C., and Knoeferle, K. (2020). Timing is everything: onset timing moderates the crossmodal influence of background sound on taste perception. *Journal of Experimental Psychology: Human Perception and Performance*, 46(10), 1118–26. Available from: https://doi.org/10.1037/xhp0000820

Welch, R. B. and Warren, D. H. (1980). Immediate perceptual response to intersensory discrepancy. *Psychological Bulletin*, 3, 638–67.

Welch, R. B. and Warren, D. H. (1986). Intersensory interactions. In *Handbook of Perception and Performance: Vol. 1. Sensory Processes and Perception*, edited by K. R. Boff, L. Kaufman, and J. P. Thomas, 25-1–25-36. New York, NY: Wiley.

Wilson, K. A. (2021). Individuating the senses of 'smell': orthonasal versus retronasal olfaction. Synthese, 199, 4217–42. Available from: https://doi.org/10.1007/s11229-020-02976-7

Winkielman, P., Ziembowicz, M., and Nowak, A. (2015). The coherent and fluent mind: how unified consciousness is constructed from cross-modal inputs via integrated processing experiences *Frontiers in Psychology*, 6, 83. Available from: https://doi.org/10.3389/fpsyg.2015.00083

Wozny, D. R., Beierholm, U. R., and Shams, L. (2008). Human trimodal perception follows optimal statistical inference. *Journal of Vision*, 8(3), 24.1–11.

Young, N. (2017). Hearing spaces. *Australasian Journal of Philosophy*, 95(2), 242–55. Available from: http://dx.doi.org/10.1080/00048402.2016.1164202

17
Multimodal binding as mereological co-constituency

Jonathan Cohen

> Inflation is as violent as a mugger, as frightening as an armed robber and as deadly as a hit man.
>
> — Ronald Reagan, 1978

1 Introduction

Perceptual features are said to be bound when they are bundled together and distinguished from other features. Thus, in a canonical (unimodal, visual) example, if the stimulus contains a green triangle and a red square, ordinary visual perception organizes the *greenness* and *triangularity* together into one bundle, organizes the *redness* and *squareness* together into a second bundle, and treats the two distinct bundles as separate wholes. We know this binding occurs because, without it, visual systems would be unable to distinguish (as they obviously can) the configuration just described (*greenness* bundled with *triangularity*, *redness* bundled with *squareness*) from the distinct configuration consisting of a green square and a red triangle (*greenness* bundled with *squareness*, *redness* bundled with *triangularity*).[1]

On the dominant theoretical description (henceforth, 'the convergence conception'), binding amounts to the representation of features as convergently exemplified by one and the same object—a so-called sensory individual. This understanding of binding relies on (and provides important support for) the idea that perception does not only register distal property instances, but (in some modalities, at least some of the time) also connects with and attributes properties to individual entities.[2] These sensory individuals serve as points of contact between mind and world—distal individual *res* that perception connects us to, and that anchor our *de re* cognitive contact with things outside our own heads. Moreover, they are the loci of perceptual attribution, serving as targets for focal attention and/or perceptual demonstration. Importantly, sensory individuals' role as the loci for perceptual attribution grounds their role in explaining perceptual binding: the conception of binding as convergent attribution of features to one sensory individual does not get off the ground without the assumption that perception attributes its features to such individuals.

[1] This is an instance of Jackson's (1977) Many Properties Problem.

[2] As stated, this thesis says very little about the nature of the individual entities involved; in particular, it should not be understood as committing to the perceptibility of bare particulars.

Jonathan Cohen, *Multimodal binding as mereological co-constituency* In: *Sensory Individuals*. Edited by: Aleksandra Mroczko-Wąsowicz and Rick Grush, Oxford University Press. © Oxford University Press 2023. DOI: 10.1093/oso/9780198866305.003.0018

The convergence conception is, as noted, the canonical model of unimodal binding. It is endorsed, typically explicitly, by such disparate figures as Treisman (1998), Clark (2000), Shams *et al.* (2000), Campbell (2002), Treisman (2003), Cohen (2004), Matthen (2005), and Dickie (2010). It is fair to say that the convergence conception of binding has proven itself an important and fruitful contribution to the contemporary understanding of unimodal perception.

However, research has now moved beyond unimodal cases to focus on a broad class of multimodal cases in which features from distinct perceptual modalities are bundled in what appears to be something like the manner seen in unimodal binding. Thus, visual and auditory features appear to be bundled in such much-discussed cases as the McGurk effect (McGurk and MacDonald, 1976), the ventriloquist illusion (Pick *et al.*, 1969; Vroomen and de Gelder, 2000, 2004), the motion-bounce effect (Sekuler *et al.*, 1997), and the sound-induced flash illusion (Shams *et al.*, 2000, 2002). The cutaneous rabbit illusion (Geldard and Sherrick, 1972) and rubber hand illusion (Botvinick and Cohen, 1998) provide evidence of tactile/visual bundling. Similarly, there is evidence of bundling between olfaction and vision (Kuang and Zhang, 2014), olfaction and gustation (Dalton *et al.*, 2000; Spence *et al.*, 2014), vision and gustation (Morrot *et al.*, 2001), audition and gustation (Spence, 2015), and more. In view of these findings, it is natural to wonder how and whether to extend theoretical ideas about perceptual binding, first developed with respect to unimodal cases, to multimodal perception.[3]

Below I will survey a range of answers to this question by working through what we might think of as (with apologies to Quine) three grades of multimodal involvement. The first grade is convergence; I'll argue that, despite what its proponents have claimed, the convergence conception is not straightforwardly applicable in multimodal settings, and that recent attempts to overcome this difficulty have been unsuccessful (Section 2). Next I'll consider critically a second grade of multimodal involvement—a revisionary account in terms of associative linking (Section 3); I'll argue that, while prominent objections against this view are unconvincing, it leaves important phenomena unexplained. This will lead me to offer a novel, deflationary, third grade of multimodal involvement, which I call the mereological co-constituency view; I'll present this account of multimodal binding, and argue that it answers to the empirically motivated needs that multimodal binding has been called on to address (Section 4). Finally, I'll briefly draw some morals (Section 5).

Before I begin, a few qualifications are in order. First, I won't worry about the difficult and important question of how to individuate perceptual modalities. All examples discussed will be drawn from the traditional, Aristotelian, five modalities; I will assume that these are modalities and that they are distinct (though possibly not in all explanatory contexts). Second, though the reasons I will advance for a deflationary/non-traditional treatment of intermodal binding are, in principle, applicable to intramodal/unimodal binding as well, I think this view is much less well motivated by the empirical facts about intramodal cases; in particular, considerations about the ontological diversity of feature-bearing objects, which

[3] As we'll see in Sections 2–3, many theorists have taken the view that the bundling in these multimodal cases is sufficiently close to unimodal paradigms to merit being treated as instances of binding. Of course, in the absence of a commitment to a theoretical conception of binding that would prejudge the question of the paper, one might reasonably question that choice. Thus, one might instead reserve 'binding' for unimodal cases, and describe the multimodal cases as instances of 'multimodal integration' (arguably a more standard term for them in any case).

There's no point in fighting over the label. I propose to apply 'binding' to multimodal cases, with the caveat that this terminological choice should be understood to open, rather than close off, the substantive question of what theoretical account of the notion we should adopt.

I take to be central to motivating the move away from the convergence conception, do not arise intramodally. Third, though below I'll be engaged in advocacy for the co-constituency account as a powerful and general conception of multimodal binding, it would be inappropriate to insist at the outset that there has to be a single best model that applies to every case. If, at the end of the day, co-constituency earns its keep merely as one component of a well-motivated pluralism about multimodal binding, then so be it. Finally, nothing in what I say should be taken to deny that we multimodally perceive distal objects and events. There's no question that we have perceptually informed representations of individual objects and events bearing properties, and that we have multiple, distinct modal channels that contribute to these representations. Nothing I say below is intended to conflict with these facts.

2 Convergence

To fix intuitions, consider a canonical multimodal case of perceptual bundling, such as the McGurk effect. You see video of a speaker uttering the syllable /ga/, while simultaneously hearing an audio recording of the speaker uttering the syllable /ba/; as a result, you perceive the syllable /da/.[4] On its face, this appears to be an instance of informational interaction between bundled visual and auditory features—note that you experience things quite differently, and indeed the modification to /da/ disappears, when you apprehend them in separate, successive (unbundled) unimodal presentations.

If this and the other phenomena catalogued above are to be regarded as instances of perceptual binding, we need a model for thinking about binding in a multimodal setting. A natural starting answer presents itself: why not simply extend the convergence model, familiar from discussions of unimodal binding, to multimodal cases? Recall that the convergence account construes binding as the convergent attribution of distinct features to one and the same sensory individual. At this level of description, the convergence model seems just as applicable to cases of multimodal as unimodal binding; the difference will lie merely in whether the distinct features attributed to one sensory individual are extracted by multiple modalities or one. Applying this model to the McGurk effect, the claim would be that vision and audition convergently attribute their own features to one single individual (say, an utterance event).

There is a wrinkle. In an important paper, Nudds (2014) points out that the extension just contemplated can take two quite distinct forms, which he labels 'amodal' and 'crossmodal' integration. Though both count as robust types of integration between information derived from distinct unimodal perceptual processes, they differ in the types of representations they envisage as resulting from the integrative process. In amodal cases, integration of unimodal representations results in the construction of a brand new, amodal representation of the distal target:

> [in amodal integration ...] a number of initially distinct processing streams combine to produce a single amodal representation of an object, that represents it as having features—such as spatial and temporal features—that may have been perceived with more than one sense modality, as well as features—such as colour—that are modality specific. A single amodal object representation may represent an object as shaped, coloured, and as the source

[4] For an example, see https://youtu.be/yJ81LLxfHY8

> of a sound. It follows that the same kind of amodal-object representation plays a role in explaining our perceptual awareness of particular things perceived with any of the sense modalities. Both our visual perception of an object and our auditory perception of a sound source is explained by appeal to the same kind of amodal object representation (Nudds, 2014, p. 173).

In contrast, on the crossmodal model, the result of integration is not the creation of a brand new amodal representation, but rather a coordination relation holding between unimodal representations that all converge on the same distal target:

> with respect to features that can be perceived with more than one sense, the visual-object representation and the auditory-object representation represent the object as having the same features, for example, as being at the same location, occurring at the same time, and so on, but the visual-object representation of the object will also represent it as having features that are specific to vision, and the auditory-object representation of the object will also represent it as having features specific to audition (Nudds, 2014, p. 174).

These two models of multimodal integration are importantly different, and therefore useful to have on the table in attempting to come to grips with individual cases.[5] However, and notwithstanding their differences, it is important to see that both crossmodal and amodal integration are species of convergence: both are committed to construing binding as the attribution of features to a single distal object. The two models differ on the questions of how many representations (one or many) and what types of representation (amodal or unimodally specific) result from integration. But they agree that, whatever representations do so result have the role of attributing multiple features—namely, those features derived from the unimodal inputs to integration—to one and the same distal individual. Despite their differences, then, they agree that binding integration results in the convergent attribution of *many* features to *one* individual.

The convergence conception of multimodal binding—particularly as usefully precisified by Nudds—is a simple, straightforward, and well-articulated extension of ideas that have proven their merit in unimodal settings. It has been defended by leading theorists (e.g., Nudds, 2014; O'Callaghan, 2014, 2016, 2017; Green, 2019).[6] Despite this, the application of the convergence to multimodal cases faces a *prima facie* worry concerning the diversity of sensory individuals from distinct modalities.

The concern arises from the reasonable assumption that if features from distinct modalities are convergently bound to one common individual, then that individual must be an

[5] As Nudds points out, one needn't construe every instance of integration as conforming to the same model; moreover, both models are compatible with the existence of unimodal/unintegrated representations in perceptual systems. The point of such models is not to facilitate a once-for-all time description that will apply to all perceptual processing, but simply to make available well-formulated descriptive options.

[6] O'Callaghan's position is more nuanced than this glib attribution lets on. Though he has certainly endorsed convergence as a description of (some cases of) *property* binding, he explicitly follows Treisman in construing binding as applying to both properties and parts (e.g., O'Callaghan (2014), pp. 74–5, 78)—often using 'feature' to range over both; and his descriptions of *part* binding are more readily understood as co-constituency than convergence. (In contrast, my topic here is exclusively the binding of properties.) Moreover, once he allows a diversity of models under the broad heading of binding, it is open to him to extend a co-constituency story to individual cases of property binding as well (e.g., see the discussion of audiovisual property binding in O'Callaghan, 2011, p. 395ff). In short, O'Callaghan is not a uniform proponent of convergence. (Thanks to O'Callaghan for discussion of these matters.)

appropriate locus for feature bearing in both modalities. That is, if modalities m_1 and m_2 share one and the same sensory individual a, it must be that m_1 and m_2 overlap in the types of objects that they recognize as sensory individuals (and, of course, that a falls into the overlap). However, it's unobvious that this constraint is satisfied by pairs of modalities associated with sensory individuals, and for which there is evidence of multimodal binding. On the contrary, it seems likely that the individuals associated with different modalities often fail to overlap.

For example, consider the modalities of vision and audition, both of which have been thought to attribute qualities to sensory individuals, and involving which many theorists have claimed that there is good evidence of intermodal binding. On the one hand, many authors hold that visual sensory individuals are (at least in canonical instances) ordinary material objects—cohesive, temporally persistent objects extended and bounded in space (e.g., Cohen, 2004; Dickie, 2010; Marr, 1982; Matthen, 2005, pp. 277–82; Nanay, 2013, p. 51; Pylyshyn, 2007; Spelke, 1990; but cf. Clark, 2000). On the other hand, while many have been attracted by the idea that audition attributes qualities to sensory individuals, it hasn't seemed attractive to claim (and, as far as I'm aware, no one has claimed) that such auditory individuals are ordinary material objects. Instead, there are a number of proposals in the literature that construe auditory individuals as things other than material objects. Thus, O'Callaghan (2007, 2009) holds that auditory properties qualify medium-disturbing events (cf. Matthen, 2005); Casati and Dokic (2009) believe they qualify monadic events befalling material objects; Sorensen (2008) and O'Shaughnessy (2009) think they qualify waves; Nudds (2009, 2010a, b) thinks they qualify structures exemplified by waves; Pasnau (1999), Kulvicki (2008), Cohen (2010), and Kulvicki (2014) think they qualify dispositional properties of objects; Soteriou (2018) thinks they can qualify all of these; and Young and Nanay (2020) think they qualify temporally extended, causally composite individuals built from the types of entities cited by the others. But, given the standard conception of the sensory individuals of vision as ordinary objects, every one of these views about auditory individuals would seem to threaten the possibility of overlap between the sensory individuals of vision and those of audition. Consequently, every one of them threatens the possibility of extending the convergence conception of binding to intermodal cases involving vision and audition.

The same concern arises for putative instances of multimodal binding involving other pairs of modalities. Thus, consider audition and touch. There's reason for thinking both that there are sensory individuals for touch, and that these may, at least in some cases, overlap with the sensory individuals of vision (Fulkerson, 2014b); but if so, then (for the reasons just rehearsed), the sensory individuals of touch are unlikely to overlap with those of audition. Similarly, if there are sensory individuals for olfaction and gustation (Lycan, 1996; Matthen, 2005; Batty, 2014; Carvalho, 2014; Smith, 2015), it is unobvious that these are of the same type as either the ordinary material objects that serve as visual (and perhaps tactual) individuals or any of the candidate types proposed as auditory individuals.

It would seem, then, that the diversity of sensory individuals associated with distinct modalities poses a significant *prima facie* threat to the extension of the convergence view to multimodal instances of binding. What to do?

Defenders of the convergence conception (O'Callaghan, 2008; Nudds, 2009; O'Callaghan, 2014, 2016; Green, 2019) have attempted to answer this threat by understanding multimodal sensory individuals as structured, mereologically complex individuals that have unimodal sensory individuals as constituents, but that can (*qua* complex wholes) be shared by multiple modalities:

> Multisensory perceptual objects are mereologically complex individuals with hybrid structure. Some of their parts and features are perceptible through one sense, and some are perceptible through another. Each sense provides a partial perspective on the whole. The complex whole is perceptible as such through the coordinated use of multiple senses. The key is that perceiving a whole does not require perceiving each of its parts or features. Visuo-tactile objects include the material bodies on which vision and touch converge—a subset of visible objects. Audio-visual objects are environmental happenings that involve bodies and include sounds. Flavors are complex, and they are only fully perceptible using multiple senses; however, flavors are properties, attributed to things we ingest, rather than individuals (O'Callaghan, 2016, p. 1287).[7]

Before we can determine whether this suggestion resolves the problem about the diversity of sensory individuals, and thereby clears the way for the extension of the convergence conception to multimodal binding, I offer two clarificatory observations. The first is that, if they are to be of help in understanding binding, the mereological complexes at issue cannot be mere, unstructured bags of features (as appeals to mereological summation may evoke). Rather, these must be *structured* complexes, where the structure in question extends to part-whole relations, predicational relations, and possibly more. The second observation concerns the source and nature of this structure. As presented by O'Callaghan and others, the proposal appears to be that multimodal sensory individuals should be identified with complex mereological entities licensed by the One True Metaphysical Theory of Mereology.[8] But that's potentially misleading. For, if the mereological complexes invoked here are to serve their purpose in defending an understanding of multimodal binding as convergent psychological attribution of features to single individuals, then those individuals must be not only metaphysically real, but also available for the psychological construction of perceptual representations (and, for forms of binding of which there is awareness, also available to conscious awareness). As such, the sort of mereology required by the solution on offer is not a metaphysical mereology of parts and wholes, but a psychologized mereology of part/whole groupings made available by our perceptual-cum-cognitive endowments. Mere metaphysical mereological structure without a psychological reflection won't do.

Be this as it may, I now want to observe that the proposed solution does not, as advertised, solve the diversity puzzle, or, therefore, save the convergence conception of multimodal binding. The problem is that, while the mereological complex proposal provides a common individual accessible to distinct modalities—viz. the psychologized mereological whole—that common individual is, in many ordinary cases, not a bearer of features attributed by perceptual modalities. To see why, consider O'Callaghan's (2016, p. 1284) example of a clapping event, which he proposes to construe as a mereologically complex happening consisting of

[7] NB. The mereological conception of multimodal *sensory individuals*, *qua* objects of perception (O'Callaghan, 2016), should be distinguished from both the mereological account of conscious multimodal *experiences*, *qua* complex phenomenal states (Bayne and Chalmers, 2003; Bayne, 2010), and the idea that some *unimodal* sensory individuals are themselves mereologically complex (Young and Nanay, 2020).

[8] And, indeed, the proposal appears to depend on a mereological realism that, philosophy being what it is, is denied by some (e.g., Rosen and Dorr, 2002; Sider, 2013). As this issue goes substantially beyond the scope of this paper, I'll grant as much metaphysical realism about mereology as needed for present purposes.

visible and audible components. Suppose we accept this, and that vision and audition converge in connecting us with this mereologically complex individual in virtue of connecting us with distinct parts of this common whole. Perhaps we will also grant, therefore, that we see the whole (in virtue of seeing its visible components) and hear the whole (in virtue of hearing its audible components), hence that there is an individual that we both see and hear. Still, notwithstanding these claims, it is the ordinary material objects we see—not the mereological complex having material objects as constituents—that exemplify visible properties such as colours and shapes. Vision represents the *hands*, and not the happening of a clapping, as bearing a size and colour. Likewise, audition represents the *sound*, and not the complex visual-cum-auditory happening of a clapping, as bearing the auditory qualities of volume and timbre. As such, while the mereological complex proposal may give us a way of saying that distinct modalities connect us to one and the same individual, it does not vindicate the convergence conception of multimodal binding as perceptual attribution of features from distinct modalities to one and the same individual.[9]

It would seem, then, that our *prima facie* challenge to the possibility of multimodal binding remains unanswered and stands as a threat to the convergence conception (on any precisification).

3 Association

A different proposal for understanding multimodal binding, defended by Goldstone (1998), Fulkerson (2011), Bayne and Spence (2014), and Goldstone and Byrge (2015) (and possibly by Connolly (2014a, b, 2019)), construes multimodal binding as a relation of 'association' between distinct individuals figuring in unimodal attributions. On this conception, there is no single sensory individual associated with multiple modalities and that bears the features supplied by each of them. Rather, each modality (or, for that system, within a given modality) attributes features to its own individual; binding occurs when the numerically distinct individuals that result from such multiple perceptual attributions stand in (consciously represented) relations of association:

> Multisensory perceptual experiences do not involve the direct predication of features onto individual perceptual objects. In- stead, there is an association between experiences .... What we experience is a higher-order association between sensory experiences (Fulkerson, 2011, p. 506).

It may be helpful to think of the association view as a weakening of the crossmodal convergence model: where the latter involves coordination between distinct unimodal representations converging on the same individual target, the association view involves coordination between distinct unimodal representations that need *not* converge on a common target.

[9] Response: Properties can be inherited by mereological complexes from their constituents. If the hands have a property *F* (say, a size) and are constituents of a mereologically complex clapping, then we can attribute *F* to the complex as well.

Counter-response: That sort of inheritance, construed so as to apply to the perceptible features of all mereologically complex perceptible items, is untenable. A mereologically complex figure might have components that are circular and square; if inheritance obtained in general, it would license the conclusion that the complex figure is circular and square. Surely that cannot be right.

To see what this amounts to concretely, consider a case of multimodal binding between an auditory attribution of a pitch and a visual attribution of a spatial location. Where, on the convergent attribution view, such binding requires attribution by both modalities to a common individual, the associative proposal treats binding as an instance of psychological association holding between one auditory individual bearing a pitch and a distinct visual individual bearing a location. Since, on the latter view, the only sensory individuals required are those associated with the individual modalities, there is no need for individuals spanning separate modalities. Consequently, the observed diversity of unimodal objects is not a challenge to this conception of binding, as it is for any version of the convergence conception. This seems an important advantage.

Despite this, the association view confronts a few challenges. First, as O'Callaghan (2014, p. 84) notes, mere association won't account for the observation that subjects typically display fallible conscious awareness of whether feature attributions from distinct modalities are bound or not. After all, though subjects are sometimes fallibly aware of associations holding between their psychological states, in many cases they are not. Consequently, if any old relations of association were sufficient for intermodal binding, we should arguably expect less awareness of whether binding obtains than we observe. To answer this, the association view should require for binding not only that there be higher-order associations between unimodal perceptual representations, but that these associations are (at least) typically fallibly available to conscious awareness.

But O'Callaghan objects that even this is insufficient—that even consciously available relations of association are too weak to capture the phenomenon of experienced binding:

> ... if seeming to be associated or to tend to co-occur does not guarantee seeming to share a common object or source, then appearing merely as being associated or as tending to co-occur is too permissive to capture the relevant distinctions among the cases discussed above. For instance, a sound and an image may seem merely to be associated or to tend to co-occur without seeming perceptually to share a common source. A rough surface and a red surface may seem to be associated without their seeming perceptually to be one surface or to belong to one object. Mere associations thus do not suffice for an account of that to which one may be multimodally perceptually sensitive, and they do not suffice for an account of multimodal perceptual awareness (O'Callaghan, 2014, pp. 84–5).

O'Callaghan's worry is well taken: we can represent feature exemplifications as co-occurring (or even tending to co-occur) without thereby representing them as bound, so a mere co-occurrence account of binding overgenerates instances of binding.[10] However, this is a point that the proponent of the association view can accommodate, so long as she understands the notion of association in a way that distinguishes it from mere co-occurrence. Even on a classical construal of association, spatio-temporal co-occurrence is neither necessary (today's olfactory experience of Grandma's brand of perfume is now associated with the remembered, long past, experience of that perfume, or with the also long past gustatory experience of the cookies Grandma used to bake) nor sufficient (if a subject is conditioned to associate the sound of a bell with an electric shock, the equally proximate and co-occurring colour of the

[10] On the other hand, O'Callaghan's formulation of the point so as to demand that an account of binding amounts to a representation of common sourcehood, in particular, seems unjustified. This demand goes well beyond the data, and is tantamount to begging the question against weaker (non-convergence) accounts of binding.

walls in the room need not be so associated) for association.[11] (As Fulkerson (2011, p. 506) writes, 'The coordination involved in the auditory-visual case is often (though not always) sensitive to temporal and spatial continuity.')

For that matter, the association proposal need not be tied to such a narrowly classical associationist conception of psychological association. Thus, in developing his version of the proposal, Fulkerson (2011) endorses a more flexible understanding of association:

> The idea of an associative relation is intended as a general concept that can explain a wide variety of multisensory interactions . . . (p. 494).

> This notion of an associative relation is meant to be a general means of characterizing the structure of a range of distinct sensory mechanisms relating perceptual experiences . . . (p. 497).

> . . . multisensory experiences involve some higher-level relation between separate experiences . . . (p. 504).

Though it would be nice to have a more substantive account of the liberalized conception of association than what emerges from these quotations, they give us reason for thinking that a version of the association view might evade the problem we are considering.[12]

A potentially more worrisome overgeneration objection to the association account comes from Green (2019, pp. 15–16). Green notes that the association relation does not, by itself, come with numerical limitations: it can apply as easily between two items as between three, four, or even more (subject to limitations on working memory). Consequently, he suggests, an association account of binding, unless supplemented by further constraints (that would require their own motivation), would seem to predict more binding than we observe. Thus, for example, in a case where one sound is presented with multiple visible flashes, an association account should allow binding between the one sound and many of the presented flashes. In contrast, if we thought of binding as constrained to apply to features of one and the same individual (as on a convergence view of the type Green favours), we would expect that the logic of identity would effectively constrain the binding of the sound to just the visible flash that is an aspect of the very same sensory individual. And he points to results of van der Burg *et al.* (2013) suggesting that there are, indeed, one-to-one constraints on at least some cases of audible/visible binding. In these experiments, subjects were shown a circular arrangement of 24 discs. Every 150 ms, a random subset of the discs changed colour from black to white, or vice versa. At an unpredictable point during the trial, one of the change events was accompanied by a tone. The subject was told beforehand that the discs that changed colour alongside the tone were the targets, and the task was simply to remember them. At the end of a trial, the subject was directed to a particular disc and asked whether it had been one of the targets . . . They found that capacity limits never

[11] Indeed, some psychotherapeutic conditioning techniques (e.g., extinction therapy (Marks, 1979), applied behavioural analysis (Lovaas, 1987), exposure-based therapy (Huppert and Roth, 2003)) specifically aim at gradually increasing the spatio-temporal distance between associated items. This aim would be incoherent if association relations were restricted to items in narrow spatio-temporal proximity.

[12] In conversation, Fulkerson assures me that he always intended a more flexible construal than that found in classical associationists. While I still do not understand Fulkerson's version of the association view as fully as I would like (my fault, not his), I suspect that his position will ultimately share more with the view defended in Section 4 than with that of other association theorists.

exceeded 1 (the average across experiments was 0.75). In other words, given a single auditory cue, at most, one visible disc could be remembered (Green, 2019, pp. 15–16).

To be clear, I do not believe that this challenge is completely decisive (nor does Green). For one thing, the challenge, as posed, rests on a single study and concerns only a single form of multimodal binding. For another, even if adopting an association conception of binding leaves us without an explanation of the observed numerical limitations on binding in terms of the binding relation itself, we cannot rule out the possibility of independent explanations of these facts in terms of properties of working memory, attention, or something else. Alternatively, we might even accept these limitations as brute facts about perceptual systems. Still, I think it is fair to say that the association account of binding leaves us without an explanation of an important limitation on the phenomenon and that, other things equal, it would be preferable to have such an explanation.

4 Binding and co-constituency

4.1 The mereological co-constituency view

I propose that the way forward lies in bringing together insights from two views already discussed—the mereological complex answer to our puzzle about the diversity of sensory individuals (Section 2), and the associative account of multimodal binding (Section 3).

In Section 2, I argued that mereological complexes built from unimodal constituents are unsuitable as targets of convergent feature attribution because these complexes (as opposed to their unimodal mereological components) do not bear perceptual features contributed by individual modalities—*a fortiori*, cannot bear *multiple* features contributed by *distinct* perceptual modalities. But this dissatisfaction gives us no reason for denying that there are such mereological complexes, that perception represents them, or that they have an important role to play in our understanding of multimodal perception.

On the contrary, there is ample reason for accepting that we are perceptually connected to such complexes, quite independently of one's views about sensory individuals and binding. It should be relatively uncontroversial (barring the mereological nihilism set aside earlier) that the objects and events we perceive are mereological constituents of larger, more complex objects and events. An apple is a mereologically complex object composed of (among other constituents) the surface facing you and the surfaces not facing you. So, too, a car crash is a mereologically complex event composed of (among other constituents) both visible and audible components. Just as O'Callaghan points out, we perceive the wholes by virtue of perceiving their parts. You see the apple, the complex object, by seeing one of its constituents (its facing surface), even though you do not see others of its constituents (e.g., its non-facing surfaces, its insides). Similarly, you see the complex event of the car crash by seeing its visible aspects, even though you do not see its non-visible aspects; and you hear it by hearing its audible aspects, even though you do not hear its non-audible aspects. Moreover, a distinguished history of research in perceptual psychology dating from at least the Gestalt period makes clear that perception organizes units at various levels into complex arrangements of parts and wholes—for example, those that comprise objects (e.g., Biederman, 1987; Spelke, 1990; Tarr and Bülthoff, 1998), groups (Koffka, 1935; Wertheimer, 1938; Palmer, 1999; Chang *et al.*, 2007; Wagemans, Elder, *et al.*, 2012; Wagemans, Feldman, *et al.*, 2012b; Goldstone and

Byrge, 2015), and events (Michotte, 1946/63; Johansson, 1973; Gibson, 1979; Zacks and Tversky, 2001; Zacks *et al.*, 2007). This is to say that these forms of mereological organization are not only real, but also psychologically represented.

On the other hand, while we saw in Section 3 that the association account of multimodal binding may underconstrain/overpredict the phenomenon of binding, there was much to like about this proposal as well. In particular, it offered the hope of avoiding the need for convergence, and so sidestepping the challenge posed to convergence by the diversity of unimodal individuals.

The solution is to join these ideas—to recognize the reality of (psychologically represented) multimodal mereological complexes, and to understand multimodal binding as the coordinated representation of co-constituency, but without requiring that bound features convergently apply to any one individual.[13] I'll call the resulting position the (MEREOLOGICAL) CO-CONSTITUENCY view of binding.

To understand the view, consider a paradigmatic instance of multimodal binding—the visual-*cum*-auditory perception of a car crash. A car crash is a complex mereological event, which contains visible and audible aspects. When we perceive this complex event, perception builds a complex, multimodal representation of this structured target. The current proposal is that what binds visible and audible aspects in perception is not the convergent attribution of visible and audible features to any single entity, but the fact that perception coordinates its representations of visible and audible aspects, treating them as sisters/co-constituents (i.e., nodes dominated by a single node) of the complex, hierarchically organized event of the car crash. What we have been calling 'visible aspects' of the car crash are exemplifications of visual features (say, colours, forms) by specifically visual sensory individuals (say, ordinary material objects). Likewise, the crash's 'audible aspects' amount to exemplifications of auditory features (say, pitches, loudnesses) by specifically auditory sensory individuals (say, medium-disturbing events). The co-constituency view says that these visual and auditory aspects are bound in this sense: both the exemplification of visual features by visual individuals and the exemplification of auditory features by auditory individuals are represented by perception as mereological constituents of one and the same complex event of the car crash. That is, visual and auditory features are bound by figuring in coordinated representations, making attributions to parts of the same (represented) complex whole.

The co-constituency view has a number of attractive features.

First, it is ontologically and representationally conservative. It is committed only to the existence and representation of (i) unimodal features and unimodal individuals already motivated by consideration of unimodal binding, together with (ii) the mereologically complex events built from the latter that all parties to the debate have reason to accept. It is worth emphasizing that, for reasons discussed in Section 2, the co-constituency theorist's reliance

[13] Qualification: Perhaps there are cases in which binding results in the attribution of novel features not exemplified in any unimodal perceptual episode; O'Callaghan (2017, p. 174) offers the example of *mintiness*, which plausibly requires binding between olfactory, gustatory, and oral somatosensory features, but is not itself attributed in any single modality. The co-constituency account can treat such novel features as qualifying the complex—and not the unimodal individuals of any individual modality. (In contrast, and as Fulkerson (2014a) urges, such cases are hard to understand for association views on which binding relates attributions from individual modalities but never contributes novel features.) Even so, on this treatment, such cases are not helpfully construed as instances of convergence. Here it is not that unimodal features from distinct modalities convergently qualify the complex. Rather, unimodal features from distinct modalities non-convergently qualify the distinct individuals of modalities contributing to the complex, and this leads to the downstream attribution of numerically distinct, and novel, features to the complex itself.

on a psychologically represented structure of part-whole relations is shared by convergence theorists, who invoke mereology as a response to concerns about the diversity of unimodal individuals; hence, this commitment is not a special burden for the co-constituency view.[14]

Second, though it allows that distinct perceptual modalities relate to a common entity (the mereological complex), the co-constituency view is not committed to construing binding as convergence—it does not require for binding that distinct modalities convergently attribute features to any single entity. As such, if distinct modalities predicate their features to distinct entities, this is no obstacle to the view's treating such features as bound. The diversity problem is not a problem for the co-constituency view.

Third, and unlike (the classical associationist construal of) the association view, the co-constituency view can allow for numerical limitations on binding arising from numerical limitations on the psychological representation of mereological structure. The crucial insight here is that, whatever restrictions one takes to constrain merely metaphysical mereological summation, there are substantive psychological restrictions on the organization of represented parts into represented wholes. These psychological restrictions are complicated and multifarious, and span a range of levels at which parts can be organized into wholes (as attested by the vast literature on perceptual grouping). However, it does seem, as a matter of empirical fact, that the grouping operations at issue do not permit arbitrary/unrestricted recombination. Whether or not the scattered collection of the Queen's nose and the Eiffel Tower constitutes a metaphysically legitimate mereological sum, these two objects won't be constituents of a perceptually represented whole (in ordinary cases). (This is something that can be tested by standard psychological grouping criteria such as object-specific preview effects, preferential looking times, etc.). Consequently, the features of these objects won't ordinarily enter into perceptual binding relations. Against this backdrop, the one-to-one constraints uncovered by van der Burg *et al.* (2013) are naturally understood as further limitations on psychological grouping. Their results support the idea that, when the psychological mechanisms underlying grouping build a whole connecting an auditory tone with one visually presented event, this operation inhibits an alternative grouping of that same tone with earlier or later visually presented events. Though it might be that classical association (and metaphysical mereology) lacks analogous constraints, the co-constituency view has no trouble accommodating the finding that relevant psychologically mediated grouping operations are, as a matter of fact, subject to numerical limitations.

Finally, the co-constituency view can be viewed as a conservative extension of the convergence conception: the former can explain the attractions of, and inherit the successes of, the convergence view, while succeeding in cases where the latter fails. Specifically, convergence can be seen as a special, degenerate case of co-constituency—a case where co-constituents of a single represented whole happen to stand in the identity relation (as well as the mereological co-constituency relation). In such cases, perceptual feature F_1 applies to a_1, and perceptual feature F_2 applies to a_2, and whether these features and individuals are contributed by the same or different modalities, it turns out that $a_1 = a_2$; hence F_1 and F_2 convergently qualify the very same individual. However, given that a_1 can be counted a trivial constituent of any complex of which it itself is a constituent, it is (in a good, if trivial, sense) its own

[14] One possible account of such a generalized psychological representation of this structure might invoke event files (Hommel, 2004); other options include appeals to a general notion of indexing, predictive coding, and more. My defence of co-constituency need not take sides on these (important) implementational matters.

co-constituent. Consequently, such cases can just as easily be described within the co-constituency framework.

4.2 Co-constituency and the ties that bind

I now want to argue that the account of binding sketched in Section 4.1 explains the phenomena that have provided the strongest support for believing in multimodal binding, and so is plausibly up to the job of answering the theoretical needs for which that notion has been invoked.

A first point to make is that, on the co-constituency view, multimodal binding arises from a perceptual, rather than a wholly post-perceptual, representation of co-constituency, and so should be counted a substantively perceptual phenomenon. This is important because, while it might initially seem tempting to think of intermodal binding in exclusively post-perceptual terms—say as a judgement that features *F* and *G* are linked, there are persuasive reasons for rejecting this proposal. One reason is that there are illusions of intermodal binding (as distinguished from illusory representations of any of the features bound), in which perception erroneously represents binding, even though post-perceptual judgement does not (Cinel *et al.*, 2002). Another is that illusory representations of binding can, in many cases, survive the addition of cognitively represented information that would undercut a fully post-perceptual representation of binding: for example, the McGurk illusion persists even after subjects learn about the effect and experience the auditory stimulus without the visual stimulus, and vice versa.[15] A third, though perhaps weaker, reason is that perceptual judgements about intermodal binding and the perceptual guidance of action on the basis of intermodal binding seem to have an immediacy that some have thought incompatible with a post-perceptual understanding of the phenomenon (O'Callaghan, 2014). In any case, all of this is just what we should expect on a perceptual understanding of intermodal binding such as the co-constituency view.

A second consideration is that the co-constituency account allows for a wide range of intermodal informational interactions going beyond the aggregation of information from multiple modalities, including those that many authors have appealed to explicitly to motivate the convergence conception (and, since these are often not distinguished, for accepting intermodal binding itself).[16] One paradigm of such an interaction occurs in the McGurk effect, where it is not only true that a visual feature is bundled together with an auditory feature, but that the fact of bundling results in modulation or modification of the auditory feature representation: bundling the visual representation of /ga/ with the auditory representation of /ba/ results not just in an experience combining co-present contributions from two modalities, but also modification of the content of the auditory representation (from /ba/ to /da/).

[15] That is, this particular binding effect seems to be encapsulated from cognition in just the sense that is true of some other illusory perceptual states—say, that produced by the Müller–Lyer configuration (Fodor, 1983). (This is not to say that all instances of binding are so encapsulated; on the contrary, and as discussed in footnote 17, evidence involving the motion-bounce illusion and other cases suggests that at least some instances of the representation of binding are penetrated by—that is, are not encapsulated from—cognition.)

[16] Thus: 'The main source of empirical evidence for intermodal binding is that sensory systems interact and share information' (O'Callaghan, 2014, p. 81); see also Nudds (2014), O'Callaghan (2015, 2017), and Green (2019). Theorists have used a variety of labels to mark out these interactions. In the parlance of Ernst and Bülthoff (2004), these are instances of *cue integration*, as opposed to mere *cue combination*. De Vignemont (2014) makes a similar distinction between *integrative* and *additive* forms of binding.

In another type of interaction, a feature representation in one modality cues or disambiguates a feature representation in another. Thus, for instance, in the motion-bounce (/sound-induced bouncing) illusion (Sekuler *et al.*, 1997), a sound played at the moment when two visually perceived moving objects cross trajectories leads observers to disambiguate the visually ambiguous crossing event as a collision rather than as one object streaming through another: the proposal is that binding the auditory feature together with the visual feature results in a modification or reconstrual of the latter.[17] A third paradigm of such interactions is manifest in the finding that there are intermodal object-specific preview advantages and penalties (Zmigrod *et al.*, 2009; Jordan *et al.*, 2010) mirroring the object-specific preview advantages and penalties for unimodally perceived objects (Kahneman *et al.*, 1992): following recognition of a feature in modality m_1, reaction times are lower for the recognition of a second and congruent feature in modality m_2, but higher for the recognition of a second and incongruent feature in m_2, when the second feature is bound to the first, whether or not $m_1 = m_2$.

The lesson from these cases is that the operation of grouping itself can have significant psychological consequences for grouped features. However, and crucially for our purposes, this lesson is agnostic between the convergence and co-constituency conceptions of binding: we can take the lesson on board whether we think of the features in question as applying to one and the same individual (per convergence) or not (per co-constituency). Proponents of convergence will explain the data by saying that distinct feature representations can influence one another in interesting ways when they involve convergent attribution to a single individual: that the value assigned to one such feature can modify the value assigned to another, that there is facilitation/suppression between them, and so on. But none of this depends essentially on convergence, as such. Proponents of the co-constituency account will claim that such influence occurs between distinct feature representations when the latter stand in the relevant sort of co-constituency relation—that the perceptual attribution of F_1 to o_1 and F_2 to o_2 can exert these forms of mutual psychological influence even when $o_1 \neq o_2$.[18]

All of this is to say that the central features of multimodal binding—features that not only have convinced many writers to believe in multimodal binding, but that many have thought can only be understood as resulting from multimodal convergence—are equally explicable in terms of mereological co-constituency.

5 Conclusion

The co-constituency account of multimodal binding has much to recommend it. It avoids problems that beset competing views, provides illuminating accounts of phenomena that have compelled theorists to embrace multimodal binding, and highlights the importance of continuing research into the psychological mechanisms underlying various forms of perceptual grouping.

[17] Results of Grassi and Casco (2010) show that this type of binding is penetrated by the presumably cognitive representation of feature congruence: the bounce interpretation is favoured only when the auditory feature is consistent with an impact event (cf. Vatakis and Spence, 2007).

[18] Note that, on this view, such interaction is no 'mere causal influence' between otherwise independent entities (O'Callaghan, 2014, p. 81; cf. Nudds, 2014, p. 179); it is causal influence *between items that stand in psychologically significant co-constituency relations*.

Despite these advantages, and as mentioned at the outset, there is no reason for insisting in advance that all multimodal binding must conform to the co-constituency model. It may be that the best approach involves the application of different models to different cases. This is, of course, a broadly empirical question that should not be prejudged from the armchair. My hope is that spelling out the co-constituency model, and comparing it with rivals, will aid in this assessment.

Finally, it is worth acknowledging that the co-constituency view has deflationary, revisionist consequences regarding binding and sensory individuals. Perhaps the view's most revisionary aspect (shared with the association view, but not with the convergence view) lies in its recognition of two different levels of sensory individuals, corresponding to two different roles the latter have been called upon to play. On the one hand, the co-constituency view allows that there are whatever unimodal sensory individuals that are contributed by individual perceptual modalities, and that bear the perceptual features of those modalities. On the other hand, the view also allows that perception connects us with multimodal, complex entities, themselves built from unimodal constituents. These mereologically complex entities won't, in general, bear perceptual features themselves, though their mereological constituents will; and, of course, these complex entities are at the heart of the view's explanation of binding relations between perceptual features.

Accepting these two distinct levels of sensory individuals amounts to recognizing that, in the multimodal setting (apart from degenerately unimodal special cases), the role of bearing perceptual features comes apart from the role of accounting for binding. If accepting this lesson is the price of an adequate account of intermodal binding, we should pay it.

Acknowledgements

Thanks for discussion and comments from Kevin Connolly, Casey O'Callaghan, Will Davies, Matthew Fulkerson, Mohan Matthen, Bence Nanay, Ayoob Shahmoradi, Gedalyahu Wittow, and co-participants in a 2020 seminar on perceptual interactions.

References

Batty, C. (2014). Olfactory objects. In *Perception and its Modalities*, edited by S. Biggs, D. Stokes, and M. Matthen, 222–45. Oxford: Oxford University Press.

Bayne, T. (2010). *The Unity of Consciousness*. Oxford: Oxford University Press.

Bayne, T. and Chalmers, D. J. (2003). What is the unity of consciousness? In *The Unity of Consciousness*, edited by A. Cleeremans, 23–58. Oxford: Oxford University Press.

Bayne, T. and Spence, C. (2014). Is consciousnes multi-sensory? In *Perception and Its Modalities*, edited by D. Stokes, S. Biggs, and M. Matthen, 95–132. New York, NY: Oxford University Press.

Biederman, I. (1987). Recognition-by-components. *Psychological Review*, 94, 115–47.

Botvinick, M. and Cohen, J. (1998). Rubber hands' feel'touch that eyes see. *Nature*, 391, 756.

van der Burg, E., Awh, E., and Olivers, C. N. L. (2013). The capacity of audiovisual integration is limited to one item. *Psychological Science*, 24, 345–51.

Campbell, J. (2002). *Reference and Consciousness*. Oxford: Oxford University Press.

Carvalho, F. (2014). Olfactory objects. *Disputatio*, 6(38), 45–66.

Casati, R. and Dokic, J. (2009). Some varieties of spatial hearing. In *Sounds and Perception: New Philosophical Essays*, edited by M. Nudds and C. O'Callaghan, 97–110. Oxford: Oxford University Press.

Chang, D., Nesbitt, K., and Wilkins, K. (2007). The Gestalt principles of similarity and proximity apply to both the haptic and visual grouping of elements. In *Conferences in Research and Practice in Information Technology*, Vol. 64, edited by W. Piekarski and B. Plimmer, 79–86. Ballarat: The Australian Computer Society.

Cinel, C., Humphreys, G., and Poli, R. (2002). Cross-modal illusory conjunctions between vision and touch. *Journal of Experimental Psychology: Human Perception and Performance*, 28, 1243–66.

Clark, A. (2000). *A Theory of Sentience*. New York, NY: Oxford University Press.

Cohen, J. (2004). Objects, places, and perception. *Philosophical Psychology*, 17(4), 471–95.

Cohen, J. (2010). Sounds and temporality. *Oxford Studies in Metaphysics*, 5, 303–20.

Connolly, K. (2014a). Making sense of multiple senses. In *Consciousness Inside and Out: Phenomenology, Neuroscience, and the Nature of Experience*, edited by R. Brown, 351–64. Dordrecht, Springer.

Connolly, K. (2014b). Multisensory perception as an associative learning process. *Frontiers in Psychology*, 5, 1095.

Connolly, K. (2019). *Perceptual Learning: The Flexibility of the Senses*. New York, NY: Oxford University Press.

Dalton, P., Doolittle, N., Nagata, H., and Breslin, P. A. S. (2000). The merging of the senses: integration of subthreshold taste and smell. *Nature Neuroscience*, 3(5), 431–2.

Dickie, I. (2010). We are acquainted with ordinary things. In *New Essays on Singular Thought*, edited by R. Jeshion, 213–45. Oxford: Oxford University Press.

Ernst, M. and Bülthoff, H. (2004), Merging the senses into a robust percept. *Trendsincognitivesciences*, 8, 162–9.

Fodor, J. A. (1983). *The Modularity of Mind*. Cambridge, MA: MIT Press.

Fulkerson, M. (2011). The unity of haptic touch. *Philosophical Psychology*, 24(4), 493–516.

Fulkerson, M. (2014a). Explaining multisensory experience. In *Consciousness Inside and Out: Phenomenology, Neuroscience, and the Nature of Experience*, edited by R. Brown, 365–73. Dordrecht: Springer.

Fulkerson, M. (2014b). *The First Sense: A Philosophical Study of Human Touch*. Cambridge, MA: MIT Press.

Geldard, F. A. and Sherrick, C. E. (1972). The cutaneous 'rabbit': a perceptual illusion. *Science*, 178, 178–9.

Gibson, J. J. (1979). *The Ecological Approach to Visual Perception*. Boston, MA: Houghton Mifflin.

Goldstone, R. L. (1998). Perceptual learning. *Annual Review of Psychology*, 49(1), 585–612.

Goldstone, R. L. and Byrge, L. A. (2015). Perceptual learning. In *The Oxford Handbook of the Philosophy of Perception*, edited by M. Matthen, 812–32. Oxford: Oxford University Press.

Grassi, M. and Casco, C. (2010). Audiovisual bounce-inducing effect: when sound congruence affects grouping in vision. *Attention, Perception, and Psychophysics*, 72, 378–86.

Green, E. J. (2019). Binding and differentiation in multisensory object perception. *Synthese*, 198(5), 4457–91. doi: 10.1007/s11229-019-02351-1.

Hommel, B. (2004). Event files: feature binding in and across perception and action. *Trends in Cognitive Sciences*, 8(11), 494–500.

Huppert, J. and Roth, D. (2003). Treating obsessive-compulsive disorder with exposure and response prevention. *The Behavior Analyst Today*, 4, 66–70 (January 2003). doi: 10.1037/h0100 012.

Jackson, F. (1977). *Perception: A Representative Theory*. New York, NY: Cambridge University Press.

Johansson, G. (1973). Visual perception of biological motion and a model for its analysis. *Perception and Psychophysics*, 14, 201–11.

Jordan, K. E., Clark, K., and Mitroff, S. R. (2010). See an object, hear an object file: Object correspondence transcends sensory modality. *Visual Cognition*, 18, 492–503.

Kahneman, D., Treisman, A., and Gibbs, B. J. (1992). The reviewing of object files: object-specific integration of informa-tion. *Cognitive Psychology*, 24, 175–219.

Koffka, K. (1935). *Principles of Gestalt Psychology*. New York, NY: Harcourt, Brace, and World.

Kuang, S. and Zhang, T. (2014). Smelling directions: Olfaction modulates ambiguous visual motion perception. *Scientific Reports*, 4, 5796.

Kulvicki, J. (2008). The nature of noise. *Philosophers' Imprint*, 8(11), 1–16.

Kulvicki, J. (2014). Sound stimulants: defending the stable disposition view. In *Perception and Its Modalities*, edited by D. Stokes, S. Biggs, and M. Matthen, 205–21. Oxford: Oxford University Press.
Lovaas, O. I. (1987). Behavioral treatment and normal educational and intellectual functioning in young autistic children. *Journal of Consulting and Clinical Psychology*, 55, 3–9.
Lycan, W. G. (Ed.) (1996). *Consciousness and Experience*. Cambridge, MA: MIT Press.
Marks, I. (1979). Exposure therapy for phobias and obsessive-compulsive disorders. *Hospital Practice*, 14, 101–8. doi: 10.1080/21548331.1979.11707486.
Marr, D. (1982). *Vision: A Computational Investigation into the Human Representation and Processing of Visual Information*. San Francisco, CA: W. H. Freeman.
Matthen, M. (2005). *Seeing, Doing, and Knowing: A Philosophical Theory of Sense Perception*. Oxford: Oxford University Press.
McGurk, H. and MacDonald, J. (1976). Hearing lips and seeing voices. *Nature*, 264, 746–8.
Michotte, A. (1946/63). *The Perception of Causality*. London: Methuen.
Morrot, G., Brochet, F., and Dubourdieu, D. (2001). The color of odors. *Brain and Language*, 79, 309–20.
Nanay, B. (2013). *Between Perception and Action*. Oxford: Oxford University Press.
Nudds, M. (2009). Sounds and space. In *Sounds and Perception: New Philosophical Essays*, edited by M. Nudds and C. O'Callaghan, 69–96. Oxford: Oxford University Press.
Nudds, M. (2010a). What are auditory objects? *Review of Philosophy and Psychology*, *1*, 105–22.
Nudds, M. (2010b). What sounds are. In *Oxford Studies in Metaphysics, Vol. 5*, edited by D. Zimmerman, 279–302. Oxford: Oxford University Press.
Nudds, M. (2014). Is audio-visual perception 'amodal' or 'crossmodal'? In *Perception and Its Modalities*, edited by S. Biggs, M. Matthen, and D. Stokes, 166–88. New York, NY: Oxford University Press.
O'Callaghan, C. (2007). *Sounds*. Oxford: Oxford University Press.
O'Callaghan, C. (2008). Seeing what you hear: cross-modal illusions and perception. *Philosophical Issues*, 18, 316–38.
O'Callaghan, C. (2009). Sounds and events. In *Sounds and Perception: New Philosophical Essays*, edited by M. Nudds and C. O'Callaghan, 26–49. Oxford: Oxford University Press.
O'Callaghan, C. (2011). XIII—hearing properties, effects or parts? *Proceedings of the Aristotelian Society*, 111, 375–405.
O'Callaghan, C. (2014). Intermodal binding awareness. In *Sensory Integration and the Unity of Consciousness*, edited by D. J. Bennett and C. S. Hill, 73–103. Cambridge, MA: MIT Press.
O'Callaghan, C. (2015). The multisensory character of perception. *Journal of Philosophy*, 112, 551–69.
O'Callaghan, C. (2016). Objects for multisensory perception. *Philosophical Stud-ies*, 173, 1269–89.
O'Callaghan, C. (2017). Grades of multisensory awareness. *Mind and Language*, 32, 155–81.
O'Shaughnessy, B. (2009). The location of a perceived sound. In *Sounds and Perception: New Philosophical Essays*, edited by M. Nudds and C. O'Callaghan, 111–25. Oxford: Oxford University Press.
Palmer, S. E. (1999). *Vision Science: Photons to Phenomenology*. Cambridge, MA: MIT Press.
Pasnau, R. (1999). What is sound? *Philosophical Quarterly*, 49, 309–24.
Pick Jr., H. L., Warren, D. H., and Hay, J. C. (1969). Sensory conflict in judgements of spatial direction. *Attention, Perception, and Psychophysics*, 6, 203–5.
Pylyshyn, Z. W. (2007). *Things and Places: How the Mind Connects with the World*. Cambridge, MA: MIT Press.
Rosen, G. and Dorr, C. (2002). Composition as a fiction. In *The Blackwell Companion to Metaphysics*, edited by R. Gale, 151–74. Oxford: Blackwell.
Sekuler, R., Sekuler, A. B., and Lau, R. (1997). Sound alters visual motion perception. *Nature*, 385, 308.
Shams, L., Kamitani, Y., and Shimojo, S. (2000). What you see is what you hear. *Nature*, 408, 788.
Shams, L., Kamitani, Y., and Shimojo, S. (2002). Visual illusion induced by sound. *Cognitive Brain Research*, 14, 147–52.
Sider, T. (2013). Against parthood. *Oxford Studies in Metaphysics*, 8, 237–93.
Smith, B. C. (2015). The chemical senses. In *Oxford Handbook of Philosophy of Perception*, edited by M. Matthen, 314–52. Oxford: Oxford University Press.
Sorensen, R. (2008). Hearing silence: the perception and introspection of absences. In *Sounds and Perception: New Philosophical Essays*, edited by M. Nudds and C. O'Callaghan, 126–45. Oxford: Oxford University Press.

Soteriou, M. (2018). Sound and illusion. In *Perceptual Ephemera*, edited by T. Crowther and C. MacCumhaill, 31–49. Oxford: Oxford University Press.

Spelke, E. (1990). Principles of object perception. *Cognitive Science*, 14, 29–56.

Spence, C. (2015). Eating with our ears: assessing the importance of the sounds of consumption to our perception and enjoyment of multisensory flavour experiences. *Flavour*, 4, 3.

Spence, C., Smith, B., and Auvray, M. (2014). Confusing tastes and flavours. In *Perception and Its Modalities*, edited by D. Stokes, M. Matthen, and S. Biggs, 247–74. Oxford: Oxford University Press.

Tarr, M. J. and Bülthoff, H. (1998). *Object Recognition in Man, Monkey, and Machine*. Cambridge, MA: MIT Press.

Treisman, A. M. (1998). Feature binding, attention and object perception. *Philosophical Transactions of the Royal Society of Lon- don, Series B: Biological Sciences*, 353, 1295–306.

Treisman, A. M. (2003). Consciousness and perceptual binding. In *The Unity of Consciousness*, edited by A. Cleeremans, 95–113. Oxford: Oxford University Press.

Vatakis, A. and Spence, C. (2007). Crossmodal binding: evaluating the 'unity assumption' using audio-visual speech stimuli. *Perception and Psychophysics*, 69, 744–56.

de Vignemont, F. (2014). Multimodal unity and multi-modal binding. In *Sensory Integration and the Unity of Consciousness*, edited by D. J. Bennett and C. S. Hill, 125–50. Cambridge, MA: MIT Press.

Vroomen, J. and de Gelder, B. (2000). Sound enhances visual perception: cross-modal effects of auditory organization on vision. *Journal of Experimental Psychology: Human Perception and Performance*, 26, 1583–90.

Vroomen, J. and de Gelder, B. (2004). Ventriloquism and the freezing phenomenon. In *The Handbook of Multisensory Processes*, edited by G. A. Calvert, C. Spence, and B. E. Stein, 141–50. Cambridge, MA: MIT Press.

Wagemans, J., Elder, J. H., Kubovy, M., *et al.* (2012a). A century of Gestalt psychology in visual perception: I. Perceptual grouping and figure-ground organization. *Psychological Bulletin*, 138, 1172–217.

Wagemans, J., Feldman, J., Gepshtein, S., *et al.* (2012b). A century of Gestalt psychology in visual perception: II. Conceptual and theoretical foundations. *Psychological Bulletin*, 138, 1218–52.

Wertheimer, M. (1938). Laws of organization in perceptual forms. In *A Source Book of Gestalt Psychology*, edited by W. D. Ellis, K. Paul, 71–88. London: Trench, Trubner, and Company.

Young, N. and Nanay, B. (2020). Audition and somposite sensory individuals. In *Sensory Individuals*, edited by A. Mroczko-Wasowicz and R. Grush. Oxford: Oxford University Press.

Zacks, J. M., Speer, N., Swallow, K., Braver, T., and Reynolds, J. (2007). Event perception: a mind/brain perspective. *Psychological Bulletin*, 133, 273–93.

Zacks, J. M. and Tversky, B. (2001). Event structure in perception and conception. *Psychological Bulletin*, 127, 3–21.

Zmigrod, S., Spapé, M., and Hommel, B. (2009). Intermodal event files: integrating features across vision, audition, taction, and action. *Psychological Research*, 73, 674–84.

18

Visuo-haptic object processing in the multisensory brain

Simon Lacey and K. Sathian

1 Introduction: the multisensory brain

Our perception of individual events and objects in the environment is invariably through multisensory processes, but, until 30 or so years ago, sensory research was overwhelmingly concentrated on vision. In recent decades, however, greater prominence has been given to non-visual modalities, particularly the haptic and auditory modalities, and to crossmodal and multisensory interactions such as those involved in audiovisual speech perception and visuo-haptic object recognition. For example, our laboratory was the first to show that an extrastriate visual region near the parieto-occipital fissure, known to be involved in visual discrimination of grating orientation (Sergent *et al.*, 1992) and corresponding to V6 (Pitzalis *et al.*, 2006), is also active during tactile discrimination of grating orientation (Sathian *et al.*, 1997). Furthermore, this parieto-occipital region is functionally involved in touch, since applying transcranial magnetic stimulation (TMS) to it disrupts tactile orientation discrimination (Zangaladze *et al.*, 1999). Since then, many other brain regions exhibiting selectivity for specific visual tasks have also been shown to process non-visual versions of these tasks. As a result, the idea that the brain is organized around discrete sensory channels providing inputs into higher-order multisensory regions has been replaced by a broad consensus around the idea of the 'metamodal brain' (Pascual-Leone and Hamilton, 2001)—that is, a brain with a largely modality-independent, task-based organization (Lacey *et al.*, 2009a; Pascual-Leone and Hamilton, 2001). In this chapter, we review how vision and touch contribute to perception and recognition of physical objects. We first review behavioural studies before examining the neural basis of the reported behavioural effects. In addition to the relevant multisensory perceptual interactions, we draw attention to the interaction between haptic perception and visual imagery. Finally, we pull these lines of inquiry together in a model of visuo-haptic object representation and discuss the evidence for this model.

First, though, a word about terminology. A variety of labels have been applied where more than one sensory modality is involved (see Stein *et al.*, 2010); for example, apart from 'multisensory', these include 'modality-independent', 'metamodal', 'intermodal', 'amodal', 'intersensory', and more. We choose 'multisensory' because it is intuitively easy to understand and, following Stein *et al.* (2010), we define it as 'any neural or behavioral process associated with multiple sensory modalities'. In particular, the use of 'amodal' (i.e., 'without' modality) seems to us to be better suited to information that is primarily represented linguistically or

Simon Lacey and K. Sathian, *Visuo-haptic object processing in the multisensory brain* In: *Sensory Individuals*. Edited by: Aleksandra Mroczko-Wąsowicz and Rick Grush, Oxford University Press. © Oxford University Press 2023.
DOI: 10.1093/oso/9780198866305.003.0019

propositionally (Lacey and Sathian, 2008). When we refer to a 'multisensory' representation, we mean a representation that can be encoded via multiple sensory modalities, and that retains the modality 'tags' of the associated sensory inputs, thus enabling retrieval by any of the associated modalities (Sathian, 2004). By contrast, a unisensory representation is encoded and generally retrieved via a single sensory modality, although there may be circumstances when it could be activated by another (Lacey *et al.*, 2009a); for example, the smell of the sea might evoke visual imagery of the sea.

2 Haptic and visuo-haptic object processing

2.1 Accuracy and asymmetry

It is well established that visual object processing and recognition are both fast and accurate. For familiar objects at least, haptic recognition is also highly accurate, with 68% of items correctly identified in less than 3 seconds and 94% within 5 seconds (Klatzky *et al.*, 1985). Although these haptic responses are obviously slower than their visual counterparts, they likely reflect the longer time required to manually explore larger items, compared to essentially parallel visual processing across the retinal image (Klatzky *et al.*, 1985); nonetheless, a 'haptic glance' of less than 1 second can be sufficient in some contexts (Klatzky and Lederman, 1995). Haptic perception proceeds via a series of property-specific hand movements, known as 'exploratory procedures' (EPs; Lederman and Klatzky, 1987); for example, texture information is extracted by lateral motion, while precise shape information is provided by contour following (Lederman and Klatzky, 1987). However, the salience of these properties is modulated by context—when instructions emphasized haptic processing, salience progressively decreased in the order texture > shape > hardness; but under neutral instructions, this became hardness > texture > shape (Klatzky *et al.*, 1987). Relevant to this chapter, in both simultaneous visual and haptic perception, and in haptic perception under instructions to use visual imagery, the salience order changed to shape > texture > hardness/size (Klatzky *et al.*, 1987).

Crossmodal visuo-haptic object recognition is generally less accurate than within-modal recognition (Bushnell and Baxt, 1999; Casey and Newell, 2007; Lacey *et al.*, 2007), and is generally better when visual encoding is followed by haptic recognition rather than the reverse (Easton *et al.*, 1997a, b; Jones, 1981; Lacey and Campbell, 2006; Nabeta and Kawahara, 2006; Reales and Ballesteros, 1999; Streri and Molina, 1994). Why this asymmetry occurs is not clear: in infant studies, it might result from constraints imposed by different stages of motor development (Streri and Molina, 1994), but since asymmetry is also observed in adulthood (Easton *et al.*, 1997a, b; Lacey and Campbell, 2006), this cannot be a complete explanation. One possibility is that crossmodal shape might be encoded less efficiently in the haptic-visual crossmodal condition because, as noted above, hardness and texture information are more salient to haptics than shape. This effect could be reduced if haptic perception is accompanied by visual imagery when shape becomes most salient, because shape is common to vision and touch. Note too, that in simultaneous visual and haptic perception, properties that have different weights for vision and haptics could be optimally combined on the basis of maximum likelihood estimates (Ernst and Banks, 2002; Helbig *et al.*, 2012; Helbig and Ernst, 2007; Takahashi and Watt, 2014).

Crossmodal asymmetry might also be explained by differences in visual and haptic memory capacity. Haptic working memory capacity appears to be more limited and variable than visual working memory, and consequently may be more error-prone (Bliss and Hämäläinen, 2005). Haptic representations also decay faster and more abruptly than visual representations: the haptic decay function appears to occur entirely in a band of 15–30 seconds post-stimulus rather than decaying gradually (Kiphart *et al.*, 1992); see also Craddock and Lawson (2010), who found no decline in performance at 15 seconds, although longer intervals were not tested. Decreased accuracy in the haptic-visual condition might simply be because haptically encoded representations have substantially decayed in the interval. Other studies, however, have shown little difference between haptic-visual and visual-haptic recognition after delays of up to 30 seconds (Garvill and Molander, 1973; Woods *et al.*, 2004) or even a week (Hutmacher and Kuhbandner, 2018; Pensky *et al.*, 2008; see also Ferreira *et al.*, 2021). Thus, an explanation of the asymmetry in terms of a simple function of haptic memory properties is likely insufficient.

2.2 Visuo-haptic view-(in)dependent recognition

Changes in the orientation of an object change the related sensory input, for example, the retinal pattern, and therefore constitute a challenge to successful object recognition. Achieving perceptual constancy so that objects can be recognized independently of such changes is a major goal of sensory systems. Visual object recognition is considered view-dependent if changing an object's orientation impairs subsequent recognition, and view-independent if not (Peissig and Tarr, 2007). Intuitively, one would think that haptic recognition should be view-independent because the hands can contact an object from several sides simultaneously, thus acquiring information about different 'views'. Surprisingly, numerous studies show that haptic object recognition is instead view-dependent (Craddock and Lawson, 2008, 2010; Lacey *et al.*, 2007, 2009b; Lawson, 2009, 2011; Newell *et al.*, 2001; Ueda and Saiki, 2007, 2012). Interestingly, different types of change in orientation affect vision and touch in different ways. Rotation about the x- and y-axes, that is, in depth (Figure 18.1), results in slower, less accurate visual recognition than rotation about the z-axis, that is, in the picture plane, for both 2D (Gauthier *et al.*, 2002) and 3D stimuli (Lacey *et al.*, 2007). By contrast, rotation about any axis impairs haptic recognition (Lacey *et al.*, 2007), suggesting that visual and haptic view dependence have a different basis. One possibility is that vision and haptics differ in whether or not a surface is occluded by rotation. For vision, rotating an object can involve not only a transformation in perceptual shape, but also the occlusion of one or more surfaces—unless the observer physically changes position relative to the object (Bülthoff and Newell, 2006; Tarr and Pinker, 1989). Comparing Figure 18.1A and C, rotation about the x-axis turns the object upside down so that the former top surface is no longer visible—that is, it becomes occluded. In haptic exploration, no surface is necessarily occluded because the hands are free to move over all surfaces of an object and to explore different orientations relative to the hand. Thus, no single axis of rotation should be more or less disruptive than another due to surface occlusion, so that haptic recognition only has to deal with a shape transformation. Further work is required to examine whether this explanation is, in fact, correct.

Recognition of unfamiliar objects is typically view-dependent, transitioning to view independence as object familiarity increases in both vision (Bülthoff and Newell, 2006; Tarr and

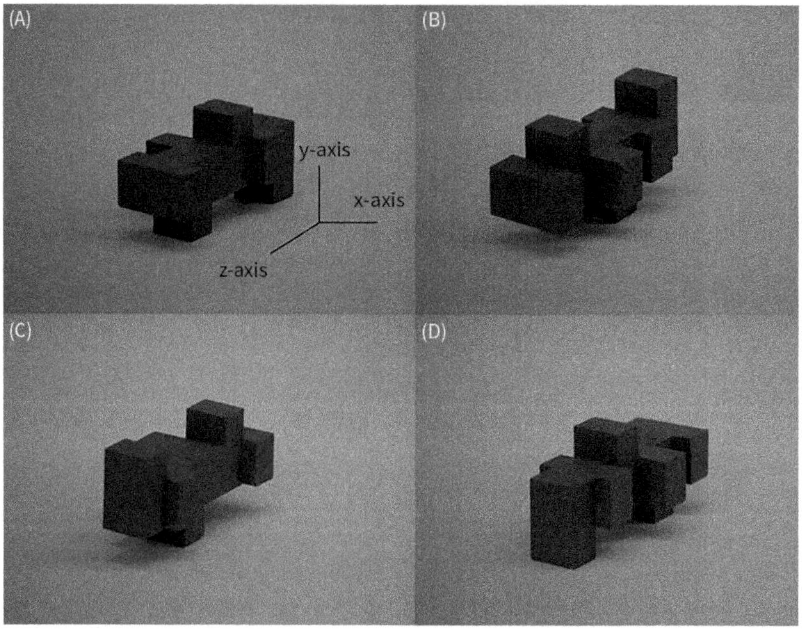

Figure 18.1 An example of a 3D unfamiliar object shown (A) in its original orientation, and in 180° rotations in depth about the *z*-axis (B), *x*-axis (C), and in the picture-plane about the *y*-axis (D).
Reproduced from Lacey *et al.*, 2007.

Pinker, 1989) and haptics (Craddock and Lawson, 2009). An exception occurs when a familiar object has a prototypical orientation, known as a canonical view, such as the front view of a house (Palmer *et al.*, 1981). Recognition may still be view-independent for a limited range of orientations around the canonical view, but can revert to view dependence for radically non-canonical views, for example, a teapot seen from directly above (Bülthoff and Newell, 2006; Palmer *et al.*, 1981; Tarr and Pinker, 1989). The same is true for non-canonical orientations in haptic recognition (Craddock and Lawson, 2008), but vision and haptics differ in what constitutes a canonical view. The preferred view in vision is one in which the object is aligned at 45° to the observer (Palmer *et al.*, 1981), while objects are generally aligned either parallel or orthogonal to the body midline for the haptic canonical view (Woods *et al.*, 2008). Such canonical views may facilitate view-independent recognition either because they provide the most structural information about an object or because they most closely match a stored representation.

Visuo-haptic crossmodal recognition involves a change in modality from study to test and should presumably be disrupted by an additional change in orientation. Again surprisingly, crossmodal recognition is view-independent even for unfamiliar objects that are highly similar (Figure 18.1), whether visual study is followed by haptic test, or vice versa, and whatever the axis of rotation (Lacey *et al.*, 2007, 2010a; Ueda and Saiki, 2007, 2012). For familiar objects, crossmodal view independence has only been shown in the haptic-visual condition (Lawson, 2009). However, this study used both scale-model (e.g., bed, shark) and actual-size objects (e.g., jug, pencil); since scale-model objects at their actual size would typically be

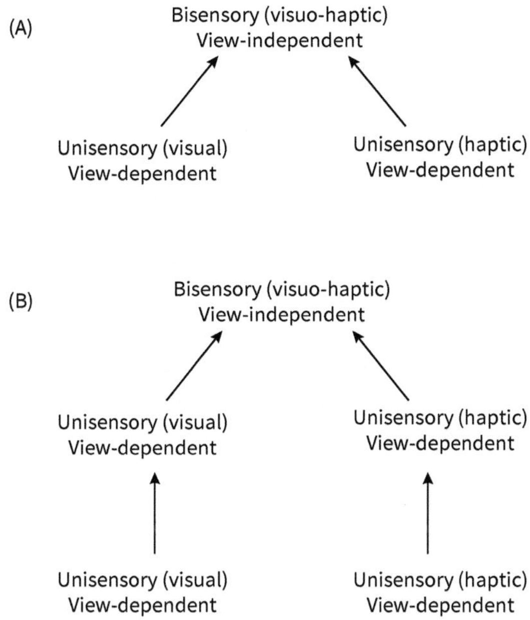

Figure 18.2 Competing models of visuo-haptic view-independence. (A) Direct integration of the unisensory view-dependent representations into a multisensory view-independent representation. (B) Bisensory view-independence gated by separate, unisensory view-independent representations.
Reproduced from Lacey *et al.*, 2009b.

more familiar visually than haptically, increased error rates in the visual-haptic condition might be due to visually familiar objects having to be recognized by touch. Further research is required on potential effects of differential familiarity.

The simplest way to achieve crossmodal view independence would be direct integration of the unisensory visual and haptic view-dependent representations into a multisensory view-independent representation (Figure 18.2A); alternatively, it might be necessary to achieve unisensory view independence first (Figure 18.2B). In a perceptual learning study, we showed that view independence acquired in one modality transferred completely and symmetrically to the other without additional training (i.e., the untrained modality did not need to achieve view independence); thus, both visual and haptic within-modal view independence rely on the same view-independent representation (Lacey *et al.*, 2009b). Crossmodal training (whether haptic-visual or visual-haptic) also produced both visual and haptic within-modal view independence (Lacey *et al.*, 2009b). Consistent with models proposed for vision (Riesenhuber and Poggio, 1999), visuo-haptic view independence is therefore likely supported by a single multisensory representation that directly integrates the unisensory view-dependent representations (Figure 18.2A) and appears to support both crossmodal recognition and view independence (whether within-modal or crossmodal). Thus, at least in these circumstances, unisensory visual and haptic representations appear to converge on a common multisensory representation.

2.3 Imagery preferences and visuo-haptic object representations

Successful object recognition depends on what information is contained in the representations that support it. Although visual shape, colour, and texture are processed in different cortical areas (Cant *et al.*, 2009; Cant and Goodale, 2007), these structural (shape) and surface (colour, texture, etc.) properties are integrated into visual object representations: shape recognition is impaired if object colour or part-colour combinations change between study and test (Nicholson and Humphrey, 2003). Since recognition is unaffected by changes to the background colour against which objects were presented, this impairment could be isolated to the object representation, which must therefore contain both shape and colour information (Nicholson and Humphrey, 2003). We extended these findings by showing that both visual and haptic within-modal object discrimination are impaired by changes in surface texture (Lacey *et al.*, 2010a). Thus, not only do haptic representations also integrate structural and surface properties, but also surface property information in visual representations is not limited to modality-specific properties like colour. In addition, we tested whether the multisensory representation supporting crossmodal object recognition also integrates structural and surface properties. Thus, we tested object discrimination across changes in texture, orientation (thus requiring access to the view-independent multisensory representation discussed in the previous section), or both. Crossmodal discrimination was view-independent, provided there was no change in texture, consistent with prior studies (Lacey *et al.*, 2007; Ueda and Saiki, 2007, 2012); but when texture changed, performance was reduced to chance levels, whether orientation was also changed or not (Lacey *et al.*, 2010a). Interestingly, texture changes produced considerable variability in individual performance: we hypothesized that this arose from individual differences in object representations related to individual preferences for different kinds of imagery.

Visual imagery preferences can be for either 'object imagery' (defined as involving vivid, detailed pictorial images conveying the literal appearance of an object's shape, colour, brightness, etc.) or 'spatial imagery' (defined as more schematic images focusing on the spatial relationships between the component parts of an object and spatial transformations; Blajenkova *et al.*, 2006; Kozhevnikov *et al.*, 2002, 2005). Importantly, while both encode the structural property of shape, object imagery includes information about surface properties like colour and texture, while spatial imagery does not. We investigated whether object and spatial imagery differences also occur in haptic imagery, using tasks requiring discrimination of shape across changes in texture, and vice versa (Figure 18.3), in both visual and haptic within-modal conditions (Lacey *et al.*, 2011). Object imagers were impaired at discriminating shape across changes in texture, but not texture across changes in shape, indicating that their object representations integrate surface texture. By contrast, spatial imagers could discriminate shape across changes in texture, but not the reverse, presumably because they tend not to encode surface properties (Lacey *et al.*, 2011). Importantly, performance largely reflected both self-reports of visual imagery preference and scores on the Object and Spatial Imagery Questionnaire (OSIQ; Blajenkova *et al.*, 2006). Thus, object and spatial imagery dimensions exist in both haptic and visual imagery, and are reflected in the extent to which surface properties are integrated into object representations (Lacey *et al.*, 2011). We also showed that the variability in the texture change condition of our earlier crossmodal study referred to above

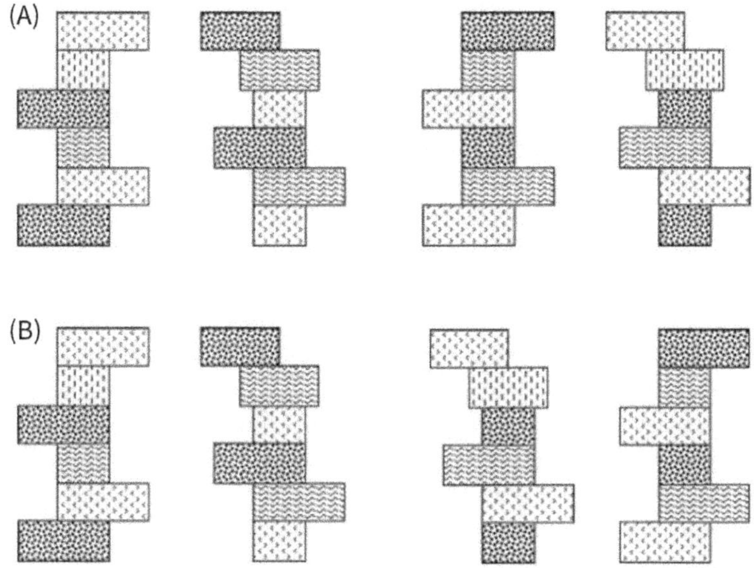

Figure 18.3 (A) Schematic example of shapes 1 and 2, with original texture schemes (left pair) and with the texture schemes exchanged (right pair). (B) Example of textures 1 and 2, with original shapes (left pair) and with the shapes exchanged (right pair).
Reproduced from Lacey *et al.*, 2011.

(Lacey *et al.*, 2010a) was indeed related to imagery preference: although object and spatial imagers showed crossmodal view independence, object imagers were impaired by texture changes whereas spatial imagers were not (Lacey *et al.*, 2011). The degree of the impairment was correlated with OSIQ scores: stronger preference for object imagery led to greater impairment by texture changes, suggesting that only object imagers integrate surface properties into the multisensory representation (Lacey *et al.*, 2011). Thus, the multisensory representation has some features that are stable across individuals, like view independence, and some that vary across individuals, such as integration of surface property information, correlating with individual differences in imagery preference. Finally, object and spatial imagery preferences likely generalize across vision and haptics within individuals since visual and haptic performance were highly correlated in both the shape and texture change conditions (Lacey *et al.*, 2011, 2013).

3 The neural basis of visuo-haptic object processing

3.1 Segregated ventral and dorsal cortical pathways

Visual object processing has long been divided along a ventral 'what' pathway dealing with object identity and recognition, and a dorsal 'where/how' pathway dealing with object location and action, for example, reaching and grasping (Goodale and Milner, 1992; Ungerleider and Mishkin, 1982; see Freud *et al.*, 2017 for a discussion of more distributed shape processing in these pathways). A similar functional division is seen in auditory ventral and dorsal pathways mediating perception of object identity and location (Alain *et al.*, 2001; Arnott *et al.*, 2004;). In the somatosensory system, a similar segregation was proposed, with a dorsal 'action' stream targeting the posterior parietal cortex and a ventral 'perception' stream terminating in the insular cortex (Dijkerman and de Haan, 2007); the analogy between the latter and the ventral visual pathway was originally suggested by Mishkin (1979).

A functional magnetic resonance imaging (fMRI) study found that frontal cortical areas and inferior parietal regions were activated during haptic object recognition, while a haptic object location task activated superior parietal regions (Reed *et al.*, 2005). In a later fMRI study from our laboratory (Sathian *et al.*, 2011), we proposed that texture would be a better index of haptic object identity because of its salience to touch (Klatzky *et al.*, 1987). We found that both visual and haptic location judgements involved a similar dorsal pathway comprising multisensory regions in the intraparietal sulcus (IPS) and frontal eye fields (FEFs) bilaterally (Sathian *et al.*, 2011), while haptic texture perception recruited the ventrally located parietal operculum in the higher-order somatosensory cortex (Sathian *et al.*, 2011; Stilla and Sathian, 2008). In addition, both visual and haptic texture perception engaged parts of the extrastriate visual cortex (Sathian *et al.*, 2011; Stilla and Sathian, 2008). Moreover, activation magnitudes for visual and haptic perception of either texture or location were correlated in several of these bisensory areas, suggesting shared, task-specific cortical processing across modalities (Sathian *et al.*, 2011). These findings were further extended by studies showing that activation magnitudes in the early visual cortex were related to the inter-dot spacing of dot patterns and were also modulated by matching haptic input (Eck *et al.*, 2013). However, multivoxel pattern analyses (MVPA) showed that roughness could only be decoded in the higher-order somatosensory (parietal opercular) and visual (inferotemporally in the collateral sulcus) cortex (Eck *et al.*, 2016).

3.2 Cortical regions processing multisensory shape

Both the ventral and dorsal visual pathways are functionally involved during analogous haptic tasks (for reviews, see Amedi *et al.*, 2005; Lacey and Sathian, 2011, 2014; Sathian and Lacey, 2007). Haptic shape perception may engage the entire ventral stream since very early areas in the human visual pathway that project to both dorsal and ventral streams respond to haptic and tactile stimuli (Gaglianese *et al.*, 2020; Snow *et al.*, 2014). Many studies of visuo-haptic object processing have focused on the lateral occipital complex (LOC), a higher-order, object-selective region in the ventral visual pathway (Malach *et al.*, 1995). A sub-region of the LOC is selective for objects in both touch and vision (Amedi *et al.*, 2001, 2002; Stilla and Sathian, 2008), whether haptic 3D (Amedi *et al.*, 2001; Stilla and Sathian, 2008; Zhang *et al.*, 2004) or tactile 2D stimuli (Prather *et al.*, 2004; Stoesz *et al.*, 2003) are used. Whether the LOC is engaged during auditory object processing is less clear. On the one hand, object recognition cued by object-specific sounds does not engage the LOC (Amedi *et al.*, 2002). However, the shape of objects (either rods or balls) could be reliably inferred from the impact sounds made when they were dropped onto a hard surface, and the LOC was more activated during that task than when participants had to categorize the sounds by their material (wooden or metallic; James *et al.*, 2011). Moreover, the LOC responds to auditory shape information conveyed by a sensory substitution device (SSD; Amedi *et al.*, 2007). SSDs convert visual information into an auditory 'soundscape' in which auditory duration and stereo panning represent the horizontal axis, tone frequency represents the vertical axis, and loudness equates to brightness (Amedi *et al.*, 2007). Thus, the soundscapes represent shape in a principled way, enabling both blind and sighted individuals to recognize objects; importantly, the LOC only responds to soundscapes if they adhere to these principles, and not if they are arbitrarily associated with shapes (Amedi *et al.*, 2007). More recently, Dormal *et al.* (2018) showed that object sounds, but not voices, evoked activity in the LOC in both early blind and sighted participants.

In the dorsal pathway, several parietal regions are involved in visuo-haptic object processing. Both object shape and location activate the anterior IPS, with co-activation of the LOC for shape and the FEF for location (Saito *et al.*, 2003; Sathian *et al.*, 2011; Stilla and Sathian, 2008); the role of the IPS may be to compute the spatial relations between object parts in order to construct a complete object representation (Lacey *et al.*, 2009a; Sathian *et al.*, 2011). The post-central sulcus (PCS), corresponding to Brodmann's area 2 of the primary somatosensory cortex (S1; Grefkes *et al.*, 2001) and traditionally considered exclusively somatosensory, also shows visuo-haptic shape selectivity (Stilla and Sathian, 2008) consistent with earlier neurophysiological studies that suggested visual responsiveness in parts of S1 (Iriki *et al.*, 1996; Zhou and Fuster, 1997). In these and many other studies, however, visual cortical recruitment might be explained by visual imagery; but even if haptic perception were mediated by visual imagery, this implies the existence of polysynaptic cortical pathways between somatosensory and visual cortices in humans, as shown in network analyses of the macaque brain (Négyessy *et al.*, 2006). We will discuss the potential role of visual imagery in a later section. Returning to the idea of crossmodal asymmetry discussed in the opening section, neural activity in each crossmodal condition may be asymmetric. In a match-to-sample task, visual samples followed by haptic targets resulted in greater response in the bilateral LOC, fusiform gyrus (FG), and anterior IPS (aIPS), compared to unimodal visual matching (Kassuba *et al.*, 2013). However, these regions showed no such increase either when haptic samples were followed by visual targets, or in unimodal haptic matching (Kassuba *et al.*, 2013). This asymmetric increase in activation in the visual-haptic condition suggests that the haptic modality integrates previously presented visual information more readily than vision integrates previous haptic information (Kassuba *et al.*, 2013).

The multisensory responses discussed above could reflect integration by either multisensory neurons or separate, but interdigitated, unisensory neuronal populations. Tal and Amedi (2009) attempted to distinguish between these alternatives in the case of visuo-haptic responses using functional magnetic resonance adaptation (fMR-A). This technique exploits the repetition suppression effect—that is, attenuation of the blood-oxygen level-dependent (BOLD) signal when the same stimulus is repeated, thus revealing neuronal selectivity profiles (Grill-Spector *et al.*, 2006; Krekelberg *et al.*, 2006). The LOC, aIPS, ventral premotor cortex, and right anterior insula all showed crossmodal adaptation for stimuli presented visually, but repeated haptically, suggesting that these areas integrate multisensory inputs at the neuronal level (Tal and Amedi, 2009). Other regions, however, failed to show crossmodal adaptation, implying that their multisensory responses likely resulted from interdigitated unisensory neurons. Mur *et al.* (2010) argued that fMR-A effects may not reliably index neuronal selectivity; thus, the findings of Tal and Amedi (2009) will need to be corroborated using other methods.

A critical question is whether activation of visual cortical areas during haptic or tactile tasks is functionally relevant—that is, *necessary* for task performance, or merely a by-product of other processes. Several case studies suggest that the LOC is necessary for both haptic and visual shape perception. A patient with a left occipito-temporal lesion, likely including the LOC, had impaired tactile and visual object recognition (agnosia) while somatosensory cortex and basic somatosensory functions were spared (Feinberg *et al.*, 1986). Another patient with bilateral LOC lesions could not learn new objects in either vision or touch (James *et al.*, 2006). These findings are consistent with a shared multisensory representation in the LOC. In contrast, bilateral occipito-temporal lesions including the LOC resulted in visual, but not haptic, agnosia (Snow *et al.*, 2015), but since the patient with these lesions was tested 12 years post-stroke, there may have been some degree of cortical reorganization. TMS can be used to create 'virtual lesions' by temporarily disrupting specific, functionally defined cortical areas (Sack, 2006), as in the case of V6 and tactile orientation discrimination described above (Zangaladze *et al.*, 1999). Applying repetitive TMS (rTMS) to the LOC suggests a critical role in visual shape processing (Chouinard *et al.*, 2017; Mullin *et al.*, 2011). While rTMS over the left aIPS impaired visual-haptic, but not haptic-visual, shape matching using the right hand, rTMS over the right aIPS had no effect on left hand performance in either crossmodal condition (Buelte *et al.*, 2008). Thus, much work remains to be done in order to fully describe the precise roles of the IPS and LOC in multisensory shape processing.

3.3 View-(in)dependent cortical regions

Evidence for the neural basis of the multisensory view-independent representation is mixed. Studies of visual recognition show that LOC responses are view-dependent for unfamiliar objects (Gauthier *et al.*, 2002) and view-independent for familiar objects (Eger *et al.*, 2008; Pourtois *et al.*, 2009; Valyear *et al.*, 2006), as would be expected. But other studies have shown view dependence in the LOC even for familiar objects (although there was position independence; Grill-Spector *et al.*, 1999), or view independence regardless of object familiarity (James *et al.*, 2002). A developmental fMRI study found view-independent LOC responses for unfamiliar objects in adults, but not in adolescents or children (Nishimura *et al.*, 2015). Application of TMS to the LOC suggests functional involvement in view-independent recognition of 2D shape (Silvanto *et al.*, 2010), but only at a rotation of 20°—no TMS effects were seen at 70°. FG responses are less sensitive to orientation changes on the left than on the right (Andresen *et al.*, 2009; Harvey and Burgund, 2012). In the parietal cortex, visual view-dependent responses have been observed in the IPS (James *et al.*, 2002) and a parieto-occipital area (Valyear *et al.*, 2006). The superior parietal cortex is view-dependent during mental rotation, but not during visual object recognition (Gauthier *et al.*, 2002; Wilson and Farah, 2006). However, these regions are in the dorsal 'action' pathway (see above; Goodale and Milner, 1992; Ungerleider and Mishkin, 1982), and so view-dependent responses are perhaps to be expected: for example, reaching and grasping will be affected if changes in object orientation require a different or unusual grip. Consistent with this suggestion, the lateral parieto-occipital cortex shows view-dependent responses for graspable, but not for non-graspable, objects (Rice *et al.*, 2007). To our knowledge, there are no neuroimaging studies of haptic or crossmodal processing of stimuli across changes in orientation to date. Although James *et al.* (2002) varied object orientation, this study concentrated on haptic-to-visual priming rather than on the crossmodal response to same vs different orientations per se.

4 Perception and imagery in visuo-haptic object representation

LOC activity during tactile or haptic perception could arise from either 'bottom-up' or 'top-down' processes. Activity in S1 during tactile shape discrimination propagates to the LOC as early as 150 ms after stimulus onset, approximately the same time frame in which S2 activity appears and consistent with feedforward (i.e., bottom-up) projections to the LOC (Adhikari *et al.*, 2014; Lucan *et al.*, 2010). This suggests that the LOC can be activated directly from bottom-up somatosensory input, although the existence of polysynaptic pathways cannot be ruled out. A patient with bilateral ventral occipito-temporal lesions, but likely excluding the dorsal multisensory sub-region of the LOC, showed visual, but not haptic, agnosia with associated activation of the intact dorsal LOC, also suggesting direct activation by somatosensory input (Allen and Humphreys, 2009). Top-down activation of the LOC during haptic perception might be associated with visual imagery of the felt object (Sathian *et al.*, 1997). Consistent with this suggestion, the left LOC is active during mental imagery of familiar objects in both blind and sighted participants, although object information is derived predominantly from touch in blind people and vision in those who are sighted (De Volder *et al.*, 2001). Furthermore, measures of vividness of visual imagery were strongly correlated with activation magnitudes in the right LOC during haptic shape processing (Zhang *et al.*, 2004).

We formulated a model of visuo-haptic multisensory object representation, in which object representations in the LOC can be flexibly accessed either bottom-up or top-down, depending on object familiarity, and modulated by object and spatial imagery processes (Lacey *et al.*, 2009a; Figure 18.4). Since unfamiliar objects have no stored representation, these must be fully explored during haptic shape perception to relate the component parts to each other, and thus compute the global shape. We proposed that this involves a bottom-up pathway from the somatosensory cortex to the LOC, facilitated by spatial imagery processes occurring in the IPS. By contrast, familiar objects do have pre-existing representations; during haptic exploration, enough information is acquired to trigger a stored visual image and generate a hypothesis about object identity (Bar, 2007). We suggested that this involves a top-down pathway from the prefrontal cortex to the LOC, facilitated by object imagery processes (though spatial imagery may have a role in processing familiar objects, for example, in view-independent recognition).

To test this model, we analysed inter-task correlations of activation magnitude between haptic shape perception and both visual object (Lacey *et al.*, 2010b) and spatial imagery (Lacey *et al.*, 2014) tasks, as well as effective connectivity in these different tasks (Deshpande *et al.*, 2010; Lacey *et al.*, 2014). If the tasks rely on similar processes, task-specific activation magnitudes should be correlated across participants and there should be similar patterns of effective connectivity across tasks. Participants performed haptic shape discrimination tasks using familiar and unfamiliar objects. In contrast to previous studies, we ensured that participants engaged in imagery throughout each scan by requiring an explicit, recorded response on tasks that reflect object and spatial imagery. As predicted by the model, object familiarity modulated inter-task correlations between haptic shape perception and visual object imagery (Lacey *et al.*, 2010b). Relatively fewer regions showed inter-task correlations between spatial imagery and haptic perception of either familiar or unfamiliar shape (Lacey *et al.*, 2014). This

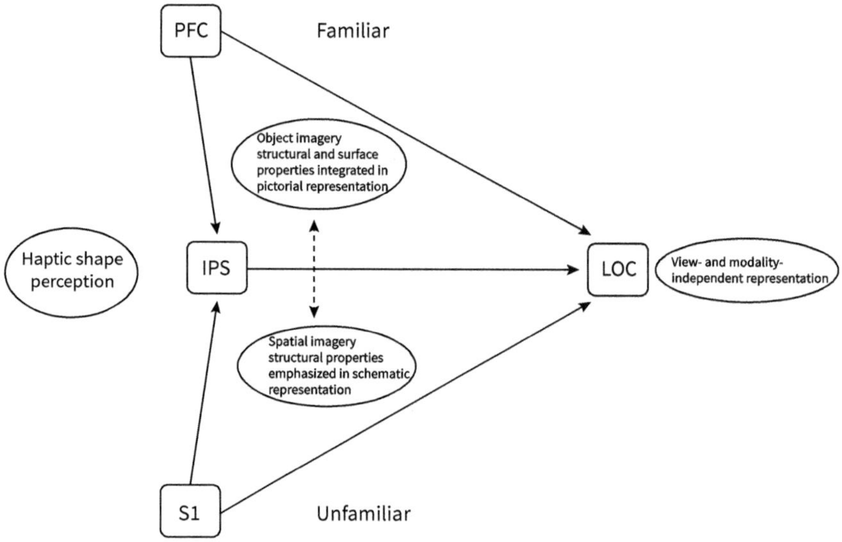

Figure 18.4 Schematic model of haptic object representation in the LOC modulated by object familiarity and visual imagery type. For unfamiliar objects, the LOC is driven bottom-up from the somatosensory cortex (S1) supported by spatial imagery processes in the intraparietal sulcus (IPS). For familiar objects, the LOC is driven top-down from the prefrontal cortex (PFC) via object imagery processes. Thus, the LOC houses an object representation that is flexibly accessible, via both vision and touch, both bottom-up and top-down, and which is additionally view-independent.
Reproduced from Lacey & Sathian, 2014.

suggests that spatial imagery is relevant to haptic shape perception, independently of object familiarity, whereas object imagery is more strongly associated with haptic perception of familiar, compared to unfamiliar, objects (Lacey *et al.*, 2010b).

Converging evidence from effective connectivity analyses showed that the spatial imagery network was highly similar to the network associated with unfamiliar shape perception, while the object imagery network was highly similar to the network involved in familiar shape perception (Lacey *et al.*, 2014). While these findings supported the overall architecture of the model, more detailed aspects were also supported. For example, in the network common to spatial imagery and unfamiliar haptic shape perception, the LOC was driven by multiple parietal foci, with cross-talk between posterior parietal and somatosensory foci, consistent with the proposal that global shape representations for unfamiliar objects are primarily computed via a bottom-up pathway facilitated by spatial imagery processes (Lacey *et al.*, 2009a, 2014). These IPS and somatosensory interactions were absent from the much smaller network common to spatial imagery and haptic perception of familiar shape. Conversely, in the network shared by object imagery and haptic perception of familiar shape, bilateral LOC activity was driven by a focus in the left inferior frontal gyrus, consistent with the prediction that this network is characterized by top-down pathways from prefrontal areas (Lacey *et al.*, 2009a): these pathways were absent from the sparser network common to object imagery and unfamiliar haptic shape perception (Lacey *et al.*, 2014).

5 Conclusion

In this chapter, we have shown that the visual and haptic modalities are interconnected at almost every level of object processing, sharing common object representations in crossmodal and view-independent recognition, and common dimensions in imagery preferences. These commonalities are supported by multisensory neural substrates and complex interactions between bottom-up and top-down processes, as well as between object and spatial imagery. However, much remains to be done in providing a detailed account of visuo-haptic multisensory behaviour and its neural basis. An important goal will be to apply this understanding in the service of neurorehabilitation, especially for those with sensory deprivation of various kinds.

Acknowledgements

Support to K. S. from the National Eye Institute at the National Institutes of Health, the National Science Foundation, and the Veterans Administration is gratefully acknowledged.

References

Adhikari, B. M., Sathian, K., Epstein, C. M., *et al.* (2014). Oscillatory activity in neocortical networks during tactile discrimination near the limit of spatial acuity. *NeuroImage,* 91, 300–10.

Alain, C., Arnott, S. R., Hevenor, S., *et al.* (2001). 'What' and 'where' in the human auditory system. *Proceedings of the National Academy of Sciences of the United States of America,* 98, 12301–6.

Allen, H. A. and Humphreys, G. W. (2009). Direct tactile stimulation of dorsal occipito-temporal cortex in a visual agnosic. *Current Biology,* 19, 1044–9.

Amedi, A., Jacobson, G., Hendler, T., *et al.* (2002). Convergence of visual and tactile shape processing in the human lateral occipital complex. *Cerebral Cortex,* 12, 1202–12.

Amedi, A., Malach, R., Hendler, T., *et al.* (2001). Visuo-haptic object-related activation in the ventral visual pathway. *Nature Neuroscience,* 4, 324–30.

Amedi, A., Stern, W. M., Camprodon, J. A., *et al.* (2007). Shape conveyed by visual-to-auditory sensory substitution activates the lateral occipital complex. *Nature Neuroscience,* 10, 687–9.

Amedi, A., von Kriegstein, K., van Atteveldt, N. M., *et al.* (2005). Functional imaging of human crossmodal identification and object recognition. *Experimental Brain Research,* 166, 559–71.

Andresen, D. R., Vinberg, J., and Grill-Spector, K. (2009). The representation of object viewpoint in human visual cortex. *NeuroImage,* 45, 522–36.

Arnott, S. R., Binns, M. A., Grady, C. L., *et al.* (2004). Assessing the auditory dual pathway model in humans. *NeuroImage,* 22, 401–8.

Bar, M. (2007). The proactive brain: using analogies and associations to generate predictions. *Trends in Cognitive Sciences,* 11, 280–9.

Blajenkova, O., Kozhevnikov, M., and Motes, M. A. (2006). Object-spatial imagery: a new self-report imagery questionnaire. *Applied Cognitive Psychology,* 20, 239–63.

Bliss, I. and Hämäläinen, H. (2005). Different working memory capacity in normal young adults for visual and tactile letter recognition task. *Scandinavian Journal of Psychology,* 46, 247–51.

Buelte, D., Meister, I. G., Staedtgen, M., *et al.* (2008). The role of the anterior intraparietal sulcus in crossmodal processing of object features in humans: an rTMS study. *Brain Research,* 1217, 110–18.

Bülthoff, I. and Newell, F. N. (2006). The role of familiarity in the recognition of static and dynamic objects. *Progress in Brain Research,* 154, 315–25.

Bushnell, E. W. and Baxt, C. (1999). Children's haptic and cross-modal recognition with familiar and unfamiliar objects. *Journal of Experimental Psychology: Human Perception and Performance,* 25, 1867–81.

Cant, J. S., Arnott, S. R., and Goodale, M. A. (2009). fMR-adaptation reveals separate processing regions for the perception of form and texture in the human ventral stream. *Experimental Brain Research,* 192, 391–405.

Cant, J. S. and Goodale, M. A. (2007). Attention to form or surface properties modulates different regions of human occipitotemporal cortex. *Cerebral Cortex,* 17, 713–31.

Casey, S. J. and Newell, F. N. (2007). Are representations of unfamiliar faces independent of encoding modality? *Neuropsychologia,* 45, 506–13.

Chouinard, P. A., Meena, D. K., Whitwell, R. L., *et al.* (2017). A TMS investigation on the role of lateral occipital complex and caudal intraparietal sulcus in the perception of object form and orientation. *Journal of Cognitive Neuroscience,* 29, 881–95.

Craddock, M. and Lawson, R. (2008). Repetition priming and the haptic recognition of familiar and unfamiliar objects. *Perception and Psychophysics,* 70, 1350–65.

Craddock, M. and Lawson, R. (2009). Do left and right matter for haptic recognition of familiar objects? *Perception,* 38, 1355–76.

Craddock, M. and Lawson, R. (2010). The effects of temporal delay and orientation on haptic object recognition. *Attention, Perception, and Psychophysics,* 72, 1975–80.

De Volder, A. G., Toyama, H., Kimura, Y., *et al.* (2001). Auditory triggered mental imagery of shape involves visual association areas in early blind humans. *NeuroImage,* 14, 129–39.

Deshpande, G., Hu, X., Lacey, S., *et al.* (2010). Object familiarity modulates effective connectivity during haptic shape perception. *NeuroImage,* 49, 1991–2000.

Dijkerman, H. C. and de Haan, E. H. F. (2007). Somatosensory processes subserving perception and action. *Behavioral and Brain Sciences,* 30, 189–239.

Dormal, G., Pelland, M., Rezk, M., *et al.* (2018). Functional preference for objects sounds and voices in the brain of early blind and sighted individuals. *Journal of Cognitive Neuroscience,* 30, 86–106.

Easton, R. D., Greene, A. J., and Srinivas, K. (1997a). Transfer between vision and haptics: memory for 2-D patterns and 3-D objects. *Psychonomic Bulletin and Review,* 4, 403–10.

Easton, R. D., Srinivas, K., Greene, A. J. (1997b). Do vision and haptics share common representations? Implicit and explicit memory within and between modalities. *Journal of Experimental Psychology: Learning, Memory, and Cognition,* 23, 153–63.

Eck, J., Kaas, A. L., and Goebel, R. (2013). Crossmodal interactions of haptic and visual texture information in early sensory cortex. *NeuroImage,* 75, 123–35.

Eck, J., Kaas, A. L., Mulders, J. L., *et al.* (2016). The effect of task instruction on haptic texture processing: the neural underpinning of roughness and spatial density perception. *Cerebral Cortex,* 26, 384–401.

Eger, E., Ashburner, J., Haynes, J.-D., *et al.* (2008). fMRI activity patterns in human LOC carry information about object exemplars within category. *Journal of Cognitive Neuroscience,* 20, 356–70.

Ernst, M. O. and Banks, M. S. (2002). Humans integrate visual and haptic information in a statistically optimal fashion. *Nature,* 415, 429–33.

Feinberg, T. E., Rothi, L. J., and Heilman, K. M. (1986). Multimodal agnosia after unilateral left hemisphere lesion. *Neurology,* 36, 864–7.

Ferreira, C. D., Gadelha, M. J. N., da Fonsêca, E. K. G., *et al.* (2021). Long-term memory of haptic and visual information in older adults. *Aging, Neuropsychology, and Cognition,* 28, 65–77.

Freud, E., Culham, J. C., Plaut, D. C., *et al.* (2017). The large-scale organization of shape processing in the ventral and dorsal pathways. *eLife,* 6, e27576.

Gaglianese, A., Branco, M. P., Groen, I. I. A., *et al.* (2020). Electrocorticography of tactile responses in visual cortices. *Brain Topography,* 33, 559–70.

Garvill, J. and Molander, B. (1973). Effects of standard modality, comparison modality and retention interval on matching of form. *Scandinavian Journal of Psychology,* 14, 203–6.

Gauthier, I., Hayward, W. G., Tarr, M. J., *et al.* (2002). BOLD activity during mental rotation and view-dependent object recognition. *Neuron,* 34, 161–71.

Goodale, M. A. and Milner, A. D. (1992). Separate visual pathways for perception and action. *Trends in Neuroscience,* 15, 20–5.

Grefkes, C., Geyer, S., Schormann, T., *et al.* (2001). Human somatosensory area 2: observer-independent cytoarchitectonic mapping, interindividual variability, and population map. *NeuroImage,* 14, 617–31.

Grill-Spector, K., Henson, R. and Martin, A. (2006). Repetition and the brain: neural models of stimulus-specific effects. *Trends in Cognitive Sciences,* 10, 14–23.

Grill-Spector, K., Kushnir, T., Edelman, S., *et al.* (1999). Differential processing of objects under various viewing conditions in the human lateral occipital complex. *Neuron,* 24, 87–203.

Harvey, D. Y. and Burgund, E. D. (2012). Neural adaptation across viewpoint and exemplar in fusiform cortex. *Brain and Cognition,* 80, 33–44.

Helbig, H. B. and Ernst, M. O. (2007). Optimal integration of shape information from vision and touch. *Experimental Brain Research,* 179, 595–606.

Helbig, H. B., Ernst, M. O., Ricciardi, E., *et al.* (2012). The neural mechanisms of reliability weighted integration of shape information from vision and touch. *NeuroImage,* 60, 1063–72.

Hutmacher, F. and Kuhbandner, C. (2018). Long-term memory for haptically explored objects: fidelity, durability, incidental encoding, and cross-modal transfer. *Psychological Science,* 29, 2031–8.

Iriki, A., Tanaka, M., and Iwamura, Y. (1996). Attention-induced neuronal activity in the monkey somatosensory cortex revealed by pupillometrics. *Neuroscience Research,* 25, 173–81.

James, T. W., Humphrey, G. K., Gati, J. S., *et al.* (2002). Differential effects of view on object-driven activation in dorsal and ventral streams. *Neuron,* 35, 793–801.

James, T. W., James, K. H., Humphrey, G. K., *et al.* (2006). Do visual and tactile object representations share the same neural substrate? In Touch and *Blindness: Psychology and Neuroscience,* edited by M. A. Heller and S. Ballesteros, 139–55. Mahwah, NJ: Lawrence Erlbaum Associates.

James, T. W., Stevenson, R. W., Kim, S., *et al.* (2011). Shape from sound: evidence for a shape operator in the lateral occipital cortex. *Neuropsychologia,* 49, 1807–15.

Jones, B. (1981). The developmental significance of cross-modal matching. In *Intersensory Perception and Sensory Integration,* edited by R. D. Walk and H. L. Pick Jr., 108–36. New York, NY: Plenum Press.

Kassuba, T., Klinge, C., Hölig, C., *et al.* (2013). Vision holds a greater share in visuo-haptic object recognition than touch. *NeuroImage,* 65, 59–68.

Kiphart, M. J., Hughes, J. L., Simmons, J., *et al.* (1992). Short-term haptic memory for complex objects. *Bulletin of the Psychonomic Society,* 30, 212–14.

Klatzky, R. L. and Lederman, S. J. (1995). Identifying objects from a haptic glance. *Perception and Psychophysics,* 57, 1111–23.

Klatzky, R. L., Lederman, S. J., and Metzger, V. A. (1985). Identifying objects by touch: an 'expert system'. *Perception and Psychophysics,* 37, 299–302.

Klatzky, R. L., Lederman, S. J., and Reed, C. L. (1987). There's more to touch than meets the eye: the salience of object attributes for haptics with and without vision. *Journal of Experimental Psychology: General,* 116, 356–69.

Kozhevnikov, M., Hegarty, M., and Mayer, R. E. (2002). Revising the visualiser-verbaliser dimension: evidence for two types of visualisers. *Cognition and Instruction,* 20, 47–77.

Kozhevnikov, M., Kosslyn, S. M., and Shephard, J. (2005). Spatial versus object visualisers: a new characterisation of cognitive style. *Memory and Cognition,* 33, 710–26.

Krekelberg, B., Boynton, G. M., and van Wezel, R. J. (2006). Adaptation: from single cells to BOLD signals. *Trends in Neuroscience,* 29, 250–6.

Lacey, S. and Campbell, C. (2006). Mental representation in visual/haptic crossmodal memory: evidence from interference effects. *Quarterly Journal of Experimental Psychology,* 59, 361–76.

Lacey, S., Feng, H., Caesar, E., *et al.* (2013). Preferences for integrative versus schematic sensory imagery across modalities. Abstract, *Society for Neuroscience.*

Lacey, S., Flueckiger, P., Stilla, R., *et al.* (2010b). Object familiarity modulates the relationship between visual object imagery and haptic shape perception. *NeuroImage,* 49, 1977–90.

Lacey, S., Hall, J., and Sathian, K. (2010a). Are surface properties integrated into visuo-haptic object representations? *European Journal of Neuroscience,* 31, 1882–8.

Lacey, S., Lin, J. B., and Sathian, K. (2011). Object and spatial imagery dimensions in visuo-haptic representations. *Experimental Brain Research,* 213, 267–73.

Lacey, S., Pappas, M., Kreps, A., *et al.* (2009b). Perceptual learning of view-independence in visuo-haptic object representations. *Experimental Brain Research,* 198, 329–37.

Lacey, S., Peters, A., and Sathian, K. (2007). Cross-modal object representation is viewpoint-independent. *PLoS One,* 2, e890.

Lacey, S. and Sathian, K. (2008). Haptically evoked activation of visual cortex. *In Human Haptic Perception: Basics and Applications,* edited by M. Grunwald, 251–7. Basel: Birkhäuser.

Lacey, S. and Sathian, K. (2011). Multisensory object representation: insights from studies of vision and touch. Progress in Brain Research, 191, 165–76.

Lacey, S. and Sathian, K. (2014). Visuo-haptic multisensory object recognition, categorization and representation. *Frontiers in Psychology,* 5, 730.

Lacey, S., Stilla, R., Sreenivasan, K., *et al.* (2014). Spatial imagery in haptic shape perception. *Neuropsychologia,* 60, 144–58.

Lacey, S., Tal, N., Amedi, A., *et al.* (2009a). A putative model of multisensory object representation. *Brain Topography,* 21, 269–74.

Lawson, R. (2009). A comparison of the effects of depth rotation on visual and haptic three-dimensional object recognition. *Journal of Experimental Psychology: Human Perception and Performance,* 35, 911–30.

Lawson, R. (2011). An investigation into the cause of orientation-sensitivity in haptic object recognition. *Seeing and Perceiving,* 24, 293–314.

Lederman, S. J. and Klatzky, R. L. (1987). Hand movements: a window into haptic object recognition. *Cognitive Psychology,* 19, 342–68.

Lucan, J. N., Foxe, J. J., Gomez-Ramirez, M., *et al.* (2010). Tactile shape discrimination recruits human lateral occipital complex during early perceptual processing. *Human Brain Mapping,* 31, 1813–21.

Malach, R., Reppas, J. B., Benson, R. R., *et al.* (1995). Object-related activity revealed by functional magnetic resonance imaging in human occipital cortex. *Proceedings of the National Academy of Sciences of the United States of America,* 92, 8135–9.

Mishkin, M. (1979). Analogous neural models for tactual and visual learning. *Neuropsychologia,* 47, 139–51.

Mullin, C. R. and Steeves, J. K. E. (2011). TMS to the lateral occipital cortex disrupts object processing but facilitates scene processing. *Journal of Cognitive Neuroscience,* 23, 4174–84.

Mur, M., Ruff, D. A., Bodurka, J., *et al.* (2010). Face-identity change activation outside the face system: 'release from adaptation' may not always indicate neuronal selectivity. *Cerebral Cortex,* 20, 2027–42.

Nabeta, T. and Kawahara, J. (2006). Congruency effect of presentation modality on false recognition of haptic and visual objects. *Memory,* 14, 307–15.

Négyessy, L., Nepusz, T., Kocsis, L., *et al.* (2006). Prediction of the main cortical areas and connections involved in the tactile function of the visual cortex by network analysis. *European Journal of Neuroscience,* 23, 1919–30.

Newell, F. N., Ernst, M. O., Tjan, B. S., *et al.* (2001). View dependence in visual and haptic object recognition. *Psychological Science,* 12, 37–42.

Nicholson, K. G. and Humphrey, G. K. (2003). The effect of colour congruency on shape discriminations of novel objects. *Perception,* 32, 339–53.

Nishimura, M., Scherf, K. S., Zachariou, V., *et al.* (2015). Size precedes view: developmental emergence of invariant object representations in lateral occipital complex. *Journal of Cognitive Neuroscience,* 27, 474–91.

Palmer, S. E., Rosch, E., and Chase, P. (1981). Canonical perspective and the perception of objects. In *Attention and Performance IX,* edited by J. Long and A. Baddeley, 135–51. Hillsdale, NJ: Lawrence Erlbaum Associates.

Pascual-Leone, A. and Hamilton, R. H. (2001). The metamodal organization of the brain. Progress in Brain Research, 134, 427–45.

Peissig, J. J. and Tarr, M. J. (2007). Visual object recognition: do we know more now than we did 20 years ago? *Annual Review of Psychology,* 58, 75–96.

Pensky, A. E. C., Johnson, K. A., Haag, S., *et al.* (2008). Delayed memory for visual-haptic exploration of objects. *Psychonomic Bulletin and Review,* 15, 574–80.

Pitzalis, S., Galletti, C., Huang, R. S., *et al.* (2006). Wide-field retinotopy defines human cortical visual area V6. *Journal of Neuroscience,* 26, 7962–73.

Pourtois, G., Schwarz, S., Spiridon, M., *et al.* (2009). Object representations for multiple visual categories overlap in lateral occipital and medial fusiform cortex. *Cerebral Cortex,* 19, 1806–19.

Prather, S. C., Votaw, J. R., and Sathian, K. (2004). Task-specific recruitment of dorsal and ventral visual areas during tactile perception. *Neuropsychologia,* 42, 1079–87.

Rauschecker, J. P. and Tian, B. (2000). Mechanisms and streams for processing of 'what' and 'where' in auditory cortex. *Proceedings of the National Academy of Sciences of the United States of America,* 97, 11800–6.

Reales, J. M. and Ballesteros, S. (1999). Implicit and explicit memory for visual and haptic objects: cross-modal priming depends on structural descriptions. *Journal of Experimental Psychology: Learning, Memory, and Cognition,* 25, 644–63.

Reed, C. L., Klatzky, R. L., and Halgren, E. (2005). What vs. where in touch: an fMRI study. *NeuroImage,* 25, 718–26.

Rice, N. J., Valyear, K. F., Goodale, M. A., *et al.* (2007). Orientation sensitivity to graspable objects: an fMRI adaptation study. *NeuroImage,* 36, T87–93.

Riesenhuber, M. and Poggio, T. (1999). Hierarchical models of object recognition in cortex. *Nature Neuroscience,* 2, 1019–25.

Sack, A. T. (2006). Transcranial magnetic stimulation, causal structure-function mapping and networks of functional relevance. *Current Opinion in Neurobiology,* 16, 593–9.

Saito, D. N., Okada, T., Morita, Y., *et al.* (2003). Tactile-visual cross-modal shape matching: a functional MRI study. *Cognitive Brain Research,* 17, 14–25.

Sathian, K. (2004). Modality, quo vadis? Comment. *Behavioral Brain Sciences,* 27, 413–14.

Sathian, K. and Lacey, S. (2007). Journeying beyond classical somatosensory cortex. *Canadian Journal of Experimental Psychology,* 61, 254264.

Sathian, K., Lacey, S., Stilla, R., *et al.* (2011). Dual pathways for somatosensory information processing. *NeuroImage,* 57, 462–75.

Sathian, K., Zangaladze, A., Hoffman, J. M., *et al.* (1997). Feeling with the mind's eye. *NeuroReport,* 8, 3877–81.

Sergent, J., Ohta, S., and MacDonald, B. (1992). Functional neuroanatomy of face and object processing. A positron emission tomography study. *Brain,* 115, 15–36.

Silvanto, J., Schwarzkopf, D. S., Gilaie-Dotan, S., *et al.* (2010). Differing causal roles for lateral occipital complex and occipital face area in invariant shape recognition. *European Journal of Neuroscience,* 32, 165–71.

Snow, J. C., Goodale, M. A., and Culham, J. C. (2015). Preserved haptic shape processing after bilateral LOC lesions. *Journal of Neuroscience,* 35, 13745–60.

Snow, J. C., Strother, L., and Humphreys, G. W. (2014). Haptic shape processing in visual cortex. *Journal of Cognitive Neuroscience,* 26, 1154–67.

Stein, B. E., Burr, D., Constantinidis, C., *et al.* (2010). Semantic confusion regarding the development of multisensory integration: a practical solution. *European Journal of Neuroscience,* 31, 1713–20.

Stilla, R. and Sathian, K. (2008). Selective visuo-haptic processing of shape and texture. *Human Brain Mapping,* 29, 1123–38.

Stoesz, M., Zhang, M., Weisser, V. D., *et al.* (2003). Neural networks active during tactile form perception: common and differential activity during macrospatial and microspatial tasks. *International Journal of Psychophysiology,* 50, 41–9.

Streri, A. and Molina, M. (1994). Constraints on intermodal transfer between touch and vision in infancy. In *The Development of Intersensory Perception: Comparative Perspectives,* edited by D. J Lewkowicz and R. Lickliter, 285–307. Hove: Lawrence Erlbaum Associates.

Takahashi, C. and Watt, S. J. (2014). Visual-haptic integration with pliers and tongs: signal 'weights' take account of changes in haptic sensitivity caused by different tools. *Frontiers in Psychology,* 5, 109.

Tal, N. and Amedi, A. (2009). Multisensory visual-tactile object related network in humans: insights gained using a novel crossmodal adaptation approach. *Experimental Brain Research,* 198, 165–82.

Tarr, M. J. and Pinker, S. (1989). Mental rotation and orientation dependence in shape recognition. *Cognitive Psychology*, 21, 233–82.

Ueda, Y. and Saiki, J. (2007). Viewpoint independence in visual and haptic object recognition. *Japanese Journal of Psychonomic Science*, 26, 11–19.

Ueda, Y. and Saiki, J. (2012). Characteristics of eye movements in 3-D object learning: comparison between within-modal and crossmodal object recognition. *Perception*, 41, 1289–98.

Ungerleider, L. G. and Mishkin, M. (1982). Two cortical visual systems. In *Analysis of Visual Behavior*, edited by D. J. Ingle, M. A. Goodale, and R. J. W. Mansfield, 549–86. Cambridge, MA: MIT Press.

Valyear, K. F., Culham, J. C., Sharif, N., *et al.* (2006). A double dissociation between sensitivity to changes in object identity and object orientation in the ventral and dorsal streams: a human fMRI study. *Neuropsychologia*, 44, 218–28.

Wilson, K. and Farah, M. J. (2006). Distinct patterns of viewpoint-dependent BOLD activity during common-object recognition and mental rotation. *Perception*, 35, 1351–66.

Woods, A. T., Moore, A., and Newell, F. N. (2008). Canonical views in haptic object perception. *Perception*, 37, 1867–78.

Woods, A. T., O'Modhrain, S., and Newell, F. N. (2004). The effect of temporal delay and spatial differences on crossmodal object recognition. *Cognitive, Affective, and Behavioral Neuroscience*, 4, 260–9.

Zangaladze, A., Epstein, C. M., Grafton, S. T., *et al.* (1999). Involvement of visual cortex in tactile discrimination of orientation. *Nature*, 401, 587–90.

Zhang, M., Weisser, V. D., Stilla, R., *et al.* (2004). Multisensory cortical processing of object shape and its relation to mental imagery. *Cognitive, Affective, and Behavioral Neuroscience*, 4, 251–9.

Zhou, Y.-D. and Fuster, J. M. (1997). Neuronal activity of somatosensory cortex in a cross-modal (visuo-haptic) memory task. *Experimental Brain Research*, 116, 551–5.

19
Crossmodal identification

Casey O'Callaghan

1 Introduction

In crossmodal identification, a subject token identifies an individual thing or feature that is perceived in one sensory modality with an individual thing or feature that is perceived in a distinct sensory modality. You may surmise that the tennis ball you feel in your hand is the tennis ball you see. You may register that you hear and see it bounce. Sometimes, when you perceive something with each of two or more different senses, you can tell it is the same thing.

Crossmodal identification takes place just in case a subject s perceives an individual o_1, in a sensory modality m; s perceives an individual o_2, in a distinct sensory modality n; and s token identifies o_1 with o_2.

This paper is about the nature of crossmodal identification. To simplify, it focuses on perceptible individuals, and not on attributes. It concerns the capacity to token identify an individual object or event that a subject perceives using distinct senses. Focusing on attributes perceived using distinct senses raises thorny additional issues.

What kind of psychological phenomenon is crossmodal identification? Which capacities does it comprise? In particular, does crossmodal identification always implicate thought or cognition, or is crossmodal identification sometimes wholly perceptual?

In the first place, crossmodal identification can be cognitive. One way to token identify an item perceived using distinct senses is to work it out. Using perceptual clues, a subject might infer that what is perceived through each of two separate senses is the same thing. Or a subject might just recognize it to be one thing, with past experience or stored knowledge as a guide. If crossmodal identification relies on inference or recognition, and each of these is extraperceptual, then so too is crossmodal identification.

It is tempting to think crossmodal identification can be perceptual. If so, a subject does not just recognize as one thing or identify in thought what is perceived using each of two distinct senses. Instead, one perceives the sameness, the numerical identity, of an individual which one perceives through distinct senses. Perceptibly, the thing you see is the thing you touch; you perceive the event you hear and the event you see to be one event. If so, one may conclude that crossmodal identification takes place in perception.[1]

So, we have two views about crossmodal identification. According to one, crossmodal identification always takes place in thought or extraperceptual cognition. According to the other, crossmodal identification sometimes occurs in perception. This paper negotiates the merits and drawbacks to each approach, and it offers a third way. It resists the claim that

[1] See Bayne, 2014; Green, 2021; and O'Callaghan, 2014. Spence and Bayne (2015) are sceptical.

Casey O'Callaghan, *Crossmodal identification* In: *Sensory Individuals*. Edited by: Aleksandra Mroczko-Wąsowicz and Rick Grush, Oxford University Press. © Oxford University Press 2023. DOI: 10.1093/oso/9780198866305.003.0020

crossmodal identification takes place in perception. Strictly speaking, crossmodal identification is not a perceptual capacity because it is too demanding. Perception does not represent individuals perceived with one sense as being identical to individuals perceived with another sense. However, perception is not unalive to crossmodal identity. Perception in typical human subjects is differentially sensitive to the identity or sameness of individuals presented to distinct senses. This relies on distinguishing crossmodal identification from perceiving crossmodal identity. Perception registers crossmodal identity but does not represent it as such.

2 Crossmodal identification in thought

The first option says crossmodal identification takes place outside perception. Philosophy has focused on token identification in thought. Testimony, rather than night sky viewing, taught me Hesperus is Phosphorus. Students understand that saying Hesperus is Hesperus differs in informativeness from saying Hesperus is Phosphorus. In the paradigms, an object is thought of in different ways, on multiple occasions, or using distinct words, and it is recognized, appreciated, or said to be one thing.

Suppose a subject perceives an object by means of one sense and also perceives it by means of another sense, such that one could continue to perceive it independently using each sense. Each way of perceiving enables thoughts about the object. Given differences between senses, a subject may not recognize or appreciate the identity. An uninitiated perceiver could think there are two distinct objects. Nevertheless, an experienced subject facing the scene could think there is one perceptible object. Tradition says the main difference is what initiated and uninitiated subjects think, not what they perceive.

Williamson (1990) suggests a model. In *Identity and Discrimination*, he says to discriminate is to activate knowledge of distinctness. When a subject discriminates *a* from *b*, the subject knows *a* is distinct from *b*, and that knowledge becomes active.

By analogy, identification may be understood as activating knowledge of identity or sameness. When a subject identifies *a* with *b*, the subject knows *a* is numerically identical with *b*, and that knowledge becomes active.

Notice two things. First, if knowledge requires belief and belief is extraperceptual, then both discrimination and identification are cognitive and extraperceptual.[2] Second, the approach just sketched treats identification as factive. Just as one cannot discriminate something from itself, it is not possible to token identify distinct items. This factive requirement may be waived to focus on a fallible mental counterpart. The target then is roughly the activation of warranted occurrent belief in identity.

In this picture, crossmodal identification takes place in cognition. A subject perceives an individual object or event with one sense, perceives it with another sense, and activates knowledge or belief that it is one and the same thing. A subject may see a tennis ball, touch that tennis ball, and think the item seen is identical to the item touched.

[2] If perception entails belief, this does not follow. However, if so, the perceptual account of crossmodal identification may collapse into the cognitive account.

3 Appreciating identity

Crossmodal identification requires token identifying an item perceived using one sense with an item perceived using another sense. Token identification requires appreciating identity. Typical human subjects can appreciate crossmodal identity in thought. To determine whether crossmodal identification occurs in perception, it helps to say what it is to appreciate identity across senses.

Consider three types of cases. The first evidently involves just one presentation of an object *a*. For instance, a subject may at once ascribe distinct attributes to an individual, as when observing that a perceptible object is yellow and spherical, or yellow and fuzzy. Using a term in language to refer to one thing on an occasion offers an example, as in, 'Hesperus is a planet visible in the evening.' Regarding it once reflects that one appreciates the identity of that which bears distinct features.

The second involves regarding one thing presented in the same way or manner in distinct episodes as the same. Seeing one tennis ball on two occasions offers an example. You may think the tennis ball is yellow, then later think it is spherical, while thinking of it in the same particular visual way. This reflects a way to appreciate token identity. Recanati (2012) would say it involves a single mental file. Using distinct tokens of one proper name in language offers an example, as in, 'Hesperus is visible' and 'Hesperus is a planet.' Fine (2007) calls this *coordination*, or *representing as the same*. He says (p. 40):

> An object is represented as the same in a piece of discourse only if no one who understands the discourse can sensibly raise the question of whether it is the same.

It is a constraint that the term '*a*' refers to the same thing in each use in a discourse. That is not true for '*a*' and '*b*'. Not appreciating this betrays misunderstanding.

The third involves regarding something presented in differing ways or manners to be one thing. Using distinct Fregean modes of presentation to express an identity offers an example. So, too, can seeing and feeling a tennis ball. You may think what you see is yellow and think what you touch is fuzzy. Vision and touch enable you to think of it in differing ways. Perhaps each gives rise to a distinct mental file. Nonetheless, you may recognize that what you see is what you touch and thus appreciate the identity. In language, distinct proper names can serve to express an identity claim, as in, 'Hesperus is Phosphorus.' Fine calls this *representing as being the same*, which is incompatible with representing as the same (coordination). Representing as being the same, but not representing as the same, is informative (p. 40):

> The idea of representing objects *as* the same is to be distinguished from the idea of representing the objects as *being* the same. The sentences 'Cicero = Tully' and 'Cicero = Cicero' both represent the objects as being the same but only the second represents them as the same. And, in general, one cannot informatively represent objects as being the same compatibly with representing them as the same.

In general, it is informative to be told that *a* is *b* because, typically, it is not a constraint in a discourse that '*a*' and '*b*' refer to one thing.

Crossmodal identification belongs to the third type. Using Fine's terms, for one to token identify an individual perceived by means of distinct senses relies on representing it as being the same.

The argument proceeds by elimination. Crossmodal identification does not just belong to the first type of case, in which an object or event evidently is presented just once, in a given manner, while bearing one or more features (as in, *a is F and G*). In crossmodal identification, a subject perceives an item and its features by means of distinct senses. You see the tennis ball, and you touch the tennis ball. Moreover, these dissociate. You can see the tennis ball without touching it, and you can touch the tennis ball without seeing it. So, sight and touch each present the tennis ball, and neither presentation evidently depends on the other for its existence. Accordingly, to appreciate the identity across senses requires either that the item presented in sight and in touch is represented as the same (*a is F* and *a is G*), or that it is represented as being the same by means of an informative identity ($a = b$).

Crossmodal identification also does not just belong to the second type of case, in which an object or event evidently is presented more than once but as the same, or in the same manner. The appearance of identity across senses can be misleading or mistaken. You might just be wrong that the tennis ball you see is the tennis ball you touch. Fine says representing as the same does not allow such errors (p. 40):

> Finally, what are in fact two distinct objects can be represented as being the same, as with the sentence 'Cicero = Caesar.' But two distinct objects cannot ever be represented as the same—or, at least, not without taking two names to be one or committing some other error of this kind.

More to the point, it need not be evident that what you see is what you touch. A subject may fail to discern the identity. And one can sensibly raise the question of whether what is seen is what is touched. Indeed, it can be rational for a subject who sees and touches one object to doubt the identity claim or ascribe incompatible features. So the identification is not trivial. It is informative to learn a crossmodal identity.[3]

Crossmodal identification thus belongs to the third type of case, in which an object or event presented with distinct senses in differing ways or manners is taken to be one thing. This means regarding what is perceived with one sense as being the same as what is perceived with another sense. Using Fine's terms, crossmodal identification is akin to representing as being the same.

Identification in such a case requires more than appreciating each of two distinct presentations. To see and to touch one thing does not suffice to appreciate its identity. One must take what is seen and what is touched to be the same. It must be regarded as being a single thing. Fine says (p. 40):

> A further difference is that only a single sentence (such as 'Cicero = Tully' can represent its objects as being the same but two different sentences (e.g., 'Cicero is Roman', 'Cicero is an orator') can represent their objects as the same.

[3] For discussion of co-reference in distinct modalities, see Rescorla, 2020.

Maybe no explicit identity expression is required. What Campbell (1997, p. 59) calls *trading on identity* may suffice. If one appreciates that *a* is *F* and that *b* is *G*, appreciating that something is both *F* and *G* can be evidence one has traded on identity.

4 Crossmodal identification in perception

A typical subject can think something seen and something touched are one thing. In thought, one can represent as being the same something that is perceived in each of two distinct sensory modalities. A lot of philosophy since Frege addresses how. Thus, assuming common sensibles, perception can activate knowledge of crossmodal identity.

The view that crossmodal identification takes place in perception faces obstacles. There are reasons to doubt that perception represents items perceived with distinct senses *as being the same*. And it is not clear perception appreciates crossmodal identity in the way required for token identification. Perception does not fit the model of crossmodal identification in thought.

Notice first that to perceive numerical identity across senses is not trivial. One cannot perceive the identity of what is seen with what is touched just by independently seeing and touching an item. To perceive crossmodal identity requires not only using two senses in parallel, but also perceiving jointly with those senses. It relies on registering individuals, spatio-temporal characteristics, and relations among them across senses. This is a form of ineliminably multisensory perception.

Next consider what crossmodal identification requires. According to the previous section, crossmodal identification requires representing an item perceived using one sense as being identical with an item perceived using another sense. To represent an individual *a* (perceived with *m*) as being the same as an individual *b* (perceived with *n*) means representing *a*, representing *b*, and representing *a* to be identical with *b*. One must represent identity, as such.

It is not obvious contents structured in this way figure elsewhere in synchronic perception. To see an object and its mirror image at once is not evidently perceptually to represent what is doubly perceived as being the same. Nor is seeing double perceptually representing as being the same.

How about diachronic reidentification? Consider finding your car in a lot or your eyeglasses in the dark. Clear cases in which a subject appreciates identity are not evidently the ones in which the item is perceived in differing manners and represented perceptually as being the same. Instead, in visual or tactual reidentification, given how perception presents the item, it is plausible that what is perceived is represented as the same, or in the same manner. The high-level content, *being the same as*, does not obviously find a place in perception.

This is not conclusive. Consider an example involving perceptual indexicals. Suppose you see your car, see it later, and appreciate the identity. Both the indexical manners and their targets seem to match. But distinct uses of an indexical need not entail representing as the same. One can sensibly ask whether *here* is the same as *here*. Since you appreciate your car's identity, two uses of an indexical are compatible with representing a target as being the same. And this might occur within perception.

Of course, this presumes without argument that visual perception represents the identity as such. Still, even if we find no unisensory example of perceptually representing as being the same, crossmodal perceptual identification could be unique.

A perceptual approach to crossmodal identification faces additional obstacles. Crossmodal identification entails somehow appreciating the identity of what is perceived using distinct senses. However, it is always sensible to ask if what you perceive in one modality is identical to what you perceive in another. A subject rationally can doubt whether what is seen is the same as what is touched. Considering the question is not enough to settle it. Accordingly, to appreciate crossmodal identity requires further considerations beyond just perceiving one item using each of two senses. Crossmodal identification calls for evidence, warrant, or reasons supporting the identity.

It is questionable whether perception provides evidence by representing identity across senses. On one hand, perception may directly present a subject with crossmodal identity as such. If so, perception supplies the needed warrant by revealing crossmodal identity. One could then grasp immediately that a seen item is the same as a touched item. However, it is doubtful that perceiving subjects are directly presented with crossmodal identity *as such*. It is not evident phenomenologically that one perceives something seen and something touched as being numerically the same. Earnestly introspecting subjects may wonder. In contrast with *redness*, *texture*, or *motion*, there is ample room for scepticism (see Spence and Bayne, 2015).

On the other hand, perception may present a subject with indirect evidence of crossmodal identity, such as spatio-temporal coincidence and correlation. This offers the support a subject needs to appreciate identity across senses. However, it enfeebles crossmodal identification in perception. Suppose crossmodal identification in thought finds perceptual support from features such as spatio-temporal relations. What does perceiving identity across senses add? Without an answer, crossmodal identification in perception is superfluous.

This challenge, too, is answerable. Multisensory perception may represent as such the identity of an item perceived with each of two senses. However, a subject may fail to appreciate it, or may appreciate it for other reasons. Whether and how a subject appreciates the identity in thought need not couple tightly with the fact that perception reveals or represents it.

Still, if crossmodal identification requires appreciating identity, a further obstacle remains. To perceptually represent something seen as being identical with something touched requires an appreciative response to the identity. Seeing a tennis ball and touching a tennis ball are compatible with seeing and touching distinct tennis balls. Crossmodal identification in perception thus calls for support from further features, such as spatio-temporal coincidence and correlations.

Suppose crossmodal identification requires appreciating such considerations—in effect, treating them as warranting identification. Appreciation, in this conception, entails reason responsiveness. However, it is not clear perception responds appreciatively to considerations as favouring a proposition or to reasons as such. If perception does not respond rationally or appreciatively to factors supporting identity, crossmodal identification is not perceptual.

Put another way, crossmodal identification requires evidence that what is perceived using each of two senses is the same. That is because crossmodal identification belongs to the third type of cases in which a subject appreciates identity. In such cases, an item is presented in differing ways that are compatible with distinctness. Appreciating identity requires responding to reasons for identity as such. If perception does not engage in what Siegel (2017) dubs *reckoning*, in which considerations are *taken* to support an upshot, as described by Boghossian (2014), then perception does not identify individuals across sensory modalities. If so, crossmodal identification does not take place in perception.

This argument sets the bar high for an appreciative response. However, perception could respond to factors that support identification, even if a subject is not aware of its doing so. This echoes debates about whether perception involves inferences (Orlandi, 2014; Siegel, 2017). Transitions that are, in several respects, like inferences may occur in perception, even if they are unlike inferences subjects perform. Accordingly, identification could take place in perception, even if a subject does not identify *a* with *b*.

The clear upshot is that crossmodal identification in perception must differ in character from crossmodal identification in thought. That, in principle, is no problem. What matters is that crossmodal identification requires representing something perceived with one sense as being the same as something perceived with another sense. Given differing presentations across senses, it is not obvious perception has the resources to do so.

Each obstacle could be answered, in the ways I have suggested. However, the lesson is that doing so takes work, and success is not guaranteed. Any account is bound to meet opposition. Some I have described, but others run deeper. For instance, many naïve realists will resist any proposal that requires perception constitutively to involve representing something perceived with one sense as being the same as something perceived with another sense. How to translate this idiom into friendlier terms is not evident. The hope was an account of which capacities crossmodal identification comprises, not a contentious partisan description of identification within perception. Given agreement that crossmodal identification occurs in thought, why go there?

5 Sensitivity to identity without crossmodal identification

Here is where things stand. Clear paradigms of token identification occur in thought. There are reasons to doubt that crossmodal identification takes place in perception. It is not evident what the cognitive approach leaves out. Thus, according to a reasonable line of thought, crossmodal identification is cognitive and not perceptual. Adapting Williamson's account of discrimination, we might say crossmodal identification is activating knowledge or warranted belief in identity across senses.

One response is to vindicate the perceptual approach to crossmodal identification. For instance, dig in and defend the claim that perceiving involves representing as being the same an item perceived using each of two distinct senses. I am not hopeful. Let us pursue another approach. I would like to consider the evidence that crossmodal identification takes place in perception and to offer an explanation that avoids the obstacles in the previous section.

According to the account I propose, perception is differentially sensitive to situations in which one thing is perceived using distinct senses. Perception is sensitive to sameness across senses. It registers crossmodal identity. My proposal does not say perception represents items perceived with distinct senses as being the same. It is compatible with saying that appreciating crossmodal identity requires extraperceptual cognition. Thus, it does not say crossmodal identification takes place in perception. This captures all we should ask from an account of perception's role in crossmodal identification. Perceiving crossmodal identity does not entail crossmodal identification.

Before turning to the evidence, it helps to describe what it takes for perception to register identity across senses and consider an objection. In the first place, while perceiving often is a way of coming to know, perceiving need not activate knowledge or warranted belief. One might perceive *p* without believing *p*; one might perceive *p* without being justified in believing *p*, due to defeaters or lack of awareness; one might perceive *p* but lack concepts to think *p*; or one might seem to perceive *p* while *p* is false. More simply, one might perceive *p*, know *p*, but fail to activate one's knowledge that *p*. Therefore, perceptually registering the identity of an item presented to each of two senses does not entail activating knowledge or warranted belief in crossmodal identity. One could register perceptually the identity of something seen with something touched without cognitively appreciating it.

As I understand it, the bar for registering identity perceptually is low. It does not require thought. Indeed, perception can register the identity of *a* and *b* without representing *a* as being identical to *b*. And one can register identity perceptually without awareness of identity as such.

Perception therefore can register a crossmodal identity, even if a subject fails to appreciate it. How a subject perceives can be sensitive to the fact that what is presented to one sense is the same as what is presented to another sense, even if the subject does not recognize it. Most notably, perception can register crossmodal identity without representing an item perceived using one sense and an item perceived using another sense as being the same. Perceptually registering crossmodal identity does not imply perceptually representing anything *as being the same*. To insist that it does just renames the puzzle.

Which perceptual capacities does registering identity across senses comprise? In short, perception must be selectively sensitive to crossmodal identity. Perception at minimum ought to be differentially responsive to scenarios in which what is presented to one sense is token identical to what is presented to another sense.

To be differentially sensitive to a feature requires the capacity perceptually to detect it and to differentiate it from other features. Three conditions must be met. First, to detect it, one must be capable of responding to the relevant feature. So, one must be affected by instances in which a single thing is perceived using each of two distinct senses. Since one may respond to diverse features in just the same way, mere detection does not suffice for selective sensitivity. Second, one must be capable of differentiating the relevant feature from others. This means responding differently to it from how one responds to some range of other features. So, one must respond differently to crossmodal identity from how one responds to shapes, textures, or kinds. The most relevant contrast is between one's response to crossmodal identity and one's response to distinct items perceived with distinct senses. Without such a difference, there is little reason to think perception singles out crossmodal identity.[4] Finally, such differential sensitivity must be perceptual. Thus, one's responsiveness to crossmodal identity must rely on extracting information from sensory stimulation, and it must play the right role in one's psychology, especially with respect to cognition, belief, and action. If these conditions are met, perception is differentially sensitive to the token identity of an item perceived using each of two distinct senses. This is a way perceptually to register crossmodal identity.

Indiscriminable counterpart scenarios pose an objection. Suppose a subject perceives the same item using distinct senses. An indiscriminable counterpart is a scenario a subject cannot differentiate but which contains two distinct objects or events. For all a subject can tell, the scenarios match. But, in one case, two things are before the senses rather than one.

[4] Cf. Burge's (2010, p. 466), 'Principle of Relevant Representational Alternatives'.

Being differentially sensitive to crossmodal identity requires responding differently to its presence and to the presence of other features. If one cannot tell the difference between crossmodal identity and its indiscriminable counterparts, why say perception registers crossmodal identity rather than some other feature, such as spatio-temporal coincidence?

This objection concerns the specificity or determinacy of sensitivity claims. These are some of the trickiest issues in philosophy of mind, perception, and psychosemantics, and I will not try to resolve them here. Instead, I hope to sidestep and to blunt their impact on this paper's project. Indeed, in three ways, sensitivity to a feature such as spatio-temporal coincidence helps support this paper's conclusions.

First, to appreciate crossmodal identity requires being sensitive to considerations favouring identity over distinctness. Perceptually registering crossmodal identity relies on cues to identity across senses. Spatial and temporal alignment are key evidence. Thus, being sensitive to spatio-temporal coincidence shows perception satisfies a necessary condition on registering crossmodal identity.

Second, being sensitive to either crossmodal identity or spatio-temporal coincidence differs from failing to be sensitive to any such relation between individuals perceived using distinct senses. Suppose perception is differentially sensitive to crossmodal coincidence, but not identity. That remains a significant feature of perception, when compared with being sensitive only to spatio-temporal locations perceived using individual senses. Thus, in this project, perceptual sensitivity to spatio-temporal alignment might stand in for perceptual sensitivity to token identity across senses.

Third, spatio-temporal coincidence is being in the same place and time. Being sensitive to spatio-temporal coincidence is being sensitive to the identity of place and of time across senses. This may not imply perceptual sensitivity to the token identity of an individual perceived using distinct senses. Nonetheless, it is perceptual sensitivity to identity across senses.

6 Evidence for sensitivity to identity

This section describes evidence that perception is differentially sensitive to scenarios in which an item presented to one sense is token identical to an item presented to another sense. Where a subject perceives o_1 in m, and perceives o_2 in distinct n, sometimes the subject is perceptually sensitive to the oneness of o_1 with o_2. Results from psychophysics support the claim that perceptual systems are selectively responsive to cases in which the same item is before each of two distinct senses. This is less fancy than identification in thought. And it does not entail representing *as being the same*. Still, it is a way perceptually to register crossmodal identity.

Evidence for sensitivity to identity stems from studies of *feature binding* and *object files*. Feature binding occurs when distinct features are treated perceptually as bundled or bound—to belong together—despite being registered separately by sensory systems. If visual systems extract separately information about shape and colour, something accounts for a tennis ball's looking at once both yellow and round. Perceptually binding *yellow* and *round* reflects sensitivity to the token identity of that which bears each feature. Object files are temporary, episodic representations posited to explain how distinct features at a time are associated with a particular item and how perceivers keep track of an item over time. Tracking a

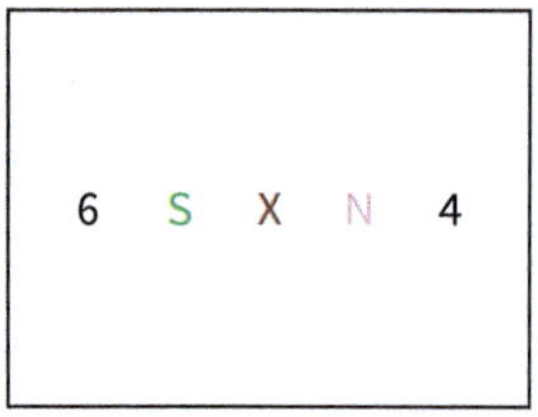

Error type	*Example*	*Answers per trial*
Correct	Middle, brown X	0.44
Illusory conjunction	Left, green N	0.33
	Right, brown S	0.06
Single feature correct	Right pink	0.49
Single feature wrong	Something blue	0.06

Figure 19.1 Visual illusory conjunction.
Treisman and Schmidt (1982) (Ptak, 2008, Figure 14).

tennis ball from moment to moment, even if it changes colour, reflects sensitivity to its identity over time.[5]

The *Many Properties Problem* illustrates what is at stake in feature binding (Jackson, 1977). Perceivers differentiate seeing a red square beside a yellow circle from seeing a red circle beside a yellow square. But, in each case, just the same basic features are present. Something must account for the difference. What explains why *red*, *square*, and *left* go together in one case, and not in the other? Plausibly, a subset of features in a scene is bundled or bound. In this case, binding signals a common feature-bearing individual.

Translated to multisensory contexts, this argument does not offer such clear support for sensitivity to identity across senses. Consider how perceiving *colour*$_1$ and *pitch*$_1$ on the left with *colour*$_2$ and *pitch*$_2$ on the right differs from perceiving *colour*$_1$ and *pitch*$_2$ on the left with *colour*$_2$ and *pitch*$_1$ on the right. It is not evident *colour*$_1$ and *pitch*$_1$ are bundled or bound together in one case, but not in the other. Colour and pitch might belong to distinct individuals with similar locations. Moreover, it is not obvious introspectively that perception itself identifies locations across senses. Thus, it begs the question to appeal to the introspective exercise described in the *Many Properties Problem* to establish perceptual sensitivity to identity across senses.

6.1 Illusory conjunctions

Illusory conjunctions outside focal attention are evidence for feature binding in unisensory contexts (Treisman and Schmidt, 1982). If presented with a field of red circles and blue squares while doing another task, it is not uncommon to report a red square. Presented with a green 'S' and a pink 'N', Treisman and Schmidt's participants sometimes perform as if they have seen a green 'N' or a pink 'S' (Figure 19.1). Indeed, subjects have high confidence they have experienced such figures. According to Treisman and Schmidt, features are first coded separately, then bound together during object perception. The illusion reveals the capacity perceptually to bind distinct features that belong to the same object. Feature binding thus shows subjects are perceptually sensitive to the identity and distinctness of token feature bearers.

In a multisensory setting, Cinel *et al.* (2002) argue that illusory conjunctions occur between vision and touch. They first build objects with distinctive surface textures by gluing carpeting, Legos, fake fur, beans, and bubble wrap to 12 × 3 cm cardboard rectangles. They

[5] Green and Quilty-Dunn (2021) is a state-of-the-art treatment of object files.

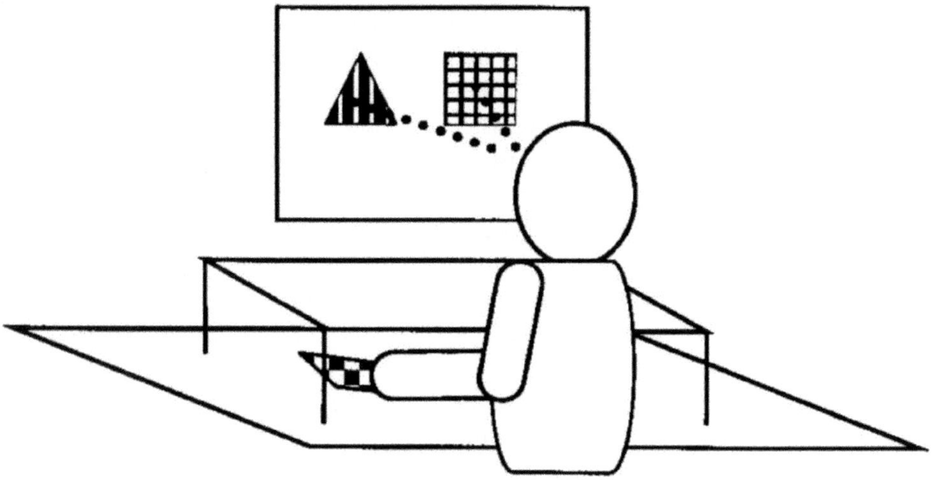

Figure 19.2 Crossmodal illusory conjunction.

photograph each texture and use it to fill three visual figures: a square, a triangle, and a circle. The first experiment presents a tactual object, hidden beneath a screen, in one of two orientations, and two visual objects, for 75–300 ms (Figure 19.2). The task is to report the orientation of the tactual bar, and the texture and shape of each visual object. When the tactual and visual textures differ, participants on average accurately report the visible texture 76% of the time. However, 8% of the time, participants report the felt texture as the visible texture. In illusory conjunction cases, subjects misattribute a texture they felt to a shape they saw.

Cinel *et al.* conduct further experiments using forced-choice matching, increased attention demands, memory loading, reversed vision and touch, and performance on a person with parietal damage. They argue that illusory conjunctions of visible and tactual features are 'perceptual in nature' (p. 1261) and 'preattentive' (p. 1244), rather than effects of memory or cognition. Consistent with unisensory results, the effect persists in a target-matching task, and it is enhanced with greater memory demands, reduced attention, and same-hemifield stimuli. They conclude, 'These results demonstrate that ICs are possible not only within the visual modality but also between two different modalities: vision and touch.' (p. 1245).

If a tactual feature can be ascribed illusorily to a visible figure, this suggests that in the good case, perception is sensitive to the fact that what is touched is what is seen. In that case, the object seen does possess the tactual feature attributed to it. If so, one is selectively sensitive to token identity between vision and touch. If not, what explains attributing felt texture to a visible figure?

Here, there are reasons to pause. Studies supporting illusory conjunction rely on subjective reports, recall, and prior target matching, rather than on more direct psychophysical measures. Indeed, Cinel *et al.* prompt reports and forced choices to probe feature conjunctions. This methodology is subject to extraperceptual interference and confounds, such as snap judgement, recognition, and recollection. So it is valuable to have further evidence.

Cinel *et al.*'s touch–vision paradigm raises further doubt, when compared with Treisman and Schmidt's unisensory paradigm. Treisman and Schmidt aim to test whether illusory conjunctions are part of visual phenomenology (especially experiments I and II). Based on error patterns, confidence judgements, and their own impressions, Treisman and Schmidt (1982, p. 121) report, 'The phenomenal descriptions subjects gave made it clear that at least some of the errors were experienced as genuine illusions.' They say, 'No conclusive answer can be given, but the evidence seems consistent with our own conviction, from the pilot experiment, that on some trials at least, illusory conjunctions have the character of perceptual experiences' (p. 120).

Cinel *et al.* do not purport to examine perceptual experience. Their studies do not mention 'experience', 'phenomenology', or 'confidence'. Since their crossmodal paradigm does not test for experience, and it does not describe subjective reports or confidence judgements, it does not demonstrate that an illusory perceptual experience has occurred. So these results do not establish a true illusory conjunction.

There is a further disanalogy. In the unisensory paradigm, participants report visually experiencing colour and shape jointly to qualify an apparent visible object. One figure appears both 'S'-shaped and pink, just as a true pink 'S' looks. What seems pink seems also 'S'-shaped. Evidently, one item appears to bear both visible features.

The touch–vision paradigm, however, does not show that a tactual feature and a visible feature are bound perceptually in this way as features of a single object. The felt texture does not evidently infuse visible shape like a colour or fill it like a pattern. The visible figure looks to have visible shape and pattern, and the cardboard bar feels like it has texture and orientation. The tactual and visible features do not seem commingled or directly to be coinstantiated by one individual. The crossmodal paradigm does not demonstrate that felt texture (a texture as felt) seems to qualify any object that is seen. For all it shows, what is touched, which perceptibly has texture and orientation, may or may not appear to be one with what is seen, which perceptibly has shape and pattern.

Thus, in a key respect, the tactual–visual case is unlike the unisensory visual case. Because it does not demonstrate that tactual and visible features illusorily appear to belong to a single perceptible item, it does not show that some tactual feature bearer perceptibly is identical to a visual feature bearer. The crossmodal paradigm does not establish crossmodal identification in perception.

It does show that felt textures interfere with reporting and matching conflicting visual patterns. Consider a standard crossmodal illusion, such as ventriloquism, the McGurk effect, or the sound-induced flash. These illusions demonstrate that stimulating one sense can reshape experience associated with another. Each stems from attempting to resolve a conflict concerning features perceived using distinct senses, such as locations, phonemes, or quantities. Ventriloquism, for instance, reconciles a spatial discrepancy between sights and sounds.

In the crossmodal illusory conjunction paradigm, touch does not trigger an illusory visual pattern. Touch does not make the bubble wrap look like fur. So it is not a standard crossmodal illusion. But touch does affect performance in visual tasks. Roughly one in 12 times when they conflict, touch impacts a subject's ability to report, recall, or match visual texture. This, however, does not require an illusory experience.

What then explains the interference? Handling conflict can reshape performance without causing illusion. What matters here is the performance, not the experience. The reason is that conflict presupposes a common subject matter. Handling conflict, being sensitive to it, means being sensitive to identity or sameness, for nothing prevents distinct things from having conflicting features, including distinct textures. Cinel *et al.*'s crossmodal paradigm therefore illustrates that performance in perceptual tasks is differentially responsive to the fact that what bears felt texture typically also yields visible pattern. Felt texture sometimes overrides the visual pattern.

According to this interpretation, the crossmodal illusory conjunction paradigm shows typical human subjects are differentially sensitive to identity across senses. However, these results do not demonstrate crossmodal identification has taken place. Crossmodal conflict resolution does not require representing what is touched and what is seen as being the same. It does not require appreciating crossmodal identity.

6.2 Object-specific preview effects

Turn to the *object-specific preview effect*, which is evidence for object files in visual perception (Kahneman *et al.*, 1992). Suppose several objects at once are presented visually. Following a brief interval, the same objects reappear. On the second viewing, subjects more quickly report a feature of one of those objects if that feature is unchanged from the preview to the target. Participants in one experiment (Figure 19.3) on average are 35 ms faster reporting which letter appears in a small box when it matches the letter that appeared earlier in that box (Same Object) than if it is an entirely new letter (No Match) or if that letter appeared earlier in a different box (Different Object)—roughly a 7% advantage for reappearing on the same object. Belonging to the same object, not simply having been viewed earlier, drives the speeded response. So the effect is object-specific, unlike non-specific feature priming.[6]

According to Kahneman *et al.*, attention to the target triggers *reviewing*, in which one current object is selected and assigned to a corresponding object from the preview field. If the target and preview match, it facilitates recognition. If the target and preview mismatch, it interferes with recognizing the feature. The effect reveals the capacity to differentiate and assign features to distinct objects at a time and the capacity to keep track of an individual object viewed at distinct times. If so, the object-specific preview effect shows that subjects are perceptually sensitive to the situation in which a single object has distinct perceptible features and to the situation in which an object perceived at one time is token identical to an object perceived at another time.

Object-specific preview effects have been reported in multisensory settings. Zmigrod *et al.* (2009) describe audiovisual and audio-tactile experiments. First, they present a colour–pitch preview, followed by an interval, then a colour–pitch target (Figure 19.4). Participants in distinct trials report either the pitch or the colour of the target. Subjects are quicker and more accurate in reporting a target's pitch or its colour when both pitch and colour match the preview than when either pitch or colour differs from the preview. Being paired with the same colour facilitates one's response to a pitch, and vice versa. Similar results hold in a pitch–vibration study (Experiment 2).

This study is not a direct parallel to the unisensory visual paradigm introduced by Kahneman *et al.* (1992). It does not present multiple objects at once during the preview or target stage. And it collapses the distinction between No Match and Different Object conditions. So it does not distinguish an object-specific effect from non-specific priming. To

[6] In a study using whole words, the authors observed both non-specific priming and object-specific preview effects (Kahneman *et al.*, 1992, p. 188).

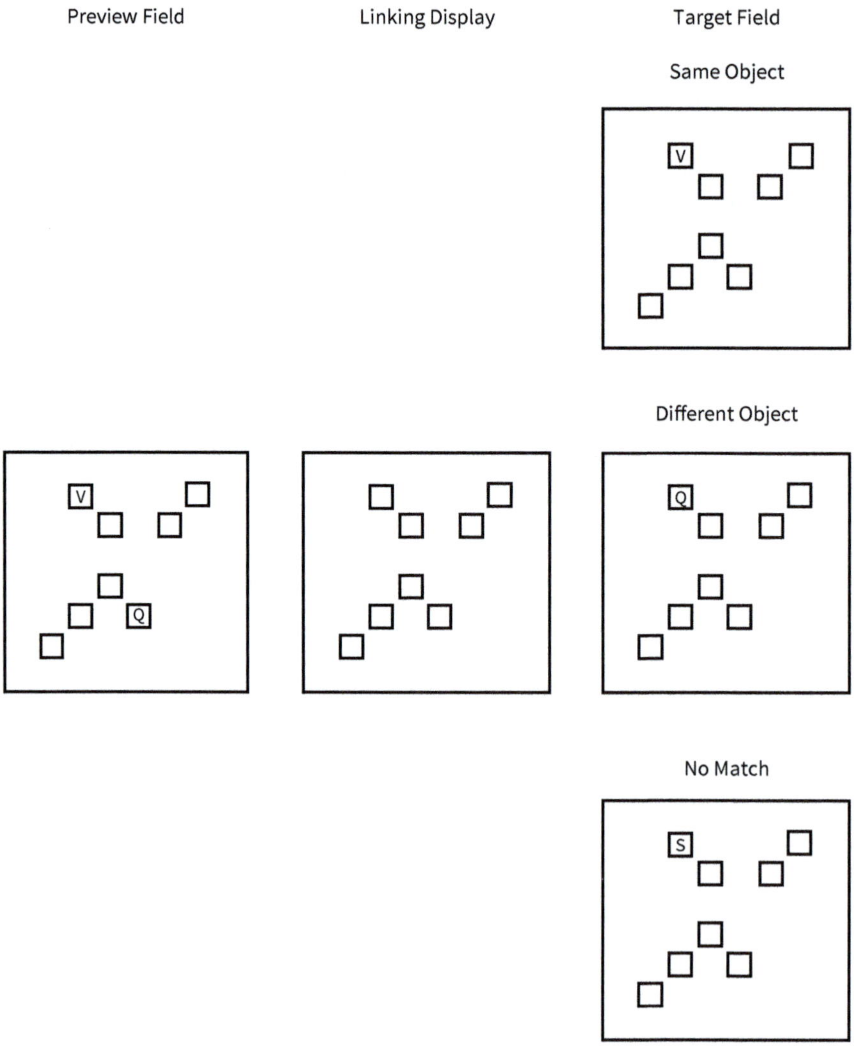

Figure 19.3 Visual object-specific preview effect.
Reprinted with permission from Elsevier (Kahneman *et al.*, 1992).

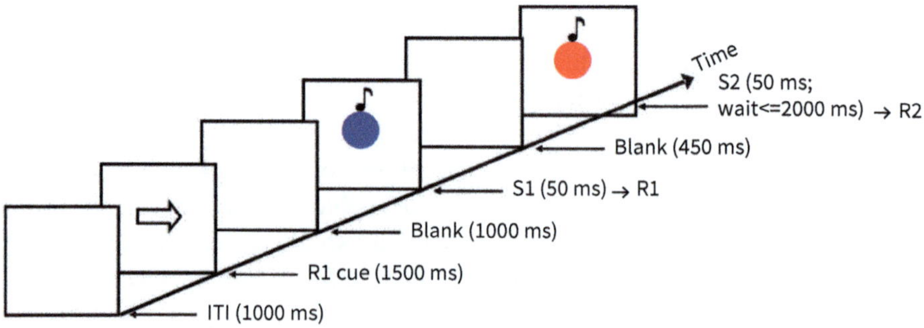

Figure 19.4 Audiovisual object-specific preview.
Zmigrod *et al.*, 2009.

explain the effect does not require object files or same-object perception. The result does show that features perceived with distinct senses interact in a task for which one sense is not relevant. This means distinct features are associated across senses and compared over time. However, this experiment does not guarantee that colour and pitch are perceived to belong to a single individual.

Nonetheless, if features perceived with distinct senses belong to one thing, this result makes sense. Objects tend not to change, so even task-irrelevant feature differences may slow you down. The study does not establish that the associated features are perceived to belong to one object. However, it does support perception's sensitivity to real-world situations in which one thing has features perceived using distinct senses.

Jordan *et al.* (2010) describe an intriguing variant to the original unisensory paradigm. This does demonstrate a multisensory object-specific preview effect. The authors present a brief preview picturing two objects framed in separate boxes (Figure 19.5). The pictures then disappear from the frames. After 1 s, the target display plays a sound at the location of one of the two frames. The sound either matches the object that appeared in that frame during preview, matches the object that appeared in the other frame during preview, or matches neither previewed object. The task is to report 'match' if the sound fits *either* of the two previewed pictures, and 'no match' if not. Subjects on average are 18 ms faster in congruent than in incongruent trials—a 2.5% advantage. They are 54 ms faster in incongruent than in no match trials, reflecting a 7% general, non-specific priming advantage, even for the incongruent feature. While the advantage is half that observed by Kahneman *et al.* in visual tasks, Jordan *et al.* nevertheless record a significant multisensory object-specific preview effect.

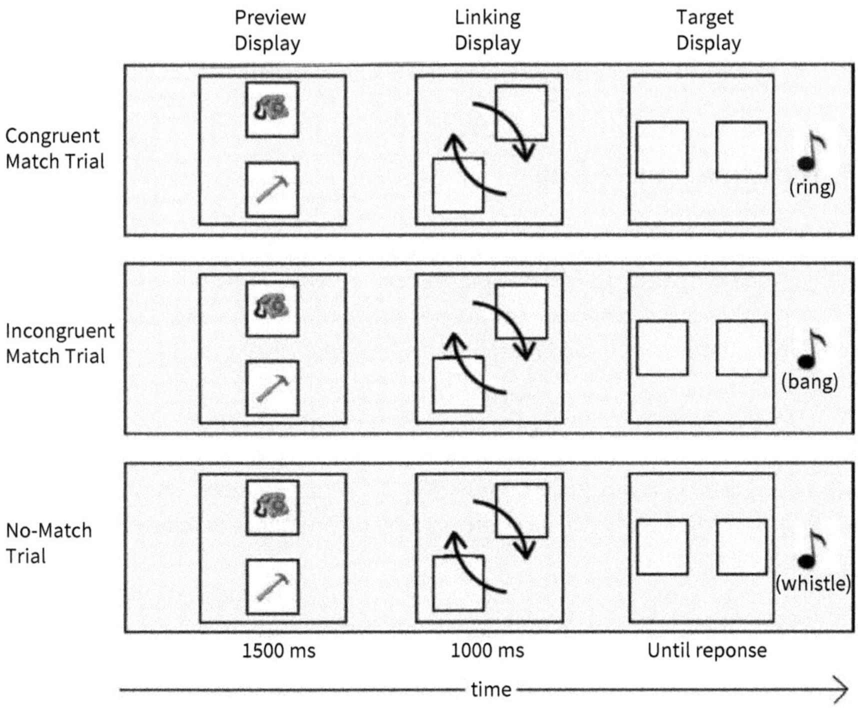

Figure 19.5 Multisensory object-specific preview effect.
Reprinted with permission from Taylor & Francis (Jordan *et al.*, 2010).

What follows? Notice first that this study does not present the same feature in the preview and the target. In that respect, it is unlike the visual paradigm. It displays a picture and later plays a corresponding sound. A phone is not the same feature as a ringing sound. Thus, the effect must not require perceptually identifying a feature seen with a feature heard.

Instead, the effect relies on discerning what makes the picture and sound congruent or incongruent. They are pictures and sounds of artefacts and animals—'a dog, whistle, train, hammer, piano, and phone' (Jordan *et al.*, 2010, p. 496). But one may doubt that perception registers such kinds, natural or artefactual. Plausibly, perception reveals at most sense-specific gestalts. Thus, if the effect relies on recognizing a common kind shared by preview and target, that may signal an advantage in extraperceptual cognition rather than in perception.

However congruence is explained, this multisensory preview advantage remains object-specific. Object-specific preview effects require reidentifying something that persists through time. In the unisensory visual paradigm, this is a visible object and it bears visible features. The object-specific advantage holds for features of that object that match—they are identical—between the preview and target. In Jordan *et al.*'s multisensory paradigm, the advantage also relies on reidentifying a visible frame in the target display as a visible frame from the preview. For the advantage to occur, the previewed picture and the target sound each must be associated with the same visible object. But this does not require representing what is seen and what is heard as being the same.

It follows that an audible feature must be associated with a visible object (frame). When a sound (ringing) and a congruent image (phone) each are associated with a single persisting visible object (or location), typically that means one thing in the environment is responsible for each. Sight and hearing then converge on one individual. The target display mirrors scenes in which one thing is responsible for the sights and the sounds. The multisensory object-specific preview effect therefore illustrates that perception is differentially sensitive to situations in which one thing is presented to vision and audition.

6.3 Attentional spreading

Finally, consider *attentional spreading*, which supports sensitivity to identity across senses. Attending to one feature of an object enhances the salience of its other features. It does so more than it enhances salience for features of other objects. Unattended features do not have this effect. Attentional spreading relies on object-oriented attention, not simple spatial attention.

> An attentional spreading view of object-based attention proposes that the rate and efficiency of perceptual processes are improved by spreading attention through an attended object. This spreading likely enhances the representation of the attended object relative to unattended objects.
>
> Richard *et al.*, 2008, p. 843.

Richard *et al.* present features in fixed locations on distinct objects or the same object (Figure 19.6A). The target is the feature at the centre of each bar—a circular or rectangular bite. Compatible flankers on the same object speed one's response to the target (middle stimulus, Figure 19.6A). Incompatible flankers interfere, slowing one's response (bottom stimulus).

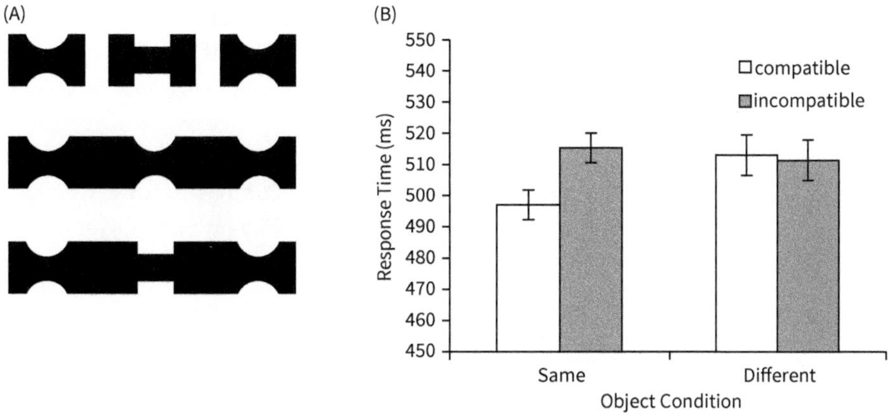

Figure 19.6 Visual object-based attentional spreading.

Flankers on distinct objects have no effect (top stimulus). Attentional spreading illustrates sensitivity to situations in which distinct visible features belong to one object.

Attentional spreading also occurs across senses. When a sound accompanies an image, attending to the image enhances not only the image's salience, but also the sound's salience. In a study measuring event-related potentials associated with auditory perception, Busse *et al.* (2005) ask participants to focus on a central cross while covertly attending to one side. On either the attended or the unattended side, an image is shown (Figure 19.7). The task is to detect a target feature that appears on the image in roughly one in seven trials (in Figure 19.7, the two small squares). When a brief sound plays with an attended visual image, event-related potentials demonstrate enhanced auditory responsiveness to the sound itself, as compared to when the sound occurs with an unattended image. The sound does not affect accuracy

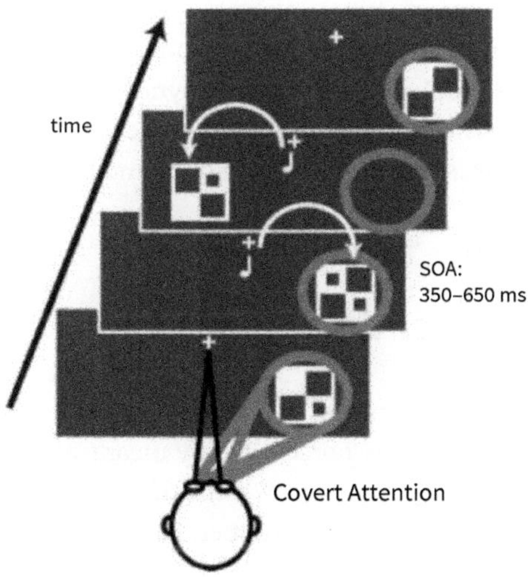

Figure 19.7 Audiovisual object-based attentional spreading.

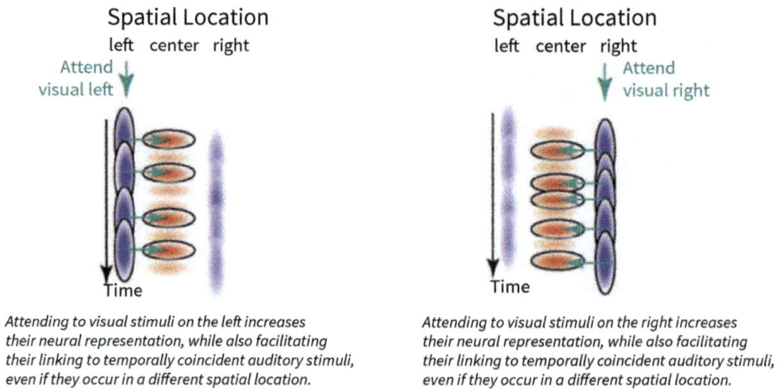

Figure 19.8 Audiovisual object-based attentional spreading.
Reprinted with permission from Elsevier (Talsma *et al.*, 2010).

or response time for the attended visual target. Subjects unsurprisingly are slower and less accurate in detecting the visual feature on an unattended image. This study does not include behavioural measures regarding the sound.

In virtue of the parallel, it is tempting to conclude that this study reveals object-based attentional spreading. Attending the visual feature enhances one's response to the auditory feature. Since their locations differ, this is not just spatial attention. And the auditory feature is irrelevant, so it is not a task effect. According to this interpretation, attention spreads from visual feature to auditory feature because the two perceptibly belong to a single object (Figure 19.8).

This is too quick. Synchronous sights and sounds typically do stem from one source. So, crossmodal attentional spreading indicates sensitivity to situations in which one object or event is responsible for visible and audible features. However, to explain attentional spreading does not require that perception token identifies what is heard with what is seen. Perception need not represent an audible individual as being identical to a visible individual. It need not identify a single object or an event as the bearer of both the visible feature (two small squares) and the audible feature (a tone). It is enough that synchronous visible and audible features are targets of special interest. What is seen and what is heard become associated because they occur at one time. Nevertheless, being associated or bound in this way does not require being token identified.

7 Binding without identification

The empirical evidence I have described shows that perception is differentially sensitive to circumstances in which one thing is perceptible by means of distinct senses. Features that typically stem from a common source receive enhanced processing, facilitate speedier reporting, and sometimes are taken for one another. Perceptual responsiveness thus is selectively sensitive to crossmodal identity.

This operationalizes feature binding and object files across senses. But the evidence does not guarantee that crossmodal identification takes place within perception. Being differentially sensitive to crossmodal identity does not require co-ascribing features to a single individual or token identifying an individual perceived using one sense with an individual perceived using another. It does not entail representing crossmodal identity as such.

Feature binding is too permissive. It does not require a unified perceptual process or mechanism (cf. Treisman, 2003, p. 98). And it does not specify a relation between things in the environment. As operationalized, feature binding is compatible with a variety of relations, not each of which entails perceptually identifying one feature bearer with another feature bearer.

Consider some options. First, one might think feature binding encodes *co-instantiation*. This is plausible in certain unisensory examples. If a tennis ball looks yellow and round, one thing perceptibly bears both visible features at once. That item which one sees to be yellow, one sees also to be round. *Yellow* and *round* seem perceptibly to be co-instantiated by one individual.

However, co-instantiation is not plausible for other examples of feature binding. When one sees a red circle and hears a pitched sound (as in Zmigrod *et al.*, 2009), it is not necessary that colour, shape, pitch, and loudness perceptibly are, or seem to be, co-instantiated by one item. The visible individual that seems perceptibly to have colour and shape need not also seem audibly to have pitch and loudness. The red circle is not the sound. No single item seen and heard evidently bears colour, shape, pitch, and loudness at once.

Second, one might instead think feature binding indicates *co-constituency* or *co-parthood*. Touch conveys the fuzzy felt underside of a tennis ball, while vision reveals its yellow upper hemisphere. The undersurface is a tangible part, the upper surface is a visible part, and each perceptibly is a part of a whole. Binding in this understanding occurs partwise. One perceives bound tactual parts and visible parts to form a common whole. This suggests a model for crossmodal perceptual identification, if the whole that is felt is the whole that is seen. In that case, something perceptibly bears parts perceived with distinct senses.

The parthood model even illuminates audiovisual binding. Audible sounds and visible objects perceptibly belong to an encompassing event, such as a collision or an explosion, that one perceives multisensorily. This approach is especially promising, and I have developed it elsewhere (O'Callaghan, 2016; see also Green, 2019).

In the present context, however, there are two noteworthy caveats. One is that nothing in the studies described suggests binding across senses ought to be understood to specify co-constituency. In Zmigrod *et al.* (2009), what is the whole to which a red circle and a C-sharp tone perceptibly belong? In Jordan *et al.* (2010), is it plausible that subjects see a picture of a phone, represent it to be part of a broader event in which that phone is caused to ring, later hear a ringing sound, represent it to be part of a broader event, and perceptually token identify the event seen with the event heard? These interpretations are not obvious or mandatory. The experimental evidence thus does not show that feature binding requires co-constituency. The theory of binding is one thing; a theory of perceptible objects is another.

Even so, there is another caveat. Perceptible co-constituency does not demonstrate crossmodal identification has taken place. Suppose one sees a red phone while hearing a loud ringing sound. And suppose binding occurs, such that the phone and the ringing perceptibly are parts of a whole. If so, a red visible object and a loud audible sound perceptibly are parts of some further object or event.

Nevertheless, the whole need not be perceptible with any single sense. Perceiving the whole, a composite, may require joint or cooperative use of multiple senses. If so, it is perceived only through multisensory perception. But such a novel perceptual act need not token identify what is seen with what is heard. The visible object may remain numerically distinct, according to multisensory perception, from the audible sound. The same can hold for each visible object and each audible event. If so, it need not be the case that, for an individual o_1 that *s* perceives with *m*, and for an individual o_2 that *s* perceives with *n*, perception token identifies o_1 with o_2. Where o_1 and o_2 are distinct perceptible proper parts of o_3, and perceiving o_3 requires the joint use of both *m* and *n*, partwise binding can occur without token identifying what is visible with what is audible. Therefore, partwise multisensory binding does not entail crossmodal identification.

Third, one might think binding reflects numerical identity *across time*. For instance, the object-specific preview effect relies on tracking one thing between two times. Something heard may be bound temporally with something seen at another time, thus implicating crossmodal identity over time.

Current evidence does not show that token identification across times and senses occurs in perception. For instance, in Jordan *et al.*'s multisensory object-specific preview paradigm, subjects see a phone, follow the visible frame, then hear a ringing sound. The object-specific advantage relies on visual tracking plus synchronous binding between the frame and sound. The previewed phone and later sound need not be bound. Crossmodal reidentification may rely on extraperceptual resources, such as recognition, long-term memory, concepts, or inference.

Fourth, one might think binding conveys a common *causal source*. Rather than co-instantiation, co-constituency, or identity over time, binding could signal features perceived using distinct senses with the same causal origin. Felt heat and flickering light each come from the stove. Binding implicates their sharing a source. This is one way to interpret Bayesian causal source analyses proposed by empirical researchers (see Shams and Beierholm, 2010).

Causal source analysis is a powerful way to explain crossmodal binding effects. But it is one thing to explain perceptual processes in Bayesian terms; it is another to say perception represents causal sources of features perceived using distinct senses as such. Perception may be sensitive to the token identity of causal sources in a multisensory setting, thanks to Bayesian processes, but binding does not establish that perception token identifies the cause of a feature perceived using one sense with the cause of a feature perceived using another sense (see also Green, 2019; Rescorla, 2020).

Finally, one might think empirical evidence is compatible with treating binding across senses as a kind of *association* (Connolly, 2014; Fulkerson, 2011). Association is very permissive. Since it is so non-committal, association allows a range of differing relations in which audible and visible features may stand. Understood this way, binding could reflect perceptual sensitivity to identity across senses. However, even regular crossmodal association does not show perception token identifies an item perceived using one sense with an item perceived using another sense.

Perception is sensitive to situations in which one item is presented to distinct senses. However, studies of feature binding do not establish that perception involves token identification across senses. Binding nonetheless reflects a way to be sensitive to crossmodal identity.

8 Registering identity across senses

Empirical studies of binding and object files do not establish crossmodal identification in perception. The result of binding is compatible with a variety of relations in which perceptible items may stand. Binding as operationalized in such studies does not require a perceptual process in which an individual perceived in one sensory modality is token identified with an individual perceived in a distinct sensory modality.

Studies of binding and object files do provide evidence that perception is differentially sensitive to crossmodal identity. Illusory conjunctions, object-specific preview effects, and attentional spreading demonstrate that perception typically registers something presented at once to distinct senses. The lesson is that perception is selectively responsive to identity across senses, even if perception does not represent identity across senses as such.

This captures all we should ask from an account of perception's role in crossmodal identification. In the first place, perception's pattern of sensitivity is an apt response to identity across senses. Perception responds selectively to cues that are evidence for identity, such as spatial proximity and temporal coincidence. And the way it responds makes sense in the face of crossmodal identity. For instance, resolving conflict between felt texture and visible pattern, taking more time to report barking when it accompanies an object with a history as a phone, and devoting greater attention to sounds that are synchronous with items commanding visual attention are appropriate when one thing is presented to distinct senses.

Furthermore, perception's pattern of sensitivity to crossmodal identity redounds outside perception. For instance, it benefits perception-based recognition. The effects described reorganize perception in ways that enhance perceptual cues to crossmodal identity: conflict resolution reduces feature misalignment; same-object advantages prioritize continuity and consistency over time; attention spreading highlights multisensory clusters or configurations of features. Each positions perceivers better to recognize one thing that is presented to distinct senses.

Sensitivity to crossmodal identity also improves action, making it easier to coordinate complex behaviour informed by distinct senses. With better perceptual information, one can act more fluently and more effectively. There is less to work out, less evidence to gather, and less need to sort out whether to trust one's ears or one's eyes. Being differentially sensitive to crossmodal identity therefore is not idle or superfluous in relation to cognition and action.

One may conclude that typical human subjects perceive token identity across senses. That is alright. However, this sets the bar low for perceiving crossmodal identity. It does not require crossmodal identification in perception, representing an item perceived with one sense as being the same as an item perceived using another sense, or awareness of sameness. The existence of a rationalizing account appealing to identity across senses does not imply representing or appreciating crossmodal identity. Thus, it is a mistake to model perceiving crossmodal identity on crossmodal identification in thought.

This approach recognizes that multisensory perception is sensitive to identity in what is presented to distinct senses. But the evidence does not establish that perception token identifies what is perceived using one sense with what is perceived using another sense. So this approach does not claim that crossmodal identification takes place in perception. In this respect, it remains a sceptical approach. This is a Goldilocks position. It appreciates the evidence, but it does not overintellectualize perception. My contention is that this does all the work a theorist needs. We should not ask more of perception.

9 Conclusion

This paper describes two opposing views about crossmodal identification. According to the first, crossmodal identification occurs only in thought or extraperceptual cognition. According to the second, crossmodal identification also occurs in perception. Strictly speaking, it endorses the first and denies the second.

In this picture, crossmodal identification involves appreciating identity across senses. It may consist in activating knowledge or warranted belief in identity. However, existing evidence does not establish that perception itself token identifies an individual perceived in one sensory modality with an individual perceived in another sensory modality.

Perception is not unalive to crossmodal identity. Empirical studies show multisensory perception is differentially sensitive to situations in which distinct senses are presented with a single individual. Studies of feature binding and object files pioneered in vision research have been adapted to demonstrate multisensory effects that parallel unisensory ones. Illusory conjunctions, object-specific preview effects, and attentional spreading have been observed in multisensory settings. Multisensory perception responds, and it responds in ways that are appropriate, when a single item is presented to distinct senses. It resolves conflicts, speeds responsiveness, and spreads attention. This, in turn, enhances one's capacity to recognize and to act on multisensory objects.

But binding is very permissive. Binding does not require a unified type of perceptual process, and it does not yield a single, uniform perceptual structure. The binding observed in multisensory contexts is compatible with a variety of relations in which perceptible features and objects may stand. Not each such relation requires a single object or feature bearer. Moreover, binding does not guarantee representing what is perceived using one sense as being identical to what is perceived using another sense. So, as operationalized in empirical studies of multisensory perception, binding can occur across senses even if crossmodal identification as such does not take place in perception.

In this approach, typical human perception is differentially sensitive to crossmodal identity. This is distinct from the capacity to token identify an individual across senses. But it is enough to serve the needs of recognition and action. Perception registers crossmodal identity, but crossmodal identification as such belongs to thought.

Acknowledgements

Many thanks to Austin Andrews, Kevin Connolly, E. J. Green, Jake Quilty-Dunn, Jeff Speaks, and an anonymous reviewer for generous written comments.

References

Bayne, T. (2014). The multisensory nature of perceptual consciousness. In *Sensory Integration and the Unity of Consciousness*, edited by D. Bennett and C. Hill, 15–36. Cambridge, MA, MIT Press.

Boghossian, P. (2014). What is inference? *Philosophical Studies*, 169, 1–18.
Burge, T. (2010). *Origins of Objectivity*. Oxford, Oxford University Press.
Busse, L., Roberts, K., Crist, R., Weissman, D., and Woldorff, M. (2005). The spread of attention across modalities and space in a multisensory object. *Proceedings of the National Academy of Sciences of the United States of America*, 102, 18751–6.
Campbell, J. (1997). Sense, reference, and selective attention. *Proceedings of the Aristotelian Society*, 71, 55–74.
Cinel, C., Humphreys, G., and Poli, R. (2002). Cross-modal illusory conjunctions between vision and touch. *Journal of Experimental Psychology: Human Perception and Performance*, 28, 1243–66.
Connolly, K. (2014). Making sense of multiple senses. In *Consciousness Inside and Out*, edited by R. Brown, 351–64. Dordrecht: Springer.
Fine, K. (2007). *Semantic Relationism*. Malden, MA: Blackwell.
Fulkerson, M. (2011). The unity of haptic touch. *Philosophical Psychology*, 24, 493–516.
Green, E. J. (2019). A theory of perceptual objects. *Philosophy and Phenomenological Research*, 99, 663–93.
Green, E. J. (2021). Binding and differentiation in multisensory object perception. *Synthese*, 198, 4457–91.
Green, E. J. and Quilty-Dunn, J. (2021). What is an object file? *British Journal for the Philosophy of Science*, 72, 665–99.
Jackson, F. (1977). *Perception: A Representative Theory*. Cambridge: Cambridge University Press.
Jordan, K., Clark, K., and Mitroff, S. (2010). See an object, hear an object file: object correspondence transcends sensory modality. *Visual Cognition*, 18, 492–503.
Kahneman, D., Treisman, A., and Gibbs, B. (1992). The reviewing of object files: object-specific integration of information. *Cognitive Psychology*, 24, 175–219.
O'Callaghan, C. (2014). Intermodal binding awareness. In *Sensory Integration and the Unity of Consciousness*, edited by D. Bennett and C. Hill, 73–103. Cambridge, MA, MIT Press.
O'Callaghan, C. (2016). Objects for multisensory perception. *Philosophical Studies*, 173, 1269–89.
Orlandi, N. (2014). *The Innocent Eye: Why Vision Is Not a Cognitive Process*. New York: Oxford University Press.
Ptak, R. (2008.). Saliency, Selection, and Spatial Neglect. Habilitation thesis: University of Geneva.
Recanati, F. (2012). *Mental Files*. Oxford: Oxford University Press.
Rescorla, M. (2020). Perceptual co-reference. *Review of Philosophy and Psychology*, 11, 569–89.
Richard, A., Lee, H., and Vecera, S. (2008). Attentional spreading in object-based attention. *Journal of Experimental Psychology: Human Perception and Performance*, 34, 842–53.
Shams, L. and Beierholm, U. (2010). Causal inference in perception. *Trends in Cognitive Sciences*, 14, 425–32.
Siegel, S. (2017). *The Rationality of Perception*. Oxford: Oxford University Press.
Spence, C. and Bayne, T. (2015). Is consciousness multisensory? In *Perception and Its Modalities*, edited by D. Stokes, M. Matthen, and S. Biggs, 95–132. New York: Oxford University Press.
Talsma, D., Senkowski, D., Soto-Faraco, S., and Woldorff, M. (2010). The multifaceted interplay between attention and multisensory integration. *Trends in Cognitive Sciences*, 14, 400–10.
Treisman, A. (2003). Consciousness and perceptual binding. In *The Unity of Consciousness*, edited by A. Cleeremans, 95–113. Oxford: Oxford University Press.
Treisman, A. and Schmidt, H. (1982). Illusory conjunctions in the perception of objects. *Cognitive Psychology*, 14: 107–41.
Williamson, T. (1990). *Identity and Discrimination*. Oxford: Basil Blackwell.
Zmigrod, S., Spapé, M., and Hommel, B. (2009). Intermodal event files: integrating features across vision, audition, taction, and action. *Psychological Research*, 73, 674–84.

20
The multisensory perception of persistence

E. J. Green

1 Introduction

We perceive objects as persisting through time. As I watch my cat dash across the room, it is visually apparent that the animal I see now is the same one I saw a moment ago, even if he passes briefly behind a barrier. While discussion of the perception of persistence has focused mostly on vision, the capacity generalizes to other modalities. If I pass a pencil between my hands, I experience the object now in my left hand as the same one I earlier felt in my right. Nonetheless, while we can perceive objects as persisting *within* more than one modality, this does not settle how, if at all, the modalities *interact* in the perception of persistence. The present chapter examines this issue.

While many philosophers have considered crossmodal interactions in object perception, the primary focus has been on multimodal *binding*—that is, whether perception binds features apprehended through separate modalities to a single individual (Deroy, 2014; de Vignemont, 2014; O'Callaghan, 2014). However, ordinary episodes of object perception recruit many capacities beyond binding. The perception of objects as persisting over time is one example. Multisensory contributions to the perception of persistence are ripe for investigation.

This chapter explores two questions about the multisensory perception of persistence.[1] The first concerns perceptual processing. I ask whether perceptual computations of persistence are *multimodally penetrable*—roughly, whether they ever receive representations from multiple modalities as input. I argue that research on the audiovisual bounce effect makes a compelling case for multimodal penetration.

The second question concerns perceptual experience. I ask whether experiences of persistence ever exhibit *constitutively multisensory* phenomenal character—roughly, whether the phenomenal character associated with perceiving an object as persisting over time ever outstrips that which is uniquely associated with just one modality or another. The clearest examples would involve *multisensory diachronic grouping*—experiencing an object as persisting over time, even though it is alternately perceived through separate modalities. While there is suggestive evidence of this, the case remains inconclusive. I consider ways of resolving the issue.

[1] Caveats: I will limit my discussion to vision, audition, and touch, omitting the chemical senses. Nor will I address the apparent persistence of dynamic, temporally extended auditory streams like melodies or speech (but see Green, 2019a). The latter cases may differ from the perception of ordinary material objects in notable respects. According to O'Callaghan (2016) (although see Skrzypulec, 2020), sound streams seem to *perdure*: they seem to be composed of a series of temporal parts. Conversely, material objects seem to *endure*: my cat strikes me as wholly present at each moment I see him.

E. J. Green, *The multisensory perception of persistence* In: *Sensory Individuals.* Edited by: Aleksandra Mroczko-Wąsowicz and Rick Grush, Oxford University Press. © Oxford University Press 2023. DOI: 10.1093/oso/9780198866305.003.0021

2 Crossmodal processing of persistence

Certain perceptual processes receive inputs from multiple modalities. Consider multisensory cue combination. When two modalities produce conflicting estimates of a variable—for example, size or shape—our sensory systems often integrate these representations to form a single combined estimate of that variable. Typically, the output estimate is a reliability-weighted average of the unimodal estimates (van Dam *et al.*, 2014).

The processes responsible for cue combination receive distinct visual and, say, haptic representations of a variable as input, and produce a combined estimate of that variable as output. When a process thus receives representations from multiple modalities as input, let us say it is *multimodally penetrable.*

A sensory process might be affected by processing in multiple modalities without being multimodally penetrable. Suppose an auditory representation of sound location is affected by the visual representation of the location of its source. And suppose that another process computes audible motion based on that auditory representation of sound location, but receives no further information as input. Then the motion computation is affected by both vision and audition, but is not multimodally penetrable, since its inputs would consist of auditory representations alone.

The multimodal penetration issue mirrors familiar discussions of the cognitive penetration of perception (Green, 2020; Macpherson, 2012; Pylyshyn, 1999). Several have argued that directness is a necessary condition for cognitive penetration. There is, on this view, an important distinction between the claim that a perceptual process is merely affected by cognition and the claim that a perceptual process receives cognitive representations as input and computes over them (Gross, 2017; Quilty-Dunn, 2020). In the latter case, perceptual processing is, in an important sense, *continuous* with cognition (Pylyshyn, 1999). A similar distinction applies to crossmodal effects. Some processes are merely affected by processing in multiple modalities; others receive representations from multiple modalities as input.

This section considers whether perceptual processes that compute object persistence are multimodally penetrable. I will focus on two cases that have received extensive investigation: crossmodal influences on apparent motion, and the audiovisual bounce effect.

2.1 Crossmodal influences on apparent motion

Suppose you see two objects flashed in alternation on opposite sides of a computer screen. With the proper timing and spacing, you will experience a single object moving back and forth. The problem of determining persistence through apparent motion is called the *correspondence problem*. When objects are perceptible at times T1 and T2, perception needs to determine correspondences (persistence relations) among them.

The direction of apparent motion in one modality can *capture* the direction of apparent motion in another. When subjects are required to judge the direction of an apparent motion stimulus in one modality, they are less accurate if an irrelevant stimulus is presented to another modality moving in the opposite direction (Soto-Faraco *et al.*, 2004). Here I will focus on a case where auditory motion influences ambiguous visual motion.

Alink *et al.* (2012) showed participants a stimulus where several columns of dots were visible through an aperture. This was followed by a second frame in which the columns shifted some distance rightward. When the rightward shift was exactly half the distance between columns, subjects were equally likely to see the columns moving rightward or leftward. This display was presented alongside an auditory stimulus (four sounds in succession) that produced an impression of either rightward or leftward auditory motion. Critically, auditory apparent motion influenced reports of visual apparent motion. Ambiguous visual stimuli were more often reported as moving in the same direction as the auditory stimulus.

The perception of motion does not imply the perception of persistence.[2] However, supposing in this case that either the columns or the dots within them were seen as persisting between apparent motion frames, then audition affected the apparent persistence of visually perceived objects.

I turn to another case where audition influences visual apparent motion.

Suppose you see two objects flashed in alternation, one on the right and one on the left. If the temporal spacing between flashes is constant, you typically experience a single object hopping back and forth. However, if spacing is uneven—if, say, the time from left to right flash is shorter than the time from right to left flash—the percept shifts. Instead, you typically see an object moving from left to right and then vanishing, followed by a new object appearing on the left and moving in the same direction (von Gruneau, 1986).

Freeman and Driver (2008) employed an apparent motion stimulus with equal temporal spacing between flashes. However, they played sounds that either slightly led or slightly lagged some of the flashes. For example, one sound could be played just after the left flash, and another just before the right flash. Because the perceived timing of sounds attracts the perceived timing of visible events (Morein-Zamir *et al.*, 2003), Freeman and Driver predicted that this display would induce a unidirectional percept of objects moving from one side to the other and vanishing. This prediction was confirmed through both subjective reports and adaptation after-effects. Thus, correspondence solutions for visually perceived stimuli were causally influenced by audition. In the presence of auditory input, subjects perceived an object *ceasing to persist*. Without auditory input, they would have perceived it as *continuing to persist*. Again, we have a crossmodal effect on perceived persistence.

Both of these cases exhibit causal influences of processing in one modality on persistence computations for stimuli apprehended through another modality. But do they show that these persistence computations are multimodally penetrable? I suggest not.

Consider the Freeman and Driver (2008) study. In this case, audition affects perceived persistence in visual apparent motion, but it plausibly does so *by way* of affecting the visual representation of temporal properties. Thus, the correspondence computation might have received only visual representations as input. It is just that some of these representations (representations of temporal features) were modified by audition.[3]

What about crossmodal capture? I suggest that this also need not involve multimodal penetration of persistence computations. It could be mediated by crossmodal influences on the allocation of attention, which in turn affects the perception of persistence.

[2] Perceived motion *typically* involves perceived persistence (Paul, 2010; Scholl, 2007), but not always. Motion energy signals can induce an impression of movement without any perceptually differentiable object appearing to move (Cavanagh, 1992).

[3] Another possibility is that audition affects not the perceived *timing* of visual stimuli, but their apparent *perceptual grouping* (Roseboom *et al.*, 2013).

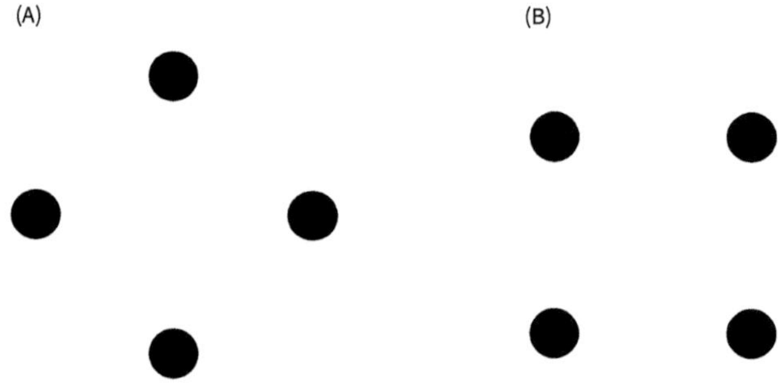

Figure 20.1 Two frames of an apparent motion sequence. The dots may be seen as moving either clockwise or counterclockwise.

Suppose you see four dots forming the vertices of an imaginary tilted square (Figure 20.1A). In the next frame, the dots have shifted, forming an imaginary square resting on its base (Figure 20.1B). If you see these frames in alternation, you can experience the dots moving either clockwise or counterclockwise. In displays of this sort, we have some control over the perceived direction of movement (Verstraten *et al.*, 2000). A leading explanation is that the computations underlying apparent motion are sensitive to shifts in attention (Cavanagh, 1992; Xu *et al.*, 2013). On the 'attentional pointer' hypothesis (Cavanagh *et al.*, 2010), for example, the perception of persistence in high-level, object-based apparent motion sometimes results from shifting spatial attention. When you shift attention from one object's location in frame 1 to another object's location in frame 2, this '[links] the two locations together as the changing location of a single target' (Cavanagh *et al.*, 2010. p. 151). By shifting attention in one direction vs another, you can influence the perceived direction of motion in an ambiguous stimulus.

Such attentional effects could be involved in crossmodal capture. There is independent evidence that the allocation of either endogenous (Driver and Spence, 2004) or exogenous (Spence and Driver, 1997) attention in one modality biases the allocation of attention in other modalities. Suppose that in Alink *et al.*'s (2012) ambiguous motion display, perceived motion direction was biased by subjects' sequential allocation of visual attention. If visual attention was first directed towards locations or dots on the left, and then to locations or dots on the right, this biased the system towards left–right motion. Suppose, further, that the concurrent unambiguous auditory motion stimulus attracted attention sequentially to the locations through which the sound appeared to move. If the direction of *auditory* attention biased the direction of *visual* attention, it may have also biased the perception of visual apparent motion, and thus visual persistence.[4]

Consistent with the hypothesis that crossmodal capture effects go via attention, attended motion streams in one modality are more effective in capturing apparent motion in another modality (Oruc *et al.*, 2008). There is also evidence that louder, more attention-grabbing auditory stimuli exert stronger capture effects on tactile motion (Occelli *et al.*, 2009). Finally, there is evidence that attention-grabbing static sounds can bias the perception of an

[4] This story is even simpler if there is a single multimodal system of spatial attention, since there would be no need for a mediating influence of auditory attention on visual attention.

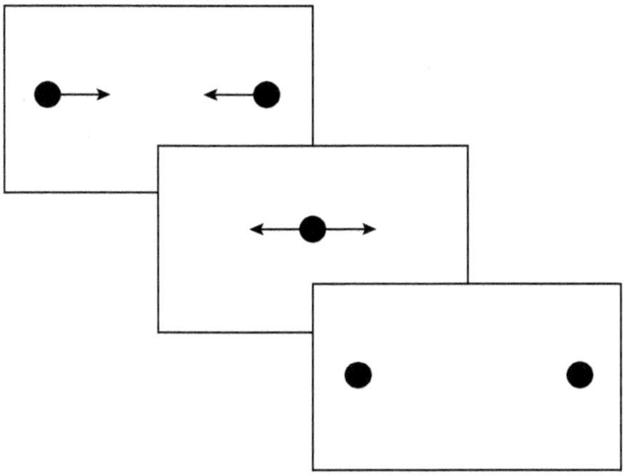

Figure 20.2 Bouncing/streaming display.

ambiguous visual apparent motion stimulus, even without any auditory motion (McBeath *et al.*, 2019).[5]

If crossmodal capture effects are mediated by attention, this raises the possibility that crossmodal capture does not involve multimodal penetration. Instead, persistence computations within a modality might receive *attention-weighted unimodal representations* as input. Thus, the computation of persistence in visual apparent motion might receive purely visual representations of objects, properties, and locations, alongside information about the amount of attention allocated to each. Processing in other modalities could influence the allocation of attention, and thus modify these inputs to visual persistence computations, but auditory and haptic representations would not themselves constitute inputs to those computations.

Thus, while extra-modal factors *affect* perceived persistence through apparent motion within a modality, it is unclear whether the relevant persistence computations literally integrate and compute over representations from multiple modalities. I note that there is a striking analogy here with the literature on cognitive penetration, where it is often disputed whether purported cognitive effects on perception are direct or rather mediated by attention.

2.2 The audiovisual bounce effect

I turn to another candidate case of multimodal penetration. Suppose two objects start on opposite sides of a computer screen and approach each other until they overlap. Afterward, two objects emerge following the original motion trajectories (Figure 20.2). Two percepts are possible: the objects can appear either to *stream past* each other or to *bounce off* each other. Typically, streaming percepts dominate. However, if a sound is played at the moment of overlap, the pattern reverses, with bouncing perceived on nearly 80% of trials (Sekuler *et al.*, 1997).

[5] McBeath *et al.* created a visual stimulus that could be seen as either starting on the left and moving rightward or starting on the right and moving leftward. They found that static auditory cues presented on the left promoted percepts of rightward visual motion originating on the left, and vice versa for auditory cues presented on the right.

The objects at the beginning of the bouncing/streaming display can be seen as persisting in two incompatible ways, and audition helps settle which interpretation dominates. But is this a case where persistence computations are multimodally penetrated? It depends on what the computations receive as input. We can distinguish two models.

First, it is possible that the audiovisual bounce effect is mediated by auditory influences on visual attention. Perhaps the sound attracts attention away from the motion event, causing the visual system to 'miss' the moment of complete overlap between the objects. Accordingly, the visual system non-veridically represents what Scholl and Nakayama (2004) call an 'illusory crescent'—a slice of one object that is never fully covered by the other. The attentional diversion reduces the perceived overlap between the objects, and in the absence of full overlap, bouncing becomes a more likely interpretation than streaming. Meyerhoff and Scholl (2018) found that perceivers are indeed more likely to perceive illusory crescents in the bouncing/streaming display on sound-present trials than on sound-absent trials. They suggest that attentional diversion causes illusory crescents, which promote bouncing percepts (2018, pp. 92–3).[6] Call this the *attentional model*. This model is consistent with (but does not entail) the absence of multimodal penetration. Perhaps the inputs to visual persistence computations are exhausted by visual representations of motion trajectories and spatial relations. By diverting attention, audition modifies the visual representation of spatial relations (viz. whether a crescent is represented), and so modifies those inputs.

Second, it is possible that the inputs to persistence computations include representations of relations between events apprehended through separate modalities. When a sound occurs at the moment of overlap, perhaps the perceptual system determines that the visible event *caused* the sound.[7] This information is passed to the persistence computation, which determines that the objects bounced rather than streamed. After all, bouncing normally causes sound, while passing normally does not. Call this the *causal model*. On this model, persistence computations receive both visual and auditory representations as input. The representation of the causal relation delivered to the computation contains representations from both modalities as constituents (*visible event V caused audible event A*).[8]

Some evidence appears to support the attentional model. Watanabe and Shimojo (1998) found that bounce percepts are promoted not only by sounds, but also by visible flashes, which would also be expected to distract attention from the bouncing/streaming event. Further studies suggest that other abrupt perceptible events—also known as 'transients'—also facilitate bouncing. A brief tactile pulse promotes bouncing and also induces illusory crescents (Meyerhoff *et al.*, 2018; Watanabe and Shimojo, 2005). Meyerhoff and Suzuki (2018) found that sudden *offsets* of sound have a similar result. Meyerhoff *et al.* (2018, p. 2236) suggest that the attentional model might explain the generality of these effects: '[I]t is possible that any coinciding transient distracts attention from the bouncing/streaming display. [This] might result in missing the central frame which in turn induces illusory crescents as well as bouncing impressions.' Such evidence seems to favour the attentional model over the causal model. Collisions normally produce sounds but rarely produce visible flashes, tactile vibrations, or

[6] One concern with the attentional model is that it is unclear why diverting processing resources away from the overlapping objects would lead the system to infer an illusory crescent, instead of simply representing the objects as continuing their original trajectories. (Thanks to Casey O'Callaghan for this point.)

[7] The proposal that perceptual systems represent causal relations raises many interesting issues, such as whether sensory cues to causation are innate, learnt, or both. Unfortunately, I cannot address these issues here.

[8] Alternatively, one might hold that some non-causal relation between visual and auditory events is represented and drives the bounce effect. This would also be a form of multimodal penetration.

sound offsets. It is unclear why the perceptual system would opt for a causal interpretation in the latter cases.

However, other evidence favours the causal model. First, not all sounds promote bouncing with equal effectiveness. Real-world impact sounds possess a characteristic amplitude profile: abrupt 'attack' followed by gradual decay. Grassi and Casco (2009) generated artificial sounds that could be either impact-consistent or impact-inconsistent in this respect. One sound had a loud onset and decayed gradually, while the other had a gradual onset and abrupt decay. Critically, the impact-consistent sounds generated more bounce percepts than impact-inconsistent sounds. Furthermore, Grassi and Casco (2010) found that billiard ball collision sounds generate a stronger bounce effect than water drops or fireworks, even though all three sounds are equally attention-grabbing. Together, these findings suggest that the processes responsible for distinguishing bouncing from streaming *are* sensitive to whether the transient is a viable effect of the motion event.

Another study provides particularly strong evidence against a pure attentional model. Adams and Grove (2018) reasoned that if transients induce the bounce effect simply because they distract attention, then their effect should be stronger when they occur farther from the moving objects, since this diverts attention farther from the overlap event. However, in the case of visible flashes, precisely the *opposite* occurred. Visible flashes produced more bounce responses when they occurred at the same location as the objects' overlap than when they occurred on the opposite side of the screen. This result conflicts with the attentional model's predictions but comports well with the causal model. Perhaps the perceptual system is wired to treat abrupt transients at or near the location of the overlap as indicating a causal transaction has occurred.

So there is evidence supporting both the attentional and causal models. What could break the impasse? I suggest that both models capture aspects of the perceptual computation of persistence in bouncing/streaming displays.

Recall that Grassi and Casco (2009) found that impact-consistent sounds generate more bounce percepts than impact-inconsistent sounds. However, they also discovered an important distinction *within* the class of impact-inconsistent sounds. In one experiment, they examined impact-inconsistent sounds that had the same average intensity as impact-consistent sounds: the latter sounds began at 87 dB and softened to 47 dB, while the former sounds did the opposite. In this case, impact-inconsistent sounds produced *no* bounce effect—bouncing percepts were no more frequent than in the silent display. However, an interpretive difficulty here is that when a sound is softer at onset, it may also be less attention-grabbing. Thus, a proponent of the attentional model could reply that impact-inconsistent sounds were simply less distracting. To address this concern, Grassi and Casco ran another experiment where impact-inconsistent sounds had the same intensity as impact-consistent sounds at onset, but gradually got louder rather than decaying. In this case, the impact-inconsistent sound *did* induce a bounce effect (nearly 50% bounce responses vs 20% in the silent display), but the effect was nonetheless weaker than that produced by impact-consistent sounds (nearly 80% bouncing).

A hybrid model could explain these results. When sounds have a louder onset, they are more attention-grabbing (Grassi and Casco 2009, exp. 3). Consistent with the attentional model, this attentional distraction promotes illusory crescents, which bias the system towards bouncing. Beyond this, however, the persistence computation also consults information about whether the sound is a credible effect of the motion event. When the motion event is represented as causing the sound, this information is sent as input to the persistence computation, which is then even more likely to opt for bouncing. On this view, attentional and causal factors are mutually reinforcing. Similar remarks apply to the Adams and Grove (2018) study. Although visible flashes near the overlapping objects resulted in a stronger bounce effect, the effect was not eliminated with distant flashes. The hybrid model could claim that while distant flashes were less likely to be represented as caused by the visible motion event, they effectively diverted attention and induced illusory crescents. Finally, the hybrid model can explain why abrupt transients with no obvious causal connection to the motion event (e.g., brief silences) also promote bouncing: they still draw attention.

If this analysis is correct, then the answer to our first question—whether persistence computations are multimodally penetrable—is yes. At a computational level, the perception of persistence is sometimes a multimodal operation. In determining how objects apprehended through one modality persist through time, the perceptual system sometimes consults representations from other modalities. Such representations constitute inputs to persistence computations.

3 Multisensory experience of persistence

So far I have discussed crossmodal interactions in the subpersonal computation of persistence. This section turns to the question of how crossmodal interactions might reshape our perceptual experience of persistence. I ask whether such experiences are ever *constitutively multisensory*.

3.1 Preliminaries

Some perceptual capacities are multisensory at the level of perceptual processing, but not at the level of experience. In the McGurk effect, our experience of an uttered syllable arises from computations that combine visual information about mouth movement with auditory information about acoustic features, but arguably the experience of the syllable is purely auditory (McGurk and MacDonald, 1976). Likewise, although persistence computations sometimes integrate information from multiple modalities, the experiences that result from these interactions might not be multisensory in any deep respect. In the audiovisual bounce effect, the experience of the objects bouncing may be purely visual despite issuing from crossmodal interactions.

The present question is whether perceptual experiences of persistence are ever *constitutively multisensory*. To say that an experience is constitutively multisensory is, roughly, to say that its phenomenal character cannot be fully factored into components, each of which is uniquely associated with just one modality. However, applying this principle presupposes some conception of when an aspect of phenomenal character is 'associated' with one modality vs another. Clarifying this idea is notoriously difficult (Macpherson, 2011). In recent years, Casey O'Callaghan (2015, 2019) has offered the most refined approach to the issue.

On O'Callaghan's account, an aspect of phenomenal character is associated with a modality when it could be instantiated by a *corresponding mere experience of that modality*. A *mere* experience of modality A is an experience that belongs to A, and not to any other modality. Thus: '[A] merely visual experience is visual but not auditory, tactual, olfactory, or gustatory. To get a fix on this, consider the other sense organs as blocked or anesthetized' (O'Callaghan, 2015). So let us say a merely visual experience is an experience of the sort that is normally produced, absent sensory malfunctions, when other sense organs receive no stimulation. A *corresponding* mere experience of a modality on an occasion is 'a perceptual experience merely of that modality under equivalent stimulation' (2015). Suppose that a subject actually undergoes a perceptual experience owing to stimulation S1 of her retina and S2 of her cochlea. Then a corresponding merely visual experience is one that the subject could have undergone on that occasion had she received only S1, and not S2. Call this conception of constitutively multisensory experience the *unisensory correspondence model.*

One concern should be flagged. O'Callaghan proposes that a corresponding mere experience of a modality is one that the subject could have had under *equivalent* stimulation. However, Wadle (2020) argues that this rule yields implausible consequences. Our experiences of uttered syllables during the McGurk effect are plausibly purely auditory, not multisensory, even though they arise from processes that integrate visual and auditory information. However, the unisensory correspondence model appears to deem such experiences constitutively multisensory. Consider someone who receives auditory input consistent with /ba/ and visual input consistent with /ga/, but experiences /da/ instead. If they had received only the equivalent auditory input, they would have experienced /ba/, not /da/. Thus, the model implies that the actual phenomenal character of experiencing /da/ is constitutively multisensory. Thus, the unisensory correspondence model has trouble distinguishing constitutively multisensory experiences from unisensory experiences that are only producible through subpersonal crossmodal interaction.

One response to this problem is to relax the requirement that corresponding unimodal experiences be producible under *equivalent* stimulation. Perhaps they need only be producible under some *appropriately similar* stimulation. Even if the McGurk subject could not have experienced /da/, given only the acoustic input she, in fact, received, there is a related acoustic stimulus that could have produced an experience with that phenomenal character without concurrent inputs to other sense organs. It is unclear just what it means for two conditions of stimulation to be 'appropriately similar', and certain construals raise further puzzles (Wadle, 2020, pp. 14–15). However, as I will explain below, my application of the unisensory correspondence model does not hinge on how exactly we construe the notion of appropriately similar stimulation, so we need not settle the issue here.[9]

3.2 Multisensory diachronic grouping

We can distinguish the phenomenal character associated with an object's apparent *persistence* from that associated with its other properties. Suppose you see a red ball pass behind a barrier

[9] For those doubtful that any version of the unisensory-correspondence model can succeed, a weaker principle would suffice for my purposes. Specifically: if someone perceptually experiences a relation between two objects, or two time slices of an object, that are wholly apprehended through separate modalities, then the phenomenal character associated with experiencing this relation is constitutively multisensory. Suppose you experience a visible event as *simultaneous* with an audible event. Then the experience of simultaneity is neither uniquely visual nor uniquely auditory, but constitutively audiovisual (O'Callaghan, 2017, pp. 166–7). Likewise, if you perceive an object as persisting from T1 to T2, but it is perceived wholly through vision at T1 and wholly through audition at T2, then the experience of persistence is neither uniquely visual nor uniquely auditory, but constitutively audiovisual.

and then emerge. You have a strong impression that the emerging object is the *same one* that passed behind the barrier. This impression of persistence plausibly possesses a distinctive phenomenal character separate from that associated with, say, the ball's apparent colour and shape. Studies suggest that by slightly increasing the interval before an object emerges, we can remove the impression of persistence while leaving the phenomenology of other properties unchanged (Flombaum and Scholl, 2006).

My present interest concerns whether this distinctive phenomenology of persistence is ever constitutively multisensory. Given O'Callaghan's unisensory-correspondence criterion, the issue becomes more precise: consider the phenomenal character of persistence exhibited by an experience of an object moving through the scene. Could that very phenomenal character—abstracting away from the phenomenology associated with other features—be instantiated by a corresponding experience merely of a single modality? If it could not, then it is constitutively multisensory.

What could establish constitutively multisensory phenomenology of persistence? Let us say *multisensory diachronic grouping* occurs if an object is perceived wholly through one modality at time T1, then wholly through a separate modality at time T2, but is experienced as persisting from T1 to T2. For example, you might see a ball roll behind a barrier at T1, then feel the ball behind the barrier at T2 (without seeing it). If you experience the ball as persisting from T1 to T2, then your experience exhibits multisensory diachronic grouping. This would establish constitutively multisensory experience of persistence. No corresponding unimodal experience could share the phenomenal character associated with perceiving the ball as persisting from T1 to T2. Any experience produced under equivalent visual stimulation, but without haptic stimulation, would omit the object at T2, and vice versa for haptic stimulation without vision. This holds even if we relax the requirement of equivalent unimodal stimulation. No remotely similar pattern of visual stimulation could produce an experience of the object as persisting at T2.

Section 2.1 examined cases where apparent motion in one modality influences apparent motion in another modality. While these are not candidate instances of multisensory diachronic grouping, other crossmodal stimuli are—namely, displays where stimuli are alternately presented to different modalities and the subject is asked whether they perceive motion between them.

Consider visual–tactile apparent motion. Harrar *et al.* (2008) had subjects place their index fingers in a pair of cups that could emit brief taps to the skin. Both cups had LED displays mounted on top that emitted brief flashes. This set-up allowed comparison of visual–visual, tactile–tactile, and visual–tactile apparent motion sequences by presenting flashes and/or taps in rapid succession. Subjects rated the quality of apparent motion across various inter-stimulus distances and temporal intervals. Crucially, for inter-stimulus delays of 200–300 ms, the quality of visual–tactile apparent motion was equivalent to visual–visual and tactile–tactile motion.

There is debate about whether reports of crossmodal apparent motion are genuinely perceptual (O'Callaghan, 2017), or instead result from post-perceptual inference or compliance with task demands (Spence, 2015; Spence and Bayne, 2014). I believe the non-perceptual hypothesis is unlikely in the present case. Suppose that subjects reported experiencing visual–tactile motion simply because they believed that the purpose of the experiment was to document its existence. This account provides no obvious explanation of why the rated quality of visual–tactile motion was systematically governed by the temporal interval

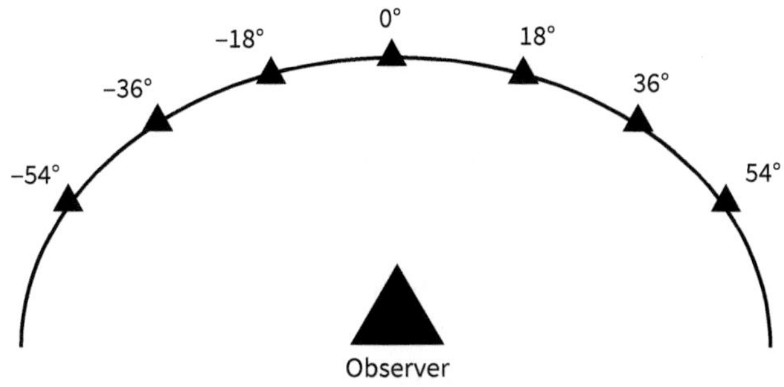

Figure 20.3 Design of the Kluss *et al.* (2012) study.

between stimuli. Further support for the perceptual account derives from evidence that crossmodal apparent motion has the propensity to capture unimodal auditory apparent motion (Jiang and Chen, 2013).

Sceptics about crossmodal apparent motion have observed that in another study, Huddleston *et al.* (2008) tested for the presence of audiovisual apparent motion and found negative results. Subjects were shown an array of two loudspeakers and two LED displays arranged in a circle. Lights and white noise bursts occurred in either clockwise and counterclockwise order. The authors found that while subjects could reliably distinguish the implied direction of audiovisual motion, they reported no genuine percepts of audiovisual motion. However, another study investigated audiovisual apparent motion, with more promising results.

Kluss *et al.* (2012) had subjects sit in a chair surrounded by loudspeakers arranged in a semicircle, with an LED attached to each speaker cone (Figure 20.3). They compared three conditions: a *unimodal–auditory* condition where, say, the speakers at –54°, –18°, 18°, and 54° emitted noise bursts in rapid succession; a *coherent–bimodal* condition, where the sounds were presented in the same order, but flashes were spatio-temporally interpolated between the sounds; and an *incoherent–bimodal* condition, where the sounds were presented in the same order, but flashes were temporally interpolated between them at random locations. Relative to the other conditions, subjects reported perceiving continuous motion at longer temporal intervals between sounds in the coherent–bimodal condition. It is as though flashes at the appropriate spatial locations helped 'fill the gaps' between sounds, promoting a percept of seamless motion. The authors take this to support an 'amodal interpolation subsystem accepting unimodal activity from both the auditory and the visual domain' (2012, p. 64; cp. Stiles *et al.*, 2018).

Suppose, then, that studies of crossmodal apparent motion are unearthing a real perceptual phenomenon. Do they establish constitutively multisensory experience of persistence? I argue that, at present, they do not. While crossmodal apparent motion involves constitutively multisensory experiences of *some* kind, it is unclear whether these are experiences of persistence, or of some other relation.

Consider Harrar *et al.*'s (2008) study of visual--tactile apparent motion. I argued that the systematic dependence of quality ratings on inter-stimulus interval suggests that subjects were reporting their perceptual experiences at face value. At longer intervals, the subjects experienced the stimuli as standing in some relation—a relation they did not experience at shorter intervals. Consider the phenomenology associated with perceiving this relation. On the unisensory-correspondence model, this phenomenology is constitutively multisensory if it is not shared by any corresponding experience merely of a single modality. Plausibly this is true. Any merely visual experience would omit the tap, while any merely tactual experience would omit the flash. And if one of the relata is not experienced at all, then, plausibly, neither is the relation between them. Thus, experiences of crossmodal apparent motion have constitutively multisensory phenomenology.

But do they have constitutively multisensory phenomenology of *persistence*? When subjects see a flash and then feel a tap, do they experience the tap as a *continuation* of the flash? While this may seem like a reasonable interpretation, subjects' self-reports in the Harrar *et al.*'s (2008) study suggest otherwise. Thus (p. 810):

> [S]ubjects in the visuotactile condition reported perceiving some type of multimodal apparent motion, but they often described it as being 'more causal' than the unimodal apparent motion. Our participants mainly interpreted their perception like a switch flicking on a light or like a cannon firing that was felt on one hand and then the flash from the landing explosive was seen on the other hand.

As I have argued elsewhere (Green, 2019b), such reports raise the possibility that in crossmodal apparent motion, subjects do not experience persistence at all, but rather a causal relation between distinct objects or events. Compare Michotte-style causal perception displays (Kominsky *et al.*, 2017), wherein one disc approaches another and contacts it, after which the second disc begins moving in the same direction. Here subjects report a strong impression that the first disc *launched* the second. But while they arguably experience the discs as causally related, they do not experience them as stages of a single persisting object. Crossmodal apparent motion may fit the same mould.[10]

How could we settle whether crossmodal apparent motion involves experience of persistence? One option would be to apply paradigms thought to reveal the perception of persistence in unimodal cases. One such paradigm is the *object-reviewing* task (Kahneman *et al.*, 1992): subjects perceive a pair of objects, and features briefly appear on the objects before vanishing. Then the objects move to new locations, and a feature appears on one of them. Subjects' task is to report whether the feature matches either of those encountered earlier. Typically, responses are faster when there is a match, but faster still when the feature appears in the object in which it initially appeared—an *object-specific preview benefit* (OSPB). The standard explanation of the OSPB is that the perceptual system maintains object representations—*object files*—over time and stores information about an object's currently and recently perceived features in its file (Green and Quilty-Dunn, 2021). When an object's feature matches information already in its file, responses to the feature are speeded. To perceptually represent an object as persisting from time T1 to T2 just is to maintain a file that sustains reference to the object from T1 to T2.

Suppose this story is correct. Then a natural idea would be to examine object-reviewing in a crossmodal context. Suppose an object is first seen, then moves behind an occluder where it can only be touched. And suppose that the features to be compared are perceptible both

[10] This story could also apply to audiovisual apparent motion in Kluss *et al.* (2012). Rather than experiencing a single moving object, perhaps subjects experienced an uninterrupted causal chain linking the lights and sounds, and simply reported this as continuous motion for lack of a better description.

visually and haptically (e.g., a simple shape or oriented line segment). If an OSPB emerges in reidentifying these features, then the perceptual system must have deemed the object to persist from one time to the next despite being perceived wholly through separate modalities.

Notably, there is evidence that OSPBs are not modality-specific. Jordan *et al.* (2010) found that when subjects saw a picture of a telephone appear within an object, there was an OSPB for reidentifying the feature by its audible ring (see also Zmigrod *et al.*, 2009). However, while Jordan *et al.*'s experiment provides evidence for multisensory *binding*, it does not directly bear on the question of multisensory *persistence*, since the objects (square-shaped wireframes) to which the features were bound could have been selected and reidentified through vision alone. To address the persistence issue, we need a case where the *objects*, and not merely the features bound to them, are picked out and then reidentified using separate modalities.

Unfortunately, there are problems with using object-file maintenance as a guide to the experience of persistence. Some evidence has been taken to show that object files sometimes diverge from the experience of persisting objecthood (Mitroff *et al.*, 2005). In general, however, the perceptual experience of persistence is at least reliably yoked to the maintenance of object files. Odic *et al.* (2012) examined a variety of subtly different apparent motion displays biased towards different correspondence solutions and found that object-file maintenance, as indexed by the OSPB, did systematically track the conscious experience of persistence in apparent motion. Divergence between object files and perceptually experienced persistence is probably the exception, not the rule.

I suggest that while studies of object-file maintenance do not offer an *experimentum crucis* regarding whether crossmodal apparent motion involves the experience of persistence, they offer a key source of evidence. Objects in crossmodal apparent motion are experienced as 'linked' in some way—the question is what sort of link they are experienced as exhibiting. So suppose an OSPB is produced during crossmodal apparent motion, indicating that an object-file is maintained throughout. Then we would have evidence that the perceptual system is representing persistence between objects apprehended through separate modalities. A prima facie plausible hypothesis would be that the associated perceptual experience represents persistence as well. If so, that experience would be constitutively multisensory.

Summing up: it remains an open question whether we enjoy constitutively multisensory experiences of persistence. This section has described the type of phenomenon (multisensory diachronic grouping) that would establish this convincingly. However, while crossmodal apparent motion supplies suggestive evidence for multisensory diachronic grouping, alternative interpretations remain viable. The issue is not whether crossmodal apparent motion is *perceptual*, but whether it involves the perception of *persistence*.

4 Conclusion

This chapter has explored multisensory interactions in the perception of persistence. This issue can be framed at the level of either perceptual processing or perceptual experience. I have argued that research on the audiovisual bounce effect offers compelling evidence that certain perceptual persistence computations are directly penetrated by representations from more than one modality. I then considered the issue of whether perceptual experiences of persistence are ever constitutively multisensory. I argued that the available evidence does not settle this question, but that it is not beyond empirical resolution.

Acknowledgements

Thanks to Tyler Brooke-Wilson, Alex Byrne, Casey O'Callaghan, and an anonymous reviewer for helpful comments that greatly improved this chapter.

References

Adams, K. L. and Grove, P. M. (2018). The effect of transient location on the resolution of bistable visual and audiovisual motion sequences. *Perception*, 47(9), 927–42.

Alink, A., Euler, F., Galeano, E., Krugliak, A., Singer, W., and Kohler, A. (2012). Auditory motion capturing ambiguous visual motion. *Frontiers in Psychology*, 2(391), 1–8.

Cavanagh, P. (1992). Attention-based motion perception. *Science*, 257(5076), 1563–5.

Cavanagh, P., Hunt, A. R., Afraz, A., and Rolfs, M. (2010). Visual stability based on remapping of attention pointers. *Trends in Cognitive Sciences*, 14(4), 147–53.

Deroy, O. (2014). The unity assumption and the many unities of consciousness. In *Sensory Integration and the Unity of Consciousness*, edited by D. J. Bennett and C. Hill, 105–24. Cambridge, MA: MIT Press.

de Vignemont, F. (2014). Multimodal unity and multimodal binding. In *Sensory Integration and the Unity of Consciousness*, edited by D. J. Bennett and C. Hill, 125–50. Cambridge, MA: MIT Press.

Driver, J. and Spence, C. (2004). Crossmodal spatial attention: evidence from human performance. In *Crossmodal Space and Crossmodal Attention*, edited by C. Spence and J. Driver, 179–220. Oxford: Oxford University Press.

Flombaum, J. I. and Scholl, B. J. (2006). A temporal same-object advantage in the tunnel effect: facilitated change detection for persisting objects. *Journal of Experimental Psychology: Human Perception and Performance*, 32(4), 840–53.

Freeman, E. and Driver, J. (2008). Direction of visual apparent motion driven solely by timing of a static sound. *Current Biology*, 18(16), 1262–6.

Fulkerson, M. (2014). Rethinking the senses and their interactions: the case for sensory pluralism. *Frontiers in Psychology*, 5, 1426.

Grassi, M. and Casco, C. (2009). Audiovisual bounce-inducing effect: attention alone does not explain why the discs are bouncing. *Journal of Experimental Psychology: Human Perception and Performance*, 35(1), 235–43.

Grassi, M. and Casco, C. (2010). Audiovisual bounce-inducing effect: when sound congruence affects grouping in vision. *Attention, Perception, and Psychophysics*, 72(2), 378–86.

Green, E. J. (2019a) A theory of perceptual objects. *Philosophy and Phenomenological Research*, 99(3), 663–93.

Green, E. J. (2019b). Binding and differentiation in multisensory object perception. *Synthese*, doi: https://doi.org/10.1007/s11229-019-02351-1

Green, E. J. (2020). The perception–cognition border: a case for architectural division. *Philosophical Review*, 129(3), 323–93.

Green, E. J. and Quilty-Dunn, J. (2021) What is an object file? *British Journal for the Philosophy of Science*, 72(3), 665–99.

Gross, S. (2017). Cognitive penetration and attention. *Frontiers in Psychology*, 8, 1–12.

Harrar, V., Winter, R., and Harris, L. R. (2008). Visuotactile apparent motion. *Perception and Psychophysics*, 70(5), 807–17.

Huddleston, W. E., Lewis, J. W., Phinney, R. E., and DeYoe, E. A. (2008). Auditory and visual attention-based apparent motion share functional parallels. *Perception and Psychophysics*, 70(7), 1207–16.

Jiang, Y. and Chen, L. (2013). Mutual influences of intermodal visual/tactile apparent motion and auditory motion with uncrossed and crossed arms. *Multisensory Research*, 26(1–2), 19–51.

Jordan, K. E., Clark, K., and Mitroff, S. R. (2010). See an object, hear an object file: object correspondence transcends sensory modality. *Visual Cognition*, 18(4), 492–503.

Kahneman, D., Treisman, A., and Gibbs, B. J. (1992). The reviewing of object files: object-specific integration of information. *Cognitive Psychology*, 24, 175–219.

Kluss, T., Schult, N., Schill, K., Fahle, M., and Zetzsche, C. (2012). Investigating the in-between: multisensory integration of auditory and visual motion streams. *Seeing and Perceiving*, 25, 45–69.

Kominsky, J. F., Strickland, B., Wertz, A. E., Elsner, C., Wynn, K., and Keil, F. C. (2017). Categories and constraints in causal perception. *Psychological Science*, 28(11), 1649–62.

Macpherson, F. (2011). Cross-modal experiences. *Proceedings of the Aristotelian Society*, 111(3), 429–68.

Macpherson, F. (2012). Cognitive penetration of colour experience: rethinking the issue in light of an indirect mechanism. *Philosophy and Phenomenological Research*, 84(1), 24–62.

McBeath, M. K., Addie, J. D. and Krynen, R. C. (2019). Auditory capture of visual apparent motion, both laterally and looming. *Acta Psychologica*, 193, 105–12.

McGurk, H. and MacDonald, J. (1976). Hearing lips and seeing voices. *Nature*, 264(5588), 746–8.

Meyerhoff, H. S., Merz, S., and Frings, C. (2018). Tactile stimulation disambiguates the perception of visual motion paths. *Psychonomic Bulletin and Review*, 25(6), 2231–7.

Meyerhoff, H. S. and Scholl, B. J. (2018). Auditory-induced bouncing is a perceptual (rather than a cognitive) phenomenon: evidence from illusory crescents. *Cognition*, 170, 88–94.

Meyerhoff, H. S. and Suzuki, S. (2018). Beep, be-, or–ep: the impact of auditory transients on perceived bouncing/streaming. *Journal of Experimental Psychology: Human Perception and Performance*, 44(12), 1995–2004.

Mitroff, S. R., Scholl, B. J., and Wynn, K. (2005). The relationship between object files and conscious perception. *Cognition*, 96(1), 67–92.

Morein-Zamir, S., Soto-Faraco, S., and Kingstone, A. (2003). Auditory capture of vision: examining temporal ventriloquism. *Cognitive Brain Research*, 17(1), 154–63.

O'Callaghan, C. (2014). Intermodal binding awareness. In *Sensory Integration and the Unity of Consciousness*, edited by D. Bennett and C. Hill, 73–103. Cambridge, MA: MIT Press.

O'Callaghan, C. (2015). The multisensory character of perception. *Journal of Philosophy*, 112(10), 551–69.

O'Callaghan, C. (2016). Objects for multisensory perception. *Philosophical Studies*, *173*, pp. 1269–1289.

O'Callaghan, C. (2017). Grades of multisensory awareness. *Mind and Language*, 32(2), 55–81.

O'Callaghan, C. (2019). *A Multisensory Philosophy of Perception*. Oxford: Oxford University Press.

Occelli, V., Spence, C., and Zampini, M. (2009). The effect of sound intensity on the audiotactile crossmodal dynamic capture effect. *Experimental Brain Research*, 193(3), 409–19.

Odic, D., Roth, O., and Flombaum, J. I. (2012). The relationship between apparent motion and object files. *Visual Cognition*, 20(9), 1052–81.

Oruc, I., Sinnett, S., Bischof, W. F., Soto-Faraco, S., Lock, K., and Kingstone, A. (2008). The effect of attention on the illusory capture of motion in bimodal stimuli. *Brain Research*, 1242, 200–8.

Paul, L. A. (2010). Temporal experience. *Journal of Philosophy*, 107(7), 333–59.

Pylyshyn, Z. (1999). Is vision continuous with cognition? The case for cognitive impenetrability of visual perception. *Behavioral and Brain Sciences*, 22(3), 341–65.

Quilty-Dunn, J. (2020). Attention and encapsulation. *Mind and Language*, 35(3), 335–49.

Roseboom, W., Kawabe, T., and Nishida, S. Y. (2013). Direction of visual apparent motion driven by perceptual organization of cross-modal signals. *Journal of Vision*, 13(1):6.1–13.

Scholl, B. J. (2007). Object persistence in philosophy and psychology. *Mind and Language*, 22(5), 563–91.

Scholl, B. J. and Nakayama, K. (2004). Illusory causal crescents: misperceived spatial relations due to perceived causality. *Perception*, 33(4), 455–69.

Sekuler, R., Sekuler, A. B., and Lau, R. (1997). Sound alters visual motion perception. *Nature*, 385, 308.

Skrzypulec, B., 2020. Visual endurance and auditory perdurance. *Erkenntnis*, *85*, pp. 467–488.

Soto-Faraco, S., Spence, C., and Kingstone, A. (2004). Congruency effects between auditory and tactile motion: extending the phenomenon of cross-modal dynamic capture. *Cognitive, Affective, and Behavioral Neuroscience*, 4(2), 208–17.

Spence, C. (2015). Cross-modal perceptual organization. In *The Oxford Handbook of Perceptual Organization*, edited by J. Wagemans, 639–54. Oxford: Oxford University Press.

Spence, C. and Bayne, T. (2014). Is consciousness multisensory? In *Perception and Its Modalities*, edited by D. Stokes, M. Matthen, and S. Biggs, 95–132. Oxford: Oxford University Press.

Spence, C. and Driver, J. (1997). Audiovisual links in exogenous covert spatial orienting. *Perception and Psychophysics*, 59(1), 1–22.

Stiles, N. R., Li, M., Levitan, C. A., Kamitani, Y., and Shimojo, S. (2018). What you saw is what you will hear: two new illusions with audiovisual postdictive effects. *PLoS One*, 13(10), e0204217.

van Dam, L. C. J., Parise, C. V., and Ernst, M. O. (2014). Modeling multisensory integration. In *Sensory Integration and the Unity of Consciousness*, edited by D. J. Bennett and C. Hill, 209–29. Cambridge, MA: MIT Press.

Verstraten, F. A., Cavanagh, P., and Labianca, A. T. (2000). Limits of attentive tracking reveal temporal properties of attention. *Vision Research*, 40(26), 3651–64.

von Grünau, M. W. (1986). A motion aftereffect for long-range troboscopic apparent motion. *Perception and Psychophysics*, 40(1), 31–8.

Wadle, D. C. (2020). Sensory modalities and novel features of perceptual experiences. *Synthese*, doi: https://doi.org/10.1007/s11229-020-02689-x

Watanabe, K. and Shimojo, S. (1998). Attentional modulation in perception of visual motion events. *Perception*, 27(9), 1041–54.

Watanabe, K. and Shimojo, S. (2005). Crossmodal attention in event perception. In *Neurobiology of Attention*, edited by L. Itti, G. Rees, and J. Tsotsos, 538–43. New York, NY: Elsevier Academic Press.

Xu, Y., Suzuki, S., and Franconeri, S. L. (2013). Shifting selection may control apparent motion. *Psychological Science*, 24(7), 1368–70.

Zmigrod, S., Spapé, M., and Hommel, B. (2009). Intermodal event files: integrating features across vision, audition, taction, and action. *Psychological Research PRPF*, 73(5), 674–84.

21
Faces as multisensory individuals

Matthew Fulkerson

1 Introduction

We are surrounded by objects. Nature herself supplies objects with an abundance of colours, textures, shapes, and arrangements of form. The built world contributes an immense store of ever-changing artefacts. We are directly connected to these objects via perception. Some objects we can eat. Others are dangerous. Some fade into the background, almost never drawing any attention or interest from creatures like us. My focus in this chapter is on an especially interesting perceptual object, one that creatures like us find immensely important: the human face. No matter what room or environment we may find ourselves in, the faces around us will almost always be among the most interesting, important, and salient objects. Faces convey essential information about individuals: their identity, their current emotional state, and their present state of mind. They separate friend from stranger, warn of those with dangerous or deceptive intentions, and convey enough detail about another's inner feelings to elicit empathy and engagement. Faces provide numerous important social cues about a person's identity, age, race, character traits, direction of eye gaze, interest, vocal properties, and so much more. For social creatures like us, faces are among the most important perceptual objects we can imagine.

It might seem strange to talk about faces as an *object*, even restricted to a *perceptual object*. I do not here presume that all perceptual objects are material particulars or middle-sized dry goods. As I am using the term, music or spoken words can constitute perceptual objects, as do smells, even if the most plausible metaphysical account of these items treats them as (say) events and molecular clouds, respectively.[1] They are what Green (2019, p. 665) describes as 'the individuals that our perceptual systems function to pick out'. As O'Callaghan (2016, p. 1270) describes them, 'Objects are constituents of the conscious perceptual manifold, they are targets of perceptual attention, and they are subjects of perception-based demonstrative thought.' Faces are objects in this sense: we perceive them and their features, we attend to them, and these perceptual interactions ground our subsequent thoughts and beliefs about those around us.

To the truism that faces are objects of special perceptual significance, we can add the truism that our experience of these faces is mediated primarily through vision. It is thus not surprising (or in itself problematic) that most research on the perception of faces has been focused exclusively on vision. It is no exaggeration to say that the visual awareness of human faces is among the most studied of visual processes. Interestingly, this intense research focus in psychology and neuroscience did not significantly spill over into related work in the

[1] See, for example, O'Callaghan (2007) on sounds, and Batty (2009) on smells.

Matthew Fulkerson, *Faces as multisensory individuals* In: *Sensory Individuals*. Edited by: Aleksandra Mroczko-Wąsowicz and Rick Grush, Oxford University Press. © Oxford University Press 2023. DOI: 10.1093/oso/9780198866305.003.0022

philosophy of perception. These decades of investigation strongly suggested that the visual awareness of faces was mediated through dedicated, highly specialized mechanisms. Some of these areas and mechanisms have been extensively studied, especially the fusiform face area (FFA; Kanwisher *et al.*, 1997; Kanwisher and Yovel, 2006). For this reason, faces might seem like paradigm instances of a unimodal sensory individual, a perceptual object of unique significance to a single modality. Consider that the 900-page *Oxford Encyclopedia of Face Perception*, published in 2011, contains 45 excellent and authoritative chapters on the perception of faces, exactly one of which discusses the perception of faces through a non-visual modality. The words 'touch' and 'haptic perception' have no entries in the index, despite the fact that Kilgour and Lederman (2002) had demonstrated clearly that humans are extremely good at the haptic identification of faces, and extensive subsequent work revealed that haptic perception showed significant specialization for the perception of faces (both identifying them and perceiving emotion state from them) (Lederman *et al.*, 2007).

In this chapter, I argue for the claim that faces are best understood as *multisensory* objects of special significance. I mean by this two distinct, but interconnected, claims. First, there is evidence that faces are preferentially processed by dedicated mechanisms in modalities other than vision. There seem to be dedicated haptic processing areas and preferential patterns of awareness involved in the perception of faces through touch. Second, *despite* these dedicated unisensory mechanisms, we also see extensive evidence of coordinated interactions between other modalities and the visual awareness of faces. Combined, these claims converge on the surprising conclusion that faces, whatever specialized processes they may involve within each modality, are best understood as multisensory perceptual individuals.

2 Faces in vision

Faces are objects of immense significance, and their features are readily available to vision. It is thus not surprising that there has been an incredible amount of research in recent decades on the visual awareness of faces. This work provides an essential foundation before understanding the importance of faces in other modalities.

In vision, faces seem to be unique perceptual objects that are perceived with highly specialized mechanisms. These mechanisms doubly dissociate. That is, subjects who can perceive and recognize other ordinary objects can fail to perceive and recognize faces (Wada and Yamamoto, 2001). In addition, subjects with deficits for ordinary objects can still perceive faces (Moscovitch *et al.*, 1997; see also McNeil and Warrington, 1993).[2] Unlike many other perceptual objects, faces seem to be processed holistically, rather than by their parts (these supposed specialized functions are all, for the moment, restricted to the *visual* perception of faces) (Kanwisher and Yovel, 2006). Faces are also extremely sensitive to orientation, leading to the so-called 'inversion effect', where perception of faces in ordinary orientations is facilitated compared to inversely presented faces (Haxby *et al.*, 1999; Yin, 1969). In addition, faces have special emotional salience, especially when they convey emotional information in the form of facial expressions (Ekman, 1993). Emotional faces are especially capable of breaking

[2] This fact is not unique to faces: there is ample evidence of specific perceptual deficits for some artefacts, tools (functional kinds), and various elements of spatial content. Together these dissociations suggest that perception in general may involve many more specific dedicated areas of processing.

through masking; for instance, subjects are more likely to see a face masked via continuous flash suppression (CFS) when the face is fearful (Yang *et al.*, 2007).

Additional evidence supports the identification of dedicated neural machinery for face perception. Early work focused on individual neurons, and single-celled recordings in primates revealed neurons that were especially sensitive to particular faces (so-called 'face cells') (Rolls, 2012). These early advances in our understanding eventually led researchers to focus on more specific brain regions. As noted by Kanwisher and Yovel (2006, p. 2109), 'Considerable evidence from behavioural, neuropsychological and neurophysiological investigations supports the hypothesis that humans have specialized cognitive and neural mechanisms dedicated to the perception of faces (the face-specificity hypothesis).' This specialized brain region seems essential for the specialized perception of faces and their properties.[3] One of the most important advances in our understanding of face perception concerned the discovery by Nancy Kanwisher and her collaborators (Kanwisher *et al.*, 1997) of a dedicated brain region in the fusiform gyrus for the preferential processing of faces, the so-called FFA. As they note in their review article summarizing the extensive evidence for the functional specialization of the FFA, Kanwisher and Yovel (2006, p. 2110) write, 'substantial evidence supports the hypothesis that at least one complex cognitive function—face perception—is implemented in its own specialized cortical network that is not shared with many if any other cognitive functions'. In particular, they review their own previous work and corroborating evidence that strongly suggests that the FFA is dedicated to the perception of faces, and that this functional architecture is seemingly incompatible with any of the alternative hypotheses. As they summarize (Kanwisher and Yovel, 2006, p. 2116): '[E]xisting data support the hypothesis that the FFA is selectively engaged in the processing of faces *per se*.'

I do not wish here to weigh in on the ongoing debate in vision science about whether the *face-specialization* or the alternative *domain-general* hypotheses are correct (though my own sensitivities favour the former). That is, I wish to remain neutral about whether the FFA is an area dedicated to *faces* or one that is involved in more general recognitional capacities. Either way, the visual awareness of faces is behaviourally and functionally distinct, with strong evidence that such awareness is grounded in specialized neural machinery. These unique elements show up in our daily lives, where faces, especially emotionally charged faces and the expressions and features of those we care most about, often take centre stage in our experiences of the world. The experience of faces is central to nearly everything we do with others, guiding our capacity for mindreading, empathy, communication, joint attention, and emotional intelligence, etc. As an educator, recent months of virtual teaching to a sea of mostly blank screens have reinforced the importance of faces to our human interactions. These many features have guided immense swaths of recent research on the specialized functional architecture that makes these special abilities possible. This is a rich research field, one that is yielding immense progress in our understanding of perception and the nature of face awareness. I do not think this focus on vision has been a mistake. And much remains to be worked out and understood in the visual awareness of faces. These investigations, of course, are proceeding in both psychology and neuroscience. More importantly for my own interests, these investigations of the visual experience of faces raise immensely important questions

[3] The claims of specificity to follow have been strongly disputed by some, with some evidence that the FFA is a general purpose area (see, for example, Tarr and Gauthier, 2000). Nothing in what follows hangs on the *exact* interpretation of FFA function, since all parties agree that it is an area that plays a special role in the processing of faces. In addition, work on these issues continues to accumulate (Schalk *et al.*, 2017).

for ongoing work in the philosophy of perception, from the content of visual and perceptual experience to the proper way of theorizing about the interactions between perception, cognition, and emotion. There are also many important developmental, computational, Artificial Intelligence, and social questions of relevance for better understanding the visual experience of faces that should interest all researchers in numerous fields.

Despite these many advances and outstanding questions, faces should not be considered perceptual objects of interest to vision alone. Indeed, even within vision, there has been an overemphasis on *static* faces, with relative less focus on facial dynamics (see, for example, Dobs *et al.*, 2019). Faces are important objects for multiple modalities, and for our understanding of *perception* in general. Indeed, as I will survey below, there is ample evidence that faces are important perceptual objects in other modalities, and should be treated and investigated, especially by philosophers of perception, as a significant target of multisensory theorizing.

3 Faces in haptic touch

As we have seen, faces are visual objects of immense importance. This importance has supported extensive visual specialization for the perception of faces and their features. But the importance of faces goes beyond vision. This fact has not, to my mind, been adequately appreciated in the broader work on the perception of faces. In this section, I want to argue that faces also have special significance for another modality: haptic touch.

In the last two decades, work by psychologists, especially by Susan Lederman and her many collaborators (e.g., Klatzky *et al.*, 1993; Klatzky and Lederman, 2008; Lederman and Loomis, 2003), has provided evidence of specialization for faces in haptic touch. This evidence suggests that faces are salient objects for touch, and that we should not think of their theoretical or practical importance solely in visual terms. In this section, I will briefly survey the evidence that, if nothing else, we should think of faces as perceptual objects with significant bimodal significance.

We start with some of the basics of human haptic touch. Haptic touch is also known as 'active touch' and involves the combined operations of tactile surface receptors on the skin along with kinaesthetic and proprioceptive awareness of our own exploratory movements. There have been many works dedicated to exploring haptic touch in the last two decades (Fulkerson, 2014b; Gallace and Spence, 2014; Jones and Lederman, 2006). Of importance in this context is the fact that human haptic perception involves immense specialized capacities for identifying ordinary material objects. Work by Susan Lederman and Roberta Klatzky (Klatzky *et al.*, 1993; Klatzky and Lederman, 2008; Lederman and Klatzky, 2004) demonstrated that the identification of objects through touch requires specialized exploratory actions that they call *exploratory procedures* (EPs). These included such specialized movements as *unsupported holding* and *contour following*. These movements involved the integration of features sampled over time, combined with the distinctive sensory inputs provided by the exploratory activity itself (especially important for the experience of heft and the use of dynamic inputs (Turvey, 1996)). As I noted in my own work on touch (Fulkerson, 2014b, Chapter 5), these aspects of touch suggest a much more prominent role for *intensive* features such as weight and thermal properties, which vary in intensity along some dimension, as compared to more holistic geometrical features. Still we seem to be quite good at recognizing objects through touch: 'Contrary to both scientific and lay thinking at the time, our initial investigations showed that humans are highly skilled at haptically recognizing common objects' (Lederman *et al.*, 2007, p. 230). The relative importance of intensive features in haptic object identification, combined with their requirement of specialized exploratory activities, suggests important differences between haptic and visual processing. These differences justify the strongly held assumption that perceptual processing of features in touch and vision are fundamentally distinct and separate. And yet, this assumption fails to account for the systematic and pervasive ways in which vision and touch do interact. These interactions are especially important for faces, as we will see.

I have previously maintained that there are theoretically important and interesting differences between modalities, and that we should not assume that some feature of vision automatically carries over to the others.[4] And yet, in general, we should be thinking of perceptual experience as genuinely multisensory, revealing a set of distinctly multisensory perceptual individuals. As O'Callaghan (2008, p. 317) argued in the early days of the now pervasive multisensory perspective: '[T]heorizing about perception is not just a matter of assembling independently viable stories about vision, audition, olfaction, and the rest. Considering the relationships and interactions among perceptual modalities sheds light on what is most striking about perception: its capacity to furnish a sense of awareness of a world of things and happenings independent from oneself.' The almost total focus on the visual awareness of faces can be a pernicious impediment to advancing our knowledge of perception. That said, a strong thread in my previous work on touch was to emphasize the deep *continuities* between touch and vision, of the many features they share (see especially Fulkerson, 2011, 2012). Thus, while they operate with distinct, proprietary elements, touch and vision often end up serving many of the same fundamental functions. This is not surprising, since both modalities support awareness of the same sets of ordinary material objects (whereas smell, audition, and gustation, on many accounts, have their own metaphysically distinct objects).

Since haptic touch provides rich awareness and recognitional capacities for material objects, and faces are after all ultimately a material object (albeit a very special one), it is natural to accept that haptic touch provides cues for recognition of faces. More importantly, as in vision, there also appears to be extensive specialization in haptic touch for faces. Kilgour and Lederman (2002) were the first to experimentally demonstrate that subjects could identify and recognize faces through haptic touch. Since even in the conditions where they were using 3D face masks with uniform material properties, they inferred that subjects *were* using geometric features to identify the faces through touch. That said, there was also a significant drop in accuracy in these conditions, suggesting that, as for other objects of touch material, features were playing an important role.

Lederman *et al.* (2007, p. 231) summarize their earlier work, establishing evidence for specialized face processing in touch:

> [W]e present the results of a multidisciplinary research program that has been guided by three converging methodologies: behavioural experimental studies with neurologically intact participants, neuropsychological behavioural research with prosopagnosic individuals,

[4] In particular, I have defended a 'moderate pluralism' about sensory modalities (Fulkerson, 2014a), according to which there are multiple equally good ways of carving up modalities. For the purposes of this paper, I am going to remain neutral on the best ways to individuate the modalities and their interactions.

> and neuroimaging studies using fMRI techniques. Our results clearly reveal that face processing is in fact a bimodal phenomenon that can involve haptic or visual processing.

The evidence includes findings of a haptic inversion effect for faces (Kilgour and Lederman, 2006), and functional magnetic resonance imaging (fMRI) evidence for specialized brain circuits for the haptic processing of faces (located especially in the left fusiform gyrus) (Kilgour *et al.*, 2005). These brain regions overlap somewhat with the visual areas dedicated to faces, but there is also evidence of significant divergence (Lederman *et al.*, 2010, p. 282):

> The occurrence of strong activation in the left hemisphere coupled with no significant right hemisphere activation suggests, rather, that the neural systems that mediate haptic and visual face recognition diverge. Because manual exploration is so often sequential, perhaps activation in the left fusiform gyrus was greater than in the right because haptically derived inputs about the facial configuration must be integrated over time. Conversely, visually derived inputs about the face may be simultaneously integrated over space.

More precise localization supports evidence for a 'haptic face region' (HFR). This area, located close to the FFA, nevertheless seems distinct (Kitada *et al.*, 2009, p. 2043):

> We conclude, therefore, that although subareas within the fusiform gyrus are sensitive to faces regardless of presentation modality, the subarea that is most sensitive to haptically presented faces is functionally distinct from the subarea that is most sensitive to visually presented faces. It is reasonable to speculate that both visual and haptic face recognition recruit distributed regions within the fusiform gyrus (including the HFR and FFA), but that the degree to which these areas are recruited depends on the modality.

More recent work by Matsumiya (2012, 2013) further supports the idea that haptic touch and vision share some important functional relationships. Matsumiya (2013) used the well-studied phenomenon of 'face adaptation effects' (FAEs) to show strong interaction effects between touch and vision. Face adaption occurs when exposure to a face with one salient property (such as emotional valence of an expression) causes a bias in favour of the *opposite* feature when exposed to a neutral stimuli. For instance, if a subject visually attends to a *happy* face, then they are more likely to experience a subsequent neutral face as *sad*. Since there is ample evidence that we can experience faces in both vision and touch, using dedicated circuits, Matsumiya (2013) hypothesized that there may be evidence of crossmodal transfer of face adaptation from touch to vision, and vice versa. The study demonstrated: '[T]hat cross-modal FAEs operate not only from haptics to vision but also from vision to haptics, which suggests the existence of shared facial representations between vision and haptics' (Matsumiya, 2013, p. 2095). These findings, combined with the evidence surveyed above from Kitada *et al.* (2009) (and also Kitada *et al.*, 2010), strongly suggest that face perception involves extensive shared circuits between touch and vision (Matsumiya, 2013, p. 2097):

> [T]he present results demonstrate that haptics can influence the perception of visual faces, and also demonstrate that vision can influence the perception of haptic faces. These findings suggest that face processing can show bidirectional cross-modal interactions. Together with the view that face processing depends on shared neural processing between visual and

> haptic faces, these findings suggest that even face processing is essentially multisensory, which is consistent with the emerging view that much of neocortex is multisensory.

Additional evidence comes from Dopjans *et al.* (2009), who used a series of experiments to show that there was extensive crossmodal transfer of facial information from vision to touch, and back again. As they summarized (p. 240):

> [W]e found cross-modal transfer in face recognition in both directions, H-V and V-H, with a cost in cross-modal (V-H) relative to within-modal (V-V) recognition of a set of faces learned visually, but no such cost for cross-modal (H-V) compared to within-modal (H-H) recognition of the same faces learned haptically. The fact that participants were able to accurately recognize faces in the cross-modal conditions, despite the novelty of the task, suggests that haptic and visual face recognition share commonalities that allow for efficient, 'default' information transfer about faces.

The facts concerning these transfers suggest some interesting work yet to be done on understanding the exact nature and structure of these interactions. The cost of transfers suggests that there is not a single, amodal representation of faces (which would suggest more symmetric information transfer). This supports the initial observations that vision, for instance, seems to have some proprietary unimodal machinery for facial perception. But the fact of transfer itself suggests that the unimodal elements are not the whole story, and that there are significant areas of overlap and interaction between the modalities of touch and vision when it comes to the perception of faces.

More recent evidence comes from a study by Murty *et al.* (2020). They showed that visual experiences were not necessary at all for the development of face selectivity in the FFA. This was demonstrated by activations in these regions in congenitally blind subjects who perceived faces through touch. As they noted, 'Our finding of robust face-selective responses in the lateral fusiform gyrus of most congenitally blind participants indicates that neither preferential foveal input, nor perceptual expertise, nor indeed any visual experience at all is necessary for the development of face-selective responses in this region' (Murty et al., 2020, p. 23015). This evidence suggests that face selectivity, even in these brain regions, is not a purely visual affair.

The evidence supports the claim that even for touch, faces are objects of special significance, exhibiting preferential processing and dedicated recognitional mechanisms. More importantly, these special features of haptic face perception carry over into the substantial interaction effects observed between vision and touch, suggesting that touch and vision are providing redundant, overlapping, and mutually-reinforcing information about faces and their features. Together, these facts reveal that the perception of faces is at least fundamentally bimodal, with shared neurological and functional areas of activation, deep interconnections between the processing of faces in both touch and vision, and evidence for similar shared specializations (especially in identification and emotional sensitivity). We should not think of faces as of special interest just in vision.

Can we go even further? Is there any evidence that faces can also be perceived through audition or smell? (I am going to assume that faces, however important, are not interesting objects of taste.) I want to suggest that there is plausible evidence for faces being significant objects of perceptual interest for other modalities.

4 Faces in hearing and smell?

Faces are objects of special significance in both vision and touch. There is, as far as I know, nothing equivalent to the FFA or HFR for the auditory perception of faces, and it is not clear that the other modalities supply direct awareness of faces per se. There is evidence, however, for dedicated, preferential processing of *voices*, and many of these regions overlap and interact strongly with regions implicated in the visual awareness of faces (Belin *et al.*, 2004, 2011; Latinus and Belin, 2011). More importantly, there is significant evidence of specialized multisensory interactions in the processing of faces.

This reveals an important difference that should be noted between vision and touch and hearing and smell. While the former two bring awareness of the same set of material objects, hearing seems to provide awareness primarily of *sounds*, and smell brings awareness primarily of *odours*.[5] On no theoretical accounts do we *directly* perceive faces through sounds or smells, and certainly not via the extraction of higher-level geometric features or from distributions of characteristic material properties (like the distinctive noses and mouths that play an important role in perception of faces in touch). Nevertheless, both hearing and smell provide important information connected to faces. In particular, the perception of vocalized speech can carry important information about facial expressions and features of vocal articulation, and contribute to visual information about identity and emotional state (a topic discussed in Fulkerson 2020a).

There are many important kinds of multisensory interactions. Sometimes one sense boosts the signal of another, leading to *facilitation*. Sometimes one modality weakens the effect of another, leading to sensory *domination*. Even stronger effects can lead to total sensory *suppression*. Sometimes the joint operation of two or more modalities can lead to novel forms of emergent awareness, not available in either modality alone. In all such cases, we see one sensory modality interact with another, producing some novel effect. These interactions are not automatic or universal, but operate according to strict spatio-temporal overlap and joint functional operations. Some involve *redundant* features, the ones available in multiple modalities. Others involve *reinforcing* features.[6] In all such cases, the interaction among the senses indicates some important functional coordination between distinct modalities.

As noted by Campanella and Belin (2007, p. 535):

> Most of our social interactions involve combining information from both the face and the voice of other persons (with the notable exception of telephone conversations). It is advantageous for the brain to integrate these two sources of information, both by exploiting redundancies between face and voice to increase the reliability of sensory estimates and by combining nonredundant, complementary cues to maximize information gathered from the two modalities.

[5] Of course, this is still a contested claim, with some holdouts for sounds being the secondary properties of material objects. My point is only that there are no such debates or fundamental metaphysical differences between touch and vision. The prevalence of these other views is a significant factor in assumptions about the role of hearing and smell in the perception of faces.

[6] This terminology initially comes from Fulkerson (2020b).

In the remainder of their review, they summarize the evidence already available at that time concerning the many ways in which voice and face perceptions are intertwined (there *is* a single entry in the *Oxford Handbook* on the perception of speech, which also nicely summarizes much of this evidence) (Campbell, 2012). In particular, they note evidence that voices provide extensive information about the articulation properties of the face (p. 535). Like faces, voices provide information about a person's age, gender, and state of mind. In addition, vocal properties can carry information about facial expression (p. 536):

> Facial movements involving the mouth region affect the resonant properties of the vocal tract in a way that can often be heard in the voice. For example, smiling causes a decrease in the vocal tract length and a resulting increase in formant frequencies that is often clearly noticeable. During phonation, it is only the upper extremity of the vocal tract, the 'filter' part of the vocal apparatus, that is accessible to vision. The larynx, or 'source' part, remains largely hidden from sight.

These features work together to provide strong cues to personal identity (Anzellotti and Caramazza, 2017; Blank *et al.*, 2014; Tsantani *et al.*, 2019).[7] That perception of voices and faces should overlap and interact so strongly is not in itself surprising or controversial. Mounting evidence in a number of fields supports the view that object perception in general is inherently multisensory (Naumer and Kaiser, 2010). But the interaction effects between faces and voices go beyond this simple fact. The sound source for voices after all is the face, and articulated via the lips and lower jaw. As Nudds (2001) and O'Callaghan (2015) have argued, such correspondences are paradigm examples of a special multisensory overlap. And yet, even so, faces seem to be even more intertwined with voices than simply serving as their sources. Many of the most celebrated and classic case studies in multisensory perception—the McGurk effect and the ventriloquism effect—demonstrate audiovisual integration of vocal speech and visual faces.[8] There seem to be strong connections here in the information provided by both modalities. Many of the central cases used to urge for a more multisensory conception of perceptual experience were actually built on the claim that faces and voices are deeply integrated and intertwined. The facilitation and suppression effects in congruent vs incongruent presentations suggest that vocal information is also used to inform about the state of a subject's face (and vice versa). If this is right, then we should not think of faces just as important bimodal targets (between touch and vision), but also as inherently multisensory individuals.

When we add the contributions made by vision to the perception of faces, in particular to the features of expression and articulation generated by the facial movements involved in the production of speech, we can now properly think of faces as genuine multisensory individuals. Faces are complex objects that are the perceptual target of multiple overlapping sensory modalities. That touch plays an important role in the perception of faces is not an implausible leap, since the objects of touch and vision so strongly overlap and the perception of voices is obviously connected in various ways with facial information. But why think that smell plays any role in the awareness of faces?

[7] I am grateful to an anonymous referee for pointing out this work on identity.

[8] While many studies use faces for both sets of examples, there is ample evidence for distinct mechanisms involved in these two case studies, with McGurk significantly more attuned to faces. See Bertelson, 1999 and Bertelson *et al.*, 1994 for discussion.

For one, both face perception (but especially the emotional perception of faces) and smell are involved in the processing of affective valence and emotional significance. In this, they play a similar functional role. In addition, indications of disgust and bad smells on the faces of others provide extremely important information for the multisensory detection of disgust elicitors. There is some evidence that the perception of expressions of disgust activates the same regions of the brain as disgusting smells (Haxby *et al.*, 2000, p. 229). In addition, there is evidence for priming (facilitation) effects for the perception of disgusted faces when co-presented with odours (independently of the valence of the odour) (Seubert *et al.*, 2010). The active areas involved in this study strongly overlapped with the active areas involved in audiovisual speech integration. Similarly, Leleu *et al.* (2015) found evidence for facilitation effects from odours. They found that odours enhanced the perception of ambiguous faces in congruent conditions (p. 12):

> A major finding of this study is that the odor context significantly facilitated the perception of ambiguous low-intensity expressions of faces, especially when the emotional meaning of the odorant was congruent with the emotional meaning of the facial expression. Specifically, happiness was perceived at a lower intensity of facial expression when the participants were in the pleasant odor context. Conversely, disgust—but also anger—was perceived at a lower intensity of expression in the aversive odor context. Thus, the perception of facial expressions is not a purely visual process but also integrates cues from other sensory modalities, including olfaction.[9]

These behavioural findings have been known for some time (see, for example, Leppänen and Hietanen, 2003) and were confirmed by Cook *et al.* (2015), who also showed (via event-related potential, or ERP, data) that activations elicited by the odours were tied to activations in the perception of faces.[10] As we can see, there is, in fact, evidence that the perception of faces is influenced by, and connected with, several modalities, including (somewhat surprisingly) smell.

5 Faces as multisensory individuals

We have seen that there is significant evidence that in both vision and touch, faces are perceived in highly specialized ways. In addition, we have seen evidence for extensive multisensory interactions in the perception of faces through audition and olfaction. These interactions seem domain-specific for the perception of faces. In the case of audition, these centre on the specific articulation properties of vocalizing faces and the auditory qualities generated by expressive faces. In the case of smell, they centre primarily on the joint function of faces and odours in conveying salient emotional information (especially concerning

[9] This study involved a perceptual matching task that showed facilitation in odour contexts above the baseline established in non-odour pilot studies. The authors describe this, plausibly, as a perceptual difference in visual acuity generated by the odour. They assess several hypotheses for *why* there is such an effect and suggest some combination of top-down influence through verbal categories, priming of shared visual representations by emotional and multisensory processes, and motor priming. Whatever the ultimate mechanisms, the kind of matching task used in this study strongly suggests a genuine perceptual effect (and not merely a downstream judgement or appraisal).

[10] Alas, as for touch, there are also no chapters or index entries for 'smell' or 'odour' in *The Oxford Handbook of Face Perception*.

elicitors of disgust). These specialized and idiosyncratic interactions reveal faces as objects of special sensory significance. The perception of faces is not merely multisensory in the mundane way in which any form of object perception might be multisensory. Instead, the perception of faces is a unique and special domain of uniquely multisensory interactions.

What does it mean to say that faces are uniquely multisensory individuals? We can now return to our initial conception of perceptual objects. Unsurprisingly, the focus of many discussions of perceptual objects have focused on *visual* objects. In discussing *multisensory* objects, O'Callaghan (2016, p. 1271) proposed the following: '[P]erceptual objects are structured mereologically complex individuals. Roughly, objects for perception are items that bear perceptible features and have perceptible parts arranged to form a unified whole.' I do not myself think about multisensory individuals in this way, as complex mereological objects. Nor do I think that we should think of faces as multisensory individuals in the sense that they—the faces—are the entities or individuals to which each modality assigns features (see also Cohen, this volume, Chapter 17). Instead, the evidence strongly supports the idea that the individual modalities each treat faces as objects of special significance for themselves, using their own proprietary mechanisms and features. While several modalities provide overlapping (redundant) information about faces, they also do so in their own proprietary ways. Haptic touch and vision do assign features to the same faces—they deeply influence and interact with one another and this only makes sense if the senses are ultimately targeting the very same individuals—but they also have their own unique proprietary mechanisms and formats for assigning features to faces (see the late engagement of ERP signals discussed in Cook *et al.*, 2015). This suggests that faces can be genuine and distinctly multisensory individuals, while also being targets of specialized unisensory mechanisms.

This perspective offers some advantages in our understanding of face perception. As Griffiths and Warren (2004, p. 887) state (in the context of auditory objects):

> Objects in any sensory domain can be considered to be perceptual entities that depend on the brain mechanisms available to represent and analyse sensory information. Therefore, in neuroscientific terms, the concepts of an object and of object analysis can be regarded as inseparable.

They go on to highlight four principles of such object analysis: analysis of information obtained from the sensory object; separation of information about the object from background information; abstraction of sensory information about the object; and finally, generalizing of the information about the object for use by other senses. At each stage of analysis, the brain is using what it knows about the scene and the context and the present bodily state to provide useful information about the object to make that information available for use by other systems. This is, in part, the essential functional role of perception. On this understanding of object awareness, it is essential that we understand the nature of the processes that help perform these critical functions. And when we do, we can see that faces are ideal objects for such awareness in multiple modalities: faces are extremely important for us along many dimensions—they are complex items that generate useful information made available to multiple modalities, and this rich, overlapping, but not identical, information needs to be reconciled and integrated to be of maximal usefulness. These features all suggest strongly that faces are multisensory individuals par excellence. This perspective is nicely described by Dopjans *et al.* (2009, p. 236), who note that multisensory engagement can be useful when each modality supplies incomplete information: 'This incompleteness might be resolved by combining information from different sources that complement each other, thus rendering recognition less sensitive, for example, to ambiguous cues.'

A further point to be emphasized: it is unclear what the multisensory nature of face perception means for our understanding of face perception generally or of multisensory interactions in general. For instance, many theories of face perception in vision are squarely described as 'cognitive theories', which hold that the special status afforded to the perception of faces is partly the result of extraperceptual (cognitive or emotional) processing. Trapp *et al.* (2018) attempt to show how such apparent influences can be explained with predictive coding models. If faces are genuine multisensory objects, this might have some implications for which elements are genuinely part of perception and which recruit domain-general cognitive or emotional elements. In addition, even in multisensory contexts, perceptual learning and patterns of development might play a role in mediating face awareness. As Bermúdez (2007, p. 47) has argued about the perception of objects:

> To perceive a body is to perceive a compact clustering of visually detectable features and to recognize that that cluster is associated with further features detectable in different sensory modalities. But perceiving objects is not just a matter of how one parses the perceptual array. It is also a matter of the expectations one has about how those compact clusters of visually detectable features will behave, both when one is perceiving them and when one is not perceiving them.

As he later elaborates (p. 47):

> Genuine objects obey certain principles and behave in certain regular ways. A world articulated in terms of genuine objects is a world that obeys certain basic physical regularities. Consequently, any creature that perceives such a world must be sensitive to these physical regularities that circumscribe what it is for something to count as a particular or a body at all.

These regularities, for visual objects, include defined edges, uniform movement of connected parts, relative spatial and shape stability, and so on. Faces exhibit these same features, but they also connect with our expectations about other people and their inner states and emotional lives. Faces in this way play a special role and carry immense social and interpersonal information.[11] Consider the role that eye gaze direction can play in mediating joint attention, for instance. This suggests that the multisensory experience of faces will be *especially* complex, and likely to implicate significant cognitive and emotional resources beyond those used in the perception of more mundane objects. When we see an angry face, it can generate immediate emotional responses like fear, in turn drawing attention and biasing a range of distinct

[11] Here and above, one might worry that it is the *person* that is the target of awareness, rather than the face itself. After all, it is the person who has the emotion. While obviously connected, I do think there is strong reason to distinguish awareness of persons from awareness of faces. Consider an amateur actress in a play. We might recognize an expression of sadness on her face, without thinking she is actually sad. Or consider prosopagnosics, who can be aware of persons (often through body language cues and voice) but who cannot experience faces as such. These sorts of dissociations suggest that faces can function as unique targets of perceptual awareness, though, of course, it is true that our experiences of faces and persons are deeply intertwined (indeed, this is one of the main reasons perception of faces is so important and interesting). I am grateful to an editor for pushing me on this point.

behavioural inclinations (many of these roles for face perception are discussed in Fulkerson, 2020a).

6 Conclusion

Faces are extremely important features in human environments. As social creatures, it is essential that we are able to reliably identify our friends and family, and assess their current moods, emotional states, and cognitive states. Faces have played an essential role in empirical work on vision and the identification of dedicated functional neuroanatomy. Similar work has occurred in haptic touch. But faces are not merely visual or haptic objects. Many of our senses provide overlapping and interacting information about the faces around us, and by looking more carefully at the nature of these interactions, we can better reveal some of the central defining features of perception as a whole.

References

Anzellotti, S. and Caramazza, A. (2017). Multimodal representations of person identity individuated with fMRI. *Cortex*, 89, 85–97. Available from: https://doi.org/10.1016/j.cortex.2017.01.013

Batty, C. (2009). What's that smell? *Southern Journal of Philosophy*, 47(4), 321–48.

Belin, P., Bestelmeyer, P. E. G., Latinus, M., and Watson, R. (2011). Understanding voice perception. *British Journal of Psychology*, 102(4), 711–25. Available from: https://doi.org/10.1111/j.2044-8295.2011.02041.x

Belin, P., Fecteau, S., and Bédard, C. (2004). Thinking the voice: neural correlates of voice perception. *Trends in Cognitive Sciences*, 8(3), 129–35. Available from: https://doi.org/10.1016/j.tics.2004.01.008

Bermudez, J. L. (2007). The object properties model of object perception: between the binding model and the theoretical model. *Journal of Consciousness Studies*, 14(9–10), 43–65.

Bertelson, P. (1999). Ventriloquism: a case of crossmodal perceptual grouping. *Advances in Psychology*, 129, 347–62.

Bertelson, P., Vroomen, J., Wiegeraad, G., and De Gelder, B. (1994). Exploring the relation between McGurk interference and ventriloquism. *ISCA Archive*, 94, 559–62. Available from: https://www.isca-speech.org/archive/pdfs/icslp_1994/bertelson94_icslp.pdf

Blank, H., Wieland, N., and von Kriegstein, K. (2014). Person recognition and the brain: merging evidence from patients and healthy individuals. *Neuroscience and Biobehavioral Reviews*, 47, 717–34. Available from: https://doi.org/10.1016/j.neubiorev.2014.10.022

Campanella, S. and Belin, P. (2007). Integrating face and voice in person perception. *Trends in Cognitive Sciences*, 11(12), 535–43. Available from: https://doi.org/10.1016/j.tics.2007.10.001

Campbell, R. (2012). Speechreading: what's MISS-ing? In *Oxford Handbook of Face Perception*, edited by Calder *et al.*, Oxford University Press, 605–29. Available from: https://doi.org/10.1093/oxfordhb/9780199559053.013.0031

Cook, S., Fallon, N., Wright, H., *et al.* (2015). Pleasant and unpleasant odors influence hedonic evaluations of human faces: an event-related potential study. *Frontiers in Human Neuroscience*, 9, 1–10. Available from: https://doi.org/10.3389/fnhum.2015.00661

Dobs, K., Isik, L., Pantazis, D., and Kanwisher, N. (2019). How face perception unfolds over time. *Nature Communications*, 10(1), 1–10. Available from: https://doi.org/10.1038/s41467-019-09239-1

Dopjans, L., Wallraven, C., and Bülthoff, H. H. (2009). Cross-modal transfer in visual and haptic face recognition. *IEEE Transactions on Haptics*, 2(4), 236–40.

Ekman, P. (1993). Facial expression and emotion. *American Psychologist*, 48(4), 384–92. Available from: https://doi.org/10.1037/0003-066X.48.4.384

Fulkerson, M. (2011). The unity of haptic touch. *Philosophical Psychology*, 24(4), 493–516. Available from: https://doi.org/10.1080/09515089.2011.556610

Fulkerson, M. (2012). *Touch without Touching. Philosophers Imprint*, 12(5), 1–15.

Fulkerson, M. (2014a). *Rethinking the Senses and Their Interactions: The Case for Sensory Pluralism*, 5(1426), 1–15.

Fulkerson, M. (2014b). The first sense: a philosophical study of human touch. Cambridge, MA: MIT Press.

Fulkerson, M. (2020a). Perception, emotion, and the interconnected mind. *Journal of Consciousness Studies*, 27(7–8), 7–30.

Fulkerson, M. (2020b). Sensory interactions and the epistemology of haptic touch. In *The Epistemology of Non-Visual Perception*, edited by B. Brogaard and D. Gatzia, 53–76. New York: Oxford University Press. Available from: https://doi.org/10.1093/oso/9780190648916.003.0003

Gallace, A. and Spence, C. (2014). *In Touch with the Future: The Sense of Touch from Cognitive Neuroscience to Virtual Reality*. New York: Oxford University Press.

Green, E. J. (2019). A theory of perceptual objects. *Philosophical and Phenomenological Research*, XCXI(3), 663–93. Available from: https://doi.org/10.1111/phpr.12521

Griffiths, T. D. and Warren, J. D. (2004). Opinion: what is an auditory object? *Nature Reviews Neuroscience*, 5(11), 887–92.

Haxby, J., Hoffman, E., and Gobbini, M. I. (2000). The distributed human neural system for face perception. *Trends in Cognitive Science*, 4(6), 223–33.

Haxby, J. V., Ungerleider, L. G., Clark, V. P., Schouten, J. L., Hoffman, E. A., and Martin, A. (1999). The effect of face inversion on activity in human neural systems for face and object perception. *Neuron*, 22(1), 189–199. Available from: https://doi.org/10.1016/S0896-6273(00)80690-X

Jones, L. A. and Lederman, S. J. (2006). *Human Hand Function*. New York: Oxford University Press.

Kanwisher, N., McDermott, J., and Chun, M. M. (1997). The fusiform face area: a module in human extrastriate cortex specialized for face perception. *Journal of Neuroscience*, 17(11), 4302–11. Available from: https://doi.org/10.1523/jneurosci.17-11-04302.1997

Kanwisher, N. and Yovel, G. (2006). The fusiform face area: a cortical region specialized for the perception of faces. *Philosophical Transactions of the Royal Society of London B: Biological Sciences*, 361(1476), 2109–28.

Kilgour, A. R., Kitada, R., Servos, P., James, T. W., and Lederman, S. J. (2005). Haptic face identification activates ventral occipital and temporal areas: an fMRI study. *Brain and Cognition*, 59(3), 246–57.

Kilgour, A. R. and Lederman, S. J. (2002). Face recognition by hand. *Perception and Psychophysics*, 64(3), 339–52. Available from: https://doi.org/10.3758/BF03194708

Kilgour, A. R. and Lederman, S. J. (2006). A haptic face-inversion effect. *Perception*, 35(7), 921–31. Available from: https://doi.org/10.1068/p5341

Kitada, R., Johnsrude, I. S., Kochiyama, T., and Lederman, S. J. (2009). Functional specialization and convergence in the occipito-temporal cortex supporting haptic and visual identification of human faces and body parts: an fMRI study. *Journal of Cognitive Neuroscience*, 21(10), 2027–45. Available from: https://doi.org/10.1162/jocn.2009.21115

Kitada, R., Johnsrude, I. S., Kochiyama, T., and Lederman, S. J. (2010). Brain networks involved in haptic and visual identification of facial expressions of emotion: an fMRI study. *NeuroImage*, 49(2), 1677–89. Available from: https://doi.org/10.1016/j.neuroimage.2009.09.014

Klatzky, R. L. and Lederman, S. (2008). Object recognition by touch. In *Blindness and Brain Plasticity in Navigation and Object Perception*, edited by J. J. Rieser, D. H. Ashmead, F. F. Ebner, and L. L. Corn, 185–207). Lawerence Erlbaum.

Klatzky, R. L., Loomis, J. M., Lederman, S. J., Wake, H., and Fujita, N. (1993). Haptic identification of objects and their depictions. *Perception and Psychophysics*, 54, 170–8.

Latinus, M., and Belin, P. (2011). Human voice perception. *Current Biology*, 21(4), 143–45. Available from: https://doi.org/10.1016/j.cub.2010.12.033

Lederman, S. and Loomis, J. (2003). Tactual perception. *Handbook of Perception and Human Performances*, 2(2), 1–41.

Lederman, S. J., Kilgour, A., Kitada, R., Klatzky, R. L., and Hamilton, C. (2007). Haptic face processing. *Canadian Journal of Experimental Psychology*, 61(3), 230–41.

Lederman, S. J. and Klatzky, R. L. (2004). Haptic identification of common objects: effects of constraining the manual exploration process. *Perception and Psychophysics*, 66(4), 618–28.

Lederman, S. J., Klatzky, R. L., and Kitada, R. (2010). Haptic face processing and its relation to vision. In *Multisensory Object Perception in the Primate Brain*, edited by M. J. Naumerand J. Kaiser, 273–300. New York: Springer. Available from: https://doi.org/10.1007/978-1-4419-5615-6_15

Leleu, A., Demily, C., Franck, N., Durand, K., Schaal, B., and Baudouin, J. Y. (2015). The odor context facilitates the perception of low-intensity facial expressions of emotion. *PLoS One*, 10(9), 1–19. Available from: https://doi.org/10.1371/journal.pone.0138656

Leppänen, J. M. and Hietanen, J. K. (2003). Affect and face perception: odors modulate the recognition advantage of happy faces. *Emotion*, 3(4), 315–26. Available from: https://doi.org/10.1037/1528-3542.3.4.315

Matsumiya, K. (2012). Haptic face aftereffect. *I-Perception*, 3(2), 97–100. Available from: https://doi.org/10.1068/i0496sas

Matsumiya, K. (2013). Seeing a haptically explored face: visual facial-expression aftereffect from haptic adaptation to a face. *Psychological Science*, 24(10), 2088–98. Available from: https://doi.org/10.1177/0956797613486981

McNeil, J. E. and Warrington, E. K. (1993). Prosopagnosia: a face-specific disorder. *The Quarterly Journal of Experimental Psychology Section A*, 46A(1), 1–10. Available from: https://doi.org/10.1080/14640749308401064

Moscovitch, M., Winocur, G., and Behrmann, M. (1997). What is special about face recognition? Nineteen experiments on a person with visual object agnosia and dyslexia but normal face recognition. *Journal of Cognitive Neuroscience*, 9(5), 555–604. Available from: https://doi.org/10.1162/jocn.1997.9.5.555

Murty, N., Teng, S., Beeler, D., Mynick, A., Oliva, A., and Kanwisher, N. (2020). Visual experience is not necessary for the development of face-selectivity in the lateral fusiform gyrus. *Proceedings of the National Academy of Sciences*, 117(37), 23011–20. Available from: https://doi.org/10.1073/pnas.2004607117

Naumer, M. J. and Kaiser, J. (2010). *Multisensory Object Perception in the Primate Brain*. New York: Springer.

Nudds, M. (2001). Experiencing the production of sounds. *European Journal of Philosophy*, 9(2), 210–29. Available from: https://doi.org/10.1111/1468-0378.00136

O'Callaghan, C. (2007). *Sounds: A Philosophical Theory*. New York: Oxford University Press.

O'Callaghan, C. (2008). Seeing what you hear: cross-modal illusions and perception. *Philosophical Issues*, 18(1), 316–38.

O'Callaghan, C. (2015). Intermodal binding awareness. In *Sensory Integration and the Unity of Consciousness*, edited by D. Bennet and C. Hill. Cambridge MA: MIT Press. Available from: https://doi.org/10.7551/mitpress/9780262027786.003.0004

O'Callaghan, C. (2016.) Objects for multisensory perception. *Philosophical Studies*, 173(5), 1269–89.

Rolls, E. T. (2012). Face neurons. In *Oxford Handbook of Face Perception*, edited by Calder *et al.*, New York: Oxford University Press. Available from: https://doi.org/10.1093/oxfordhb/9780199559053.013.0004

Schalk, G., Kapeller, C., Guger, C., *et al.* (2017). Facephenes and rainbows: causal evidence for functional and anatomical specificity of face and color processing in the human brain. *Proceedings of the National Academy of Sciences of the United States of America*, 114(46), 12285–90. Available from: https://doi.org/10.1073/pnas.1713447114

Seubert, J., Kellermann, T., Loughead, J., *et al.* (2010). Processing of disgusted faces is facilitated by odor primes: a functional MRI study. *NeuroImage*, 53(2), 746–56. Available from: https://doi.org/10.1016/j.neuroimage.2010.07.012

Tarr, M. J. and Gauthier, I. (2000). FFA: a flexible fusiform area for subordinate-level visual processing automatized by expertise. *Nature Neuroscience*, 3, 764–9. Available from: https://doi.org/10.1038/77666

Trapp, S., Schweinberger, S. R., Hayward, W. G., and Kovács, G. (2018). Integrating predictive frameworks and cognitive models of face perception. *Psychonomic Bulletin and Review*, 25(6), 2016–23. Available from: https://doi.org/10.3758/s13423-018-1433-x

Tsantani, M., Kriegeskorte, N., Mcgettigan, C., and Garrido, L. (2019). NeuroImage faces and voices in the brain: a modality-general person-identity representation in superior temporal sulcus. *NeuroImage*, 201, 116004. Available from: https://doi.org/10.1016/j.neuroimage.2019.07.017

Turvey, M. T. (1996). Dynamic touch. *American Psychologist*, 51(11), 1134–52. Available from: https://doi.org/10.1037/0003-066X.51.11.1134

Wada, Y. and Yamamoto, T. (2001). Selective impairment of facial recognition due to a haematoma restricted to the right fusiform and lateral occipital region. *Journal of Neurology, Neurosurgery & Psychiatry*, 71(2), 254–7.

Yang, E., Zald, D. H., and Blake, R. (2007). Fearful expressions gain preferential access to awareness during continuous flash suppression. *Emotion*, 7(4), 882–6. Available from: https://doi.org/10.1037/1528-3542.7.4.882

Yin, R. K. (1969). Looking at upside-down faces. *Journal of Experimental Psychology*, 81(1), 141–5. Available from: https://doi.org/10.1037/h0027474

22
Multimodal structure of painful experiences

Błażej Skrzypulec

1 Introduction

Authors working on the topic of pain often claim that experiences of pain have two aspects: a sensory aspect, which determines descriptive, non-evaluative aspects of pain, for instance, that it is 'pinching' or 'cramping'; and an evaluative aspect, which determines that pain feels bad (e.g., Aydede and Güzeldere, 2002; Bain, 2013; Corns, 2018). An important problem in analysing the sensory aspects of pains is that they are characterized by such a great variety that it is difficult to find any experiential quality shared by all painful experiences. Nevertheless, while sensory aspects of pains do not constitute a uniform set, they are also not completely random. For instance, the widely used McGill Pain Questionnaire distinguishes ten groups of descriptors used by people in characterizing the sensory character of pains. Many such descriptors characterize pains in touch-related terms, like 'cutting', 'pressing', 'sharp', and 'pulsing', or temperature-related terms, like 'hot' or 'burning'.

This suggests that at least some pains, in addition to pain-specific terms, like 'aching' or 'hurting', are characterized by terms strongly related to tactile or thermal modality. Because application of such terms is likely to happen due to the specific phenomenal character of painful experiences, it is plausible to assume that there are pains which are experienced as having some tactile-like or thermal-like character. Hence, not only are pain experiences 'physiologically multimodal', as mechanisms involved in producing painful sensations process various types of sensory inputs (see Julius and Basbaum, 2001 for a review), but some of them are also 'phenomenally multimodal'. The goal of this paper is to investigate the structure of such phenomenally multimodal pain experiences—and by structure, I mean the way in which various elements presented in an experience of pain, like its unpleasantness, intensity, or tactile-like character, are experientially organized.

In recent years, philosophers of perception have become increasingly interested in analysing structures of multimodal experiences (e.g., Briscoe, 2016; Green, 2019b; O'Callaghan, 2015; Richardson, 2014; Skrzypulec, 2021). In particular, it has been recognized that the perceived elements of multimodal experiences are not presented simply as co-occurring, but also as organized into complex wholes. For instance, when having an audiovisual experience of a barking dog, one does not merely experience that something has a dog-like appearance and that something is barking, but that both visual and auditory elements somehow characterize the same object. In consequence, a question arises of how to characterize the way in which the content of multimodal experiences is organized.

These investigations have given rise to two main ideas regarding the structure of multimodal experiences. According to the first, the structure of multimodal experiences should be

Błażej Skrzypulec, *Multimodal structure of painful experiences* In: *Sensory Individuals*. Edited by: Aleksandra Mroczko-Wąsowicz and Rick Grush, Oxford University Press. DOI: 10.1093/oso/9780198866305.003.0023

characterized in terms of property instantiation (see Macpherson, 2011; O'Callaghan, 2014, 2015). In this case, when one perceives a multimodal object, what is presented is that a single object instantiates properties related to distinct modalities—for instance, visual and auditory properties. The second, alternative proposal is to analyse the multimodal experiential structure in mereological terms (see Matthen, 2010; O'Callaghan, 2016), which holds that the structure of multimodal experiences should be characterized in terms of parts and wholes. For instance, when a multimodal, audiovisual object is perceived, the experience presents it as a whole composed of a visual proper part and an auditory proper part.

However, I argue that these two accounts of multimodal structure cannot be plausibly applied to multimodal tactile-like and thermal-like experiences of pain. I set forth two major reasons. First, the mechanisms underlying pain experiences are unlike those which underline multimodal experiences, whose structure can be characterized in terms of instantiation and parthood. Second, the considered accounts entail that many adaptive tactile-like and thermal-like pain experiences are illusory. Instead, I propose to characterize the structure of tactile-like and thermal-like experiences of pain by referring to the notion of crossmodal correspondences. Crossmodal correspondences are associations which people experience between elements belonging to distinct modalities (e.g., Parise, 2016; Spence, 2011; Spence and Deroy, 2013). For instance, ascending visual movement is judged as corresponding to transition from lower to higher pitch and brighter colours are experienced as corresponding to louder sounds (see Deroy and Spence, 2016).

In the subsequent investigations, I do not rely on any specific theory of painful tactile and thermal experiences. However, I assume that such experiences have some intentional content which concerns bodily states. This content does not necessarily have to be merely indicative content characterizing a bodily state in some way. In particular, it has been proposed that evaluative aspects of pains may be grasped by the notion of imperative, command-like content (e.g., Klein, 2007). Furthermore, while I do not endorse any particular view on individuation of experiences, I assume that a thermal-like or a tactile-like painful experience is a single experience, and not some combination of two experiences—one thermal or tactile, and the second a 'pure' painful experience. While arguing for this thesis goes beyond the scope of the paper, I believe that it is intuitive that while having, for instance, a burning pain in one's foot, we recognize it as a single, and not as two, distinct experiences.

More specifically, to account for the multimodal pains by using the notion of crossmodal correspondences, I argue that tactile-like and thermal-like painful experiences have such content that they present painful bodily states as corresponding to tactile or thermal bodily states of certain types. My proposed analysis can avoid problems of instantiation and parthood because the mechanisms underlying experiences of pain are able to sustain crossmodal correspondences, and analyses in terms of crossmodal correspondences do not entail the illusory character of many adaptive multimodal pains.

The paper starts by presenting the major mechanisms responsible for pain perception (see Section 2). Subsequently, in the Sections 3 and 4, I argue that the mechanisms of pain perception are of a distinct type than those underlying experiences with a structure organized by relations of instantiation and parthood. Furthermore, I show that analyses in terms of instantiation and parthood lead to an implausible conclusion that many adaptive multimodal pains are illusory (see Section 5). Finally, in the Section 6, I analyse the experiential structure of tactile-like and thermal-like pains in terms of crossmodal correspondences and show that such an analysis avoids the problems of other theories.

2 Mechanisms of pain perception

The modern paradigm in scientific investigations regarding pain has emerged due to the formulation of the gate control theory of pain by Melzack and Wall (1965) in their seminal paper (see also Mendell, 2014 and Moayedi and Davies, 2013 for contemporary reviews). This theory postulates pain-gates, implemented in the spinal cord, which receive input from various afferent fibres. The activities of some of these fibres 'open' a gate, while those of other fibres 'close' a gate. The opening of a gate leads to the occurrence of a conscious painful sensation. In consequence, an experience of pain happens when there is a certain combination of activities of various afferent nerve fibres, such that 'closing' activities are weaker than 'opening' activities.

The gate control theory has been developed in opposition to two earlier major theories: the specific system theory and the intensity theory (see Moayedi and Davies, 2013). The specific system theory claimed that there is a specialized system for processing noxious stimuli whose activities lead to experiences of pain. On the other hand, proponents of the intensity theory believe that painful sensations arise not due to activities of a specialized system, but rather as a result of a detection of intense stimulation made by any sensory system. The success of the gate theory has demonstrated the shortcomings of these two approaches.

First, while there are fibres, such as Aδ and C fibres, which are specialized in transferring noxious stimuli, pain, contrary to the specific system theory, is experienced not due to the functioning of a separate system, but in virtue of certain (gate-opening) combinations of activities of afferent fibres related to various sensory systems. Second, in contrast to the intensity theory, intense stimulation is neither necessary nor sufficient for the occurrence of pain. In some circumstances, even intense stimulation may be countered by certain gate-closing activities and thus does not lead to pain. Similarly, pain may occur even without intense stimulation due to a lack of usual gate-closing inputs. Nevertheless, the gate control theory preserves some insights of earlier theories. In particular, as observed above, while there is no specialized pain system, there are certain nociceptive receptors and related afferent fibres whose activities contribute to the opening of pain-gates. Furthermore, while there is no strict relationship between intense stimulation and experiences of pain, in a typical situation, an intense stimulus is likely to cause pain gate opening.

Over the years, the gate control theory has undergone various modifications and extensions (see Wall, 1996). For instance, while in their initial model, Melzack and Wall strictly differentiated between gate-closing large-fibre activities and gate-opening small-fibre activities, the actual interactions between activities of afferent fibres are more complex (see Peláez and Taniguchi, 2016). Furthermore, while the gate control theory mainly concerns bottom-up processing, nowadays it is well recognized that there are important top-down contributions that determine the occurrence of sensations of pain (e.g., Tsao *et al.*, 2011). Melzack himself has proposed a more general theory of pain by introducing the concept of the 'neuromatrix' (Melzack, 1993). The neuromatrix is a distributed neural network, shaped both by genetic and developmental factors, which processes information about the state of the body provided by various bottom-up and top-down systems (e.g., the neuromatrix processes sensory data, information about emotional states, and inputs from stress regulation system; see Melzack, 1999). When a particular combination of activities within the neuromatrix occurs, known as a pain 'neurosignature', conscious pain is experienced.

Though the gate control theory does not explain all pain-related phenomena, and the original model proposed by Melzack and Wall is not considered fully accurate, it provides a general idea concerning mechanisms underlying pain perception that is crucial in contemporary investigations of pain. This idea is that mechanisms of pain perception are, to an important extent, mechanisms that detect particular combinations of inputs. A pain-gate itself is such a detector, which, by virtue of having 'open' and 'closed' states, differentiates combinations of various types of inputs which lead to experiences of pain from combinations which do not lead to such a sensation. Similarly, the neuromatrix is a mechanism which processes combinations of bottom-up and top-down inputs and, as an output, recognizes whether the input satisfies a pain or non-pain neurosignature. Such mechanisms can be called 'pattern classification mechanisms': as an input, they take a pattern of elements, such as a certain combination of sensory stimulations, and as an output, they classify this pattern as being of a certain type.

In the next two sections, I argue that experiential structures, which can be characterized in terms of instantiation or parthood, arise as a result of mechanism types that differ from pattern classification mechanisms. Hence, experiences of pain, including tactile-like and thermal-like pains, are not likely to have content organized by instantiation and parthood relations.

3 Mechanisms of experiential instantiation

An influential idea in investigations concerning the structure of multimodal experiences is to characterize their structure in terms of an instantiation relation between a property and a subject possessing this property (see Macpherson, 2011; O'Callaghan, 2014, 2015). For example, in the case of an audiovisual experience presenting a barking dog, the structure of an experience consists of a common subject which instantiates both vision-related properties (concerning how the dog looks) and audition-related properties (concerning how the dog sounds). Due to this structure, an audiovisual experience presents not only that there is something that looks a certain way and something that sounds in a certain way, but also that there is a single entity which instantiates both of these visual and auditory characteristics. If the analogous idea were applied to the multimodal tactile-like and thermal-like experience of pain, it would mean that such experiences present a common subject, like a bodily state, as possessing various properties, including those related to activities of the tactile and thermal sensory systems.

Mechanisms underlying the experiential attribution of properties to a common subject are often characterized by referring to the notion of 'object-files' (e.g., Green and Quilty-Dunn, 2017; Mitroff *et al.*, 2005). According to object-file theories, a perceptual system organizes information about perceived objects by storing them in 'files' such that a single file contains information about exactly one object. The content of an object-file can be updated over time, which allows representation of the diachronic sameness of the perceived entities. For instance, an object may change colour from red to green but still be visually represented as numerically the same entity due to being represented using the same object-file. In the case of multimodal experiences, an object-file combines information provided by different modalities, and serves as a basis for an experiential structure in which an entity is presented as instantiating multimodal properties (Green, 2019b).

The object-files framework is not the only one used to characterize mechanisms underlying experiential presentations of individual entities as subjects simultaneously possessing a variety of properties. For instance, Pylyshyn (2007) famously proposed a notion of preattentive 'visual indices', by virtue of which the visual system is able to simultaneously refer to several objects, treat them as distinct, and track them through time. A visual index functions like a demonstrative term, such as 'this' or 'that', which refers to an entity without attributing any properties to it. As a result of combining indices with qualitative information, perceived entities can be presented as instantiating properties. Another classic approach, the Feature Integration Theory (Treisman, 1998), proposes an attentional mechanism in virtue of which focusing attention on an element of a visual scene allows the integration of information from distinct feature-maps and representation of the attended entity as possessing various properties. Similarly, according to Rensink's coherence theory of attention (Rensink, 2000), attentional mechanisms combine short-lived representations of properties into a stable 'nexus' which represents a persisting object instantiating properties. In terms closer to the neural implementation level, a neural synchronization mechanism has been proposed, which allows representation of a single entity as having many properties and differentiating between properties instantiated by distinct entities because the neurons coding properties of a single entities fire in synchrony (e.g., O'Reilly *et al.*, 2003).

These theories describing mechanisms which are likely to be responsible for perception of individuals as subjects instantiating properties differ in many respects and they describe the relevant mechanisms using terminology from distinct levels of description, sometimes closer to a specific neural implementation, and sometimes postulating abstract structures which could be implemented in many ways. Nevertheless, despite these differences, all of them characterize mechanisms of the same type. These mechanisms are 'unification mechanisms' in virtue of which initial, separate input elements become constituents of some higher-order whole. For instance, in object-files accounts, separate pieces of information about perceived features are gathered within a single object-file; similarly, in the visual indices theory, a complex representation is formed that combines a visual index with representations of properties. The same general pattern can be identified in other theories, as in the coherence theory of attention where feature-representations are unified within a nexus; in the feature integration theory, an attentional mechanism combines information from separate feature-maps; and in neural synchronization accounts, separate neural representations start to form a synchronized whole.

However, the fact that experiential instantiation structures arise from the function of unification mechanisms poses a problem for interpreting tactile-like and thermal-like pains in terms of instantiation. As described in the Section 2, the mechanisms underlying painful sensation are pattern classification mechanisms, and not unification mechanisms. In particular, pattern classification mechanisms do not add any new element to the input elements but merely recognize that the arrangement of input elements satisfies a certain pattern. For instance, pain-related mechanisms recognize whether the pattern of inputs is 'gate-opening' or satisfies a certain neurosignature. On the other hand, unification mechanisms modify the input by adding a new element, such as an object-file, an index, or a nexus, in virtue of which other input elements are unified into a higher-order whole. By contrast, mechanisms such as pain-gates and the neuromatrix do not create any such higher-order structures by adding new elements to the input elements. As a result, it is not plausible to characterize the structure of multimodal pains in terms of property instantiation, as the major mechanisms of pain are distinct from those usually associated with experiential instantiation structures. This is not to deny that mechanisms responsible for creating structures such as object-files can also be described in terms of pattern recognition, especially when characterized using low-level terms concerning their neural implementation. However, they are some specific types of pattern recognition mechanisms which can serve a unificatory function associated with object-files. On the other hand, the mainstream description of pain mechanisms does not provide grounds for believing that such unificatory functioning is also present in the case of pattern classification mechanisms that underlie pain perception.

4 Mechanisms of experiential parthood

Structures of multimodal experiences are characterized not only in terms of instantiation, but also in terms of parthood. In particular, O'Callaghan has argued that in the case of audiovisual experiences, it is implausible to describe experiential structure of multimodal objects by referring solely to the notion of instantiation, as sounds are not experienced as properties of objects, but themselves are experienced as individuals instantiating properties (O'Callaghan, 2016; see also Green, 2019a). Instead, he has proposed that an audio-visual object is presented as a mereological whole with visual and auditory proper parts, each instantiating separate visual and auditory properties. In the case of multimodal tactile-like and thermal-like pains, a mereological account entails that such experiences of pain present a bodily state divided into proper parts, some of which are tactile or thermal. For instance, it may be that when a tactile-like pain is felt, what is presented are two proper parts: a tactile state of a body (e.g., that a pressure is inflicted on the skin) and a purely painful state of the body (e.g., that a bodily disorder is present), which jointly constitute a mereological whole (i.e., a tactile-like painful bodily state). Nevertheless, I believe that such a solution faces an analogous problem to characterization in terms of instantiation, as the mechanisms responsible for part perception are of a distinct type from those underlying painful sensations.

Studies on part perception are most extensively developed in regard to the visual modality, which usually present objects as composed of spatial parts. Mechanisms of visual part perception are mainly concerned with detecting qualitative discontinuities and analysing the structure of edges to identify specific arrangements, like concavity, which serve as a cue for determining where one parts ends and another starts (e.g., Hoffman and Richards, 1984; Palmer and Rock, 1994; Xu and Singh, 2002). For instance, an hourglass shape will be perceptually divided into two parts due to an arrangement of edges creating a concavity. While vision was the primary target of part perception studies, it is likely that analogous mechanisms also operate in other modalities. In particular, it has been proposed that sounds are perceived as having temporal parts designated by detecting discontinuities in pitch (see O'Callaghan, 2008).

While there are significant differences between mechanisms of part perception in various modalities, the relevant mechanisms are of the same type. Such mechanisms, which may be called 'differentiation mechanisms', detect differences and similarities between input elements and, as an output, arrange them so similar elements are combined with each other and separated from dissimilar elements. For instance, in the case of visual part perception, the relevant mechanisms arrange perceived elements into spatially and qualitatively coherent fragments separated from each other by qualitative discontinuities and specific arrangements of edges. Similarly, in the auditory modality, part perception mechanisms divide the stream of auditory information into fragments relying on temporal and qualitative properties such as pitch.

This observation leads to an analogous problem for applying the mereological notions in analysing the structure of tactile-like and thermal-like pains, as in the case of analysis in terms of instantiation. While, at the neural level of description, mechanisms of part perception may also be characterized as engaged in pattern recognition, their specific function is to serve as differentiation mechanisms. On the other hand, it does not seem to be true in the case of major mechanisms of pain perception. In particular, there is nothing in the way in which pain-gates and the pain neuromatrix function that suggests they detect and process qualitative differences between input elements to divide them into some coherent arrangements. In consequence, it is not plausible to characterize the experiential structure of multimodal pains in terms of parthood, as the mechanisms responsible for part perception are significantly distinct from those underlying pain perception.

5 Illusory pains

The previous sections demonstrated that characterizing the structure of multimodal tactile-like and thermal-like pains in terms of instantiation or parthood is implausible, given the type of mechanisms that are responsible for pain perception. Below, I argue that there is an additional problem in applying these accounts to multimodal experiences of pains. Accepting them entails that various adaptive multimodal experiences of pain are, in fact, illusory. Both considered frameworks are representational frameworks, as they assume that multimodal experiences present perceived entities as having certain characteristics, and in doing so, they may be accurate or inaccurate, leading to a veridical or an illusory experience. In particular, when applied to a tactile-like pain experience, an analysis in terms of instantiation states that the bodily state is represented as having some touch-related properties. For instance, it may be presented that the concerned state instantiates properties associated with being under pressure or being in contact with an external entity. Similarly, in an analysis in mereological terms, a bodily state is represented as having a tactile proper part. For example, such a theory may postulate that in a mereological structure of the whole bodily event, there is a tactile proper part concerning being under external pressure.

Such a representational approach to multimodal pains is problematic, as there are types of tactile-like and thermal-like painful experiences which do not intuitively seem illusory but are illusory from the perspective of instantiation and parthood frameworks. One such example can be presented by referring to the distinction between so-called 'first pain' and 'second pain' (Price and Aydede, 2005). When one cuts one's finger, at the first moment, a sharp, short-lasting first pain is felt, which, in an important way, is due to information carried by Aδ fibres. However, after a moment, due to information carried by C fibres, a second pain occurs which is longer-lasting and more spatially diffuse. Such second pains are commonly described in thermal-like terms (specifically, as 'burning'), despite the fact they happen without any actual thermal stimulation or significant change in body temperature. Because of that, they are illusory from the perspective of an instantiation framework; according to this framework, the considered painful bodily state is inaccurately represented as having some thermal-related properties. Similarly, experiences of second pains are inaccurate from the perspective of a mereological framework, as the represented bodily states do not have any actual thermal component that can serve as a thermal proper part. Nevertheless, treating second pains as illusory is problematic because they have adaptive value as they facilitate prolonged protection of injured area and do not involve any malfunction of pain mechanisms. In fact, they are one of the most common responses for typical noxious stimuli. What is particularly important is that the adaptiveness of second pains is not independent from the presence of thermal-like aspects. Quite the opposite—second pains are, in an important way, adaptive precisely due to their thermal-like character: a second pain has a qualitative character which is plausibly described as 'burning' and the presence of this thermal-like character determines that a proper behaviour consists in a prolonged care and not, for instance, a rapid withdrawal movement. In consequence, instantiation and parthood accounts that treat the thermal-like character of second pains as illusory face problems in explaining the adaptiveness of such experiences of pain.

Another example, this time concerning tactile-like pains, comes from experimental studies in which painful sensations are evoked by applying laser stimuli (Haggard *et al.*, 2013). As a result, the occurring pain is described in tactile terms as a 'pinprick pain', even though the laser stimuli are used specifically to avoid activation of tactile Aβ fibres. Similar to the case of second pain, the pinprick sensation is inaccurate from the perspective of instantiation and parthood frameworks, as the 'pinprick' description suggests the presence of a bodily state which has properties, or a proper part, concerning the occurrence of a sudden increase in pressure inflicted on a small skin fragment, though no such event happens after laser stimulation. However, also in this case, it is far from obvious that the pain experience is illusory. Again, the painful sensation does not happen because of some malfunction of pain mechanisms and is adaptive because its tactile-like, 'pinprick' character motivates a proper behaviour. In the case of 'pinprick' pain, such a behaviour is a swift withdrawal of a bodily part, which is likely to break contact with a noxious stimulus.

It should be noted that it is not inconsistent to interpret each case as a form of pain illusion. In particular, it may be the case that some experiences are both illusory and adaptive. However, it is a disadvantage of a representational approach which holds that a significant number of common experiences is inaccurate, so such an approach should not be adopted unless there is no other viable theory. In the subsequent sections, I argue that such an alternative theory can be formulated using the notion of crossmodal correspondences.

6 Crossmodal correspondences

Crossmodal correspondences are associations experienced between elements presented by distinct modalities (Parise, 2016; Spence, 2011; Spence and Deroy, 2013). Some elements are phenomenally presented in such a way that people judge them as matching some other elements, presented by a distinct sensory system. For instance, transition from lower to higher pitch is experienced as matching ascending visual movement, and various vivid correspondences occur between olfactory and taste perception (e.g., odours are often characterized as 'sweet'). One of the most famous demonstrations is the kiki/bouba effect, showing that people feel that the sound of the name 'kiki' is more appropriate for angular shapes, while the 'bouba' sound is associated with shapes without sharp corners (see Köhler, 1947 for the classic study). The presence of crossmodal correspondences has been demonstrated between virtually all modalities, including associations between sounds and tastes, shapes and tactile hardness, or colours and temperatures (Deroy and Spence, 2016). Studies on crossmodal correspondences constitute a fruitful area of investigations in which psychologists consider, inter alia, the influence of correspondences on behavioural performance (Brunetti *et al.*, 2017), cultural differences in experienced correspondences (Wan *et al.*, 2014), and the emergence of correspondences in ontogenetic development (Asano *et al.*, 2015). It is believed that crossmodal correspondences occur mainly due to similarities in the way corresponding elements are represented, the learnt statistical associations between corresponding elements, and possession of common semantic labels, as in the case of 'high' sounds and 'high' visual elements (Deroy and Spence, 2016; Parise, 2016; Spence, 2011).

I propose that multimodal tactile-like and thermal-like pains should be understood as cases of crossmodal correspondences. More specifically, I claim that tactile-like pain consists in experiencing that a painful bodily state corresponds to a tactile bodily state of a certain type, and that having a thermal-like pain consists in experiencing that a painful bodily state corresponds to a thermal bodily state of a certain type. This correspondence is a relation in virtue of which an element is experienced as associated with another element, such that they tend to be described in similar terms. For instance, a visual stimulus and a sound may be characterized as being 'high' and some painful bodily states may be described as 'pressing' or 'pulsing' in analogy to certain tactile states.

As observed above, there are several types of correspondences, and so the specific character of the correspondence relation may vary depending on a particular type of crossmodal correspondence. For instance, experiencing element *A* as corresponding to *B* may consist in experiencing *A* as likely co-occurring with *B*, as being associated with the same external entity as *B*, as motivating similar actions as *B*, or as having a similar significance for the organism as *B*. The correspondence occurs between a bodily state and a bodily state of a certain type because one element typically corresponds to many elements with a common characteristic. For instance, the sound 'kiki' is not experienced as corresponding to a particular angular shape, but to many shapes with a property of being angular.

The approach described above does not require postulating that bodily states presented by multimodal experiences of pain are experienced as having any internal structure organized by instantiation or parthood relations. The tactile-like or thermal-like character of pain does not consist in presenting some combination of a subject and properties or in presenting proper parts composing a mereological whole. Instead, tactile-like and thermal-like pains occur when a presented painful bodily state, which may be atomic in respect of an experiential internal structure, is experienced as standing in a correspondence relation to a tactile or a thermal state of a certain type. It should be noted that this position is coherent even if one believes that correspondence is a form of similarity, and that similarity between any entities *A* and *B* entails the presence of a property *F* instantiated both by *A* and *B*. This is because representing introduces an intentional context—that is, even if similarity between *A* and *B* entails that a property *F* is instantiated both by *A* and by *B*, it may still be the case that *A* and *B* are *represented* as similar without being *represented* as instantiating the property *F*. Hence, it can be maintained that a painful bodily state is represented as similar to a tactile or thermal bodily state without representing that this state has the structure of a subject instantiating properties.

Further, I argue that the approach formulated in terms of crossmodal correspondences is free from difficulties that threaten the instantiation and parthood theories. First, I show that mechanisms of pain perception are such that they can sustain crossmodal correspondences between painful, tactile, and thermal bodily states. Second, I show how interpreting multimodal pains as cases of crossmodal correspondences allows us to avoid interpreting various adaptive multimodal pains as inaccurate representations.

6.1 Mechanisms of crossmodal correspondences

The huge variety of crossmodal correspondences makes it unlikely that there is a single mechanism responsible for all of them. Nevertheless, scientists investigating crossmodal correspondences have proposed three non-exclusive types of correspondences which are likely to happen due to distinct relations between mechanisms responsible for experiencing corresponding elements (Deroy and Spence, 2016; Parise, 2016; Spence, 2011). The first type is the structural correspondence in which an element *A* is experienced as corresponding to an element *B* because a mechanism in virtue of which *A* is experienced is similar to a mechanism in virtue of which *B* is experienced. For instance, the correspondence between bright and loud stimuli is likely to be structural, as both higher levels of brightness and loudness are coded by an increase in neuron firing. The second type is the statistical correspondence arising from learning that some elements are likely to occur together in the environment. In this case, an element *A* is experienced as corresponding to an element *B* because it has been learnt that activities of mechanisms in virtue of which an element *A* is experienced frequently co-occur with activities of mechanisms in virtue of which an element *B* is experienced. It has been proposed that the correspondence between smaller visual size and higher pitch is a statistical correspondence, as smaller objects are more likely to emit higher-pitched noises. Finally, there are semantic correspondences, which occur due to common, higher-order (e.g., linguistic) classification of experienced elements. In this case, the correspondence between an element *A* and an element *B* is likely to be due to a common top-down influence on the outputs or functioning of mechanisms in virtue of which *A* and *B* are experienced. It is believed that correspondence between higher pitch and higher visual elevation is at least partially a result of the use of the same predicates ('high' and 'low') in both cases.

In the Sections 3 and 4, I argued that multimodal tactile-like and thermal-like pains should not be analysed in terms of instantiation or parthood. Such phenomenal structures occur due to the function of unification and differentiation mechanisms, not due to pattern classification mechanisms which underline experiences of pain. Now, it can be asked whether such pattern classification mechanisms are suitable for sustaining experiences which present an element as corresponding to a distinct element. I believe that the answer is positive no matter which type of correspondences is considered.

The structural correspondence occurs when an element *A* is experienced as corresponding to an element *B* because the experience of *A* and the experience of *B* happen in virtue of similar mechanisms. The question is whether pattern classification mechanisms responsible for pain experiences are such that structural correspondences can happen in virtue of them. I believe that it is so because pain mechanisms, such as pain-gates, process input from various types of afferent fibres. Some of these fibres are specialized in providing information concerning noxious stimuli, but others are specialized in transferring information concerning tactile or thermal stimulation. In consequence, input elements, in virtue of which pain experiences arise, may be, in some instances, partially similar to input elements which lead to tactile or thermal experiences. Thus, the functioning of pain mechanisms in cases of some tactile-like and thermal-like pains is likely to be similar to the functioning of mechanisms processing non-painful tactile or thermal stimulation, which may lead to structural correspondences.

Similarly, statistical correspondences can be sustained by pattern classification mechanisms because frequent co-occurrence, and so co-computing, of certain stimuli may shape the function of such mechanisms in a process we may broadly label as 'learning'. For instance, it may be the case that due to frequent co-occurrence of certain nociceptive and tactile inputs, the neurosignatures provided by the neuromatrix undergo modifications. Then, even when a pinprick pain is caused by laser stimulation without activating the fibres that usually transfer tactile information, the correspondence occurs because, in virtue of the usual co-occurrence, the neuromatrix provides a neurosignature typically produced by processing both nociceptive and tactile information.

Semantic correspondences can also arise from interactions between pattern classification mechanisms and some higher-order mechanisms—for instance, those responsible for the application of linguistic categories. In this case, a correspondence between an element *A* and an element *B* may occur because the output of pain mechanisms processing *A* and the output of tactile or thermal mechanisms processing *B* are further classified in the same way by a higher-order system. For example, the second pain that persists after injury may correspond to some thermal sensations partially because both sensations are linguistically categorized as 'hot' or 'burning'.

The above considerations demonstrate that while experiential structures organized by instantiation and parthood relations are unlikely to be sustained by pattern classification mechanisms, such problems do not occur in the case of crossmodal correspondences. In consequence, the framework of crossmodal correspondences is suitable for analysing tactile-like and thermal-like multimodal pains. It should be noted that due to the variety of crossmodal correspondences, I do not postulate that all tactile-like and thermal-like pains are purely sensory experiences. While a purely sensory character is plausible in the case of structural correspondences, it is likely that statistical and semantic correspondences also occur in virtue of input provided by mechanisms responsible for memory, imagery, and language.

6.2 Content of crossmodal correspondences

Proponents of instantiation theories and parthood theories of multimodal experiences adopt representationalist positions which lead to the conclusion that many multimodal pains, despite being adaptive and resulting from the normal function of pain mechanisms, are inaccurate. In consequence, it should be asked whether characterization in terms of crossmodal correspondences can provide an approach to multimodal pains that does not render adaptive pains illusory.

When a crossmodal correspondence is experienced, an element *A* is presented as corresponding to an element *B* of a certain type. In virtue of the occurrence of correspondence relations, elements are associated with each other and are likely to be described in similar terms. As stated earlier, the specific character of the correspondence relation may vary depending on the particular type of crossmodal correspondence. For instance, small visual stimuli may be experienced as corresponding, in the sense of likely co-occurrence, to a high-pitched, but not to a low-pitched, sound. Similarly, sudden, bright visual stimuli may be experienced as corresponding, in the sense of leading to a similar defensive reaction, to a sudden loud sound. From the perspective of the current investigation, the crucial characteristic of such experiences is that when *A* is experienced as corresponding to *B*, it is not the case that properties of *B* are attributed to *A* or that parts of *B* are presented as parts of *A*. Instead, *A* is represented as standing in a correspondence relation to *B*. For example, when a bright visual stimulus is experienced as corresponding to a loud noise, it is not represented as having a certain loudness or pitch, but as standing in a correspondence relation to a noise of a certain type.

When applied to multimodal pain experiences, this position states that in the case of tactile-like or thermal-like pains, a bodily state is represented as corresponding to a tactile or thermal experience of a certain type. From this perspective, a painful sensation is described as 'pressing' because the represented bodily state is experienced as corresponding to a tactile state of a type which is likely to be described in such terms. However, in the case of crossmodal correspondence, the painful bodily state is not represented as instantiating some tactile properties or having some tactile parts. In consequence, even when a multimodal experience of pain happens without co-occurring tactile or thermal stimulation, such an experience of pain can still be accurate. It is so because what is represented is merely that a bodily state corresponds (e.g., in the sense of frequently co-occurring) to some tactile or thermal state of a certain type. It should be noted that a painful bodily state can be accurately represented as corresponding to a different bodily state, even if the corresponding element is not currently present. This can happen because the information concerning the corresponding tactile or thermal states of the relevant types can be available to the cognitive system due to the functioning of memory, imagery, and semantic knowledge, without the actual occurrence of a given bodily state. Thus, it may be accurately represented that a pinprick pain corresponds to a tactile state of a certain type, even if no such tactile state is currently occurring.

On the other hand, painful experiences are often preceded by a tactile or thermal stimulation that increases in intensity and finally causes a painful sensation. Relying on this common pattern, one may doubt whether there is a strong analogy between phenomenally multimodal pains and other cases of crossmodal correspondences. In usual cases of crossmodal correspondences, the correspondence concerns elements presented by two experiences belonging to distinct modalities. However, when a painful experience is the result of a tactile or thermal stimulation, it may seem that there is a sequence of experiences belonging to the same modality—for instance, at first a non-painful tactile experience, and then a painful tactile experience, as each experience is caused by the same type of stimulus. Nevertheless, even if such analysis is correct. it is not inconsistent with my main proposal (even if it is no longer strictly a *crossmodal* correspondence). I argue that phenomenally multimodal pains present painful bodily states as corresponding to tactile or thermal bodily states of certain types. It may be the case that an experience, at first, presents a non-painful tactile or thermal bodily state *S* and subsequently starts to present a painful bodily state *P* as corresponding to the earlier state *S*. Furthermore, while there is no philosophical agreement regarding the individuation of sensory modalities, there are reasons to believe that when a painful sensation is caused by a tactile or thermal stimulation, experiences from two modalities are present. This is because when an earlier non-painful experience leads to a subsequent painful experience, the underlying combinations of inputs change what causes a difference in sensory processing: Only the later combination of inputs opens a pain-gate and leads to recognition of a pain neurosignature.

Furthermore, characterization in terms of crossmodal correspondences can easily account for the fact that tactile-like and thermal-like pains often lead to adaptive reactions, even when no actual tactile or thermal stimulus is present. It is so because in virtue of representing a painful bodily state as corresponding to some type of tactile or thermal bodily state, a multimodal painful sensation may motivate behaviours that are suitable, given the content of an occurring correspondence. For instance, a laser-induced pinprick pain represents a bodily state as corresponding to a bodily state evoked by a rapid, finely localized tactile stimulation, and so evokes a swift withdrawal reaction. On the other hand, a thermal-like second pain represents a bodily state as corresponding to a longer-lasting bodily state caused by thermal stimulation, and so leads to actions aimed towards prolonged care concerning the bodily fragment.

7 Conclusions

I have argued that in the case of phenomenally multimodal tactile-like and thermal-like pains, the usual frameworks referring to the notion of instantiation and parthood are not plausible. A more promising approach is to characterize multimodal pains in terms of crossmodal correspondences. According to this proposal, tactile-like pain consists in experiencing that a painful bodily state corresponds to a tactile bodily state of a certain type, and thermal-like pain consists in experiencing that a painful bodily state corresponds to a thermal bodily state of a certain type.

Acknowledgements

The author would like to thank the anonymous reviewers for their comments. The work was supported by the National Science Centre (Poland), grant 2018/31/D/HS1/00363.

References

Asano, M., Imai, M., Kita, S., Kitajo, K., Okada, H., and Thierry, G. (2015). Sound symbolism scaffolds language development in preverbal infants. *Cortex*, 63, 196–205.

Aydede, M. and Güzeldere, G. (2002). Some foundational problems in the scientific study of pain. *Philosophy of Science*, 69, S265–83.

Bain, D. (2013). What makes pains unpleasant?. *Philosophical Studies*, 166, 69–89.

Briscoe, R. E. (2016). Multisensory processing and perceptual consciousness: part I. *Philosophy Compass*, 11(2), 121–33.

Brunetti, R., Indraccolo, A., Mastroberardino, S., Spence, C., and Santangelo, V. (2017). The impact of cross-modal correspondences on working memory performance. *Journal of Experimental Psychology: Human Perception and Performance*, 43(4), 819–31.

Corns, J. (2018). Recent work on pain. *Analysis*, 78(4), 737–53.

Deroy, O. and Spence, C. (2016). Crossmodal correspondences: four challenges. *Multisensory Research*, 29, 29–48.

Green, E. J. (2019a). A theory of perceptual objects. *Philosophy and Phenomenological Research*, 99(3), 663–93.

Green, E. J. (2019b). Binding and differentiation in multisensory object perception. *Synthese*, 198, 4457–4491.

Green, E. J. and Quilty-Dunn, J. (2017). What is an object file? *British Journal for the Philosophy of Science*, 72(3), 665–699.

Haggard, P., Iannetti, G. D., and Longo, M. R. (2013). Spatial sensory organization and body representation in pain perception. *Current Biology*, 23, R164–76.

Hoffman, D. D. and Richards, W. A. (1984). Parts of recognition. *Cognition*, 18(1–3), 65–96.

Julius, D. and Basbaum, A. I. (2001). Molecular mechanisms of nociception. *Nature*, 413(6852), 203–10.

Klein, C. (2007). An imperative theory of pain. *Journal of Philosophy*, 104, 517–32.

Köhler, W. (1947). *Gestalt Psychology: An Introduction to New Concepts in Modern Psychology*. New York, NY: Liveright.

Macpherson, F. (2011). Cross-modal experiences. *Proceedings of the Aristotelian Society*, 111, 429–68.

Matthen, M. (2010). On the diversity of auditory objects. *Review of Philosophy and Psychology*, 1, 63–89.

Melzack, P. (1993). Pain: past, present and future. *Canadian Journal of Experimental Psychology*, 47(4), 615–29.

Melzack, P. (1999). From the fate to the neuromatrix. *Pain Supplement*, 6: S121–6.

Melzack, R. and Wall, P. D. (1965). Pain mechanisms: a new theory. *Science*, 150, 971–9.

Mendell, L. M. (2014). Constructing and deconstructing the gate theory of pain. *Pain*, 155, 210–16.

Mitroff, S. R., Scholl, B. J., and Wynn, K. (2005). The relationship between object files and conscious perception. *Cognition*, 96, 67–92.

Moayedi, M. and Davis, K. D. (2013). Theories of pain: from specificity to gate control. *Journal of Neurophysiology*, 109, 5–12.

O'Callaghan, C. (2008). Object perception: vision and audition. *Philosophy Compass*, 3(4), 803–29.

O'Callaghan, C. (2014). Intermodal binding awareness. In *Sensory Integration and the Unity of Consciousness*, edited by D. J. Bennett, C. S. Hill, 73–103. Cambridge, MA: MIT Press.

O'Callaghan, C. (2015). The multisensory character of perception. *Journal of Philosophy*, 112(10), 551–69.

O'Callaghan, C. (2016). Objects for multisensory perception. *Philosophical Studies*, 173, 1269–89.

O'Reilly, R., Busby, R., and Soto, R. (2003). Three forms of binding and their neural substrates: alternatives to temporal synchrony. In *The Unity of Consciousness: Binding, Integration, and Dissociation*, edited by A. Cleeremans, 168–90. Oxford: Oxford University Press.

Palmer, S. and Rock, I. (1994). Rethinking perceptual organization: the role of uniform connectedness. *Psychonomic Bulletin and Review*, 1(1), 29–55.

Parise, C. V. (2016). Crossmodal correspondences: standing issues and experimental guidelines. *Multisensory Research*, 29, 7–28.

Peláez, F. J. R. and Taniguchi, S. (2016). The gate theory of pain revisited: modeling different pain conditions with a parsimonious neurocomputational model. *Neural Plasticity*. Available from: http://dx.doi.org/10.1155/2016/4131395

Price, D. D. and Aydede, M. (2005). Painfulness is not a auale. In *Pain: New Essays on its Nature and the Methodology of its Study*, edited by M. Aydede, 315–24. Cambridge, MA: MIT Press.

Pylyshyn, Z. W. (2007). *Things and Places*. Cambridge, MA: MIT Press.

Rensink, R. A. (2000). The dynamic representation of scenes. *Visual Cognition*, 7(1/2/3), 17–42.

Richardson, L. (2014). Non sense-specific perception and the distinction between the senses. *Res Philosophica*, 91(2), 215–39.

Skrzypulec, B. (2011). The structure of audio–visual consciousness. *Synthese*, 198, 2101–27.

Spence, C. (2011). Crossmodal correspondences: a tutorial review. *Attention, Perception, and Psychophysics*, 73, 971–95.

Spence, C. and Deroy, O. (2013). How automatic are crossmodal correspondences?. *Consciousness and Cognition*, 22, 245–60.

Treisman, A. M. (1998). Feature binding, attention and object erception. *Philosophical Transactions of the Royal Society of London. Series B. Biological Sciences*, 353(1373). 1295–306.

Tsao, H., Danneels, L. A., and Hodges, P. W. (2011). ISSLS Prize Winner: smudging the motor brain in young adults with recurrent low back pain. *Spine*, 36(21), 1721–7.

Wall, P. D. (1996). Comments after-30 years of the gate control theory. *Pain Forum*, 5(1), 12–22.

Wan, X., Woods, A. T., van den Bosch, J. J. G., McKenzie, K. J., Velasco, C., and Spence, C. (2014). Cross-cultural differences in crossmodal correspondences between basic tastes and visual features. *Frontiers in Psychology*, 5(1365), doi: 10.3389/fpsyg.2014.01365.

Xu, Y. and Singh, M. (2002). Early computation of part structure: evidence from visual search. *Perception and Psychophysics*, 64(7), 1039–54.

23
Pictures as supramodal sensory individuals

Fabrizio Calzavarini and Alberto Voltolini

1 Introduction

It is unclear whether pictures may be taken as *sensibilia*—that is, merely sensory individuals that are ascribed only perceptual properties, like rainbows and tricks of the light. For even holograms, the paradigmatic cases of pictorial *visibilia* (Martin, 2012), may be seen as having *merely surface properties* (Lopes, 2005)—that is, properties affecting the physical constitution of their *vehicles*, their physical bases. The famous scene in *The Terminator* series, in which the Terminator character, played by Arnold Schwarzenegger, deceives his enemies by letting them fight against a hologram of him they mistake for him, clearly shows this. For in the movie, his hologram clearly shows it is a hologram that merely presents the Terminator insofar as it vanishes in a blaze of lights.

Yet independently of whether the above idea is correct, pictures can be taken as *sensory individuals* in a weaker sense—namely, ordinary individuals to which properties are ascribed perceptually (Nanay, 2013).

In a sense, this is quite trivial. For, qua physical objects, pictures surely instantiate perceptual properties like any other such objects. Yet this is obviously not the sense in which one is interested in classifying a picture as a sensory individual in the previous characterization. For the interesting question is whether, qua picture—that is, in displaying its pictorial character qua figurative representation or *depiction*, a picture is a sensory individual.

This question can be hard to answer if one considers the properties that are ascribed not to the picture's vehicle, but to the picture's *subject*—that is, what a picture presents (Voltolini, 2015a). For some people (first of all, Wollheim, 1980, 1987, 1998), one must take this for granted. For in the so-called *recognitional fold* of the sui generis twofold perceptual experience of *seeing-in* that concerns a picture qua picture—that is, the fold that is addressed to grasp the picture's subject—one basically perceives the properties ascribed to that subject (Wollheim, 2003). Yet for some other people (Walton, 1990, 2008), even if pictorial experience can be analogously conceived as twofold, the properties ascribed to that subject are basically imagined, and not perceived.

Fortunately enough, however, there is no need to address this controversy, for the vehicle's perceptual properties that, for Wollheim, are captured in the other fold of the seeing-in experience—that is, the *configurational fold*, which is precisely addressed to grasp the picture's vehicle—are constituted not only by the aforementioned merely surface properties, but also by the so-called (Lopes, 2005) *design properties*—that is, the properties responsible for the fact that a certain subject is seen in that picture. Thus, pictures are sensory individuals qua pictures in virtue of their instantiating design properties in their pictorial vehicles.

Fabrizio Calzavarini and Alberto Voltolini, *Pictures as supramodal sensory individuals* In: *Sensory Individuals*. Edited by: Aleksandra Mroczko-Wąsowicz and Rick Grush, Oxford University Press.
DOI: 10.1093/oso/9780198866305.003.0024

Primarily, these properties are the vehicle's colours, shape, and texture. But these *low-level* perceptual properties are not the only design properties that there are. For, as Lopes himself (1996, p. 3) admits and Kennedy (2000, pp. 82–4) stresses, also *high-level* perceptual properties bear the same responsibility, as the cases of the so-called aspect dawning pictures, such as the picture of a Dalmatian and of perceptually ambiguous pictures prompting a multistable perception such as the duck–rabbit picture, clearly show. Consider *grouping properties*—that is, the properties for the vehicle's elements to be arranged in a certain way along a certain direction and under a certain dimension (Voltolini, 2015a). Such properties are high-level perceptual properties, since they merely generically depend on the vehicle's colours and shapes (Voltolini, 2020). In the picture of a Dalmatian, one can only see a Dalmatian if one groups the black-and-white spots of its vehicle in a certain way; in the duck–rabbit picture, one can only see either a duck or a rabbit once one groups the vehicle's lines either in one way or in another one.[1]

Now, our main claim in this paper is that many at least of such design properties are also *multimodal* (Kennedy, 1997, p. 81), or even *supramodal*, high-level perceptual properties. By 'supramodal' we mean properties that can be indifferently accessed via different sense modalities, since they are not specifically related to any such modality. We will try to show this by appealing in particular to so-called *tactile pictures*—that is, pictures that are not only objects of sight, but also and primarily objects of touch, for such pictures show that, among their design properties, not only the shape properties, but also the grouping properties of their vehicles, are supramodal high-level perceptual properties. Or so we claim. Indeed, this might provide strong support for the conjecture that pictures are not *either* visual *or* tactile, but supramodal sensory individuals (as regards their vehicles).

The architecture of this paper is as follows. In Section 2, we sum up the philosophical debate about tactile pictures and provide some theoretical reasons as to why there can be such pictures. In Section 3, we gather some empirical evidence showing that congenitally blind people can tactilely grasp 3D shapes, in order to adequately sustain the hypothesis that the perception of shapes is neither intrinsically visual nor intrinsically tactile, but supramodal. In Section 4, we focus on what enables people to have a pictorial perception, even of tactile pictures—namely, to be able to grasp 3D groupings of 2D elements that physically constitute a picture. In Section 5, we appeal to the extant empirical evidence pointing to the idea that congenitally blind people are also able to gain such a grasp, and suggest how further empirical evidence could be found in order to prove that pictorial perception itself is supramodal as well.

2 The philosophical debate on tactile pictures

From an empirical point of view, there exists a vast body of literature in the field of cognitive psychology stating that when pictures are shown to congenitally blind people in forms

[1] Some people (Brewer, 2011; Kriegel, 2011) question that grouping properties are perceptual properties. Indeed, according to some (Pylyshyn, 2003), grasping such properties is merely a matter of attention, not of perception. Yet this overlooks that attention here plays a holistic, not a focalizing, role (Jagnow, 2011); hence, it yields a specifically perceptual contribution to the relevant experience (Martina and Voltolini, 2017; Voltolini, 2020), as it does whenever it is so involved (Burnston, 2017; Stokes, 2018). We will later see that on the empirical side, this claim is supported by the extant psychological, as well as neurological, evidence. For this idea, see also Block, 2014 and Kennedy, 2000.

Figure 23.1 Tactile pictures.
Reproduced with permission of Lea Ferro.

of bas-relief (paintings, drawings, sketches), they are able to recognize the figurative value of the pictures by touch—that is, to recognize what makes them *pictures* or *depictions*, in other words a particular type of representation, namely *figurative*. What is proven on the side of recognition corresponds on the side of production. The same people are able to produce standard 2D drawings based on raised-line drawings. As with any other ordinary picture, these pictures depict subjects in a 3D scenario (Kennedy, 1993; see also D'Angiulli *et al.*, 1998; Kennedy and Bai, 2002; Pawluk *et al.*, 2011). In Figure 23.1, one can see, on the left, a raised-line drawing recognizable as a picture by a congenitally blind, and on the right, a standard 2D drawing with a similar pictorial subject made by such a person.

In the philosophical debate on the figurative value of pictures, and thus the debate on what makes a picture a *depiction*, a *figurative* type of representation, some philosophers have made full use of these empirical data. In the framework of his theory of depiction based on the ability to recognize objects, Lopes (1996) has argued that, just as there are visual pictures, there are also *tactile pictures*—that is, pictures whose figurative value consists in the fact that the subjects interacting with them by grasping this value are able to recognize by touch the same objects and 3D scenes they would have been able to if they had had real life perception of them. In other words, tactile pictures, too, have a prosthetic value, as they make it possible to extend our ability to recognize objects and 3D scenes. This makes them bona fide pictures (Lopes, 1997).

Other philosophers, instead, are more sceptical about this. Hopkins, for example, has argued that, despite appearances, tactile pictures are not genuine pictures, unlike visual ones. Unlike vision, touch does not activate a *point of origin* from which the scene perceived can be apprehended. There is no 'point of touch', as instead exists with a point of view. Indeed, touch tends to allow full contact by the perceiver, and not perspectival contact, or rather contact between certain parts of her body and the objects tactilely perceived. Thus, a 3D scene is not tactilely rendered in perspective. Yet, perspective is vital precisely since the picture's figurative value, as the 3D scene presented by a picture, is obtained by starting out from a perspective (which, incidentally, does not change, as Wollheim (1980) noted, with any change in the position of the picture's perceiver) (Hopkins, 2000, pp. 156–8; 2004, pp. 155–6).

But Hopkins' theory is controversial. For one thing, let us think of the Frege (1980) problem, the problem that, (when exposed) on a mental, and not linguistic, level (as was originally the case for Frege himself), an individual may not recognize that what in her mental states prove to be different aspects of the same object actually are one and the same object. This problem paradigmatically occurs in visual mental states: the ancient Babylonians did not recognize that the object that presented itself to them visually in the evening as the first celestial body to appear in the heavens—the evening star—was actually the same object Venus that presented itself to them as the first object to disappear from the heavens—the morning star. But there is nothing to prevent a similar problem from also occurring in the tactile mode: an individual may fail to recognize that the object she is groping on the bedside table is the same object she was tactilely searching for a moment before. Now, the problem on the visual level is clearly one of perspective: a subject does not recognize that an object seen from a certain visual perspective is the same object as seen from another. But the same is true on the tactile level: the object perceived by touch in one of its aspects is not recognized as the same object perceived by touch in another of its aspects. This is indirectly confirmed by the fact that different blind, or merely blindfolded, individuals may have different tactile experiences of the same object, since it is given to them by different perspectives (Gallace and Spence, 2014, p. 117). Indeed, many experiments in psychology have stressed that there is a vantage point in touching objects on the blind people's part, since they grasp occlusions and directions in what they touch, possibly different ones if orientation changes, just as in sight (Kennedy, 1997, p. 78; Kennedy, 2000, p. 80; Kennedy and Juricevic, 2006, pp. 80–3). What may make it seem to Hopkins (2000, pp. 157–9) that there is no 'point of touch' is that, since touch is phenomenologically based on contact between the toucher and the touchee, it appears arbitrary to select a point of origin for touch from among the different parts of the toucher's body in contact with the touchee. But this simply shows that unlike sight, in touch, a vantage point is not anchored at a specific distance from the touchee (Kennedy and Juricevic, 2006, p. 90). Indeed, even in the case of vision, there is something that should not be called arbitrary, but rather an idealization. The point of origin giving rise to the vision of a scene is geometric, not subjective, as binocular vision demonstrates; if another individual were at the same point of geometrical origin at which I am now, she would see the scene that I perceive from the same perspective.

3 The supramodal perception of a three-dimensional shape

Hopkins' thesis is controversial not only at the theoretical level, but also at the level of empirical research. A series of behavioural studies carried out on neurotypical subjects does, in fact, suggest clearly that tactile recognition, too, like visual, depends on perspective or, rather, is influenced by the position of the object with respect to the observer (Lacey and Sathian, 2014). For example, it has been demonstrated that the subjects' performance in a tactile recognition task regarding unfamiliar objects (combinations of Lego blocks assembled in various ways) is dramatically less successful if the object to be recognized is rotated through 180° along any one of its axes, with respect to its original position (Figure 23.2; Newell *et al.*, 2001). Or again: experimental research suggests that objects presented in tactile modality, just like those presented in visual modality, have a privileged perspective or a *canonical* perspective. When objects are presented in this perspective, tactile recognition is easier (Woods *et al.*, 2008). Moreover, a long tradition of studies using the graphesthesia task has revealed that individuals tend to adopt a *natural* (self-centred) perspective when recognizing ambiguous tactile symbols, such as the letters b, d, p, and q (e.g., Arnold *et al.*, 2016), although a number of factors might influence the adoption of this spatial perspective (Arnold *et al.*,

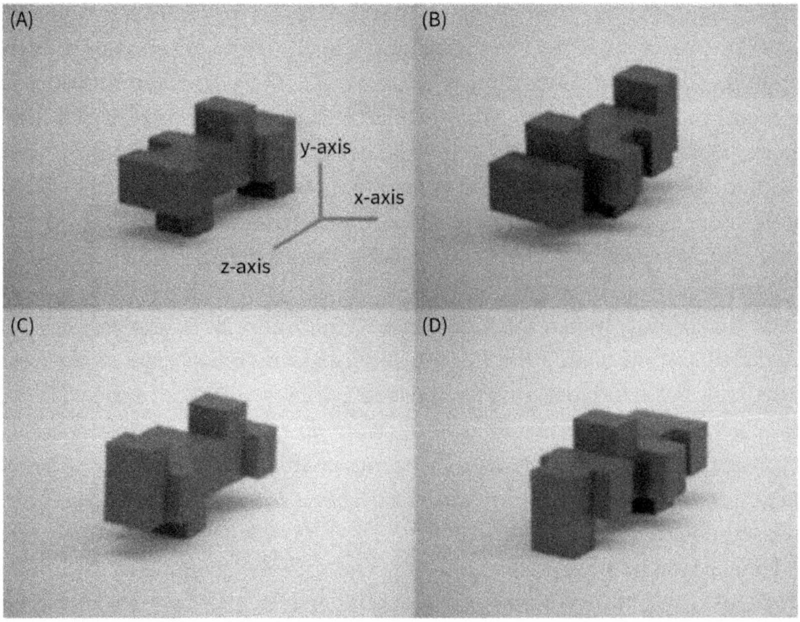

Figure 23.2 Sample of unfamiliar objects as those used in the study of Newell *et al.* (2001).

2017). These behavioural data seem to confirm the intuition that there exists a 'point of touch', in the same way there exists a point of view; or, contrary to Hopkins' affirmations, touch, too, has a perspectival dimension, at least as far as 3D objects are concerned.

On the basis of these findings, some psychologists have even argued that, as regards the 3D apprehension of the shape of objects, the visual system and the tactile system are functionally equivalent. The thesis of the functional equivalence of sight and touch is also supported by other data from behavioural studies, such as the fact that tactile recognition, like visual, is influenced by the *size* of objects. In other words, a change in the size of the objects presented produces very similar effects in both visual and tactile recognition tasks, causing a significant drop in success in both cases (Craddock and Lawson, 2009). More important still, functional equivalence is supported by a growing number of experimental studies based on widely varying techniques, such as functional magnetic resonance imaging (fMRI) and transcranial magnetic stimulation (TMS), or experimental studies on subjects with brain damage, all of which converge to suggest the existence of significant overlap of the neural structures serving both visual and tactile modalities (Lacey and Sathian, 2014; Ricciardi *et al.*, 2014).

The overlap between visual and tactile neural structures seems to be at its height at the so-called *lateral occipital complex* (LOC), a group of brain areas localized in the ventral visual pathway, between the occipital gyrus and the inferior temporal gyrus. For a long time, the LOC was considered a specifically visual structure, devoted to visual apprehension of the shapes of objects (Grill-Spector *et al.*, 2001). From the early 2000s onwards, however, research in neuroimaging has clearly shown that in subjects with normal eyesight, this structure responds selectively to the shape of objects, even if they are touched only (Amedi *et al.*, 2002). One immediate explanation may be that, when we explore an object by touch, we tend, either consciously or unconsciously, to form a mental visual image of the object. This hypothesis is supported by the observation that the LOC is normally activated in visual *imagery* tasks (De Volder *et al.*, 2001) or in tasks in which subjects must recover from memory information on the shape of objects (Newman *et al.*, 2005). In addition, the activation of this area seems to be directly correlated with how vivid the subjects reported their imagery experience to be (Zhang *et al.*, 2004). Nevertheless, it has been observed that the LOC is activated both in subjects who are unable to form visual mental images of objects, such as the congenitally blind, and in those with early blindness.[2] These findings lead us to suppose that activation of the LOC by tactile stimuli is 'direct'—that is, it does not necessarily require the mediation of visual imagery.

One of the first studies in neuroimaging to yield this information was carried out by Pietro Pietrini *et al.* (2004), who used the fMRI technique to investigate the neural responses of subjects with normal eyesight during a visual and tactile recognition task, and those of sight-impaired subjects (congenitally or early blind; from now onwards, we take this specification for granted) in a tactile recognition task. The stimuli in this study were 3D objects divided into three categories: bottles, shoes, and masks representing human faces. During the tactile recognition task, subjects were asked to explore the objects with both hands, attempting to form an overall representation of their shape. Subjects with normal eyesight were blindfolded. The study's results show that in subjects with normal eyesight, the experience of the shapes of objects in both visual modality and tactile modality activates an area of the brain included in the LOC, namely the inferior temporal cortex (IT), with different patterns of activation for the specific categories of objects (bottles, shoes, faces). What is more, this same region of the brain, like other 'visual' areas of the ventral temporal cortex (e.g., the fusiform gyrus), was also activated by tactile stimuli in sight-impaired subjects, showing activation patterns very similar to those observed in the brains of subjects with normal eyesight (Figure 23.3).

For this paper's purposes, what must be stressed is that these findings suggest substantial convergence of the neural structures involved in the recognition of objects, in particular the LOC, both in those with normal eyesight and in the congenitally blind. It is interesting to observe that involvement of the LOC in the perceptual recognition of 3D shapes does not seem to be limited to visual and tactile modalities. The LOC, for example, is also activated by 'sensory-substitution' systems able to use specific algorithms to convert visual information on the shape of objects into auditory information (Amedi *et al.*, 2007). For some authors, the multisensorial profile of the LOC demonstrates that, even in those with normal eyesight, this region supports processes of elaboration of the 3D shape of objects operating indifferently in multiple modalities of perception (Amedi *et al.*, 2002; Lacey and Sathian, 2014; Ricciardi *et al.*, 2014). In other words, for this hypothesis, perception of the shape of objects is intrinsically *neither* visual *nor* tactile, but rather *supramodal* (or amodal). This hypothesis reflects the growing consensus that the so-called *meta-modal* or *supramodal* paradigm (Pascual-Leone and Hamilton, 2001; Ricciardi *et al.*, 2014; Heimler *et al.* 2015) is gaining in the field of neuroscience, for which the architectural principle guiding the development and operation of the human brain is not subdivision according to perceptual (visual, tactile, auditory, olfactory, gustatory) and motor modalities, but rather the (type of) task the brain is called upon to perform. If this paradigm is accepted, there is obviously one more reason

[2] Granted, this does not mean that congenitally blind individuals are unable to engage in other forms of mental imagery, such as tactile or spatial imagery (Marmor and Zaback, 1976). Also note that, according to some scholars (Nanay, 2020), a portion of blind people, including congenitally blind individuals, are able to experience vivid visual mental images.

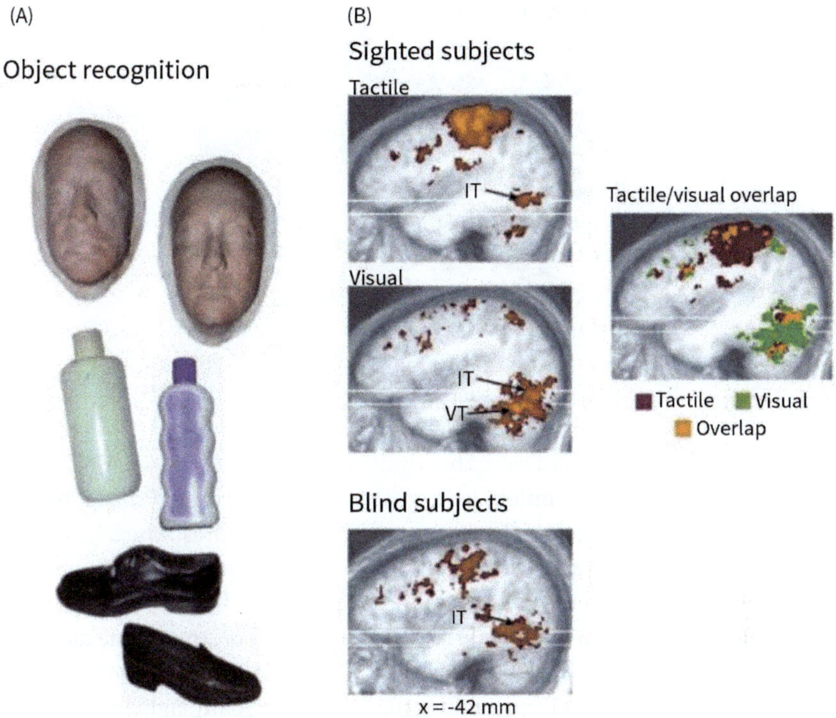

Figure 23.3 Stimuli (left) and results (right) of the study by Pietrini *et al.* (2004).
Adapted from Pietrini *et al.* (2004). Copyright (2004) National Academy of Sciences, U.S.A.

to believe that the brain of a subject with normal eyesight and that of a congenitally blind person operate on the same level when the recognition of an object's 3D shape is at stake.

4 Picture perception depends on the supramodal apprehension of 3D groupings

Thus far, the results from neuroscience allow us to postulate that, being given indifferently by both visual and tactile modalities, perceptual apprehension of the shapes of objects and their 3D depth ratios is supramodal.[3] But at most this tells us that the congenitally blind are able to grasp *sculptures*, 3D objects used to represent other objects or 3D scenes. On this, too, cognitive psychology can already offer confirmation. Congenitally blind subjects are able to recognize by touch certain objects or 3D scenes rendered in perspective by their 3D models (e.g., models of city squares and monuments) (Tinti *et al.*, 2018). This is not astounding. In general, sculptures were originally conceived to elicit tactile reactions towards them, and some have been specifically thought to be tactilely addressed by blind people (Gallace and

[3] There is no room for addressing the question here, but if shape recognition is supramodal, then plausibly enough, one can positively answer to the famous Molyneux's question of whether, if a congenitally blind were able to suddenly recover sight, she would able to visually recognize the same shapes she was able to tactilely grasp. For a recent debate on this, cf. Ferretti and Glenney (2020).

Figure 23.4 A group of horses.
Reproduced with permission of Paola Tosti.

Spence, 2014, pp. 109–10). Yet, unless we argue that sculptures are *pictures*, but simply in 3D—another controversial thesis in the literature regarding pictures, which has its defenders (Walton, 2008) and its opponents (Hopkins, 2004)—this tells us nothing about whether the congenitally blind are able to grasp pictures in terms of their figurative value, particularly ordinary 2D pictures like paintings, and whether therefore there are genuine tactile pictures.

In order to prove this, it is necessary to establish whether the congenitally blind are able to carry out the particular *grouping* operation that allows normally sighted people to understand that a 2D object is a picture of a 3D subject. This operation makes it possible to grasp the grouping properties of elements of an object arranged in one direction, in terms of a certain dimension. Grouping properties are, in fact, the properties of organization of an object that allow its elements to be grouped according to a certain directional order (related to coordinates of reference) along a certain dimension. The import of grouping operations in perceiving pictures has been already acknowledged in psychological research (Kennedy, 2000, p. 84; Kennedy and Juricevic, 2006, pp. 83–4). In particular, the decisive factor for grouping that makes it possible to grasp the figurative value of a 2D object is the grouping in front → behind (or even behind → in front) along the dimension of depth, because this is what enables the object to present a 3D scene, and thus to count as a picture (Voltolini, 2015a). Now, if this sort of grouping operation may occur also as regard tactile pictures, then there is no further reason to deny them a pictorial status.[4]

The decisive nature of this type of grouping for figurative purposes can clearly be seen in the case of so-called aspect-dawning pictures, those very pictures whose figurative value is not immediately grasped until a similar grouping operation has been carried out. Consider the Figure 23.4. For a certain time lapse, all we can make out in this figure are black-and-white patches. But then, after a while, by distinguishing certain subjective contours (Lopes, 2005, pp. 41–2), we group some of the patches in order to characterize something standing in the *forefront* against the *background* of something else. This is when we are able to see the

[4] A further proof on this concern would be given if one were able to show that, just as with visual pictures (Ferretti, 2018a; Voltolini, 2015a), the pictorial perception one has of a tactile picture lacks a feeling of presence as regards the picture's subject, so that even the affordance properties that are ascribed to that subject are neutralized. We leave this point for another investigation. The same must be said as regards whether this pictorial perception is inflected in the tactile case, like in the visual case (Voltolini, 2015b).

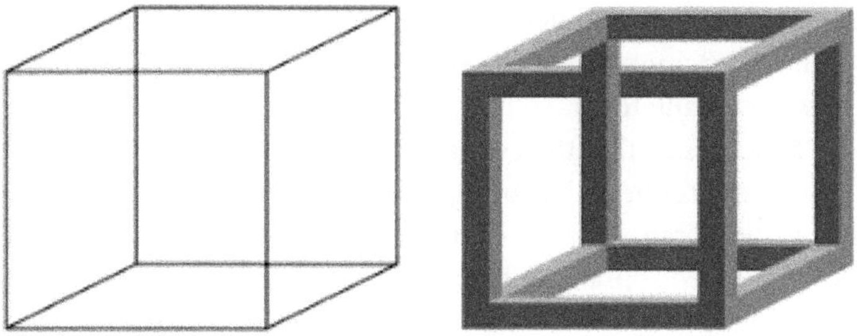

Figure 23.5 The Necker cube in 3D and 2D.
https://en.wikipedia.org/wiki/Necker_cube

figure as showing a group of horses against a certain background. For other examples of such groupings centred on performing subjective contours, see Kennedy, 2000.

Now, are the aforementioned grouping properties grasped visually only, or also tactilely? Everything leads us to believe that the latter is correct. In the previously quoted raised-line drawings or those done directly by congenitally blind people (Figure 23.1), what is captured by these people are precisely the same grouping properties of elements ordered in an in front → behind direction along a dimension of depth that also makes it possible for normally sighted people to grasp the closed thumb in front of, and partly concealing, the other fingers of a hand against a background. If this is so, then it is quite plausible that these properties are perceived supramodally. Thus, they are responsible to the same degree for the figurative value of the corresponding visual and tactile pictures.

More ambitious proof of this point may be had from the aforementioned case of perceptually ambiguous pictures leading to multistable experiences of a figure, in which the difference in the grouping properties of elements arranged in a certain direction along a certain dimension determines the ambiguity of the picture—namely, the fact that a figure actually activates two pictures with different figurative values.[5] This ambiguity is not picture-specific but also occurs in ordinary 3D objects whose ambiguity can be grasped both on the visual and on the tactile level; thus, it cannot be ascribed to either of the two sensorial modalities. The ambiguity can actually be seen on a pictorial level in 2D objects interpreted as presenting different 3D scenes, thus determining a pictorial difference, which again may present itself either visually or tactilely, hence not essentially modally.

Consider the so-called Necker cube (Figure 23.5). As Bruno *et al.* (2007) have emphasized, 3D examples of the perceptual ambiguity of this cube are given by a 3D cubical object which, according to how its elements are grouped in an in front → behind direction along the dimension of depth, is perceived at times as having one side at the front and one at the back, and at times as having a different side in the front and a different one at the back. More famously, however, 2D examples are also provided, in which, according to how its elements are grouped in the in front → behind direction along a dimension of depth, a 2D figure is perceived at

[5] Such ambiguities are implicitly admitted by Kennedy himself (2000, pp. 84–5) as also graspable by blind people, even if he focuses on cases in which, in virtue of Gestalt psychological laws, only one reading is actually performed. Given that the elements in a 2D object are nonetheless interpreted as presenting a 3D scene, the ambiguity in question does not necessarily reside in a direction following the dimension of depth, as the case of the famous duck–rabbit figure shows (here the ambiguity mobilizes the left → right and the right → left direction along the dimension of width). We can, however, omit these complications.

times as presenting a cube with a certain side in the front and another side at the back, at times as presenting a different cube with a different side at the front and a different side at the back. This results in two different pictures.

Now, we can hypothesize that the congenitally blind are able to recognize by touch the same ambiguity of the 3D cube as the normally sighted can visually; hence, the same grouping properties of elements arranged in an in front → behind direction along the dimension of depth prove to be grasped supramodally. It is a matter of picking up relations of occlusion that change according to which side of the Necker cube is perceived as more prominent. This, as we have seen in the previous paragraph, is related to an ability to grasp orientations that the congenitally blind are also capable of; here it is simply activated first in one way and then in the other. Likewise, it is just as arguable that the congenitally blind are able to recognize by touch, when pictorially grasping a 2D figure raised in relief, the same ambiguity that the normally sighted would be able to recognize visually; hence, once again, the same grouping properties of elements arranged in an in front → behind direction along the dimension of depth prove to be grasped supramodally. Indeed, the congenitally blind do seem to be able to grasp ambiguities of grouping along the first two dimensions (Carter *et al.*, 2008). It is thus easy to believe that they may be able to grasp this ambiguity in cases where the figure-background relations of occlusion are arranged alternatively, but such that groupings along the third dimension make it possible to perceive, as originally hypothesized by Shopland and Gregory (1964) and then taken up again by Kennedy and Domander (1984, p. 259). In this latter case, these distinct perceptions would determine the recognition of *two* tactile pictures of different cubes.

5 The neural bases of supramodal pictorial perception

The hypothesis that grouping properties—as involved in pictorial perception—are supramodal would enjoy robust empirical confirmation if it were possible to demonstrate that the neural structures selectively involved in the recognition of visual pictures and those involved in the recognition of tactile pictures were more or less the same, and that they were active both in the normally sighted and in the congenitally blind. Unfortunately, the data available from neuroscience up to now are not sufficient to confirm this hypothesis. As far as we know, there is only one neuroimaging (fMRI) study that has investigated neural activity during the recognition of seemingly tactile pictures (Stoesz *et al.*, 2003). An interesting piece of information resulting from this study is that the elaboration of shapes originating from a tactile pictorial stimulus selectively activates a bilateral network of areas in the brain that include the LOC, the structure most involved in the supramodal perception of 3D objects[6] (see Section 3). This seems to suggest that the LOC also plays a critical role in supporting the grouping operation that makes it possible to grasp the figurative value of a picture (visual or tactile). Nevertheless, the tactile stimuli used in this study were aluminium plates raised in relief with extremely modest figurative value, as they simply consisted of two letters for recognition ('V' and 'U'). This sort of stimulus does not involve the in front → behind grouping

[6] For some authors (Briscoe and Schwenkler, 2015), the LOC can play a crucial role also in visually guided actions during action planning. For this reason, it has been argued that such a region might be critical also for pictorial perception (see Ferretti, 2018b).

along the dimension of depth, which, as we saw, is an essential factor for something to count as a picture. Most importantly, in this study, only normally sighted subjects were involved. It is thus impossible to exclude the possibility of the LOC activation observed here reflecting processes of visual imagery, rather than processes of grasping tactile stimuli in their pictorial import.

One simple way to empirically verify the hypothesis that pictorial perception is supramodal would consist in carrying out a variation of Pietrini and colleagues' (2004) fMRI experiment, described in Section 3, this time involving both the normally sighted and the congenitally blind (or early blind) as the subjects of the experiment and using as stimuli raised-line drawings instead of 3D objects. The experiment should involve a task of visual and tactile recognition for the normally sighted and a task of tactile recognition for the sight-impaired. To obtain behavioural responses as accurate and rapid as possible from the participants, the tactile stimuli should be kept extremely simple. They might, for example, consist of raised-line drawings depicting single objects presented in perspective: a hand standing out from a background, a cup seen in profile, the reproduction of a human face perceived from the front, and so on. In a more demanding variation on the same experiment, the tactile stimuli might be categorized into three types, according to their figurative complexity: (1) raised-line drawings of single objects seen in perspective; (2) raised-line drawings of simple scenes seen in perspective (e.g., two children in front of a house); and (3) raised-line drawings marked by perceptual ambiguity (e.g., the Necker cube) (Figure 23.6). For both versions of this potential experiment, the hypothesis regarding supramodal pictorial perception makes two general predictions. First, the performances of the normally sighted and the sight-impaired in tasks of tactile recognition are more or less the same, demonstrating that the grouping processes in both groups of subjects are functionally equivalent. The second prediction concerns instead the neural structures activated in tasks of pictorial recognition which, as we have said, should at least partially overlap in the various experimental circumstances (visual–tactile) and in the comparison between normally sighted and sight-impaired people. Concerning the specific neural structures involved in the previously imagined experiment, that prediction says that the LOC may be activated in all the experimental circumstances and in all types of subjects, demonstrating that this structure constitutes the neural substrate of the grouping operation distinguishing supramodal pictorial perception. This result would correlate not only with the role of the LOC in the supramodal elaboration of the shape of objects, but also with those studies that suggest the existence in this region of vision-dependent representations of objects—that is, codified in accordance with a certain directional order (Grill-Spector *et al.*,

Figure 23.6 Samples of stimuli that might be used in the experiment described in this section.

1999; for proof to the contrary, see James *et al.*, 2002). Lastly, this result would be coherent with the role that the LOC seems to play in supramodal pictorial *production*. The latter hypothesis emerges from an interesting study by Amedi and colleagues (2008) carried out on the subject EA, an early blind artist able to produce detailed drawings of objects or scenes that prove perfectly recognizable to normally sighted subjects. The study used the fMRI technique to investigate the neural correlates associated with EA's ability to transform 3D objects explored by touch into 2D drawings, using a series of control conditions (e.g., a condition in which EA was requested to make senseless doodles, in order to remove from the analysis those areas of the brain associated with hand movements). The neural activity recorded during EA's production of the drawings involved various 'visual' areas, including the LOC.

Of course, it is quite reasonable to expect that the LOC is not the *only* area of the brain involved in understanding tactile stimuli with figurative value. If tactile stimuli require deep comprehension of the spatial relations between the elements depicted (as in the case of the raised-line drawings representing scenes including several objects arranged in perspective), then some neural structures distributed in the frontal lobes and the 'visual' dorsal pathway might also be involved, just as the bilateral intraparietal sulcus and the posterior parietal cortex may be.[7] Indeed, in a previous study carried out by Emiliano Ricciardi and colleagues (2006), it was observed that these regions operate supramodally (both visually and tactilely, both in the normally sighted and in the congenitally blind) in tasks of spatial memory involving 3D and 2D stimuli. Moreover, the frontal regions and, above all, the inferior frontal lobe, just as some areas of the parietal cortex, may be involved in the perception of tactile stimuli marked by perceptual ambiguity, such as a raised-line drawing of the Necker cube. Indeed, these structures seem to play a critical role in processes of multistable visual perception, like perception of the Necker cube, in both a 3D and a 2D version (Brascamp *et al.*, 2018).

To sum up, grasping tactile pictures may involve a fairly extended network of distinct cerebral structures, each with relative functional autonomy. Nonetheless, to the extent that one may demonstrate that this network is the same in both normally sighted and sight-impaired people, and that it coincides at least partially with the one involved in visual perception of pictures, one can reasonably conclude that pictorial perception really is supramodal and that tactile pictures, to the contrary of what Hopkins argues, are bona fide pictures.

6 Conclusions

In this paper, we first argued that both at a theoretical and at an empirical level (both in the fields of cognitive psychology and the neurosciences), congenitally blind subjects using touch to perform tasks involving 3D objects seemingly manage to grasp their perspectival arrangement, as normally happens for the normally sighted by means of sight. Not only does this fact contribute to supporting the idea that the perception of shapes and 3D relations between objects is supramodal, but it also challenges the assumption underlying the philosophical theory denying that there are properly speaking tactile pictures—namely, a

[7] Note that the brain regions included in the 'visual' dorsal pathway have also been associated with action processing and spatial awareness (Gallese, 2007), two cognitive processes that, according to some scholars, have a role in picture perception (Ferretti, 2018b).

lack of any perspective in tactile perception. In order to legitimately state that there really are tactile pictures just as there are visual ones, and that consequently pictorial perception is ultimately supramodal too, it would, however, be necessary to demonstrate something further. That is, what is decisive from a figurative point of view (i.e., the determination of grouping typically in the in front → behind direction along the dimension of depth with respect to 2D objects—what pictures are like from a physical point of view) can be performed by starting out from the tactile perception of these objects that the congenitally blind may have, just as these groupings can be performed by normally sighted subjects starting out from their visual perception. From a theoretical point of view, we have many reasons to believe that this is how things stand. For experiments in psychology tell us that the congenitally blind are able to grasp the same relations of occlusion that are also essential to the normally sighted for performing this grouping, and thus have pictorial perception. Granted, from the point of view of neuroscience, proof is still lacking that the areas of the brain underlying these perceptual groupings are, at least partially, the same both for the congenitally blind and for the normally sighted. Yet, on the basis of the present evidence, it is reasonable to suppose that this proof may be forthcoming.

Acknowledgements

Previous versions of this study were presented at the 14th International Symposium of Cognition, Logic and Communication, Linking Senses: Cross-Modality in Perceptual Domains across Cultures, 7 December 2019, University of Latvia, Riga, and at the Associazione Italiana di Scienze Cognitive (AISC) Conference 2019, University of Rome 3, 11–13 December, Rome. Although the work was jointly conceived and discussed by both authors, Fabrizio Calzavarini is responsible in particular for Sections 3 and 5, and Alberto Voltolini for Sections 2 and 4.

References

Amedi, A., Jacobson, G., Hendler, T., Malach, R., and Zohary, E. (2002). Convergence of visual and tactile shape processing in the human lateral occipital complex. *Cerebral Cortex*, 12, 1202–12.

Amedi, A., Stern, W. M., Camprodon, J. A., *et al.* (2007). Shape conveyed by visual-to-auditory sensory substitution activates the lateral occipital complex. *Nature Neuroscience*, 10, 687–9.

Amedi, A., *et al.* (2008). Neural and behavioral correlates of drawing in an early blind painter: a case study. *Brain Research*, 1242, 252–62.

Arnold, G., Spence, C., and Auvray, M. (2016). Taking someone else's spatial perspective: natural stance or effortful decentring? *Cognition*, 148, 27–33.

Arnold, G., Spence, C., and Auvray, M. (2017). A unity of the self or a multiplicity of locations? How the graphesthesia task sheds light on the role of spatial perspectives in bodily self-consciousness. *Consciousness and Cognition*, 56, 100–14.

Block, N. (2014). Seeing-as in the light of vision science. *Philosophy and Phenomenological Research*, 89, 560–72.

Brascamp, J., Sterzer, P., Blake, R., and Knapen, T. (2018). Multistable perception and the role of the frontoparietal cortex in perceptual inference. *Annual Review of Psychology*, 69, 77–103.

Brewer, B. (2011). *Perception and Its Objects*. Oxford: Oxford University Press.

Briscoe, R. and Schwenkler, J. (2015). Conscious vision in action. *Cognitive Science*, 39, 1435–67.

Bruno, N., Jacomuzzi, A., Bertamini, M., and Meyer, G. (2007). A visual-haptic necker cube reveals temporal constraints on intersensory merging during perceptual exploration. *Neuropsychologia*, 45, 469–75.

Burnston, D. C. (2017). Cognitive penetration and the cognition–perception interface. *Synthese*, 194, 3645–68.

Carter, O., Konkle, T., Wang, Q., Hayward, V., and Moore, C. (2008). Tactile rivalry demonstrated with an ambiguous apparent-motion quartet. *Current Biology*, 18, 1050–4.

Craddock, M. and Lawson, R. (2009). Size-sensitive perceptual representations underlie visual and haptic object recognition. *PLoS One*, 4, e8009.

D'Angiulli, A., Kennedy, J. M., and Heller, M. A. (1998). Blind children recognizing tactile pictures respond like sighted children given guidance in exploration. *Scandinavian Journal of Psychology*, 39, 187–90.

De Volder, A. G., Toyama, H., Kimura, Y., *et al.* (2001). Auditory triggered mental imagery of shape involves visual association areas in early blind humans. *NeuroImage*, 14, 129–39.

Ferretti, G. (2018a). Visual feeling of presence. *Pacific Philosophical Quarterly*, 99, 112–36.

Ferretti, G. (2018b). The neural dynamics of seeing-in. *Erkenntnis*, 84, 1285–324.

Ferretti, G. and Glenney, B. (eds). (2020). *Molyneux's Question and the History of Philosophy*, London: Routledge.

Frege, G. (1980). On sense and reference. In *Translations from the Philosophical Writings of Gottlob Frege*, edited by P. T. Geach and M. Black, 56–78. Oxford: Blackwell.

Gallace, A. and Spence, C. (2014). The neglected power of touch: what the cognitive neurosciences can tell us about the importance of touch in artistic communication. In *Sculpture and Touch*, edited by P. Dent, 107–24. Farnham: Ashgate.

Gallese, V. (2007). The conscious dorsal stream: embodied simulation and its role in space and action conscious awareness. *Psyche*, 13(1), 1–20.

Grill-Spector, K., Kushnir, T., Edelman, S., Avidan, G., Itzchak, Y., and Malach, R. (1999). Differential processing of objects under various viewing conditions in the human lateral occipital complex. *Neuron*, 24, 187–203.

Grill-Spector, K., Kourtzi, Z., and Kanwisher, N. (2001). The lateral occipital complex and its role in object recognition. *Vision Research*, 41, 1409–22.

Heimler, B., Striem-Amit, E., and Amedi, A. (2015). Origins of task-specific sensory-independent organization in the visual and auditory brain: neuroscience evidence, open questions and clinical implications. *Current Opinion in Neurobiology*, 35, 169–77.

Hopkins, R. (2000). Touching pictures. *British Journal of Aesthetics*, 40, 149–67.

Hopkins, R. (2004). Painting, sculpture, sight, and touch. *British Journal of Aesthetics*, 24, 149–66.

James, T. W., Humphrey, G. K., Gati, J. S., Menon, R. S., and Goodale, M. A. (2002). Differential effects of view on object-driven activation in dorsal and ventral streams. *Neuron*, 35, 793–801.

Kennedy, J. M. (1993). *Drawing and the Blind: Pictures to Touch*. New Haven, CT: Yale University Press.

Kennedy, J. M. (1997). How the blind draw. *Scientific American*, 276, 60–5.

Kennedy, J. M. (2000). Recognizing outline pictures via touch: alignment theory. In *Touch, Representation, and Blindness*, edited by M. A., Heller, 67–98. Oxford: Oxford University Press.

Kennedy, J. M. and Bai, J. (2002). Haptic pictures: fit judgments predict identification, recognition memory, and confidence. *Perception*, 31, 1013–26.

Kennedy, J. M. and Domander, R. (1984). Pictorial foreground/background reversal reduces tactual recognition by blind subjects. *Journal of Visual Impairment and Blindness*, 78, 215–16.

Kennedy, J. M. and Juricevic, I. (2006). Form, projection and pictures for the blind. In *Touch and Blindness: Psychology and Neuroscience*, edited by M. A. Heller and S. Ballestreros, 73–93. Mahwah: Lawrence Erlbaum Associates.

Kriegel, U. (2011). *The Sources of Intentionality*. Oxford: Oxford University Press.

Jagnow, R. (2011). Ambiguous figures and the spatial contents of perceptual experience: a defense of representationalism. *Phenomenology and the Cognitive Sciences*, 10, 325–46.

Lacey, S. and Sathian, K. (2014). Visuo-haptic multisensory object recognition, categorization, and representation. *Frontiers in Psychology*, 5, art. 730.

Lopes, D. (1996). *Understanding Pictures*. Oxford: Oxford University Press.

Lopes, D. (1997). Art media and the sense modalities: tactile pictures. *The Philosophical Quarterly*, 47, 425–40.
Lopes, D. (2005). *Sight and Sensibility*. Oxford: Oxford University Press.
Marmor, G. S. and Zaback, L. A. (1976). Mental rotation by the blind: does mental rotation depend on visual imagery? *Journal of Experimental Psychology: Human Perception and Performance*, 2, 515–21.
Martin, M. G. F. (2012). Sounds and images. *British Journal of Aesthetics*, 52, 331–51.
Martina, G. and Voltolini, A. (2017). Perceiving groupings, experiencing meanings. Rivista di estetica, 58, 22–46.
Nanay, B. (2013). *Between Perception and Action*. Oxford: Oxford University Press.
Nanay, B. (2020). Molyneux's question and interpersonal variations in multimodal mental imagery among blind subjects. In *Molyneux's Question and the History of Philosophy*, edited by G. Ferretti and B. Glenney, 257–263. London: Routledge.
Newell, F. N., Ernst, M. O., Tjan, B. S., and Bülthoff, H. H. (2001). Viewpoint dependence in visual and haptic object recognition. *Psychological Science*, 12, 37–42.
Newman, S. D., Klatzky, R. L., Lederman, S. J., and Just, M. A. (2005). Imagining material versus geometric properties of objects: an fMRI study. *Cognitive Brain Research*, 23, 235–46.
Pascual-Leone, A. and Hamilton, R. (2001). The metamodal organization of the brain. *Progress in Brain Research*, 134, 427–47.
Pawluk, D., Kitada, R., Abramowicz, A., Hamilton, C., and Lederman, S. J. (2011). Figure/ground segmentation via a haptic glance: attributing initial finger contacts to objects or their supporting surfaces. *IEEE Transactions on Haptics*, 4, 2–13.
Pietrini, P., Furey, M. L., Ricciardi, E., *et al.* (2004). Beyond sensory images: object-based representation in the human ventral pathway. *Proceedings in the National Academy of Sciences of the United States of America*, 101, 5658–63.
Pylyshyn, Z. (2003). *Seeing and Visualizing*. Cambridge, MA: MIT Press.
Ricciardi, E., Bonino, D., Gentili, C., Sani, L., Pietrini, P., and Vecchi, T. (2006). Neural correlates of spatial working memory in humans: a functional magnetic resonance imaging study comparing visual and tactile processes. *Neuroscience*, 139, 339–49.
Ricciardi, E., Bonino, D., Pellegrini, S., and Pietrini, P. (2014). Mind the blind brain to understand the sighted one! Is there a supramodal cortical functional architecture? *Neuroscience and Biobehavioral Reviews*, *41*, 64–77.
Shopland, C. and Gregory, R. L. (1964). The effect of touch on a visually ambiguous three-dimensional figure. *Quarterly Journal of Experimental Psychology*, 16, 66–70.
Stoesz, M. R., Zhang, M., Weisser, V. D., Prather, S. C., Mao, H., and Sathian, K. (2003). Neural networks active during tactile form perception: common and differential activity during macrospatial and microspatial tasks. *International Journal of Psychophysiology*, 50, 41–9.
Stokes, D. (2018). Attention and the cognitive penetrability of perception. *Australasian Journal of Philosophy*, 96, 303–18.
Tinti, C., Chiesa, S., Cavaglià, R., Dalmasso, S., Pia, L., and Schmidt, S. (2018). On my right or on your left? Spontaneous spatial perspective taking in blind people. *Conscious Cognition*, 62, 1–8.
Voltolini, A. (2015a). *A Syncretistic Theory of Depiction*. Basingstoke: Palgrave.
Voltolini, A. (2015b). Why, as responsible for figurativity, seeing-in can only be inflected seeing-in. *Phenomenology and the Cognitive Sciences*, 14, 651–67.
Voltolini, A. (2020). Different kinds of fusion experiences. *Review of Philosophy and Psychology*, 11, 203–22.
Walton, K. (1984). Transparent pictures: on the nature of photographic realism. *Critical Inquiry*, 11, 246–77.
Walton, K. (1990). *Mimesis As Make-Believe*. Harvard, MA: Harvard University Press.
Walton, K. (2008). Seeing-in and seeing fictionally. In *Marvelous Images. On Values and the Art*, edited by J. Hopkins and A. Savile, 133–42. Oxford: Oxford University Press.
Wollheim, R. (1980). Seeing-as, seeing-in, and pictorial representation. In *Art and Its Objects*, edited by R. Wollheim, 205–26. Cambridge: Cambridge University Press.
Wollheim, R. (1987). *Painting as an Art*. Princeton, NJ: Princeton University Press.
Wollheim, R. (1998). On pictorial representation. *Journal of Aesthetics and Art Criticism*, 56, 217–26.

Wollheim, R. (2003). What makes representational painting truly visual? *Proceedings of the Aristotelian Society*, 77, 131–47.

Woods, A. T., Moore, A., and Newell, F. N. (2008). Canonical views in haptic object perception. *Perception*, 37, 1867–78.

Zhang, M., Weisser, V. D., Stilla, R., Prather, S. C., and Sathian, K. (2004). Multisensory cortical processing of object shape and its relation to mental imagery. *Cognitive, Affective, and Behavioral Neuroscience*, 4, 251–9.

Index

For the benefit of digital users, indexed terms that span two pages (e.g., 52–53) may, on occasion, appear on only one of those pages.

Pakistan’s Foreign Policy

Pakistan's Foreign Policy

Security, State, and Society

Wali Aslam

OXFORD
UNIVERSITY PRESS

OXFORD
UNIVERSITY PRESS

Great Clarendon Street, Oxford, OX2 6DP,
United Kingdom

Oxford University Press is a department of the University of Oxford.
It furthers the University's objective of excellence in research, scholarship,
and education by publishing worldwide. Oxford is a registered trade mark of
Oxford University Press in the UK and in certain other countries.

Published in the United States of America by Oxford University Press
198 Madison Avenue, New York, NY 10016, United States of America

British Library Cataloguing in Publication Data
Data available

Library of Congress Control Number: 2026933574

ISBN 9780198909521

DOI: 10.1093/9780198909552.001.0001

Printed and bound by
CPI Group (UK) Ltd., Croydon, CR0 4YY

The manufacturer's authorized representative in the EU for product safety is
Oxford University Press España S.A. of Parque Empresarial San Fernando de Henares,
Avenida de Castilla, 2 – 28830 Madrid (www.oup.es/en or product.safety@oup.com).
OUP España S.A. also acts as importer into Spain of products made by the manufacturer.

FSC www.fsc.org
MIX
Paper | Supporting
responsible forestry
FSC® C013604

For Jeevan—the thread that binds us all

Acknowledgements

I extend my sincere gratitude to friends and colleagues whose intellectual and practical contributions were essential to this scholarly work. My interviewees were generous with their time, allowing for frank exchanges. Some of them were available for multiple sessions, which proved very useful. I am grateful for their valuable insights.

I am particularly indebted to four anonymous reviewers, whose incisive feedback significantly refined the manuscript, enhancing the rigour and nuance of its arguments. The theoretical guidance provided by Jim Fearon, Valerie Hudson, and Juliet Kaarbo was instrumental in shaping the book's conceptual framework. Katharine Adeney and Maya Tudor advised on matters of data collection, for which I remain deeply appreciative.

Thanks are due to Rajalakshmi Ezhumalai for her excellent project management and for her understanding of my time constraints during different phases of the book's final production. Her colleagues, Saravanan Rajendran and Ashwini Shanmugam, were responsible for typesetting, proof checks, and quality control. I am especially thankful to all three for their assistance during the review of the final manuscript prior to publication. Sharon Laverick assisted with the indexing and helped correct certain errors in the text, which was greatly appreciated.

Karen Bunn at Oxford University Press extended support during the initial phase of production. Her successor, Phoebe Aldridge-Turner, ensured a seamless transition and provided helpful advice in the later stages of production. I remain indebted to them for their support and guidance. I am particularly grateful to Dominic Byatt at Oxford University Press for commissioning this work and for his expert oversight of the rigorous peer-review process, which greatly strengthened the book's academic quality.

I celebrate the legacy of my great-grandfather, Jeevan, who was known for his humility, selflessness, and compassion. His ethos of service and generosity continues to inspire my academic and personal life.

Contents

1

Why and how to study Pakistan's foreign policy

1.1 Introduction

Like most other postcolonial states, Pakistan has faced a variety of institutional challenges throughout its chequered history that have severely hampered the country's ability to provide a meaningful existence to its society. The endemic civil–military imbalance is among the foremost of such challenges colouring the country's politics since the 1950s. This imbalance, and its implications for the domestic politics of Pakistan, has been widely studied by a range of scholars for different objectives.[1] Most of these studies have examined the causes and consequences of the civil–military imbalance on a chronological basis. These works, however, have primarily analysed the impact of this imbalance on the domestic politics of Pakistan.[2] There is still a gap in the literature regarding what this imbalance means for Pakistan's foreign policy. To that point, certain key questions remain unanswered, such as: how does the civil–military imbalance impact upon Pakistan's foreign policy? To what extent is a less-than-optimal foreign policy of Pakistan a result of the military's dominance of the decision-making process? How can the external implications of the internal imbalance within Pakistani institutions be studied? How do certain ruling elites use the country's foreign policy to remain dominant domestically? These are some of the pertinent questions that this work is going to answer.

This book is not a straightforward descriptive history of Pakistani foreign policy.[3] Though there is ample space for further scholarly work on this

[1] See, for example, Hasan Askari Rizvi, *The Military and Politics in Pakistan 1947–1997* (Lahore: Sang-e-Meel publishers, 2000); Husain Haqqani, *Pakistan: Between Mosque and Military* (Washington, D.C.: Carnegie Endowment for International Peace, 2010).

[2] Shuja Nawaz, *Crossed Swords: Pakistan, Its Army, and the Wars Within* (Karachi: Oxford University Press Pakistan, 2008).

[3] The works of Abdul Sattar and Shahid Amin might be more useful on that topic, as they engage with almost all major aspects of the country's foreign policy since independence, albeit from an official angle. See Abdul Sattar, *Pakistan's Foreign Policy 1947–2019* (Karachi: Oxford University Press Pakistan, 2020); Shahid Amin, *Pakistan's Foreign Policy: A Reappraisal* (Karachi: Oxford University Press Pakistan, 2021).

Pakistan's Foreign Policy. Wali Aslam, Oxford University Press. © Wali Aslam (2026).
DOI: 10.1093/9780198909552.003.0001

topic, it is not the primary objective of this contribution. This book goes one step further and presents an analytical assessment to understand Pakistan's less-than-optimal foreign policy choices of the past and the policy's negative outlook for the future. Instead of detailing the key episodes of the state's history, this book employs a novel theoretical framework rooted in post-structuralism[4] and ontological security theory of International Relations[5] to outline the way the self-serving foreign policy elites of Pakistan, which primarily include the country's military, use 'fantasmatic logic'[6] to take over the decision-making in the name of security. The concept of 'fantasy'[7] is invaluable in giving us insights into the decision-making processes in different settings, and foreign policy is one such domain where this concept can be usefully applied. These fantasmatic endeavours result in the state machinery serving the interests of a tiny minority of Pakistani society while the rest of the populace remains uneducated and malnourished, hunting for scraps to live each day. This book is a study of the process through which these foreign policy choices are made and justified by those in power. It is also a study of the deleterious implications of such choices—not only for Pakistan but also for its neighbours and the wider international community.

Conducting an academic analysis of this subject is of importance for the international community for a number of reasons. First, Pakistan has played a significant role in shaping contemporary global security dynamics in various ways, ranging from its nuclear cooperation with North Korea and Iran[8] to its double games in Afghanistan—which contributed to a great extent to the Western withdrawal from the country and the eventual takeover of the Taliban in August 2021.[9] The Pakistani component of China's Belt and Road Initiative (BRI), known as the China Pakistan Economic Corridor (CPEC), is of great relevance for the emerging dynamics of global geopolitics.[10] The recent Pakistani leadership of the 'loss and damage fund' to raise money for the developing countries affected by climate change also has significant

[4] Linda Ahall, 'Poststructuralism', in *Security Studies: An Introduction*, edited by Paul Williams and Matt McDonald (Abingdon: Routledge, 2023), pp. 86–99.

[5] Jennifer Mitzen, 'Ontological Security in World Politics: State Identity and the Security Dilemma', *European Journal of International Relations* 12, no. 3 (2006): pp. 341–370.

[6] Jason Glynos and David Howarth, *Logics of Critical Explanation in Social and Political Theory* (Abingdon: Routledge, 2007).

[7] Jakub Eberle, 'Narrative, Desire, Ontological Security, Transgression: Fantasy as a Factor in International Politics', *Journal of International Relations and Development* 22, no. 1 (2019): pp. 243–248.

[8] Gordon Corera, *Shopping for Bombs: Nuclear Proliferation, Global Insecurity, and the Rise and Fall of the A.Q. Khan Network* (London: Hurst, 2006).

[9] Sahar Khan, *Double Game: Why Pakistan Supports Militants and Resists U.S. Pressure to Stop*. Cato Institute, 2018. JSTOR, http://www.jstor.org/stable/resrep23042.

[10] Andrew Small, *The China-Pakistan Axis: Asia's New Geopolitics* (New York: Oxford University Press, 2015).

repercussions going forward as far the politics of climate change is concerned.[11] Given the outsized role played by a financially impoverished state like Pakistan in shaping the contours of international politics, it is more than obvious that a study of international politics and security matters, great power competition (especially the rise of China and the American pivot to Asia), and the challenges facing the developing world can benefit from an informed analysis of the country's foreign policy. This book performs the task of studying the country's external relations and their impact on contemporary global political processes through its domestic political landscape.

Second, scrutinizing Pakistan's foreign policy is helpful for those studying the domestic political developments in the country. As is argued below, the book follows the perspective that emphasizes the linkage between the two: internal and the external. In this case, both play a critical role in consolidating the dominant position of the military within the state and in shaping its domestic and foreign policies.

Third, an in-depth study on this topic would be valuable for the international community in learning about how to engage with a state like Pakistan.[12] As stated, the country does perform a significant role in influencing the international politics of Central Asia, the Persian Gulf, and the Indo-Pacific, and challenges concerning these regions can be better addressed through a focus on this subject.

After outlining the case for conducting this analysis, it is now important to outline how this chapter will proceed. Section 1.2 presents a glimpse into the history of civil–military relations in Pakistan and its attendant imbalance. This will show both that this imbalance is relatively uncommon compared to the rest of the world and that the way the state currently operates can neither be defined as a democracy or a dictatorship. I am sympathetic to the idea of seeing the case of Pakistan as a 'hybrid state'[13] that lies somewhere in between the two extremes on the spectrum.

Section 1.3 conducts a detailed literature review in three areas. First, it examines the body of literature on Pakistan's civil–military imbalance from its inception to the present day, critically appraising its strengths. It also outlines how this literature either neglects the foreign policy implications of the imbalance or, when it engages with them, does so in a limited way,

[11] Nina Lakhani, '"We Couldn't Fail Them": How Pakistan's Floods Spurred Fight at COP for Loss and Damage Fund', *The Guardian*, 20 November 2022.

[12] That has been a pressing concern among various policy elites in the West, especially the United States, over the last four decades since the Soviet invasion of Afghanistan. See United States Institute for Peace, *How to Deal with Pakistan*, Washington, D.C., Seminar, 18 October 2017.

[13] Katharine Adeney, 'How to Understand Pakistan's Hybrid Regime: The Importance of a Multidimensional Continuum', *Democratization* 24, no. 1 (2017): pp. 119–137.

often as a secondary focus or by-product of analysing the domestic context. The second set of literature examined in Section 1.3 covers the foreign policy of Pakistan. It illustrates that this subject is generally presented as a descriptive history without a deeper analysis driven by a specific theoretical approach. Specifically, and more relevant to this book, works in this domain make almost no attempt to study the linkages between the internal and external dimensions of Pakistani politics, a gap in the literature that will be filled by this book. The third type of literature consulted in Section 1.3 examines the emerging research on hybrid states, some of which has Pakistan at its centre. Falling within the domain of the literature studying regime types and foreign policy,[14] this research is relatively new and asserts that the politics of such states varies from other regime types. While analysing that research, I will, again, emphasize how research on hybrid states (like Pakistan) often focuses on the domestic dimensions, and there remain gaps covering the foreign policies of hybrid states. If one agrees with the typical Liberal assumptions[15] that a country's foreign policy is a reflection of its domestic make-up, it goes without saying that a hybrid state would have a different foreign policy compared to a democracy or a dictatorship.[16]

After outlining these gaps in the literature, Section 1.4 will study how this book contributes to the literature in the three domains listed above while answering the project's research questions. This section will also put forward the details of the theoretical background that informed my broad outlook while writing the book. The objective of the theoretical underpinning is to help me develop a unique insight into the topic at hand and to say something new, which might not be possible without employing that framework. However, this book utilizes the theory in a relatively light-touch way to ensure as wide an appeal as possible among academic scholars, university students, policy circles, journalists and beyond. Not utilizing a theoretical framework might deprive me of the opportunity of accessing insights that its ideas provide. On the other hand, leaning too heavily on the use of such a framework can limit the scope of the readership and risks alienating parts of the intended audience.

[14] Susan Hyde and Elizabeth Saunders, 'Recapturing Regime Type in International Relations: Leaders, Institutions, and Agency Space', *International Organization* 74, no. 2 (2020): pp. 363–395.

[15] Charles Lipson, *Reliable Partners: How Democracies Have Made a Separate Peace* (Princeton, NJ: Princeton University Press, 2003).

[16] Jack Levy, 'Domestic Politics and War', *The Journal of Interdisciplinary History* 18, no. 4 (1988): pp. 653–673.

Section 1.5 puts forward the theoretical framework utilized in this study. It starts by defining the key ideas of ontological security theory, post-structuralist discourse theory (PDT), and the concept of 'fantasy'. After defining them, the section discusses the mechanism through which the ideas from the three areas are combined to build the theoretical framework for the examination of Pakistan's foreign policy. Section 1.5 also outlines the deductive methodology used in this work, as well as the different sources of data used in this book. Section 1.6 will conclude by presenting a brief outline of each chapter of the book, to introduce the reader to what will be studied in each chapter, for what purpose it is studied, and the main lessons learnt from specific chapters.

1.2 The context of civil–military imbalance in Pakistan

Since its inception, Pakistan has experienced three types of political systems: military rule, emerging or fragile democracy, and hybrid regime. That is in addition to Pakistan's first decade—which was marked by turmoil for the country, the two wings of which (East and West Pakistan) struggled to deal with the aftermath of Partition and to chart a path together within one union. Civilian and military bureaucrats were in charge of running the state during that period while the country went through several prime ministers and governors general.[17] The instability of this interim period gave way to a military takeover of the country in 1958, marking the start of the first military regime under General Ayub Khan. During his tenure, the country fought a major war with India in September 1965. Facing popular protests on the streets and a vacuum at the foreign-policy level, Ayub Khan resigned as the president of Pakistan in March 1969, paving the path to the rule of General Yahya Khan for another two years.[18] The country's third military ruler was General Zia-ul-Haq, who governed the country between 1977 and 1988. The fourth military regime came in the form of General Pervez Musharraf's takeover in 1999, which lasted until 2008. As we will see throughout the book, each of Pakistan's military regimes invariably saw a closer alliance with the United States and significantly benefited from American largesse.

As concerns democratic dispensation within Pakistan, it is generally considered that the government of President (later, Prime Minister) Zulfikar Ali

[17] Bashir Ahmad and Maawra Salam, 'Bureaucracy in Pakistan Between Delivery of Civil Services and Politics', *Journal of Political Studies* 23, no. 2 (2016): p. 421.

[18] Altaf Gauhar, 'Pakistan: Ayub Khan's Abdication', *Third World Quarterly* 7, no. 1 (1985): pp. 102–131.

Bhutto came the closest to a democratic set-up that the country has seen in its history since 1947.[19] During that period, the military was recovering from the shock of the 1971 war, which led to the separation of East Pakistan to become Bangladesh. Zulfikar Ali Bhutto enjoyed substantially greater autonomy in decision-making than was afforded two decades later to the governments of his daughter Benazir Bhutto or that of her political rival, Nawaz Sharif, during the 1990s. The decade between 1988 and 1999 was also democratic, but was marred by the again-ascendent military's efforts to weaken the system. Opinions differ regarding that period: some describe it as a decade of a hybrid regime,[20] while others consider it a time when Pakistan was a 'feeble democracy', with the military remaining relatively separate from governance and playing the role of arbiter among politicians.[21] According to the latter view, civilians asserted themselves more and fought back during the 1990s in response to the military's encroachment upon their turf—especially compared to how events unfolded after 2008.[22]

Pakistan emerged from the shadows of Musharraf's military rule in 2008, again, as a feeble democracy which soon transformed into a hybrid state: the military expanded its role in the governance of the state with the cooperation of elected leaders of the country. Prime Minister Nawaz Sharif's invitation to the military to take part in the proceedings of the Cabinet Committee of National Security in 2013 facilitated that move.[23] The military gradually grew its share in civil sector governance, with Pakistan being described as an archtypical hybrid regime: neither a military dictatorship nor a democracy.[24]

Studying these types of regimes is critical to understanding the country's foreign policy, as the latter is primarily an external manifestation of the internal imbalance. Given the preponderance of the military in running the affairs of the state, Pakistan's foreign policy can be effectively studied through the military's narratives about the external threats facing Pakistan. That is especially pertinent at a time when direct military rule is not acceptable to the international community, on which Pakistan relies for regular economic and financial support. This dependence leads the military to tolerate some degree of civilian rule while continuing to influence the policy from behind the scenes through a hybrid system. Such a hybrid system benefits the military

[19] Interview, Politician 1, Islamabad, Pakistan, January 2024.

[20] Interview, Columnist, Islamabad, Pakistan, January 2024.

[21] Interview, Academic 2, London, United Kingdom, March 2024.

[22] Interview, Politician 1.

[23] See 'DCC to be Reconstituted as Committee on National Security', *Dawn*, 22 August 2013, Interview, Academic 2.

[24] Ian Talbot, 'Pakistan's Hybrid Regime: Growing Democratization, Or Increased Authoritarianism?', in *Routledge Handbook of Autocratization in South Asia*, edited by Sten Widmalm (Abingdon: Routledge, 2021), pp. 141–150.

because civilians are forced to take the blame for negative policy outcomes while the military claims credit for the success of its own vision when the outcome is positive.[25] A hybrid regime is also hard for the 'masses' to resist, as the chains of power and responsibility are diffused—with the military generally exercising power behind the scenes.[26] A focus on the hybrid nature of the state also allows for the study of social forces that are the permanent decisive actors in the country, including the civil and military bureaucracy, landed aristocracy, and urban big businesses.[27] This represents a coalition of various dynasties in which the 'institutional dynasty' of the military top brass commands the upper hand.[28] The military dynasty guards its economic interests jealously, and effectively uses other dynasties for its benefit, though these benefit from this arrangement as well.

Here a question arises: if the foreign policy of Pakistan is mainly the purview of the military, why does studying the contours of different regime types within the state matter? It is true that the military guards foreign and security policies jealously, but there have been instances in which civilian leaders have been able to assert themselves and put forward a more positive foreign policy than that of the military. As we will note throughout the book, the country's civilian leadership has occasionally sought to diversify Pakistan's foreign policy and minimize the state's dependence on external rent by presenting several promising visions for the welfare of the people. Though civilians have also made mistakes while in or out of power, their opportunity to learn from these errors has been limited due to the lack of a stable political process, which has been disrupted by persistent military interference in the system. A focus on different regimes can help pick out those nuances that characterize the policy and must be given due attention to study the potential for change.

1.3 Minding the gap

In this section, I conduct a broad and critical review of the literature to appraise scholarship on Pakistan's foreign policy and highlight how the current volume examines the topic from a different angle. In other words, this review will outline the current gaps that this work will bridge. Books written on cognate topics can be divided into three broad areas. The first set of

[25] Interview, Politician 1.
[26] Interview, Journalist 3, Lahore, Pakistan, January 2024.
[27] Interview, Journalist 1, Islamabad, Pakistan, January 2024.
[28] Interview, Columnist.

writings relates to the military's interference in Pakistani politics. The second group concerns the literature on Pakistan's foreign policy. The third section looks at the highly relevant research on 'hybrid states', with Pakistan being an archetypal example of such a state.

1.3.1 Pakistan's military, politics, and society

This is a very well-studied area of research given that civil–military imbalance characterizes a variety of states across the postcolonial world. In the case of Pakistan, this research can be divided roughly into two time periods. The first category of research studies the dynamics from independence in 1947 through the military coup in 1999. The second set of the literature covers the period from 1999 through today. Other than just ease of analysis, a justification for division lies in the fact that pre-1999 literature considers the country's civil–military imbalance to be some kind of anomaly. These works suggest that Pakistan might emerge from the shadow of the military's influence in the near future. They consider the military's role as part of the 'growing pains' that any newly independent state goes through. The second group of the literature considers the imbalance to be a more or less ongoing phenomenon, and contributing authors consider the military's dominance as a regular feature of the country's identity. Simply put, the literature in the first strand broadly regards the military interference as an interruption in Pakistan's journey to become a democracy, while that in the second strand treats it as the destination and not something that the society will be able to easily dislodge.

The writings of Husain Haqqani, Asghar Khan, Hasan Askari Rizvi, and Saeed Shafqat are among the landmark contributions in the first category above. Haqqani's work, *Pakistan: Between Mosque and Military*,[29] broaches the very topical subject of the Pakistani military's relationship with Islamists since independence. The detailed analysis looks at the multiple ways in which the former has nurtured that relationship and exploited it for internal and external objectives. Internal objectives range from using Islamist groups to destabilize civilians' attempts to assert themselves,[30] to helping strengthen the military's control over society when the military is directly in power—such as during General Zia's dictatorship,[31] whereas external objectives include promoting an Islamist 'jihad without borders'.[32] Some of Haqqani's conclusions

[29] Haqqani, *Pakistan: Between Mosque and Military*; see, on the same subject, 'Pakistan: The Mullahs and the Military', *International Crisis Group*, Asia Report 49, Islamabad/Brussels, 20 March 2003.
[30] Haqqani, *Pakistan*, p. 177.
[31] Ibid., pp. 118–141.
[32] Ibid., p. 234.

concern the impact of the internal civil–military imbalance on the country's foreign policy, though the book does not deliberately study the linkages between the former and the latter from the perspective of a conceptual perspective.

Asghar Khan, who served in a variety of significant positions in Pakistan, including the Chief of the Pakistan Air Force between 1957 and 1965, has authored several volumes in Urdu and English studying the military's involvement in Pakistani politics.[33] These works are valuable personal chronicles of Pakistan's turbulent history across different phases of democracy and dictatorship through the eyes of someone who played a significant role in opposing the generals' reluctance to limit their role to their respective domain, as defined by the state's constitution.[34]

Rizvi's work on military and politics in Pakistan,[35] conducts a detailed analysis of the issues from thematic as well as chronological angles. The thematic side of the study looks in great detail at the societal, internal military, and international factors that can help explain the military's dominance in Pakistani politics.[36] The chronological analysis covers the period from the first military takeover in 1958 through the fourth coup in 1999. The analytical depth to which the author goes to scrutinize the details of how events unfolded as well as the underlying causes and implications of these developments is laudable. However, these works remain almost completely silent on how the military coups (and subsequent military governments) and military interference in the times of the nominal civilian tenures impacted the country's foreign policy. Although there is a mention of Pakistan's relations with the international community, it is done primarily to understand how disparate military rulers used the developments at the international level to consolidate control internally or to provide justifications for continuing to cling on to power.[37]

Saeed Shafqat's work incorporates a comparative politics perspective to study the issues of military hegemony, centre–province relations, matters of federalism, and the challenges of political economy.[38] Where foreign policy

[33] See Asghar Khan, *Generals in Politics* (New Delhi: Vikas, 1983).

[34] Asghar Khan, *We've Learnt Nothing from History: Pakistan: Politics and Military Power* (Karachi: Oxford University Press Pakistan, 2005).

[35] Hasan Askari Rizvi, *Military and Politics in Pakistan* (New Delhi: Konark, 1989); Rizvi, *The Military and Politics in Pakistan 1947–1997*; Hasan Askari Rizvi, *Military, State and Society in Pakistan* (Basingstoke: Palgrave Macmillan, 2000).

[36] Rizvi, *Military, State and Society in Pakistan*, pp. 21–25.

[37] See, for example, ibid., p. 107.

[38] Saeed Shafqat, *Civil-Military Relations in Pakistan: From Zulfikar Ali Bhutto to Benazir Bhutto* (Boulder, CO: Westview, 1997).

is discussed, it is presented with reference to the varying views that different actors within the state might hold regarding that subject.[39]

The interference of the military in the politics of Pakistan continues to inspire the work of a number of authors, who devote greater attention to the second strand of our analysis: the military takeover of 1999 to the present. Mazhar Aziz, Cornelia Baciu, Christine Fair, Shuja Nawaz, and Aqil Shah feature in this set of authors. Aziz's work challenges the views of the military apologists who assert that civilian ineptitude in the face of sectarian, political, ethnic, and economic troubles forces the military to interfere in politics in order to save Pakistan from failure.[40] Instead, he argues that the military, who the author describes as a 'parallel state', interferes in Pakistani politics primarily to protect its institutional interests.[41] Applying the ideas from theories of institutionalism, this work makes a valuable contribution to alternative and critical understandings of different coups in the country.

Cornelia Baciu examines Pakistani politics from the perspective of Samuel Huntington's ideas of military professionalism in her book *Civil-Military Relations and Global Security Governance*.[42] This work asserts that most analyses in contemporary IR on the topic tend to focus on established democracies, and that it is critical to learn how the experience of hybrid states like Pakistan can inform theories on military change, transformation, and civilian control.[43]

Shuja Nawaz's well-researched work, *Crossed Swords*, includes a variety of archival and interview sources that had previously not been consulted, especially those pertaining to the military.[44] The focus, however, of these books is still primarily on the domestic politics of Pakistan, and specifically the military's perspective regarding the political and security developments concerning Pakistan.

Aqil Shah's work studies the military's own belief system or 'what is commonly known as the military mind-set or the military mentality' to explain military politics in Pakistan.[45] Shah argues that adopting such a perspective is important in order to think about how that system can be reformed.[46] Rabia Chaudhry's work examines the Pakistani military's dominance over

[39] Ibid., p. 93.
[40] Mazhar Aziz, *Military Control in Pakistan: The Parallel State* (Abingdon: Routledge, 2008).
[41] Ibid.
[42] Cornelia Baciu, *Civil-Military Relations and Global Security Governance: Strategy, Hybrid Orders and the Case of Pakistan* (Abingdon: Routledge, 2021).
[43] Ibid., pp. 4–6.
[44] Nawaz, *Crossed Swords*.
[45] Aqil Shah, *The Army and Democracy: Military Politics in Pakistan* (Cambridge, MA: Harvard, 2014), p. 8.
[46] Ibid., pp. 254–287.

the civilian sector, highlighting its control of civilian institutions and significant influence on economic decision-making to advance its own business empire.[47] However, these works have not expanded their discussion to the domain of foreign policy decision-making and have primarily concerned themselves with the domestic politics of the country.

A number of interesting studies on the institution of the Pakistan army also exist. Ayesha Siddiqa's landmark work on the Pakistani military's business empire remains the only work of its kind to meticulously detail the organization's business interests. The military has infiltrated the country's economy to such an extent that control of economic policy is key for military decision makers.[48] Brian Cloughley's history of Pakistan's army is conducted with substantial internal insight that gives the reader an idea of the organizational culture of the institution.[49] Unlike Siddiqa's work, Cloughley's history maintains a greater engagement with the external dimension of relations with the country's neighbours, but that is in the form of a military history rather than an analysis of the state's foreign policy *per se*.

Christine Fair's study of the Pakistan army, conducted over a decade ago, looks at the organization from the perspective of 'strategic culture', which in the army's case is shaped through an unending competition with India.[50] That competitive reasoning is the defining and existential feature of the army—thus abandoning it is not merely impossible, it is unthinkable. The analysis of the organization through a 'fighting-to-the-end' culture illustrates why the military is involved in so many unending conflicts with its neighbours. The work dedicates particular attention to the militant groups that the military uses for its external objectives.[51] This valuable work does not, however, extensively discuss the contours of Pakistan's foreign policy in the twenty-first century, nor does it capture the linkages between the internal dynamics of the country and its foreign policy. My book bridges both of these gaps.

1.3.2 Studies of Pakistan's foreign policy

The second obvious gap that this book seeks to bridge concerns Pakistani foreign policy. This domain remains relatively underdeveloped and neglected,

[47] Rabia Chaudhry, *The Changing Dynamics of Civil Military Relations in Pakistan: Soldiers of Development* (Abingdon: Routledge, 2024).
[48] Ayesha Siddiqa, *Military Inc.: Inside Pakistan's Military Economy* (London: Pluto, 2007).
[49] Brian Cloughley, *A History of the Pakistan Army: Wars and Insurrections* (New York: Carrel publishers, 2016).
[50] Christine Fair, *Fighting to the End: The Pakistan Army's Way of War* (Oxford: Oxford University Press, 2014).
[51] Ibid., pp. 226–260.

which is surprising given the crucial role played by Pakistan in regional and global political and security-related challenges.

Shahid Amin and Abdul Sattar, both retired Pakistani career diplomats, have written the 'go-to' works on Pakistani foreign policy studies. Sattar authored a descriptive study of Pakistani foreign policy from Partition in 1947 through 2019.[52] As the name suggests, the work (currently in its fifth edition), is a historical study of major developments concerning Pakistan and the world. Despite being the main source on the subject so far, the book focuses primarily on policy debates rather than academic analysis. Similarly, Amin's work chronicles Pakistan's foreign policy from the author's own perspective during his long career with the Pakistani Foreign Office.[53] Given that both of these authors were part of the state machinery and close to the security establishment, their works primarily present the state's official perspective, often portraying Pakistan as an innocent victim of Western 'betrayal' throughout its history.[54] Furthermore, there is little to no linkage with the domestic politics of Pakistan, and no description of how the internal informs the external or vice versa in this case.

Aparna Pande's *Explaining Pakistan's Foreign Policy: Escaping India* points towards the country's competition with India as the primary explanatory variable of Pakistani foreign policy.[55] As argued in this book, though the India factor is critical, it remains limited in helping understand the less-than-optimal Pakistani foreign policy. A focus on that variable alone is likely to have limited utility when it comes to making sense of a variety of foreign policy behaviours espoused by Pakistan.

S. Paul Kapur reflects on Pakistani alliances with militant groups in *Jihad as Grand Strategy*. He argues that the approach has had negative as well as positive dividends.[56] T. V. Paul looks towards Pakistan's 'geostrategic curse' to understand why Pakistan has lagged behind in the domains of democratic consolidation and economic progress compared to other developing nations.[57] These works, however, tend to stay away from the civil–military imbalance in Pakistan and its impact on the state's foreign policy.

Given the serious topical nature of the subject, Pakistan's relations with the great powers have also come under scrutiny. The works by Husain Haqqani

[52] Abdul Sattar, *Pakistan's Foreign Policy*.

[53] Amin, *Pakistan's Foreign Policy*.

[54] Sattar, *Pakistan's Foreign Policy*, p. 291.

[55] Aparna Pande, *Explaining Pakistan's Foreign Policy: Escaping India* (Abingdon: Routledge, 2011).

[56] S. Paul Kapur, *Jihad as Grand Strategy: Islamist Militancy, National Security, and the Pakistani State* (New York: Oxford University Press, 2017).

[57] T. V. Paul, *The Warrior State: Pakistan in the Contemporary World* (Oxford: Oxford University Press, 2014).

and Dennis Kux expand upon Pakistan–US relations. Kux benefited from a range of archival sources to write a diplomatic history of the relations.[58] In '*Magnificent Delusions*',[59] Haqqani has combed through the US National Security Archives to chart a historical evolution of Pakistan–US relations. Although both of these important works contribute to a specific area of Pakistan–US relations, they cannot be described as comprehensive evaluations of Pakistan's foreign policy. Furthermore, these diplomatic histories do not generally concern themselves with the developments of the twenty-first century.

Pakistan's relations with China are often described as an 'enduring entente'.[60] Andrew Small's investigative work on the subject remains an important resource.[61] Similar to Kux and Haqqani but with a different focus, Small outlines the contours of Pakistan's relations with China since the late 1940s. Boni[62] and Wolf[63] have also contributed to the subject, but their work primarily tackles the (relatively recent) Belt and Road Initiative and its Pakistani component, the China Pakistan Economic Corridor (CPEC), and does not engage with the broader challenges of Pakistani foreign policy beyond China.

Despite their enduring rivalry at the state level, surprisingly little has been written on the relations between India and Pakistan. Dilip Hiro[64] and Satinder Singh Lambah[65] have put forward their two historical accounts of this intricate relationship since 1947. Hiro presents an interesting and accessible history of the subject, while Lambah has provided the historical narrative based on his own work as an Indian career diplomat. Lambah's book studies the terms of six Indian prime ministers and their different styles of engagement with Pakistan—with varying degrees of success. Though valuable, such volumes only provide an understanding of the bilateral relations between the two neighbours and are of limited value for those interested in the broader study of Pakistani foreign policy in its entirety.

[58] Dennis Kux, *The United States and Pakistan 1947–2000: Disenchanted Allies* (Baltimore, MD: The Johns Hopkins University Press, 2001).

[59] Husain Haqqani, *Magnificent Delusions: Pakistan, the United States, and an Epic History of Misunderstanding* (New York: PublicAffairs, 2013).

[60] Sattar, *Pakistan's Foreign Policy*, p. 327.

[61] Andrew Small, *The China-Pakistan Axis*.

[62] Filippo Boni, *Sino-Pakistani Relations: Politics, Military and Regional Dynamics* (Abingdon: Routledge, 2019).

[63] Siegfried Wolf, *The China-Pakistan Economic Corridor of the Belt and Road Initiative: Concept, Context and Assessment* (London: Springer, 2020).

[64] Dilip Hiro, *The Longest August: The Unflinching Rivalry Between India and Pakistan* (New York: Nation Books, 2015).

[65] Satinder Singh Lambah, *In Pursuit of Peace: India-Pakistan Relations Under Six Prime Ministers* (Gurugram: Penguin India, 2023).

1.3.3 Foreign policies of hybrid states

The distinction of hybrid states from democratic and authoritarian ones has gained substantial attention since the end of the Cold War.[66] Featuring 'elections without democracy', Larry Diamond made an emphatic case to study 'pseudo-democracy' as a distinct regime type.[67] Whilst scholarly contributions have noted that it is hard to define a 'hybrid regime', there is a general consensus that it possesses the characteristics of both a democracy and a dictatorship.[68] Quoting different scholars on the subject, Mufti helpfully describes various ways that such regimes are understood when she states that '[h]ybrid regimes are variably understood as diminished subtypes of democracy, diminished subtypes of authoritarianism, transitional "situations" that are expected to revert back to either democracy or authoritarianism, a residual category of regimes that fit neither democracy nor authoritarianism, or as clear-cut instances of authoritarianism.'[69]

Despite the widespread presence of hybrid states in the world, limited detailed analyses of the regimes and their behaviour exist. Steven Levitsky's work stands as an exception to that rule as it goes on to look at thirty-five cases from Asia, Africa, Latin America, and post-Soviet states in Europe to helps us understand what factors play a role in countries' transitions from autocracies to hybrid states or democracies.[70] There are also some useful single-n studies focused on understanding the democratic processes (or lack thereof) in Uganda,[71] Bangladesh,[72] Lebanon,[73] and Turkey.[74] However, the discussion on the subject has primarily focused on how to understand hybrid states as a category, and has failed to move beyond this to shed more light on decision-making processes underlying hybrid states' foreign and domestic

[66] Leonardo Morlino, 'Are There Hybrid Regimes? Or Are They Just an Optical Illusion?', *European Political Science Review* 1, no. 2 (2009): pp. 273–296.

[67] Larry Diamond, 'Elections Without Democracy: Thinking About Hybrid Regimes', *Journal of Democracy* 13, no. 2 (2002): pp. 21–35.

[68] Mariam Mufti, 'What Do We Know About Hybrid Regimes After Two Decades of Scholarship?' *Politics and Governance* 6, no. 2 (2018): p. 113.

[69] Ibid., p. 113; in-text references omitted. On the issue of definitions, see also, Leah Gilbert and Payam Mohseni, 'Beyond Authoritarianism: The Conceptualization of Hybrid Regimes', *Studies in Comparative International Development* 46, no. 3 (2011): pp. 270–297. Andrea Cassani, 'Hybrid What? Partial Consensus and Persistent Divergences in the Analysis of Hybrid Regimes', *International Political Science Review* 35, no. 5 (2014): pp. 542–558.

[70] Steven Levitsky, *Competitive Authoritarianism: Hybrid Regimes After the Cold War* (New York: Cambridge University Press, 2010).

[71] Aili Mari Tripp, *Museveni's Uganda: Paradoxes of Power in a Hybrid Regime* (Boulder, CO: Lynne Rienner, 2010).

[72] Ali Riaz, *Voting in a Hybrid Regime: Explaining the 2018 Bangladeshi Election* (Singapore: Palgrave, 2019).

[73] Nora Stel, *Hybrid Political Order and the Politics of Uncertainty: Refugee Governance in Lebanon* (Abingdon: Routledge, 2020).

[74] Magdalene Antreou, *The Ailing Turkish Democracy: The Transformation and Perpetuation of a Hybrid Competitive System* (Cambridge: Cambridge Scholars, 2019).

policies. A limited number of works introduce some discussion regarding civil society activism in hybrid states, but such discussion remains far from the subject of foreign policy studies.[75]

Pakistan as a quintessential hybrid state has featured prominently in such debates.[76] Though hybridity has been a feature of Pakistan's politics since the late 1980s, it started gaining academic attention in the late 1990s[77] as it became clear that, short of outright coups, the military was able to employ a variety of tactics to constrain the authority of democratically elected political leaders to make and implement decisions.[78] In the 1990s, the military realized that the costs of clinging on to power were higher than letting a democratic transition take place.[79] Often described as a period of 'temporary democracy',[80] that decade offers insights into how the military occasionally worked actively to constrain democratic consolidation in Pakistan.[81]

Though the interest in the Pakistani regime's hybridity is justified, this literature continues to espouse the same feature as the literature on hybrid regimes generally: the discussion does not move far beyond the issue of definitions and classifications, and the focus remains on transitions between hybridity on one side and democracy or dictatorship on the other. As mentioned earlier, if the Liberal assumption regarding the foreign policy of a country reflecting its internal identity is accepted as holding some weight, the assertion that hybrid states' policies will have distinctly different features from those of states in other categories is justified. An interesting question to explore would be whether the foreign policy of a hybrid state differs from that of a democratic state or a dictatorship. Furthermore, are there different processes involved in foreign policy decision-making in the former category that distinguish them from the latter categories? This book contributes to these debates. As mentioned above, though its main focus is the study of Pakistani foreign policy, a secondary contribution of this work is in the conceptual domain of the foreign policies of hybrid states. We now turn to this discussion.

[75] Rustamjon Urinboyev, *Migration and Hybrid Political Regimes: Navigating the Legal Landscape in Russia* (Oakland, CA: University of California Press, 2020); Gabriele Wilde, et al., eds., *Civil Society and Gender Relations in Authoritarian and Hybrid Regimes: New Theoretical Approaches and Empirical Case Studies* (New York: Columbia University Press, 2018).

[76] Adeney, 'How to Understand Pakistan's Hybrid Regime'.

[77] Rizvi, *Military, State and Society in Pakistan*, p. 32.

[78] Aurel Croissant, David Kuehn, Paul Chambers, and Siegfried O. Wolf, 'Beyond the Fallacy of Coupism: Conceptualizing Civilian Control of the Military in Emerging Democracies', *Democratization* 17, no. 5 (2009): pp. 950–975.

[79] Steven Wilkinson, 'Democratic Consolidation and Failure: Lessons from Bangladesh and Pakistan', *Democratization* 7, no. 3 (2007): pp. 203–226.

[80] Michael Hoffman, 'Military Extrication and Temporary Democracy: The Case of Pakistan', *Democratization* 18, no. 1 (2011): pp. 75–99.

[81] As mentioned, the country returned to military rule in 1999, which continued until 2008.

1.4 Theoretical contribution

1.4.1 Domestic politics and foreign policy: studying the linkages

After situating the topic of the book in the broader literature on the subject, I will now turn towards outlining a theoretical framework to help conduct the above-outlined analysis. This section begins by looking at the academic literature on International Relations (IR) and Foreign Policy Analysis (FPA) that is reflective of a country's identity. Structural Realism, one of the conventional theories of IR, considers states to be 'black-boxes': what happens inside them has little impact on the external domain. For scholars in that tradition, like Kenneth Waltz, 'a systemic IR theory is not a theory of foreign policy because a systemic theory explains "why states similarly placed behave similarly despite their internal differences". By contrast, he suggested, a theory of foreign policy "would explain why states similarly placed in a system behave in different ways".'[82] That aligns with the thinking that 'states often react differently to similar systemic pressures and opportunities, and their response may be less motivated by systemic-level factors than domestic ones.'[83] Despite this being a common understanding, it is surprising that FPA still lacks robust theoretical literature on 'how domestic factors influence foreign policy and under what conditions these factors become more important.'[84] Although there exists a decent body of literature on the 'democratic peace thesis', which asserts that liberal democratic states do not wage war against each other, such discussion does not stretch to states which are non-democracies—especially hybrid states. The field in that regard remains generally open; addressing this could answer a variety of questions, including, for example: how do decision-making elites use foreign policy to enhance their control over the levers of power domestically? How do certain domestic imperatives lead to a specific type of foreign policy reflecting those domestic requirements?[85]

Given the increased interest in a variety of regimes, as listed above, scholars like Hudson have argued that future research in this domain, including on 'the nature of the regime itself, with all its particular strengths and weaknesses and

[82] Binnur Ozkececi-Taner, 'Domestic Politics and Foreign Policy', *Oxford Research Encyclopedia of Politics*, 26 September 2017; in-text references omitted.

[83] Randall Schweller, 'Unanswered Threats: A Neoclassical Realist Theory of Underbalancing', *International Security* 29, no. 2 (2004): p. 164.

[84] Ozkececi-Taner, 'Domestic Politics and Foreign Policy.'

[85] Jeffrey Knopf, 'Domestic Sources of Preferences for Arms Cooperation: The Impact of Protest', *Journal of Peace Research* 35, no. 6 (1998): pp. 677–695.

predispositions, must be made a central part of any analysis of the domestic roots of foreign policy'.[86] The focus on regime types has the potential to help us bring into the discussion a range of factors, including 'legitimation of rule, structure, scope and way of rule'.[87] Analysing the foreign policies of various totalitarian states[88] such as Nazi Germany, Fascist Italy, and the Soviet Union, scholars have asserted that a focus on domestic decision-making explains such countries' expansionist actions and, by extension, their foreign policies.[89] Referring to Bracher,[90] Kneuer argues that 'removing all controls and hurdles in the domestic realm constituted the precondition for the expansionist policy' with such states.[91] Pertinently, 'without an outward projection against a real or imaginary enemy, these regimes could not marshal the fanatical devotion the system requires for survival'.[92] This process has a mutually reinforcing mechanism, in which not only certain regime dispositions inform or shape specific foreign policies, but also the implementation of certain foreign policies requires a 'comprehensive system of control over all areas of political, economic, and societal life, the elimination of any opposition or veto player and the reliance on the mass movements or mass support reached by agitation and propaganda'.[93]

The FPA literature in the 1970s and 1980s concerned itself with studying specific countries' foreign policies but, according to Kaarbo, the rise of interest in Constructivist and Liberal scholarship led scholars to link the FPA literature to the broader IR debates.[94] That interest has also been accompanied by the studies of regime types and attendant foreign policies.[95] However, the regimes studied continue to be primarily democracies (as opposed to hybrid regimes) and a 'regime-sensitive differentiation' has not been employed while conducting such studies.[96] A range of FPA scholarship covers the differences

[86] Valerie Hudson, *Foreign Policy Analysis: Classic and Contemporary Theory* (Lanham, MD: Rowman & Littlefield, 2013), p. 142.
[87] Marianne Kneuer, 'Autocratic Regimes and Foreign Policy', *Oxford Research Encyclopedia of Politics*, 26 September 2017.
[88] Gerhard Weinberg, *The Foreign Policy of Hitler's Germany: Diplomatic Revolution in Europe, 1933–36* (Chicago, IL: University of Chicago Press, 1971).
[89] Kneuer, 'Autocratic Regimes'.
[90] Karl Bracher, *Stages of Seizing Power* (Berlin: Ullstein, 1960).
[91] Kneuer, 'Autocratic Regimes'.
[92] Carl Joachim Friedrich and Zbigniew Brzezinski, *Totalitarian Dictatorship and Autocracy* (Cambridge, MA: Harvard University Press, 1965), p. 353.
[93] Kneuer, 'Autocratic Regimes'.
[94] Juliet Kaarbo, 'A Foreign Policy Analysis Perspective on the Domestic Politics Turn in IR Theory', *International Studies Review* 17, no. 2 (2015): pp. 189–216.
[95] Joe Hagan, 'Regime Type, Foreign Policy, and International Relations', *The International Studies Encyclopedia*, 9, 2010, pp. 6153–6178; Kneuer, 'Autocratic Regimes'.
[96] Kneuer, 'Autocratic Regimes'.

within democracies 'and how these differences influence foreign policy-making processes and outcomes'.[97] In other words, despite these significant developments mentioned above, in-depth studies of non-democracies' foreign policies are still lacking, primarily because of the Anglo-Saxon-centric origins of the discipline of International Relations itself, in which scholars from these countries simply end up studying their own systems—which are, uncoincidentally, primarily liberal and democratic. As Kinne asserts, most studies examining the linkage between domestic and foreign policies 'have focused on decision making in democratic states; while some do explore autocratic decision making, they tend to do so on a case-by-case basis, without speculating on the potential systematic differences that might separate autocracies from democracies, or autocracies from other autocracies'.[98] He goes on to say that

> We know, or can at least make a reasonable guess, that the foreign policy apparatus of a non-democratic state functions much differently from that of a democratic state, and, further, that the political interests of dictators are most likely different from those of democratic presidents. It is therefore appropriate to more fully investigate the nature of non-democratic decision making.[99]

To summarize, where Structural Realism provides little scope for domestic-level factors to explain foreign policy decision-making, Liberalism considers it to be almost an automatic outcome of the state's domestic regime type. Constructivist scholarship, on the other hand, puts a different twist on this argument, asserting that where a state's identity shapes its foreign policy interests, that identity is 'constituted primarily through state-society dynamics, discursively constructed and, most importantly, habitualized—that is, implicit, solidified, and taken-for-granted'.[100] However, as mentioned above, Constructivist literature also agrees with the assertion made earlier that in a two-way process identity is also shaped by foreign policy interests of states. Referring to David Campbell's seminal work on the politics of identity with regard to the United States' foreign policy,[101] Vucetic asserts that according to this strand of thinking, 'U.S. national identity ... is not something that shapes U.S. foreign policy, but rather that which is constituted by it'.[102]

[97] Kaarbo, 'A Foreign Policy Analysis Perspective', p. 196.
[98] Brendon Kinne, 'Decision Making in Autocratic Regimes: A Poliheuristic Perspective', *International Studies Perspectives* 6, no. 1 (2005): p. 115.
[99] Kinne, 'Decision Making in Autocratic Regimes', p. 115.
[100] Srdjan Vucetic, 'Identity and Foreign Policy', *Oxford Research Encyclopedia of Politics*, 26 April 2017.
[101] David Campbell, *Writing Security: United States Foreign Policy and the Politics of Identity* (Saint Paul, MN: University of Minnesota Press, 1998).
[102] Vucetic, 'Identity and Foreign Policy'.

In this section, I have made a case to look at the complicated scenario in which the domestic and foreign are linked in international politics. The picture is far more complex, however, than some Structural Realist writings might lead us to believe. As Kaarbo argues, 'how decision makers interpret and respond to their domestic and international environments is then subject to a number of factors: psychological, societal, ideational, political, institutional, and material.'[103]

1.4.2 Scholarly literature linking the internal with the external

Despite the significance and contemporary relevance of studying the internal identity-related matters and their link with a state's foreign policy, the literature on that topic is still generally sparse and does not do justice to the level of importance the matter holds. Robert Putnam's work on domestic politics and diplomacy remains one of the touchstone pieces of work on this topic, as it studies the role of domestic politics in states' diplomacy.[104] The same can be said about the studies looking at the external relations of Japan[105] and Brazil[106] and how they are affected by the respective politics of these states. However, this literature includes more general references to domestic politics instead of examining the issues of identity, or how certain groups of elites manipulate that identity for specific foreign policies and how some foreign policy agendas are projected to foster a certain kind of state identity with the intent of serving those elites' vested interests.[107] Ted Hopf's theoretically oriented work, in which he looks at the role of societal-level dynamics in the Soviet Union in shaping the country's foreign policy during the Cold War,[108] comes closest to addressing these points.

The dearth of literature on this topic is compounded by the even greater scarcity of studies on contemporary developments in postcolonial (or developing world) states from that perspective. A notable exception is Turkey,

[103] Kaarbo, 'A Foreign Policy Analysis Perspective', p. 207.

[104] Robert Putnam, 'Diplomacy and Domestic Politics: The Logic of Two-Level Games', *International Organization* 42, no. 3 (1988): pp. 427–460.

[105] Leonard Schoppa, 'Two-level Games and Bargaining Outcomes: Why *Gaiatsu* Succeeds in Japan in Some Cases but Not Others', *International Organization* 47, no. 3 (1993): pp. 353–386.

[106] Howard Lehman and Jennifer McCoy, 'The Dynamics of the Two-Level Bargaining Game: The 1988 Brazilian Debt Negotiations', *World Politics* 44, no. 4 (1992): pp. 600–644.

[107] Authors writing in the domain of Neoclassical Realism have considered the linkages between states' domestic politics and their foreign policies, but that scholarship has paid limited attention to these questions.

[108] Ted Hopf, *Reconstructing the Cold War: The Early Years, 1945–1958* (New York: Oxford University Press, 2012).

which has attracted some attention in that regard. Ciftci has assessed how the military's economic interests in Turkey lead it to aspire for a certain type of foreign policy.[109] Despite the obvious parallels with Pakistan, Ciftci's study is more about economic factors influencing foreign policy than the elites' tinkering with identity—which is the subject of this book. Hakan has analysed how the nostalgia in Turkey regarding a specific identity from the country's imperial past shapes the country's foreign policy today.[110] Oğuzlu's valuable contribution studies foreign policy change with reference to Turkey's domestic identity, with its focus shifting away from the West and towards the Middle East.[111] India is another state that has been the subject of literature in this area. For example, Priya Chacko looks at how India's postcolonial identity has shaped its foreign policy and how we might be witnessing a change in the former which, in turn, has affected the latter.[112]

Though valuable, this collection of works falls short of helping with the type of analysis conducted in my book, which focuses on how a section of the ruling elite in Pakistan, specifically the military's senior command, justifies its total control over the foreign policy narrative of an ever-present (manufactured) national security 'threat'. The presence of such a threat helps the military consolidate control domestically (because, according to military leaders, the civilians cannot be trusted to deal with such a 'serious' situation). With typical Orwellian undertones, the internal and external domains in Pakistan are linked and support one another in a circular fashion: domestic control is essential to fight wars abroad—and wars abroad require that the populace at home be kept under tight control.

Hintz has looked at the processes of contestation between different societal groups in Turkey regarding the type of identity the state should have, and how that identity impacts upon its foreign policy. It is a valuable contribution linking the internal with the external, but the focus is on contesting certain types of Turkish identities and how consensus regarding them might change. In the case of Pakistan, a decades-old foreign policy outlook shaped by a top-down manufactured identity is the only paradigm and is never contested.[113] There

[109] Sabri Ciftci, 'Military Might: A Domestic Economy Explanation of Turkish Foreign Policy', *Turkish Studies* 24, no. 5 (2023): pp. 764–787.

[110] M. Hakan Yavuz, *Nostalgia for the Empire: The Politics of Neo-Ottomanism* (New York: Oxford University Press, 2020); see also M. Hakan Yavuz, 'The Motives Behind the AKP's Foreign Policy: Neo-Ottomanism and Strategic Autonomy', *Turkish Studies* 23, no. 5 (2022): pp. 659–680.

[111] Tarik Oğuzlu, 'Middle Easternization of Turkey's Foreign Policy: Does Turkey Dissociate from the West?' *Turkish Studies* 9, no. 1 (2008): pp. 3–20.

[112] Priya Chacko, *Indian Foreign Policy: The Politics of Postcolonial Identity from 1947 to 2004* (Abingdon: Routledge, 2012).

[113] Lisel Hintz, *Identity Politics Inside Out: National Identity Contestation and Foreign Policy in Turkey* (New York: Oxford University Press, 2018).

has been little change in those identity narratives and foreign policy outlooks in Pakistan (for example, those centred on a 'victim' identity or Pakistan as a 'fortress of Islam'), and challenging them is possible neither for the country's civilian elites nor for society in general.

1.5 Theoretical Framework: Ontological Security, Post-Structuralist Discourse Analysis, and the Concept of 'Fantasy'

In order to fill the above-defined gap in the literature, I seek guidance from the emerging literature combining ontological security theory, post-structuralist discourse theory, and the concept of 'fantasy'. In this section, I will first outline the main contours of these ideas and then describe the method through which these approaches can be combined to build a theoretical framework. The section will also provide a rationale for using this particular theoretical framework for conducting the said analysis.

1.5.1 Defining the three concepts

The emerging interest in a post-structuralist discourse analysis takes inspiration from the theoretical works of Michel Foucault, Ernesto Laclau, and Chantal Mouffe, which collectively challenge the distinction between a discursive, or linguistic, and a non-linguistic realm and point towards different sets of reality. According to this thinking, effective discourses are 'relational and differential systems of signification' that operate in a field of intelligibility and have the capacity to present certain realities and can also lead the recipients of those realities to act in a certain way.[114] That means that discourses do not just describe the identities of subjects and objects in a world detached from human agency and experience. These identities are the product of an 'articulatory practice' which creates a 'relational structure through which subjects, objects, words and actions become meaningful'.[115] However, because by their definitions discourses have a relational and differential character that is constantly in flux, identities shaped through them can never have fixed meanings and 'are thus contingent and open for political contestation'.[116] While post-structuralist discourse theory studies the way structures are made

[114] Thorsten Wojczewski, 'Identity and World Order in India's Post-Cold War Foreign Policy Discourse', *Third World Quarterly* 40, no. 1 (2019): p. 181.
[115] Ibid., p. 182.
[116] Ibid.

and unmade through discursive practices, it also makes an emphatic case for critically reflecting on the matters of power and (in)equality.[117] Foucault's invaluable contribution in this regard is to study the productive dimensions of power which 'does not repress but makes possible certain practices'.[118]

A post-structuralist discourse analysis argues that identities and meanings are not 'given' or 'stable' in the way Newton's laws of physics might be. They are discursively constructed and are constituted against the difference of an 'Other'.[119] Using this approach gives us room to analyse how states' foreign policy discourses help us understand how those employing them attempt to create an idea of 'Self' and the 'Other' and for what purpose. 'Foreign policy and identity mutually constitute each other in that the formation of the identity is achieved through this boundary-drawing practice, but the identity is simultaneously constructed as the legitimisation for the proposed policy.'[120]

Discourses play a role in using foreign policies (and associated world orders) as sites for the production and reproduction of representations of certain identities of an actor (like Pakistan) by defining 'what it is'.[121] State elites reference certain identities (such as 'the castle of Islam' in the case of this writing) that emphasize affective dimensions which obscure 'the absence of an extra-discursive foundation on which national identities could be grounded by endowing the Self with an imaginary essence and seemingly unique qualities'.[122] Given that meanings can never be permanently stable, they provide openings for struggles of 'discursive hegemony' over how certain meanings could prevail 'as the "natural" or "normal" understanding of the world and become "common sense"'. Those in powerful positions can construct meaning in line with their interests and demands, and they try to naturalize and universalize certain identities that suit those interests and demands.[123] Those in positions of power might talk about their definitions of identities as 'normal' and 'common-sensical', but such definitions are primarily the masked forms of their interests and demands. That means that PDT does not regard 'social' as an objective ground, but instead sees discursive practices as shaping, and being shaped by, the social. Before moving on to the next concept, PDT can be summarized as follows: first, this approach is interested in a discursive construction of subjectivity, perceiving agency to be an effect of

[117] Johannes Angermuller, 'Poststructuralist Discourse Studies: From Structure to Practice', in *The Cambridge Handbook of Discourse Studies*, edited by Anna de Finna and Alexandra Georgakopoulou (Cambridge: Cambridge, 2020), p. 240.
[118] Ibid., p. 240.
[119] Wojczewski, 'Identity and World Order', p. 181.
[120] Ibid., p. 183.
[121] Ibid., p. 180.
[122] Ibid.
[123] Ibid., p. 183.

discursive practices. Second, it believes that practices constitute the social. Third, through discourses, the social is 'subject to "policing" practices which flatten and smoothen the social out and make it governable'.[124]

The next concept that needs to be defined is ontological security. This concept starts with individual-level identities and their associated insecurities and employs them to understand the roots of states' insecurities.[125] It can be defined as an individual's desire

> to feel oneself as a whole and comprehend one's sense of self. Agents therefore need to formulate a stable way to exist to feel secure vis-à-vis their unique autobiographical narratives and self-identity. Articulating this self-identity requires a certain cognitive setting from which agents formulate and persist in their behavioral routines in light of their self-narratives.[126]

In order to feel secure, individuals like to have cognitive certainty, which they achieve through routinized behaviour in their interaction with significant others.[127] Wars and instances of terrorism are 'dislocatory moments' which disrupt actors' sense of ontological security as they do not know how to 'go on' during those moments.

The next concept that requires definition is 'fantasy', which remains an underutilized concept for the analysis of world politics. For Eberle, fantasies are stories we tell ourselves and one another in order to make sense of the ambiguous and multifaced world; subjects co-construct and relate to social reality through these stories.[128] The stories are created by a country's elites through certain discourses to hide the 'essential incompleteness of identities' and try to root identities in 'an imaginary, transcendental essence' by placing them into 'a seemingly coherent narrative and projecting everything that challenges this secure and stable sense of Self onto the Other'.[129]

> (F)antasy captures the process whereby subjects (social actors) relate to and reproduce reality (social structures) by outlining the relationship between subjectivity, social order and desire. Fantasies connect subjects to social orders by arousing

[124] Angermuller, 'Poststructuralist Discourse Studies', p. 241.
[125] Mitzen, 'Ontological Security in World Politics', p. 343.
[126] Umut Can Adisonmez and Recep Onursal, 'Governing Anxiety, Trauma and Crisis: The Political Discourse on Ontological (In)Security After the July 15 Coup Attempt in Turkey', *Middle East Critique* 29, no. 3 (2020): p. 294.
[127] Mitzen, 'Ontological Security in World Politics', p. 342.
[128] Eberle, 'Narrative, Desire, Ontological Security', p. 245.
[129] Wojczewski, 'Identity and World Order', p. 182; the root of this idea is in the Lacanian psychoanalytic theory. For more, see Ty Solomon, *The Politics of Subjectivity in American Foreign Policy Discourses* (Ann Arbor, MI: University of Michigan Press, 2015).

> desire and channelling it to socially constructed 'objects' like commodities, partners or ideological goals.[130]

Within a post-structuralist method of thinking, discourses are powerful tools in the hands of policy elites because state identities are in perpetual need of reproduction by differentiating 'Self' from a set of 'Others;' a post-structuralist discourse analysis helps one identify the underlying dynamics and methods employed to generate and sustain that reproduction.[131] The absence of a stable foundation in which one's identities are grounded 'creates anxieties, insecurities and desires within the subject and motivates it to identify with a discourse that promises to fill this lack and provide the Self with a sense of wholeness, certainty and stability'.[132] This is where the concept of 'fantasy' comes in. Contemporary scholarship on the subject makes a case for studying it as a mechanism to capture the missing essence of the subject, as opposed to an 'illusion' or 'false consciousness'.[133] Hence, fantasies construct

> stable, natural or transcendental foundation on which the imaginary essence of the Self can be grounded (e.g., nature, religion or ancient epics) and places the Self in a linear, coherent story that is often characterised by an imaginary origin, a moment of purity, grandeur and perfection, which has been lost and must be recovered, or an ideal of a glorious, consummate future, whose realisation is blocked by something: the Other.[134]

Fantasies have two faces, and both of them work in tandem to impart a sense of ontological security to a subject. The first aspect is described as the beatific side, defined as 'a narrative structure involving some reference to an idealized scenario promising an imaginary fullness or wholeness' and the second aspect, the horrific side, refers to a disaster scenario.[135]

Fantasies can be recognized for having a '"black and white" character' that leaves little space for ambiguity and their substructural discourses come with '"good" and "bad" political objects that are very one-dimensional'.[136] As Eberle argues,

[130] Eberle, 'Narrative, Desire, Ontological Security', p. 245.
[131] Wojczewski, 'Identity and World Order', p. 1.
[132] Ibid.
[133] Ibid., p. 1.
[134] Ibid.,
[135] Jason Glynos, 'Ideological Fantasy at Work', *Journal of Political Ideologies* 13, no. 3 (2008): p. 283.
[136] Jelle Behagel and Ayşem Mert, 'The Political Nature of Fantasy and Political Fantasies of Nature', *Journal of Language and Politics* 20, no. 1 (2021): p. 6.

> fantasies are those types of narratives, which shelter us from anxiety by denying any ambiguity or complexity of the given situation. They feed our 'hunger for certainty' by offering us the false safety of a clear-cut choice between two—and only two—options, which are constructed as excessively beatific and horrific respectively.[137]

Given their very nature, fantasies are 'structurally entwined with discourse and meaning, desires and emotions that drive political choices, as well as political ontologies and subjectivities'.[138] Fantasy can be rightfully described as narrative imbued with desire.[139] They do not only manage desires, they also make worlds. Understanding fantasy helps one grasp why subjects cling to unfeasible or unrealistic hopes that give them certainty in an increasingly damaged world.[140] The role of fantasy can thus be central to explaining the ideological force of certain discourses. This helps one understand how and why certain discourses grip the political subject: '"the imaginary promise of recapturing our lost/impossible enjoyment provides the fantasy support" for our political and social choices'.[141] The elites strategically mobilize political projects and ideologies that promise the delivery of this desire by first identifying an obstacle—an 'other' that stands in the way—and then suggesting its removal in a certain way.[142] That is done through various ritualistic practices and with the expert deployment of both dystopian and utopian narratives.[143]

It is important not to regard fantasies as thoughts that are incompatible with reality. In fact, they stabilize our sense of reality 'in the face of the ultimately uncertain world and our own subjective incompleteness'.[144] By doing so, fantasies provide subjects a feeling of 'security of being',[145] 'security as being',[146] 'identity security',[147] and 'the subjective sense of who one is'.[148] Given that these feelings are at the heart of the literature on ontological security, it can be stated that fantasies provide subjects with a feeling of this security. Furthermore, stability provided by a routinized existence that is enabled by fantasies plays a critical role in imparting that feeling of ontological security to the subject.

137 Eberle, 'Narrative, Desire, Ontological Security', p. 249.
138 Behagel and Mert, 'The Political Nature of Fantasy', p. 80.
139 Eberle, 'Narrative, Desire, Ontological Security', p. 248.
140 Behagel and Mert, 'The Political Nature of Fantasy', p. 81.
141 Ibid., p. 8.
142 Ibid.
143 Ibid., p. 8
144 Eberle, 'Narrative, Desire, Ontological Security', p. 249.
145 Catarina Kinnvall, 'Globalization and Religious Nationalism: Self, Identity, and the Search for Ontological Security', *Political Psychology* 25, no. 5 (2004): p. 746.
146 Brent Steel, *Ontological Security in International Relations* (Abingdon: Routledge, 2008), p. 51.
147 Mitzen, 'Ontological Security in World Politics', p. 342.
148 Ibid., p. 344.

The search for ontological security is a political practice with clear political consequences and cannot be detached from the issues of power and exclusion.[149] In other words, the policy elites' attempts to make subjects feel ontologically secure depends on the power they enjoy; the process to achieve that security is based on including some while excluding others from the collective. Given that fantasies help to impose closure on the ambiguities and complexities of the social world, 'someone is always excluded from the community in whose name closure is performed, and there are unintended consequences that we may not even be aware of'.[150]

Jason Glynos and David Howarth identify multiple 'logics of critical explanation' and, according to their understanding, a 'fantasmatic logic' helps us study how certain types of discourses are used to construct specific practices and 'regimes of truth', and to understand *why* specific practices and regimes 'grip' subjects.[151] In other words, such logics help us understand why subjects desire to identify with a discourse and take it to be 'real'.[152] By doing so, this type of logic helps us understand the resistance to change of social practices.[153] Seen this way, the notion of a 'fantasmatic logic' asserts that fantasies should not just be treated as fictions detached from reality, but that they have the power to project one's sense of lacking onto the other and can ultimately promise the 'self' a state of wholeness once the 'other' is overcome.[154]

By accepting the elites' presentations of certain meanings and identities, subjects are also complicit in concealing or covering over radical contingency—'the impossibility of fixing social meaning and identity'[155] in a given context—namely that of social relations.[156] By pushing the radical contingency of social relations to the background, fantasmatic narratives provide simple and easy-to-digest answers. However, while doing this, they also move the contestation of the political dimensions of the practices through which these meanings are articulated and accepted to the back burner, thereby consolidating the hegemony of those behind such meaning-making.[157] As PDT does away with the idea of all forms of essentialism and rejects the notion that social phenomena can be rooted in a single underlying principle, for elites the

[149] Eberle, 'Narrative, Desire, Ontological Security', p. 251.
[150] Ibid., p. 252.
[151] Jason Glynos and David Howarth, *Logics of Critical Explanation in Social and Political Theory* (Abingdon: Routledge, 2007), p. 145.
[152] Wojczewski, 'Identity and World Order', p. 184.
[153] Glynos and Howarth, *Logics of Critical Explanation*, p. 145.
[154] Wojczewski, 'Identity and World Order', p. 181.
[155] Umut Can Adisonmez and Recep Onursal, '"Strong, but Anxious State": The Fantasmatic Narratives on Ontological Insecurity and Anxiety in Turkey', *Uluslararası İlişkiler/International Relations* 19, no. 73 (2022): p. 65.
[156] Glynos and Howarth, *Logics of Critical Explanation*, p. 15.
[157] Ibid., p. 145.

social becomes 'an open and clear field that can be hegemonized in an infinite number of ways, free of prior determinations'.[158]

1.5.2 Linking the three concepts

Where 'dislocatory moments' cause a deep upset to actors' sense of security, they also open up space for 'political practices (which can) shape, modify, or reorder the regime itself'.[159] However, the hegemonic social groups within states can use that opportunity to 'actively reproduce the existing power relations, norms, values and institutions by keeping the public contestation' of alternatives at bay.[160] In states like Pakistan, such moments are often manufactured by the state elites to rekindle the masses' desire for ontological security by giving them a 'sense of identity, order, and a future direction'.[161] Such practices help to reinforce the hegemony of these elites over the lives of the masses and further reduce the space for other voices to present an alternative vision for those involved. Where complex realities confronting individuals might cause anxiety, they yearn for 'unitary, consistent and singular' identities which 'cancel out political, social, and cultural differences and demands within the society to facilitate a *sense of order* and maintain a *coherent sense of self*'.[162]

Ontological security-seeking behaviour yearns to tackle anxieties more than fears. Here, anxiety can be defined as 'a sensation, an emotion, and a thought—(which) prevails through all stages of being and must therefore be central for any political analysis'.[163] Anxiety is different from fear, because fear concerns threat or risk that is visible and known. The former, on the other hand, concerns situations in which the object of threat is not visible and identifiable.[164] Anxiety's links to ontological security can be understood through the lenses of time and space. Anxiety is 'a future-oriented and long-acting response to an unknown threat, fear is a present-oriented and short-lived response to a clearly identifiable and tangible threat'. In moments of dislocations, state elites mobilize fantasmatic narratives that shelter them from anxiety and, by doing so, the elites feed the masses' hunger for

[158] Thomas Jacobs, 'The Dislocated Universe of Laclau and Mouffe: An Introduction to Post-Structuralist Discourse Theory', *Critical Review* 30, nos. 3–4 (2018): p. 297.
[159] Adisonmez and Onursal, 'Governing Anxiety, Trauma and Crisis', p. 296.
[160] Ibid., p. 296.
[161] Ibid., p. 299.
[162] Ibid., emphasis in the original.
[163] Adisonmez and Onursal, 'Strong, but Anxious State', p. 64.
[164] Ibid., p. 64.

security, 'showing them their place in the world—a relatively whole, stable, and complete identity'.[165]

Fantasmatic narratives are presented in an oversimplified way with the aim of removing any room for ambiguity and uncertainty.[166] Their proponents highlight only two options: a beatific frame and a horrific frame, intentionally steering individuals towards choosing the former and rejecting the latter. No other alternatives are allowed to enter into the frame, and public contestations that might present a different diagnosis of any dislocatory moment are not allowed. As a result, 'alternative becomings and articulations'[167] are discarded without consideration, and the hegemon's answer to a problem is presented as the 'only solution' to it. In other words, policy choices suiting the elites are mobilized without any contestation and debate. This is how fantasies (and fantasmatic narratives associated with them) emerge as a tool to consolidate hegemony of those in powerful positions. Because an individual's 'need to experience oneself as a whole person in time can never be fully satisfied', the elites deploy fantasmatic narratives to impart the former with a degree of ontological security. As we will see in the case of Pakistan, manufactured perpetual conflicts on different fronts produce anxiety and make subjects feel ontologically insecure. That in turn opens space for those in power to articulate certain fantasmatic narratives to make the subjects feel whole and secure again. Such narratives protect subjects from 'the anxiety stemming from a confrontation with the radical contingency',[168] and given their simpler and clear-cut solutions, they impart individuals with a sense of 'epistemological peace'.[169] Such narratives require constant reinforcement, and elites perpetually manufacture dislocatory moments which continue to open up space for them. Where they highlight a 'beatific scenario [that] promises an anxiety-free future, a great state, and a peaceful society, this imaginary fullness-to-come (being "one") is also complemented by a horrific scenario. The horrific narrative foregrounds a disaster to come (a divided country) if implied or named obstacles (internal/external enemies) cannot be overcome'.[170] These fantasmatic narratives are one-dimensional, as they offer the same solution for a range of diverse problems and often lack creativity and imagination, with homogeneity presented as security.[171] They present the vision of a society that has no political diversity and hence no ontological insecurity, and any public

[165] Adisonmez and Onursal, 'Governing Anxiety, Trauma and Crisis', p. 301.
[166] Ibid., p. 302.
[167] Ibid., p. 304.
[168] Adisonmez and Onursal, 'Strong, but Anxious State', p. 66.
[169] Felix Berenskötter, 'Anxiety, Time, and Agency', *International Theory* 12, no. 2 (2020): p. 280.
[170] Adisonmez and Onursal, 'Strong, but Anxious State', p. 70.
[171] Ibid., p. 73.

contestation of such a colourless ontological security regime is presented as 'a threat to the survival of the state'.[172]

1.5.3 Justification and rationale

Why was a Constructivist analysis of foreign policy not used here, since it provides a more straightforward framework to study how the external behaviour is shaped by the identity of a state? The answer to this question is exactly that: it presents 'nationalistic, ethnocentric accounts of world politics', which mirror official foreign policy narratives of countries.[173] That means that such an approach does not really do justice to capturing the internal as effectively as the above-defined approach does by linking the three concepts and by adopting a critical methodology instead of the one emphasizing description only. My chosen framework affords me the possibility to analyse fantasmatic narratives as 'simplified and mythical stories ... that order experiences by placing actors and events into a seemingly linear, coherent story in which everything that contradicts this linear story is excluded or projected onto the Other'.[174] Dislocatory moments are helpful because they enable the continuation of myths through social and political practices and rituals. A Constructivist analysis does not have space for capturing the nuances of narratives that this approach provides—hence the choice of this framework.

1.5.4 Information on methodology and data

The *modus operandi* of my engagement with the theory is deductive in nature, as it performs a theoretically informed analysis.[175] I take the conceptual ideas mentioned above in their current form and utilize them to gain unique insights into the formation and evolution of Pakistani foreign policy. The study benefits from four main sources of primary data, namely: the United States Foreign Relations Archives; Pakistani military resources (including the monthly magazine *Hilal* and the speeches of its army chiefs such as Pervez Musharraf and Qamar Bajwa); US Embassy Cables; and interviews.

[172] Ibid.

[173] Wojczewski, 'Identity and World Order', p. 181.

[174] Ibid., p. 184.

[175] Nicole Schadewitz and Timothy Jachna, 'Comparing Inductive and Deductive Methodologies for Design Patterns Identification and Articulation', in: International Design Research Conference IADSR 2007 Emerging Trends in Design Research, 12–15 November 2007, Hong Kong, p. 2, available at https://oro.open.ac.uk/15257/

The documents from the US Foreign Relations Archives have been critical to the study of Pakistani foreign policy through the end of the Cold War.[176] Empirical details covering the engagement between two countries during that period are rich, and no other source contains as much information on Pakistani narratives regarding the country's domestic and foreign policies. The US Embassy Cables were obtained from the WikiLeaks data (released in 2010), which consisted of 250,000 cables from various American embassies around the world to Washington, D.C. *The Guardian* newspaper's sorting of the cables and organizing them according to different topics and countries was helpful in collating the relevant documents for this study.[177] These cables primarily cover the first ten years of the current century and have provided insights into Pakistani foreign policy at the start of the War on Terrorism in 2001.

The Pakistani military publishes various outputs that provide an insight into the military's raw thinking. These publications include *the Green Book*, the monthly magazine *Hilal*, *Margalla Papers*, *The Review*, and *The Citadel*, among others.[178] Out of these, *Hilal*[179] and *Margalla Papers*[180] are accessible online. I have chosen to primarily consult *Hilal*, which is the military's flagship monthly magazine and reaches all rank-and-file members of the organization. It provides timely and undiluted insights into the military's thinking, concerning a wide variety of issues ranging from regional and global politics to climate change and economic challenges facing Pakistan. The authors of *Hilal* are primarily serving and retired military officers, but also include retired diplomats, journalists, and academics who are sympathetic to the military standpoint. A number of authors writing for *Hilal* also contribute to the *Margalla Papers*, but the latter are published once a year and are broadly academic in nature. *Hilal*, on the other hand, provides raw and unadulterated insights into the message the organization is looking to convey to both its foot soldiers and the common person. Given its monthly frequency, this source is tremendously valuable to analyse the military's narrative on day-to-day events. *Hilal* is also a credible source for such an analysis because its writings closely align with, and support, the official military pronouncements. Importantly, these writings almost never criticize the military's vision regarding the country or its preponderant role in Pakistan's affairs.

[176] https://history.state.gov/
[177] https://www.theguardian.com/us-news/the-us-embassy-cables
[178] https://cscquetta.gov.pk/publications/
[179] https://www.hilal.gov.pk/
[180] https://margallapapers.ndu.edu.pk/site/issue/archive

Hilal is a bilingual magazine (in Urdu and English) that has been in publication for over seventy-five years, and has had a 'significant impact on shaping public opinion and fostering a sense of unity among [the] Armed Forces and the citizens of Pakistan'. The editors proudly declare that the magazine 'fosters a sense of unity, instilling pride and a shared purpose among its readers' through what it describes as 'accurate and unbiased reporting'.[181] It is important to note that citations to earlier issues of the magazine include page numbers, as these editions were originally published in print format, and were later scanned for online access. More recent issues, however, are available online and do not contain fixed pagination. Accordingly, page numbers are provided where available and omitted where the digital format does not permit them.

I also conducted a range of interviews with noted analysts, academics, security and civilian officials, and foreign observers over a period of five years. These interviews informed the main contours of the book and were instrumental in putting forward a balanced study of Pakistani foreign policy. I also wish to add two notes regarding the text of the book. First, I have used varying forms of the phrase war on terrorism/War on Terrorism. I have generally used capitalization when I refer to a specific policy that was initiated after the attacks of 11 September 2001, thereby treating the phrase as a proper noun. In contrast, I have used the lower case 'war on terrorism' when I refer to broader counterterrorism efforts more generically. The choice therefore reflects whether the phrase is being used to refer to a historically bounded policy construct or to a broader descriptive category. Second, I have extensively relied on English-language sources originating in Pakistan and have done my best to introduce additional punctuation to ensure a smooth flow for the reader. However, I have kept my interventions to the bare minimum so that the reader is also able to glean the true essence of the text being consulted.

1.6 Book structure

The civil–military imbalance has characterized Pakistan's domestic politics since the mid-1950s, and a number of scholars have written extensively on the topic. However, there is no book-length study that analyses how that imbalance shapes and influences Pakistan's foreign policy—that is something this work achieves. Through this book, I argue that the foreign policy of the country is very much representative of the military's dominance over the

[181] Asif Raja, 'A Beacon of Hope: 75 Years of Hilal's Journey to Zenith', *Hilal*, July 2023.

system. The make-up of Pakistan's internal identity is reflected in its external behaviour. However, that does not imply that the military has been the most influential actor domestically from the country's very inception. As we saw earlier, there have been occasions where the military was not in the driving seat (such as in the early 1950s) or was on the backfoot (such as after the separation of Pakistan's Eastern wing in 1971). As concerns the former, the civilian leadership's incompetence in the 1950s opened the way for the military to consolidate its control domestically and on Pakistan's foreign policy. On the other hand, though there were immense opportunities for Zulfikar Ali Bhutto to recalibrate the civil–military imbalance in Pakistan in the 1970s, he adopted increasingly authoritarian means and used the military's own nationalist rhetoric to gain popularity. However, doing so strengthened the country's beleaguered military and paved the way for Bhutto's eventual overthrow at the hands of General Zia-ul-Haq in 1977.

A study of the discourse of Pakistan's ruling elite, primarily the military, reveals that various threats facing the country—some real, others imagined—are addressed through specific discourses that foster anxiety and a sense of perpetual insecurity among the masses. In order to cope with that anxiety, a number of fantasies are put forward to make sense of that alternative reality. The official discourse on Pakistani foreign policy is truly a 'factory of fantasies', which itself belongs in an alternative reality. Its main objective is to consolidate the hegemony of those articulating these fantasies and to keep them permanently relevant to the business of the state. This chapter has prepared the foundations for that discussion.

The second chapter of the book conducts a study of Pakistan's foreign policy from the creation of the state in 1947 through its civil war and eventual dismemberment in 1971, leading to the creation of Bangladesh as an independent state. It argues that since its inception, Pakistan's foreign policy has been based on two pillars. The first is a rentier outlook that renders the state reliant on external financial assistance as the primary means of sustaining its economic activities. The second pillar is the use of militants as state proxies to achieve Pakistan's foreign policy objectives. Since Pakistan's creation, these policies have been justified by those in power on the basis of the threat posed by India. The state has used militant proxies to foster perpetual instability along Pakistan's eastern and western borders, which in turn strengthens the military dominant position within the country.

The third chapter examines Pakistani foreign policy from the civil war of 1971 through the start of the fourth military dictatorship in 1999. Pakistan experienced democracy for the first time between 1971 and 1977, which gave the civilian Prime Minister Zulfikar Ali Bhutto a rare opportunity of course correction. However, Bhutto chose to continue to rely on foreign financial assistance and the use of militancy as instruments of foreign policy in Afghanistan. The Afghan jihad of the 1980s deepened Pakistan's reliance on external rent, which led to severe economic challenges after the Soviet withdrawal from the region and the end of American aid shortly thereafter. Soon after the country's return to democracy in 1988, Pakistan's military escalated the use of militant proxies in the Indian-occupied Kashmir, which was soon followed by its support for the Afghan Taliban. The violence in Kashmir sabotaged the peace overtures to India made by the civilian governments of Benazir Bhutto and Nawaz Sharif during the 1990s and perpetuated the military's hegemony domestically.

The fourth chapter examines the case of Pakistan's policy towards Afghanistan from the emergence of the Taliban in the mid-1990s through the American withdrawal from the country in August 2021. I argue that the policy of playing 'double games' represented a missed opportunity on the part of the state's ruling elites to raise the country's profile and put it on a path of sustainable development. The civilians were sidelined while following this policy in the face of serious international recrimination and opposition. The post-withdrawal scenario can only be seen as a bounce back of that flawed approach, in which the Afghan Taliban are now colluding with the Pakistani Taliban to attack the Pakistani security forces in the northwest of the country. However, Pakistan cannot claim that it was caught off guard by that approach. I assert that there were ample warnings of that eventuality long before the US withdrawal in 2021, but the Pakistani state chose to ignore them. A number of mistakes by the United States administration soon after the overthrow of the Taliban in 2001 resulted in Pakistan siding with the Taliban as an insurance against a Northern Alliance–controlled Afghanistan, as the Northern Alliance was favoured by the international community at the time. However, Pakistan had over two decades to make a change in the policy, especially after the Bush administration realized its mistakes and vowed to work with Pakistan to help address its threat perceptions.

The fifth chapter studies the period from the fourth military takeover of 1999 through the US–Pakistan Strategic Dialogue of 2010. Pakistan faced international isolation as a result of the coup by Pervez Musharraf in 1999, but there was a shift in Western policy after Pakistan was asked to join the American-led War on Terrorism. The initial bonhomie, however, quickly gave way to a confrontational relationship between the two partners as Pakistan was not willing to sever its links with the Afghanistan Taliban at the time. The military justified its support for the group by citing the perceived

Indian threat on Pakistan's eastern border, prompting US efforts to facilitate rapprochement between India and Pakistan. However, I argue in this chapter that there was limited appetite for serious engagement with India on the part of the Pakistani military despite the civilian leadership emphasizing its need. Pakistan was presented with many chances of obtaining international guarantees regarding the threat from India in order to put the two countries on a path towards amicability but, as the evidence shows, the former continued to use militant proxies to ensure that India–Pakistan relations remained unstable. The straining of the relations by the end of the first decade of the twenty-first century led the US and Pakistan to reset their relationship through a strategic dialogue which emphasized the need for a longer-term engagement. However, the Pakistani ruling elite viewed this dialogue as an opportunity to convince the United States of Pakistan's stance and to value the sacrifices it had made instead of the former reflecting on its own approach.

The sixth chapter studies the period from 2011 to the present. The first year of this period is often described as the *annus horribilis* of relations between Pakistan and the United States, with the killing of Osama Bin Laden and the Salala attack (both in 2011) causing a major rift between the two states. These developments led Pakistan to look towards alternative power centres for alliance-building and to move out of the American orbit. Though a diversified foreign policy could have been beneficial for Pakistan, the country lacked the economic wherewithal to engage with its new partners (China and Russia). Given its reliance on external rent, in the form of American aid and funding from international financial institutions like the International Monetary Fund (IMF), it had little choice but to revert back to the American side for its continued survival. Since the early 2020s, the military leadership has been emphasizing the need for a balanced and diverse foreign policy. It has also been highlighting the importance of a resilient economy, which itself is linked to domestic political stability. Importantly, the military has also been reassessing the policy of using militants as tools of foreign policy, as this approach has adversely impacted Pakistan's standing with its key partner, China. The country sorely lacks competent civilian leadership which could put the Pakistani economy back on track. Additionally, there is no desire in the military to cede control of the reins of power to the civilian population and return to the barracks. While such a move would stabilize Pakistani politics, it would significantly diminish the military's influence, which it remains unwilling to accept.

The seventh chapter concludes the study of Pakistan's foreign policy and its justifying narratives, centred on the concept of Pakistan as 'the fortress of Islam'. I argue that this idea perfectly encapsulates the country's plight. Its rulers have used a narrative of all-encompassing threats throughout its history to cause anxiety and insecurity among the populace. The ruling elite, primarily the military, fosters a 'siege mentality' to shape Pakistan's foreign policy, thereby strengthening its domestic hegemony. This hegemony secures a disproportionate share of the country's budget for the military elite, while much of society remains impoverished, malnourished, and uneducated, with little prospect of change. A long history of poorly managed foreign policy has painted the military, and the country, into a corner, from which it is unable to escape the consequences of the policy of 'controlled chaos' that it has condoned and even promoted for decades.

2

Born insecure

Pakistan's quest for ontological security (1947–1971)

2.1 Introduction

The Partition of the subcontinent in 1947 was a major 'dislocatory moment'[1] in the history of the region as it was accompanied by incidents of ethnic cleansing on both sides of the border.[2] The creation of Pakistan stemmed from a 'siege mentality', which asserted that the existence of Muslims in the subcontinent would be endangered without a separate country of their own. The state was created on the basis of the Two-Nation Theory, which held that Hindus and Muslims were two separate nations and could not live peacefully within the same political entity. The Two-Nation Theory contradicted centuries of generally peaceful coexistence among diverse communities in the region—though minor skirmishes were not uncommon, as in any multi-faith society. This approach also clashed with the long-standing historical evidence that 'precolonial India was a mosaic of ethnic and linguistic variety and diversity which coexisted peacefully under normal circumstances'.[3]

Pakistan faced genuine challenges to its ontological security from its inception, fostering deep anxiety among its citizens, but such precarious insecurity is not unique to that one country. Many states in the world (such as Israel and Singapore) started their lives in uncertain geopolitical environments and feared their survival to be under serious threat from the beginning. However, Pakistan's leadership deliberately reinforced the nation's identity as a perpetually insecure state due to its geostrategic location, sustaining a constant sense of anxiety among citizens. The ruling elites decided to overplay the role of

[1] Umut Can Adisonmez and Recep Onursal, 'Governing Anxiety, Trauma and Crisis: The Political Discourse on Ontological (In)Security After the July 15 Coup Attempt in Turkey', *Middle East Critique* 29, no. 3 (2020): p. 296.

[2] Ishtiaq Ahmed, *Jinnah: His Successes, Failures and Role in History* (London: Penguin, 2019), p. 533. See also Ishtiaq Ahmed, *The Punjab: Bloodied, Partitioned and Cleansed* (Karachi: Oxford University Press Pakistan, 2012), p. 166.

[3] Ahmed, *Jinnah*, p. 61.

Pakistan's Foreign Policy. Wali Aslam, Oxford University Press. © Wali Aslam (2026).
DOI: 10.1093/9780198909552.003.0002

the country's geographic location in determining its engagement with the rest of the world while neglecting the importance of a robust economy and prosperous citizenry. From its inception, Pakistan lacked a foreign-policy vision that prioritized strategic resource allocation to thrive globally. Rather than overcoming their insecurities, Pakistan's ruling elite established the nation's foreign policy on that foundation. An exaggerated emphasis on the security threats from all sides led to the creation of 'an anxious state'. In order to mitigate these insecurities, the country's rulers decided to shape Pakistan's foreign policy on two key elements: foreign assistance to support the country's economic and defence needs, and the use of non-state actors as proxies to enhance its power. I study both of these elements in detail below. The chapter concludes with an assessment of Pakistan's foreign policy between 1947 and 1971 through an examination of different crises faced by the state, such as the India–Pakistan War of 1965 and the Civil War of 1971. By prioritizing insecurity as the state's central focus, Pakistan's ruling elites enabled the supremacy of the institution tasked with addressing security challenges: the Pakistan army. The army's dominance over policy led to foreign-policy adventurism, exemplified by Operation Gibraltar, the primary cause of the 1965 war. It also contributed to the 1971 war with India that resulted in the secession of East Pakistan. The policy of competing with India consolidated the military's control domestically while also isolating the country from the wider Muslim world, which was more sympathetic to India's non-aligned stance.

As argued below, India also contributed to establishing an antagonistic relationship between the two nations, yet Pakistan's leadership—both civilian and military—chose to base its foreign policy on this negative foundation rather than overcoming these adverse dynamics. While some challenges, such as the border dispute with China in the early 1960s, were resolved diplomatically, Pakistan's reliance on non-state actors and foreign assistance set an undesirable trajectory for its subsequent history.

2.2 The role of internal and external actors in forming Pakistan's insecure identity

Although a siege mentality shaped Pakistan's identity from its inception, this mindset became more entrenched after Partition because of various factors. The biggest challenge concerned India which, the Pakistani leadership believed had refused to come to terms with the 1947 Partition of the subcontinent and was stalling in delivering Pakistan its share of the military equipment and other resources.[4] Political leaders associated with the Indian National Congress publicly doubted the viability of the Pakistan project. For example, Maulana Abul Kalam Azad, who went on to become India's first education minister, wrote that Indian leaders like Sardar Vallabhbhai Patel believed that 'the new state of Pakistan was not viable and could not last. [Patel] thought [that] the [creation] of Pakistan would teach the Muslim League a bitter lesson [and] Pakistan would collapse in a short time.'[5] Amin writes that '*Akhand Bharat*, or united India, remained the conviction of many top Indian Hindu leaders in the early years after independence'.[6] He quotes a senior Congress leader who said '[n]either the Congress nor the nation has given up its claim of a united India'.[7] Accordingly, it was not surprising that the new nation's siege mentality worsened yet further. Ever since independence, similar hardline statements by the leadership of parties on all points of the political spectrum in India (including the Indian National Congress and the Bharatiya Janata Party) have played a role in further consolidating the insecurities of Pakistan. Indian military exercises near the Pakistani border in 1950 and 1951 did not help to ameliorate the feelings of vulnerability, though some have argued that these exercises were conducted to send a signal to Pakistan to refrain from interfering in the Indian-occupied Kashmir after the war of 1947–1948.[8] Kashmir had become a major irritant between the two nations within weeks of Partition, as the Maharaja had acceded to India to seek its help in thwarting the advance of the Pakistani tribesmen who had entered the territory on behalf of the Pakistani state. Developments like these led the Pakistani leadership to believe that India was bent upon 'strangling the new state at its very birth'.[9]

Kashmir was not the only territorial dispute facing Pakistan. Along with India, Pakistan had potential territorial disputes with all of its neighbouring countries, including China, Afghanistan, and Iran. These problems surfaced in the 1950s, not long after Partition. The issue with Afghanistan, often dubbed 'the Pashtunistan issue', dated back to the 1893 demarcation of the Durand Line, which created the present-day border between Afghanistan and Pakistan. The government of Afghanistan maintained that the line was forced upon it by the British Empire, dividing the Pashtun people on both

[4] S. Paul Kapur, *Jihad as Grand Strategy: Islamist Militancy, National Security, and the Pakistani State* (New York: Oxford University Press, 2017), p. 34.

[5] See Kalam cited in Shahid Amin, *Pakistan's Foreign Policy: A Reappraisal* (Karachi: Oxford University Press Pakistan, 2000), pp. 25–26.

[6] Amin, *Pakistan's Foreign Policy*, p. 28.

[7] Ibid.

[8] Ahmed, *Jinnah*, p. 471. I will discuss that war below.

[9] Kapur, *Jihad as Grand Strategy*, p. 34.

sides of the line. It insisted that the arrangement was only meant to last for one hundred years.[10] As a result, Afghanistan has traditionally laid claim to significant territory in the north-west of Pakistan.

China also claimed 3,400 square miles of territory within Pakistani Kashmir, including the principality of Hunza.[11] The rulers of Hunza had traditionally recognized Chinese suzerainty over their territory. The Kuomintang, the Chinese ruling party at the time of the Partition of the Indian subcontinent, expected Hunza to remain an independent state under Chinese fealty before the rulers of Hunza decided to accede to Pakistan. Several Chinese incursions in Pakistan were reported in the early 1950s, adding to the insecurities of the incipient Pakistani state.[12]

The Iranians also asserted claims to the territory within Pakistani Balochistan,[13] although these claims were never officially given support from the Shah in Iran or the country's senior politicians.[14] According to declassified CIA reports from 1954, certain Iranian foreign ministry officials made a case to their superiors that an area of around '200 kilometres' belonged to Iran that was unfairly extorted from the country by the British through the treaties of 1871 and 1896.[15] The report did acknowledge that these claims were not as serious as those raised by Afghanistan, but they further added to the insecurity of the Pakistani state. Pakistanis suspected that the government of the Shah of Iran had 'designs' on Balochistan, a belief that reinforced Pakistan's perception of Iran's pursuit of close ties with India.[16]

These external anxieties were compounded by internal challenges. Pakistan's division into two wings, separated by a thousand miles, posed significant difficulties for a resource-scarce nation striving to maintain cohesion between East and West Pakistan. While India inherited the British Raj's administrative infrastructure, most of West Pakistan encompassed the empire's frontier territory, primarily serving as its external defence rather than a hub for 'effective domestic administration'.[17] In other words, India had inherited 'the best of the Raj' and Pakistan had inherited all the problems that

[10] Bijan Omrani, 'The Durand Line: History and Problems of the Afghan-Pakistan Border', *Asian Affairs* 40, no. 2 (2009): pp. 177–195.

[11] Andrew Small, *The China-Pakistan Axis: Asia's New Geopolitics* (New York: Oxford University Press, 2015), p. 21.

[12] Ibid.

[13] Muhammad Akbar Notezai, 'Trouble at the Pakistan-Iran Border', *The Diplomat*, 30 October 2018.

[14] 'Iran-Pakistan Border Dispute: An Evaluation of the Iranian Claims', Intelligence Report number 6564, 16 September 1954, *Central Intelligence Agency*, United States of America, available at https://www.cia.gov/library/readingroom/docs/CIA-RDP08C01297R000500130001-4.pdf

[15] Ibid., p. 3.

[16] Aparna Pande, *Explaining Pakistan's Foreign Policy: Escaping India* (Abingdon: Routledge, 2011), p. 146.

[17] Kapur, *Jihad as Grand Strategy*, p. 34.

the Raj had faced (e.g. boundary issues with Afghanistan and uncertainties concerning the country's tribal areas where the state's writ did not extend).

Pakistan also had myriad challenges concerning its heterogeneous population. A disparate group of people were brought together in the name of Islam, and they shared precious little with each other apart from their religion. As the Islamic identity was given a priority, other identities (such as the Baloch or Bengali identities) were supressed, leading to these groups' resentment from the very beginning.[18]

Pakistan's civilian political leadership, including its founder Muhammad Ali Jinnah, lacked a clear understanding of the nation's place in the region. This absence of imagination hindered the development of a coherent vision for Pakistan's future in the world. In fact, the story of Pakistan's foreign policy is a history of competing, often contradictory visions, regarding the country's positioning in the world and how it wants to achieve its objectives. On the one side, the leadership championed a relatively neutralist vision that stressed Pakistan's goodwill towards its neighbours. On the other hand, there was a clear tilt towards the United States in the geopolitical rivalry of the Cold War. The expressions of goodwill towards neighbouring India also went hand in hand with the desire to compete with it at every forum, owing to what Pakistan perceived to be India's hegemonic tendencies.

Soon after independence, Jinnah proclaimed Pakistan's interest in 'friendliness and goodwill towards all the nations of the world.'[19] The first prime minister, Liaquat Ali Khan, famously asserted that Pakistan would not have any 'narrow and special commitments' or 'any prejudices in international sphere.'[20] The country was 'neither tied to the apron strings of the Anglo-American bloc nor was it a camp-follower of the communist bloc.'[21] Instead, it would champion the principles of 'friendship towards all states, but with regard to each individual question, standing on the side of fairness and subject to that, to help and succour the weak.'[22] However, in the very first meeting of the cabinet, Jinnah stated that the defence of Pakistan in the north-west (referring to the Soviet threat) was not just a Pakistani concern but a concern of the

[18] Nasir Islam, 'Islam and National Identity: The Case of Pakistan and Bangladesh,' *International Journal of Middle East Studies* 13, no. 1 (1981): pp. 55–73.

[19] Mohammad Ali Jinnah, *Quaid-i-Azam Mohammad Ali Jinnah: Speeches as Governor-General of Pakistan, 1947–1948* (Karachi: Ferozsons, 1949), p. 65; see also Mehrunnisa Ali, 'Jinnah's Perception of Pakistan's Foreign Policy and the Kashmir Issue,' *Pakistan Horizon* 43, no. 2 (1990): pp. 57–70.

[20] See also Khwaja Sarwar Hasan, 'The Foreign Policy of Mr. Liaquat Ali Khan,' *Pakistan Horizon* 4, no. 4 (1951): pp. 181–199.

[21] See also Naveed Ahmad, 'The Non-Aligned Movement and Pakistan,' *Pakistan Horizon* 32, no. 4 (1979): pp. 79–91.

[22] Abdul Sattar, *Pakistan's Foreign Policy: 1947–2005* (Karachi: Oxford University Press Pakistan, 2007), p. 12.

rest of the world. On the basis of this 'fantasy', the country made a request for generous financial assistance from the United States, which was swiftly turned down.[23]

Similarly, the leadership repeatedly expressed its desire to 'live in peace and let others live in peace'.[24] It reiterated that Pakistan would 'endeavour to create and maintain goodwill and friendship with Britain and our neighbourly dominion, Hindustan, along with other sisterly nations [to] make [its] greatest contribution for the peace and prosperity of the world'.[25] Confusingly, this pronouncement accompanied a fear of India shown in the assertions that the country was bent upon destroying the new state of Pakistan from the very beginning.[26]

Divided into factions and preoccupied with infighting, Pakistan's civilian leadership failed to unite in establishing robust foundations for the new state or developing a coherent foreign-policy vision. According to Ishtiaq Ahmed, these 'internal squabbles and intrigues amongst the Muslim Leaguers greatly undermined the party and weakened its prestige'.[27] The leadership's commitment to democracy was eclipsed by its own desire to stay in power. For example, the first prime minister Liaquat Ali Khan 'evaded elections, primarily because he had no constituency from which to ensure his election—that fear and diffidence became an excuse for his lesser successors to also postpone general elections'.[28] Where India had adopted a Constitution by November 1949, Pakistani leadership took nearly ten years after Partition to agree on one. Perpetual infighting among incompetent and irresponsible politicians occasionally culminated in physical violence. In one such incident on 23 September 1958, the Speaker of East Pakistan's Provincial Assembly disqualified six members of the ruling party, affecting the party's thin majority. *Time* magazine wrote on 6 October 1958, '[T]hough the eleven-year-old republic of Pakistan has yet to hold its first general election, its politicians staged some of the fiercest parliamentary battles of the British Commonwealth. Last week in Dacca, the evenly matched government and opposition forces of East Pakistan waged the biggest brawl in the young country's brief history'.[29] After the

[23] Dennis Kux, *Disenchanted Allies: The United States and Pakistan 1947–2000* (Maryland, MD: The Johns Hopkins University Press, 2001), p. 20.
[24] Jinnah, *Speeches as Governor-General*, p. 62.
[25] Ibid., p. 11.
[26] Kapur, *Jihad as Grand Strategy*, p. 34.
[27] Ishtiaq Ahmed, *Pakistan: The Garrison State* (Karachi: Oxford University Press Pakistan, 2013), p. 111.
[28] Ibid.
[29] 'Pakistan: Death in the Chair', *Time*, 6 October 1958, available at https://time.com/archive/6888304/pakistan-death-in-the-chair/

Speaker's ruling, 'outraged government deputies laid down a barrage of paperweights, desk panels and curtain rods, chased him out the door, voted him "insane"'.[30]

> When parliament met again, the new speaker (Shahid Ali) readmitted the six deputies. Opposition members exploded with fury. They tore their desks from the floor[,] ripped their microphones out of their stands, and charged. Steel microphone stands whipped at Ali's face, a desk panel struck him full on the head, and he went down in a pool of blood. After steel-helmeted cops arrived to break up the melee, sergeants-at-arms bore Mr. Speaker off to a hospital on a stretcher. He died two days later, the first presiding officer of any parliament in the history of the British Commonwealth to perish of injuries received while occupying the chair.[31]

The lack of serious political leadership gave the country's civilian bureaucracy the opportunity to take control of the policymaking of the nascent state from the early 1950s.[32] The ascendent army quickly eclipsed the civilian bureaucracy and emerged as perhaps the only disciplined institution in the new state. The lack of leadership from the civilian politicians and their perpetual infighting led many Pakistanis to welcome the first military takeover in 1958. One of my interviewees recalled seeing people dancing in the street welcoming the military's control of the state, given the incompetence of the civilians.[33] The role of the civilian politicians in helping exacerbate Pakistan's early sense of insecurity cannot be understated. In fact, it was the incompetence of the civilians that made space for the military's domination in the formative years of the country's history, a trend which continues until this day. As a prestigious Pakistani newspaper noted:

> The new country's early politicians, religious figures, intellectuals, artistes and society as a whole, spent most of their time and energies groping and grappling with various conflicting and vague ideas about the Pakistani nationhood; and/or what sort of an economic system Pakistan should have, how much of a role religion should play in the matters of the state and government, and what was a person's ethnic identity vis-à-vis his or her national (Pakistani) identity.[34]

[30] Ibid.
[31] Ibid.
[32] Ahmed, *Pakistan: The Garrison State*, p. 113.
[33] Interview, Academic 4, Location withheld, August 2024.
[34] Nadeem F. Paracha, 'The Pakistan Zeitgeists: A Nation Through the Ages', *Dawn*, 29 May 2014.

2.3 Security through alliances

In the absence of a clear vision from dedicated civilian leadership, Pakistan's foreign policy was unsurprisingly based on a rent-seeking paradigm first proposed by Jinnah. Pakistan's reliance on foreign rent was a logical outcome of overplaying the role of its geographical location in its destiny.[35] After interviewing Jinnah in September 1947, Margaret Bourke-White wrote that he firmly believed the West would fund Pakistan's defence for its own security. Jinnah told Bourke-White: 'America needs Pakistan more than Pakistan needs America ... Pakistan is the pivot of the world, as we are placed... [on] the frontier on which the future position of the world revolves'.[36] Jinnah believed that if the United States was ready to defend Greece and Turkey against Russia, it would be 'much more interested in pouring money and arms into Pakistan'.[37] He said, 'Russia is not so very far away ... If Russia walks in the whole world is menaced'.[38] Summarizing her own impressions of the newly formed country's foreign policy vision, Bourke-White wrote:

> In the weeks [after interviewing Jinnah, his] thesis [was] echoed by government officials throughout Pakistan. 'Surely America will build our army' they would say to me. 'Surely, America will give us loans to keep Russia from walking in.' But when I asked whether there were any signs of Russian infiltration, they would reply almost sadly, as though sorry not to be able to make more of the argument. 'No, Russia has shown no signs of being interested in Pakistan.'[39]

According to Bourke-White, the entire Pakistani state machinery was intent on securing US Treasury funds, and this idea was 'voiced so persistently that one wonders whether the purpose was to bolster the world against Bolshevism or to bolster Pakistan's own uncertain position as a new political entity'.[40] She goes on to write:

[35] Small sections of this and subsequent chapters have also been published in Wali Aslam, 'Misaligned Expectations and Broken Promises: United States Foreign Policy Approach Towards Pakistan (1947–2021)', in *The Routledge Handbook of US Foreign Policy in the Indo Pacific*, edited by Oliver Turner, Nicola Nymalm, and Wali Aslam (Abingdon: Routledge, 2022), pp. 330–342. They are reproduced here with the permission of Taylor and Francis (Books) Limited UK (Licence number: 1504349-1, 12 July 2024).

[36] Margaret Bourke-White, *Halfway to Freedom* (New York: Simon & Schuster, 1949), p. 93. See also Bourke-White cited in Ahmed, *Jinnah*, p. 401.

[37] Quoted in T. V. Paul, *The Unfinished Quest: India's Search for Major Power Status from Nehru to Modi* (New York: Oxford University Press, 2024), p. 140.

[38] Bourke-White, *Halfway to Freedom*, p. 93.

[39] Ibid.

[40] Ibid.

> Actually, I think it was more nearly related to the even more significant bankruptcy of ideas in the new Muslim state—a nation drawing its spurious warmth from the embers of an antique religious fanaticism, fanned into a new blaze. Jinnah's most frequently used technique, during the struggle for his new nation, had been the playing of opponent against opponent. Evidently, this technique was now to be extended into foreign policy.[41]

This section examines Pakistan's foreign policy strategy of maximizing US financial aid since 1947. Even if one were to grant that a foreign policy built on such dependence on allies is a useful approach, the story of Pakistan's relations with the US appears to be a poor handling of these relations. The country had several opportunities to use its relations with the US to enhance its own security, but they were not utilized.[42] The 'discursive hegemony'[43] of that vision remained unchallenged throughout the country's history.

Despite asserting that Pakistan would not favour one Cold War bloc over the other, the leadership placed the country very firmly in the Western bloc directly after independence. The country was in serious need of cash, which realistically could only have come from the United Kingdom or the United States. Additionally, Pakistan's refusal to compromise with India necessitated external support to counter its regional rival. The founding fathers put forward several rationales for aligning with the West. For example, chairing a cabinet meeting a month after independence, Jinnah said, 'Pakistan [is] a democracy and communism [does] not flourish in the soil of Islam. It [is] clear therefore that our interests [lie] more with the two great democratic countries, namely, the U.K. and the U.S.A. rather than with Russia.'[44] The idea was that the country's strategic location (at the confluence of South East Asia, the Middle East, and Central Asia) was such that it would be obligatory on the Western powers to pay for Pakistan's defence. In a meeting on 11 September 1947, Muhammad Ali Jinnah referred to the Soviet threat to South Asia along the lines of the 'Great Game' of the nineteenth century. Jinnah asserted that 'the safety of the North West Frontier [was] of world concern and not merely an internal matter for Pakistan alone.'[45] On that basis, Pakistan swiftly sent a request to the

[41] Ibid.
[42] Adisonmez and Onursal, 'Governing Anxiety, Trauma and Crisis', p. 296.
[43] Thorsten Wojczewski, 'Identity and World Order in India's Post-Cold War Foreign Policy Discourse', *Third World Quarterly* 40, no. 1 (2019): p. 183.
[44] Quoted in Kux, *Disenchanted Allies*, p. 20.
[45] Ibid., p. 20.

United States for a $2 billion aid package for military and economic assistance.[46] The idea behind that fantasy was that the enormity of the sums being requested was justified given the potential threat to Pakistan from the Soviet Union.[47] The country's policy elites thought that the American leadership would meet the country's demands given the threat Pakistan faced and the role the country could play as a bulwark against a Communist encroachment. The US government denied this request immediately, offering Pakistan a $10 million relief grant instead.[48] This symbolic episode highlights how Pakistan's policy elites, over decades, have overstated the nation's geopolitical significance, evidenced by unrealistic demands. The US officials would often highlight that Pakistanis continually presented requests for military assistance to every American official visiting Pakistan in the 1950s, with the hope of achieving their ends.[49] Significantly, when these demands went unfulfilled, Pakistan viewed the US as a barrier to its desired sense of security.[50]

Pakistan eagerly sought to capitalize on its geostrategic position for US financial aid, but the Truman administration was unreceptive. Though interested in Pakistan's potential to counter Soviet ambitions, US policymakers avoided a deep military relationship, fearing it might 'ensnare' them in an India–Pakistan dispute.[51] They also believed that a cosy relationship with Pakistan would affect the US's inclination to work with India, which had much greater economic potential. The US hoped that if the Indian government was able to quickly lift its citizens out of poverty, that would minimize the chances of a Communist-inspired revolution in the country.[52] Given India's demographic and economic potential, this strategy seemed sensible, but Pakistan's foreign-policy elites did not share that understanding. The Truman administration's tilt towards India was clear from the fact that it invited the Indian prime minister Nehru to visit the US in October 1949. Around six months after Nehru, the Pakistani prime minister, Liaquat Ali Khan, was invited for a visit to the US in the summer of 1950.

[46] Michael Rubin, 'When Realities Collide: Differing US-Pakistan Threat Perceptions', in *Pakistan: The US, Geopolitics and Grand Strategies*, edited by Usama Butt and Julian Schofield (London: Pluto, 2012), p. 46. Kux asserts that the request was for a loan instead, see Kux, *Disenchanted Allies*, p. 20.

[47] Rubin, 'When Realities Collide,' p. 46.

[48] Kux, *Disenchanted Allies*, p. 21.

[49] Editorial Note, Foreign Relations of the United States (1958–1960), South and Southeast Asia, Volume XV.

[50] Jelle Behagel and Ayşem Mert, 'The Political Nature of Fantasy and Political Fantasies of Nature', *Journal of Language and Politics* 20, no. 1 (2021): p. 8.

[51] Mariam Mufti, 'The Influence of Domestic Politics on the Making of US-Pakistan Foreign Policy', in *Pakistan: The US, Geopolitics and Grand Strategies*, p. 67.

[52] Ibid.

Pakistan's relations with the US strengthened after the election of President Eisenhower as the new administration was more sold on the role that Pakistan could play against the Communist threat. The new president was more willing to buy the Pakistani narrative regarding its geostrategic location. The Eisenhower administration believed Pakistan could play a critical role in defending Asia against communism, a goal it pursued through alliances and defence pacts with Asian and Middle Eastern states. The new Secretary of State John Foster Dulles envisioned working with three states (Turkey, Pakistan, and Iran) to build a northern tier of defence to counter the potential Communist threat. Washington believed that Pakistan's land and naval bases could be used by the former to place the US at a significant advantage vis-à-vis the Soviet Union during the Cold War. The administration also believed that Pakistan could provide troops for securing the Middle East in case of a Soviet invasion.[53]

With the US open to support, Pakistan's leaders prioritized securing military assistance to address their defence needs. The country's army chief, General Muhammad Ayub Khan, visited the United States in September 1953 and told the American officials that Pakistan's army could be 'your army if you want' it to be.[54] He asserted that Pakistan could 'serve as the West's eastern anchor in an Asian alliance structure' for the '*right price*'.[55]

As before, the US was cautious about how aiding Pakistan militarily and economically might affect its relations with India. In a meeting with the visiting US Vice President Richard Nixon in December 1953, the Governor General, Ghulam Muhammad, assured him that it would be 'absurd to think of Pakistan attacking India with 40 million Muslims' in it.[56] Muhammad also assured the US ambassador to Pakistan that if the US were to comply with the Pakistani request for military and economic assistance, he would personally 'reassure [the Indian Prime Minister Nehru] on any fears he might have of the intent of Pakistan in its use of [the US supplied] military equipment'.[57] The Governor General also offered to help Washington in its endeavour to build the northern tier, but he 'had nothing to talk about that was of any use' until Pakistan was assured of military assistance by the US.[58] However,

[53] National Intelligence Estimate, 30 June 1953, Foreign Relations of the United States (1952–1954), Volume XI, Part II: 1083–1089.

[54] Husain Haqqani, *Magnificent Delusions: Pakistan, the United States and an Epic History of Misunderstanding* (New York: PublicAffairs, 2013), p. 61.

[55] See Shirin Tahir-Kheli, *The United States and Pakistan—The Evolution of an Influence Relationship* (New York: Praeger, 1982), p. 3, emphasis added; also see Kux, *Disenchanted Allies*, p. 57.

[56] Memorandum of Conversation, by the Ambassador in Pakistan, 7 December 1953, Foreign Relations of the United States (1952–1954), Volume XI, Part II.

[57] Ibid.

[58] Ibid.

when military assistance was indeed assured, Pakistan 'could be very helpful in bringing Iran into an agreement and could go further by being helpful in lining up support of Saudi Arabia because the new King of Saudi Arabia was a very good friend of [Ghulam Muhammad]'.[59]

Pakistan sought US aid by linking it to resolving the Kashmir dispute, with Ghulam Muhammad arguing that India would not negotiate with a weaker Pakistan. The country's fantasy of self-importance was clear from Ghulam Muhammad's belief that not providing aid to Pakistan, 'in view of all the publicity', would be akin to 'taking a poor girl for a walk and then walking out on her, leaving her only with a bad name'.[60] In a separate meeting with the visiting American Vice President, Pakistan's Prime Minister pushed that idea further by asserting that it was only a matter of time until Russia would 'strive for access to the Persian Gulf through Iran, Afghanistan and Pakistan. [The problem] was of vital urgency which the free world should appreciate'.[61] He emphasized that the 'aid to GOP [Government of Pakistan] must and will do all it can to prevent such action both on [the] ground of preventing physical destruction of GOP and on the ground that all true Muslims will always have to fight[,] putting [the] all powerful state ahead of importance and well-being of individual'.[62]

Exaggerating the symbolic importance of the US treatment of Pakistan, Prime Minister Muhammad Ali Bogra said to the Vice President in 1953 that a denial of aid to Pakistan would send a negative message to all of America's allies. Discursively constructing the case of the country as that of a canary in a coal mine, he contended that 'if one so close to [the] United States [could] not get help when willing to stand up and be counted, what hope [was] there for others who [were] willing to stand up and be counted'.[63] Bogra also dismissed the 'silly' idea that assistance to Pakistan would enable it to fight India. He asserted that if the US were to provide arms to Pakistan, the latter could 'never maintain an offensive without continuing support of industrial powers, so merely giving equipment would not make possible war against India'.[64] If anything, military aid would better equip Pakistan for its defence against an India that was dominating the entire subcontinent. In other words, US assistance would help stabilize the region. The Prime Minister contended that US military assistance would help Pakistan to focus on the 'economic problems

[59] Ibid.
[60] Ibid.
[61] Memo from the Ambassador in Pakistan (Hildreth) to the Department of State, 8 December 1953, Foreign Relations of the United States (1952–1954), Volume XI, Part II.
[62] Ibid.
[63] Ibid.
[64] Ibid.

[at home] which now ha[d] to take second place to defense'.[65] This fantasmatic narrative was premised on the idea that the American side would readily accept the logic presented by the Pakistanis. The narrative oversimplified the scenario and did not leave room for any ambiguity or uncertainty.[66] The Prime Minister even tried to convince the Americans that Nehru used 'neutralism' 'purely for bargaining purpose of playing East against West'.[67] Once again exaggerating its sense of self-importance, the Prime Minister asserted that a rejection of the aid would be a diplomatic defeat not only of Pakistan but a 'diplomatic victory for Russia and India and blow to [the] United States'.[68] Such fantasmatic narratives presented Pakistan as 'a strong bulwark against communism; that was in fact the reason why it was the victim of most vicious communist and neutralist propaganda' and if it 'should go down, American influence *in all of Asia* would diminish or disappear'. In order to avoid such a 'horrific scenario', the narrative underlined that the rest of the world needed to do what they could to support Pakistan economically and militarily and not expose it to 'unnecessary dangers'.[69] As history showed, Pakistan obtained substantial US military and economic aid, yet this did not resolve the Kashmir dispute or promote South Asian stability. When it came to helping the US with troops for the Vietnam War, Pakistanis refused to contribute, citing defence needs at home.

While Pakistan's leaders privately sought US aid in exchange for base access, they were less transparent with their public about this 'arms-for-bases' deal. According to Haqqani, Ghulam Muhammad firmly denied in public that 'Pakistan intended to join US-led alliances or had offered to provide military bases' to an outside power.[70] A different narrative was presented to the audience at home, as Ghulam Muhammad asserted in an interview that 'reports that my government [was] negotiating with the US Government for military assistance in return for American bases in Pakistan [were] absolutely unfounded and baseless'.[71] Pakistani foreign minister Zafrullah Khan also denied that the country was engaged in negotiating a longer-term alliance with the US and asserted that it was only making inquiries for the purchase of weaponry.[72] This public stance was a far cry from the private conversations in

[65] Ibid.
[66] Adisonmez and Onursal, 'Governing Anxiety, Trauma and Crisis', p. 302.
[67] Ibid.
[68] The Ambassador in Pakistan (Hildreth) to the Department of State, 8 December 1953, Foreign Relations of the United States (1952–1954), Volume XI, Part II.
[69] Memorandum of a Conversation, Karachi, 8 December 1959, 9 a.m., Foreign Relations of the United States (1958–1960), Volume XV, emphasis added.
[70] Haqqani, *Magnificent Delusions*, p. 64.
[71] Quoted in ibid.
[72] Ibid., p. 67.

which Pakistanis keenly suggested that the Pakistani army could be America's army at the 'right price'. Since its inception, Pakistan has adopted contrasting foreign and domestic stances on its US alliance, a defining feature of their relations. As shown further on in this book, this approach later caused complications when Pakistan pursued a 'double game' with the international community after 11 September 2001.

Apparently convinced by the reasoning of Pakistan's government, the US announced in February 1954 that it would start a military aid programme for the country.[73] It also helped Pakistan in its food shortages at the time by providing $74 million worth of wheat for free.[74] The US also signed a Mutual Defense Assistance Agreement with Pakistan in May 1954, which sought to 'make available equipment, materials, services or other assistance with such terms and conditions' as may be agreed among the two parties.[75] Washington was to supply Pakistan with American weapons 'exclusively to maintain its internal security, its legitimate self-defense or to permit it to participate in the defense of the area, or in the United Nations collective security arrangement and measures'. A key clause of the agreement entailed that Pakistan 'would not undertake any act of aggression against any other nation'.[76] This was the basis on which the US would assist Pakistan militarily, but the former did not clarify the nature and scope of that assistance.[77]

Despite the agreement, bureaucratic delays hindered US aid to Pakistan. By mid-1954, Pakistan's leaders grew frustrated with Washington's failure to deliver promised support. Pakistan's foreign minister Zafrullah Khan met Harold Stassen, Director of the United States Foreign Operations Administration, in June 1954 to emphasize the Pakistani need for economic help given the decline in its main exports, namely jute and cotton.[78] Foreign Minister Khan said that Pakistanis did not believe in concealing their 'beggar's bowl', insisting their plight required American mercy and not justice.[79] Though willing to assist Pakistan, the US believed it could not foster a 'well balanced expanding economy' in Pakistan through 'massive financial donations'.[80] This American view led to resentment in Pakistan, which warned that without

[73] Rubin, 'When Realities Collide', p. 47.
[74] Sattar, *Pakistan's Foreign Policy*, p. 43.
[75] Haqqani, *Magnificent Delusions*, p. 69.
[76] Ibid.
[77] Kux, *Disenchanted Allies*, p. 67.
[78] Ibid., p. 68.
[79] Memorandum of Conversation, by the Officer in Charge, Economic Affairs, Office of South Asian Affairs (Fluker), 22 June 1954, Foreign Relations of the United States (1952–54), Volume XI, Part II.
[80] The Memorandum by the Secretary of State to the Embassy in Pakistan, Washington DC, 20 August 1954, Foreign Relations of the United States (1952–1954), Volume XI, Part 2: 1865–1867; Haqqani, *Magnificent Delusions*, p. 73.

adequate military assistance, it would avoid a 'defence arrangement with the United States'.[81]

In 1954, the US invited Pakistan to join the Southeast Asia Treaty Organization (SEATO) which was a significant development. The treaty sought to support Pakistan's defence against Communist aggression in the region.[82] Pakistan, however, also sought protection from non-communist states, particularly India,[83] but the US did not pay heed to that demand. For the latter, the treaty was aimed at the threat of communism and targeted China (along with the Soviet Union), with which Pakistan was to have cordial relations later on. On American persuasion, Pakistan also joined a British-led Pact of Mutual Cooperation (first named as the Baghdad Pact and then the Central Treaty Organization or CENTO) which had the same purpose as SEATO. Pakistan showed less enthusiasm for the Baghdad Pact than SEATO, as the latter lacked US involvement, which Pakistan considered essential for defending its sovereignty.[84] Though neither of these pacts held much military value for Pakistan, joining them helped it in other ways.[85] For example, they gave Pakistan a 'strengthened claim on US resources'.[86] Similarly, by involving Pakistan in these arrangements, the United States 'acquired an even larger stake in Pakistan's well-being'.[87]

Pakistan's motives for joining SEATO and the Baghdad Pact did not fully align with the aim of countering Communist expansion. The Americans were 'clear' that the foundational treaty of SEATO did not oblige them to consider 'any [local] dispute in the area' as 'a threat to the peace and security of the U.S.'.[88] In October 1954, US Secretary of State Dulles told Pakistani prime minister Muhammad Ali Bogra that a non-communist dispute, like one between Burma and Thailand, would not concern the US. Pakistan's leaders, dissatisfied with this stance, believed their relations with the US were significant enough to influence American Cold War policy. They complained that the American view 'tended by implication to condone aggression from a non-communist country'.[89] Pakistani prime minister Bogra voiced frustration

[81] Pervaiz Cheema, *Pakistan's Defence Policy 1947–58* (London: Palgrave, 1990), p. 148.

[82] The United States convinced India that the weapons supplied to Pakistan would only be used by the latter in order to deter any Soviet expansionism. However, Pakistan did use the US-supplied weapons against India in the war of 1965. See Sattar, *Pakistan's Foreign Policy*, p. 45 and Amin, *Pakistan's Foreign Policy*, p. 44.

[83] Sattar, *Pakistan's Foreign Policy*, p. 47.

[84] See ibid., pp. 48–49.

[85] Kux, *Disenchanted Allies*, p. 74.

[86] Ibid., p. 74.

[87] Ibid.

[88] Memorandum of Conversation, by the Acting Officer in Charge of Pakistan-Afghanistan Affairs (Thacher), Foreign Relations of the United States (1952–1954), Volume XI, Part II.

[89] Ibid.

that the US had promised only $30 million in military aid. He deemed this inadequate for the risks Pakistan faced by aligning with the West against India, Afghanistan, and the USSR.[90] Washington responded by increasing economic assistance to $109 million for the year. In October 1954, the US pledged, through an aide-memoire, to increase Pakistan's military aid to $50 million annually and outline future goals. This included equipping 5.5 army divisions, supplying twelve naval vessels, and providing new aircrafts for Pakistan's air force. The US estimated the military aid programme's cost at $170 million, with a 3.5-year duration from the agreement's signing.[91] This commitment, however, ran into trouble soon afterwards, as both sides believed that the promised assistance to Pakistan would cost substantially more than the sums promised. Following heated US–Pakistani talks and given Pakistan's reliance on US support, Secretary of State Dulles raised the military aid ceiling in December 1955:[92] from 1954 to 1959, Pakistan secured $425 million in US military aid. The package supplied by the US included Patton tanks, F-86 fighter jets, transportation equipment, communication devices, and naval vessels. In addition, Pakistan received $855 million in economic assistance over this period. Pakistani officials nevertheless argued that US aid was inadequate given Pakistan's geopolitical importance and the risks from aligning with the West. Meanwhile, the US began questioning whether its large aid expenditures on Pakistan were worthwhile.[93]

The Americans agreed that Pakistan's military needed an upgrade, but they were beginning to doubt whether investing in Pakistan's military would eventually help American defence. A visiting American official sent a cable home in February 1955 stating that '[t]here [was] no doubt that Pakistan need[ed] direct forces support'. He said that he had 'seen military encampments where the men [were] living in tattered and torn tents which compare[d] very unfavorably with some of the ramshackle huts built, particularly in the Middle East, by the various Arab and Pakistani refugees'. However, the US officials did not have a 'clear idea of the part Pakistan was expected to play in the defense of the Middle East or whether that role would be developed into an important one'.[94]

[90] Ibid.

[91] Kux, *Disenchanted Allies*, p. 69.

[92] See Saqib Qureshi, *US Foreign Policy to Pakistan, 1947–1960: Re-Constructing Strategy*, PhD Thesis, Department of International Relations, London School of Economics and Political Science, United Kingdom, September 2001, pp. 278–281, available at http://etheses.lse.ac.uk/3055/

[93] Haqqani, *Magnificent Delusions*, p. 83.

[94] Memorandum by the Assistant Secretary of Defense for International Security Affairs (Hensel), 17 February 1955, Foreign Relations of the United States (1955–1957), Volume VIII.

The Americans noted that it was increasingly becoming 'quite clear that Pakistan regard[ed] the Indian threat as much more serious to Pakistan than the Russian or Communist China threats. The Pakistani Army [was] now deployed along the Indian border and all tactical and strategic planning by the Pakistani[s] center[ed] on Indian problems as having first priority.'[95] The assessment concluded that Pakistan was unlikely to 'commit any more than token forces outside Pakistan territory unless its armed forces [were] considerably strengthened, its economy improved, and its fear of India greatly reduced'.[96] Officials also worried that the US was spending money on Pakistan faster than 'it could be absorbed'.[97]

Contrary to Pakistani rulers' logic, reliance on the US ended up making Pakistan weaker vis-à-vis India instead of shoring up its defences. The inflow of foreign aid precluded the country from using its foreign policy to improve its economy. Its overtures to the United States also encouraged India to build close relations with the Soviet Union by the mid-1950s. Nehru visited Moscow in 1955, a gesture reciprocated by visits from Nikita Khrushchev, the Soviet leader, to India, Afghanistan, and Burma—all of these countries surrounded Pakistan's eastern or western wings.[98] Speaking in Srinagar, Khrushchev criticized the Pakistanis for allowing American bases in the area. The Pakistani government denied that it would allow any country to establish bases in Pakistan.[99] This, in turn, was seen by the Americans as a betrayal, as they had been discussing the very issue of bases with the Pakistanis. In private, Pakistan had expressed openness to allowing American bases in the region as long as certain prerequisites were met, including American 'support for Pakistan on Kashmir'.[100] This led the US to openly question Pakistan's utility in American defence, yet the Pakistanis were still unable to understand the other side's reservations. These reservations held that the country could simply not be trusted to move its focus away from India in order to help defend against communism or to increase investment in the economy to ensure self-reliance and self-sufficiency.

By the late 1950s, well-placed Americans had started asserting that, although Pakistan could act as the 'eastern stanchion for defense of ME [Middle East] against overt Communist aggression or Communist subversion

[95] Ibid.

[96] National Intelligence Estimate, 15 March 1955, Foreign Relations of the United States (1955–1957), Volume VIII.

[97] Memorandum From the Operations Coordinator (Bishop) to the Assistant Secretary of State for Near Eastern, South Asian, and African Affairs (Allen), 28 July 1955, Foreign Relations of the United States (1955–1957), Volume VIII.

[98] Haqqani, *Magnificent Delusions*, p. 77.

[99] Ibid.

[100] Ibid., p. 78.

comparable to stanchion represented by Turkey in [the] West', in order to do so it 'must overcome serious hurdles'.[101] The biggest hurdle, in that regard, included Pakistan's 'continuing difficulties with India', which appeared to have 'few prospects for improvements'. The Governor of West Pakistan told the US ambassador in March 1957 that 'when Pakistanis [spoke] of "defense" they meant defense against India'.[102] Americans, on the other hand, believed that 'dispute with India cause[d] diversion of resources and attention from internal problems and Soviet enemy'. The ambassador cautioned that 'Pakistan now carr[ied] crushing burden of defense expenditures which it consider[ed] necessary for protection against India'.[103]

In a letter to the State Department in December 1957, America's new ambassador to Pakistan, James Langley, seriously criticized the fantasy-based narrative of the Pakistani officials. He wrote that the idea that US military assistance to Pakistan would help counter the Soviet threat was based on a 'hoax'.[104] Langley believed that Pakistan's military forces were 'unnecessarily large' and needed to be 'appreciably trimmed'. It was not a question of 'whether' but by 'how much' the armed forces should be cut. Langley went on to say that, by that time, 'Pakistan military expenses absorb[ed] approximately 65% of the government's tax revenues. The drain on the economy [was] such that U.S. aid, as important as it [was], only serve[d] to maintain precarious living standards, and, while this [was] important, it [was] difficult for the man in the street to appreciate a benefit which ha[d] to be measured in almost negative terms!'[105] The diplomat was also perturbed by the attitude in the 'highest political places', which suggested that 'the United States must keep up and increase its aid to Pakistan, and conversely, that Pakistan [was] doing the U.S. a favor in accepting aid, in addition to the Pakistan pro-western posture in the Baghdad Pact and SEATO and the UN, when actually these postures [were] in part dictated by Pakistan[i] hatred for India'. Deriding Pakistan's sense of entitlement, Langley asserted that 'the U.S. [was] entitled to a reasonable degree of respect' in its dealings with Pakistan.[106]

As before, American criticisms did not prompt reflection from the Pakistani leadership, which were reluctant to accept that reductions in the size of the army were necessary. Instead, the Pakistani ambassador to the US told the

[101] Telegram From the Embassy in Pakistan to the Department of State, 30 March 1957, Foreign Relations of the United States (1955–1957), Volume VIII.

[102] Ibid.

[103] Ibid.

[104] Letter From the Ambassador in Pakistan (Langley) to the Assistant Secretary of State for Near Eastern, South Asian, and African Affairs (Rountree), 27 December 1957, Foreign Relations of the United States (1955–1957), Volume VIII.

[105] Ibid.

[106] Ibid.

US Secretary of Defense in July 1959 that 'ordinary Pakistanis' were unable to understand the American insistence that Pakistan should reduce the size of its army.[107] The ambassador also worried that the Indians maintained 'a strength much greater than that of Pakistan and so disposed as to be able to mount an offensive against Pakistan in an estimated period of 10 days, it would be almost impossible to convince those Pakistanis living near the Indian border that any reduction in the strength of the Pakistan Army was a reasonable course of action.' Referring to the possible threats to Pakistan's domestic stability, the ambassador argued that 'it was even probable that if the Pakistan border garrisons were reduced significantly, there might develop a sudden flight of Pakistanis from the border area which could develop into panic proportions.'[108]

Pakistani officials often referred to the 'common' man and woman when addressing their US counterparts, as if ordinary Pakistanis' views were a primary consideration in their foreign policy. They also emphasized that their people's anxiety would rise if Western governments did not 'recognize their needs, including needs in [the] security field'[109] while omitting that the state often manufactured this anxiety to control the narrative. By the late 1950s, Pakistani officials also pointed to the public pressures 'that Pakistan should endeavor to follow policies less hostile to Communist bloc countries.'[110] They would often assert that the public reaction to a foreign policy that did not embody their wishes could make things 'unpredictable,' and that the country's leadership might not be able to resist the masses. In reality, the state, under dictatorship or democracy, largely ignored the people's views on foreign policy. Pakistani officials regularly used the reference to 'ordinary Pakistanis' to extract more aid from the US and express resentment towards the latter when aid was not forthcoming. In the decades following its independence, Pakistan's sense of strategic importance led it to believe that it was entitled to influence US foreign policy in South Asia, particularly towards India. Pakistani officials repeatedly complained that they could not understand the US military assistance to India and the American desire to guarantee India's territorial integrity.[111] They also criticized the American press, which was often critical of Pakistan while emphasizing

[107] Memorandum of a Conversation, Washington, 17 July 1959, Foreign Relations of the United States (1958–1960), Volume XV.

[108] Ibid.

[109] Airgram From the Embassy in Pakistan to the Department of State, 23 September 1959, Foreign Relations of the United States (1958–1960), Volume XV.

[110] Telegram From the Embassy in Pakistan to the Department of State, 16 September 1959, Foreign Relations of the United States (1958–1960), Volume XV.

[111] Memorandum of a Conversation, Washington, 17 July 1959, Foreign Relations of the United States (1958–1960), Volume XV.

that 'the way to Free World security in Asia lay in promoting the welfare of India through the granting of massive economic aid'. Pakistani officials wanted the US to be more sympathetic to their country compared to India given the Pakistani 'support of the Western concept of collective security'. They told their American counterparts that 'Nehru harbour[ed] a personal animosity toward Pakistan because he regard[ed] the Pakistani alignment with the West as the principal cause of his failure to achieve leadership of a large neutral Asian Bloc'.[112] This view clearly exaggerated Pakistan's importance in American eyes. It also overestimated Pakistan's influence in Asia. By this time, US policymakers increasingly viewed Pakistan as a liability rather than an asset. They also did not consider Pakistan to be a reliable partner against communism. Pakistan failed to recognize India's greater significance to the US in global geopolitics, particularly as a counterweight to China. This misguided foreign policy would later cost Pakistan dearly.

Overstating their influence on US policy towards India, Pakistani officials asked the Americans to not be 'apologetic', especially to India, in supporting Pakistan. They warned of the 'Indian pressure for increased economic assistance [to India] in order to compensate them for military expenditures which they would claim were the result of US military aid to Pakistan'.[113] The Pakistani leadership believed that the US could and should use its status to influence India in Pakistan's favour.[114] Pakistani officials argued that India risked falling to communism and that only good relations with Pakistan 'could save India from Communism', warning that failing to cooperate weakened India's defences against China.[115] In other words, Pakistani officials urged their US counterparts that the spread of communism in India could only be prevented by increasing US aid to Pakistan.

The American officials, on the other hand, often asserted that they had limited control over how India ran its policies. They argued that if 'they could not even bring Chiang Kai-shek, whom [they] had helped more than anyone, to do what [they] saw was in his own interest', how could they pressure India to do its bidding, given the minimal role they played in India's independence?[116] The irony was lost on Pakistani leadership that, as a smaller, weaker state reliant on US aid, it still disregarded American advice to downsize its military

[112] Ibid.

[113] Telegram From the Embassy in Pakistan to the Department of State, Foreign Relations of the United States (1958–1960), Volume XV.

[114] Haqqani, *Magnificent Delusions*, p. 98.

[115] Ibid., p. 97.

[116] Memorandum of Conversation, 11 July 1961, Foreign Relations of the United States (1961–1963), Volume XIX.

and improve relations with India. How was India—a larger, potentially more significant US ally in global geopolitics—expected to heed US advice?

Just as Pakistan–US relations differed significantly during the administrations of Truman and Eisenhower, the scenario changed again when Kennedy came to power in early 1961. Indeed, relations took a turn for the worse when he arrived: Kennedy was significantly more pro-India and was very impressed by the country's diplomatic credentials.[117] When the Kennedy administration showed a clear preference for India, Pakistan's state-controlled media decried that 'a neutral nation was being given preference over an American ally'.[118] Pakistan struggled to manage challenges to its relationship with the US, stemming from established diplomatic and strategic interactions, which heightened its ontological insecurity. The country's officials mocked America in not wanting to be 'left behind in the race to woo Pandit Nehru'.[119] Pakistan even warned that it might turn to China or the Soviet Union for security cooperation. When Vice President Lyndon Johnson asked the Pakistani leadership if they would be willing to help the US in Laos, they responded that an 'American victory against communism depended on American willingness to help Pakistan against India'.[120] Though the Americans tried to convince the country that India did not want to 'march on Pakistan', Pakistanis believed that the way the Indian army was 'deployed [indicated] that they regarded Pakistan as enemy number one'.[121] Pakistanis did not comprehend that the US did not buy their narrative regarding the India 'threat'.

Washington prioritized international communism and domestic communist movements in East Pakistan, which it viewed as 'a serious threat to the continued unity of Pakistan'.[122] The divergence of threat perception between the two sides was clear.

In 1959, Pakistan agreed to a US request for the establishment of a base in Badaber (near Peshawar in north-west Pakistan). Though it was meant to be a 'communications facility', the US used it to conduct electronic surveillance of the USSR.[123] The Badaber airbase hosted American U-2 spy planes for surveillance of the Soviet Union and China. From there, a U-2 piloted by Gary Powers was shot down by the Soviets in May 1960. The US kept these operations secret, and Pakistan later discovered the base was used to

[117] Roby Barrett, *The Greater Middle East and the Cold War: US Foreign Policy Under Eisenhower and Kennedy* (London: I.B. Tauris, 2009), pp. 278–292.
[118] Haqqani, *Magnificent Delusions*, p. 97; Kux, *Disenchanted Allies*, p. 118.
[119] Haqqani, *Magnificent Delusions*, p. 97.
[120] Ibid.
[121] Ibid., p. 99.
[122] National Intelligence Estimate, 5 May 1959, Foreign Relations of the United States (1958–1960), Volume XV.
[123] Ibid., p. 91; Sattar, *Pakistan's Foreign Policy*, pp. 49–50.

spy on both nations. Consequently, allowing US use of the base significantly strained Pakistan's relations with the USSR. The Pakistani leadership used the Badaber airbase as a bargaining chip to maximize US financial support, exaggerating its strategic importance in Cold War surveillance operations. They argued that the facility had 'aroused the suspicions of the Russians and the Indians', threatening Pakistan's security and thus requiring increased US aid.[124] Relying on roundabout logic, the Pakistani leadership argued that its strategic location was a critical entry point to the Middle East for the USSR. In turn, the Middle East was 'a natural "jumping off point" for Africa'. Pakistan's military dictator General Ayub Khan, who had come to power in a coup in 1958, declared that the 'free world should certainly not lower its guard. Indeed, if anything, it needed to put more arms and aid into the hands of those willing to use them [e.g. Pakistanis] to defend the free world'.[125] This narrative held that it was obviously 'in [the] US interest to have well-trained and equipped local forces about the world who were willing to contribute to collective defense'. The two scenarios were clear to the military leader. He asserted that it would be 'much less expensive and politically more palatable to have friendly forces abroad than to feel compelled to send US forces to meet requirements for local hostilities as they arose'.[126] However, the Pakistani leadership argued that before they could cooperate with the US, Western powers must assist Pakistan in addressing its perceived external threats, claiming it was 'surrounded' by enemies.[127] The ultimate objective of such narratives, as ever, was to make the West recognize Pakistan's defence needs.[128] Given Pakistan's inflated sense of its geopolitical importance, its leadership struggled to understand why the US would not 'insist that India be reasonable in its attitude concerning disputes with its neighbors'.[129]

As post-structuralist discourse theory suggests, the discursive construction of a certain Pakistani geopolitical identity by its elites led them to propose policies they assumed the US would follow. However, this logic, while convincing to Pakistani leadership, had clear flaws: first, there was little evidence

[124] Memorandum of a Conversation, Department of State, Washington, 8 May 1959, 3 p.m., Foreign Relations of the United States (1958–1960), Volume XV.

[125] Telegram From the Embassy in Pakistan to the Department of State, 17 August 1959, Foreign Relations of the United States (1958–1960), Volume XV.

[126] Airgram From the Embassy in Pakistan to the Department of State, 23 September 1959, Foreign Relations of the United States (1958–1960), Volume XV.

[127] Telegram From the Embassy in Pakistan to the Department of State, 16 September 1959, Foreign Relations of the United States (1958–1960), Volume XV.

[128] Airgram From the Embassy in Pakistan to the Department of State, 23 September 1959, Foreign Relations of the United States (1958–1960), Volume XV.

[129] Telegram From the Embassy in Pakistan to the Department of State, 17 August 1959, Foreign Relations of the United States (1958–1960), Volume XV.

that the USSR planned a large-scale invasion of the Middle East in the 1950s. Second, it assumed the international community would do nothing in the event of such an invasion. Third, the Pakistani leadership claimed it could withstand USSR power with full US support, implying that an economically weak Pakistan, reliant on US aid, was the sole bulwark against Soviet influence in the Middle East and Africa. With this mindset, they complained that 'some Americans did not fully appreciate the importance to the Free World of having a strong Pakistan',[130] while believing the path to 'conquer' Moscow went via Delhi and expecting US blind acceptance of this logic. Fourth, the Pakistani leadership failed to recognize that the US doubted their commitment to fighting communism, as their officials had admitted India was their primary threat. The US, viewing Pakistan's India concerns as exaggerated, had urged Pakistan to address this fixation. In the early 1950s, Pakistan sought US anti-communist support without preconditions, downplaying tensions with India to secure aid. Over time, however, it shifted its stance, expecting the US to adopt its anti-India worldview. Pakistan's persistent anti-communist rhetoric to gain foreign funds, without acknowledging distrust from others, reflected limited self-reflection. Fifth, the Pakistani leadership assumed the West would accept its narrative of being 'surrounded by enemies', despite there being no evidence that any of Pakistan's neighbours (China, India, Iran, and Afghanistan) had an intention to invade it. A State Department Telegram from January 1960 noted that 'it seems unlikely Pakistan would be subjected in near future to overt military aggression on scale justifying revision [of] U.S. military aid policy and program [of] accelerated modernization [of] Pakistan[i] forces'.[131] Like before, the US view was that the 'best way for Pakistan to acquire additional increment[al] political, economic and military strength' would be by finding a 'rapprochement' with India which would allow it to focus 'on internal development and face [the] communist threat'.[132] Washington believed that further military aid to Pakistan would more likely 'hinder than help' Pakistan's ability to reach a suitable understanding with India[133] as additional aid would 'stimulate' Pakistan's arms race with India.[134] Importantly, President Eisenhower had told General Ayub in the past that 'if Pakistan should attack India,

[130] Airgram From the Embassy in Pakistan to the Department of State, 3 December 1960, Foreign Relations of the United States (1958–1960), Volume XV.

[131] Telegram From the Department of State to the Embassy in Pakistan, 26 January 1960, Foreign Relations of the United States (1958–1960), Volume XV.

[132] Ibid.

[133] Ibid.

[134] Letter From the Officer in Charge of Pakistan-Afghanistan Affairs (Poullada) to the Special Assistant at the Embassy in Pakistan for Mutual Security Affairs (Linebaugh), 14 March 1960, Foreign Relations of the United States (1958–1960), Volume XV.

the United States would be on India's side'.[135] However, Pakistan's leadership dismissed views challenging their perceived threats, ignoring whether the US or Western powers shared their perspective. Interestingly, Pakistanis agreed that the 'Indian subcontinent [could] not be adequately defended militarily nor [could] it progress economically and politically so long as India and Pakistan are at swords' point'. However, the country was only interested in resolving the issue in its own way, rather than adopting the approaches suggested by the rest of the world.[136] Unsurprisingly, the US decision makers were often more sympathetic to a 'neutral India' than their Pakistani 'ally'.[137]

By the early 1960s, Pakistan had largely lost the sympathy and goodwill of the United States due to its questionable foreign policy mindset. Americans highlighted that, whereas they had 'predicated [US] military aid [to Pakistan] on the communist threat, the Pakistanis were anxious to receive it principally in order to strengthen their position against India. Thus the program from its inception was based on somewhat divergent motivations by the parties to the contract'.[138] Though the US continued to provide aid to Pakistan despite realizing that it was dealing with an 'unruly horse',[139] it did so to maintain leverage over Pakistan's leaders.[140] It was judged that aid could help ensure political stability in the country, but the belief in Pakistan's ability to fight communism ceased to be the primary motive for aid-giving, with the Americans noting that Pakistan would be of 'limited deterrence' against Communist aggression.[141] They agreed that Pakistan faced a threat from the 'Soviet and Chinese Communist expansion', but they did not believe that it affected 'Pakistan to a greater degree than it affect[ed] a number of other countries'.[142]

[135] Memorandum of a Conversation, Karachi, 8 December 1959, Foreign Relations of the United States (1958–1960), Volume XV.

[136] Letter From the Officer in Charge of Pakistan-Afghanistan Affairs (Poullada) to the Special Assistant at the Embassy in Pakistan for Mutual Security Affairs (Linebaugh), 14 March 1960, Foreign Relations of the United States (1958–1960), Volume XV.

[137] Telegram From the Embassy in Pakistan to the Department of State, 17 August 1959, Foreign Relations of the United States (1958–1960), Volume XV.

[138] Letter From the Officer in Charge of Pakistan-Afghanistan Affairs (Poullada) to the Special Assistant at the Embassy in Pakistan for Mutual Security Affairs (Linebaugh), 14 March 1960, Foreign Relations of the United States (1958–1960), Volume XV.

[139] Letter From the Ambassador in Pakistan (Langley) to the Assistant Secretary of State for Near Eastern, South Asian, and African Affairs (Rountree), 27 December 1957, Foreign Relations of the United States (1955–1957), Volume VIII.

[140] Letter From the Officer in Charge of Pakistan-Afghanistan Affairs (Poullada) to the Special Assistant at the Embassy in Pakistan for Mutual Security Affairs (Linebaugh), 14 March 1960, Foreign Relations of the United States (1958–1960), Volume XV.

[141] Ibid.

[142] Memorandum From the Assistant Secretary of State for Near Eastern and South Asian Affairs (Jones) to the Deputy Under Secretary of State for Political Affairs (Hare), 2 December 1960, Foreign Relations of the United States (1958–1960), Volume XV.

Diplomatic communications from the period reveal that American officials were frequently alarmed by Pakistan's misguided foreign policy, which undermined its own interests and weakened its position, resulting in significant domestic and international repercussions. They contended that the 'economic burden on Pakistan, not only from direct expenditures in support of its military budget but in wasteful competition with its partner in the sub-continent, was underestimated' by the rest of the world.[143] The emerging American view was that Pakistan could not continue to be an American 'pensioner' and it needed to achieve 'a substantial measure of economic sufficiency'.[144] Washington hoped that 'Pakistan in its own major self-interest will see the wisdom of' such an approach.[145]

When India and China had a border clash in 1959 and a limited border war in 1962, the US made no secret of its support for India, helping it with weapons and financial aid. Washington viewed New Delhi as a potential ally against a developing China. This American stance towards India incensed Pakistan, which viewed this as the betrayal of an ally or 'jilting'[146] of a lover. Pakistan worried that the arms supplied to India for use against China 'will undoubtedly be used against [Pakistan] at the very first opportunity'.[147] That was one of the reasons for Pakistan strengthening its relations with China, fostering a closer partnership in subsequent years. The two neighbours signed a boundary agreement in March 1963 to ward off the chances of any friction. Pakistan also established aviation links with China in August 1963. Interestingly, Pakistan had been wary of Chinese incursions into Pakistan throughout the 1950s and the country's military leader had declared in 1959 that any further 'Chinese incursions into Pakistan would be repelled by Pakistan with all the force at her command'.[148] In a meeting with the visiting President Eisenhower in December 1959, Pakistanis cited the threat posed by China to secure American aid. When Eisenhower urged India and Pakistan to reduce tensions in Kashmir, President Ayub Khan argued that withdrawing from Pakistan-controlled Kashmir would make it 'almost certain that the Chinese would simply move in and take over. Other points of Kashmir [would] fall to the communists'.[149] US officials were frustrated by the Pakistani leadership's use

[143] Ibid.
[144] Ibid.
[145] Ibid.
[146] Interview, Politician 2, Islamabad, Pakistan, January 2024.
[147] Telegram From the Department of State to the Embassy in Pakistan, 13 November 1962, Foreign Relations of the United States (1961–1963), Volume XIX; see also Kux, *Disenchanted Allies*, p. 146.
[148] Quoted in Small, *The China-Pakistan Axis*, p. 21; Ayub Khan also suggested that India and Pakistan create a 'joint defence union' to counter 'the inexorable push of the north in the direction of the warm waters of the Indian Ocean,' see Small, *The China-Pakistan Axis*, p. 21.
[149] Memorandum of a Conversation, Karachi, Pakistan, 8 December 1959, Foreign Relations of the United States (1958–1960), Volume XV.

of the Chinese threat to gain aid. They reminded them that Pakistan had adopted an 'anti-communist stand because it was the right one not just to make itself eligible for certain sums of dollar aid.'[150] The Pakistani leadership's pivot to China further distanced Pakistan from the US, deepening its isolation during the 1965 India–Pakistan War. Ironically, it was Pakistan that used American-supplied weapons against India in that war, rather than the other way around.[151]

By the mid-1960s, Pakistanis worried about the growing role of India in the US's geopolitical calculations. Pakistan's President Ayub Khan wrote to his American counterpart Lyndon Johnson in July 1964 that the 'US aid [to India] imperil[ed] the security of Pakistan, your ally; it prevent[ed] an Indo-Pakistan rapprochement over Kashmir which immobilise[d] the bulk of their armed forces in a dangerous confrontation.'[152] He went on to say that the US aid to India risked an 'arms race between India and Pakistan', placing a 'crushing burden on their economies'. The Pakistani President seemed to overlook that this was the basis for the US decision to reduce military assistance to Pakistan over several years. By the mid-1960s, Pakistanis had also begun 're-examining' their ties to the American-led alliance commitments. The country's President asserted that 'Pakistan had never had a deep intrinsic interest of its own in SEATO anyway. Pakistan had joined [the organisation] in 1954 only as a cooperative gesture to the US'. Ayub Khan wanted Pakistan to focus on its 'liabilities nearer home and therefore it might be the most honorable and realistic course for Pakistan to withdraw.'[153] In the early 1950s, the Pakistani leadership worked to assure the US of its anti-communist commitment, deeming the notion of using US military aid against India 'absurd', despite Pakistan's later actions in the 1965 India–Pakistan War.

Washington believed that Pakistan could have its 'territorial integrity protected from large-scale aggression given the umbrella of [the] various security pacts and by the forthright commitments they ha[d] already obtained from the U.S.'[154] Being assured by this 'shield [Pakistan] could then concentrate [its] energies and resources on the paramount tasks of achieving economic development and political stability.'[155] US officials noted that other newly

[150] Memorandum of Conversation, by the Acting Officer in Charge of Pakistan-Afghanistan Affairs (Thacher), 18 October 1954, Foreign Relations of the United States (1952–1954), Volume XI, Part II.

[151] Sattar, *Pakistan's Foreign Policy*, p. 45.

[152] Memorandum From Robert Komer of the National Security Council Staff to the President's Special Assistant for National Security Affairs (Bundy), Washington, 2 June 1964, Foreign Relations of the United States (1964–1968), Volume XXV.

[153] Telegram From the Embassy in Pakistan to the Department of State, 11 August 1964, Foreign Relations of the United States (1964–1968), Volume XXV.

[154] Letter From the Officer in Charge of Pakistan-Afghanistan Affairs (Poullada) to the Special Assistant at the Embassy in Pakistan for Mutual Security Affairs (Linebaugh), 14 March 1960, Foreign Relations of the United States (1958–1960), Volume XV.

[155] Ibid.

independent nations, despite insecurities, devised effective foreign policies and achieved self-reliance. They urged Pakistan to follow suit. That is 'what the U.S. did in its early history, developing its resources protected by two oceans and defended by the protective screen of the British fleet'. Similarly, 'Japan's spectacular economic recovery [after the Second World War was] also based on this idea of letting the U.S. carry its military burden directly'.[156] The Pakistani leadership, however, chose not to leverage US-led alliances to achieve self-reliance. Instead, it did not capitalize on the goodwill of Western allies, clinging to misguided threat perceptions that failed to appreciate mid-1960s geopolitical realities.

2.4 Defence through non-state actors

Pakistan emerged as a significantly weaker state than India from the Partition of the subcontinent in 1947. Although detailed plans were made for dividing the British Indian army's resources, Pakistan received only a 'tiny' fraction of the allocation intended for it.[157] Cloughley writes that 'India received a high proportion of base installations (stores depots and the like), because most of these were located within India itself. Wrangles took place concerning apportionment of equipment, and Pakistani officers complain to this day about the failure of the former Viceroy (then Governor General of India), Lord Mountbatten, to ensure the fair division that had been agreed'. He goes on to say that '170,000 tons of equipment and stores for Pakistan were intended to be despatched in three hundred trainloads from India; only three railway wagons arrived, and these contained either obsolete items or rubbish'.[158]

Another historian of the Pakistan army writes that it received a fair share of officers, but that was insufficient to fill approximately 4000 vacant posts. Due to this mismatch in manpower and resources, Pakistan turned to non-state actors to bolster its capabilities, rather than adopting alternative strategies to address its insecurity, as states like Singapore have done by strengthening their economic capacity.

Reliance on non-state actors for its external policies remains a major cornerstone of Pakistan's foreign policy. The roots of this approach go back to October 1947 when India and Pakistan engaged in their first clash over Kashmir. When these two countries came into being, the Maharaja of Kashmir

[156] Ibid.

[157] Brian Cloughley, *A History of Pakistan Army: Wars and Insurrections* (New York: Carrell Books, 2016), p. 4.

[158] Ibid., pp. 3–4; Anwesha Sengupta, 'Breaking up: Dividing Assets Between India and Pakistan in Times of Partition', *The Indian Economic and Social History Review* 51, no. 4 (2014): pp. 529–548.

remained indecisive regarding which state Kashmir might accede to. Pakistan believed that Kashmir should become part of its territory given its overwhelming Muslim population. Considering it to be an easy catch, the country decided to acquire Kashmir by force through a two-phase strategy. In the first phase, Pakistan was to provide material assistance to those protesting against the Maharaja's rule in Kashmir, and in the second phase, several thousand tribesmen from Pakistan's north-west would be assisted to move into Kashmir. The strategy was devised with Jinnah's tacit approval and it was based on the premise that these two tactics would undermine the Maharaja's control, compelling him to announce Kashmir's accession to Pakistan.[159] Jinnah believed Pakistan had the right to intervene in Kashmir and directed General Douglas Gracey, the Acting Chief of the Pakistan army, to deploy troops there. However, Gracey, viewing Pakistan as a weak state, argued that risking war with India so soon after independence was not in the country's interest.[160]

Though the architects of this plan envisaged a quick accession, the outcome proved markedly different. The Maharaja appealed to the Indian government for military assistance, which the Indian government readily provided on the condition that Kashmir accede to the Indian Union.[161] The accession of Kashmir to India, signed by Maharaja Hari Singh on 26 October 1947, required ratification by a plebiscite after hostilities ended. Accepting India's terms, the Maharaja secured immediate military assistance, with Indian forces deployed to stop the advance of Pakistani tribesmen, who were engaging in 'looting, plundering, and rape'.[162] According to the accounts of the then-NWFP governor Sir George Cunningham, Jinnah was aware of the activities of the tribesmen, but said to his advisors: 'Don't tell me anything about it. My conscience must be clear'.[163] The fighting ended when the two sides reached a stalemate and the United Nations sponsored a ceasefire on 1 January 1949.

Where the decision to 'liberate' Kashmir was taken by then-Prime Minister Liaquat Ali Khan, Colonel (later General) Akbar Khan was the mastermind of the idea of using tribal militants to try to provoke an uprising inside Kashmir.[164] Khan was given instructions by Mian Iftikharuddin, a prominent Muslim League leader, to prepare a plan which ensured that 'any action by [Pakistan] was to be of an unofficial nature, and no Pakistani troops or

[159] Kapur, *Jihad as Grand Strategy*, p. 44.
[160] Ahmed, *Pakistan: The Garrison State*, p. 85.
[161] Kapur, *Jihad as Grand Strategy*, p. 45.
[162] Ahmed, *Pakistan: The Garrison State*, p. 83.
[163] Mitchell Norval, *Sir George Cunningham: A Memoir* (Edinburgh: Blackwood, 1968), pp. 140–141.
[164] Kapur, *Jihad as Grand Strategy*, pp. 43, 45; Alastair Lamb, *Incomplete Partition: The Genesis of the Kashmir Dispute* (Oxford: Oxford University Press, 2002), p. 125; for an instructive, detailed account of the war, see Shuja Nawaz, 'The First Kashmir War Revisited', *India Review* 7, no. 2 (2008): p. 119.

officers were to take an active part in it'.[165] The decision was kept secret from the British officers serving in the Pakistan army at that time. Initially, Indian leaders were willing to let Pakistan claim Kashmir if it avoided interfering in India's negotiations with other princely states, such as Hyderabad.[166] However, Jinnah chose to use militant proxies to capture Kashmir, believing the strategy would secure the region while keeping open the possibility of persuading other states to join Pakistan.

In 1965, Pakistan used tribal militants in Operation Gibraltar to spark an insurgency in Indian-administered Kashmir, believing it would trigger a localized uprising. This strategy, rooted in unrealistic expectations, aimed to secure Kashmir while avoiding broader conflict with India.[167] The leadership believed that people in Kashmir would join the uprising and India would not retaliate by attacking Pakistan-occupied Kashmir or the international boundary.[168] The planners of the Operation were left stunned when India launched an attack on the international boundary across the major Pakistani cities of Lahore and Sialkot.

India also played a role in the Kashmir dispute becoming a major irritant between the two states over the decades. The hardline Indian leadership at the time showed reluctance towards finding a relatively fair resolution of the problem. There is some merit to the Pakistani argument that, if the country had ceded all of the Kashmiri territory to India that it had occupied with the help of the tribal militia, the major Pakistani cities of Rawalpindi, Jhelum, and Sialkot would be too vulnerable to an Indian invasion.[169] Where the Indian city of Amritsar was given due protection during the demarcation of Punjab's borders, key Pakistani cities, including Lahore, were left vulnerable. It was only natural for Pakistan to aspire to attain some or all of Kashmir as part of its territory in order to achieve greater security. Given this scenario, the stance of the hardliner Indian statesmen enabled the Pakistani state to divert valuable resources towards defence to counter the Indian threat.

The Pakistani state used proxy fighters, again, during the civil war of 1971 to help supress dissent in East Pakistan.[170] Dubbed as *razakars* (or

[165] Shuja Nawaz, *Crossed Swords: Pakistan, Its Army, and the Wars Within* (Karachi: Oxford University Press Pakistan, 2008), p. 44; for the details on the planning of the war, see pp. 45–48.

[166] Ishtiaq Ahmed, 'Not Easily Forgotten: Reading An Indian Diplomat's View of Kashmir', *The Friday Times*, 2 August 2022, available at https://thefridaytimes.com/02-Aug-2022/not-easily-forgotten-reading-an-indian-diplomat-s-view-of-kashmir

[167] See also Ahmed, *Jinnah*, pp. 455–456, Julian Schofield and Reeta Tremblay, 'Why Pakistan Failed: Tribal Focoism in Kashmir', *Small Wars and Insurgencies* 19, no. 1 (2008): pp. 23–38.

[168] Sattar, *Pakistan's Foreign Policy*, p. 94.

[169] Ahmed, *Pakistan: The Garrison State*, p. 80.

[170] Suzannah Linton, 'Completing the Circle: Accountability for the Crimes of the 1971 Bangladesh War of Liberation', *Criminal Law Forum* 21, no. 2 (2010): pp. 191–311.

collaborators), they engaged in looting, plundering, and rape of the locals. It is estimated that the West Pakistani army and the *razakars* were responsible for the deaths of around 3 million people and the rape of around 200,000 women in East Pakistan.[171] Their use and tactics further angered Bengalis against West Pakistan, sowing the seeds for the eventual break-up of the country.

The historical injustices against East Pakistan by the West dated back to 24 March 1948, when the country's founding father, Muhammad Ali Jinnah, visited Dhaka and declared Pakistan's national language to be Urdu (and not Bengali). That laid the foundations of East Pakistan's resentment of the Western wing, which persisted until the eventual separation of the two in 1971. In the 1950s and 1960s, the majority-Bengali population of East Pakistan was treated as a colony by West Pakistan. The unjust treatment was most evident in the refusal to transfer power to Mujibur Rehman after the 1970 elections. Instead, the West Pakistan military launched Operation Searchlight in March 1971, which was blamed for human rights abuses by the country's military and the *razakars* at a magnitude unseen since the Second World War. In reaction to Operation Searchlight, the Bengali nationalists came together under the banner of a pro-independence militia called Mukti Bahini. India, of course, benefited from the situation and started supporting Mukti Bahini; it was a great opportunity for the former to cut Pakistan down to size given the latter's revisionist tendencies, as displayed in Operation Gibraltar and Operation Desert Hawk. As Christine Fair contends:

> West Pakistan's abusive policies toward the Bengali citizens in East Pakistan devolved into a civil war. As refugees began streaming into India, India had numerous concerns to evaluate. India began providing extensive political and military support to the insurgents, including training the Mukti Bahini, who were fighting to liberate East Pakistan. By summer 1971, India saw the insurrection as an opportunity to intervene and break up Pakistan. While India was preparing its war plans, Pakistan started the war in December when it began air strikes on Indian positions.[172]

Fair goes on to write:

> While India had prepared war plans, the war formally commenced on December 3, 1971, with an 'Israeli-style pre-emptive air attack by Pakistan on India's northern air bases. ... The attack failed miserably on all counts' (Ganguly 2001, 67). The

[171] Nayanika Mookherjee, '"Remembering to Forget": Public Secrecy and Memory of Sexual Violence in the Bangladesh War of 1971', *Journal of the Royal Anthropological Institute* 12, no. 2 (2006): p. 436.

[172] Christine Fair, *Fighting to the End: The Pakistan Army's Way of War* (Oxford: Oxford University Press, 2014), online version.

> war was short. On December 15, Gen. A. A. K. Niazi, the commander of the Pakistani forces in the east, sought an unconditional ceasefire, a suggestion his Indian counterpart, Lt. Gen. Jagjit Singh Aurora, rebuffed. The war finally ended on December 17, when Gandhi ordered a unilateral ceasefire; [Yahya] Khan immediately ordered Pakistani forces to reciprocate. The war ended with the emergence of independent Bangladesh.[173]

2.5 Ignoring the Muslim world

Pakistan's choice to not seek diplomatic accommodation with India after independence curtailed its ability to gain influence among Muslim countries in the Middle East and beyond. Eager to unite the Muslim world in condemning India's perceived hegemonic ambitions in South Asia, Pakistan underestimated the broader priorities of these nations. In the decades following colonial rule, countries like Egypt and Syria were more focused on anti-imperialist struggles against Western powers than on supporting Pakistan's rivalry with India. India's non-aligned stance, emphasizing sovereignty, and resistance to neocolonial influences, resonated more deeply in Arab capitals, overshadowing Pakistan's appeals and limiting its diplomatic outreach. As Rizvi writes, 'Islam did not play the same strategic role in the nationalist movements of most Middle Eastern states as was the case in Pakistan. For most of them it was a straight struggle against a colonial ruler'. He goes on to say that these countries 'were not a minority community and did not feel that their social, political and economic interests were threatened by a non-Muslim majority. Thus, Islam did not appear prominently in their struggle for nationhood. It was not the *raison d'etre* of their state and nation'.[174]

Pakistan was established in 1947 as a result of the Partition of British India, which separated its Muslim minority from the Hindu majority through colonial divide-and-rule policies. Shaped by this postcolonial identity, Pakistan framed its foreign policy through an Islamic lens, seeking solidarity with Muslim-majority nations. However, this approach drew criticism from countries like Egypt, whose own postcolonial struggles against British and French imperialism prioritized anti-colonial unity over religious rhetoric. In 1956, Egypt's King Farouk expressed this frustration, stating that 'Pakistanis believed Islam was born on 14 August 1947',[175] highlighting the disconnect

[173] Ibid.
[174] Hasan Askari Rizvi, 'Pakistan: Ideology and Foreign Policy', *Asian Affairs* 10, no. 1 (1983): p. 50.
[175] Pande, *Explaining Pakistan's Foreign Policy*, p. 148.

between Pakistan's Islamist narrative and the broader anti-imperialist priorities of Arab states navigating their own colonial histories.

Pakistan showed little enthusiasm for building alliances with Muslim countries to diversify its foreign policy. Though it vocally supported issues like the Palestinian cause, it did not prioritize forging economic or defence ties with these nations. Two main factors drove this stance: first, the economies of most Muslim countries, weakened by postcolonial challenges, could not provide the financial aid Pakistan needed, and second, no single Muslim state had the capacity to support Pakistan's defence against India's perceived regional ambitions. Iran, Pakistan's nearest middle power, had a weaker economy, still recovering from the Anglo-Soviet occupation during the Second World War. This limited its potential as a strategic partner for Pakistan.[176]

The second reason an Islamic alliance proved unworkable was the competition for leadership among Muslim countries, with none willing to accept a subordinate position. Saudi Arabia, as the birthplace of Islam and guardian of its holiest mosques in Mecca and Medina, claimed primacy over the Muslim *Ummah*.[177] In contrast, Turkey presented itself as the modern, secular representative of the *Ummah*, shaped by its post-Ottoman identity. This rivalry hindered Pakistan's efforts to build a cohesive Islamic bloc.

Iran also saw itself as the natural leader of the Muslim world, drawing on its ancient Persian civilization. It valued the widespread prestige of the Persian language in regions like Central and South Asia and considered its culture more sophisticated than that of neighbouring states. This self-perception fuelled its leadership ambitions but clashed with competing claims in the Muslim world.[178] Egypt valued its own anti-imperialist stance, which gained prominence after the Suez Crisis of 1956, when it successfully challenged British and French control over the Suez Canal. This triumph likewise led Egypt to see itself as a natural leader of the Muslim world, particularly in advocating pan-Arab and anti-colonial causes.[179]

The rise of Arab nationalism in the 1950s sidelined Pakistan due to its non-Arab identity, aligning it more closely with Iran, another non-Arab state often distanced from Arab movements. Leading Arab nationalist states like

[176] Shirin Tahir-Kheli, 'Iran and Pakistan: Cooperation in an Area of Conflict', *Asian Survey* 17, no. 5 (1977): p. 474.

[177] *Ummah* refers to the global community of Muslims bound together by a shared faith in Islam, transcending territorial, ethnic, and national boundaries. In contemporary political discourse, appeals to the Muslim *Ummah* often refer to the ideals of transnational solidarity among Muslim-majority societies.

[178] Khalida Qureshi, 'Pakistan and Iran; A study in Neighbourly Diplomacy', *Pakistan Horizon* 21, no. 1 (1968): p. 33.

[179] Amin, *Pakistan's Foreign Policy*, p. 40.

Syria and Egypt, under figures like Gamal Abdel Nasser, criticized Pakistan for joining the British-led Baghdad Pact in 1955. Nasser's objections, rooted in anti-imperialist rhetoric, were more related to domestic agendas than the country's international priorities.[180]

Pakistan's reluctance to unite with Muslim states contrasted with strong public support for Muslim unity, particularly the Palestinian cause. Inspired by Pakistan's creation as a Muslim homeland in 1947, many Pakistanis questioned why Palestinians could not achieve similar statehood in the Middle East, viewing the Palestinians' struggle as akin to their own post-colonial aspirations.[181] Public support for Muslim unity in Pakistan led only to organizing conferences of Muslim countries, rather than deeper diplomatic or economic ties. When a journalist asked a former prime minister of Pakistan, Hussein Shaheed Suhrawardy (1956–1957), about why Muslim countries did not 'band together instead of getting tied to the West', his apt reply was: 'zero plus zero plus zero is after all equal to zero'.[182] Delvoie states that 'it [was] difficult to discern any influence of Islamic ideology in the main elements of Pakistani foreign policy over the first 25 years of its existence as an independent state. ... Pakistan's forthrightly pro-Arab stance on the Palestine question was not essentially different from that of many non-Muslim Third World countries, including India'.[183] Similarly, the Pakistani support of the liberation movements of several North African states could be seen through an Islamic as well as a postcolonial lens.[184]

In contrast to Pakistan, India built strong ties with Muslim-majority states in the Middle East and South East Asia, where its non-aligned ideals, rooted in postcolonial autonomy, gained traction. Countries like Egypt and Indonesia embraced India's stance, enhancing its regional influence.[185] Instead of Pakistan, leaders like Nasser and Seokarno 'found greater identity of views with Nehru and developed closer ties with India than with the leaders of Pakistan'.[186]

Pakistan sought support from the Muslim world for its fears of Indian expansionism, which it believed threatened the first state founded

[180] Sattar, *Pakistan's Foreign Policy*, p. 53.
[181] Ibid., p. 18.
[182] Amin, *Pakistan's Foreign Policy*, p. 41; Sattar, *Pakistan's Foreign Policy*, p. 57.
[183] Louis Delvoie, 'The Islamization of Pakistan's Foreign Policy', *International Journal* 51, no. 1 (1995/6): p. 132; an alternative opinion holds that the state used Islam quite early on to tackle its insecurities. See Ziad Haider, *The Ideological Struggle for Pakistan* (Stanford, CA: Hoover Press, 2010), p. 7.
[184] Interview, Politician 2.
[185] Rizvi, 'Pakistan: Ideology and Foreign Policy', p. 51.
[186] Ibid., p. 51.

on Islam. However, several Arab states, prioritizing their own interests, refused to back Pakistan in 'its dispute with India—a non-Muslim state'.[187]

2.6 Reliance on foreign assistance and non-state actors: Assessing the policy foundations

Since its formation in 1947, Pakistan's foreign policy relied on two pillars: foreign assistance and non-state proxies. These elements have shaped the contours of the country's foreign policy through today, and I will critically appraise the effectiveness of each approach in this section. Pakistan's decision to align with the United States from early on in its history yielded significant negative foreign policy consequences. The United States, valuing Pakistan's geostrategic location at the mouth of the Persian Gulf and near the Soviet border, welcomed this alignment through alliances like the Baghdad Pact and SEATO. However, this reliance on Western aid alienated Pakistan from non-aligned Muslim states like Egypt and limited its influence in the Muslim world, reinforcing its diplomatic isolation and dependence on external powers.

The Pakistani decision to play the 'location card' to obtain foreign economic and military assistance thwarted the country from developing a stable economy or a resilient democracy, unlike neighbouring India. Paul writes that the '[t]he military and political elites who gained materially from this interaction had little reason to innovate in either the economy or the political sphere. Nor did they have to reform the semi-feudal landlord system in order to extract resources from it'. He goes on to say that 'foreign support was largely sufficient to pay the costs of waging war externally and buttress the warrior state internally. In essence, Pakistan became a rentier state, living off the rents provide by its external benefactors for supporting their particular geostrategic goals'.[188]

Paul likens Pakistan's reliance on foreign aid to the 'oil curse' in Gulf states, where dependence on external resources hindered economic reforms, technological innovation, and global trade. Pakistan's civil and military bureaucracy dominated rent extraction, with little incentive to reform.[189] Consequently, alternative income sources were discouraged, preserving the power

[187] Ibid.

[188] T. V. Paul, *The Warrior State: Pakistan in the Contemporary World* (Oxford: Oxford University Press, 2014), p. 18.

[189] Ibid., p. 19.

of those benefiting from the system. Over time, the military became the main beneficiary of foreign financial assistance, which it procured by brandishing Pakistan's geostrategic location.

The military's dominance in Pakistan's domestic politics and its foreign policy was connected with political upheavals in the first decade of the country's existence. Pakistan had seven heads of government and four heads of state during that period.[190] Amid Pakistan's political instability, the United States and its Western allies engaged directly with the country's armed forces. Pakistan's army became the primary interlocutor with the outside world, dominating defence and security policies. This role marginalized civilian politicians, who had little influence over national security policy.[191] As national security became Pakistan's primary focus, its leadership reshaped the entire social and political orientation of the country along the lines of a national security state. That meant prioritizing security over citizens' welfare, resulting in Pakistan's low rankings in global social and economic indicators.[192]

On the other hand, Pakistan's use of militant proxies in India and Afghanistan yielded limited success, such as territorial gains in Kashmir, but its long-term negative impacts included regional instability and strained diplomatic relations. The policy lowered warfare cost for Pakistan, and it helped the country secure 'one third of the territory of Kashmir'[193] in the first Indo–Pakistan War of 1947–1948. The military knew that it was a useful way through which a conventionally inferior Pakistan could challenge a dominant India.[194] Kapur argues that states benefit from using non-state proxies by reducing warfare costs, limiting adversaries' military options, and strengthening bargaining with the opponent through principal–agent dynamics.[195]

Given that the approach of using militant proxies required secretive decision-making, the lack of transparency led to blunders like the Pakistani incursion, Operation Gibraltar, into the Indian-occupied Kashmir. As the planners believed that the insurgency would remain limited to Kashmir, they considered it unnecessary to involve the other branches of the military in

[190] India's first prime minister, Jawaharlal Nehru, was in office for nearly seventeen years from 15 August 1947 until 27 May 1964.

[191] The army does call upon civilians when things get out of hand, as was the case in the aftermath of the 1971 civil war and the Kargil conflict.

[192] See Anwar Iqbal, 'Pakistan's Human Development Ranking Declines', *Dawn*, 14 March 2024; Mahnoor Sherazee and Hammad Abbasi, 'Health Indicators Project Bleak Future for Pakistani Children', *Dawn*, 26 February 2014.

[193] Kapur, *Jihad as Grand Strategy*, p. 49.

[194] Ibid.

[195] Ibid., pp. 46–48.

the planning.[196] As a result, Pakistan's air force was unprepared when India launched an attack on the international boundary between the two states. An open decision-making process would have likely prepared for alternative scenarios to avoid the fallout of the Kashmir incursion.

Along with adverse foreign policy consequences, this approach has also had significant costs domestically. Myriad militant groups have huge domestic presence, and they operate welfare arms to provide relief in areas such as health and education—arenas in which the state has failed miserably. The result of these groups' involvement in such activities has led to a rise in extremism in the country.[197]

Pakistan's use of militant proxies in Kashmir enhanced the army's influence in politics. Proposed by Mian Iftikharuddin and autonomously executed by the army, based in Rawalpindi, 900 miles from the country's then-capital Karachi, this strategy further strengthened the army's dominance in the country's politics. Shuja Nawaz argues that the army's autonomy in executing the Kashmir proxy strategy limited civilian input, worsened by poor communication between Karachi's political capital and Rawalpindi's military headquarters.[198] In his words, 'the intellectual distance between the army and the politicians was heightened by the physical separation of the two: the political leadership in Karachi and the army in Rawalpindi at a time of uncertain communications. This led to a greater autonomy of thought and action in the army leadership'[199]—a trend that became institutionalized through the course of history.

During Pakistan's first decade, marked by instability, the country's civil and military bureaucracy often colluded to stop the prime ministers from completing their terms, which helped the former to assume a disproportionately large role in the affairs of states.[200] It suited the military's interest if the civilian political leadership lost credibility in the public's eyes.[201] Pakistan's civil bureaucracy was initially dominant, but the army, under Ayub Khan's leadership, rose to prominence by the mid-1950s and assumed control via a 1958 coup, reinforcing military dominance.[202]

Pakistan's military had not played any role in the freedom struggle. In fact, it was part of the imperialist British army, which was employed to crackdown

[196] 'Nur Khan Reminisces '65 war', *Dawn*, 6 September 2005.
[197] Christine Fair, *In Their Own Words: Understanding Lashkar-e-Tayyaba* (Oxford: Oxford University Press, 2018), pp. 149–197.
[198] Nawaz, 'The First Kashmir War revisited', p. 150.
[199] Ibid.
[200] Interview, Academic 1, Lahore, Pakistan, January 2024.
[201] Interview, Journalist 3, Lahore, Pakistan, January 2024.
[202] Interview, Journalist 2, Lahore, Pakistan, January 2024.

on nationalists and was familiar with using the coercive apparatus of the state from the beginning.[203] According to an astute observer of the civil–military relations in Pakistan, the country's army has never had respect for the sanctity of the Constitution. Instead, it made a 'deliberate and conscious attempt to remove democracy from people's minds' and from the common parlance.[204] For example, the country celebrated 'Republic Day' on every 23rd of March until the 1958 takeover, when the day was renamed 'Pakistan Day'. Unlike the Indian army, the Pakistan army continued recruiting primarily from five districts of northern Punjab, which meant that the military personnel were not familiar with the culture, geography, and economy of other parts of Pakistan. As a result, the officers could not develop a sense of 'fellowship' with the nation it ruled over and had no problem being instruments of foreign power at the right price.[205]

On the other hand, the main party leading the freedom struggle, the Muslim League, did not have as much experience of politics as the Indian National Congress because its leadership primarily included landed elites who did not have roots among the masses.[206] The leadership had an authoritarian mindset, which was visible in Jinnah's presiding over cabinet meetings despite being the constitutional head of state as Governor General.[207] It has therefore been observed that, owing to an underdeveloped political class and an overdeveloped civil and military bureaucracy,[208] Pakistan lost its balance from the very beginning, which it has never been able to recover.[209]

Jinnah championed the idea that Pakistan could capitalize on its strategic location, which became a key pillar of the country's foreign policy over time.[210] That foreign policy vision of the country's founding father meant 'surrendering his nation to an over-developed military which would consume the majority of the financial resources of Pakistan'.[211] The use of non-state actors was also a policy that represented 'Jinnah's vision', as it was developed and deployed in the first Kashmir War of 1947–1948 when he was the most powerful leader in the country. The policy has been used repeatedly since then, as seen in the cases of Operation Gibraltar and the 2016 attack on an airbase in the Indian city of Pathankot, which was widely believed to have been

[203] Interview, Journalist 3.
[204] Ibid.
[205] Ibid.
[206] Interview, Journalist 3.
[207] Interview, Journalist 2.
[208] Hamza Alavi, 'The State in Post-Colonial Societies: Pakistan and Bangladesh', *New Left Review* 1, no. 74 (1972): pp. 59–81.
[209] Interview, Journalist 3.
[210] Ibid.
[211] Ibid.

conducted by a group aligned with the Pakistani state.[212] In brief, the pillars of Pakistan's foreign policy studied in this chapter emanate directly from the personality of Jinnah and continue to be part of the country's foreign policy tradition, with severe, deleterious consequences.

2.7 Conclusion

Pakistan's adoption of the role of a 'warrior state'[213] resulted in the domination of the military in the foreign policy decision-making from the outset. The state's joining of SEATO and CENTO left the country isolated in the world while tying it closely with the US. These alliances were useless in its competition with India. Pakistan's foreign policy, dependent on Western alliances and proxies, failed to secure the support of America in the 1971 Indo–Pakistan War. While militant proxies offered short-term gains, the longer-term negative consequences of their use continue to unfold. The defeat in the war of 1965 and the separation of East Pakistan show that this foreign policy approach stopped the country from achieving security or territorial integrity.[214]

From the beginning, the policy elite were unable to move away from self-serving policies to the policies that served the entire country. The decision-making elites made competition with India the central feature of foreign policy, in preference to policies such as improving the economy. During a meeting with US National Security Advisor Henry Kissinger in August 1969, the Governor of then-West Pakistan stated that attempting to seize an advantage in Kashmir at that point in time would not be possible and Pakistan would prefer to wait.[215] Kissinger asked whether Western powers should 'appease' India to gain concessions agreeable to Pakistan. The Governor replied that such an approach would not be acceptable because the West appeasing India might lower Pakistan to a 'second class status in the subcontinent'. In Pakistan's foreign policy fantasy, parity with India counted more than resolving specific disputes with that country.[216]

[212] Interview, Academic 1.
[213] Paul, *The Warrior State.*
[214] Amin, *Pakistan's Foreign Policy*, p. 45; Mussarat Jabeen and Muhammad Saleem Mazhar, 'Security Game: SEATO And CENTO as Instrument of Economic and Military Assistance to Encircle Pakistan', *Pakistan Economic and Social Review* 49, no. 1 (2011): p. 129.
[215] Memorandum of Conversation, Lahore, 1 August 1969, Foreign Relations of the United States (1969–1972), Volume E.
[216] Haqqani, *Magnificent Delusions*, p. 137.

Another regular feature of Pakistan's foreign policy is its portrayal of itself as a victim, regardless of whether this aligns with the rest of the world's perceptions of Pakistan. Those sympathetic to this view assert that the United States 'used' Pakistan on four occasions: in its fight against Communism during the Cold War, in its opening to China in the late 1960s, the Afghan Jihad of the 1980s, and the War on Terrorism after 2001. According to this view, when their purpose was served, Pakistan was ruthlessly abandoned, as could be seen in the war of 1965.[217] However, this narrative is clearly biased towards the official narrative in Pakistan. For example, it does not account for the fact that the 1965 war started as a result of Pakistan's 'forays in Kashmir',[218] as Ayub Khan put it. The official narrative continues to portray the war as Indian aggression that was successfully deterred by Pakistan.[219]

Pakistan was isolated on the international stage by 1971, shunned by all major powers. The state had enjoyed a close relationship with China from the early 1960s, and these relations improved further after the US refusal to side with Pakistan in the 1965 war. Ayub Khan wrote to the US that their refusal to help Pakistan appeared to be 'designed to push [Pakistan] towards the Chinese'. He said: 'I don't want to sit in the Chinese lap, and I won't do so if it can be avoided. But if [the] U.S. can't give me any help, I'll have no choice'.[220] During the Pakistani civil war, however, China (like the US) did not offer any material help[221] because it 'would not pull Pakistan out of the holes it insisted on digging for itself'.[222]

Pakistan had played a key role in the Sino–US rapprochement in the late 1960s and it hoped to be suitably 'rewarded' for its services.[223] The country's officials fantasized about a 'Washington-Islamabad-Beijing Axis' against the 'evil designs' of India.[224] However, when the Soviet Union declared its support for India in 1971, neither China nor the United States considered Pakistan to be valuable enough to risk a war with the Soviet Union.[225] The West Pakistani army's human rights violations in East Pakistan were also a major factor motivating the world to distance itself from the country.

Instead of reflecting on its brutal treatment of East Pakistanis during the civil war, the military in Pakistan continued to believe in the country's

[217] Interview, Politician 2.
[218] Haqqani, *Magnificent Delusions*, p. 113.
[219] Sattar, *Pakistan's Foreign Policy*, p. 95.
[220] Haqqani, *Magnificent Delusions*, p. 114.
[221] Ibid., p. 165.
[222] Small, *The China-Pakistan Axis*, p. 15.
[223] Hasan Zaheer, *The Separation of East Pakistan: The Rise and Realization of Bengali Muslim Nationalism* (Oxford: Oxford University Press, 1994), p. 296; Haqqani, *Magnificent Delusions*, p. 160.
[224] Haqqani, *Magnificent Delusions*, p. 160.
[225] Small, *The China-Pakistan Axis*, pp. 10–11.

importance for the great powers.[226] In a clear indication of what can only be described as this fantasy-based mindset, only two days before the historic Pakistani surrender in Dhaka, the military high command was informing the soldiers to stay put, as 'yellow and white help' was on its way from the north and the south—but no such help was forthcoming.[227] The country's global isolation was mirrored in the isolation felt by the common Pakistani foot soldiers. Its security forces were totally 'isolated' in the stranger territory of East Pakistan 'amidst a hostile population', and struggled with 'physical and mental exhaustion'.[228] There was limited appetite shown by any country in the world to side with it in that scenario, given the human rights violation of Bengalis.[229] In the next chapter, I examine to what extent the major debacle of losing its eastern wing led to a re-evaluation in Pakistan's foreign policy thinking.

[226] Christopher Van Hollen, 'The Tilt Policy Revisited: Nixon-Kissinger Geopolitics and South Asia', *Asian Survey* 20, no. 4 (1980): p. 360.

[227] Husain Haqqani, *Pakistan: Between Mosque and Military* (Washington, D.C.: Brookings, 2005), p. 84.

[228] John Gill, *An Atlas of the 1971 India-Pakistan War: The Creation of Bangladesh* (Washington, D.C.: NDU Press, 2003), p. 12.

[229] Sarmila Bose, 'Anatomy of Violence: Analysis of Civil War in East Pakistan in 1971', *Economic and Political Weekly* 40, no. 41 (2005): pp. 4463–4471.

3

'New Pakistan' and resetting Pakistan's foreign policy (1971–1999)

The civil war of 1971 and secession of East Pakistan into the state of Bangladesh was devastating for Pakistan in more ways than one. Apart from the obvious loss of the country's eastern wing, Pakistan was stuck with an 'image problem' due to its army's extensive human rights violations in East Pakistan that earned it worldwide notoriety.[1] With 90,000 of its soldiers and civilians held as prisoners of war in India, the army's morale was also at an all-time low. An independent Bangladesh, closely aligned to India, had dealt a sharp blow to Pakistan's foundational narrative that emphasized 'Muslim separateness' with a firm belief in Islam's ability to hold the country together. It is no wonder that the then prime minister of India, Indira Gandhi, famously remarked that 'we have drowned the Two-Nation Theory in the Bay of Bengal'.[2]

Pakistan's leadership believed that the country's foreign policy until 1971 had fundamental flaws. Pakistan had positioned itself close to the US and China, but such a 'bipolar'[3] approach had not ensured the country's defence and territorial integrity. Pakistan's new democratic leader, Zulfikar Ali Bhutto, argued that 'a certain correction' in the foreign policy direction was needed.[4] He contended that getting too close to the West had impaired Pakistan's ability to defend itself. Bhutto wanted Pakistan to follow a more 'non-aligned' path which would bring more independence in its foreign policy orientation. Bhutto's 'correction' envisioned a multi-vector foreign policy that was based on the vision of non-alignment, bilateralism, and pragmatism. Along with being close to the West, Pakistan would also be close to the Muslim countries, the developing world, as well as China.[5] An 'independent' Pakistan would

[1] David Dunbar, 'Bhutto—Two Years On', *The World Today* 30, no. 1 (1974): pp. 23–24.

[2] Syed Jazib Shamim, *Nuclear Proliferation in South Asia—Towards World War III*, Social Sciences Research Network, 27 January 2018, p. 49, available at https://ssrn.com/abstract=3111513 or http://dx.doi.org/10.2139/ssrn.3111513

[3] Vladimir Moskalenko, 'Pakistan's Foreign Policy', *Asian Survey* 14, no. 3 (1974): p. 278.

[4] Ibid.

[5] A. G. Noorani, 'Bhutto and Nonalignment', *Economic and Political Weekly* 11, no. 51 (1976): pp. 1957–1958.

Pakistan's Foreign Policy. Wali Aslam, Oxford University Press. © Wali Aslam (2026).
DOI: 10.1093/9780198909552.003.0003

chart its own path on the basis of its needs and interests. Bhutto was keen to 'move with the times', as he believed that Pakistan had been left behind because it refused to change its outlook.[6] This intended adjustment in foreign policy proposed that 'Pakistan must normalize relations with all the great powers, lessen its dependence on the United States, and develop more intimate connections with the Arab states'.[7] It was thought that this pragmatic approach would hopefully serve the country better than the previous approach. Bhutto thought that the idea of 'new' Pakistan closer to the Muslim countries as well as the developing world would not only benefit the country externally, but would also help boost morale at home.[8] In a speech on 20 December 1971, Bhutto stated:

> My brothers and sisters, in this moment of anguish and sorrow please do not have any doubt in your mind. Please stand up and face the world. It is a bitter world but we have to face the bitter truth. I can assure you that we will succeed. There can be no doubt that we will succeed. Pakistan was made for a great cause, for a great ideal, that ideal is imperishable, that ideal is living. So let us pledge together. We shall see to it that this stigma (of defeat) is wiped out even if it has to be done by our children's children.[9]

Influential foreign policy voices of the 1970s emphasized that Pakistan needed to acknowledge India's military and economic superiority. The two countries did not have to become best friends, but Pakistan needed to accommodate India instead of being in perpetual competition with it. For example, Pakistan's ambassador to the United States at the time, Sahabzada Yaqub Khan, remarked that '[t]he disproportion of [Pakistan's] forces [with regard to India] is so large that (Pakistan's acquisition of weapons) could not possibly be conceived of as an arms race'.[10] An eminent analyst of Pakistan in the 1970s, Lawrence Ziring, also expressed hope with regard to Pakistan's future outlook in the comity of nations by saying: '[f]or more than two decades, Pakistan harbored the notion that it was the equal of India, despite that country's vast numerical superiority'. He contended that 'as a result, successive Pakistani governments assumed an aggressive posture toward India that made any kind of meaningful accommodation impossible. Apparently, a new realism has now over-taken the Pakistani leadership'.[11]

[6] Lawrence Ziring, 'Recent Trends in Pakistan's Foreign Policy', *Asian Affairs* 2, no. 5 (1975): p. 303.
[7] Ibid., p. 305.
[8] Ziring, 'Recent Trends in Pakistan's Foreign Policy'.
[9] Robert La Porte, Jr., 'Pakistan in 1972: Picking Up the Pieces', *Asian Survey* 13, no. 2 (1973): p. 198.
[10] Ziring, 'Recent Trends in Pakistan's Foreign Policy', p. 303.
[11] Ibid.

This chapter studies Pakistan's foreign policy from 1971 through 2001. As mentioned above, the country's leadership tried to provide a new foreign policy vision which bore significant promise. They believed that the country's foreign policy needed a 'correction'. However, I argue that Pakistani ruling elite did not stay on course and instead made mistakes that resulted in its further isolation in the international community by the end of the century.

Section 3.1 looks at Pakistan's opening to the Muslim world. Doing so resulted in immediate rewards, but Pakistan quickly lost balance as it tilted disproportionately towards Saudi Arabia. A close relationship with Saudi Arabia–that continues to this day–bore certain economic benefits, but its negative consequences for Pakistan domestically (in both societal and political ramifications) and externally (in terms of its relations with the rest of the Islamic world) have outweighed the positive consequences.

Section 3.2 focuses on the further warming of Pakistan's relations with China.[12] Similar to the situation with the Muslim countries, the Pakistani vision had promise, but relations with China went through a difficult time due to Pakistan's stance on Kashmir and the country's support of the Taliban. China found it hard to support the country in both cases, as doing so would have counted against its own interest. Its relations with Pakistan did not, of course, exist in a vacuum, detached from the wider geopolitical setting.

Section 3.3 outlines how Pakistan continued to use non-state actors to interfere in Kashmir and Afghanistan over this time. Again, the country did reap limited benefits from such an approach, but it also brought several costs, including severe damage to its reputation. Section 3.4 studies Pakistan's relations with the United States. Where Pakistan's leadership wanted close relations with the US, it did not significantly change the country's previous pattern of engagement. Most of its dealings with the US remained India-centric, leading to frustration on the American side by the end of 1990s. Though both countries had worked together to repel the Soviet invasion of Afghanistan, things quickly returned to how they were before the invasion. Tensions over Pakistan's nuclear programme also resurfaced. Pakistan lost several opportunities of winning the trust of various US administrations due to its unwillingness to be open about its nuclear programme and its support of terrorist groups after the Soviet withdrawal from Afghanistan.

The conclusion (Section 3.5) will summarize how a 'new' Pakistan, boosted by its new-found allies, committed mistakes and missed several opportunities to follow a foreign policy course that could ensure its security and prosperity.

[12] Aparna Pande, 'Friends of Last Resort: Pakistan's Relations with China and Saudi Arabia', in *Pakistan's Enduring Challenges*, edited by Christine Fair and Sarah Watson (Philadelphia, PA: University of Pennsylvania Press, 2015), p. 258.

It had many chances to look after its interest and to position itself better vis-à-vis neighbouring India, but it made little use of those opportunities. By the beginning of 2001, Pakistan was not only isolated among fellow Muslim countries, it had lost the sympathies of its closest allies of the past: China and the USA. Post–civil war Pakistan started its journey in 1971 as an isolated country and ended it in 2001 as the same: a lonely country in the world. More than successes, the story of Pakistan in this period is a story of mistakes and missed opportunities.

3.1 Pakistan and the opening to the Muslim world

By the beginning of 1972, the Pakistani political leadership wanted to rectify the previous mistake of not enjoying warm relations with the Muslim world due to its close ties with the West. Pakistan also wanted to capitalize on the opportunity to find new allies among Muslim countries and beyond.[13] Bhutto believed that the failure to embrace authentic pan-Islamism led to the break-up of the country a year earlier, and he pursued a political and rhetorical strategy to align Pakistan more closely to the countries of the Middle East.[14] He also believed that strengthening links with the Middle Eastern countries would help 'diversify Pakistan's sources of financial and political support at a time when the country had become overly dependent on the United States and precisely when the Gulf states were beginning to deploy the wealth accumulated as a result of spectacular increases in the price of oil'.[15]

Bhutto's rationale for courting the Arabs was presented through an Islamic lens, but it primarily bore an economic angle.[16] That explains numerous joint economic ventures between Pakistan and Islamic countries in the 1970s. Given that economic focus, Pakistan received 'soft credits, interest-free loans and financial grants from a number of oil-rich Middle Eastern states, especially Saudi Arabia, Iran (pre-revolution), Kuwait, the U.A.E. and Libya'.[17] These countries also invested in several industrial projects in Pakistan. Most importantly, Pakistan sent nearly a million of its citizens to the countries of

[13] Zubeida Mustafa, 'Recent Trends in Pakistan's Foreign Policy Towards the Middle East', *Pakistan Horizon* 28, no. 4 (1975): pp. 1–17.

[14] Christine Fair, 'Islam and Politics in Pakistan', in *The Muslim World After 9/11*, edited by Angel Rabasa, Cheryl Benard, Peter Chalk, Christine Fair, Theodore Karasik, Rollie Lal, Ian Lesser, and David Thaler (Santa Monica, CA: Rand, 2004), p. 273.

[15] Author interview with a close confidante of Zulfikar Ali Bhutto, Lahore, 2 January 2020; see also Louis A. Delvoie, 'The Islamization of Pakistan's Foreign Policy', *International Journal* 51, no. 1 (1996): p. 132.

[16] Niloufer Mahdi, *Pakistan's Foreign Policy, 1971–1981: The Search for Security* (Lahore: Ferozsons, 1999), p. 170.

[17] Hasan Askari Rizvi, 'Pakistan: Ideology and Foreign Policy', *Asian Affairs* 10, no. 1 (1983): p. 54.

the Gulf as labour, attracting lucrative remittances in the process. The Gulf Kingdoms were always sceptical of opening up their societies to outsiders in case they might threaten their regimes, but Pakistanis were welcomed because they did not involve themselves in any activities undermining these countries' stability.[18] With the loss of East Pakistan, the country looked towards the Middle East as a new market for its exports.[19] These countries also helped Pakistan tremendously in dealing with the burden of Afghan refugees after the Soviet invasion. Bhutto used Pakistan's relations with Islamic countries to champion its role as a leader of the developing world.[20] The Pakistani effort to find closer identification with the Muslim countries was also part of an effort to overcome the trauma caused by the separation of East Pakistan.[21] This 'fitted in squarely with the renewed interest in Islam in the truncated and demoralized Pakistan to address the question of national identity, and thus, enjoyed popular support'. Rizvi believes that 'at international level, it aimed at rehabilitating Pakistan's image with support from the Muslim states. In other words, the use of Islam in foreign policy did not merely satisfy the ideological dimensions of Pakistani identity, it was also conducive to the achievement of Pakistan's political, economic and security goals.'[22]

The adoption of this approach by the country's democratic leadership made much sense as the support of Muslim countries restored Pakistan's 'crisis of confidence' and also strengthened its position in the developing world, providing it space on several international forums.[23] Rizvi believes that these ties also helped Pakistan tremendously in resisting the American pressure to restrict its nuclear programme in the 1970s.[24] Pakistan also knew that the countries of the Gulf would value it to be a 'a reliable and predictable force on the eastern marches of the Gulf', whereas 'disintegration in Pakistan would profoundly worsen the politico-strategic environment in the Gulf'.[25]

Pakistan's closer alignment with the Muslim countries began as a success. This new vision held much promise for the country's future. However,

[18] Gawdat Bhagat, 'Pakistan and Saudi Arabia Relations: An Assessment', in *Pakistan: The US, Geopolitics and Grand Strategies*, edited by Usama Butt and Julian Schofield (London: Pluto, 2012), pp. 201–202.

[19] Ibid., pp. 54–55; John Bray, 'New Directions in Pakistan's Foreign Policy', *The World Today* 48, no. 4 (1992): pp. 65–68.

[20] Mahdi, *Pakistan's Foreign Policy, 1971–1981*, p. 268.

[21] T. V. Paul, *The Warrior State: Pakistan in the Contemporary World* (Oxford: Oxford University Press, 2014), p. 14.

[22] Hasan Askari Rizvi, *Pakistan and the Geostrategic Environment: A Study of Foreign Policy* (New York: St. Martin's Press, 1993), p. 73.

[23] Ibid., p. 75.

[24] Ibid.; also see Shirin Tahir-Kheli, 'In Search of an Identity: Islam and Pakistan's Foreign Policy', in *Islam in Foreign Policy*, edited by Adeed Dawisha (Cambridge: Cambridge University Press, 1983), pp. 68–83.

[25] W. Howard Wriggins, 'Pakistan's Search for a Foreign Policy After the Invasion of Afghanistan', *Pacific Affairs* 57, no. 2 (1984): p. 292.

Pakistan quickly lost its sense of 'balance' by the late 1970s. It placed a disproportionately high emphasis on its relations with Saudi Arabia, costing its relations with Iran and other countries sympathetic to Iran in the Iran–Saudi rivalry, such as Syria and Algeria.[26] Disregarding the economic side, General Zia-ul-Haq, who overthrew Bhutto in 1977 in a military coup, cultivated Arabs 'purely in an Islamic context', as explained below.[27] Some have argued that this was as much to do with strategic calculations as with personal convictions of Zia.[28] He considered Syria and Libya to have a soft spot for the Bhutto family and as a result the 'enthusiasm and cordiality that marked Pakistan's relations with these states, cultivated by the democratic leadership of Zulfikar Ali Bhutto in the early seventies, was replaced with business-like functional interaction in the early and mid-eighties'.[29] General Zia's regime wanted to mould Pakistan along 'Islamic' lines, but in the absence of a single definition of what an Islamic polity would look like, it was not clear whether such a state would represent the Iranian brand of Islam or the Saudi brand.[30] Choosing one type was sure to alienate Pakistan in the eyes of the other—as eventually happened after it sided with Saudi Arabia when Pakistan placed a disproportionately high emphasis on its relations with that country.

Pakistan's unusually warm relations with Saudi Arabia did bring some initial benefits (such as a large-scale moving of Pakistani white-collar workers to work in the country and other destinations), but it ended up worsening Pakistan's problems of rentierism.[31] Instead of establishing relations on the basis of equality, Pakistan ended up developing dependence on Saudi loans and, of course, on cheap oil.

From the start of General Zia's term in 1977, Pakistanis also saw a major increase in 'Saudi expansionism' in their country, which was manifested in both economic and politico-ideological ways.[32] Bredi asserts that Saudi Arabia was keen to mould social and political dynamics in Pakistan based on the concept of Islamic revival, which was logical given the context of the Iranian revolution. As Ziring has put it, 'with Khomeini preaching Islamic revolution and seemingly capable of influencing Muslims everywhere, the Saudis were

[26] Rizvi, *Pakistan and the Geostrategic Environment*, p. 83; Rizvi, 'Pakistan: Ideology and Foreign Policy', p. 57.

[27] Mahdi, *Pakistan's Foreign Policy, 1971–1981*, p. 170.

[28] Delvoie, 'The Islamization of Pakistan's Foreign Policy', p. 134.

[29] Rizvi, *Pakistan and the Geostrategic Environment*, p. 83.

[30] Wriggins, 'Pakistan's Search for a Foreign Policy After the Invasion of Afghanistan', p. 292.

[31] Paul, *The Warrior State*, p. 18.

[32] Daniela Bredi, 'Notes on Saudi Arabia's Economic and Ideological Expansionism—The Case of Pakistan', in *Pakistan in the 80s: Ideology, Regionalism, Economy, Foreign Policy*, edited by Wolfgang Zingel and Stephanie Lallemant (Lahore: Vanguard, 1985), p. 541.

determined not to lose the initiative', exporting their own brand of Islam into Pakistan, Afghanistan, and beyond.[33]

In the absence of any checks and balances, the military dictatorship of General Zia was a willing recipient of that brand of Islam in the region and welcomed it with open arms. It oversaw a 'mammoth' influx of Arab (primarily Saudi) financial, cultural, and political capital into the country, transforming the society's moderate outlook.[34] A number of Muslim seminaries (called '*madrassas*') were established which imported the conservative, Saudi brand of Islam to a relatively more pliant Pakistani society, which traditionally followed a mystic version of Islam rooted in the country's subcontinental heritage.[35] Pande details how Pakistan welcomed donations from wealthy Saudi individuals and charities for these seminaries. This '*madrassa* boom' from the late 1970s was linked with the militarization of Pakistan's society (that was aimed at providing the foot soldiers of the Afghan jihad) and the later rise of ethnic and sectarian tensions in the country. It also enabled the military to follow more aggressively Islamist policies in Kashmir and Afghanistan from the late 1980s onward, which also helped it strengthen its dominance internally at the time of the country's return to democracy in 1988.[36]

The Saudis (and to some extent the Emiratis) also started interfering in Pakistan's internal affairs, which would continue through the 1990s and beyond. After the country's return to democracy, these states started supporting certain political parties that matched their hard-line Islamist inclinations.[37] That support came in the form of aid, deferred loans, or oil payments and helped enhance the clout of such parties in Pakistani politics. That, in turn, boosted the popularity of conservative politicians in Pakistan at the cost of those who championed progressive causes. As some had warned in the early 1980s, Pakistan's tilt towards the conservative monarchies of the Gulf would embroil it in the broader Middle Eastern conflicts.[38] That is what happened in the 1990s, when Pakistan found itself at the centre of an Iran–Saudi proxy war, fought on Pakistani streets in the guise of violent sectarian conflict.[39] The stage for that war had been set during the military dictatorship of General Zia in the 1980s.

[33] Lawrence Ziring, *Pakistan at the Crosscurrent of History* (London: Oneworld, 2003), p. 181.
[34] Fair, 'Islam and Politics in Pakistan', p. 273.
[35] Ziring, *Pakistan at the Crosscurrent of History*, p. 187.
[36] Fair, 'Islam and Politics in Pakistan', p. 274.
[37] Pande, 'Friends of Last Resort', pp. 260–261.
[38] Rizvi, 'Pakistan: Ideology and Foreign Policy', p. 57.
[39] See Frédéric Grare, 'The Evolution of Sectarian Conflicts in Pakistan and the Ever-Changing Face of Islamic Violence', *South Asia* 30, no. 1 (2007): pp. 127–143.

Where Bhutto sought to set Pakistan on a course to establish sound economic relations with many Muslim countries, the military dictatorship that followed him squandered that chance. An over-reliance on economic aid from these countries (primarily Saudi Arabia) had several negative consequences. Investment and assistance from Muslim states in Pakistan declined as their resources 'dwindled' in the 1980s due to the declining oil prices, as well as the Iranian Revolution and the outbreak of the Iran–Iraq War in 1980—all of which left Pakistan hanging in the middle.[40] Where Pakistan did not build robust economic relations with these countries, neighbouring India's economic revival during that period opened doors for its relations with the same countries. Saudi Arabia, in particular, started warming its relations with India from the 1990s in the fields of foreign investment, defence, and counterterrorism.[41] Since that time, it has been hard for Pakistan to reconcile with the idea that Saudis would not be able to ignore 'economic interests in India in favor of an ideological affinity for Pakistan'.[42] Saudi Arabia has also supported an Indian presence at the Organisation of Islamic Countries (OIC) due to its large Muslim population, but that proposal is unacceptable to Pakistan.[43]

Pakistan's increased focus on closer ideological alignment with the countries in the Middle East (as opposed to an economic relationship) led it to cognitively dissociate itself from South Asia, where it had far stronger cultural, historical, and religious linkages.[44] As Harrison asserts, Pakistani Muslims and Indian Muslims 'have more in common with each other' than Pakistani Muslims and Middle Eastern Muslims.[45] This resulted in the transformation of a pliant Pakistani culture into a Saudi-style intolerant one, as opposed to it being reflective of the more tolerant culture of the 'Indian civilization'.[46] Pakistan's feelings of 'misplacement' were further worsened because its Arab counterparts never

[40] Rizvi, *Pakistan and the Geostrategic Environment*, p. 79; See Joseph Nye, 'Energy and Security in the 1980s', *World Politics* 35, no. 1 (1982): pp. 121–134.

[41] Pande, 'Friends of Last Resort', p. 269.

[42] Ibid., pp. 269–270.

[43] Rajendra Abhyanker, *West Asia and the Region: Defining India's Role* (New Delhi: Academic Foundation, 2007), pp. 237–256.

[44] Wali Aslam, 'Pakistan's Hard Misplacement and the Politics of Regional Identity', *Cambridge Review of International Affairs* 33, no. 4 (2020): pp. 542–554; see also Wali Aslam, Leslie Wehner, Kei Koga, Janis van der Westhuizen, Cameron Thies, and Feliciano de Sá Guimarães, 'Misplaced States and the Politics of Regional Identity: Towards a Theoretical Framework', *Cambridge Review of International Affairs* 33, no. 4 (2020): pp. 505–526.

[45] Selig Harrison, 'Fanning Flames in South Asia', in *Pakistan and South Asian Environment*, edited by Ejaz Aslam Qureshi (Lahore: Sang-e-Meel, 1991), p. 30.

[46] See Pervez Hoodbhoy, 'Saudizing Pakistan: How Pakistan is Changing and What this Means for South Asia and the World', in *Routledge Handbook of Contemporary Pakistan*, edited by Aparna Pande (Abingdon: Routledge, 2016), pp. 466–478.

accepted it as one of their own, exacerbating its isolation and dislocation.[47]

3.2 The warming of Pakistan's relations with China

Pakistan's relations with China warmed significantly because of the Sino–Indian War of 1962.[48] Further consolidating Pakistan's relations with China was part of Zulfikar Ali Bhutto's vision of having strong ties with more than one great power. He believed that '[t]he idea of conducting and developing [Pakistan's] relations with each of the Great Powers on a bilateral basis identifying areas of co-operation with one without repudiating an alliance with another and thus evolving an internally consistent and integrated policy requires no justification and implies no moral pretence'.[49]

Pakistan's befriending of China during the 1960s was a remarkably positive move, as doing so diversified Pakistan's sources of acquiring weapons from the 1970s onward. Its closeness with China also made sense given the global shift in the balance of power—especially bearing in mind the Indo-Soviet treaty of friendship. China was economically and militarily far weaker than the Soviet Union in the 1970s and it was looking for allies in South Asia. The warming of relations between Pakistan and China took place at the time when 'the Soviet Union was subjecting Beijing to its own variety of containment, and Pakistan once again was found on the other side of the political and diplomatic equation'.[50] Reflecting a broad convergence of interests, China and Pakistan (under Bhutto) shared a common perspective regarding the problems facing the developing world and presented similar (though not identical) policies to deal with them.[51] As Ahmad has stated, 'Pakistan required China's military equipment, technology, and political support to deny India hegemonic influence over it, and Beijing ... extended valuable economic and military assistance to Pakistan at times when the latter found this hard to obtain from its traditional Western allies'. He goes on to say that

> Pakistan's concern about India's military threat converged with China's interests in South Asia. Given the interlinkages of South Asian states with the external powers,

[47] Aslam, 'Pakistan's Hard Misplacement', p. 2.
[48] As mentioned earlier, Pakistan pinned high hopes on Chinese military support in the war of 1971, yet none was forthcoming. See Andrew Small, *The China-Pakistan Axis: Asia's New Geopolitics* (New York: Oxford University Press, 2015), p. 15.
[49] Noorani, 'Bhutto and Nonalignment', p. 1957.
[50] Rasul B. Rais, 'Pakistan in the Regional and Global Power Structure', *Asian Survey* 31, no. 4 (1991): p. 390.
[51] Naveed Ahmad, 'Sino-Pakistan Relations', *Pakistan Horizon* 34, no. 3 (1981): p. 64.

> two types of parallel, if not outright hostile alignments emerged—Indo-Soviet and Sino-Pakistani—shaped by Sino-Soviet rivalry and Indo-Pakistani antagonism. The patterns of diplomatic support, positions on disputes, and military supply relations followed the lines drawn by the rivalries.[52]

The Pakistani ruling elite, however, used the warming of relations with China primarily to secure longer-term Chinese diplomatic support to counteract what it perceived as India's hegemonic designs instead of utilizing the opportunity to strengthen Pakistan's economic foundation. India's nuclear tests of 1974 added to Pakistan's insecurity, as they were conducted soon after the separation of Pakistan's eastern wing. Initially, Pakistan worked with China to call for the establishment of a nuclear-free zone in South Asia.[53] China was fully supportive of such a stance—not because of its deep-seated affection for Pakistan but because a nuclear-free South Asia would have ensured that India (a Chinese rival) did not develop nuclear weapons. It is no wonder that, upon his return from a trip to China in May 1980, General Zia-ul-Haq stated that 'we have perfect understanding in all fields'.[54]

The Chinese and Pakistani defence cooperation became very close from the 1970s onward, and by 1982, Chinese weapons formed the 'backbone' of the Pakistani arsenal: they comprised 75 per cent of the tank force and 65 per cent of Pakistan's air force.[55] China also assisted Pakistan in the nuclear field (in the 1980s) and missile development (in the 1990s).[56] It further supported the country in providing weapons to the Afghan mujahidin (after the Soviet invasion) through Pakistan.[57] Relations with Pakistan also served Chinese interest tremendously as Communist China was diplomatically isolated in the early 1970s and Pakistan was a gateway for Beijing to access Europe (especially Great Britain), Africa, and the Middle East (particularly the Gulf monarchies).[58] China also benefited from its alliance with Pakistan once it became a captive buyer of certain Chinese equipment.[59] Pakistan also granted China access to Western technology and equipment, which it received from the producers of highly sophisticated weapons such as the United States.[60]

[52] Ibid., p. 391.
[53] Ibid., p. 65.
[54] Ibid., p. 71.
[55] Pande, 'Friends of Last Resort', p. 263.
[56] Amitabh Mattoo, 'Shadow of the Dragon: Indo-US relations and China', in *Engaging India: U.S. Strategic Relations with the World's Largest Democracy*, edited by Gary K. Bertsch, Seema Gahlaut and Anupam Srivastava (Abingdon: Routledge, 1999), pp. 219–220.
[57] These weapons were funded by Saudi and American money but the use of indirect routes allowed these states the pretence that they were not actively involved in confronting the Soviet Union.
[58] Ahmad, 'Sino-Pakistan Relations', p. 66.
[59] Pande, 'Friends of Last Resort', p. 263.
[60] Ibid., p. 258.

Although China–India relations followed a negative trajectory in the 1960s, both sides decided to change course by the mid-1970s to better serve their mutual interests. China believed that normalizing relations with India would ensure that the latter did not align too closely with the Soviet Union (or, later, the US). For India, opening to China would ensure that China–Pakistan relations did not harm it. The warming of relations between both sides started in 1976, when they exchanged ambassadors after a fourteen-year hiatus.[61] The relations between the two further strengthened with the visit of Rajiv Gandhi to China in 1988.[62] In the late 1980s, the United States also started building a 'symbiotic relationship' with India in the fields of trade and technology, necessitating China to continue having good relations with the latter, despite occasional hiccups.[63]

Like India, Pakistan was unable to adjust the framework of its relations with China to keep up with the changing reality. It believed that China would always support it in its competition with India, which remained Pakistan's primary concern. In the new geopolitical scenario, Pakistani ruling elites' preoccupation with Kashmir became a source of exasperation for its Chinese counterparts, who wanted amicable relations with India. Garver has conducted a thorough analysis of the change in the discourse of high-ranking Chinese officials on Kashmir.[64] Where China's stance on Kashmir was fully supportive of Pakistan during the 1965 war between India and Pakistan, there was a visible thinning of that support in the 1980s and 1990s after the Indian Foreign Minister, A. B. Vajpayee, told his Chinese counterpart in February 1979 that China's post-1964 stance on Kashmir was 'an additional and unnecessary complication to the prospects of [healthy] Sino-Indian relations'.[65] The message was well received by Deng Xiaoping, who asserted in 1980 that the 'Kashmir issue was a bilateral dispute between India and Pakistan that should be solved peacefully'.[66] Garver asserts that 'after 1980, Chinese officials no longer mentioned the Kashmiri people's right to self-determination, much less endorsed it, as they had done consistently throughout the 1960s and 1970s'.[67] When Pakistan and India came close to war in 1990, China did not refer to the UN resolutions but talked 'only of bilateral negotiations

[61] Ibid., p. 269.

[62] John W. Garver, 'Sino-Indian Rapprochement and the Sino-Pakistan Entente', *Political Science Quarterly* 111, no. 2 (1996): p. 325.

[63] Rasul B. Rais, 'Pakistan in the Regional and Global Power Structure', p. 386.

[64] See Garver, 'Sino-Indian Rapprochement and the Sino-Pakistan Entente'; see also Rizvi, *Pakistan and the Geostrategic Environment*, p. 154.

[65] Garver, 'Sino-Indian Rapprochement and the Sino-Pakistan Entente', p. 327.

[66] Ibid., pp. 327–328.

[67] Ibid.

leading to a peaceful settlement of the issue'.[68] The rationale for doing so was clear: Beijing did not want to annoy New Delhi at the cost of Pakistan.[69] By the end of 1990, Beijing suggested that both countries resolved the Kashmir issue via 'consultations', leading the President of the Indian Council of World Affairs to assert that the Chinese 'formulation was essentially [the] same as India's'.[70] To Pakistan's great dismay, Beijing also declined to support Pakistan's efforts in 1990 to organize a special session of the United Nations on Kashmir because 'there was simply too little international interest in the issue'.[71] An economically impoverished Pakistan had much less to offer to a rapidly developing China by the late 1990s compared to India—the other rising power.

Addressing Pakistan's parliament in December 1996, then Chinese President Jiang Zemin stated that 'while it was natural for neighbors to have "differences or disputes", keeping the "larger picture" in mind, "if certain issues cannot be resolved for the time being, they may be shelved temporarily so that they will not affect the normal state-to-state relations."'[72] It was clear that Pakistani rulers' constant fixation on Kashmir was bothersome for China.[73] Pande thinks that

> Jiang wanted to convey the message that, just as China had developed economic and cultural ties with India and [had] 'temporarily shelved' Sino-Indian border issues because such disagreements could not 'be resolved for the time being', it was time for Pakistan to do the same. Pakistan, however, chose not to receive the message.[74]

As mentioned, the Chinese message was clearly not heeded by Pakistan, and it continued with its attempt to 'liberate' Kashmir via the Kargil invasion of May 1999. When its paramilitary forces were badly stuck there, Pakistan looked towards China (and the US) to gain support for its ill-fated adventure—but the Chinese were not ready to provide that support and publicly distanced themselves from Pakistan.[75] China was focused on avoiding 'unnecessary

[68] Ibid., p. 328.

[69] John Calabrese, 'Balancing on "the Fulcrum of Asia": China's Pakistan Strategy', *Indian Journal of Asian Affairs* 27/28, no. 1/2 (2014–2015): p. 2.

[70] Garver, 'Sino-Indian Rapprochement and the Sino-Pakistan Entente', pp. 328–329.

[71] Ibid.

[72] Quoted in Pande, 'Friends of Last Resort', p. 264.

[73] Mohan Malik, 'South Asia in China's Foreign Relations', *Pacifica Review: Peace, Security & Global Change* 13, no. 1 (2001): p. 86.

[74] Pande, 'Friends of Last Resort', p. 264.

[75] Nasim Zehra, *From Kargil to the Coup: Events That Shook Pakistan* (Lahore: Sang-e-Meel, 2018), p. 193; see also Jingdong Yuan, 'Beijing's Balancing Act: Courting New Delhi, Reassuring Islamabad', *Journal of International Affairs* 64, no. 2 (2011): p. 41.

entanglement' from Pakistan in the intractable Kashmir conflict.[76] It believed that a more neutral approach suited China's own geostrategic interest in the 1990s. Pakistan was diplomatically isolated, again, due to its own misadventure and the military's ill-thought-out handling of the country's foreign and security policy. The country transitioned from one democratic regime to another between 1988 and 1999, but the military remained the main arbiter of Pakistani foreign policy.

Interestingly, the reference to the UN resolutions with regard to Kashmir returned in Chinese statements only after India referred to the 'China threat' for its decision to conduct nuclear tests in May 1998.[77] Quite evidently, China's stance on the Kashmir dispute was dictated by its global position and not by its purported closeness to Pakistan. This point was lost on Pakistan's military, which was unable to adjust its historical antagonism with India to adapt to these new geopolitical realities. As John Garver said, '[d]emand for ... a plebiscite had long been and continued to be Pakistan's position regarding Kashmir. Between 1964 and 1990 it was also China's position'. He believes that,

> [i]n 1990, however, Beijing shifted course: it stopped referring to the United Nations and its resolutions in the context of Kashmir (except when Beijing wanted to needle New Delhi, as, for example, in the aftermath of India's 'China threat' justification of its May 1998 nuclear tests). Beijing instead began extolling peaceful settlement of the issue via talks between India and Pakistan.[78]

China's support for the Kashmir dispute was not the only hit Pakistan took for its military's unpragmatic competition with India in the 1990s. It also led to the weakening of China's verbal deterrent support for the country in its rivalry with India. In Pakistan's previous conflicts with India, China had threatened, explicitly or implicitly, that it might enter the conflict on Pakistani side.[79] Such support was an important feature of the two countries' relations up until the mid-1980s. For example, the Chinese Foreign Minister declared in 1983 that 'Pakistan is China's exceptionally friendly neighbor. If there is a war and Pakistan suffers foreign armed attack, the Chinese government and people will, of course, stand on the side of Pakistan.'[80] However, this stance

[76] Yuan, 'Beijing's Balancing Act', p. 41.
[77] Ibid., p. 44.
[78] John Garver, 'The Future of the Sino-Pakistan Entente Cordiale', in *South Asia in 2020: Future Strategic Balances and Alliances*, edited by Michael Chambers (Walcott, NY: Scholar's Choice Publishers, 2002), p. 399.
[79] Ibid., p. 330.
[80] Quoted in ibid.

had significantly weakened by 1987 when Chinese Premier Zhao Ziyang said on a visit to Pakistan that 'the Chinese government and people will, as always, firmly support the Pakistani government and people in their just cause of safeguarding national independence, upholding stated sovereignty and promoting economic development'.[81] Similarly, when a Chinese official visiting Pakistan in 1989 was asked if China would continue to stand behind Pakistan if it faced (Indian) aggression, he replied that it was a purely hypothetical situation and that China hoped that there would be no such conflict—pointedly not giving a direct answer.[82] Given the boost in Chinese trade with India, it made sense for China to want a peaceful South Asia and to not take one side over the other.

3.3 Pakistan's support for non-state actors

Though not officially declared a pillar of its foreign policy, Pakistan continued to use non-state actors to further its security interests in the region throughout the period under examination. When President Daoud Khan overthrew his cousin Zahir Shah in a coup in Afghanistan in 1973, Pakistan immediately decided to make use of that opportunity to pre-empt the revival of the Pashtunistan issue.[83] With its morale boosted through links with the Gulf states as well as a newfound alliance with China, the new democratic leadership in Islamabad decided to support Islamist actors in Afghanistan. From 1973 onward, its government 'supplied and controlled' Gulbuddin Hekmatyar, who was opposed to President Daoud Khan, to shape the dynamics in Afghanistan in its favour.[84] Hekmatyar led a force of 5,000 Afghans trained, supplied, and controlled by Pakistan to carry out anti-government activity in Afghanistan in 1975.[85] Islamabad believed that the support of Islamists would promote stability in the country's north-west. Husain Haqqani argues that Pakistan's policies not only destabilized Afghanistan, they provoked the Soviet Union into invading the country in December 1979.[86] Indira Gandhi, the former Indian Prime Minister, also believed that the world had misconstrued the reasons for the Soviet invasion of Afghanistan. She believed

[81] Quoted in ibid.
[82] Ibid., p. 331.
[83] Ahmad, 'Sino-Pakistan Relations', p. 63. King Zahir Shah had generally been considerate of Pakistani concerns and did not threaten the country from the north-west even when Pakistan faced crises like the wars of 1965 and 1971.
[84] Mahdi, *Pakistan's Foreign Policy, 1971–1981*, p. 134.
[85] Ibid.
[86] Husain Haqqani, *Magnificent Delusions: Pakistan, the United States and an Epic History of Misunderstanding* (New York: PublicAffairs, 2013), p. 232.

that the invasion was more defensive in nature, 'motivated by the move that resurgence of Islam in Afghanistan would endanger the loyalty of the Central Asian republics to the cause of communism'.[87] That goes against the conventional logic that the invasion was a *fait accompli* for Pakistan, which had no choice but to 'initiate' the Afghan jihad to reverse the Soviet assault. As Harrison argues, Pakistan had long held neo-imperial designs on Afghanistan and wanted to mould functions of the state to its liking. Pakistan's approach towards Afghanistan was thus inherently offensive rather than reactive.[88] It had envisioned Afghanistan as space of 'strategic depth'[89] that would ensure Pakistani security above all else.

After the Soviet withdrawal in 1989, Afghanistan endured a brutal civil war in which the Pakistani military primarily supported the Islamist faction of Hekmatyar, among others. Its vision of shaping the government of Afghanistan in Pakistan's image led it to drop Hekmatyar in 1994 when he was considered to be ineffective. Instead, the Pakistan military started supporting the Taliban, who went on to declare their rule over Afghanistan in late 1996, something that would not have been possible without Pakistani support.[90]

By late 1987, the Pakistani military had also started training and arming Kashmiri militants to launch attacks against India.[91] The country started the policy of supporting militant disruption in Indian-occupied Kashmir with the hope of 'bleeding India into concessions', but such an approach failed to bring India to the negotiating table.[92] Pakistan's strategic decision to use militants as proxies generally served the country negatively, though there were some minor successes, such as temporarily boosting its capacity vis-à-vis India.[93] Seen through a broader foreign policy lens, the strategy's numerous flaws far outweighed its meagre benefits.

This use of militant proxies contributed to Pakistan's isolation in the 1990s. The country even came close to being designated a state sponsor of terrorism in the early 1990s by the United States.[94] In 1992, the US ambassador to Pakistan, Nicholas Platt, warned the Nawaz Sharif government of the dire

[87] Ahsen Chaudhri, 'Geopolitical Factors in Pakistan India Relations', in *Pakistan and South Asian Environment*, edited by Ejaz Aslam Qureshi (Lahore: Sang-e-Meel, 1991), p. 107.

[88] Harrison, 'Fanning Flames in South Asia', p. 22.

[89] Aparna Pande, *Explaining Pakistan's Foreign Policy: Escaping India* (Abingdon: Routledge, 2011), p. 59. 'Strategic depth' in Pakistan refers to a security doctrine that emerged during the Cold War, reflecting concerns about Pakistan's limited geographic depth vis-à-vis India. Initially framed as a defensive military concept to avoid strategic encirclement, it later evolved into a broader political strategy aimed at ensuring a friendly or non-hostile Afghanistan and limiting Indian influence there.

[90] S. Paul Kapur, *Pakistan: Jihad as Grand Strategy* (Oxford: Oxford University Press, 2017), p. 99.

[91] Ibid., p. 84.

[92] Christine Fair, 'Pakistan's Ability to Mitigate Sources of Insecurity', in *Pakistan: Can the United States Secure an Insecure State?* edited by Christine Fair et al. (Santa Monica, CA: Rand, 2010), p. 110.

[93] See S. Paul Kapur and Sumit Ganguly, 'The Jihad Paradox: Pakistan and Islamist Militancy in South Asia', *International Security* 37, no. 1 (2012): pp. 111–141.

[94] Kapur, *Pakistan: Jihad as Grand Strategy*, p. 100.

consequences for Pakistan's relations with the US if it continued to support terrorism. Platt wrote in a memo to the Pakistanis: 'Please consider the serious consequences [to] our relationship if this support continues. If this situation persists, the Secretary of State may find himself required by law to place Pakistan on the state sponsors of terrorism list.'[95]

China was also unhappy with Pakistani support for the Taliban regime in Afghanistan between 1996 and 2001, which was spearheaded by Pakistan's Inter-Services Intelligence (ISI) agency. Beijing worried about the increasing presence of Uighur militants in Pakistan as well as Afghanistan and their destabilising effect on China's far west. China refused to accept Pakistani military's logic of 'strategic depth'. It considered, very briefly, establishing relations with the Taliban, but eventually chose not to do so. By September 2001, it still hosted the embassy of the Islamic State of Afghanistan, which was headed by Burhanuddin Rabbani of the Northern Alliance.[96] Beijing's rupture with Islamabad on the latter's Afghanistan policy was at its starkest when it voted in favour of US- and Russian-supported UN sanctions on Afghanistan in 1999.[97] The Pakistani military's deployment of militants in Kashmir also concerned China. In the late 1990s, Beijing expressed 'private disapproval and public non-endorsement' of Pakistan's policy to challenge India in Kashmir through militants.[98]

According to the interviews conducted by Andrew Small with Chinese officials, the Pakistani military might have tolerated the Uighur presence on Pakistani soil to keep Pakistan relevant for China.[99] On the other hand, some Chinese experts worried that Pakistan's tolerance of Uighurs was more than a 'double game' that Pakistan might be playing. Instead, it might be linked to the religious sympathies that mid-ranking intelligence officers might have for the Uighur militants.[100] These complications with China led a former Pakistani Foreign Secretary to say in the year 2000:

> Regionally, China will continue to support Pakistan, but much depends upon how Pakistan itself manages to come out of its current problems. The old intimacy and warmth, which once were a hallmark of Sino-Pakistan friendship, is a part of history . . . China has some valid grounds for being wary concerning our policy in Afghanistan and Kashmir. It is up to us to remove these obstacles.[101]

[95] Jeffrey Goldberg and Marc Ambinder, 'The Ally from Hell', *The Atlantic*, December 2011.
[96] Garver, 'The Future of the Sino-Pakistan Entente Cordiale', p. 400.
[97] Ibid., p. 434.
[98] Ibid., p. 399.
[99] Small, *The China-Pakistan Axis*, p. 90.
[100] Ibid., pp. 90–91.
[101] Quoted in Yuan, 'Beijing's Balancing Act', pp. 44–45.

The Taliban government in Afghanistan was only recognized by three countries: Pakistan, the United Arab Emirates, and Saudi Arabia. However, the Saudis were also increasingly unhappy with the Taliban by the late 1990s for their refusal to handover Osama bin Laden, who had at one point declared jihad on Saudi Arabia.[102] This demonstrates Pakistan's international isolation resulting from its foreign policy mistake of using militants as instruments of policy in Afghanistan and Kashmir. That policy necessitated a close relationship with the Taliban—a group that was actively shunned by the rest of the international community.

Pakistan's interference in Kashmir through Islamist proxies lost it support in the eyes of ordinary Kashmiris, too, who did not share the extremely violent methods adopted by groups such as Lashkar-e-Tayyaba (LeT) and Jaish-e-Muhammad. These hardline Islamist organisations were supported by the Pakistani state to conduct militant operations in Indian-occupied Kashmir against civilian and military targets.[103] Pakistan's involvement had transformed a nationalist struggle into an Islamic jihad.[104] Yoginder Sikand wrote in 2001 that 'the war-weary Kashmiris have not responded enthusiastically to jihadist appeals of [the] groups'.[105] He went on to say that the 'experience of Afghanistan after the Soviet withdrawal, and the mounting economic problems, political instability, and sectarian and ethnic violence in Pakistan, in which the Islamists have had no small role to play, have convinced many Kashmiris of the futility of the jihadist rhetoric'.[106] Pakistani military-supported fighters of LeT followed a strict *wahabi* brand of Islam; this did not marry well with the Kashmiri brand of Islam, which was more liberal in nature.

There was another problem with this type of Islam: it represented an 'external agenda' which the locals felt was being imposed on them by Pakistan.[107] The situation was not dissimilar to that of India's hegemonic designs in the region, which the Kashmiris had resented from 1947. Sikand wrote that Pakistan was 'hegemonising the terms of discourse in which the Kashmiri liberation struggle has sought to express itself [and it] has been only partially successful, at best'.[108] In other words, Pakistan was involved in

[102] Steve Coll, *Ghost Wars: The Secret History of the CIA, Afghanistan and Bin Laden* (London: Penguin, 2005), p. 402.

[103] Yoginder Sikand, 'Changing Course of Kashmiri Struggle: From National Liberation to Islamist Jihad?' *Economic and Political Weekly* 6, no. 3 (2001): pp. 218–227.

[104] George Garner, 'Chechnya and Kashmir: The Jihadist Evolution of Nationalism to Jihad and Beyond', *Terrorism and Political Violence* 25, no. 3 (2013): pp. 419–434.

[105] Sikand, 'Changing Course of Kashmiri Struggle', p. 227.

[106] Ibid., p. 227.

[107] Ibid.

[108] Ibid.

similar hegemonic practices involving extreme violence, which the majority of mainstream Kashmiris did not support.

The military's support of jihad abroad also had a profound, negative impact on the country's domestic sociopolitical landscape. It gradually radicalized Pakistani society, which would make accommodating India or good relations with the US more difficult further down the line. The approach would also contribute to the rise of domestic terrorism in Pakistan in the late 2000s in the form of Tehrik-e-Taliban Pakistan (TTP).[109] The brutal sectarian war of the 1990s between Shias and Sunnis was also directly linked to the military's policy of promoting jihad outside its borders.[110] A variety of Islamist sectarian groups were patronized by the state in the 1980s and the 1990s for domestic politico-religious purposes, for parity with India, and to help the country promote its external policies in Afghanistan.[111] The extent of Pakistani involvement in the Kashmir jihad is clear from the fact that by the late 1990s, there were more Pakistanis than Kashmiris among those fighting in Kashmir.[112] The list of 'martyred commanders' of LeT in Kashmir included names from all over Pakistan. Ironically, the place where Pakistan's 'Kashmir policy had the most far-reaching impact was within Pakistan itself'.[113] Kumar believes that '[t]he free rein given [to] the ISI to mobilize the Afghan and Kashmir jihad led to the rapid spread of militant Islamic schools across the country. Jihad rallies began to attract crowds of [up to] 200,000 people, and 50,000 volunteers might sign up at a single rally (not necessarily to be called up).' According to her, '[b]y summer 2000, Islamic militias were collecting funds for jihad in every shop in Lahore, Islamabad, Rawalpindi, Karachi, and Peshawar'.[114] Ziring also says that

> [t]he long war in Afghanistan [and its attendant consequences] had transformed the new Pakistan into an entity that could not have been forecast at the time of the 1971 civil war .... Pakistan was very different from the Pakistan of 1947, and appeared more devoted to its religious heritage, but it was hardly a polity unified in its search for a coherent destiny.[115]

Pakistani support for Islamists in Afghanistan since 1973 had been to ensure the territorial integrity of north-west Pakistan with the rest of the country and

[109] Kapur, *Pakistan: Jihad as Grand Strategy*, p. 118.
[110] Jessica Stern, 'Pakistan's Jihad Culture', *Foreign Affairs* 79, no. 6 (2000): p. 116.
[111] For a comprehensive overview of these dynamics, see 'The State of Sectarianism in Pakistan', *International Crisis Group*, Asia Report 95, 18 April 2005.
[112] Radha Kumar, 'Untying the Kashmir Knot', *World Policy Journal* 19, no. 1 (2002): p. 16.
[113] Ibid., p. 17.
[114] Ibid., p. 17.
[115] Ziring, *Pakistan at the Crosscurrent of History*, p. 200.

to stop the Pashtunistan issue from rearing its head. In a classic 'tail wagging the dog' scenario, the hope was to pre-empt an (imagined) separatist insurgency in Pashtun territories at the cost of very real—and current—instability, at that time, in the entire country.

Pakistan's use of terrorism for its statist aims contributed to the country developing an 'image problem' with the international community in the late 1990s. That problem persisted for decades afterwards and had negative consequences for Pakistani citizens travelling abroad. The economic repercussions were also real as international business stayed away from the country. Militants trained at Pakistani *madrassas* came from as far afield as Chechnya, Bangladesh, Nepal, Burma, Afghanistan, Yemen, Mongolia, Kuwait, Uzbekistan, Tajikistan, Russia, and Turkey, and they hoped to return home after completing their education to fight jihad in their homelands.[116] This helped Indian writers to describe Pakistan as 'the jihad factory'.[117] Writing in 2002 about Pakistan's 'jihad culture', Stern summarized the flaws with the Pakistani approach as follows:

> [b]y facilitating the activities of the irregulars in Kashmir, the Pakistani government is inadvertently promoting internal sectarianism, supporting international terrorists, weakening the prospect for peace in Kashmir, damaging Pakistan's international image, spreading a narrow and violent version of Islam throughout the region, and increasing tensions with India—all against the interests of Pakistan as a whole.[118]

Pakistan also faced problems with militants when their interests were not aligned with those of the Pakistani state. For example, when the Taliban came to power in Afghanistan in 1996, they refused to acknowledge the Durand Line as an international boundary (a cherished dream of Pakistan and the primary reason for its support of militants in Afghanistan).[119] Furthermore, the military in Pakistan believed that it could achieve parity with India through militant proxies, which made the country increasingly confident in its approach—albeit with dangerous consequences. Mohan Malik states that it was the Chinese provision of nuclear and missile capability to Pakistan in the 1980s and 1990s that 'emboldened Pakistan to wage a "proxy" war in

[116] Kumar, 'Untying the Kashmir Knot'.
[117] Sushant Sareen, *The Jihad Factory: Pakistan's Islamic Revolution in the Making* (New Delhi: Hindustan Publishing, 2005).
[118] Stern, 'Pakistan's Jihad Culture', p. 116.
[119] Kapur, *Pakistan: Jihad as Grand Strategy*, p. 99.

Kashmir without fear of Indian retaliation'.[120] By 2001, Pakistan-supported militants were targeting as far afield as New Delhi (and Mumbai in 2008) leading to serious India–Pakistan confrontations.

3.4 Pakistan's foreign policy towards the United States

Pakistan–US relations during the period under examination tell a story of ups and downs. Although Zulfikar Ali Bhutto proclaimed that he wanted to start a 'new era in US-Pakistan relations'[121] after the civil war of 1971, Pakistan continued the strategy of the past, which aimed at securing as much American aid and military assistance as possible in its competition with India. Senior US officials were asking Pakistan to focus on domestic reforms, but the Bhutto government continued to make its objective of parity with India a central plank of its relations with the US. After the civil war, the country increasingly presented itself as a victim of Indian aggression and expected the US to help it—a narrative that was not bought by the latter. For example, Bhutto wrote in a letter to President Nixon in January 1972:

> While the world watched, India set about to destroy with predatory cynicism the established frontiers of our homeland. India continues to occupy our territories in spite of the overwhelming verdict of the General Assembly and the demand of the Security Council for the withdrawal of all armed forces to their respective territories and to positions on the ceasefire line in Jammu and Kashmir. We seek a durable peace in the sub-continent, a just and honourable peace, but there can be no peace unless the armed forces of India first vacate their occupation of our territories.[122]

The post-civil-war Pakistani discourse neglected to reflect on the atrocities committed in the former East Pakistan by the army of then West Pakistan. Showing their fantasy-based mindset, Pakistani leaders asserted that the breakup of Pakistan was tantamount to the fall of a bulwark against 'India's military adventurism in South Asia'. They believed that 'Pakistan's dismemberment [after the civil war] [was] not only a most tragic event for [its] people but it carrie[d] with it far reaching implications far beyond the pale of [its]

[120] Mohan Malik, 'The China Factor in the India-Pakistan Conflict,' *Parameters* 33, no. 1 (2003): p. 36, available at https://press.armywarcollege.edu/cgi/viewcontent.cgi?article=2134&context=parameters

[121] Telegram 26 from the Consulate General in Karachi to the Department of State, 3 January 1972, Foreign Relations of the United States (1969–1976), Volume E-7, 1154Z.

[122] Letter from the Pakistani Ambassador (Raza) to President Nixon, Washington, 18 January 1972, Foreign Relations of the United States (1969–1976), Volume E-7.

borders'.[123] They contended that given the scenario, India, 'backed by a neighbouring superpower' (the Soviet Union), might 'go on the rampage from Burma to Iran and from the Persian Gulf to the Indian Ocean'.[124] According to this view, despite its division in 1971, Pakistan was a major 'barrier to India's territorial ambitions'.[125] Bhutto asserted that Pakistan, supported by both the US and China, could prevent what he portrayed as the 'nightmare' scenario regarding the Indian dominance of the vast swathes of West Asia.[126] Ironically, however, it was Pakistan under Bhutto, not India, that was engaging in expansionist policies in Afghanistan from 1973 onward.

The Bhutto administration also tried hard in the early 1970s for the resumption of American arms supplies to Pakistan so that it could counter India. As before, it also offered land and sea bases to the US to facilitate it in the Cold War.[127] Pakistan's Defence Secretary at the time, Ghias Uddin Ahmed, told the Americans in March 1972 that the latter might consider developing 'Jiwani, Gwadar, Sonmiani Bay, Karachi and area south and east of Karachi' as naval bases.[128] Continuing to see Pakistan's relations with the US through the Indian lens, he stated that the US 'needed to bolster [Pakistani] defenses in order to provide some credible deterrent'.[129] Despite having done so in the past, Ahmed said that 'Pakistan could not contemplate attacking India but it needed some assurance about its defense'.[130]

Ahmed also referred to India's close collaboration with the Soviet Union in developing two of India's seaports. Pakistanis often complained that the US did not understand the sensitive nature of challenges facing their country and the region. They asserted 'that the US fail[ed] to appreciate the magnitude of the Soviet threat to US interests in South Asia'.[131] Instead of moving on from it, the Bhutto government continued with the old narrative regarding Pakistan's 'geographic position'. Like leaders before him, he asserted that 'Pakistan [was] situated at the mouth of the Persian Gulf. Any state that has access to the Karachi coast [could] dominate the Gulf. That is why the Soviet Union [was]

[123] Telegram 1622 from the Embassy in Pakistan to the Department of State, 18 February 1972, Foreign Relations of the United States (1969–1976), Volume E-7, 1800Z.

[124] Ibid.

[125] Memorandum of Conversation, Islamabad, 31 October 1974, Foreign Relations of the United States (1973–1976), Volume E-8.

[126] Telegram 1622 from the Embassy in Pakistan to the Department of State, 18 February 1972, Foreign Relations of the United States (1969–1976), Volume E-7, 1800Z.

[127] Telegram 1581 from the Embassy in Pakistan to the Department of State, 17 February 1972, Foreign Relations of the United States (1969–1976), Volume E-7, 1330Z.

[128] Telegram 2213 from the Embassy in Pakistan to the Department of State, 10 March 1972, Foreign Relations of the United States (1969–1976), Volume E-7, 0725Z.

[129] Ibid.

[130] Ibid.

[131] Intelligence Note RNAN-9 Prepared in the Bureau of Intelligence and Research, Washington, 15 February 1973, Foreign Relations of the United States (1969–1976), Volume E-8.

so interested in that coast.'[132] The country was positioning itself as having a purely defensive approach, but its interference in Afghanistan and Kashmir later on would show how its approach was more offensive than defensive in nature.

Ironically, as someone who wanted to introduce balance into Pakistan's foreign policy after the civil war of 1971, Bhutto was keen to try anything to get the Americans to provide military assistance to Pakistan. He presented Pakistan's closeness to the US in terms of 'favour' to the latter that cost the former place among non-aligned countries. Bhutto said:

> Pakistan has been committed to Western civilization. We have been committed to the US. In earlier days, there was a simplistic approach to world affairs in which the choice was between God and Satan, and we chose God. Nehru used to say that Pakistan was the most allied ally of the US. Pakistan supported the UK in the Suez crisis, and the Egyptians say they have not forgiven us yet. We have always tried to make our contribution. We kept away from Third World non-alignment sentiments.[133]

Bhutto knew that the US was not happy with the size of the Pakistani army. He agreed that '[Pakistan] would like to be able to strip our army of its power and put it in its place. But we must protect our borders. If we are stronger, it will enable us to do more in negotiating. But the Indians are so stupid and arrogant they cannot negotiate.'[134] Contrary to what the events of late 1980s and 1990s will show, Bhutto strongly stated that Pakistan had 'no grandiose ideas'.[135] While Bhutto's outlook on domestic politics differed from that of Pakistan's previous leaders, his fixation on India bore a strong resemblance to the past. He believed that 'India [wanted] total domination through total security. She [wanted] Pakistan totally insecure. [Pakistan] [could] accept partial insecurity but not total.'[136]

Where the Pakistani narrative had not changed much with the arrival of Bhutto, the US response was also not dissimilar from the response it had delivered during the 1960s. It believed that Pakistan (and other countries in the region) needed to 'devote their fullest resources to programs of economic and social improvement which they wanted for their people'.[137] Instead

[132] Memorandum of Conversation, Washington, 18 September 1973, 11 a.m.–12:45 p.m., Foreign Relations of the United States (1969–1976), Volume E-8.

[133] Ibid.

[134] Memorandum of Conversation, Islamabad, 31 October 1974, Foreign Relations of the United States (1973–1976), Volume E-8.

[135] Ibid.

[136] Ibid.

[137] Telegram 1834 from the Embassy in Pakistan to the Department of State, 25 February 1972, Foreign Relations of the United States (1969–1976), Volume E-7, 1117Z.

of buying the 'Soviet threat theory', the US believed that 'the main threat to Pakistan at [that] point [seemed] to be in terms of its own internal cohesiveness', endangered by economic mismanagement and poor nation building.[138] The US chargé d'affaires to Pakistan, Sidney Sober, wrote in February 1972: '[The] [Government of Pakistan or GOP] had not yet come to grips with problem of its future military posture. There had been some talk that army would remain [of the] same size as prewar. That involved problem of conflicting needs from limited resources, of military budget and of economic programs to which Bhutto and his party were committed.' He went on to say that 'there was also the serious problem of very scarce foreign reserves to be divided between military and nonmilitary needs. GOP, like other countries, face[d] difficult choices. The best decisions ... could be made in the framework of efforts to arrive at a durable peace in the subcontinent.'[139]

While Pakistani leadership continued to think in terms of future wars, the American approach was more focused on reconciliation between India and Pakistan, as one State Department Policy Review proposed in 1972: 'We could not and should not seek to build up Pakistan as any kind of strategic counterweight to India. As we see it, our basic policy objective in South Asia should now be to encourage movement toward a broad political settlement which would replace the sharp political-military confrontation that has plagued the Subcontinent for more than 20 years.' According to this Review,

> In Pakistan this would require, in addition to our continued support for its territorial integrity and economic growth, that we encourage Bhutto in every way open to US to move in the direction of a basic settlement with India and that we avoid any action in the military field that would encourage Pakistan again to postpone the difficult decisions it must make if it is to reach a basic accommodation with its stronger neighbour.[140]

Americans acknowledged that Bhutto had the ability to look beyond the India threat. They believed that the latter could see the 'futility of resuming confrontation, and necessity of reaching some sort of accommodation with India if Pakistan [was] to endure.'[141] Americans thought that while Pakistan was slowly more willing to 'recognize Indian pre-eminence ... it will strenuously

[138] Memorandum from the President's Assistant for National Security Affairs (Kissinger) to President Nixon, Washington, 7 March 1973, Foreign Relations of the United States (1969–1976), Volume E-8.

[139] Telegram 1834 from the Embassy in Pakistan to the Department of State, 25 February 1972, Foreign Relations of the United States (1969–1976), Volume E-7, 1117Z.

[140] Memorandum from Secretary of State Rogers to President Nixon, Washington, 17 March 1972, Foreign Relations of the United States (1969–1976), Volume E-7.

[141] Telegram 32 from the Embassy in Pakistan to the Department of State, 2 January 1973, Foreign Relations of the United States (1969–1976), Volume E-8, 1048Z.

resist Indian domination or hegemony'. They asked 'whether this willingness to accommodate to new "realities" of subcontinent will be sufficient to overcome past history of hostility, distrust and suspicion remains open question, however'.[142] The US officials also pointed to several opportunities Pakistan could benefit from if it were to shift its focus away from India by having a 'more active role ... in world affairs in general'.[143]

Washington also spent the decade of the 1970s refusing to acknowledge Pakistani requests for armament and for the use of Pakistani bases.[144] It believed that it did not have any 'vital security interests in South Asia anymore', though developments in this region could have negative repercussions for the US interest in the Persian Gulf area and in the wider Middle East.[145] If anything, Kissinger was more inclined to describe India 'as a major power in South Asia'.[146]

The Soviet invasion of Afghanistan in 1979 provided an opportunity for Pakistan to repair those relations, as the US needed Pakistan at this critical juncture. The Reagan administration, from 1981, wanted a 'stronger, more self-confident Pakistan' that could 'resist direct or indirect Soviet pressure through Afghanistan', since this would be 'essential for the protection of free world interests in the region'.[147] The American embrace also helped the military dictatorship of General Zia end the isolation of his regime and raise his international esteem.[148] Furthermore, General Zia won over significant military and economic assistance for Pakistan, which allowed the country to postpone undertaking significant economic reforms.[149] Zia also used the new relationship with the US as a way to win over the civilian and military bureaucracy as well as religious political parties in Pakistan (who responded positively to his call for jihad).

Pakistan took great pains to brandish its credentials as a 'frontline state'[150] after the Soviet invasion and secured significant military and economic aid. That helped bolster Pakistan's defence as well as temporarily helping its economy, but as the Americans had feared, the aid deterred Pakistan from taking tough decisions regarding the size of its military, the state of its economy,

[142] Ibid.
[143] Ibid.
[144] Memorandum from Secretary of State Rogers to President Nixon, Washington, 17 January 1972, Foreign Relations of the United States (1969–1976), Volume E-7.
[145] Ibid.
[146] Memorandum of Conversation, Washington, 29 January 1973, 1:15–2:20 p.m., Foreign Relations of the United States (1969–1976), Volume E-8.
[147] Mahdi, *Pakistan's Foreign Policy*, p. 177.
[148] Delvoie, 'The Islamization of Pakistan's Foreign Policy', p. 135.
[149] Rais, 'Pakistan in the Regional and Global Power Structure', p. 387.
[150] M. B. Naqvi, 'South Asian Cooperation', in *Pakistan and South Asian Environment*, edited by Ejaz Aslam Qureshi (Lahore: Sang-e-Meel, 1991), p. 186.

and its future relations with India. Furthermore, the US aid made Pakistan more confident in its dealings with India and less willing to engage with or accommodate the latter.[151]

This warming in relations was to come to an end after the Soviet withdrawal from Afghanistan in 1989. That should not have come as a surprise to Pakistan, which regularly complained that the West used it at a time of need and discarded it when it was not needed anymore—just like a 'Kleenex'.[152] That impression is deeply misleading but has often been used by Pakistani leadership throughout history. This idea is employed by the Pakistani civilian and military leadership to discursively construct Pakistan's identity as a victim state in front of the country's masses. It is not clear why Pakistanis were unable to anticipate that the US would have as little interest in Pakistan after the Soviet withdrawal as Pakistan had in the Maldives.[153]

Although Pakistan and the US worked closely in the 1980s, their relationship stood on a shaky foundation during that period. However, the difficulties were not as obvious in that decade as they were in the 1970s or after the Soviet withdrawal in 1989. In April 1982, the US ambassador to Pakistan, Ronald Spiers, mentioned that there were 'four vulnerabilities' in Pakistan–US relations; these included 'the India factor, the nuclear issue, human rights and narcotics'.[154] Among these, the nuclear issue was the main cause of concern for the US. Though Pakistan had started developing nuclear weapons capability in the 1970s, it tried its best to hide its progress from the US and other countries.[155] Haqqani narrates that when the United States presented 'incontrovertible evidence' to the Pakistani ruler General Zia-ul-Haq which showed that Pakistan was buying clandestine technology to develop these weapons, Zia said that he did not have 'knowledge of any such weapons development programs'. Zia went on to say that 'Pakistan might not be a large or important country but it was an honorable one'.[156] That denial continued until the US officials confronted the Pakistani leadership in May 1990 with the proof that Pakistan had acquired nuclear weapons capability. However, the response was still one of denial, leading an infuriated Robert Gates to comment: 'If it waddles like a duck, if it quacks like a duck, then maybe it is a duck.'[157]

[151] Harrison, 'Fanning Flames', p. 23.

[152] Dennis Kux, *The United States and Pakistan, 1947–2000: Disenchanted Allies* (Baltimore, MD: The Johns Hopkins University Press, 2001), p. 365.

[153] As stated by a former Pakistani ambassador to the US between 1991 and 1993. See Haqqani, *Magnificent Delusions*, p. 285.

[154] Raziullah Azmi, 'Pakistan-United States Relations: An Appraisal,' *Pakistan Horizon* 36, no. 3 (1983): p. 237.

[155] Haqqani, *Magnificent Delusions*, pp. 171–271.

[156] Ibid., p. 226.

[157] Ibid., p. 280.

Given this context, it should not have come as a surprise to the Pakistanis that the nuclear issue would return to the front burner after the Soviet withdrawal. The 'American interest in Afghanistan was strategic in nature and primarily limited to seeking the withdrawal of the Soviet forces'.[158] The US cut off its aid to Pakistan in 1990 due to the country's nuclear programme on the basis of the Pressler Amendment (though it continued to disburse over $1 billion for assistance in ongoing projects for another three years).[159] As Pakistan–US relations took a nosedive in the 1990s, the American relations with India warmed significantly, especially in the field of economy and trade. This development was received with hostility by Pakistan's military establishment.[160] India was focused on developing its high-tech sector in the 1980s and 1990s and was encouraging multinational corporations to establish business there. Lacking innovation on that front, Pakistan's military still saw the region through a security paradigm and was keen to acquire weapons from various sources.[161] It made sense for the United States to have closer links with India for several reasons, but the Pakistani military was unable to reconcile itself with that idea.

Pakistan tested its nuclear weapons in May 1998, two weeks after India. However, the nuclear capability did not alleviate Pakistan's security concerns. Instead, it encouraged the country to employ terrorist proxies without fear of Indian retaliation. By aligning so closely with the United States, Pakistan lost much respect among non-aligned countries as well as in moderate Muslim states.[162] Pakistan had publicly claimed not to help the mujahidin fighting in Afghanistan during the 1990s,[163] but it was later revealed that it had. The nuclear tests conducted in 1998 compounded existing American mistrust of Pakistan. This was a dangerous game to play with the world's only remaining post-Cold War superpower.

3.5 Policy assessment and conclusions

The early 1970s was a time when the country's foreign policy course could be corrected. The military's morale had taken a serious blow after the secession of East Pakistan, and Bhutto enjoyed an unprecedented power

[158] Rais, 'Pakistan in the Regional and Global Power Structure', p. 387.
[159] Haqqani, *Magnificent Delusions*, p. 283.
[160] Ibid., p. 289.
[161] Ibid., p. 287.
[162] Wriggins, 'Pakistan's Search for a Foreign Policy After the Invasion of Afghanistan', p. 292.
[163] Peter Pentz, 'The Mujahidin Middleman: Pakistan's Role in the Afghan Crisis and the International Rule of Non-intervention', *Dickinson Journal of International Law* 6, no. 3 (1988): p. 384.

which no civilian leader had enjoyed up to then (or since). Despite claiming to follow the rhetoric of a diversified foreign policy, he continued to follow the foreign policy of previous military regimes. He believed that he could consolidate his control by pandering to the desire of the dominant social forces in Pakistan (which included the military, feudal landed elites, and urban businesses) to compete with India.[164] Bhutto had come into power on the shoulders of the country's powerful military and he tried his best to please them instead of charting a different path at a critical time when he had the power to do so. As one astute observer put it to me, if Bhutto had not been apprenticed under the military regime of Ayub Khan, he would have had a different conception of democracy, one that acknowledged the legitimacy of opposition.[165] Bhutto, however, did not easily tolerate opposition and considered every opponent his personal enemy.

Bhutto actively revived the anti-India rhetoric of the 1960s by asserting that Pakistan was willing to fight India for a thousand years.[166] According to another noted observer of Bhutto, he wanted to 'outsmart' the military by playing the India card to consolidate his control, believing that the military would not destabilize someone who was as opposed to India as he was.[167] Instead of using this rare opportunity to refocus India–Pakistan relations on a path to reconciliation, Bhutto even used his dealings with the United States to gain the latter's support for competition with India. His Afghan policy also laid the foundations for linking Pakistan and Afghanistan inextricably for years to come. He led the way in showing the military how to intervene in the domestic politics of Afghanistan, a move that would have disastrous consequences in the years to come. His militaristic foreign policy approach was to ensure his own survival in office without regard to the consequences for the country.[168] Ironically, despite the fact that Bhutto's tenure is considered to be the only truly democratic period in Pakistan's history, his personality had an authoritarian streak similar to that of Jinnah and other leaders before him, which led to his personal downfall later on.[169] Both Jinnah and Bhutto lacked a vision and ruled on the basis of their authoritarian charisma.[170] Bhutto could have used his dealings with the Americans for economic development, but his insistence on US weaponry inadvertently strengthened the

[164] Interview, Journalist 1, Islamabad, Pakistan, January 2024.
[165] Ibid.
[166] Interview, Academic 2, London, United Kingdom, March 2024.
[167] Interview, Academic 3, Lahore, Pakistan, January 2024.
[168] Interview, Journalist 1.
[169] Interview, Journalist 2, Lahore, Pakistan, January 2024.
[170] Interview, Journalist 3.

military, preparing it to play a major role in the country's politics in the years to come.[171]

Zulfikar Ali Bhutto functioned as the military's key ideologue on relations with India. His daughter, Benazir Bhutto, by contrast, departed from her father's approach and invited Indian Prime Minister Rajiv Gandhi to Pakistan in 1988.[172] She laid the foundations of the two countries' agreement on border demarcation in the Siachen area. However, Benazir Bhutto's peace overtures were quickly stifled by the military.[173] During his decade-long tenure, General Zia had rarely discussed Kashmir, but as soon as the country returned to democracy in 1988, the military began the policy of Kashmir jihad that further consolidated its own power while weakening civilian leaders.[174] During the 1990s, the Pakistani military refused to cede control over foreign policy to the civilian leaders, and as a result, Kashmir proved to be the 'jugular vein of Pakistan's army [rather than of Pakistan itself] as it provided the justification for their volume and influence'.[175] The Kashmir dispute ensured a permanent friction with India, causing anxiety among the Pakistani masses who, in turn, looked towards the armed forces as the saviours of the country to protect it from the hegemon next door. Based on what can only be described as a fantasy, the military leadership wanted to take over parts of Kashmir by bleeding India over time. Benazir recounts in her autobiography about a briefing that she received from the military during her second term as Prime Minister. She writes:

> In my second term I was invited to the Joint Chiefs of Staff Headquarters to get a security briefing. Director of Military Operations Major-General Pervez Musharraf (who, of course, would later become Chief of Army Staff and seize power to become President) gave me a briefing. It seemed like déjà vu as I once again heard how Pakistan would take Srinagar if only I gave the orders to do so. Musharraf concluded the briefing with the words that a ceasefire would be in place and Pakistan would be in control of Srinagar, the capital of Indian-held Kashmir. I asked him, 'And what next?' He was surprised by my question and said, 'Next we will put the flag of Pakistan on the Srinagar Parliament.' 'And what next?' I asked the general. 'Next you will go to the United Nations and tell them that Srinagar is in Pakistan's control.'
>
> . . .

[171] Ibid.
[172] Interview, Academic 2.
[173] Interview, Journalist 3.
[174] Interview, Columnist, Islamabad, Pakistan, January 2024.
[175] Interview, Journalist 3. Widely attributed to Jinnah, the phrase 'Kashmir is the jugular vein of Pakistan' is frequently cited by Pakistani policy elites to signal that the Kashmir dispute is central to the country's security and national identity. The phrase implies that control or influence over Kashmir is vital to Pakistan's survival, shaping political rhetoric and military planning, particularly vis-à-vis India.

> 'And what next?' I pushed on. I could see General Musharraf had not been prepared for this grilling and was getting flustered. He said, 'And ... and you will tell them to change the map of the world taking into consideration the new geographical realities.' 'And do you know what the United Nations will tell me?' I looked General Musharraf straight in the eye, as the army chief sat silently by and the room grew still, and pointedly said, 'They will pass a Security Council Resolution condemning us and demanding that we unilaterally withdraw from Srinagar, and we will have got nothing for our efforts but humiliation and isolation.' I then abruptly concluded the meeting.[176]

The Kashmir policy was a tool through which the military exercised control and dominance within Pakistan. The dispute justified the significant defence budget to pay for a life of prestige for its officers. Continuation of the Kashmir dispute, and not its resolution, would serve the military interests in the 1990s and beyond.[177] They sabotaged the peace overtures of Nawaz Sharif when he invited the Indian prime minister Atal Bihari Vajpayee to Lahore for a peace summit in 1999 by launching the Kargil operation. As an interviewee explained to me, the policy of proxy war in Afghanistan and Kashmir in the 1990s was deployed by the military to ensure that the pot was kept simmering so that the justification for such a large military remained. In the absence of conflicts on the borders, there was no justification for Pakistan to dedicate a major portion of its defence budget to such a large army.[178]

Pakistan was an isolated country in 1971, and it was again so by 1999, after the coup by General Pervez Musharraf to topple the Nawaz Sharif government. Throughout this period, the country missed several opportunities to better its foreign policy positioning in the world. On the other hand, India had made inroads with every country that Pakistan wanted to be close to (including Saudi Arabia, the United States, China, and Iran) through sound economic links and pragmatic foreign policies as the foundation of its relations with these states. By contrast, Pakistan relied mostly on foreign aid and loans from these countries and did not show transparency in its nuclear programme. Its policy of supporting militants also earned it the ire of the international community.

Instead of reflecting on their country's isolation, Pakistanis persistently presented themselves as victims after the Soviet departure from South Asia.[179] Delvoie says that 'conference and seminar proceedings and newspaper

[176] Benazir Bhutto, *Daughter of Destiny: An Autobiography* (London: HarperCollins, 2008), pp. 418–419.

[177] Interview, Journalist 3.

[178] Ibid.

[179] Delvoie, 'The Islamization of Pakistan's Foreign Policy', p. 139.

editorials were replete with images of Pakistan as a loyal ally of the United States and the West that had been unceremoniously discarded once it had outlived its usefulness.'[180] The eruption of tensions after the Soviet withdrawal was understandable given the very real strains between the US and Pakistan over the latter's nuclear programme, which had not gone away even in the middle of the Afghan jihad. The Pakistani elites' portrayal of their country as a 'victim' state did not help much with improving the country's relations with the United States.

A key issue with Pakistan's foreign policy between 1971 and 1999 was that the country considered its objectives in regional terms such as parity with India, a quasi-alliance with China (to deter India), an empathetic Afghanistan, and the Gulf states that supported Pakistan in its competition with India.[181] It even expected the US to discard its global outlook (that determined America's South Asia policy) to see things from a Pakistani perspective. This Pakistani worldview, however, was not shared by the country's neighbours, its Middle Eastern allies, and importantly, the United States. As studied in this chapter, even Pakistan's 'all-weather friend' China reconsidered its relations with India given the global shifts and adjusted its approach towards it. It gave more preference to sound relations with India, rather than supporting Pakistan on diplomatic and political fronts at all costs. During the period discussed in this chapter, India emerged as one of the pre-eminent Asian powers and one with which the Gulf states wanted to have warm relations. On the other hand, the George H. W. Bush and Clinton administrations largely gave up on Pakistan because of its non-cooperation over the nuclear programme and its unrealistic expectations for the American role in the region.

By the end of the 1990s, Pakistan was not only an isolated country, it was also left behind in the fast-moving world of international politics. Whereas its leadership had promised in 1971 'to move with the times',[182] Pakistan failed to do exactly that, leading Lawrence Ziring to complain that 'Pakistan [was] frozen in time.'[183] He further wrote: 'All matters considered, [Pakistan's] ... policy continues to be influenced by fear and bitterness toward India. It therefore continues to judge the U.S. by its remoteness or intimacy with its larger neighbor and sworn enemy.'[184] The failures of Pakistan's foreign policy were summed up by Ziring in his subsequent analysis: '[a]s a debtor nation it has needed more than it could possibly offer in return. More important, its stubborn struggle with India has made it difficult for even sympathetic states to

[180] Ibid.
[181] Pande, 'Friends of Last Resort', p. 270.
[182] Ziring, 'Recent Trends in Pakistan's Foreign Policy', p. 303.
[183] Lawrence Ziring, 'Pakistan and India: Politics, Personalities, and Foreign Policy', *Asian Survey* 18, no. 7 (1978): p. 726.
[184] Ibid., p. 719.

provide it with the type of assistance it sorely requires.' He stated that '[n]o country has really championed its claim to Kashmir, and despite the publicity given to the dispute inside and outside the United Nations, India would seem to have successfully incorporated the mountain state into its Union.'[185]

When it comes to lost opportunities, Pakistan missed several. The country could have used its opening up to the Gulf states for economic reforms and investment in its high-tech industry, as well as to establish stable trading links with those states. However, that was not done. The Soviet invasion of Afghanistan could also have been leveraged by Pakistan to raise its prestige if it had taken a lead to find a political settlement to the crisis. That stance would have been supported by the international community and Pakistan's neighbours.[186] However, the country used the invasion instead to brandish its status as a front-line state to amass rent from different directions. When the last round of Geneva talks to finalize the Soviet withdrawal from Afghanistan concluded in February 1988, the government of General Zia-ul-Haq refused to support it. That led the visiting Soviet First Deputy Foreign Minister, Yuli Vorontsov, to say to the Pakistanis: 'For eight years you have been asking us to leave Afghanistan. Now you want us to stay. I smell a rat!'[187] This about-turn in Pakistan's position caused confusion among the international community,[188] though a cursory glance at the situation would make it clear that the country used the invasion to amass international economic and military aid and consolidate power within the military regime of General Zia-ul-Haq. One could understand why Pakistan would want the conflict to continue as long it could. Ghori says that Zia seized the opportunity of Afghan jihad 'with both hands' as it was 'the God-sent opportunity to ingratiate himself with the Americans.'[189] Wriggins also writes that professionals in Pakistan's foreign policy circles were wary of a close alliance with the US given the (perceived) unreliability that the US displayed in the wars of 1965 and 1971.[190] That is why the country's Foreign Minister in the early 1980s described their new relationship as 'a handshake not an embrace.'[191] However, the military regime of General Zia was more keen to have a close relationship with the US for the reasons mentioned above.

[185] Ibid., p. 726.

[186] Naqvi, 'South Asian Cooperation', p. 188.

[187] Abdul Sattar, *Pakistan's Foreign Policy 1947–2005: A Concise History* (Karachi: Oxford University Press Pakistan, 2007), p. 165.

[188] Shahid Amin, *Pakistan's Foreign Policy: A Reappraisal* (Karachi: Oxford University Press Pakistan, 2000), p. 95.

[189] Karamatullah Ghori, 'Sixty Years of Pakistan's Foreign Policy,' *Pakistan Horizon* 60, no. 2 (2007): p. 16.

[190] Wriggins, 'Pakistan's Search for a Foreign Policy After the Invasion of Afghanistan', p. 290.

[191] Ibid., p. 296.

Both Saudi Arabia and China (in addition to the US) were uneasy with Pakistan's support of the Taliban.[192] Both countries were also wary of constant Pakistani demands for economic help and would have liked the country to have developed sound economic foundations. However, Pakistan was unable to do that due to persistent economic instability, itself massively exacerbated by the military's intervention in politics.[193] Despite its problems, China continued to aid Pakistan in the field of defence. It considered that the 'combined strategic and political advantages' of its relations with the country outweighed the benefits of cutting off relations.[194] Although China would not have liked Pakistan to go to war with India, it would not have favoured friendly relations between the two either. That is because 'as long as India is preoccupied with Pakistan on its western frontier, it will not stir up trouble on the Tibetan border. A secure and stable India at peace with Pakistan would, on the other hand, make New Delhi focus on China and East Asia.'[195] If Pakistan and India were to resolve their differences and India were to emerge as an uncontested power, for China that would amount to 'conceding South Asia as an Indian sphere of influence. Such a move would spell the virtual end to Chinese aspirations of being the leading Asian power and would greatly weaken China's position against Indian power.'[196] Garver also says:

> if India were to achieve dominance of the vast area between the Hindu Kush and the Mekong River, Mauritius and the Strait of Malacca—which is how some Indian leaders define the natural and historical limits of India's sphere of influence—the potential Indian threat to China would be much greater. If China is to prevent this dominance, Pakistan is perhaps its last and best bet.[197]

This chapter showed that the countries dealing with Pakistan were looking after their own national interests by developing coherent foreign policy visions regarding their positionings in the world. Pakistan, on the other hand, was unable to see the bigger picture, and its foreign policy was more concerned with the 'here and now', instead of where the country might be in the longer term. The failure to develop that vision meant that Pakistan was a lonely and insecure country by 2001.

[192] Pande, 'Friends of Last Resort', p. 270.
[193] Shuja Nawaz, *Crossed Swords: Pakistan, Its Army, and the Wars Within* (Karachi: Oxford University Press Pakistan, 2008), pp. 411–506.
[194] Malik, 'South Asia in China's Foreign Relations', p. 78.
[195] Ibid.
[196] Ibid., p. 85.
[197] John Garver, 'China and South Asia', *The Annals of the American Academy of Political and Social Science* 519, no. 1 (1992): p. 81.

4

The 'double game' as a foreign policy tool (2001–2021)

4.1 Introduction

Just as Pakistan stood isolated in the world after its civil war in 1971, it found itself isolated, again, at the start of another war in September 2001: the US-led War on Terrorism. As argued in Chapter 3, Pakistan's approach to gain 'strategic depth' in Afghanistan and its policy to support militants in the Indian-occupied Kashmir in the 1990s had already cost its standing in the world dearly. The nuclear tests of 1998 helped further exacerbate Pakistan's isolation, and by 2001, Pakistan had very few friends left. Where the United States was displeased by the country's nuclear tests and the military takeover of 1999, China had concerns about Pakistan's Afghan policy due to the fears that Afghanistan might become home to Uighur militants from the west of China. Though Saudi Arabia supported Pakistan after its nuclear tests, it was unhappy with the Taliban decision to not hand over Osama bin Laden, leaving Pakistan as the only country that continued to support the group wholeheartedly.

The attacks of 11 September 2001 changed the South Asian regional scenario dramatically. Pakistan was asked explicitly by the Bush administration whether it stood with the West or the terrorists.[1] Given Pakistan's history of support for militant proxies from the country's inception, it would have been too abrupt a switch to change tack quickly. Given the threat of imminent US bombing,[2] Pakistan proclaimed allegiance to the Western alliance. However, it continued to keep its links with the Taliban intact. These contradictory policies gave birth to the Pakistani strategy of playing a 'double game' that

[1] 'US "Threatened" to Bomb Pakistan', *BBC News*, 22 September 2006.

[2] Suzanne Goldenberg, 'Bush Threatened to Bomb Pakistan, Says Musharraf', *The Guardian*, 22 September 2006.

Pakistan's Foreign Policy. Wali Aslam, Oxford University Press. © Wali Aslam (2026).
DOI: 10.1093/9780198909552.003.0004

continued until the American withdrawal from Afghanistan in August 2021.[3] This chapter assesses this double game as a foreign policy tool. Though various scholars have featured this approach by Pakistan within their writings, no study exists that assesses its strengths and weaknesses. This chapter will perform that task.

Section 4.2 looks at the lead-up to Pakistan's more formal adoption of a double game as its foreign policy tool after 2001. It does so with reference to the historic roots of the Pakistani support of militant proxies as well as its dealings with the outside world to understand what led it to adopt this strategy. Section 4.3 examines the main features of the double game, such as courting the United States for financial support, both military and non-military, while simultaneously supporting different factions of the Afghan Taliban. Section 4.4 assesses the consequences of the Pakistani double game, such as the impact on its standing in the international community. Section 4.5 conducts a critical assessment of the double game as a foreign policy tool. The concluding section, Section 4.6, summarizes the main conclusions and briefly looks at the alternative approaches that Pakistan could have adopted.

In this chapter, I argue that, as with any foreign policy tool, the policy of the double game has deep historical roots. Pakistan's support of Islamist militants is rooted in the country's cleavage from India in 1947. Given the circumstances, it is understandable that, in the beginning, Pakistan chose to align with the United States while also keeping its alliance with terrorists intact. However, it had several opportunities to sever those links in the following twenty years which it did not utilize. Pakistan could have used its influence over the Taliban to help Afghanistan reach an internationally supported political settlement in the country upon their takeover in 1996. Given Taliban reliance on Pakistan, it could also have pressured the group to cede to the demands of the international community and become a part of the democratic process in Afghanistan. Instead, Pakistan continued to support the Taliban in the face of serious international criticism and opposition, causing itself complications for years to come.

While the Taliban takeover of Afghanistan in 2021 might have initially looked like a Pakistani victory, it later became clear that Taliban control was bad news for Pakistan. As a reputable source wrote in May 2024:

> Relations between Afghanistan's Islamist rulers and Pakistan have been tense since the Taliban returned to power in 2021. Islamabad blames the Taliban for sheltering

[3] The approach has been described by different names, including 'double game', 'double dealing', or 'double cross'. However, I will employ 'double game' as a preferred phrase for the purpose of discussion here. See Frédéric Grare, 'Pakistan's Double Dealing', *Foreign Policy*, 1 March 2006; 'BBC Documentary Reveals New Evidence of Pakistan Double Cross', *BBC Media Centre*, 26 October 2011; Sahar Khan, 'Double Game: Why Pakistan Supports Militants and Resists U.S. Pressure to Stop', *Cato Institute*, 20 September 2018.

> the Tehrik-e Taliban Pakistan (TPP), a longtime ideological and organizational ally of the Taliban.[4]

That should have come as no surprise to the Pakistani military, the primary sponsor of the Taliban, because there always existed deep-seated disagreements between the militant group and the Pakistani state. It was natural that an emboldened Taliban were not going to be a Pakistani puppet. It became clear after the Taliban takeover that the Pakistani military needed more favours from the Taliban than the other way around, given the pressure the group was able to place on the Pakistani military in the north-west by working with the anti-Pakistan militants in the area grouped under the umbrella of Tehrik-e-Taliban Pakistan (TTP). As one well-placed observed noted to me, confrontation with Pakistan suited the Taliban because they had their own battle of legitimacy vis-à-vis the Islamic State group inside Afghanistan. A tense relationship with Pakistan enabled them to claim to the Afghan people that they were not a puppet of Pakistan despite the fact that the Islamic State-Khorasan (ISK) had portrayed them that way in the past.[5]

Pakistan's persistent support of the Taliban not only kept the war going far longer than it should have, it also left the country isolated to deal with the problem, since it had burnt its bridges with the international community. At the beginning of the War on Terrorism, the Pakistani double game once again represented a policy that served the interests of a single institution—the military—at the expense of the broader interests of the state and society. As we see elsewhere, the military benefited from vast amounts of American financial assistance, whereas the adverse consequences of the policy were being faced by the entire nation by the time of writing in 2026.

4.2 Why play a 'double game'?

The strategy of supporting militants in Afghanistan and Kashmir has had deep roots in Pakistani history.[6] Their first such use by the state dates to the first Kashmir War of 1947–1948—barely months after the independence of Pakistan. This approach continued throughout the 1960s and 1970s and received a boost during the dictatorship of General Zia-ul-Haq, who further

[4] 'Intense Border Clashes Between Taliban, Pakistan Cause Deaths, Destruction', *Radio Free Europe/Radio Liberty*, 17 May 2024.

[5] Interview, Politician 1, Islamabad, Pakistan, January 2024.

[6] S. Paul Kapur, *Jihad as Grand Strategy: Islamist Militancy, National Security, and the Pakistani State* (Oxford: Oxford University Press, 2017); S. Paul Kapur and Sumit Ganguly, 'The Jihad Paradox: Pakistan and Islamist Militancy in South Asia', *International Security* 37, no. 1 (2012): pp. 111–141.

enhanced the state support of such groups at the behest of the United States, Saudi Arabia, and China as part of the Afghan Jihad of the 1980s.[7]

Islam played a central role in the creation of Pakistan and in keeping the state together after it came into being. Using all the means at its disposal, the state systematically spread a number of fantasies centred on Pakistan playing a major role in creating an Islamic caliphate.[8] The ruling elites did not shy away from weaponizing religion for their interest of ‘keeping the pot simmering’ on Pakistan’s borders by using Islamist militant proxies, which subsequently proved instrumental in creating permanent ontological anxiety among the masses throughout the country’s history.[9] The masses were told for decades about the ‘Indian threat’ and how the Muslim state was perpetually a victim of Hindu Indian aggression, boiling over into outright armed conflict on several occasions.[10] By the late 1980s and 1990s, the manufactured identity of the victim encouraged the populace to provide financial support to jihadi groups; the Pakistani public often donated generously to support jihad in Indian-administered Kashmir and elsewhere.[11] This public support for militant groups was instrumental in helping Pakistan sustain a jihad against India for decades despite severe international criticism, reprimand, and even isolation. The entire phenomenon made ideal conditions for the creation of a national security state and the militarization of society at home,[12] which relied on a ‘strong’ military to defend the nation’s frontiers. Importantly, that policy was critical in tilting the civil–military balance in the country in favour of the military. The primary actor in charge of national security affairs, the Pakistani military and its main intelligence agency, the Inter-Services Intelligence (ISI), strongly promoted the idea of the jihads in two theatres (India and Afghanistan), while denying it internationally. The institutional structures of these organizations were primed for that purpose for decades by subsequent army chiefs.[13]

The decision to pursue a double game from 2001 was closely linked to the coercive circumstances Pakistan faced after the 11 September attacks, when then U.S. Deputy Secretary of State Richard Armitage reportedly warned

[7] Olivier Roy, ‘The Taliban: A Strategic Tool for Pakistan’, in *Pakistan: Nationalism Without a Nation*, edited by Christophe Jaffrelot (London: Zed, 2002), pp. 149–160.

[8] Aparna Pande, *Explaining Pakistan’s Foreign Policy: Escaping India* (Abingdon: Routledge, 2011), p. 3.

[9] Interview, Academic 3, Lahore, Pakistan, January 2024.

[10] M. A. H. Ispahani, ‘The Alternative to the India-Pakistan Arms Race’, *Pakistan Horizon* 2, no. 2 (1967): pp. 109–119.

[11] Seth G. Jones, ‘The Rise of Afghanistan’s Insurgency: State Failure and Jihad’, *International Security* 32, no. 4 (2008): p. 14.

[12] Interview, Journalist 3, Lahore, Pakistan, January 2024.

[13] See Saroj Kumar Rath, ‘Pakistan’s Double Game: Supporting and Opposing Terrorism’, *World Affairs* 15, no. 2 (2011): p. 92.

that Pakistan must support the American war or risk being bombed 'back to the Stone Age'.[14] As Duncan Verwey asserts, in the aftermath of the twin-towers attacks, the 'Western narratives became clouded with emotionally backed binaries of "good versus evil", leading to military interventions' in which the states in the region did not have much choice but to side—at least superficially—with the interveners.[15]

The aftermath of the American invasion of Afghanistan in 2001 helps explain Pakistani concerns about security in its north-western region for several reasons.[16] First, even though the United States promised Pakistan that the Northern Alliance would not take over Kabul after the overthrow of the Taliban, the US could not prevent that outcome given its light-footprint invasion and its reliance on the Northern Alliance in the war.[17] Second, the abrupt shift of American focus onto Iraq in early 2002 also signalled to Pakistan that the US commitment to the country was short term and, given the Northern Alliance's control over Kabul, India would establish close relations with them due to the historical closeness of these two parties.[18] These developments may explain why the country did not sever its ties with the Afghan Taliban in 2001. However, as we will see below, there were many opportunities and reasons over the years for Pakistan to abandon that policy, yet its military chose not to do so.

4.3 How did Pakistan play the 'double game'?

Having placed the strategy of the double game in context, I will provide a brief overview of Pakistan's approach to this foreign policy tool and its implementation. The US invaded Afghanistan on 7 October 2001 and as the invading forces, with the support of the Northern Alliance, homed in on Kabul, several Taliban militants and commanders retreated into the mountains of Kunduz. They were accompanied by operatives of Pakistani intelligence agencies, who facilitated their moves. The Pakistani military then airlifted hundreds of these elite Taliban commanders and fighters (along with their own personnel) from Kunduz to their new, temporary, homes in Pakistan, in what became known

[14] Scott Conroy, 'Pakistanis React to "Stone Age" Claim', *CBS News*, 23 September 2006.

[15] Duncan Verwey, *How Do Orientalist Narratives Inhibit Western Understanding of Terrorism in Pakistan?* Unpublished Dissertation, Masters in International Security, University of Bath, United Kingdom, September 2019, p. 11; Verwey largely draws his arguments on Adam Curtis, *Bitter Lake* (London: BBC, 2015), available at https://www.bbc.co.uk/programmes/p02gyz6b

[16] Christine Fair, 'The US-Pakistan Relations After a Decade of the War on Terror', *Contemporary South Asia* 20, no. 2 (2012): pp. 243–253.

[17] Ibid., p. 248.

[18] Seth Jones, 'Pakistan's Dangerous Game', *Survival* 49, no. 1 (2007): pp. 17–18.

as the notorious 'Kunduz airlift'.[19] The airlift was not a secret as the Indian, American, and Northern Alliance sources came to know about it 'within minutes'.[20] For Pakistan, this airlift was one of the various ways in which it helped the Taliban in the next twenty years.

Following on from the apparent precipitous downfall of the Taliban government, Kabul was captured in 2001, and Afghanistan was apparently ready for a fresh start. The George W. Bush administration, however, soon shifted focus onto Iraq, thinking that the Northern Alliance–led state, with the help of America's new ally, Pakistan, would be in a good position to shape the contours of the future Afghanistan.[21] In the meantime, Pakistan's military government was grooming the Taliban leadership in Pakistan, which was set to start a long-running (and ultimately successful) insurgency in Afghanistan from their bases in the adjacent tribal areas of Pakistan where they had taken temporary refuge as mentioned above.[22] By 2003, they had started their attacks, and the insurgency was in full swing by 2005—taking a severe toll on the international forces, the Afghan forces, and civilians in the country.[23] At the same time, the Pakistani state was in receipt of generous American military and non-military assistance under various heads for its contribution in the War on Terrorism. To give an indication of the scale of this support, the security aid alone amounted to $1.6 billion during 2003.[24] By November 2010, the *New York Times* claimed that the US had provided Pakistan with $18 billion in aid—a significant sum, compared with past American aid to Pakistan.[25] Meanwhile, Pakistani backing for the Taliban covered a wide spectrum of political, moral, and military support,[26] ranging from intelligence and resource sharing to know-how in bomb-making, logistical support, and provision of sanctuaries in the tribal areas of Pakistan and beyond.[27] Pakistan

[19] Seymour Hersh, 'The Getaway: Questions Surround a Secret Pakistani Airlift', *The New Yorker*, 28 January 2002.

[20] Ibid.

[21] Barnett Rubin and Ahmed Rashid, 'From Great Game to Grand Bargain: Ending Chaos in Afghanistan and Pakistan', *Foreign Affairs* 87, no. 6 (2008): pp. 30–44.

[22] Ahmed Rashid, 'Musharraf's Double Game Unravels', *The New York Times*, 10 August 2005.

[23] Muhammad Akbar Notezai, 'The Dilemma of Pakistan's Foreign Policy', *The Diplomat*, 12 August 2016.

[24] Krishnadev Calamur, 'Pakistan's Week Keeps Getting Worse', *The Atlantic*, 4 January 2018.

[25] Nicolas Kristof, 'A Girl, A School and Hope', *The New York Times*, 10 November 2010; see also S. Akbar Zaidi, 'Who Benefits from Aid to Pakistan', *Economic and Political Weekly* 46, no. 32 (2011): pp. 103–109.

[26] Arwin Rahi, 'Do the Afghan Taliban and Pakistan Get on as Well as Some Think?' *Dawn*, 27 April 2019.

[27] Sandy Gall, *War Against the Taliban: Why It All Went Wrong in Afghanistan* (London: Bloomsbury, 2012), p. 315; for a damning critique of Pakistan's support of the Taliban after the start of the War on Terrorism, see Matt Waldman, 'The Sun in the Sky: The Relationship Between Pakistan's ISI and Afghan Insurgents', *Carr Center for Human Rights Policy*, Kennedy School of Government, Harvard University, June 2010.

also diverted a significant chunk of US aid to enhance its military capability against India.[28]

Fissures soon developed between Pakistan and the coalition of countries present in Afghanistan as a result of Pakistan's duplicity. In 2007, one of the US Generals in Afghanistan, Karl Eikenberry, told the US Congress that America could not win in Afghanistan as long as the Taliban enjoyed safe haven in Pakistan.[29] As will be discussed in Chapter 5 in greater detail, relations between the US and Pakistan worsened more significantly by 2011, when some US generals started more openly describing the Haqqani network, one of the Taliban groups, as 'a veritable arm' of Pakistan's ISI.[30] Unlike before, there were harsh criticisms of Pakistan's uses of 'violent extremism as an instrument of policy'.[31] The dubious Pakistani commitment to the War on Terrorism also led the United States to increase its use of drones in the tribal areas of Pakistan to target militants that the Pakistani state might be unable or unwilling to apprehend.[32] Where there were only forty-eight strikes between 2004 and 2008, there were fifty-two strikes alone in 2009 and 122 in 2010.[33] Tensions between Pakistan and the United States regarding Pakistan's support for the Taliban were among the reasons why then President Barack Obama chose not to visit Pakistan during his tenure.[34]

Relations between the two countries hit a new low when the new occupant of the White House in 2017, Donald Trump, expressed unease that the Pakistani state was not willing to relinquish its support for militants in Afghanistan, despite significant US assistance and promises of future involvement in Afghanistan. That led President Trump to tweet on 1 January 2018 that '[t]he United States has foolishly given Pakistan more than 33 billion dollars in aid over the last 15 years, and they have given us nothing but lies & deceit, thinking of our leaders as fools. They give safe haven to the terrorists

[28] Rajeev Sharma, 'US Must Wise Up on Pak Aid', *The Diplomat*, 27 April 2010. See also 'Musharraf Admits US Aid Diverted', *BBC News*, 14 September 2009.

[29] See Tricia Bacon, 'Slipping the Leash? Pakistan's Relationship with the Afghan Taliban', *Survival* 60, no. 5 (2018): p. 167.

[30] Jeff Smith, 'US Takes Gloves Off with Pakistan', *The Diplomat*, 25 September 2011.

[31] Ibid.

[32] Brian Glynn Williams, 'The CIA's Covert Predator Drone War in Pakistan, 2004–2010: The History of an Assassination Campaign', *Studies in Conflict and Terrorism* 33, no. 10 (2010): pp. 871–892; Wali Aslam, 'A Critical Evaluation of American Drone Strikes in Pakistan: Legality, Legitimacy and Prudence', *Critical Studies on Terrorism* 4, no. 3 (2011): pp. 313–329; Wali Aslam, *The United States and Great Power Responsibility in International Society: Drones, Invasion and Rendition* (London: Routledge, 2013).

[33] See 'Drone Strikes: Pakistan', New America Foundation database, available at https://www.newamerica.org/in-depth/americas-counterterrorism-wars/pakistan/

[34] See Nosheen Iqbal, 'Barack Obama and His Wild, Drug-Taking Roomie', *The Guardian*, 3 October 2012; Mohammad Shoaib Adil, 'Why Obama Won't Visit Pakistan', *Foreign Policy*, 10 April 2015; Obama visited India and Afghanistan twice in his two terms in office.

we hunt in Afghanistan, with little help. No more!' Trump reiterated similar sentiments in November 2018 when he tweeted '[o]f course we should have captured Osama Bin Laden long before we did. I pointed him out in my book just BEFORE the attack on the World Trade Center. President Clinton famously missed his shot. We paid Pakistan Billions of Dollars & they never told us he was living there. Fools!' The President of the United States went on to say '[w]e no longer pay Pakistan the $Billions because they would take our money and do nothing for us, Bin Laden being a prime example, Afghanistan being another. They were just one of many countries that take from the United States without giving anything in return. That's ENDING!'

4.4 The consequences of the 'double game'

Pakistan's perseverance in supporting the Taliban remained notable despite international efforts to dissuade it from this foreign policy strategy. The most important consequence of the double game was the strengthening of the Taliban as a militant group. While they were once thought to have been weakened permanently in 2001, they emerged as a viable opponent of the Afghan government from the mid-2000s onward with help from Pakistan. This re-emergent strength forced the United States and its allies to open avenues of negotiations with Taliban agents as early as 2010.[35]

These negotiations, however, faltered from the beginning owing to several factors, primarily the Taliban's refusal to engage with the Afghan government at the time.[36] Driven by President Trump's commitment to ending 'endless wars',[37] Washington publicly stated its desire to withdraw from Afghanistan by 2020, prompting the President to approve a February 2020 ceasefire with the Taliban, contingent on a reduction in violence.[38] The US signed a peace agreement with the Taliban in the Qatari capital, Doha, on 29 February 2020 with the intention of beginning to withdraw military personnel from Afghanistan, which was then to open 'the way for the Taliban and the Afghan government to negotiate over the future of the nation'.[39] The agreement was an American 'derecognition' of the Ashraf Ghani government and a *de facto* recognition of

[35] Mark Landler and Alissa Rubin, 'War Plan for Karzai: Reach Out to Taliban', *The New York Times*, 28 January 2010.
[36] 'Taliban Refuse to Take Part in Afghanistan Peace Talks', *The Guardian*, 5 March 2016.
[37] Patricia Zengerle, 'With Eye on Afghanistan Talks, Trump Vows to Stop "Endless Wars"', *Reuters*, 6 February 2019.
[38] 'Donald Trump "Approves Conditional Peace Deal with Taliban"', *The Telegraph*, 12 February 2020.
[39] Ibid.

the Taliban state.[40] The Taliban did not have to work hard to take over most of Afghanistan after the eventual American withdrawal the following year.

Although Pakistan played a major role in enabling the eventual Taliban return in Afghanistan, its support for the group brought serious negative consequences. Friction soon developed between Taliban-led Afghanistan and Pakistan, resulting in serious violence on both sides of the border by early 2022.[41] The violence was orchestrated by the TTP, which had conducted several attacks inside Pakistan in the first two decades of the century, often with the support of the Afghan Taliban.[42] It has been estimated that the group caused the deaths of around 80,000 Pakistani citizens between 2003 and 2015.[43] Several commentators have asserted that the terrorist attacks by the Pakistani Taliban within Pakistan, and Pakistani support for the Afghan Taliban were clearly linked.[44] Pakistan also had to face the serious economic costs of terrorism at home, as well as the curtailment of foreign aid.[45] Though the US drone strikes within Pakistan were tacitly supported (or at least condoned) by the Pakistani state,[46] American drones operating in Pakistani skies obviously undermined Pakistan's own sovereignty and territorial independence.[47]

One of the major consequences of Pakistan's duplicity was to lead the Trump administration—notably working closely with the Indian government[48]—to lobby successfully the Financial Action Task Force (FATF) to place the country again on its grey list in June 2018. FATF is an intergovernmental body established in 1989 that exists to ensure an effective implementation of a variety of measures to combat money-laundering and terrorist financing.[49] FATF uses a 'black list' for countries which it considers to have failed to implement measures against money laundering or terrorist

[40] Interview, Politician 1.

[41] Abid Hussain, 'Tensions High After Pakistan Launches Cross-Border Attacks into Afghanistan', *Al Jazeera*, 18 March 2024.

[42] Jane Perlez, 'Onetime Taliban Handler Dies in Their Hand', *The New York Times*, 24 January 2011.

[43] Anwar Iqbal, 'Terror War Killed 80,000 During 2005–13: Report', *Dawn*, 31 March 2015.

[44] Jane Perlez, 'Onetime Taliban Handler Dies in Their Hand'.

[45] Muhammad Zakaria, Wen Jun, and Haseeb Ahmed, 'Effects of Terrorism on Economic Growth in Pakistan', *Economic Research* 32, no. 1 (2019): pp. 1794–1812.

[46] Wali Aslam, 'Great-Power Responsibility, Side-Effect Harms and American Drone Strikes in Pakistan', *Journal of Military Ethics* 15, no. 2, (2016): pp. 143–162.

[47] Wali Aslam, 'Drone Opposition in Pakistan and the Issues of Post-Colonial Identity', *E-International Relations*, 26 February 2014; Wali Aslam, 'Drones and the Issue of Continuity in America's Pakistan Policy Under Obama', in *Obama's Foreign Policy: Ending the War on Terror*, edited by Michelle Bentley and Jack Holland (Abingdon: Routledge, 2013), pp. 139–161. On the societal consequences of the use of drones in Pakistan see Wali Aslam, *Terrorist Relocation and the Societal Consequences of US Drone Strikes in Pakistan* (London: Remote Control Project, 2018), available at https://researchportal.bath.ac.uk/en/publications/terrorist-relocation-and-the-societal-consequences-of-us-drone-st/

[48] Khan, 'Double Game: Why Pakistan Supports Militants and Resists U.S. Pressure to Stop'.

[49] See 'Who We Are', Financial Action Task Force, available at https://www.fatf-gafi.org/about/

financing. The 'grey list', on the other hand, is used as a warning to the countries with weak controls which might slip on to the 'black list', alerting them to adopt requisite measures to curb such activities and avoid this outcome. A country being blacklisted has obvious ramifications for foreign direct investment into that state's economy. As I will discuss in greater detail in Chapter 5, Pakistan was only removed from the grey list in October 2022.

4.5 A critical assessment of the 'double game'

The discussion so far has looked at the context in which the policy of the double game was developed, the way it took shape, and some of the key consequences of the approach. This section conducts a critical evaluation of the approach. It is divided into two parts: an analysis of the threat perceptions that drove the adoption of this approach and an examination of the measures implemented to address the perceived threat of insecurity.

Pakistan's enduring focus on shaping Afghanistan's political dynamics has persisted for decades. Pakistani interference in the country since 1973 was justified on the premise that it would secure Pakistan's border in the north-west while the country faced an 'existential' threat from India in the east.[50] The Indian leadership's reluctance to discuss disputed issues with Pakistan, its military exercises on the India–Pakistan border (such as Operation Brasstacks of 1986/7), and its nuclear programme further helped the Pakistani military to make the case for a threat from India. The military asserted that, if India assumed dominance in Afghanistan, it could squeeze Pakistan from two sides.[51] The feeling of insecurity helped the elites to lay the foundations of an anxious state. Though the Indian side was not without its flaws, the reality is that every war between India and Pakistan was started by the latter. This also includes the 1971 war, in which Pakistan's Operation Chengiz Khan, the pre-emptive airstrike on Indian targets, provided India with a pretext to intervene, directly contributing to East Pakistan's independence as Bangladesh.[52] A glance at the history of tensions between the two states could easily show that, on balance, Pakistan's posture has been more offensive in nature, while the Indian posture has been mostly reactive and, often, poorly crafted.[53]

[50] Pande, *Explaining Pakistan's Foreign Policy: Escaping India*, pp. 28–87.
[51] Khalid Iqbal, 'Fallacy of Two Front Squeeze', *Defence Journal* 20, no. 10 (2017): pp. 70–72.
[52] Saleem Samad, 'What Led to the War', *Dhaka Tribune*, 16 December 2019.
[53] T. V. Paul, *The Warrior State: Pakistan in the Contemporary World* (Oxford: Oxford University Press, 2014), p. 126.

India had several opportunities to attack Pakistan or severely weaken it throughout the two countries' chequered history, but it has continually chosen not to do so. For example, when Pakistan's 90,000 citizens (the majority of whom were the country's soldiers) were stranded in East Pakistan after the quagmire in 1971, India could have easily launched an attack on Punjab or Sindh, in an under-defended West Pakistan, but it passed up the opportunity. Similarly, when Pakistani forces were stuck on a mountaintop in Kargil in 1999, India could have refused to compromise given Pakistani aggression, but it acquiesced in the face of international, primarily American, requests. Indira Gandhi once said that India was 'not remotely desirous of territory'.[54] She found the Pakistani obsession to base the whole survival of the country on the basis of hostility to India 'irritating'.[55] In my off-the-record conversations with Pakistani security officials, they conceded that India might no longer harbour ambitions to win territory from Pakistan, but that it wanted to 'humiliate' Pakistan—an eventuality for which the country needed to be prepared.[56] However, history reveals that it is actually Pakistan that has adopted an assertive policy towards India rather than the other way around.

While promoting an anti-India worldview has not served the interests of Pakistan or its people, neither was Pakistan's sole reliance on the Afghan Taliban, rather than engaging with diverse actors, a balanced foreign policy approach.[57] The Taliban were a known entity to the Pakistani army, and the patterns of their relationship with the latter when the group was last in power (from 1996 to 2001) should have served as a guide for Pakistan. These patterns could have helped anticipate how relations between the two sides would proceed if the militant group took over Afghanistan again. Contrary to popular assumptions, Pakistan's relations with the Taliban were marked by complications from the outset, despite official rhetoric in 1994, when the then interior minister, General Naseerullah Babar, described the Taliban as 'our boys'.[58] These complications characterized their relations before the Taliban takeover of Kabul, as well as after the establishment of the Islamic Emirate of Afghanistan in late 1996.[59] In the run-up to the establishment of the Emirate, the Taliban refused to follow Pakistani requests

[54] Memorandum of Conversation Between Henry Kissinger and Indira Gandhi, 7 July 1971, Foreign Relations of the United States (1969–1976), Volume XI: 407–409.

[55] Husain Haqqani, *Magnificent Delusions: Pakistan, the United States and Epic History of Misunderstanding* (New York: PublicAffairs, 2013), p. 163.

[56] Author interviews with Pakistani security officials; locations and dates withheld.

[57] Interview, Academic 2.

[58] Jane Perlez, 'Onetime Taliban Handler Dies in Their Hand'.

[59] S. Yaqub Ibrahimi, 'The Taliban's Islamic Emirate of Afghanistan (1996–2001): "War-Making and State-Making" as an Insurgency Strategy', *Small Wars and Insurgencies* 28, no. 6 (2017): pp. 947–972.

to work with other warlords on their march north from Kandahar, the Taliban's birthplace.[60] This refusal also annoyed the warlords (Gulbuddin Hekmatyar and Abdur Rashid Dostum) whom Pakistan urged the Taliban to embrace. To the great Pakistani consternation, soon after taking over Afghanistan for the second time in 2021, the Afghan Taliban refused to respect the Durand Line as an international boundary between the two countries.[61] However, that should not have come as a surprise for Pakistan. After the first Taliban takeover of Afghanistan in 1996, Pakistan tried three times to persuade the country's new government to accept the Durand Line as an international border, but the latter refused every time.[62] The uncertainties concerning the Durand Line and the feared rise of Pashtunistan was one of the primary reasons for Pakistan's involvement in Afghanistan since 1973, but the Taliban did not show much sympathy to Pakistan on that front.[63]

The Pakistani military always looked towards Islamist Pashtuns within Pakistan to counter more secular Pashtuns. The logic behind such an approach was that Islamic Pashtuns would be more sympathetic to the cause of the 'idea' of Pakistan—a homeland of Muslims. On the other hand, in the eyes of Pakistani decision makers, secular Pashtuns would be a centrifugal force for the territorial integrity of Pakistan as they might want union with their ethnic brethren across the Afghanistan border, and their empowerment might lead to the Pashtun territories of Pakistan separating from the country. That view was based on a flawed assumption, as no secular Pashtun political party nor any civil society movement in Pakistan (from the National Awami Party of the 1970s to the Pashtun Tahafuz Movement of today) has expressed irredentist desires.[64]

The Taliban, on the other hand, had defied that logic nearly three decades ago and continued to do so after the takeover of Afghanistan in 2021. While showing allegiance to the cause of Islam, they also leaned strongly towards Pashtun nationalism, and for them to compromise on the Durand Line after coming to power in Afghanistan would, in their eyes, have been a compromise of Pashtun honour—something they could not even consider.

[60] Ahmed Rashid, *Taliban: Militant Islam, Oil and Fundamentalism in Central Asia* (London: I.B. Tauris, 2001), pp. 31–44.

[61] Kashif Imran, 'Pakistan Says Taliban Deputy FM's Rejection of Durand Line "Fanciful", Can't Change Facts', *Arab News*, 17 February 2024.

[62] 'Pakistan Thrice Tried to Formalize Durand Line: Taliban Official', *The Nation*, 29 February 2016.

[63] Patrick Seale, 'The US is Losing Pakistan', *The Diplomat*, 4 June 2011.

[64] Interview, Politician 1; Wali Aslam and Alex Neads, 'Renegotiating Societal–Military Relations in Pakistan: The Case of the Pashtun Tahafuz Movement', *Democratization* 28, no. 2 (2021): pp. 265–284.

Following the Taliban's 1996 capture of Kabul, Naseerullah Babar,[65] a retired Pakistani general and key architect of Pakistan's Taliban policy, visited Mullah Omar, the Taliban leader, in Kabul and suggested over a meal that if the latter considered recognizing the Durand Line 'all problems would be solved'.[66] Mullah Omar took serious offence at this suggestion, which he thought went against the 'Pashtun code of honour', causing him to ask Babar to stop eating and to leave at once.[67] He also described Babar, an ethnic Pashtun, as a 'traitor'. Once they consolidated their position in Afghanistan at the turn of the century, the Taliban started fostering 'Pashtun nationalism, albeit of an Islamic character, and their actions began to affect Pakistani Pashtuns'.[68] Given that background, it was no surprise that the group followed the same path again, but with greater vigour the second time.

Various notable observers had sounded warning notes four years before the 2021 Taliban takeover in Afghanistan. For example, Mohammed Ayoob wrote in 2019 that what gives 'Taliban's appeal potency is its ability to couch in religious terminology traditional Pashtun aspirations for dominance in Afghanistan as well as the tribes' aversion to foreign interference in their land'. That is because 'both these factors have been constants in Afghan politics going back at least to the nineteenth century ... [t]he resurgent Taliban are driven only partly by religion. They are motivated equally, if not more, by the search for Pashtun dignity and revenge'.[69]

During their brief control over Afghanistan, the Taliban were a source of embarrassment for Pakistan due to their abysmal human rights record and for demolishing the famous Buddha statues in Bamian. Pakistan made high-level entreaties to the Taliban regime in March 2001 not to demolish these statues, but they paid no heed. The episode led to Pakistan attracting international rebuke at the time for continuing to support the regime.[70]

The Taliban were actively participating in the drugs and goods smuggling trade from and to Afghanistan since their first takeover of the country, and the region affected by that smuggling spanned from the Persian Gulf to Central Asia. By 2001, the Taliban-controlled smuggling market had become 'the biggest smuggling racket in the world and had enmeshed the Taliban with Pakistani smugglers, transporters, drug barons, bureaucrats, politicians

[65] Naseerullah Babar has also been described as the 'Taliban architect'. See Saba Imtiaz, '"Taliban Architect" Naseerullah Babar Passes Away', *The Express Tribune*, 11 January 2011.

[66] Rahi, 'Do the Afghan Taliban and Pakistan Get on as Well as Some Think?

[67] Asking someone to do so would be considered a major mark of disrespect in Pashtun culture.

[68] Rashid, *Taliban: Militant Islam, Oil and Fundamentalism in Central Asia*, p. 187.

[69] Mohammed Ayoob, 'The Taliban and the Changing Nature of Pashtun Nationalism', *The National Interest*, 10 January 2019.

[70] Zahid Hussain, *Frontline Pakistan: The Struggle with Militant Islam* (New York: Columbia University Press, 2007), p. 42.

and police and army officers'.[71] This undercover trade contributed to the strengthening of the Taliban's position in Afghanistan but cost Pakistan 30 billion rupees ($600 million) in lost revenues during the year 1997/8, up from 3.5 billion rupees ($80 million) in 1992/3.[72] Though the government of Pakistan made several attempts to rein the group in (described by one of the most renowned experts on the Taliban in the world, Ahmed Rashid, as a 'mafia'[73]), the Taliban used a number of tactics to avoid that control, such as playing Pakistan's federal government against its provincial governments.[74] The ISI was already in a weak position in the late 1990s because some of its own officers had gone 'native' in Afghanistan and were pushing for the Taliban cause with more urgency than the Taliban themselves. As one retired ISI officer said: 'these officers became more Taliban than the Taliban'.[75] When the ISI tried to pressure the Taliban, it faced counter-pressure not only from these officers but also from various lobbies that the Taliban had infiltrated within Pakistan, including radical mullahs and drug barons.[76] With the ISI losing control and Pakistan's Foreign Ministry completely sidelined in shaping Pakistan's Afghan policy, it was not clear who had control over the policy. Ironically, the Taliban were supposed to provide 'strategic depth' to Pakistan, but in practice, it was Pakistan that appeared to be providing the Taliban with strategic depth by enhancing their chances of survival.[77]

The Taliban also sheltered various sectarian groups in Afghanistan that were involved in sectarian violence on Pakistani streets in the 1990s. With the rise of the Taliban in Afghanistan, several local Mullahs also began to copy the Taliban ideology to try to impose shariah in Pakistan. For example, Sufi Muhammad started demanding the imposition of shariah in the Malakand division of Pakistan's north-west in 1995 with the support of 'thousands' of armed supporters.[78] He occupied government offices and refused to leave until the Pakistan army launched an operation to expel him. When the United States invaded Afghanistan in October 2001, Muhammad crossed the border into Afghanistan armed with 'Kalashnikovs, rocket launchers, missiles, anti-aircraft guns, hand-grenades and swords' to fight the international

[71] Rashid, *Taliban: Militant Islam, Oil and Fundamentalism in Central Asia*, p. 189.
[72] Ibid., p. 191.
[73] Ibid., p. 193.
[74] Ibid., p. 186.
[75] Ibid., p. 188.
[76] Ibid.
[77] Ibid., p. 187.
[78] Ashok Behuria, 'Fighting the Taliban: Pakistan at War with Itself', *Australian Journal of International Affairs* 61, no. 4 (2007): p. 533.

forces.[79] The Taliban's negative influence on Pakistan was immense, and the group started asserting their independence from their Pakistani handlers long before they took charge in Afghanistan in 1996.

The Taliban recalcitrance flew in the face of the fact that the Pakistani state bent over backward to facilitate their installation in Kabul. Ahmed Rashid recounts how from 1995 onward

> [General Naseerullah] Babar ordered Pakistan Telecommunications to set up a telephone network for the Taliban, which became part of the Pakistan telephone grid. Kandahar could be dialled from anywhere in Pakistan as a domestic call using the prefix 081—the same as Quetta's prefix. Engineers from the Public Works Department and the Water and Power Development Authority carried out road repairs and provided an electricity supply to Kandahar city.[80]

He continues, 'the paramilitary Frontier Corps, directly under the control of Babar, helped the Taliban set up an internal wireless network for their commanders. Pakistan International Airlines and the Civil Aviation Authority sent in technicians to repair Kandahar airport and the fighter jets and helicopters the Taliban had captured. Radio Pakistan provided technical support to Radio Afghanistan, now renamed Radio Shariat.'[81]

Rashid goes on to say that the Pakistani efforts were intensified after January 1996 and 'officials from Civil Aviation, Pakistan Telecom, PIA [Pakistan International Airlines], Pakistan Railways, Radio Pakistan and the National Bank of Pakistan .... ministries and government corporations took on further projects to help the Taliban with budgets that were supposedly for developing Pakistan's economy'.[82] Despite this immense help from Pakistan to facilitate their rise, the Taliban 'were nobody's puppets and ... resisted every attempt by Islamabad to pull their strings'.[83] As mentioned earlier, Pakistan had accepted major domestic costs and international isolation to support the Taliban. Even after taking over Kabul, the Taliban were reliant on Pakistan for their sustenance given their own international isolation. Despite this, they were not willing to accept the Pakistani demands which they considered to be contradictory to their own agendas.

[79] Ibid.; Importantly, it was Muhammad's son-in-law, Mullah Fazlullah, who led the Pakistani Taliban later, from 2013 until he was killed in a US drone strike in 2018. Fazlullah masterminded a series of attacks in Pakistan including the tragic massacre of schoolchildren in Peshawar that killed 150 people. See 'Pakistan Taliban Chief Mullah Fazlullah "killed in Drone Attack"', *Al Jazeera*, 15 June 2018.

[80] Rashid, *Taliban: Militant Islam*, p. 185.

[81] Ibid.

[82] Ibid.

[83] Ibid.

The Taliban ambassador to Pakistan at that time, Abdul Salam Zaeef, talks about the difficulties between the two in detail in his memoir.[84] Zaeef recounts that, at the beginning of 2001, Mullah Omar wrote a letter to the then Pakistani President Musharraf calling on him to 'implement Islamic law and to give Pakistan an Islamic government'.[85] Needless to say that that letter caused much consternation to Musharraf. In fact, Zaeef's memoirs are a long list of complaints against the Pakistani army and its intelligence agencies. For example, he criticizes the ISI for 'abusing and overruling the elected government' in Pakistan![86] He writes that 'since the start of the anti-Soviet *jihad*, the ISI extended its roots deep into Afghanistan like a cancer puts down roots in the human body; every ruler of Afghanistan complained about it, but none could get rid of it'.[87] Even the personality of President Musharraf (the man who initiated the double game to protect the Taliban after 2001) is a target who Zaeef criticizes for his 'devilish hostility towards the Emirate of Afghanistan'.[88]

Although Pakistan paid a very heavy price for supporting the Taliban in the 2000s and 2010s, the latter did not shy away from showing their independence whenever they could. For example, the Taliban leader Mullah Baradar, whom the Pakistani military believed to be close to it, sought to meet then-President Karzai in 2010 to initiate negotiations between the Taliban and the Afghan government. However, Baradar did so without the knowledge or permission of the Pakistani policy elites. The Pakistani military establishment was very unhappy with him for showing independence of will and for excluding Pakistan, leading them to arrest him.[89] He was released only in late 2018 at the Americans' request and on the promise that Pakistan would be kept in the loop during future negotiations between the Taliban and the United States.[90]

Pakistan has also had to face significant isolation for its use of double game as a foreign policy tool over the first two decades of the twenty-first century. As mentioned earlier, the country was once more placed on the FATF grey list in 2018 for its inability or unwillingness to place curbs on terrorist financing. The politics surrounding this 'greylisting' are quite revealing. China did not publicly support Pakistan at this critical juncture, declining to block or oppose its placement on the FATF grey list despite earlier indications that it might

[84] Abdul Salam Zaeef, *My Life with the Taliban* (London: Hurst, 2010).
[85] Ibid., p. 120.
[86] Ibid., p. 123.
[87] Ibid.
[88] Ibid., p. 121.
[89] Dexter Filkins, 'Pakistanis Tell of Motive in Taliban Leader's Arrest', *The New York Times*, 22 August 2010.
[90] Zahid Hussain, 'Ironies of the Afghan War', *Dawn*, 29 March 2019.

defend Pakistan. It was reported that Beijing decided to withdraw its support for Pakistan after New Delhi's reassurances of a greater Chinese role in FATF-related activities in the future.[91] The episode shows clearly that for China, keeping Pakistan on its side was not as a big of a concern as reaping the significant rewards of joining institutions such as FATF. The Saudis also initially opposed the idea of placing Pakistan on the FATF grey list. However, they retracted their opposition after Washington promised Riyadh full membership of the body if they followed the American lead.[92] While Pakistan sought support from China and Saudi Arabia after a rupture in its diplomatic relations with the US earlier in 2011, both countries demonstrated limits in their ability to assist Pakistan diplomatically or financially.[93]

The FATF episode also set an unfavourable precedent for Pakistan. The country worked hard to be removed from the grey list and took a variety of measures to show its efforts against money laundering and terror financing,[94] as would be expected given the toll that being placed on the list can take on the economy. Pakistan handed down a prison sentence to the leader of Lashkar-e-Tayyaba (a Pakistan-based militant group accused of the Mumbai attacks and other attacks in Kashmir),[95] and there was surprisingly little activity from the state-aligned militant groups after India's revocation of Article 370 in August 2019,[96] which had guaranteed special autonomous status to the Indian Kashmir within the Indian union until then.[97] Such absence of these groups from Pakistani streets was remarkable, as they would have been protesting all across the country at a different time. The pressures of FATF clearly tied the hands of the country's military establishment. Interestingly, Pakistan also played a 'huge role' from the beginning of 2019 in further facilitating the US–Taliban talks—a development that could not be understood without the backdrop of FATF.[98] These developments set a precedent that no amount of financial aid could coax Pakistan to change its behaviour, but the country responded well

[91] Pranab Samanta, 'Pakistan on FATF's Grey List: How India Convinced China & Pak Shot itself in the Foot', *The Print*, 23 February 2018.

[92] Anwar Iqbal, 'Pakistan Placed on FATF "Grey List" Despite Diplomatic Efforts to Avert Decision', *Dawn*, 28 June 2018.

[93] Lisa Curtis, 'The Reorientation of Pakistan's Foreign Policy Towards its Region', *Contemporary South Asia* 20, no. 2 (2012): pp. 256–262 see also 'US Must Grow Up on Pakistan', *The Diplomat*, 26 July 2010.

[94] 'FATF Meeting', *Dawn*, 18 February 2020.

[95] However, it is doubtful that the episode reflects a complete reversal of the Pakistani policy of supporting terrorists, as the country's military elites have often used the justice system in Pakistan for their own interests as and when required. See Christine Fair, *In Their Own Words: Understanding Lashkar-e-Tayyaba* (London: Hurst, 2018), p. 74.

[96] 'India Revokes Occupied Kashmir's Special Autonomy through Rushed Presidential Decree', *Dawn*, 5 August 2019.

[97] Umair Jamal, 'Can a Negative Decision at the FATF Bolster Hardliners in Pakistan?', *The Diplomat*, 16 October 2019.

[98] Umair Jamal, 'What the Apparent Collapse of Afghan Peace Talks Means for Pakistan', *The Diplomat*, 10 September 2019.

to coercive measures.[99] These lessons will not be ignored by those engaging with Pakistan in the future.

FATF was not the only platform where Pakistan had to face embarrassment. At a BRICS meeting in 2017 in the Chinese city of Xiamen, the leaders of Brazil, Russia, China, India, and South Africa issued a joint declaration condemning the Taliban as well as a variety of other groups—three of which were closely linked to Pakistan's military establishment. As commentators noted, the declaration was particularly embarrassing for Pakistan because 'both Pakistan's "best friend" China and its "worst enemy" India were signatories'.[100]

Pakistan's isolation from China in this case was remarkable given that the country had pinned all its hopes on the latter after a rupture in relations with the US in 2011. China was not happy with the instability in Afghanistan for two reasons. It was obviously concerned about Uighur militant presence in Afghanistan (affecting its own Uighur minority in Xinjiang province) and it wanted a stable Central Asia to implement its Belt and Road Initiative project. As former Pakistani Senator Afrasiab Khattak noted in 2017, 'China was not content with looking at Afghanistan or India from Pakistan's point of view anymore'.[101] Similarly, although there was some military cooperation between Russia and Pakistan, the former remained India's top defence supplier as Russia did not want to anger India to support an isolated Pakistan.[102] Russia also did not want militant Islam to trickle up from Afghanistan into any of the Central Asian republics, given the porous nature of the border between these republics and Russia itself. That context helps in understanding the Russian warning to Pakistan via the BRICS declaration.

That BRICS declaration led the Pakistani Foreign Minister at that time, Khawaja Asif, to say: 'we need to bring our house in order to prevent facing embarrassment on the international level'.[103] Ashraf Jahangir Qazi, the former Pakistani ambassador to the United States, India, and China, wrote in 2017 that the BRICS declaration showed that Pakistan's policies in Afghanistan had been counterproductive. He asserted that '[a]lienating [Ashraf Ghani-led] Kabul and relying on the Taliban to provide leverage [was] demonstrably stupid. Pakistan's policy ha[d] also been unnecessarily India-centric. This alienate[d] Afghan political opinion and ensure[d] that Pakistan [would] lose zero-sum games with India inside Afghanistan'. He continued his statement by saying that '[Ashraf Ghani–led government's] goodwill for India [would]

99 Richard Weitz, 'How China can Fix Pakistan', *The Diplomat*, 17 May 2011.
100 Daud Khattak, 'Will Pakistan Part Ways with its Proxies', *The Diplomat*, 13 September 2017.
101 Afrasiab Khattak, 'Change or Isolation?' *The Nation*, 9 September 2017.
102 Notezai, 'The Dilemma of Pakistan's Foreign Policy'.
103 Ibid.

not translate into ill will towards Pakistan unless [Pakistani] policies [were] seen by Afghans as forcing upon them an unwanted choice between India and Pakistan'. Qazi made a strong case for a 'paradigm shift' in Pakistan's approach that would require 'policy realism and imagination'.[104]

The Taliban of 2021 held a different status in the international community compared to the Taliban of 2001. They were no longer isolated and alone in the world, primarily reliant on Pakistan.[105] Iran and Russia had started courting them before the group's takeover of Kabul, as both these countries wanted Taliban support in future to quell the Islamic State movement in Afghanistan.[106] China also started building bridges with the Taliban before the US withdrawal in 2021. There were also significant overtures from India, resulting in the Taliban announcing that they did not consider India to be their enemy.[107] It was clear before their arrival into the corridors of power that the Taliban would be less sensitive to any Pakistani pressure than in the past, given their international recognition.

The Taliban overtures towards India in the lead-up to the Doha agreement which paved the way for the US withdrawal from the country should have been a significant warning sign for Pakistan. The group announced in 2019 that the Kashmir dispute and the war in Afghanistan were not linked, and they would like India and Pakistan to resolve their disputes in a peaceful manner. When India revoked Article 370 in Kashmir, the Taliban spokesperson Zabihullah Mujahid said that '[h]aving gained bitter experiences from war and conflict, we urge peace and use of rational pathways to solve regional issues'. Mujahid also stated that '[l]inking the issue of Kashmir with that of Afghanistan by some parties [would] not help because the issue [was] not related, neither should Afghanistan be turned into a theatre of competition between countries'.[108] This statement was significant for Pakistan given that the country had historically used the Afghan territory for training militants to conduct attacks in Kashmir.

The Taliban position also resonated with India, which also did not want the future Afghanistan to be used as a training ground for militants to conduct attacks in India.[109] Mujahid's statement calling for Afghanistan not to be embroiled in disputes over Kashmir came when a prominent Pakistani

[104] Ashraf Jahangir Qazi, 'Our Afghan Policy', *Dawn*, 9 September 2017.

[105] Daud Khattak, 'The Taliban's Diplomatic Reemergence', *The Diplomat*, 25 October 2019.

[106] Barnett Rubin, 'Is Afghanistan Ready for Peace', *Foreign Affairs*, 30 July 2018.

[107] Ahmed Rashid, 'Afghan Taliban Don't See India as Their Enemy', Talk delivered at Asia Society, New York, 27 March 2012, available at https://asiasociety.org/blog/asia/ahmed-rashid-afghan-taliban-dont-see-india-their-enemy

[108] 'Don't Link J&K Issue with Kabul: Taliban', *The Tribune*, 9 August 2019.

[109] Avinash Paliwal, 'The "India Question" in Afghanistan', *Lawfare*, 6 October 2019.

politician, Shahbaz Sharif, criticized the Taliban's imminent peace deal with the United States. Sharif had said: '[w]hat kind of a deal [was] this that the Afghans enjoy[ed] and celebrate[d] peace in Kabul, but in Kashmir, blood [was being] shed? No, this is not acceptable for us.'[110] Clearly, the Taliban did not agree with Sharif on this association, which was critical for Pakistan.

Zabihullah Mujahid also asserted that, 'unlike the Lashkar [Lashkar-e-Tayyaba] which is focused on Jammu and Kashmir, the Afghan Taliban concentrate[d] on Afghanistan. We have never taken part in any attack in India, nor do we attack anyone at Pakistan's behest.' Mujahid went on to say that

> [w]e are not saying that India should get out of Afghanistan. Nor can India be completely expelled from Afghanistan. The Taliban aren't in any direct conflict with India. Indian troops aren't part of NATO forces, they haven't occupied Afghanistan. India and Afghanistan have had historic ties. If the Taliban returns to power, we would like to maintain normal relations with all countries including India. It is possible for the Taliban and India to reconcile with each other.[111]

The Taliban spokesperson asserted that whenever the group attacked the Indian Embassy in Kabul, they took responsibility and presented solid reasons for it. He also denied that there was any presence of Lashkar-e-Tayyaba in Afghanistan.[112]

Similarly, Suhail Shaheen, a key member of the political wing of the Taliban, asserted in October 2019 that there was 'no question' that the Taliban would attack Indian targets after the US were to leave the country. He said that '[a]fter the end of occupation and formation of an all Afghan-inclusive Islamic government in the country, we [would] solely focus on reconstruction and development of our ruined country. For that, we want to have good relations with all countries of the world, including the regional countries and seek their assistance for a new Afghanistan.'[113]

Referring to the fears of the Taliban attacking Indian targets, the spokesperson reiterated that

> [i]t is not a genuine fear, it is not a reality. Why [should we] turn our fighters towards India when we need to reconstruct our country after its liberation? We need to have

[110] 'Kashmir: Taliban Urges India, Pakistan to Refrain from Violence, Use "Rational Pathways"', *The Wire*, 9 August 2019.

[111] Rahimullah Yusufzai, 'The Taliban and India can be Reconciled', *Outlook*, 15 February 2020.

[112] Manoj Kumar Mishra, 'Engaging Taliban and India's Afghan Strategy', *Asia Times*, 22 October 2019.

[113] Abhinandan Mishra, 'No Question of Taliban Fighters Targeting India', *The Guardian*, 5 October 2019; see also 'Afghan Taliban Dismisses Fear of Attacks on India After US Troop Withdrawal, Says Neighbourhood's Support Vital to Rebuild Country', *Firstpost*, 15 October 2019.

> [relations] with other countries to help us in reconstructing and developing our country. We do not have any policy of interference in any other countries; rather, we want to have good relations with every global partner. That's our policy.[114]

Ahmed Rashid also reiterated in 2012 that within the last few years, the Taliban moderated their rhetoric in such a way that their 'jihadi rhetoric ha[d] been replaced by nationalist rhetoric' and they were keen to assert that they were 'Afghans first and foremost'. That view, of course, did not please their 'patrons and allies' in Pakistan.[115]

Afghanistan is a resource-starved country, and it was clear that the Taliban would need valuable resources to prop up the economy in order to legitimize their rule. India's financial muscle and its investment in Afghanistan would be valuable in that regard. Therefore, it made sense for the Taliban to mend fences with India before they took charge of Afghanistan. Pakistan, on the other hand, is often described as 'the sick man of South Asia',[116] home to one of the region's slowest growing economies. It did not have the financial wherewithal to compete with India to offer the Taliban badly needed financial resources.

Whereas Pakistan has been resented by many common Afghans for supporting terrorism in their country,[117] India has been winning over the same people since 2002 by 'building roads and dams, laying power lines, training ministry officials and army officers, and providing free scholarships to Afghan university students'.[118] According to some estimates, India spent $2 billion in aid and infrastructural development in Afghanistan since 2001.[119] India was laying the foundations for a longer-term involvement in Afghanistan. As Ahmed Rashid said in 2016, sooner or later, Pakistan will have to hold talks with India regarding Afghanistan. He asserted that 'India will be in Afghanistan, whether we like it or not'.[120] It was not a surprise that India took active part in the peace talks held in Moscow in late 2018.[121]

The Indian realignment effort, before the Taliban's takeover of Kabul, was supported by the former Indian army chief Bipin Rawat, who asked for India

[114] Mishra, 'Engaging Taliban and India's Afghan Strategy'.

[115] Ahmed Rashid, 'Afghan Taliban Don't See India as Their Enemy'.

[116] Arif Rafiq, 'Pakistan's Roller Coaster 2012', *The Diplomat*, 8 January 2012.

[117] 'Anti-Pakistan Protests Break Out Across Afghanistan', *The Business Standard*, 8 April 2017.

[118] Jerry Meyerle, 'Afghanistan-Pakistan Border: Back to Politics as Usual?' *The Diplomat*, 29 November 2013.

[119] Harsh Pant and Avinash Paliwal, 'India's Afghan Dilemma is Tougher than Ever', *Foreign Policy*, 19 February 2019.

[120] Qasim Moini, 'Carrot-and-Stick Policy Needed with Afghan Taliban, Says Ahmed Rashid', *Dawn*, 12 March 2016.

[121] Aishwarya Kumar, 'India Should Hold Backchannel Talks with Taliban, Its Return to Power in Afghanistan is Inevitable', *News 18*, 16 January 2019.

to hold talks with the group 'without any precondition'.[122] He believed that India would be an 'indispensable development partner', regardless of the government in charge in Kabul. As a former Indian diplomat put it, after the Taliban assumed power, 'national interest and not Pakistan will dictate [the group's] policies. India-Afghan ties have generally always been very good based on the fact that India has no designs in Afghanistan, has always been provider of economic support and the fact that a neighbour's neighbour is normally a good friend.'[123] The most common aspect of their relations lies in the fact that neither India nor the Taliban recognize the Durand Line, showing a major convergence of interests among the two parties.[124]

While making overtures to the Taliban from 2018, India also made an emphatic case for minimizing Pakistan's role in the future Afghanistan by arguing that any effort to stabilize Afghanistan should be 'Afghan-led, Afghan-owned, and Afghan-controlled'.[125] India had relatively quickly adapted its approach, given that it had refused to support an initiative by President Karzai in 2007 to engage with the Taliban.[126] Though India is not geographically contiguous to Afghanistan, it had been using an Iranian port—developed by India—and road facilities to trade with the country. Relations between Iran and India remained resilient despite the US pressure on the latter to curtail those links.[127] As one informed observer noted to me that after their takeover of Kabul in 2021, the Afghan Taliban were relatively close to Russia, China, and India and they did not need Pakistan anymore. It was unsurprising that the group was no longer under Pakistan's control. By the same logic, one might ask why India would not consider cooperation with the Taliban to keep Pakistan's north-western border unsettled, much as Pakistan previously used militant groups to sustain pressure in Indian-administered Kashmir. Additionally, the Pakistani Taliban had previously assisted the Afghan Taliban; it was now payback time, and the latter were paying back well.[128] According to that observer, the turmoil on that Pakistan–Afghan border after the US withdrawal was a backfiring of the two decades of narrowly conceived Pakistani policy, which was prolonged far longer than its utility for the country.[129]

122 Ibid.
123 Ibid.
124 Ibid.
125 Zeenat Saberin and Shereena Qazi, 'Afghan Peace Conference: India Shares Table with Taliban', *Al Jazeera*, 9 November 2018.
126 Ibid.
127 Curtis, 'The Reorientation of Pakistan's Foreign Policy', p. 263.
128 Interview, Journalist 3.
129 Interview, Academic 2.

Throughout the first two decades of this century, numerous warnings indicated that most Afghans, including the Afghan Taliban, resented Pakistan's interference in their country, contributing to strained relations between the two states following the international withdrawal in 2021. While conducting research in Afghanistan, Matt Waldman learnt that, although Pakistan provided immense support to the Taliban in different forms, they showed very little gratitude to Pakistan for that support, and 'deeply resented the ISI pressure'. Someone close to the Taliban leadership told him that '[i]ndeed the only people they hated more than the Americans were the ISI'.[130]

Tricia Bacon conducted a detailed study of the extent to which the Taliban resented the Pakistanis breathing down their necks during their negotiations with the US. The urge to liberate themselves from that bind was one of the reasons for the group's decision to open an office in Qatar instead of Islamabad in 2013.[131] While reducing their reliance on Pakistan during the 2010s, the Taliban also developed substantial independent revenue streams to fund their insurgency against international forces, drawing on drugs, timber, and gemstone smuggling.[132] When the Pakistani state was pushing for the policy of supporting the Taliban in the face of serious international opposition, the country was deeply unpopular in Afghanistan. In fact, a 2016 survey showed that Pakistan was even less popular in Afghanistan than the Islamic State group. Where the support for Islamic State stood at 5.8 per cent, Pakistan's support was at a meagre 3.7 per cent.[133] Bacon wrote in 2018 that the 'Taliban's increasing independence, evident in both its rhetoric and its action, diminishe[d] the importance of Pakistan's support'. She went on to say that for 'Pakistan, the Taliban ha[d] proven to be unreliable and difficult to manage. For the Taliban, Pakistan [sought] too much control and use[d] coercive measures to obtain it.' The bottom line was that 'Pakistan [could] no longer compel the Taliban to take strategic actions that [ran] counter to its own wishes. In other words, Pakistan [could] act as a spoiler but probably not as a peacemaker.'[134]

Given the innumerable warning signs, it was not a surprise that the Afghan Taliban adopted a confrontational path with Pakistan after the 2021 takeover. It was more than ironic that a Pakistani minister blamed the Afghan Taliban in 2023 for playing a 'double game'[135] with Pakistan by giving refuge to the militants attacking his country. An interviewee remarked that it took the country seventy-five years to move beyond perceiving India as a threat, only

[130] Waldman, 'The Sun in the Sky', p. 9; Gall, *War Against the Taliban*, p. 309.
[131] Bacon, 'Slipping the Leash', p. 168.
[132] Ibid., p. 172.
[133] Ibid., p. 174.
[134] Ibid., pp. 174–175.
[135] Saleem Shahid, 'Kabul Playing "Double Game", Says Minister', *Dawn*, 3 December 2023.

to confront another flashpoint along the Afghan–Pakistan border, which is likely to remain tense for years to come.[136]

4.6 Conclusion

This chapter studies the emergence of the Pakistani foreign policy tool of double game in the context of the war in Afghanistan and eventual withdrawal of international forces in 2021. This approach cost Pakistan—as well as the international community engaged in Afghanistan—greatly in lives and financial losses. Pakistan faced international opprobrium and isolation due to its support of the Taliban, which was almost a complete replay of its isolation on the matter in the late 1990s. Instead of confronting the reasons for its isolation, the country accused the rest of the world of ignoring Pakistan's human and financial sacrifices for joining the West's War on Terrorism.[137] This showed that, rather disturbingly, Pakistan had not moved on from the pattern it had adopted in the face of international criticism in the late 1990s. For example, when the UN criticized Pakistani support for the Taliban back in 1998, the country's ambassador to the organization said that it was the UN that had 'gradually marginalised itself in Afghanistan and lost credibility as an impartial mediator'.[138] The ambassador also asserted that Pakistan was not isolated in Afghanistan; rather, the rest of the world was isolated from Pakistan and would sooner or later have to accept its position on the Taliban.[139] Contrary to the ambassador's assertion, Pakistan at the time faced not only broader international isolation but also pronounced regional estrangement, as India, Iran, Russia, China, and the Central Asian states criticized its policies, with some openly accusing Islamabad of destabilizing the region.[140]

Criticizing the complete lack of vision behind Pakistan's Taliban policy in the late 1990s, Olivier Roy wrote that 'the apparent victor, Pakistan, could pay dearly for its success'.[141] Similarly, Ahmed Rashid wrote in 2001 that the Taliban takeover of Kabul appeared to be a success for Pakistan, but, rather than being the master of the Taliban, Pakistan was the group's victim.[142] Pakistan was set to reap rather unpleasant rewards for its ill-thought-out approach of

[136] Interview, Columnist, Islamabad, Pakistan, January 2024.

[137] Islamuddin Sajid, 'Pakistan: PM Recalls Sacrifices Made in US War on Terror', *Anadolu Agency*, 19 November 2018.

[138] Ahmed Rashid, 'Pakistan Undermines UN in Peace Process', *The Nation*, 23 January 1998.

[139] Ibid., see also, Rashid, *Taliban: Militant Islam, Oil and Fundamentalism in Central Asia*, p. 189.

[140] Rashid, *Taliban: Militant Islam, Oil and Fundamentalism in Central Asia*, p. 188.

[141] Olivier Roy quoted in ibid., p. 187.

[142] Rashid, *Taliban: Militant Islam, Oil and Fundamentalism in Central Asia*, p. 18.

supporting the Afghan Taliban for over two decades after the latter's diplomatic 'reemergence'.[143] As Hillary Clinton said on a visit to Pakistan in 2011: 'you can't keep snakes in your backyard and expect them only to bite your neighbors'.[144]

The policy confusion of the Bush administration after the start of the War on Terrorism is also partly responsible for Pakistan's strong alliance with the Taliban. The United States committed a number of 'tactical, operational, and strategic follies in Afghanistan' which ranged from 'Donald Rumsfeld's rejection of talks (with the Taliban), the faulty Bonn Process, and, most importantly, Washington's inability to understand Afghanistan's ethnic composition'.[145] These missteps further isolated the Taliban, who were a formidable force when it came to stabilizing Afghanistan. After overthrowing the Taliban regime in 2001, the US started the Bonn process to discuss the future of post-war Afghanistan, but the Taliban were excluded from the discussions. Instead, the Northern Alliance were prioritized as the American partner. The Northern Alliance had a historically strong relationship with India, whereas Pakistan had close links with their opponents, the Taliban.[146] The neoconservatives in the Bush administration displayed a serious resistance to engaging those with 'oppositional ideologies', which made finding a longer-term political settlement for Afghanistan even harder to achieve.[147] It could, hence, be suggested that, given the uncertain situation and American strategic mistakes, it made sense for Pakistan to not break all its ties with the Taliban.

Although it might have been unrealistic to expect Pakistan to perform an 'about-turn' immediately upon joining the War on Terrorism,[148] the

[143] Khattak, 'The Taliban's Diplomatic Reemergence'; Leonardo Jacopo Maria Mazzucco and Kristian P. Alexander, 'Saudi Arabia and Qatar are Cooperating with the Taliban. But Their Approaches to Afghanistan are Different', *The Atlantic Council*, 12 January 2023.

[144] Khattak, 'Will Pakistan Part Ways with its Proxies'.

[145] I am grateful to one of the reviewers of the manuscript for this point.

[146] Ashish Kumar Sen, *Missteps and Missed Opportunities for Peace in Afghanistan*, United States Institute of Peace, 27 October 2022.

[147] Timo Kivimäki, 'When Ideologies Became Dangerous: An Analysis of the Transformation of the Relationship Between Security and Oppositional Ideologies in US Presidential Discourse', *Global Society* 37, no. 2 (2023): pp. 225–244.

[148] It is often said that Pakistan would have faced severe backlash for turning against its own proxies. However, that claim should be regarded with some scepticism. Pakistan managed to dramatically curtail the activities of several Pakistan-based militants active in Indian-occupied Kashmir (such as Lashkar-e-Tayyaba and Jamaat-ud-Dawa) when it was placed on the FATF grey list in 2018. It is highly unusual that there were no mass killings of Indian security forces in Kashmir or suicide bombings by the Pakistan-based militants after India's revocation of Article 370 of the Indian Constitution. That shows that the Pakistani state can curb the activities of its proxies more easily than some might think. For a slightly exaggerated nature of the expected backlash, see Michael Kugelman, 'The Haqqani Threat to the US-Pakistan Détente', *The Diplomat*, 31 July 2014; for the complete absence of the Kashmiri jihadis from the scene after the revocation of Article 370, see Saeed Shah, 'Pakistan's New Plight in Kashmir: What to Do About the Jihadists', *The Wall Street Journal*, 11 August 2019.

country nevertheless had twenty years to take that turn. Despite receiving extensive international support, diplomatic reassurances, and billions of dollars, Pakistan still failed to change direction. The country could have used its leverage over Afghanistan more imaginatively. That could have involved gradually shifting its stance and pressuring the Taliban to participate in a broad-based government, recognized and supported internationally. Its active assistance in devising an internationally-recognized solution for Afghanistan would also have exalted Pakistan's status in the international community and would have brought much-needed stability in Afghanistan.[149] Given its fears of India, Pakistan could have asked for guarantees against Indian dominance in Afghanistan. Pakistan would have been internationally supported in such a scenario in the event of anything going wrong. The country could also have used that opportunity to enter into longer-term arrangements of financial assistance with the international community as a guarantor of peace in the region, in the same way that Egypt has done with regard to Israel.[150] An analyst stated that Pakistan could have put pressure on the Taliban to work with the Ghani government and to enable the American-led process succeed. Ashraf Ghani went out of his way to appease Pakistan, but the latter's military remained unresponsive. At any point during the negotiations, Pakistan could have opted to renew its support for the Taliban if it believed that its bargaining position was weakening. Instead, the country undermined all peace processes and welcomed the Taliban's rise to power in 2021 over a cup of tea.[151] The very same military that had reportedly welcomed the Taliban in 2021 was drawn into hostilities with the group just two years later.[152] Its rulers chose a path which kept the war going far too long leading the country receiving generous financial assistance from the West – funds which only served a narrow set of elites.

The events of 11 September 2001 presented a foreign policy opportunity to Pakistan to strengthen its international standing in more ways than one. Pakistan had a chance to achieve several key objectives: to win over the international community to its stance vis-à-vis its neighbours, to put its own house in order—especially economically, to cut its links with terrorist groups, and to embark on a progressive path. As Zahid Hussain asserts, 'from being an international outcast for its longstanding support for the Taliban and militant

[149] Michael Kugelman, 'How Pakistan Can Help with the Afghan Peace Process', *Herald*, 21 January 2019.

[150] Interview, Academic 1, Lahore, Pakistan, January 2024.

[151] See 'We didn't Think about Consequences of "Kabul Tea": Bilawal Bhutto-Zardari', *The Times of India*, 16 January 2024.

[152] Interview, Academic 2.

cross-border insurgents in Kashmir, Pakistan became the key strategic partner of America's war on terror' overnight.[153] However, it squandered that opportunity in the following twenty years by playing its foreign policy cards poorly. Though most actors affected by this double game (including India, Iran, China, Russia, the Taliban—even the US, since it managed to extricate itself from the Afghan quagmire) appeared to have won, one actor that *appeared to be* a winner was in fact the main loser: Pakistan.

[153] Hussain, *Frontline Pakistan: The Struggle with Militant Islam*, p. vii.

5

From the coup to a reset with the United States (1999–2010)

5.1 Introduction

In 1999, Pakistan's then Chief of Army Staff, General Pervez Musharraf, removed the civilian Prime Minister Nawaz Sharif in a military coup. In his first address to the nation on 13 October 1999, Musharraf referred to a horrific scenario of 'turmoil and uncertainty that [the] country' was experiencing at that time as justification for the takeover. Using an anxiety-inducing discourse, he blamed the Sharif government for 'playing with' and systematically destroying all the institutions of the state and bringing the economy to a state of collapse through 'the self-serving policies' which rocked 'the very foundation of the Federation of Pakistan'.[1] Further adding to the charge sheet against Sharif, Musharraf asserted that the Prime Minister tried to 'interfere with the armed forces, the last remaining viable institution' in which all Pakistanis took 'so much pride' and looked up to, 'at all times, for the stability, unity and integrity of our beloved country'. He alleged that Sharif 'tried to politicize the army, destabilize it and tried to create dissension within its ranks'.[2]

Casting the military as a saviour to address the nation's anxiety, Musharraf assured the country that the armed forces had moved in 'as a last resort, to prevent any further destabilization' and that he had done so with 'all sincerity, loyalty and selfless devotion to the country with the armed forces firmly behind' him.[3] Supporting his takeover, Musharraf presented a beatific scenario, promising to put the country back on track when he stated:

> Dear brothers and sisters, your armed forces have never and shall never let you down, Inshallah [God willing]. We shall preserve the integrity and sovereignty of our

[1] See 'General Pervez Musharraf's Address to the Nation', 13 October 1999, available at https://presidentmusharraf.wordpress.com/2007/07/10/address-nation-13-oct-1999/

[2] Ibid.

[3] Ibid.

Pakistan's Foreign Policy. Wali Aslam, Oxford University Press. © Wali Aslam (2026).
DOI: 10.1093/9780198909552.003.0005

> country to the last drop of our blood. I request you all, to remain calm and support your armed forces in the re-establishment of order to pave the way for a prosperous future for Pakistan.[4]

Five days after the takeover, General Musharraf delivered his first policy speech, in which he once again addressed the horrific scenario the country faced. He specifically outlined the dire straits the country was in and how it stood in 'darkness'.[5] 'There is despondency, and hopelessness surrounding us with no light visible anywhere around', he charged. 'The slide down has been gradual but has rapidly accelerated in the last many years.' He asserted

> Quite clearly, what Pakistan has experienced in the recent years has been hardly a label of democracy not the essence of it. Our people were never emancipated from the yoke of despotism. I shall not allow the people to be taken back to the era of sham democracy, but to a true one. And I promise you I will Inshallah.[6]

He stated that the army had two binary choices:

> The choice before us on 12th October was between saving the body—that is the nation, at the cost of losing a limb—which is the Constitution, or saving the limb and losing the whole body. The Constitution is but a part of the nation therefore I chose to save the nation and yet took care not to sacrifice the Constitution.[7]

General Musharraf promised to strengthen the Federation and hold corrupt politicians accountable. Despite being the mastermind of the Kargil War of 1999, he offered to work with India to resolve their mutual disagreements and to lift the entire South Asian region out of poverty.[8] Indeed, one of his early foreign policy forays was to visit India for the famous Agra Summit of July 2001.

As part of the discussions at Agra, both sides came close to agreeing to Kashmir's autonomy within the Indian union, with Pakistan showing flexibility on its historical demand of 'self-determination' for the Kashmiris.[9] The movements of people and goods on both sides of the Line of Control (LOC) were to be made easier, while the wider issue of a permanent resolution of

[4] Ibid.

[5] 'President's Address to the Nation: 7-Point Agenda and Future Course', 17 October 1999, available at https://presidentmusharraf.wordpress.com/2007/07/10/address-nation-7-point-agenda/

[6] Ibid.

[7] Ibid.

[8] Ibid.

[9] Uday Rana, 'When Vajpayee and Musharraf "Almost Resolved" the Kashmir Dispute', *News18*, 13 February 2018.

the dispute was to be put on hold for the time being. However, the Summit failed to secure agreement from both sides on these points. Though some have pointed to Musharraf's 'lack of intent in containing terrorism'[10] as a reason for the failure of the talk, it is widely agreed that the hardline stance of the Hindu nationalist politicians in India, such as the then Indian Minister of Defence L. K. Advani, led to the failure of the talks.[11] Like every dictator, Musharraf needed legitimacy for both his takeover as well as continuing to stay in power, and a summit with India presented him with an opportunity to help with both of these objectives.[12] The Agra summit was also a venture for Musharraf to increase his own base of support, but most of the military senior command did not support Musharraf's plan.[13] As one commentator put it, 'the military was not interested in peace with India and it has always undermined genuine peace efforts'.[14] The non-cooperative attitude of the right-wing politicians in India only strengthened those in the Pakistani military who were unwilling to make peace with the former. In other words, Indian statesmen often contributed to the feeling of insecurity that helped the Pakistani policy elite to build a 'siege mentality', which in turn enabled the country's army to portray itself as the primary saviour of the nation to combat the dangers from all sides. The Indian threat has not been an imaginary one. It has always been there, but the way the Pakistani military has used that threat to advance its own dominance internally as well as to shape the country's foreign policy is highly consequential.

A military-run Pakistan was shunned fairly quickly by the international community, but the terrorist attacks of 11 September 2001, a mere two years after Musharraf's takeover, changed things: they presented the country with an opportunity to ally itself with the US-led war on global terrorism—thereby setting the stage for a consequential phase in its foreign policy. This chapter studies the major contours of Pakistani foreign policy in the first decade of the twenty-first century. Though Pakistan joined the United States' War on Terrorism within days of the twin-tower attacks, it also continued supporting a variety of militant groups, including the Afghan Taliban. This led to the straining of relations between the two states from 2005–2006 onward, with the US regularly asking the country to take stronger action in curbing militancy across the Pakistan–Afghanistan border. Despite this, Pakistan's policy remained the same, even until the American withdrawal from Afghanistan in August 2021.

[10] 'India's Ex Diplomat Details What Led to Agra Summit Collapse in 2001', *NDTV*, 8 January 2024.
[11] Sheela Bhatt, 'Advani Destroyed Agra Summit with Pakistan: A S Dulat', *Rediff*, 6 July 2015.
[12] Interview, Journalist 3, Lahore, Pakistan, January 2024.
[13] Interview, Politician 3, Islamabad, Pakistan, January 2024.
[14] Interview, Journalist 3.

Section 5.2 studies the journey from a newly revived alliance between Washington and Islamabad after 9/11 to the American demand to Pakistan to 'do more' in both countries' common fight against terrorism. Section 5.3 details how the Pakistani military played upon the threat from India to counter the Western demands throughout the 2000s. Section 5.4 expands on the Obama administration's 'AfPak approach', in which the then US administration sought to link Afghanistan and Pakistan as one theatre of operation warranting combined attention. The Obama administration also introduced the Kerry–Lugar Bill in 2009 to bring a greater balance to the civil–military relations in Pakistan—the Bill and the Pakistani military's resistance towards it will be the focus of Section 5.5. By 2010, relations between Pakistan and the United States had reached another low point, prompting both sides to attempt a 'reset' through the Strategic Dialogue—though it ultimately yielded little improvement in their relationship. These developments will be studied in Section 5.6 which will be followed by the Conclusion (Section 5.7).

5.2 From bonhomie to 'do more'

General Pervez Musharraf addressed the nation on 19 September 2001, a week after the twin-tower attacks, preparing the ground to join the US in the latter's fight by telling Pakistani people that the country 'was facing a very critical situation. Perhaps as critical as the events in 1971', and if it took the 'wrong decisions [its] vital interests will be harmed'. Hinting towards India, Musharraf stated that 'our neighbours ... have promised [the] US all cooperation' implying that *not* siding with the United States will count against key Pakistani interests. Musharraf highlighted the need for 'prudence' (or *hikmat*) instead of 'bravery'.[15]

Soon afterwards, Pakistan's military ruler assured the Americans the country's support for 'the intelligence cooperation, use of air space and logistical support'.[16] The United States, in return, offered generous financial assistance and the lifting of US sanctions that had earlier been placed

[15] 'President's Address to the Nation', 19 September 2001, available at https://presidentmusharraf.wordpress.com/2006/07/13/address-19-september-2001/; on the issue of prudence as a 'virtue' of statecraft, generally, see Wali Aslam, 'International Interventions and Normative Prudence as a "Forgotten" Virtue of Statecraft', *Journal of Intervention and Statebuilding* 15, no. 2 (2021): pp. 181–200.

[16] 'Remarks by Secretary of State Colin L. Powell and President Pervez Musharraf in Islamabad', 15 October 2001, available at https://presidentmusharraf.wordpress.com/2007/05/12/colin-powell-press-conference-2001/

on the country due to the development of its nuclear programme. Anti-terrorist discourse became a key focus of Musharraf's speeches from that time onwards. This apparently anti-terrorist and pro-West approach won Musharraf major plaudits in Washington D.C. and other Western capitals, starting a 'Bush-Mush' bonhomie[17] which continued until 2005–2006. During this period, General Musharraf also came up with his signature narrative of 'enlightened moderation' to 'overcome terrorism and prevent a clash of civilizations'.[18] The narrative highlighted the linkage between Pakistan's 'domestic' and 'foreign' policies, as both sides were now intrinsically linked. It served the dual purposes of the General: to continue to remain in power and garner American support for himself.[19]

Despite pledging to work more closely with the United States to eradicate militancy in Pakistan and Afghanistan, Pakistan's military concurrently continued to foster robust links with a number of militant groups, including Lashkar-e-Tayyaba (LeT) and the Afghan Taliban.[20] Though the former group was more focused on India and the latter on Afghanistan, the US government was worried that the anti-India groups were also involved in supporting the insurgency in Afghanistan.[21] Pakistan's support for the militant groups in Afghanistan was purportedly to gain 'strategic depth' vis-à-vis India. As a leaked US Embassy Cable from Islamabad to the State Department noted, Pakistanis were adamant in their objective, even if that required them 'to enslave the Afghans'.[22]

General Musharraf denied that any such support existed,[23] and assured the Americans that the roots of Afghanistan's problems lay inside the country itself.[24] Musharraf assured them of his vision of becoming the new 'Muslim

[17] Ahmed Rashid, 'The Void Left Behind', *The Washington Post*, 27 December 2007.

[18] See 'President Pervez Musharraf's Speech at the United Nations General Assembly's 59th Session', 22 September 2004, available at https://presidentmusharraf.wordpress.com/2007/03/17/united-nations-59th-session/. The narrative aimed to promote economic development, political reform, and a moderate interpretation of Islam, while presenting Pakistan as a responsible partner in the global war on terrorism. The strategy emphasized balancing internal security concerns with improving Pakistan's international image.

[19] Interview, Academic 2, London, March 2024.

[20] Michelle Macander, 'Examining Extremism: Lashkar-e-Taiba', *Centre for Strategic and International Studies*, 28 October 2021.

[21] 'Terror Finance: Embassy Presses GOP on UN 1267 Committee Cases', US Embassy Cable, EO12958, 27 November 2006, available at https://www.theguardian.com/world/us-embassy-cables-documents/87074

[22] 'Musharraf Tells Boucher About Pakistan's Plans for Border Security', US Embassy Cable, EO12958, 17 January 2007, available at https://www.theguardian.com/world/us-embassy-cables-documents/92731

[23] 'Terror Finance: Embassy Presses.'

[24] 'President Musharraf Briefs Gen Schoomaker and Ambassador on Strategy for Border Security', US Embassy Cable, EO12958, 3 January 2007, available at https://www.theguardian.com/world/2010/nov/30/pervez-musharraf-taliban

voice of moderation', and argued that the American claims that Pakistan was playing a double game were 'wrong' and emboldened the common enemies of the two countries.[25] He asserted that questioning 'Pakistan's reliability as an ally of the U.S. directly damage[d] his ability to wage war on [their] common enemies'.[26]

From the start of the War on Terrorism in 2001, the Bush administration had treated Pakistan 'as an ally, Iran as an enemy and Iraq as the main threat'.[27] The Pakistani President was also keen for the United States to keep its focus on Iraq after the fall of Kabul in November 2001 and rely on him to deal with the war in Afghanistan. As the US Embassy in Islamabad wrote to Washington, Musharraf was

> worried that a premature pull-out of U.S. and coalition forces from Iraq would spread sectarian strife throughout the Gulf region. Musharraf said he understood U.S. public opinion was against prolonging U.S. presence in Iraq, but hoped U.S. leadership could communicate the importance of the mission in Iraq [to its people].[28]

Musharraf also complained that the Americans did not appreciate the 'nuances of Pakistani politics' and 'did not adequately understand the complex realities' of the threats of al-Qaeda, the Taliban, and extremism, and 'were questioning Pakistan's commitment to combat them'.[29] In brief, Pakistanis were keen for the United States to continue to support the country financially and keep its faith in their capacity to deliver.[30] They also wanted Washington to remain focused on Iraq leaving the Afghan theatre to Pakistan. One of the leaked US Embassy Cables revealed that the Americans were not confident that the funds were being used for the purpose of fighting the War on Terrorism. The US Embassy in Islamabad sent the following cable to Washington regarding those concerns in December 2007:

[25] 'President Musharraf Meets Speaker Pelosi and 110th Congressional Delegation', US Embassy Cable, EO12958, 31 January 2007, available at https://www.theguardian.com/world/us-embassy-cables-documents/94672

[26] Ibid.

[27] Barnett Rubin and Ahmed Rashid, 'From Great Game to Grand Bargain: Ending Chaos in Afghanistan and Pakistan', *Foreign Affairs* 87, no. 7 (2008): p. 36.

[28] 'Musharraf Tells McCain: Don't Pull Out of Iraq Without Tripartite Political Settlement', US Embassy Cable, EO12958, 10 April 2007, available at https://www.theguardian.com/world/us-embassy-cables-documents/103788

[29] 'Musharraf Tells Deputy Secretary Pakistan is Committed to Fight Al Qaeda, Taliban', US Embassy Cable, EO12958, 21 September 2007, available at https://www.theguardian.com/world/us-embassy-cables-documents/123103

[30] For the US support of Pakistan's education sector during the War on Terrorism, see Wali Aslam, 'Post-9/11 Higher Education Reforms and the Instruction of International Relations in Pakistan', *International Studies Perspectives* 19, no. 3 (2018): pp. 267–287.

> HELICOPTER READINESS. Pakistan received 55 million USD for helicopter operations from July 2006 to February 2007; however ... as few as 3 Cobra Helicopters were fully mission capable as recently as 10 weeks ago. Post is confident Army Aviation Command never received the 55 million.
>
> MEDEVAC ASSISTANCE TO FRONTIER CORPS. The Pakistan Army claimed 99 million USD over past 12 months for medical operations ... Yet ... the Inspector General of the Frontier Corps has repeatedly requested U.S. assistance in providing assets for medevac, obviously unaware of the resources the U.S. has provided.
>
> RADAR MAINTENANCE: *Between August 2006 and July 2007, Pakistan submitted claims for almost 70 million USD in ADA Radar Maintenance, although there is no enemy air threat related to the war on terror.*
>
> BARBED WIRE: *Between August 2006 and July 2007, we received a claim for 26 million USD in barbed wire and pickets. While these items are no doubt helpful in protecting outposts, the claim figures are highly suspect.*[31]

As before, the policy of using militants during the 2000s by the military had domestic as well as external political objectives.[32] The domestic political objectives included marginalizing moderate sections of Pakistani society and empowering religious groups against non-religious civilian political parties. The foreign policy objectives, primarily, involved using Islamist militants to launch attacks inside India, Afghanistan, and beyond. As always, the two objectives remained interlinked, with the International Crisis Group writing:

> The mullahs' usefulness for the military goes beyond domestic politics. The perpetual threat of war with India over Kashmir, a conflict coloured in religious hues, also brings the MMA[33] and the military together. The more Musharraf searches for domestic legitimacy, the more he plays up the Indian threat. The mullahs are more than willing to support the military's policies in Kashmir.[34]

A number of Taliban groups, including the Haqqani Network, had relocated to Pakistan with the help of the Pakistani security agencies after the fall

[31] See 'US Worried Taliban Battle Funds Being Wasted on Barbed Wire', *The Guardian*, 30 November 2010, emphasis in original.

[32] For details, see 'Pakistan: The Mullahs and the Military', *International Crisis Group*, Asia Report 49, 20 March 2003, available at https://www.crisisgroup.org/asia/south-asia/pakistan/pakistan-mullahs-and-military

[33] Muttahida Majlis-e-Amal (MMA) was an Islamist electoral alliance that was brought into government with the military's help in 2002.

[34] 'Pakistan: The Mullahs and the Military', *International Crisis Group*.

of Kabul in 2001, and found refuge in the tribal areas of Pakistan as well as in the settled parts of the country. That included the current Khaibar-Pakhtunkhwa province (then known as the North West Frontier Province, or NWFP), as well as Balochistan.[35] With the Pakistani military's support, the Afghan Taliban were able to mount substantial insurgency operations against the international forces in Afghanistan by 2005, leading to heavy civilian and military losses in the country.

Though the Americans were aware of the Pakistani support for these groups since the autumn of 2001,[36] it was only towards the middle of President Bush's second term (2005–2009) that Washington started to openly criticize Pakistani support for the Afghan Taliban—explicitly urging Pakistan to 'do more'[37] in their proclaimed fight against terrorism. The Western demands centred on a range of issues, from refusing safe havens for the Taliban to cutting off all military training and intelligence assistance. However, allegations of these activities were vociferously denied by the Pakistani military, with its spokesman asserting: 'To say that these militant groups were being supported by the state with the organised camps in these areas ... I think nothing could be further from the truth.'[38]

Pakistan's main intelligence agency, the Inter-Services Intelligence (ISI), was at the centre of criticism because it was the conduit through which the state corresponded with its militant proxies. Authors writing for *Hilal*, the premier magazine of the Pakistani military, pushed back in no uncertain terms:

> A volley of criticism on the performance of the Pakistan Army, particularly its ace spy set-up, the Inter Services Intelligence, [has been] unleashed from the American (mis)handlers of the war on terror ... Perhaps, what was being implied was that the control of the operation of the Taliban and Al Qaeda lies in the hands of the ISI. Should this contention be taken as a tribute to Pakistan's spy masters for leashing in successfully the powerful terrorists who the combined forces of [the] world's super-power America and its powerful allies admittedly failed to defeat over a span of eight years? Or should this contention be taken as an excuse to exonerate failure of

[35] Steve Coll, *Ghost Wars: The Secret History of the CIA, Afghanistan and Bin Laden* (London: Penguin, 2005).

[36] US official interviewed for the BBC documentary *Secret Pakistan: Double Cross*, 26 October 2011, available at https://www.bbc.co.uk/programmes/b016ltpv

[37] 'Gates: Pakistan must Do Better Job Near Border', United States Central Command Press Release, 18 July 2008, available at https://www.centcom.mil/media/news-articles/news-article-view/article/883676/gates-pakistan-must-do-better-job-near-border/

[38] See 'BBC Documentary Reveals New Evidence of Pakistan Double Cross', 26 October 2011, available at https://www.bbc.co.uk/mediacentre/latestnews/261011pakistan

the 60,000 allies' forces backed by a 70,000-strong Afghan National Army, to defeat a few thousand irregular militants?[39]

According to this view, it was not unusual for intelligence agencies to maintain contacts with multiple actors in a given theatre, but this should not be equated with having cordial relations, since

the intelligence set-up of any country serves as the eyes and ears of the government. The American official claims [regarding the] proof of the ties between the Taliban and Pakistani spies should not be taken as evidence of establishment of cordial relationship between the ISI and the Taliban. History tells us that the CIA has all along been in contact with America's sworn enemies.[40]

The supporters of the Pakistani military believed that the United States was scapegoating Pakistan for the failure in its war efforts. Bilal Shahid wrote in *Hilal* in March 2009 that 'the Bush administration ha[d] shown they [couldn't] eliminate or reduce terrorism, ... the bombing of civilians in Afghanistan and pushing the terrorists to [Pakistan's border areas was] not a solution'.[41] Shahid asserted that the United States had 'failed in Iraq as well as in Afghanistan [and the] cost of war [was] a major concern due to which the US economy ha[d] started melting. Being the superpower, America will never admit it ha[d] failed. Pakistan, a front line state ha[d] been made a scapegoat to hide the incompetence of NATO-US forces.'[42]

This view held that the West had 'more than 117,000 troops with thousands of vehicles, tanks, guns, armed personnel carriers, Cobra gunship helicopters and jet aircraft[s] ... engaged in operations against the terrorists. Thousands of rounds of major weapons and millions of bullets have been consumed to suppress the fire of terrorism.'[43] It was unfair to blame Pakistan for not succeeding in controlling the presence of the militants within Pakistan if a much stronger force on the other side of the border was unable to do the same. Pakistan was being 'forced to seal the Afghan border completely. Despite having meagre resources, Pakistan ha[d] established over 1,000 posts along the Afghan border whereas less than 100 posts [were] being manned by the coalition forces.' Shahid asked, 'why don't the Coalition Forces having all available resources

[39] Sayed G. B. Bokhari, 'Unjust Maligning of ISI', *Hilal*, May 2009, p. 13.
[40] Ibid.
[41] Bilal Shahid, 'Scourge of Extremism: Do More with What?' *Hilal*, March 2009, p. 24.
[42] Ibid.
[43] Ibid., p. 25.

seal the border from Afghanistan's side?' He went on to assert that these terrorists were the same individuals 'who were encouraged by the Western world to fight against the Soviet occupation and were abandoned later'.[44]

According to this perspective, it was, in fact, Pakistan that was the victim of infiltration from the Afghan side of the border, which the coalition forces needed to stop, as the 'unchecked and boundless influx of these militants [was] further spreading the militancy and an insurgency like situation everywhere in Pakistan besides ... tribal areas and Swat. It [was] not possible that security forces of Pakistan [could] reach everywhere in chasing these militants.'[45]

5.3 Pakistan's relations with India and the East–West Connection

By 2008, the Pakistani military high command's view of Musharraf's utility had started to change. Unhappy with his stance on Kashmir, they worried that he might sign a peace agreement with India to prolong his own rule.[46] That perception led the senior generals to overthrow him through a number of manufactured crises, such as the famous 'Lawyers' Movement' of 2007–2008.[47] Pakistan reverted to civilian rule in 2008, with the new civilian president Asif Ali Zardari leading the charge to amend the country's Constitution to its original form from 1973 through the eighteenth amendment, 'cleansing it of distortions' and recovering Pakistan's 'federal structure'.[48] According to a well-informed observer, the Pakistani military had never accepted the country's Constitution and had distorted its face in favour of their rule across several dictatorships.[49] This individual eloquently surmised, the generals 'always tried to write the civil–military equation on the military's terms, which democracy did not allow'.[50]

The new president Zardari also put forward a vision for peace with neighbouring India when talking to an audience in Delhi via video link from Islamabad on 22 November 2008:

[44] Bilal Shahid, 'The World Should Acknowledge Our Sacrifices', *Hilal*, January 2009, p. 24.
[45] Muhammad Khan, 'Curbing Militancy: The Responsibility of Allies', *Hilal*, July 2009, p. 23; Wali Aslam, 'Whose Interest? Whose Peace? An Evaluation of Pakistan's Peace Agreements with the Taliban in the Tribal Areas (2004–2015)', *Studies in Conflict and Terrorism* 46, no. 1 (2023): pp. 21–45.
[46] Interview, Politician 1, Islamabad, Pakistan, January 2024.
[47] Interview, Journalist 3.
[48] Interview, Politician 1.
[49] Interview, Politician 3.
[50] Ibid.

> I am not threatened by India, I repeat I do not feel threatened by India and India should not feel threatened by us ... We bring together peace and the message of love from Pakistan. For the first time in the history of Pakistan, today we have a Parliament, which has already pre-agreed upon a friendly relationship with India. In spite of our disputes, we have a great future together.[51]

Zardari's views on nuclear weapons were most revealing of his vision for the region. He stated: 'Most definitely, I am against nuclear warfare altogether, I do not appreciate it. Pakistan will not be the first country to use nuclear weapons ... we do not hope to ever get to that position.' He asserted that 'in every Pakistani there [was] a little Indian and in every Indian there [was] a little Pakistani' and he did not know whether it was the Indian or the Pakistani in him talking that day.[52]

Within days of those momentous words being spoken, militants linked to LeT launched the famous Mumbai attacks of 26–29 November 2008, which abruptly ended the short-lived enthusiasm for peace in the region. The timing of the attacks was critical, according to one analyst, as it happened only four days after President Zardari's landmark comments regarding Pakistan not facing a threat from India.[53] Direct evidence of the Pakistani army's links with the Mumbai attacks does not exist, though there are sources that strongly hint in that direction. Christine Fair has described LeT as Pakistan's 'most subservient proxy'.[54] She writes:

> US analysts tend to believe that the Pakistani security services maintain reasonably tight control over LeT through providing resources, monitoring group activity, and most importantly continuing to provide LeT the most important asset it enjoys: virtually unfettered access to operate (e.g., recruit, raise funds, train, plan missions) in and from Pakistan ... After the 2008 Mumbai attacks the United States granted Indian officials access to David Headley, an American citizen, after he had been arrested and charged with involvement in the attacks. According to Indian officials, Headley conceded ISI involvement. Bob Woodward reported that the previous director general of the ISI, Shuja Pasha, acknowledged that persons connected to

[51] 'Transcript of Speech and Q&A Session with President of Pakistan Asif Ali Zardari', *Hindustan Times*, 26 November 2008.

[52] Ibid.

[53] Interview, Columnist, Islamabad, Pakistan, January 2024.

[54] Christine Fair, 'The Milli Muslim League: The Domestic Politics of Pakistan's Lashkar-e-Taiba', The Hudson Institute Commentary, 3 June 2018, available at https://www.hudson.org/national-security-defense/the-milli-muslim-league-the-domestic-politics-of-pakistan-s-lashkar-e-taiba#footNote25

> the ISI were involved in the attacks but insisted that the operation was rogue. US officials have declined to endorse this claim.[55]

A senior Pakistani politician told me, 'whenever the civilian government tried to correct the "threat perception", it was always challenged by the [military's] General Headquarters'.[56] The military adopted multiple methods to redefine that perception according to their own agenda.

Despite there being substantial indication of the Pakistan army's hand in the Mumbai attacks,[57] the military in Pakistan denied any link with the event. In one of its editorials, *Hilal* criticized the 'hysteria' generated by the Indian media to deflect attention from its own incompetence:

> The whole world watched in horror as the events of November 27, 2008 unfolded on air. Day after gruelling day we were reminded of the incompetence of Indian Security System as the death toll kept rising, but nothing could have prepared us for the wave of unsubstantiated blame hurled at Pakistan. The Indian media, riding high on the hysteria generated by this mayhem, lapped up rhetoric originating from anywhere within India and thus started another standoff between the nuclear neighbours.[58]

This view held that the entire Mumbai episode was 'tailor made for vilifying Pakistan'.[59] In fact, Pakistan was 'a victim of terrorism' which was sponsored by India through its presence in Afghanistan.[60] Furthermore, India was alone in its effort to malign Pakistan because the 'US view[ed] Al Qaeda and Afghan Taliban as a greater threat' and the 'security priorities' of the regional and global powers were 'more in line with Pakistan than with India'.[61]

As well as denying any links with the November 2008 militant attacks within India, the Pakistani military regularly put pressure on the international community (and specifically the United States) to instruct India not to respond to any such attacks, as doing so would affect Pakistan's cooperation in the American fight against terrorism. The military's mouthpiece, *Hilal*, outlined a horrific scenario in which an escalation in the conflict between the two neighbours might force Pakistan 'to divert a significant portion of air effort

[55] Christine Fair, *Fighting to the End: The Pakistan Army's Way of War* (Oxford: Oxford University Press, 2014), online version; in-text references omitted.

[56] Interview, Politician 3.

[57] Stephen Tankel, 'Lashkar-E-Taiba: From 9/11 to Mumbai', *The International Centre for the Study of Radicalisation and Political Violence*, April/May 2009, available at https://icsr.info/wp-content/uploads/2009/04/Tankel_01.pdf

[58] 'National Unity', Editorial, *Hilal*, January 2009, p. 3.

[59] Ibid.

[60] Hasan Askari Rizvi, 'Complexities of Pakistan-India Relations', *Hilal*, June 2010, p. 6.

[61] Ibid.

from its current employment, along [the] western border, in support of war against terrorism, towards eastern theatre' and 'that would not only facilitate regrouping/rearming of terrorists, it would also cause a reverse gear motion with respect to already achieved objectives. Prolonged standoff warranting [the Pakistani air force's] protracted effort diversion [was] likely to affect is campaign against the terrorists quite severely, who [could] indeed get a new lease of life [*sic*].'[62]

The Indian 'threat' was also effectively deployed by the military whenever Pakistan was criticized for not doing enough in acting against militants within its territory. For example, Muhammad Khan, a Pakistani military officer, wrote in *Hilal* that 'the US and international community must realise that, [Pakistan could not] denude its eastern borders' especially given that 'the South Asian giant had thrice attacked Pakistan without an ultimatum.'[63] Linking the success of Pakistani cooperation with the West (in Afghanistan) to its fight against India, this view held that 'the world community ... must help Pakistan in the resolution of disputed issues like; Kashmir, water, Siachen glacier and Sir Creek with India. Until the resolution [*sic*], Pakistan will have to maintain its India-centric approach in its military deployment.'[64] According to Muhammad Khan, the threat from India was real, given that, 'in spite of Indian rhetoric of [a] two front war, with Pakistan and China simultaneously, its 80 percent military deployment of all three services [was] against Pakistan' with its 'defence budget being "seven times" that of Pakistan.'[65]

The fear of India among the masses in Pakistan was regularly cited as a reason to keep the country's forces focused on the eastern frontier. Unlike the West, Pakistanis did not regard the Afghan Taliban as a major threat, seeing them as far less dangerous than India. Pakistani policy elites contented that if the country diverted more forces to confront the Afghan Taliban, the state 'would invite well-deserved criticism that Pakistan [was] depleting its force levels on the Indian border at a time when India [was] once again murdering Kashmiris wanting to be rid of Indian occupation.' According to this view, 'the need of the moment' was 'for both Pakistan and the US to display maturity and wisdom; to understand their respective limitations and national sensitivities.'[66] It was repeatedly asserted that Pakistan was encircled by threats and only its military was able to understand the full extent of the problem, unlike the common masses. As *Hilal* noted:

[62] Khalid Iqbal, 'Likely Fallout of Indian Posturing', *Hilal*, March 2009, p. 16.
[63] Muhammad Khan, 'Robert Gates' Visit in the Perspective', *Hilal*, March 2010, p. 33.
[64] Ibid.
[65] Ibid.
[66] Zafar Hilaly, 'No; to Targeting Allies', *Hilal*, November 2010, p. 6.

> The adverse political circumstances have enveloped the country from all sides, either we gauge in political or economic terms [*sic*]. Drastic implications of the situation cannot be grasped by the general masses at large, who abruptly change their opinion without reasons. Hence, they become easy prey to the internal exploiters, unintentionally benefiting the external conspirators who want to weaken Pakistan by creating a rift between our general masses led by politicians[,] and the security forces.[67]

This view held that the country faced 'an obdurate, conceited, large, well armed and conniving neighbour that remain[ed] bent on undoing Pakistan' and was 'planning the encirclement of Pakistan by suborning Afghanistan to her cause. Moreover, in addition to the existential threat posed by India, [Pakistan] was confronted ... by the ideological threat posed by the Taliban [and] both threats [were] deadly. Both envisage[d] the destruction of Pakistan.'[68] In this situation, the country would rely on its saviour—its army. As the 'ably led, well educated, properly trained and imbued with courage, [Pakistani] armed forces comprise[d] the best that the nation ha[d] to offer ... they [were] the sole bulwark between [its] enemies and the abyss'. Without its armed forces, 'Pakistan would be in the hands of rustic bigots'.[69]

Pakistan also regularly sought American mediation between Pakistan and India if the US wanted to achieve victory in Afghanistan. Furthermore, it wanted the US to 'compel India to stop its anti-Pakistan subversive activities in FATA and Balochistan, implement[ed] through its consulates and covert launching pads established all along the Pak-Afghan border'.[70] The military mouthpiece *Hilal* portrayed an Indian, American, and Israeli nexus actively conspiring against Pakistan. Muhammad Khan, writing in 2010, believed that it was

> no more a mystery that [an] Indo-US-Israel-Afghanistan nexus [was] effectively operating with Afghanistan as its centre of activities. Although Indo-Afghan relationship centred on the Pakistani enmity has a history, but in the post-US attack on Afghanistan (following 9/11), India ha[d] deeply entrenched itself in that war torn country under the garb of the so called reconstruction. India ha[d] employed its combat soldiers and a very heavy composition of the elements of its spying agency RAW. These elements [were] effectively involved in the promotion of militancy in FATA, NWFP and [Pakistan's] Balochistan province.[71]

67 Atia Anees, 'National Integration', *Hilal*, June 2009, pp. 17–18.
68 Zafar Hilaly, 'Vibrant Armed Forces', *Hilal*, September 2010, p. 17.
69 Ibid., p. 16.
70 Khan, 'Curbing Militancy: The Responsibility of Allies', p. 23.
71 Muhammad Khan, 'Limited War Under Nuclear Overhang', *Hilal*, January 2010, pp. 10–11.

The enemy conspiracy to create unrest in Balochistan was linked with the territory's rich hydrocarbon and mineral reserves which, given its location, could be significant for commerce.[72]

Despite these concerns about India, the Pakistani military hindered resolution of the dispute between the two, even though the civilian leadership offered to lead the peace initiative, as will be studied below. It appeared that an unending, perpetual conflict with India provided a permanent shield against the Western demands on Pakistan to do more on the Western front, while making it harder for the West to extricate itself from Afghanistan. Prolongation of the conflict primarily advantaged the Pakistani military, as it secured ongoing foreign financial support. Moreover, the protracted conflict functioned as a source of societal anxiety, which the military systematically exploited to strengthen its domestic authority.

The leaked US Embassy Cables appear to show that Indian diplomats had more clarity on the challenges concerning India–Pakistan relations than their Western counterparts. Indian diplomats did not unquestioningly accept the Pakistani logic linking the Indian threat to the cooperation in the War on Terrorism. In the late 2000s, when the Americans were increasing their pressure on Pakistan to cooperate more with them, one Indian diplomat told the US officials that there had been 'a sharp increase in unseasonal Pakistan-inspired violence, ... incidents of cross-border shelling along the line of control and in Punjab, increased infiltration, and transfer of terrorist hardware'.[73] 'They [were] clearly trying to "stir the pot" in Kashmir', she asserted, because 'Pakistan [was] trying to deflect attention to its eastern border from the Afghanistan-Pakistan border, where the focus ought to be'.[74] The Indians were clear that supporting Pakistan's army was 'not the answer'.[75] They wanted the Americans to put pressure on Pakistan's army as 'most of the problems in Pakistan [could] be traced to the capacity and intentions of Pakistan's military'.[76]

Demonstrating insight into how a greater civil–military balance in Pakistan was also good for India, the Indian diplomats informed their American counterparts that, after the Mumbai attacks in 2008, they 'consciously [had] not

[72] Muhammad Khan, 'Towards A Better Pak-Iran Relationship', *Hilal*, December 2009, p. 16.

[73] 'SRAP Holbrooke Discusses Afghanistan and Pakistan', US Embassy Cable, EO12958, 28 January 2010, available at https://www.theguardian.com/world/us-embassy-cables-documents/245980

[74] Ibid.

[75] 'Indian Foreign Secretary: "Huge Stake" in Special Representative Holbrooke's Success', US Embassy Cable, EO12958, 17 February 2009, available at https://www.theguardian.com/world/us-embassy-cables-documents/192309

[76] Ibid.

attempted to take any measures to destabilize Pakistan's civilian government, and had made every effort to continue trade and travel relations'.[77] India 'had consciously not built up troops on the border with Pakistan, as it had following the 2001 attack on its Parliament'.[78] However, the Pakistan army 'continued to make things difficult for India, through ceasefire violations, infiltrations and continued support for terrorist groups'.[79] India believed that Pakistani fears of an Indian encirclement of the country might be genuine and it had made 'numerous efforts' over the years to 'allay Pakistan's concerns, ... [but] only to be rebuffed'.[80]

Shivshankar Menon, the Indian High Commissioner to Pakistan between 2003 and 2006, told the Americans that he even offered 'President Musharraf to sit down and explain exactly what India was doing in Afghanistan, without even asking for a reciprocal explanation from Pakistan' but 'Pakistan officials—not just Musharraf—"... avoided it in every way"'.[81]

In another meeting, Y. K. Sinha, an Indian diplomat, asked Tim Roemer, the former US Ambassador to India (2009–2011) 'to recognize and resist Islamabad's "game" of promising cooperation in Afghanistan in return for USG [US Government] pressure on India to improve ties with Pakistan, while also exaggerating India's threat to Pakistan and "trying to internationalize every bilateral issue between India and Pakistan"'.[82] He assured the Americans that India had 'no benchmarks' for talks with Pakistan on the Kashmir dispute, but the Pakistani lack of cooperation was not rooted in that dispute. Accordingly, Sinha believed that 'even if India were to lop off Kashmir and hand it on a platter to Pakistan, they would still find a reason to make trouble for us'.[83]

Such clarity, however, was lacking on the American side, whose representatives were still willing to see the region from the Pakistani perspective despite dealing with the challenge for nearly ten years. Anne Patterson, the US Ambassador to Pakistan from 2007 to 2010, believed that Pakistan was worried about Indian activities against it, which were distracting it from its fight against the Taliban. Patterson asserted that 'no amount of money' was likely to induce Pakistan to be more proactive in its counterterrorist efforts unless there was a 'change [in] the Pakistan government's own perception of

[77] Ibid.
[78] Ibid.
[79] Ibid.
[80] Ibid.
[81] Ibid.
[82] 'Current Indian Government Thinking on Afghanistan and Pakistan', US Embassy Cable, EO12958, 27 November 2009, available at https://www.theguardian.com/world/us-embassy-cables-documents/236969
[83] Ibid.

its security requirements.'[84] The American Deputy Chief of the Mission at the country's Islamabad Embassy wrote in another cable:

> If the [Mumbai] militants' plan was to force the Pakistani Army to re-focus on its eastern border and eliminate any chance (however slight) of moving forces from the Indian border to the Federally Administered Tribal Areas (FATA), then their plan may have succeeded. The Mumbai attacks likely torpedoed any prospect of Indian CBMs [confidence-building measures] on Kashmir in the immediate future.[85]

Just as the Pakistanis wanted them to do, the Americans privately worried that Pakistan would transfer more troops from its western border to its eastern front in the wake of the Mumbai attacks. They observed that, despite some cooperation on the issue of LeT, it was not clear 'if ISI ha[d] finally abandoned its policy of using these proxy forces as a foreign policy tool.'[86] Western officials appeared predisposed to accept Pakistan's claim that 'tackling the Kashmir issue remain[ed] the key to regional security.'[87]

In a cable in 2009, Patterson wrote that though the 'ending of Pakistani establishment support to terrorist and extremist groups, some Afghan-focused and some India-focused, [was] a key element for success,' there was 'no chance that Pakistan [would] view enhanced assistance levels in any field as sufficient compensation for abandoning support to these groups, which it [saw] as an important part of its national security apparatus against India. The only way to achieve a cessation of such support [was] to change the Pakistan government's own perception of its security requirements.'[88] Patterson went on to state that

> The Pakistani establishment fear[ed that] a pro-India government in Afghanistan would allow India to operate a proxy war against Pakistan from its territory. Justified or not, increased Indian investment in, trade with, and development support to the Afghan government, which the USG has encouraged, causes Pakistan to embrace Taliban groups all the more closely as anti-India allies.[89]

[84] See 'Reviewing Our Afghanistan-Pakistan Strategy', US Embassy Cable, EO12958, 23 September 2009, available at https://www.theguardian.com/world/us-embassy-cables-documents/226531

[85] 'Pakistan Reactions to Terror in Mumbai', US Embassy Cable, EO12958, 28 November 2008, available at https://www.theguardian.com/world/us-embassy-cables-documents/180496

[86] 'Scenesetter for Special Envoy Holbrooke', US Embassy Cable, EO12958, 4 February 2009, available at https://www.theguardian.com/world/us-embassy-cables-documents/190330

[87] Ibid.

[88] 'Reviewing Our Afghanistan-Pakistan Strategy'.

[89] Ibid.

Appearing to readily buy the Pakistan narrative, Patterson was keen to find a solution to the challenges at hand according to that narrative too. She believed that the Americans needed

> to reassess Indian involvement in Afghanistan and [their] own policies towards India, including the growing military relationship through sizable conventional arms sales, as all of this fe[d] Pakistani establishment paranoia and pushe[d] them closer to both Afghan and Kashmir-focused terrorist groups while reinforcing doubts about U.S. intentions. Resolving the Kashmir dispute, which [lay] at the core of Pakistan's support for terrorist groups, would dramatically improve the situation. Enhanced [United States Government] efforts in this regard should be considered.[90]

The American view held that a shift in the Pakistani focus to 'the western border and internally'[91] instead of India was key for both countries, but they also believed that shifting that focus would only be possible through resolving the India–Pakistan disputes and normalizing relations between the two. On that basis, Richard Holbrook, the US Special Representative for Afghanistan and Pakistan at that time, became 'obsessed with the idea of Pakistan-India trade'[92] during 2009/2010 despite it being something the Pakistani military had constantly opposed in the past.[93]

5.4 The AfPak approach and its critics

President George W. Bush's departure from office in January 2009 coincided with a low point in US–Pakistan relations, and the incoming Obama administration made it a priority to extricate the United States from the Afghanistan imbroglio. Soon after taking office, the new Obama administration outlined 'an AfPak approach' which made the case for treating Afghanistan and Pakistan as one area of operation.[94] This did away with the traditional India–Pakistan dyad and, instead, linked Pakistan firmly with Afghanistan. The Pakistani state's response to the development was of serious disapproval. An article in *Hilal* asserted that Afghanistan and Pakistan were so different from

[90] Ibid.

[91] 'Centcom General Petraeus Meets with Pakistan COAS Kayani', US Embassy Cable, EO12958, 24 January 2009, available at https://www.theguardian.com/world/us-embassy-cables-documents/188670

[92] Khaled Ahmed, 'Bullying Pakistan Won't Work', *Newsweek Pakistan*, 22 June 2013.

[93] Khaled Ahmed, 'Homework for Nawaz', *The Indian Express*, 24 May 2013.

[94] Wali Aslam, 'Understanding the "Pak" in "Afpak": The Obama Administration's Security Policy for Pakistan at the Mid-Term', *Journal of Policing, Intelligence and Counter Terrorism* 7, no. 1 (2012): pp. 2–21.

each other that seeing them as one theatre of operations would be a mistake. It contended that

> [t]he main purpose behind the AfPak strategy was to develop a unified approach and common strategy to hammer out terrorism, on the part of the Americans, but no matter how badly it affect[ed] geography, politics and security of the two sovereign countries of Pakistan and Afghanistan. The term represented the view that the conflicts in Afghanistan and Pakistan [were] not two separate problems, but part of a single unified issue demanding a joint approach to policy and strategy. The Obama administration thought that the strategic problems of the Afghan War playing out in both countries [meant that it was] far more useful to consider them intertwined but actually [that] was inaccurate.[95]

The author also suggested that there were racial undertones to the coinage of the phrase 'AfPak' by noting that

> the term was as derogatory as the [use of the] term 'economic animals' for the Japanese, for instance. By the same token, conversely, nobody was willing to use, for instance 'ChiJap' for 'China and Japan' and 'KorJap' for Korea and it goes on. While these terms [were] just ethnically and ethically derogatory, the term AfPak was not just an abstract [idea nor was it] trivial in nature. It was deep rooted with unforeseen consequences, [and was] going to reshape the geography, politics, security and economy of the region.[96]

The Pakistanis argued that the new idea revived their insecurities concerning the Durand Line because the AfPak approach made the Line meaningless. Sensing conspiracy, the authors of *Hilal* complained that the Americans were playing with the destiny of Pakistan in the way the British had done in the past and that lumping Pakistan and Afghanistan together was a plot to weaken Pakistan, as Americans believed that 'a divided and non-nuclearised Pakistan [would] be a lesser evil and a [lesser] threat to security in the region, willing to accept the Indian hegemony under the American flag'.[97]

The key criticism by Pakistanis of the AfPak approach centred on the fact that Pakistan and Afghanistan were too different to be viewed through a singular lens, and the policy betrayed the 'US ignorance and lack of knowledge about the distinctive features of both Afghan and Pakistani societies, people and several ... other distinctive elements that the two countries possess[ed]

[95] Ahmad Rashid Malik, 'Afpak Strategy–An Obituary', *Hilal*, April 2010, p. 15.
[96] Ibid.
[97] Ibid., p. 16.

over centuries'.[98] Afghanistan and Pakistan had 'different backgrounds, cultures and values. Except religion and limited tribal traditions, nothing [was] common between the two countries. Moreover a country under foreign occupation [could not] be equated with a sovereign nuclear power of the world.'[99]

Pakistan's never-ending historical competition with India had until then meant that the rest of the world dedicated the same policy attention to both countries, with their names often hyphenated. The AfPak approach was seen by the Pakistani military as a downgrading of their country—it was being delinked from India and lumped in with what it considered to be the failed state of 'Afghanistan'.[100] Instead of addressing international concerns regarding Pakistan's support for Afghan militants, Pakistan focused primarily on restoring the India–Pakistan 're-hyphenation'.[101] When the United States and India signed a civil nuclear agreement in 2005 to boost India's energy needs,[102] the Pakistani military was equally keen for the United States to sign a similar deal with their country. However, the chances of that happening were dim, given Pakistan's previous record of nuclear proliferation through the A. Q. Khan Network.[103] (The A. Q. Khan Network was a 'complex international network of experts, suppliers, and front companies that ... aided Iran, Libya, North Korea, and potentially others'[104] at different times in history until it was intercepted in 2004, earning Pakistan major international criticism.) The network was a major reason for American reluctance to treat Pakistan in the same manner as India in nuclear matters, which irritated Pakistan. The US Ambassador wrote in a leaked Embassy Cable in 2009: 'Islamabad has chafed over the U.S.-India 123 Agreement,[105] arguing [Pakistan] also need[ed] civilian nuclear power to meet energy demand; [however] we have repeatedly advised the [Government of Pakistan] that it should not expect a similar agreement because of AQ Khan's proliferation activities'.[106]

After the A. Q. Khan network was intercepted, Pakistani officials asserted that the country was unfairly being singled out on proliferation charges, since many other countries had engaged in similar practice. This perspective held

[98] Ibid.
[99] Khan, 'Curbing Militancy: The Responsibility of Allies', p. 23.
[100] C. Raja Mohan, 'The Great Game Folio: Obama and Pakistan', *The Indian Express*, 21 January 2015.
[101] Ninan Koshy, 'On Re-Hyphenating India and Pakistan', *Mainstream Weekly*, 21 December 2008.
[102] See 'India and US Reach Nuclear Agreement', *The Guardian*, 2 March 2006.
[103] Gordon Corera, *Shopping for Bombs: Nuclear Proliferation, Global Insecurity, and the Rise and Fall of the A.Q. Khan Network* (London: Hurst, 2009).
[104] Michael Laufer, 'A. Q. Khan Nuclear Chronology', *Carnegie Endowment for International Peace Report*, 7 September 2005.
[105] The Name of the US–India Civil Nuclear Agreement.
[106] 'Scenesetter for Special Envoy Holbrooke'.

that even the American nuclear project was made possible by European expatriates who 'were brought to the United States "with their 'stolen' nuclear secrets from Germany, Italy, Austria and the Scandinavian countries" .... While acting [as] the primary nuclear proliferator, United States later on "helped Britain and France to become the nuclear powers".'[107] Similarly, the 'Soviet Union, after becoming a nuclear power in 1949, through clandestine nuclear scientists from the Manhattan Project, helped China and North Korea in their nuclear programmes.'[108] Such assertions suggested that Pakistan's involvement in nuclear proliferation was not unusual. As *Hilal* asserted,

> India was assisted by Canada, the United States, Britain and France in the developmental stages of its nuclear programme and even [the] NPT rules were violated to assist India by [the] US through recently concluded Indo-US Nuclear Deal. Germany helped South Africa and Brazil in the achievement of their nuclear technologies. France virtually built the Israeli nuclear infrastructure; and [one] only need[ed] to ask Seymour Hersh about the ultimate destination of large quantities of fissile materials stolen from the laboratories of the United States.[109]

In fact, Pakistanis believed that India was the wrong country for the US to sign a civil–nuclear agreement with, given its 'worst' record of nuclear proliferation in 1974 and 1998, the years of its two nuclear tests: 'The US and Canadian materials were used for the Pokhran nuclear testing in the name of "peaceful uses of nuclear technology". India might use imported uranium for the civilian reactors and the indigenous uranium for nuclear bomb production again.'[110]

At the time, concerns over the security of Pakistan's nuclear weapons and the general instability of a nuclear-armed country troubled Western officials engaging with Pakistan. Analysts supportive of the military, however, went to great lengths to attribute the negative portrayal of their country to the misinformation campaigns of its enemies. As *Hilal* asserted, 'The US, Indian and the western misinformation [campaign was] baseless and self centered, spread by anti-Pakistan lobbies ... media and ill-informed scholars. Pakistan [was] neither like Somalia, nor could be compared with Afghanistan. And the western cries of Taliban marching on to Islamabad would never ever be put into practice.'[111] This perspective held that by criticizing Pakistan as a proliferator of nuclear weapons, the West sowed doubt among Pakistanis

[107] Muhammad Khan, 'Pakistani Nukes are Well Guarded', *Hilal*, September 2009, p. 20.
[108] Ibid.
[109] Ibid.
[110] Ahmad Rashid Malik, 'Nuclear Discrimination', *Hilal*, January 2012, p. 13.
[111] Khan, 'Pakistani Nukes are Well Guarded', p. 21.

regarding the 'sincerity of the US' commitment to Pakistan'.[112] Where there had been nuclear accidents in other countries, no such incident had ever taken place in Pakistan. The country had 'proved to be the most responsible nuclear power in the comity of nations' and it did not require 'a certificate for the safety and security of its nuclear arsenals from any other country'.[113] However, despite such Pakistani military protests, no nuclear deal with the United States materialized—infuriating the military, as its efforts at a 're-hyphenation' depended on the signing of such a deal.

Interestingly, while the Pakistani case for a 're-hyphenation' in *Hilal* argued that Pakistan and Afghanistan were very different, the same forum presented the opposite argument when India offered to train the Afghan security forces in 2009–2010. Muhammad Khan wrote that Pakistan should have been tasked with that responsibility instead, arguing that 'besides, sharing common values, culture and comparable stakes, both countries ha[d] a history of interdependence. Foreign interferences, influences, and imperialism [could not] force the people to think differently'.[114] Given that an Indian role in Afghanistan could be detrimental to securing the border, closer cooperation between Afghan and Pakistani security forces 'would enable both to devise common strategies to deal with the common enemy' across the 'common terrain and borders'.[115]

5.5 The Kerry–Lugar Bill, and recalibrating the civil–military equation in Pakistan

To address the challenges involving Pakistan in the AfPak region, the Obama administration introduced the *Enhanced Partnership with Pakistan Act* of 2009, commonly known as the Kerry–Lugar Bill. Its objectives included reconfiguring the civil–military balance in Pakistan and directing more aid to the country's civilian administration than to the military. It was clearly an attempt to rein in the country's powerful military, and its primary intelligence agency (ISI), weakening their control over Pakistan's foreign and security policies. The American leadership believed that Pakistan's civilian leaders supported US efforts to crack down on the Afghan Taliban, whereas the military could not be fully relied upon to do so.[116] The United States also pressured Pakistan to reduce its disproportionately large 'defense spending

[112] Ibid., p. 20.
[113] Ibid., p. 21.
[114] Muhammad Khan, 'Towards Pak-Afghan Harmonization', *Hilal*, April 2010, p. 10.
[115] Khan, 'Robert Gates' Visit in the Perspective', p. 33.
[116] Aslam, 'Understanding The "Pak" in "Afpak"'.

[to] free up money for health and education'.[117] The Kerry–Lugar Bill was to provide $7.5 billion to Pakistan between 2010 and 2014 in the areas of 'health care, education, social services, and humanitarian assistance'.[118] To increase accountability in Pakistan, the Bill also tasked the American Secretary of State to 'submit detailed monitoring reports to [US] Congress every six months, specifying such matters as how much control elected officials [would] have over military budgets and appointing senior officers'.[119]

The Pakistani military was 'angry' at the Bill and claimed that 'there were elements in the bill that would set back the bilateral relationship' with its 'critical provisions ... almost entirely directed against the [Pakistan] Army'.[120] In a meeting with the US Ambassador to Pakistan, Anne Patterson, the country's Army Chief General Ashfaq Pervez Kayani registered his protest, stressing that 'the bill refused to recognize the enormous progress which had been made bilaterally with the US military and against terrorism within Pakistan'.[121] He, and other military officials accompanying him, also stressed that 'the bill had caused a negative reaction among the Corps Commanders and younger officers'.[122] The military also cautioned that the 'bill could undermine political support for the Army's anti-terrorist effort'.[123]

In this meeting, the military also wheeled out the expected narrative that the country was facing threats from all sides, with 'huge security crises along eastern and western borders. Defence increases of neighbouring India reflect[ed] its expansionist and aggressive ambitions in the region, while Pakistan's defence [was] turning into more "self-defence" ... without the reflection of any aggressive designs.'[124] In fact, so the narrative went, Pakistan had done the international community a favour by joining in the 'War on Terror ... to ensure a secure and peaceful world' and that should have been 'acknowledged globally'.[125] Writing in *Hilal*, Ahmad Rashid Malik argued that 'there [was] no shortcut to development, defence or security. By putting the army in financial crunch, Pakistan [could not] move one extra mile toward peace and prosperity.'[126]

[117] The Interview of the United States Ambassador to Pakistan for *PBS News*, August 2009, available at https://www.pbs.org/wgbh/pages/frontline/obamaswar/interviews/patterson.html

[118] Bryan Gibel, 'U.S. Aid to Pakistan: The Kerry-Lugar Bill', *PBS News*, available at https://www.pbs.org/frontlineworld/stories/pakistan901/aid.html

[119] Ibid.

[120] 'Ambassador Meets with Kayani and Pasha About Kerry-Lugar', US Embassy Cable, EO12958, 7 October 2009, available at https://www.theguardian.com/world/us-embassy-cables-documents/228747

[121] Ibid.

[122] Ibid.

[123] Ibid.

[124] Ahmad Rashid Malik, 'Comparative Defence Perspectives', *Hilal*, September 2010, p. 21.

[125] Ibid.

[126] Ibid.

Authors supporting this narrative also objected to the idea that Pakistani defence spending was a hurdle for social sector development in Pakistan and argued that there were other causes of social deprivation in Pakistan, such as corruption in the civilian sector.[127] Sayed Rafaqat wrote in *Hilal* that

> we see around us a politico-economic apparatus erected to ensure, in violation of all principles of law, justice and decency, a massive transfer of resources daily taking place from the abysmally poor to the enormously rich. This apparatus act[ed] like a self-sustaining capillary action machine to suck the life blood of the lower [strata] to feed the inexhaustible demands of the greed and life style of the rich and still worse, the newly rich. The effects of this merciless resource transfer were visibly spread over the entire landscape of Pakistan.[128]

The military perspective also described the domestic supporters of the Kerry–Lugar Bill as 'political perverts' who had 'somehow concluded that a strong and large defence, whatever be its correlation with the threat to national security, [stood] as an obstacle in the path of their *so-called democratic ambitions*.'[129] According to this view, such groups acted irresponsibly by asking for the reduction in the defence budget at 'strategic moments.'[130] Given the internal and external security challenges faced by Pakistan, it believed that the country's 'transparent' defence budget was 'modest' and was under 'public control,'[131] but certain quarters portrayed the Pakistani armed forces 'as a "burden", "sacred cow", and a "white elephant" in the eyes of the public, which went against the ground realities, [as well as the] regional and the global scenarios. Pakistan's socio-economic development could be supported through a strong national defence structure and there [was] no other way round.'[132] Malik argued that such allegations represented the new normal or 'new reality,' and the claims portraying the army as 'the main obstacle [to] development [only served] the enemies of Pakistan.'[133]

The Pakistan military's campaign against the Kerry–Lugar Bill also blamed the Western defence cooperation with India for Pakistan's defence expenditure. Specific criticism was reserved for the US–India civil-nuclear agreement. Accordingly, a high defence budget was justified for Pakistan on the grounds of the Indian threat:

[127] Syed Rafaqat, 'Security Without Starvation,' *Hilal*, November 2008, p. 11.
[128] Ibid., p. 10.
[129] Ibid., emphasis added.
[130] Ibid., p. 11.
[131] Malik, 'Comparative Defence Perspectives,' p. 21.
[132] Ibid.
[133] Ibid.

> Unfortunately, [the] United States, Russia, France, South Korea, and now Japan, ha[d] been signing civilian nuclear power projects with the non-NPT and non-CTBT India that [was] a major concern for the non-nuclear proliferation regime. The change of non-nuclear rules specific for India could trigger a major catastrophe in the region and lead to further escalation of tensions and increases in defence budgets. The above-mentioned international actors must share the blame for the introduction of the renewed arms race and defence increases in the region.... In view of the continuous increase in defence expenditure of India and resultantly enhanced war potentials, defence budget of Pakistan must be significantly increased in the coming years.[134]

Pakistanis were also critical of the amount they received from the West in terms of military and civilian assistance, which equated to about $10 billion in the first decade of the century. They believed that the amount was too little to hold Pakistan accountable for. As Bilal Shahid wrote: '[an] assistance of $10 billion as compared to $653bn for Iraq and $172bn for Afghanistan ha[d] been given to Pakistan since 2001. Much hue and cry [was] being raised by the US and western media on this "peanut" amount. An unending blame game [had] started and the noose [was] being tightened against Pakistan.'[135] The author continued that, between 2004 and 2008, $15.3 and $20 billion were spent on the training and equipment of the Afghan and Iraqi forces, respectively. However, the Pakistani forces 'who were bearing the major brunt, who killed and captured high value targets which neither Iraq nor Afghanistan did', were only given $5.64 billion—although the cost to Pakistan was significantly more than that.[136] According to this understanding, Pakistan deserved funding equivalent to that provided to Western European countries under the Marshall Plan for post–World War II reconstruction. The idea was that

> Pakistan should [have been] given massive aid of 40 to 50 billion dollars per year as was done in the Marshall Plan ... If the US want[ed] to defeat the extremism, they ha[d] to provide money, weapons and equipment on [a] war footing. The US ha[d] already wasted trillions of dollars on GWOT, [and if even a portion of that] had been [spent] in Pakistan; the War on Terrorism would have been won much before. The time [was] running out.[137]

Extending the narrative one step further, Khan wrote in 2009 that foreign aid to Pakistan was actually a conspiracy against the country because American

[134] Ibid.
[135] Shahid, 'Do More with What?' p. 24.
[136] Ibid.
[137] Ibid., p. 25.

assistance had meant that 'Pakistan was betrayed from the passageway of self sufficiency [*sic*] and over the period of time became a foreign aid dependent state'. That was done because a Pakistan dependent on the West for financial assistance was easier to dictate to and control.[138]

Pro-military viewpoints also asserted that the small amount the US was giving them to fight their war showed the latter's lack of seriousness in bringing it to an end. If the US had been serious about ending the war, it would have provided Pakistan with substantially greater funding. As one commentator noted:

> Pakistan ha[d] lost almost everything; there [was] no peace, no safe place, no economy and still it may take more than hundreds of billions of dollars to regain what ha[d] been lost. What America ... spent in this GWOT [Global War on Terrorism] in Iraq and Afghanistan [was] an eye opener, not only for the Americans but for the Pakistanis also.... The US [didn't] seem to be serious in finishing the GWOT and want[ed] to keep the pot boiling. When America was serious in stopping Communism and Soviets' march towards warm water, arms and ammunition were sent to Pakistan on war footing. Now the only thing which [came] out from America [was] the blame game, accusing Pakistan of not doing enough. Had the Bush regime been serious in eliminating terrorism, it could have given Pakistan thousands of M-16 rifles, few predators, communication equipment, night vision goggles, IED jamming devices and gunship helicopters.[139]

Several pro-military voices also observed that from the start of its War on Terrorism, the United States provided dated weapons and equipment to Pakistan and expected the latter's security forces to fight better equipped and trained militants with those weapons, which it felt was unfair. The American 'military aid to Pakistan [was] so pathetic that it [was] not even worth mentioning. More than half of the aid [was] paid back to [the] USA in logistic issues [*sic*] [and] consultation charges.' There was 'a lot of rhetoric but no substantial arms and equipment to combat the terrorism. [Frontier Constabulary or] FC, a [Pakistani] paramilitary force [deployed for combating militancy on the Afghan-Pakistan border] ... was never trained to fight insurgency' and did not have 'the requisite arms [or] the equipment to fight militants or mercenaries who [were] trained and armed by foreign forces. Pakistan [was] being pressured to "do more", but do more with what?', asked Bilal Shahid writing in *Hilal*, 'Just 303 rifles and World War-II equipment?'[140]

[138] Khan, 'Curbing Militancy: The Responsibility of Allies', p. 22.
[139] Shahid, 'Do More with What?' p. 25.
[140] Ibid.

Pakistan's Ambassador to Saudi Arabia also told the Americans in 2009 that, although the US '"[did] a lot for Pakistan", ... it was not enough to fight this problem [on the Afghanistan–Pakistan border]. All of NATO was fighting the battle in Afghanistan ... while Pakistan alone was standing up to the extremists within its borders. For this reason, [the Ambassador] insisted that Pakistan needed more material support from the U.S. in the form of "arms, ammo and planes"'.[141]

As discussed earlier, by the end of the 2000s, Americans were keen to shift the balance of the civil–military equation in Pakistan towards the civilians, whom they trusted more than the military to help achieve their objectives. In February 2009, the US Ambassador to Pakistan wrote in a secret cable to Washington, D. C., that civilian leaders, including President Asif Ali Zardari and Prime Minister Yusuf Raza Gilani 'recogniz[ed] Pakistan's greatest threat ha[d] shifted from India to militancy concentrated on the Pak-Afghan border [which was] spreading to NWFP and beyond. The Army and ISI, however, ha[d] not turned that corner.'[142] Americans believed that while there was a need to bring the ISI under greater civilian control, whenever they asked the Pakistani military to take a step in that direction, it 'rearrange[d] the furniture.'[143] Patterson asserted in February 2009,

> After eight years of military rule under Musharraf, Zardari [was] re-shaping civilian–military relations in the shadow of Pakistan's history of repeated military coups. It [was] in USG interests for the Zardari/Gilani government to complete its full five-year term, and we should praise Kayani's efforts to support civilian democracy.... Zardari and Gilani agree[d] that Pakistan's biggest threat [came] from a growing militant insurgency on the Pak-Afghan border. The military and ISI ha[d] not yet made that leap; they still view[ed] India as their principal threat and Afghanistan as strategic depth in a possible conflict with India. They continue[d] to provide overt or tacit support for proxy forces (including the Haqqani group, Commander Nazir, [Gulbuddin] Hekmatyar, and LeT) as a foreign policy tool.[144]

[141] 'Pakistani Ambassador on Military Operations In FATA; Saudi Relations', US Embassy Cable, EO12958, 26 October 2009, available at https://www.theguardian.com/world/us-embassy-cables-documents/231326

[142] 'Focusing the U.S.-Pakistan Strategic Dialogue', US Embassy Cable, EO12958, 21 February 2009, available at https://www.theguardian.com/world/us-embassy-cables-documents/193196

[143] 'USDP Edelman's October 15 Meeting in London', US Embassy Cable, EO12958, 21 October 2008, available at https://www.theguardian.com/world/us-embassy-cables-documents/174627

[144] 'Scenesetter for General Kayani's Visit to Washington', US Embassy Cable, EO12958, 19 February 2009, available at https://www.theguardian.com/world/us-embassy-cables-documents/192895

Interestingly, although the Americans had concluded at the end of the decade that US–Pakistan relations were characterized by 'mutual mistrust',[145] they continued to accept the Pakistani military's logic regarding the country's strategic focus, which, according to the Americans, remained 'fixated on India, Afghanistan as strategic depth in the fight against India, and the core Kashmir issue'.[146] Rather than critically questioning the Pakistan army's narrative of its perception of reality, the Americans continued to embrace it and, as before, sought solutions through that lens. There remained a substantial focus on assuaging the Pakistani army's concerns about India, rather than challenging them as excuses used to prolong the war and continue receiving US financial assistance. For example, Patterson wrote in a 'Scenesetter' ahead of the Pakistani military bosses' visit to the US in February 2009 that Obama officials needed to ask them 'what political end state in Afghanistan would convince [Pakistani army] to end proxy support for militants and ... what would be required by India to allow [Pakistan] to redeploy forces from the Indian border for the fight in FATA'.[147] She recommended that '[b]oth sides should resume Composite Dialogue negotiations, re-establish back-channel negotiations, and increase trade across both the Wagah border and the Kashmir Line of Control'.[148] The Americans naïvely believed that the Pakistan army needed to focus on 'today's war against the Taliban rather than yesterday's war against India',[149] without realizing that, for the Pakistan army, the conflict with India remained a policy of the past, present, and future, ensuring its perpetually dominant position within Pakistan. It was also a key lever to ensure a continued Western presence in Afghanistan—a Western presence on whose rent the Pakistani state had come to depend heavily.

Towards the end of the 2000s, the Pakistani political and military establishment started getting worried about the American departure from the region, which would surely spell an end to the aid. In his conversations with the American diplomats in Islamabad in 2008, President Zardari expressed the Pakistani public's worry that the Americans would leave Pakistan once their objectives were achieved, instead of announcing 'a Marshall Plan for Pakistan'.[150] The President, however, sought increased funding for democracy

[145] 'Focusing the U.S.-Pakistan Strategic Dialogue'.
[146] Ibid.
[147] 'Scenesetter for General Kayani's Visit'.
[148] 'Focusing the U.S.-Pakistan Strategic Dialogue'. India and Pakistan were engaged in a dialogue process before the Mumbai attacks, but it was suspended due to the attacks.
[149] 'Special Advisor Holbrooke's Meeting with Saudi Assistant Interior Minister Prince Mohammed Bin Nayef', US Embassy Cable, EO12958, 17 May 2009, available at https://www.theguardian.com/world/us-embassy-cables-documents/207396
[150] 'CODEL Schiff Meets with PPP Leader Zardari: Political Moves and Balancing Aid', US Embassy Cable, EO12958, 30 May 2008, available at https://www.theguardian.com/world/us-embassy-cables-documents/156183

promotion, and for the civilian sector, such as the police, rather than for the military, which had been the primary focus throughout most of the 2000s. Contrary to the assertions by the Pakistani President, the country's army chief claimed in 2009 that American funds had mostly supported the civilian federal government's budget while the military's budget had 'flatlined', dropping 'sharply in real terms' at the time when it badly 'needed basic equipment like helmets, protective gear, and personnel carriers'.[151] He further asserted that 'the military had little incentive to provide the copious [audit] documentation [Americans asked for], since only 40 percent of the money had been returned to military coffers in the past'.[152]

The civil–military rift within Pakistan became serious after the country's return to democracy in 2008, as the military perceived the civilians supportive of the Kerry–Lugar Bill to be the 'enemies within' who wanted to destabilize Pakistan. It is pertinent to note that the army in Pakistan has always been wary of liberal individuals and civil society groups, owing to their demands for greater parity in the civil–military balance and for respect for the civil and political rights of Pakistanis.[153] Pointing towards the internal enemies, an author sympathetic to the military narrative wrote, 'there is an old maxim: "when a rival country demeans performance of your army: it is an excellent fighting machine". Nonetheless it becomes tragic when a segment of your own countrymen, not familiar with the psychological factor of the warfare, joins the chorus of the anti-army propaganda that emanates from a foreign source.'[154] Other such voices were critical of those civil society groups that considered the military's counter-insurgency tactics to be disproportionately violent. For example, writing in *Hilal*, Muhammad Khan condemned those who raised 'hue and cry, accusing security forces for the violation of human rights. These superfluous and unfounded criticisms put Pakistan's security forces at an enormously awkward position.'[155] It was actually the Western and Indian forces who were involved in such activities, while the 'Pakistan Army could not go for arbitrary killings [*sic*]' as was 'being done by India in occupied Kashmir or United States in Afghanistan and Iraq'.[156]

Where the military resisted the American efforts to rebalance the civil–military equation in Pakistan, its supporters also sent warnings to the civilian political leadership of Pakistan not to side with the Americans against the military, the 'saviour' of Pakistan. Doing so would be tantamount to disrupting

[151] 'Centcom General Petraeus Meets'.
[152] Ibid.
[153] Interview, Academic 2.
[154] Bokhari, 'Unjust Maligning of ISI', p. 13.
[155] Muhammad Khan, 'Convolution of Military Operations in Tribal Areas', *Hilal*, June 2009, p. 15.
[156] Ibid.

national unity. The civilians were warned in no uncertain terms that such endeavours in the past had resulted in military takeovers. *Hilal* wrote that 'most of [Pakistan's] regional and national parties [who were] divided on secular and ethnic lines had been manipulating the ongoing phenomenon of terrorism not only against one another; but also against the most honourable armed forces who ha[d] always stood to safeguard the nation.'[157] Such a 'selfish attitude [was] blocking the path of national unity. Past experience prove[d] that an unresolved crisis ultimately [led] to military rule in the country.'[158]

Referring back to the oft-rehearsed narrative of threats from all sides, the pro-military voices claimed that by 2010 Pakistan faced a delicate phase, with 'saboteurs'[159] entering the country's northwest from foreign training camps in Afghanistan. At the same time, the country was facing a barrage of Indian propaganda, alleging 'false links between Pakistan's Intelligence agency and [the] Taliban'. However, according to this narrative, this was a 'baseless and a disgusting plot against the integrity of Pakistan' and required the Pakistani nation to stand behind its military in such volatile times. It was incumbent on the political leaders to 'avoid manipulating any crisis not only against one another but also against the security forces as well whose image was deliberately being tarnished by the external plotters'. The masses were told that, although the situation was anxiety-inducing, such 'conspiracies' could be 'foiled' through the love of the 'people of Pakistan [for] their armed forces'.[160] In other words, if the masses maintained their faith in the military, and the military alone, internal or external threats could not destabilize Pakistan.

5.6 US–Pakistan Strategic Dialogue and the reset in 2010

With relations between the two sides at a low point by the beginning of 2010, the United States and Pakistan initiated a 'Strategic Dialogue Process' with a view to resetting their relationship. The objective of the Dialogue was to cultivate 'a strategic, comprehensive, and long-term partnership', with meetings taking place in March, July, and October of that year.[161] *Hilal* welcomed the Dialogue as 'a new long–lasting and interdependent strategic partnership

[157] Anees, 'National Integration', p. 17.
[158] Ibid.
[159] Ibid.
[160] Ibid.
[161] 'Joint Statement: U.S.-Pakistan Strategic Dialogue at the Ministerial Level', Washington, D.C., United States, 22 October 2010, available at https://2009-2017.state.gov/r/pa/prs/ps/2010/10/149854.htm

with the United States'.[162] The military portrayed the Dialogue as a sudden realization and acknowledgement among the American policy elites of the Pakistani counterterrorist success in the north-west of the country. In other words, those supporting the military narrative portrayed it as an opportunity for the United States to undertake a course correction, while Pakistan was seen as requiring no change. This viewpoint asserted that

> these smashing successes [had] changed the traditional US rhetoric of 'Do More' for the first time to [build] a long-lasting strategic partnership between the two allies. This led Clinton [the US Secretary of State at that time] to term this relationship the beginning of a 'new phase' in Pakistan-United States bilateral relations ... she further went on to declare at [the] start of the conference in Washington that 'Pakistan's struggles [were] American struggles'.

Ahmad Rashid Malik went on to write in *Hilal* that

> This [was a] historical acknowledgement and it turned down all the previous doubts, suspicions and mistrust prevailing since the beginning of the 9/11 and even before. This [was] a new beginning indeed. A sea change and a paradigm shift [was] being witnessed in Pak-US relations. A mere friendship and event-based relationship ha[d] been converted into a mutually beneficial long-lasting strategic partnership by covering wide arrays of areas and interests.[163]

According to this worldview, where the previous engagement was based on dealing with the day-to-day matters instead of a longer-term partnership, the new approach was to transcend that 'event-based' relationship:

> Hitherto, Pakistan-United States strategic partnership ha[d] revolved around the Cold War parameters till the end of the 1980s and then onward on stability in Afghanistan and terrorism. Their relationship had never been strategic before and remained opportunity and event-based [*sic*]. Now both Pakistan and United States needed to go several miles in order to expand the bilateral relations from the mere stability in Afghanistan and terrorism to building bilateral ties in the fields of development, trade, investment, aid, economic cooperation and energy.[164]

[162] Ahmad Rashid Malik, 'Pakistan-United States Strategic Dialogue: Shaping a New Relationship', *Hilal*, May 2010, p. 22.
[163] Ibid., p. 23.
[164] Ibid.

While the Americans sought a reset, a clear misalignment remained in the expectations[165] of the two parties, as Pakistan continued to adhere to its long-standing narrative regarding threats from all sides—including Indian nuclear domination of Pakistan. For Pakistan, a positive step forward in the Dialogue process would be a nuclear agreement with the United States, similar to the one India had secured years earlier. As *Hilal* noted:

> It [was] expected that the new strategic partnership could resolve all such crucial and outstanding issues between the two partners. The most crucial element to balance relations between the two countries depend[ed] on the signing of a similar type of nuclear deal, which United States inked with India in 2005. Analysts ha[d] noted a change of attitude of the US government over Pakistan's nuclear program.

The author went on to depict an alternative reality that made perfect sense to the Pakistani military, but which their American counterparts would find hard to comprehend:

> In last June [2009], President Barack Obama remarked, 'I have confidence that the Pakistani government has safeguarded its nuclear arsenal. It's Pakistan's nuclear arsenal'. A balanced approach by [the] US on nuclear issue could be a real deterrence to prevent both traditional rivals India and Pakistan, especially the former from nuclear blackmailing and Holocaust, and to bring them on the table [*sic*] for the resolution of outstanding disputes.[166]

In their fantasmatic mindset, the Pakistani military believed that the US–India agreement had already set a precedent, and it would not be difficult for a similar agreement to be signed with Pakistan:

> [The US] signed the nuclear deal with [India] by changing the basic guidelines of the Nuclear Suppliers Group (NSG) and International Atomic Energy Agency (IAEA) to materialize the deal [*sic*]. Against this backdrop Pakistan's signing of nuclear deal with United States would be less cumbersome and complex, as precedence [had] already been set.[167]

The American side, on the other hand, continued to harbour serious concerns that a civil-nuclear deal might be used by Pakistan to further develop

[165] Wali Aslam, 'Misaligned Expectations and Broken Promises: United States Foreign Policy Approach Towards Pakistan (1947–2021)', in *The Routledge Handbook of US Foreign Policy in the Indo-Pacific*, edited by Oliver Turner, Nicola Nymalm, and Wali Aslam (Abingdon: Routledge, 2022), pp. 330–342.
[166] Malik, 'Pakistan-United States Strategic Dialogue', p. 23.
[167] Ibid.

its nuclear weapons technology. All the Pakistani nuclear facilities were under direct military command and control, casting doubt on the true separation of the civil and military leadership in the country. The country produced a small amount of nuclear power, and the primary focus of Pakistan's nuclear activity had been the country's nuclear weapons programme.[168] Despite the persistent fantasies on the Pakistani side, the State Department's official statements regarding the Strategic Dialogue included no reference to a nuclear agreement.[169]

5.7 Conclusion

The first decade of the twenty-first century is a fascinating study of the alternative reality the Pakistani decision makers lived in and wanted everybody else to believe. It is a story of repeatedly resisting the international community's efforts to persuade the country to side with the rest of the world in the fight against terrorism and to sever its links with militant proxies. The country's ruling elites offered a variety of alternative interpretations indicating how they regarded the War on Terrorism. These interpretations are important to analyse because they can help us to make greater sense of the state's policies and attendant practices. For example, one viewpoint sympathetic to the military's logic claimed that the war was, in fact, a war against Muslims and Islam. Referring to the campaign in *Hilal*, Bilal Shahid contended that 'killing of Muslims and calling it [a] Crusade [was] nonsense. Every person ha[d] the right to live, irrespective of caste, creed and colour. So far millions of innocent people ha[d] been killed in the name of fighting terrorism.'[170] The author went on to say that

> The irony [was] that everything [was] happening in the Muslim countries. A boy in Palestine who fights using stones [was] called a terrorist, while the Israelis who use[d] gun ships [*sic*], jets and tanks against those stone-throwing children [were] named as saviors. Somalia [had] been ruined, Iraq razed to the ground, Palestinians [were] being crushed, Afghanistan [had] become a country of widows. And now Pakistan [was] being called a failed state.[171]

A postcolonial narrative was also deployed to deflect the pressures on Pakistan throughout that decade and beyond. This narrative likened the War

[168] 'N-Deal for Pak? US Denies, India Remains Anxious', *The Times of India*, 23 March 2010.
[169] See 'Joint Statement: U.S.-Pakistan Strategic Dialogue at the Ministerial Level'.
[170] Shahid, 'The World Should Acknowledge', p. 24.
[171] Ibid.

Against Terrorism to a colonial adventure. Those opposing the West were simply 'countering the USA's anti-Muslim policies'.[172] As the West had used the developing world as its colonies in the past, now the War on Terrorism was being waged with the agenda of 'crushing Muslims [, but] peace [could not] prevail by killing [and] the solution [lay] in having [a] dialogue'. '[Was] it justified to kill millions of people to take the revenge of a few thousand?' asked *Hilal*. It also issued a stern warning: '[r]emember everybody has to die and history never forgives the oppressors and murderers'.[173] This view held that the West had a dream of dominating the world:

> The dream of ruling the world must finish as there [was] no clash of civilizations. All countries, even the poorest/weakest, have the right to live according to their own faith, religion and traditions. Islam is the religion of peace, it does not encourage disorder and it gives rights to slaves, women, children, non-Muslims and even animals. So how can Islam favour terrorism ? ... [Instead] the world should acknowledge [Pakistan's] priceless contribution to war against terror and spare [it] from mudslinging.[174]

This narrative also criticized the civil society voices across the world that were silent when it came to 'those countries which [were] killing thousands of people in Iraq, Afghanistan and Palestine? They ... [had] the license to kill anybody, anywhere in the world because might is said to be right.'[175]

The victim narrative, present in the official discourses since 1947, also continued to feature in the period studied in this chapter. It was repeatedly asserted that it was time for the rest of the world to acknowledge Pakistani sacrifices, which was the price the country was paying for ensuring peace in the West. 'Pakistan became the front-line state in the war against terrorism and ... deployed more than 120,000 troops in FATA. Hundreds of soldiers ha[d] laid their lives, thousands injured ... On the other hand after 9/11, no major terrorism activity took place in America or Western countries because of Pakistan's concerted efforts', this narrative argued.[176] Accordingly, the West was unfairly critical of Pakistan, and instead of criticizing, it needed to work with the country to curb terrorism. The implication was that the West was not willing to work with Pakistan to address its problems, which, as noted above, was simply not the case. Pakistanis claimed that the West

[172] Ehsan Mehmood Khan, 'A View from the Other Side of the Planet', *Hilal*, October 2009, p. 25.
[173] Shahid, 'The World Should Acknowledge', p. 24.
[174] Ibid.
[175] Ibid.
[176] Ibid.

was creating a 'misperception' in the world about their country, whereas the nation was the victim of the West's proxies that were wreaking havoc across the country.[177]

Given the Pakistani side's distaste for the term 'AfPak', the Americans soon stopped using it altogether. By 2010, the Obama administration advised its officials not to use the term because it did not 'please' Pakistan.[178] Instead of modifying its policy of using militants as tools of foreign policy, Pakistan remained critical of the Obama administration's 'AfPak' approach as it blended 'two independent and sovereign nations and countries into one so-called identity for strategic reasons only known to decision-makers in the United States.'[179]

The Pakistani narrative also served to justify and enable the Afghan Taliban, rather than curtailing their support for the group to stabilize the Afghan government. The narrative continued to argue that the Taliban in Afghanistan were fighting for the 'righteous cause, namely their fair share of power in Kabul presently denied to them by a coalition of Tajiks and other minority groups.'[180] In that way, for one

> [t]o expect that the Taliban [would] cease fighting merely because they [were] outgunned[,] or in return for substantial sums of money[,] [was] not to understand the Taliban/Pashtun psyche! Their history or their proclivity [was] to accept stipends and continue doing what they [felt was] right. And it would be an equally great misreading of history to believe that in their fight against the other ethnic groups the Afghan Pashtuns would not have the support of their Pakistani brothers across the Pak-Afghan border.[181]

In what would be ironic words to remember in 2026, Zafar Hilaly asserted in April 2010:

> [The Afghan-Pakistan] border, for [the Pashtuns], is a notional boundary; moreover, the concept of nationality, as we understand it [, was] meaningless for them. In Afghanistan and among the tribal Pashtuns it [was] all about tribes and clans, not states or nationalities. Besides, it would be physically impossible for Pakistan to stop [the] Pashtuns crossing the border to join the Afghan Taliban.[182]

[177] Khan, 'Convolution of Military Operations', p. 16.
[178] Malik, 'Afpak Strategy', p. 15.
[179] Ibid.
[180] Zafar Hilaly, 'The Challenges Ahead', *Hilal*, April 2010, p. 6.
[181] Ibid.
[182] Ibid.

6

From *annus horribilis* to *annus miserabilis* and beyond (2011–2026)

6.1 Introduction

While 2010 ended with hopes of "resetting" Pakistan's relations with the West, 2011 proved to be an '*annus horribilis*'[1] of US–Pakistan relations. The year began with the 'Raymond Davis incident', in which an American security contractor shot and killed two Pakistani citizens in Lahore who were following Davis at the behest of the ISI. Davis left Pakistan shortly afterwards, leading to renewed strains between the two countries.[2] The killing of Osama bin Laden in the northern Pakistani city of Abbottabad in June 2011 without the prior knowledge of the Pakistani state was a clear message from the Americans that they did not trust Pakistan. The bin Laden episode was followed by the famous US Senate testimony of the then United States Chairman of the Joint Chiefs of Staff Admiral Mike Mullen, in which the Admiral openly described the Haqqani group of the Afghan Taliban as a 'veritable arm of Pakistan's Inter-Services Intelligence Agency', pushing relations to yet another low.[3] The final blow came in the form of the Salala incident in November 2011, in which NATO forces launched an incursion into Pakistan, killing 'at least 24 soldiers'.[4] By the end of 2011, the divergence in US–Pakistan relations was clear, with Pakistan openly criticizing what it regarded as the United States' unfair treatment despite its assistance in the War on Terrorism.

This divergence presented Pakistan with an opportunity to realign its foreign policy and bring greater balance to its relations with the great powers. Doing so included acknowledging the positions of Russia and China as equal to that of the United States and other powers. That determination coincided with the inauguration of the China–Pakistan Economic Corridor (CPEC) as

[1] 'Pakistan-US Relations: Annus Horribilis', *Crikey*, 9 December 2011.

[2] Declan Walsh, 'A CIA Spy, A Hail of Bullets, Three Killed and a US-Pakistan Diplomatic Row', *The Guardian*, 20 February 2011.

[3] Adam Brookes, 'Mullen Mauling Shows Patience Wearing Thin with Pakistan', *BBC News*, 23 September 2011.

[4] 'Pakistan Outrage After "NATO Attack Kills Soldiers"', *BBC News*, 26 November 2011.

Pakistan's Foreign Policy. Wali Aslam, Oxford University Press. © Wali Aslam (2026).
DOI: 10.1093/9780198909552.003.0006

the 'jewel in the crown' of China's Belt and Road Initiative (BRI) in 2015. The downward trend in US–Pakistan relations, along with Pakistan's desire to diversify its foreign policy, was reinforced in 2018 when President Donald Trump openly criticized the country for its 'lies and deceit'.[5]

The outlining of the 'Bajwa Doctrine' in 2018 was a logical culmination of the trend in Pakistani foreign policy to wean itself away from the United States and to broaden its foreign policy approach. Although the Doctrine promoted a more independent foreign policy, it soon became clear to Pakistan that closer ties with other great powers offered little benefit, given pervasive rent-seeking in its foreign policy. As a result of this realization, Pakistan sought to revive its ties with the United States toward the end of the decade. American interest in Pakistan, on the other hand, was decreasing quickly, as the former signed a withdrawal deal with the Taliban in 2020, with the actual withdrawal taking place in August 2021.

This chapter analyses Pakistan's foreign policy since 2011, a particularly momentous period, to closely examine the dynamics mentioned above. The country's relations with the major powers as well as its neighbours will be studied. I argue in this chapter that despite attempting to diversify its foreign policy, Pakistan has not been able to benefit from this vision due to the country's weak domestic foundation in commerce, services and manufacturing. By 2026, it became clear that Pakistan's economic difficulties made it more of a liability than an asset to anyone wanting to side with it.

6.2 The *annus horribilis* of US–Pakistan relations

The assassination of Osama bin Laden in May 2011 caught most public institutions in Pakistan by surprise. The incident was also a major embarrassment for the Pakistani military, as bin Laden was found living in close proximity to the Pakistan Military Academy in Abbottabad—an institution where military officers undergo training after recruitment.

The military's response was steeped in the usual victim narrative that has characterized the Pakistani foreign policy since the country's inception, portraying it as being under attack by the West. The raid was portrayed as an attack by 'the most civilized nation ... on the soil of a sovereign country by violating all the rules and laws without informing its officials'.[6] The Americans had not informed the Pakistani military first because the former did not

[5] Saba Aziz, 'Pakistan-US War of Words Over Donald Trump's Tweet', *Al Jazeera*, 2 June 2018.
[6] Saima Shabbir, 'We Stand Together', *Hilal*, June 2011, p. 11.

trust them with such privileged information—an indication of the state of the relationship between the two countries at the time.

After the assassination of bin Laden, the military-aligned voices intensified efforts to deflect blame from Pakistan's negligence (or even connivance) concerning his presence in Pakistan. '[T]he nation felt betrayed from a coalition partner', decried *Hilal*, for who Pakistan had suffered tremendously as it was 'because of [the] US that, [Pakistan] had internal instability, the weapons culture, falling apart social filament and ... a growing sense of insecurity'.[7] For Muhammad Khan, the nation was justified in asking: 'if this was the reward of what [Pakistan] did for the US and western world?' With its raid in Abbottabad, the US had sabotaged 'the sovereignty and national respect' of Pakistani people, leading to the development of hatred for the United States.[8]

The Pakistani narrative portrayed the fact that bin Laden was hiding in a strategically important city as an intelligence failure, but it also emphasized that other intelligence agencies in the world, including the American CIA, the British MI6, and the Indian Research and Analysis Wing (RAW) had also suffered various failures in the past. Accordingly, it was unfair to isolate the Pakistani intelligence agencies for this negligence.[9] As concerned the allegations of a 'double game', Pakistanis alleged that it was actually the other side (the US) who was 'playing all sides' and the 'sentiments of hurt and betrayal' were experienced by both sides and not just the Americans.[10] Western allegations made Pakistanis feel as if the two sides 'never knew each other and [were] disgusting foes'.[11]

A major Pakistani worry centred on the American abandonment of Pakistan and the drying up of the US aid.[12] If the US did 'dump' Pakistan, Rais wrote, it would be the 'third betrayal' of the country, following the United States' betrayal in the 1965 War and after the Soviet withdrawal from Afghanistan in 1989.[13] According to this view, severing ties with Pakistan would make defeating terrorists more difficult for the United States than it had been thus far. The Pakistani perspective highlighted that both countries had achieved 'remarkable successes' in the global War against Terrorism, and that even the Abbottabad raid was possible because Pakistan allowed an extensive CIA intelligence network to operate in Pakistan.[14]

7 Muhammad Khan, 'While Managing the Crises Situation', *Hilal*, July 2011, p. 23.
8 Ibid.
9 Shabbir, 'We Stand Together', p. 11.
10 Rasul B. Rais, 'Will the U.S. Walk Away Again?' *Hilal*, June 2011, p. 5.
11 'Reflection; Not Mere Reaction', Editorial, *Hilal*, June 2011, p. 3.
12 Rais, 'Will the U.S. Walk Away Again?' p. 5.
13 Ibid., p. 5.
14 Ibid., p. 10.

Hilal presented the 'humiliating intrusion' into Pakistan as an attempt to dent the 'prestige and national standing' of the Pakistani armed forces, since

> [m]any conspirators [were] looking for creating chaos [*sic*] by tarnishing the image of the high command. Let's not allow them to succeed. All times ready Pak army[,] to help the people of Pakistan[,] must continue getting nation's love and support [*sic*]. Let's tell our army and intelligence agencies that the heart of this nation and motherland is huge which can forgive their mistake and expect a secure future from them.[15]

The raid was presented as another example of the threats the country faced from all sides. Pakistanis were critical of the maligning of their intelligence agencies by Western media, the US State Department, and the Pentagon, all of whom wanted to 'pollute the global environment against Pakistan'. A number of domestic critics of the military were also characterized as American 'partner[s] in this malevolence defaming campaign'.[16] It was critical for Pakistanis to identify the 'real enemies' of the state who were 'insidious' with 'tentacles spread far and wide'.[17] The public was told not to 'play into their hands by maligning the military establishment of Pakistan' as it did not 'help to be the braggart on the street ridiculing and undermining state institutions. [I]nstead political and social dialogue should be tampered with sagacity and wisdom.'[18]

Admiral Mike Mullen's testimony to the US Senate in September 2011 was another difficult episode between the two states. Mullen asserted that 'proxies of the government of Pakistan [were] attacking Afghan troops and civilians as well as U.S. soldiers' in Afghanistan with 'the support and protection of the Pakistani government and [these groups were], in many ways, a strategic arm of Pakistan's Inter-Services Intelligence Agency'.[19] He stressed:

> In choosing to use violent extremism as an instrument of policy, the government of Pakistan—and most especially the Pakistani Army and ISI—jeopardise[d] not only the prospect of our strategic partnership, but also Pakistan's opportunity to be a respected nation with legitimate regional influence.[20]

[15] Shabbir, 'We Stand Together', p. 11.

[16] Khan, 'While Managing the Crises Situation', p. 24.

[17] Saadia Abbasi, 'Pakistan and the United States: Where Do We Go from Here?' *Hilal*, August 2011, p. 9.

[18] Ibid., p. 9.

[19] See the 'Statement of Admiral Michael Mullen, Chairman Joint Chiefs of Staff, before the Senate Armed Services Committee on Afghanistan and Iraq', 22 September 2011, available at https://www.armed-services.senate.gov/imo/media/doc/Mullen%2009-22-11.pdf

[20] 'Adm Mike Mullen's Address', *BBC News*, 22 September 2011.

For military-aligned Pakistanis, the Mullen testimony included very serious allegations 'without concrete evidence and proofs', which amounted to 'pouring oil on troubled waters at this critical juncture' when the world was getting close to finding 'a solution to the Afghan stalemate'.[21] This perspective categorically denied any support for the Haqqani Network and criticized the US for being 'the most difficult ally of Pakistan', which was 'unable to grasp the essence of Pakistan's efforts in the war on terror ... and during the Afghan war against the Soviet Union'.[22] According to this view, seemingly rooted in an alternate reality, the American allegations sought to deflect attention from the West's own failures because 150,000 international troops and 350,000 Afghan security forces could not stop the Afghan Taliban from entering Afghanistan to conduct attacks and to return to Pakistan. Despite its extensive drones and surveillance systems, the US was unable to track down the militants.[23]

The Salala incident was another serious blow for the two sides' relationship, since it was, in the Pakistani military's view, a 'planned massacre to provoke Pakistan'.[24] In late November 2011, the international forces attacked two Pakistani checkposts at the Pakistan–Afghanistan border after receiving fire from there.[25] In response to the incident, the Pakistan Army Chief issued an unusually stern command communiqué to his troops:

> I want to re-emphasise and leave no ambiguity in the rules of engagement for everyone, down the chain of command, especially the unit/sub-unit commanders [*sic*]. When under attack, you have full liberty of action to respond, employing all capabilities available at your disposal. This would require no clearance at any level. Army will continue to provide the resources as required, on the ground. All of us have resolve and faith to go down, if required, rather than be concerned into accepting fait accompli [*sic*]. Be assured that we will not let the aggressors walk away easily. No loss, I repeat, no loss, will be considered greater than upholding the national and institutional honour and unjust spilling of sacred blood of the soldiers of Pakistan army. I have full trust in your capabilities and resolve.[26]

Soon after the Salala episode, the Pakistani government closed the NATO supply line crossing its territory into Afghanistan (that had continued since the start of Operation Enduring Freedom in 2001). It also asked the United

[21] Ahmad Rashid Malik, 'An Alliance in Troubled Waters', *Hilal*, November 2011, p. 11.
[22] Ibid., p. 11.
[23] Muhammad Khan, 'Misperceiving Pakistan's Connections with Haqqanis', *Hilal*, November 2011, p. 14.
[24] Shaukat Qadir, 'In Search of a Foreign Policy', *Hilal*, June 2012, p. 5.
[25] Dion Nissenbaum, Tom Wright, Owais Tohid, and Adam Entous, 'Airstrike Ravages U.S.-Pakistan Ties', *The Wall Street Journal*, 28 November 2011.
[26] 'Together We Stand', Editorial, *Hilal*, January 2012, p. 3.

States to vacate the Shamsi airbase that the latter was leasing for its counter-insurgency operations, and ordered a review of all cooperation with the United States. The military's narrative asserted that the Pakistani checkposts were targeted by the Americans because the latter wanted to 'strengthen the Taliban' to keep the war in Afghanistan going.[27]

The year 2012 began with both sides in the US–Pakistan alliance viewing each other with suspicion. 'Pakistan was being publicly sidelined', as both the United States and India demanded 'stronger action from Pakistan on terrorism', specifically 'terrorist sanctuaries in the neighbourhood'.[28] The Pakistani rebuttal, on the other hand, noted that its relations with the US were important 'but not so important as to over-ride prime considerations of national security'.[29] Writing in *Hilal*, Brian Cloughley criticized the Americans' inflexible attitude, which sat

> poorly with that role of the world policeman that nobody asked it to assume. The war in Afghanistan was entirely of American choosing, and the disastrous consequences [could] be laid only at its door. Pakistan [was] a victim of this uncalled-for conflict, and rightly resent[ed] being blamed for the chaos in that unfortunate country. If the US want[ed] to engage with Pakistan—or with any other nation—it should moderate its aggressive approach and consider from time to time that much world turmoil could [have been] avoided had Washington listened to wise counsel rather than acting with customary bellicosity.[30]

As before, Pakistanis blamed the Americans for betraying them, in the same way they had done after the 1965 and 1971 wars, and the Afghan jihad of the 1980s.[31] They believed that Pakistan was forced to side with the Americans in their War against Terrorism under the threat of being 'bombed back to the Stone Age'.[32] The military narrative asserted that, ten years down the line, Pakistan had indeed been 'bombed back into Stone Age' and the country was 'seeking ways of wriggling out of a largely non-productive and unequal relationship [with Salala being] the final nail in the coffin' of the relationship.[33] This perspective accused American think tanks and the CIA of eyeing Pakistan's resource-rich but restive province of Balochistan, which has endured a decades-old separatist insurgency.[34] It asserted that, in collaboration with

[27] Ahmad Rashid Malik, 'The New Engagement with United States', *Hilal*, February 2012, p. 11.
[28] Brian Cloughley, 'Not Too Far', *Hilal*, October 2013, p. 8.
[29] Ibid.
[30] Ibid.
[31] Tughral Yamin, 'The Massacre at Salala and its Aftermath', *Hilal*, January 2012, pp. 9–10.
[32] 'US "Threatened" to Bomb Pakistan', *BBC News*, 22 September 2006.
[33] Yamin, 'The Massacre at Salala and its Aftermath', p. 9.
[34] Muhammad Khan, 'Balochistan Needs Attention', *Hilal*, March 2012, p. 7.

Mossad and RAW, the CIA was devising 'new nefarious methods' to commit human rights abuses in Balochistan in order to defame Pakistani security forces in the province.[35] These external enemies were joined by the internal enemies in Pakistan who were bribed by the CIA—an organization with an 'unlimited supply of dollars'. The internal enemies played an equal role in maligning the Pakistani security forces.[36] Such forces targeted the 'ideological Islamic status' of Pakistan, and the country had a duty to oppose such forces even if that meant living without US foreign assistance—because there were 'no compromises on ... sovereignty, national pride, and national integrity'.[37] Once again, the narrative advanced by Pakistan's security apparatus portrayed the country as being surrounded by threats. However, if the nation stood firmly behind its military, those dangers could be effectively contained.

6.3 Realignment of Pakistani foreign policy after 2011

The rupture of its relations with the United States in 2011 compelled Pakistan to reassess its entire foreign policy approach. *Hilal* argued that the Western security umbrella, under which it was 'forced' to take shelter after Partition, was an 'attractive trap'.[38] According to that view, Americans just wanted to exploit 'Pakistan's geo-strategic position' to 'contain the rising tide of communism'. Pakistani leadership was 'relentlessly pursued' by the Americans over several decades, who 'literally seduced it into submission' for their own benefit, only to be abandoned after their purpose had been served.[39] Mirroring this narrative, Yamin also asserted:

> These spikes in the on-again of-again [*sic*] partnership can hardly be qualified as a lovers' tiff. These were based on deliberately thought out and cold-blooded calculations. Not a moment was lost to suitably discipline Pakistanis, when their actions were in contravention to American national interests [*sic*]. These measures range[ed] from relatively mild rebukes and reprimands in closed-door sessions to more open vilification campaigns and even more painful measures like stoppage of aid and political and economic sanctions.[40]

[35] Khan, 'Balochistan Needs Attention', p. 8.
[36] Ibid.
[37] Muhammad Khan, 'A Much Awaited Pakistani Stance', *Hilal*, January 2012, p. 12.
[38] Muhammad Khan, 'Security Environment in South Asia', *Hilal*, September 2011, p. 18.
[39] Tughral Yamin, 'Pakistan–US Relationship', *Hilal*, November 2011, p. 9.
[40] Ibid.

Saadia Abbasi wrote in *Hilal* that the American approach centred on controlling Pakistan through aid packages, with conditionalities including 'greater civilian control over Pakistan's military, and a binding commitment to "democracy" goals' but that aid had only benefited the 'ruling elite'.[41] Cutting off of the aid could be a blessing in disguise, as it would 'compel Pakistan to mobilize internal resources and reduce foreign reliance to which it has got addicted over the years'.[42]

Notably, some in the Pakistan military admired India for showing greater flexibility than Pakistan in its foreign policy, as it benefited from both the Soviet and the American camps during the Cold War.[43] India diversified its foreign policy to build strong relations with the European Union, the Middle Eastern states, and the Far Eastern states (especially China). Pakistan, on the other hand, had placed all its eggs in the US basket and stood 'isolated'[44] after the US deserted it. It was high time for Pakistan to 're-think and re-set its foreign and domestic policy goals taking into account its own priorities and interests, and setting a new agenda for its future'.[45] This new approach suggested greater strategic partnership with China and reduction of tensions with India.[46]

China–Pakistan relations in the last quarter of a century have been a story of 'less than meets the eye' and have primarily remained focused on defence cooperation.[47] China has generally shown caution in its dealings with Pakistan and has not been willing to extend the blanket economic, diplomatic, and financial support that the rhetoric of the Pakistan–China relationship being 'higher than the Himalayas. ... deeper than the deepest sea in the world and sweeter than honey' would imply.[48] For example, when Pakistanis approached China to help in their Kargil venture against India, the Chinese showed little appetite for intervention or support: Beijing advised Islamabad to follow the American advice of troop withdrawal from the Indian territory.[49] China also did not side with Pakistan during the 2002 stand-off between the two nuclear-armed neighbours after Pakistan-based groups attacked the

[41] Abbasi, 'Pakistan and the United States', p. 8.
[42] Talat Masood, 'Pakistan's Membership of Shanghai Cooperation Organisation (SCO)', *Hilal*, July 2017.
[43] Interview, Academic 2, London, March 2024.
[44] Qadir, 'In Search of a Foreign Policy', p. 5.
[45] Abbasi, 'Pakistan and the United States', p. 7.
[46] Malik, 'The New Engagement with United States', p. 12.
[47] Christine Fair, 'The China-Pakistan Axis: Asia's New Geopolitics. By Andrew Small. New York: Oxford University Press, 2015. Xvi, 319', Book Review, *The Journal of Asian Studies* 75, no. 3 (2016): p. 797.
[48] 'Pakistani PM Says China Ties "Sweeter Than Honey"', *Radio Free Europe/Radio Liberty*, 5 July 2013.
[49] Siegfried Wolf, 'China's Role in Pakistan: International and Domestic Implications', in *Studies on Pakistan*, edited by Sylvia Mishra (New Delhi: Paragon, 2014), p. 173.

Indian Parliament in December 2001.[50] China's public announcement that it was 'a neighbor and friend of both countries'[51] indicated Beijing's focus on its rise as a great power, which required peace in the South Asian region, and it 'no longer consider[ed] it in China's best interest to get entangled in a conflict with two nuclear-armed neighbours'.[52]

In the context of Pakistan's newfound alliance with the US after the start of the War on Terrorism in 2001, both countries signed the 'Treaty of Friendship and Cooperation and Good Neighbourly Relations' in 2005, agreeing that '"neither party will join any alliance or bloc which infringes upon the sovereignty, security and territorial integrity" of either country, while simultaneously positing that both parties "would not conclude treaties of this nature with any third party"'.[53] More importantly, they agreed that 'neither shall allow its territory to be used by a third country to jeopardize the state sovereignty, security and territorial integrity of the other and each country will prohibit, on its own soil, the establishment of organizations or institutions which infringe upon the sovereignty, security and territorial integrity of the other'.[54]

Over the past two decades, however, China has exhibited growing concern over Pakistan's use of militant proxies as instruments of foreign policy. That is because a number of Uighur militants, challenging the state in Western China, have been found to have links with the militants in Pakistan.[55] Moreover, China fears that any gains by Kashmiri separatists (enabled by the Pakistani state) could have negative consequences for separatist movements in Xinjiang and Tibet.[56] After the Mumbai attacks of November 2008, China also supported the United Nations Security Council Resolution banning the Jamaat-ud-Dawa, a front organization for Lashkar-e-Tayyaba, even though China had vetoed similar resolutions in the past.[57] Addressing the Pakistani Parliament on 21 April 2015, Chinese President Xi Jinping encouraged Pakistan to focus on internal security, which was critical for stability and development if the country wanted to follow in China's footsteps of 'peaceful development' and 'national renewal'.[58] Conspicuous by its absence was

[50] Ibid., p. 174.
[51] Paul J. Smith, 'The China-Pakistan-United States Strategic Triangle: From Cold War to the "War on Terrorism"', *Asian Affairs* 38, no. 4 (2011): p. 209.
[52] Tilak Devasher, *Pakistan: Courting the Abyss* (New Delhi: HarperCollins, 2016), p. 324.
[53] Ibid., p. 310.
[54] Ibid., p. 310.
[55] Lisa Curtis and Derek Scissors, 'The Limits of the Pakistan-China Alliance', *The Heritage Foundation Report*, 19 January 2012.
[56] Ibid.
[57] Ibid.
[58] Devasher, *Pakistan: Courting the Abyss*, p. 311.

a mention of the Kashmir dispute in that address, hinting at the need for Pakistan to work with India directly to find a resolution to the problem and not depend on China for that purpose.[59]

China's worries about the fragility of the Pakistani state and the threat it faces from religious militants led it to veto a March 2016 UN Security Council resolution banning Jaish-e-Muhammad, another anti-India militant group based in Pakistan. The Chinese feared that militants 'would take up arms against the Pakistani state (as in the past) which [would] create further instability in Pakistan', undermining China's economic and geostrategic interests in the country.[60]

Although China was not willing to get involved in Pakistan's disputes with other countries, its military support for Pakistan remained steadfast, as it assisted with military hardware, weapons, and nuclear technology.[61] Pakistan is valuable for China because the former has experience of working closely with Western militaries—an experience which the latter values tremendously. Furthermore, China values the fact that Pakistan displays Chinese weaponry in its arsenal for the rest of the world to see.[62] Islamabad also keeps its lips tightly sealed on the sensitive issue of the persecution of the Muslims in Xinjiang despite being a large Muslim country. That provides China a diplomatic cover in the international community.

As Devasher notes, '[f]or China, militarily equipping Pakistan has been a low-cost option to keep India bogged down and threatened with a potential two-front war. By 2013, Pakistan had emerged as the largest recipient of Chinese arms export, indicating a drawing down of reliance of the Pakistan military solely on Western sources.'[63] However, there have always been limits to how much China could help Pakistan financially. China has been reluctant to provide large sums of cash that Pakistan has requested on various occasions, such as after the financial crisis of 2008 to bolster its economy.[64]

Chinese diplomatic support was particularly useful for Pakistan after the bin Laden raid, which the Pakistani military described as a clear breach of its sovereignty. Beijing asked Washington to 'respect the "independence, sovereignty and territorial integrity" of Pakistan and announced [that] it would expedite the delivery of 50 JF-17 fighter aircrafts equipped with

[59] Ibid.

[60] Raj Verma, 'China's New Security Concept: India, Terrorism, China's Geostrategic Interests and Domestic Stability in Pakistan', *The Pacific Review* 33, no. 6 (2019): p. 991.

[61] Andrew Small, *The China-Pakistan Axis: Asia's New Geopolitics* (New York: Oxford University Press, 2015), p. 29.

[62] Interview, Academic 2.

[63] Devasher, *Pakistan: Courting the Abyss*, p. 311.

[64] Jane Perlez, 'Rebuffed by China, Pakistan May Seek I.M.F. Aid', *The New York Times*, 18 October 2008.

upgraded avionics to Pakistan'.[65] Importantly though, China did not want to go overboard in its support of its South Asian partner. For example, when the Pakistani Defence Minister announced at that time that Pakistan had 'invited China to start building a naval base at Gwadar Port, Chinese officials publicly dismissed the notion'.[66]

The most discussed milestone between the two countries in the twenty-first century has been the China–Pakistan Economic Corridor (CPEC). Dubbed by some as a 'game changer',[67] CPEC was informally launched in 2013, but was formally inaugurated with much fanfare in April 2015 during the Chinese President's visit to Pakistan. China pledged to invest up to $65 billion in building a variety of infrastructural projects in Pakistan between 2015 and 2030, including power stations, roads, bridges, and railroads. The project was expected to create '2.3 million jobs in Pakistan by 2030 and provide an alternative pathway for exports and energy imports from West Asia to China, linking China's western provinces to key global sea lanes through Pakistan's Gwadar port'.[68] CPEC's two components included 'the corridor' (comprising roads and railway lines) and 'the projects' (infrastructural ventures), with roughly 25 per cent of the Chinese investment going towards the former and around 75 per cent of the funding earmarked for the latter.[69] A '10-year scorecard' of the project published in 2023 concluded

> [i]nvestment-starved Pakistan has so far reportedly attracted direct investments worth US$25.4 billion from China. Electric power projects under CPEC also fed an estimated 6000 megawatts into the grid for the energy-deficient nation, expanding the transmission network by nearly 1000 kilometres. [Pakistan] witnessed the construction of some 500 kilometres of highways, a huge expansion for the populous but underdeveloped South Asian nation.[70]

However, as with the broader China–Pakistan relationship, the benefits of CPEC have often been overstated. In practice, Pakistan's political instability, security challenges in the south-west of the country, disputes among federal units, and corruption have contributed to halting the project, and the future outlook remains unpromising.[71] According to well-informed observers, up

[65] Curtis and Scissors, 'The Limits'.
[66] Ibid.
[67] Jamil Khan, 'CPEC—A Game Changer for Pakistan', *The Business Recorder*, 12 August 2022.
[68] Ayjaz Wani, 'Pakistan: Why Has China Halted CPEC Projects', *Observer Research Foundation*, 25 November 2020.
[69] Farrukh Saleem, 'Gwadar', *The News*, 31 July 2016.
[70] Syed Fazl-e-Haider, 'China's Big Gamble in Pakistan: A 10-Year Scorecard For CPEC', *The Interpreter*, 1 August 2023.
[71] Haroon Janjua, 'Is China's Investment in Pakistan Slowing Down?' *DW*, 10 November 2023.

to 80 per cent of the investment in Pakistan can be wasted in corruption and leakages, which China is unable to avoid or benefit from.[72] A senior government official noted to me that the project stalled because Pakistanis were unable to conceive new ideas.[73] Persistent insecurity in Balochistan and Khaibar-Pakhtunkhwa (KP) provinces of Pakistan also pose a challenge, as Chinese personnel working in these provinces have frequently been targeted.[74] Some persecuted Balochis view CPEC as '"neo-colonialism" of their land by the Punjabis in collaboration with China'[75] and consider it their core mission to resist it. The Pakistan army has allowed the conflict in Balochistan to simmer over the last two decades, particularly in its latest phase, to justify tight control over the province, whose resources the state heavily relies on. Unsurprisingly, the negative effect of that policy has led to the stalling of CPEC—yet another example of two poorly thought-out policy visions coming into conflict with each other.[76] There is also a strong possibility of external involvement behind the insurgencies in Western Pakistan. In 2015, referring to the need to respond to Pakistan's long-running interference in Kashmir, the then Indian Defence Minister, Manohar Parrikar, outlined the policy of neutralizing 'terrorists through terrorists'.[77] Thus, it is often speculated that India could be one of the countries funding militants attacking Chinese targets in Pakistan.

Questions also persist over the economic viability of the project. As Small has noted, 'the cost of [shipping one barrel of oil] overland via Gwadar and Xinjiang would run at between four and five times that of the sea route through Shanghai'.[78] Farrukh Saleem calculated in 2016 that if it were to cost $10 per cubic metre for a barrel of oil to be shipped via the Malacca Straits, it would be in the range of $70 per cubic metre for it to be shipped via the overland route from Gwadar to Kashgar in China.[79] There are also tensions between the different provinces of Pakistan regarding where Chinese investment should focus, often described as the 'route controversy'.[80]

[72] Wani, 'Pakistan: Why Has China Halted CPEC'.

[73] Interview, Government official, Islamabad, Pakistan, January 2024.

[74] Manish and Prashant Kaushik, 'CPEC, Afghanistan and India's Concerns', *India Quarterly* 75, no. 2 (2019): pp. 253–261.

[75] Seema Khan and Zahid Shahab Ahmed, 'Impact of the China-Pakistan Economic Corridor on Nation-Building in Pakistan: A Case Study of Balochistan', *Journal of Contemporary China* 33, no. 150 (2024): p. 6.

[76] Interview, Journalist 3, Lahore, Pakistan, January 2024.

[77] Masahiro Kurita, 'The "Pakistan Factor" in China-Indian Relations', *The National Institute of Defence Studies Commentary Number* 48, Ministry of Defence, Japan, 29 July 2015, p. 2, available at https://www.nids.mod.go.jp/english/publication/commentary/pdf/commentary048e.pdf

[78] Small, *The China-Pakistan Axis: Asia's New Geopolitics*, p. 102.

[79] Saleem, 'Gwadar'; also see Devasher, *Pakistan: Courting the Abyss*, p. 314.

[80] Hasan Naser Khan, 'CPEC for Punjab or Pakistan: Myth and Reality', *The Express Tribune*, 17 January 2016.

The government of Balochistan worries that the dominant Punjab's political leaders prefer the eastern route (passing through Punjab and Sindh) as opposed to the western route (that goes through KP and Balochistan provinces) in order to reap greater benefits of CPEC. According to a 2015 report by the government of Balochistan, 'the provincial government claimed that "despite denials the route ha[d] been changed to pass through central Punjab"'.[81] According to *The Express Tribune*, the report asserted that '[b]y selecting the eastern route, the government [was] trading off today's security risks with provincial discord and political instability in the future'.[82] In an unprecedented move, the Chinese embassy in Islamabad issued a statement in 2016 referring to the provincial disharmony. It noted:

> We hope that relevant parties could strengthen communication and coordination, solve differences properly, so as to create favourable conditions for the CPEC. We are ready to work with [the] Pakistani side to actively promote construction of the CPEC projects, and bring tangible benefits to the peoples of the two countries.[83]

There have also been concerns regarding transparency, as the CPEC agreement has not been made public by either state, and the key details regarding the debt Pakistan will assume as part of the project remain unknown to both the Pakistani people and the wider world.[84] Additionally, there are fears of Pakistan falling into the Chinese 'debt trap'.[85] In early 2024, Pakistan's foreign debt stood at $124.5 billion, which was 42 per cent of its GDP,[86] and some estimates concluded that about one-third of that debt was owed to China.[87]

More than ten years into its construction, the CPEC route increasingly appeared to be 'a redundant transport corridor' whose value only lay in possibly protecting 'China during a naval blockade or other wartime scenarios'.[88] As a well-known Pakistani economist asserted:

> Gwadar as a trans-shipment port to China makes no economic sense. Truck cargoes from Gwadar to China travelling for 7,000 km to 8,000 km make no economic sense.

[81] Shahbaz Rana, 'Eastern CPEC Route Unfeasible: Report', *The Express Tribune*, 25 July 2015.
[82] Ibid.
[83] Obaid Abrar Khan, 'Pak Political Parties Must Hammer Out Rifts: China', *The News*, 10 January 2016; Devasher, *Pakistan: Courting the Abyss*, p. 317.
[84] Raza Khan, 'CPEC Agreement is Confidential, Ahsan Iqbal Tells Senate', *Dawn*, 28 July 2016.
[85] Keshav Padmanabhan, 'How Much Debt Is Too Much Debt? Curious Case of Pakistan's $69 Billion Chinese Debt Trap', *The Print*, 13 November 2023.
[86] Pierre van der Eng, 'Foreign Debt Will be Pakistan's Economic Touchstone in 2024', *East Asia Forum Report*, 5 March 2024, available at https://eastasiaforum.org/2024/03/05/foreign-debt-will-be-pakistans-economic-touchstone-in-2024/
[87] Fazl-e-Haider, 'China's Big Gamble in Pakistan'.
[88] Fair, 'The China-Pakistan Axis', p. 798.

> Oil to China through Gwadar makes no economic sense. Similarly, Chinese products traveling for 7,000 km to 8,000 km all the way to Gwadar make no economic sense.[89]

China appears to have realized now that 'the overland connection is beset with difficulties because of geographical, economic and security problems, and that China's long-term motivations for maintaining a presence in Pakistan are likely to be chiefly geopolitical rather than geo-economic'.[90] One South Asian defence analyst also noted:

> the primary Chinese interest in Pakistan has been regional rather than purely bilateral, and security-driven rather than economic. China has used Pakistan as a cat's paw against India. For this purpose, China has given Pakistan not merely conventional military supplies but ready-made nuclear and missile capabilities to keep India focused on the threat from Pakistan rather than on the emergence of China. These defence and political bonds are likely to remain the cornerstone of the relationship.[91]

The Pakistani military remained cautious in its engagement with China and implicitly accepted the challenges facing CPEC, reflecting its ambivalent stance towards the project. The military's premise for dealing with China was that of a 'controlled engagement', as it did not want Pakistan to become another Sri Lanka, which declared a default on its debt payments in 2022.[92] The Pakistani military did not grant China access to strategic areas near Gwadar. It was suspicious of Chinese contacts with Baloch nationalist militants, fearing that China might negotiate directly with them to halt attacks on CPEC targets. The military worried that Pakistan could become a satellite state of China, whose intelligence network was operating on its turf. These were some of the factors that led the Pakistani state to accept the slowing down of the entire venture by the end of the 2010s.[93]

On the other front, amid the downward trend in US–Pakistan relations, President Trump, in his August 2017 remarks on Afghanistan and Pakistan, accused Pakistan of providing 'safe havens to agents of chaos, violence, and terror', and stated that the US could 'no longer be silent about Pakistan's safe havens for terrorist organizations, the Taliban, and other groups that pose[d] a threat to the region and beyond'. The president asserted that 'Pakistan ha[d]

[89] Saleem, 'Gwadar'.
[90] Jeremy Garlick, 'Deconstructing the China-Pakistan Economic Corridor: Pipe Dreams versus Geopolitical Realities', *Journal of Contemporary China* 27, no. 112 (2018): p. 519.
[91] Devasher, *Pakistan: Courting the Abyss*, p. 328.
[92] Interview, Academic 2.
[93] Ibid.

much to gain from partnering with [the American] effort in Afghanistan [and] much to lose by continuing to harbor criminals and terrorists.'[94] The United States National Security Strategy (NSS) of December 2017 also added that 'the United States continue[d] to face threats from transnational terrorists and militants operating from within Pakistan.'[95] The Trump administration went one step further in January 2018, suspending 'almost all security aid to Pakistan' because the latter had 'failed to deal with terrorist networks operating on its soil.'[96]

Instead of considering a course correction, the Pakistani military narrative strongly rebutted American demands to 'do more', arguing that Pakistan had 'done enough and [... couldn't] do ... anymore for anyone. [Instead,] it [was] time Afghanistan and the U.S. d[id] more for Pakistan.'[97] The Trump approach was described by *Hilal* as a 'rehash of past failed policies' at the time when the US administration was busily working with India to give it a greater role in Afghanistan for 'the encirclement of Pakistan from the Afghan and Indian sides.'[98] As Amir Zia wrote, Pakistan was 'being painted black, defamed and criticized' for the Western failure in Afghanistan. Western and Indian propaganda was incessantly painting Pakistan as 'a "perfect villain" or a "rogue state"' to deflect blame for the West's 'failures, half-done jobs, shortcomings and poor strategy' in Afghanistan.[99] Zia declared that Pakistan could not afford to go around opening new fronts on American orders and that it needed to 'set priorities in line with its national interests.'[100]

6.4 The Bajwa Doctrine and completing the circle

Given the focus on the 'rebalance', the Pakistani army chief General Qamar Javed Bajwa put forward the 'Bajwa Doctrine' in 2018.[101] It outlined the military's vision for the domestic and foreign policies of Pakistan, with a special focus on the country's security policy in the context of its relations

[94] Zafar Jaspal, 'U.S. National Security Strategy and Pakistan: An appraisal', *Hilal*, January 2018.
[95] Ibid.
[96] 'US Cuts Pakistan Security Assistance Over Terror Groups', *BBC News*, 4 January 2018.
[97] Jaspal, 'U.S. National Security Strategy'.
[98] Zamir Akram, 'No U.S. Exit Strategy from Afghanistan', *Hilal*, January 2018.
[99] Amir Zia, 'Unrealistic Expectations', *Hilal*, February 2016.
[100] Ibid.
[101] '"Bajwa Doctrine" Will Bring Durable Peace: Ghafoor', *Dawn*, 12 January 2018; Kemal Alam, 'The Bajwa Doctrine: The Pakistani Military Has Done More Than Enough', *RUSI Newsbrief* 38, no. 1 (16 February 2018); Suhail Warraich, 'The Bajwa Doctrine: From Chauvinism to Realism', *The News*, 18 March 2018. See also the comments by General Bajwa at the Munich Security Conference, 17 February 2018, available at https://securityconference.org/assets/02_Dokumente/03_Materialien/MSC2018_speech_QamarJavedBajwa.pdf; 'General Bajwa's address at the Islamabad Security Dialogue', *Dawn*, 18 March 2021.

with the United States. The Doctrine was based on the premise that 'in the face of constant public humiliation dished out by the Americans, Pakistan desire[d] to balance its sovereignty and interests against American pressures by embracing new centers of power' such as Russia and China.[102] The Doctrine emphasized that 'the Pakistan Army ha[d] done enough, it [was] time for the world to do more'.[103] Pakistan was not willing to accept any longer 'American threats and directives' and was prepared to 'face cuts in the American military aid',[104] but wanted its own sacrifices in the War on Terrorism to be acknowledged, instead of being subjected to 'public humiliation'.[105] The Pakistani state was only responsible for stability within its own borders, while the responsibility for stability in Afghanistan rested on the shoulders of the Afghan and US security forces.[106] The Doctrine also stated that 'Pakistan ha[d] no expansionist designs about Afghanistan [and it was] openly abandon[ing] the idea ... of strategic depth [which treated] Afghanistan as the fifth province of Pakistan'.[107] It wanted both Afghanistan and Pakistan to stop using their territories against one another.

The Bajwa Doctrine advocated a reorientation of Pakistan's strategic outlook from 'geopolitics' to 'geoeconomics'. That meant prioritizing economic growth, trade enhancement with its neighbours, greater regional integration, and peace. That vision could be achieved by 'neutralizing' conflicts within Pakistan and by improving relations with its neighbours.[108] Non-interference in the affairs of other states would boost trade and connectivity within the region, while sustainable development and prosperity could be achieved by attracting investment across the wider South Asian region as a single economic hub.[109]

The Doctrine also envisaged bringing some kind of normalization to Pakistan's relations with India. The two countries had continued an off-and-on 'composite dialogue' in the first decade of the twenty-first century. The dialogue culminated in the government of Pakistan deciding at the cabinet level to grant India the status of Most Favoured Nation (or MFN) in 2011 to fully normalize trade between the two states.[110] The MFN status meant 'equal trade benefits' with 'low tariffs or nontariff barriers'[111] between the two countries.

[102] Huma Baqai, 'Pakistan-India-U.S.: The Troubled Triangle', *Hilal*, April 2018.
[103] Ibid.
[104] Ibid.
[105] Ibid.
[106] Ibid.
[107] Suhail Warraich, 'The Bajwa Doctrine'.
[108] Baqai, 'Pakistan-India-U.S.'.
[109] Huma Baqai, 'Towards Conflict Resolution and Lasting Peace', *Hilal*, April 2021.
[110] See Zeeshan Haider, 'Pakistan Grants India Most Favoured Nation Trade Status', *Reuters*, 2 November 2011.
[111] Zafar Mahmood, 'Pakistan Conferring MFN Treatment to India', *Hilal*, January 2012, p. 20.

This dialogue was carried forward by Pakistan's civilian leadership during the 2010s and received military backing, given the country's financial difficulties amid the deterioration of relations with the United States at that time. The government of Nawaz Sharif, entering office in 2013, made closer ties with India one of its key priorities, with a view to finally signing off on the MFN status of India in early 2014.[112] The military also promoted the idea at the time through its public relations outlets. This stood in stark contrast to the historical portrayal of India as a threat against which the country was willing to fight a one-thousand-year-long war. The new emphasis was on learning to live with those different from oneself. As Rizvi stated in *Hilal* at that time,

> Pakistan's foreign policy need[ed] to reduce any country specific obsession and develop relations with other states on mutually beneficial considerations. Domestic and international politics should not be viewed as a function of religion. There are no purely religious wars or purely religion based friendship [*sic*] in the present-day world. Pakistan may emphasize common religion and culture with Muslim states, especially those from the Middle East. However, this cannot be a basis of enduring relationship unless other and practical considerations are developed in terms of political, security, economic and trade interests.[113]

Rizvi went on to say that '[t]he attitude of hostility towards the states diverging from your worldview strengthens ultra-nationalism in the domestic context and results in isolationism at the global level',[114] which was not an option for Pakistan. It needed to 'cultivate friendly interaction with as many countries as possible if it want[ed] to cope with internal and external challenges'.[115] The world had moved on from a Cold War mindset and the era of 'power blocs and permanent enmity or enduring friendships'[116] had come to an end. In a world in which relationships were 'need-oriented and functional', no country could afford to have 'a permanent friend or permanent adversary'.[117] Pakistan also needed to build

> relationships with all major powers on 'need basis' rather than expecting other countries to adopt Pakistan's global and regional agenda as a pre-condition for good relations [*sic*]. Mutually advantageous relationship [could] be cultivated through astute and active diplomacy.[118]

[112] Ankit Panda, 'Pakistan to Grant India Most-Favored Nation Status', *The Diplomat*, 19 March 2014.
[113] Hasan Askari Rizvi, 'The Imperatives of Changing World Order', *Hilal*, March 2014.
[114] Ibid.
[115] Ibid.
[116] Ibid.
[117] Ibid.
[118] Ibid.

According to this new thinking, the military was popularizing the idea of trade with Pakistan's neighbours, especially India. To accomplish this objective, the country needed to be connected to the global processes of trade and commerce. These processes also had a strong regional component as, according to the military's beatific outlook, all states in a 'regional economic cooperative arrangement'[119] benefited together from the fruits of foreign investment. This meant that the 'needs of economic and trade cooperation often [impelled] ... states to defuse their political disputes.'[120] In the contemporary globalized world, regions grow and contract together, and isolation from the international system was not an option for any country, including Pakistan. *Hilal* asserted that the 'more a country is isolated the less are its chances to function effectively in the present day global order and protect its national interests. Engagement is the key to expanding a country's options at the global level.'[121]

The prime ministers of both countries at the time, Nawaz Sharif and Narendra Modi, appeared to share a cordial rapport, underscored by the latter's impromptu visit to Lahore in December 2015. However, the Pakistani military sought to maintain complete control over the dialogue process, as it had never allowed a civilian leader to engage with India by sidestepping the institution.[122] As always, the military feared losing relevance if civilians on both sides established a direct relationship. Given that background, the country's security establishment did not tolerate Nawaz Sharif taking the lead in the process, even if it meant scrapping trade normalization between the neighbours at a time when relations with the United States were practically broken and the future of American financial assistance was in doubt. Sharif's overtures to Modi were among the factors that strained relations between the military and the civilian leadership in Pakistan at the time. Another irritant between the two was a request by the then army chief, General Raheel Sharif, for an extension of his tenure—something the Nawaz Sharif government was not willing to permit.[123] These disagreements, coupled with a lack of dialogue between the civilian and military leadership in Pakistan,[124] formed the backdrop to a major terrorist attack in Pathankot, in Indian Punjab, in January 2016, which led to an immediate deterioration of relations between the two states. The attack, likely a Pakistani military venture brought about by allied militants,[125] was marked by numerous border skirmishes in subsequent years,

[119] Ibid.
[120] Ibid.
[121] Ibid.
[122] Interview, Academic 2.
[123] Interview, Journalist 3.
[124] Interview, Academic 2.
[125] Ibid.

ultimately prompting India to revoke Pakistan's MFN status in February 2019, which it had granted in the 1990s.[126]

The Bajwa Doctrine's emphasis on reviving the trade dialogue with India must be interpreted against this background. The premise was that Pakistan could model its relations with India on the trade framework that characterized India–China relations, whereby both sides set aside their border dispute in order to expand bilateral trade and derive economic benefits.[127] Similarly, if China and Taiwan also continued to trade despite serious disagreements on the matters of reunification, nothing prohibited India and Pakistan from doing the same. *Hilal* noted that Kashmir was an important issue, but the two sides could start with 'low-hanging fruits'[128] before tackling bigger challenges. In the twenty-first-century environment, where states are connected through shared economic and security interests, an exclusionary Cold War paradigm of prolonged competition and perpetual confrontation was no longer relevant. As Baqai asserted in *Hilal*

> General Bajwa's statement [was] not in any way indicative of abandonment of the Kashmiri cause. It [was] in alignment with the new thrust of Pakistan's foreign policy, which [was] pivoted on geoeconomics. Thus, the desire to normalize the relations rather than cold war situation between the two protagonists of the region [*sic*]. Pakistan and India ... tried everything, they fought wars, used other means than wars, ... indulged in propaganda war against each other, and ... also tried to talk to each other and ... failed several times. Let this thaw be sustainable and culminate into something which may change the fate of this region, and more importantly, its people.[129]

The Bajwa Doctrine was keen to reiterate that the two countries could talk peace while keeping their differences on core issues.[130] It appeared that the military took heed of the Chinese advice to Pakistan to find peaceful measures to negotiate over Kashmir instead of resorting to force.[131]

The Bajwa Doctrine also clarified that distancing itself from the United States did not mean becoming a Chinese client state. There was a need to maintain balance in Pakistani foreign policy in its dealings with the

126 'India Withdraws Most Favoured Nation Status for Pakistan', *Dawn*, 15 February 2019.
127 Rizvi, 'The Imperatives of Changing World Order'.
128 Baqai, 'Towards Conflict Resolution and Lasting Peace'.
129 Ibid.
130 Warraich, 'The Bajwa Doctrine'.
131 Curtis and Scissors, 'The Limits of the Pakistan-China Alliance'.

world's major powers.[132] Additionally, a dialogue with the United States over controversial issues was integral to the relations between the two states.[133]

Although it had been in the pipeline for years, the CPEC project acquired renewed significance within this broader context of geopolitical realignment. The supporters of the project believed that Pakistan had always tried to reap the benefits of its geography, but 'they were largely conflict dividends for which Pakistan later paid a very high cost'.[134] CPEC was the first time when the geography of Pakistan was going to be utilized 'to reap [the] real potentials of being a trade and energy corridor', which would also help it build a strategic relationship with China, an emerging global power.[135] This use of geography for greater economic benefit was described as a 'geo-economic blessing'.[136]

According to a serving military officer writing in *Hilal*, the world could be seen according to two nexuses during the mid-2010s, with the first nexus (comprising India, Israel, and the United States) vying for world domination. The second nexus included Russia, China, Iran, and 'other sufferers of [the] unnatural relationship' of the first nexus.[137] The former nexus bullied other Asian states such as Sri Lanka, Pakistan, and China in the name of human rights, and it would serve Pakistan to firmly join the Russian and Chinese bloc for its economic development.[138]

6.5 Pakistan beyond the Bajwa Doctrine

As with any other strategy or doctrine, the Bajwa Doctrine can only be analysed in the context of the time in which it was presented and with reference to the regional and global political dynamics the country faced. The Doctrine emerged at a time when the relations between the United States and Pakistan were at another historic low. A return to the level of warmth seen at the start of the War on Terrorism appeared unthinkable. Only when the Pakistani military could no longer expect to benefit from American largesse did it begin to discuss a 'rebalance' of its foreign policy, focusing on friendship and trade with its neighbours, particularly India.

[132] Baqai, 'Pakistan-India-U.S.: The Troubled Triangle'.
[133] Warraich, 'The Bajwa Doctrine'.
[134] Huma Baqai, 'Pakistan's Foreign Policy Turns a New Leaf', *Hilal*, July 2015.
[135] Ibid.
[136] Nazir Hussain, 'The Contemporary Dynamics of Middle Asia and Pakistan's Role in Peace and Development', *Hilal*, May 2018.
[137] Khan, 'Security Environment in South Asia', p. 19.
[138] Tughral Yamin, 'Sri Lanka in the Cross-Hair', *Hilal*, April 2012, p. 14.

Pakistan's desire to pursue peace with its neighbour should also be viewed in the context of the threats by the Financial Action Task Force (an international body tasked with countering money laundering and terrorist financing) to blacklist Pakistan for sponsoring militant groups, which would have devastated its economy. Pakistan was grey-listed, a step before being blacklisted, in 2008 and 2012, and removed from that list in 2010 and 2015, respectively. As mentioned previously, it was placed on the grey list again in 2018, with the latest removal taking place in 2022.[139] The pressure from FATF forced the Pakistani state to freeze its policy of using militant proxies to launch infiltration attacks inside India.[140] Being blacklisted was something that Pakistan could not afford, given the rapid drying up of international financial assistance. That meant the country had to put its key foreign policy tool of using militant proxies into cold storage and rekindle its desire to talk peace with India from 2018 onwards. The major shift in Pakistan's 'threat perception' resulted not from gentle persuasions by the international community, but rather from the threat of FATF black-listing. It was no surprise that the Pakistani state only chose to issue strongly worded press statements when India revoked Article 370 of its Constitution which had granted Jammu and Kashmir a special autonomous status within India.[141] Compared to its track record of historical violence against India, that reaction was dubbed as 'farcical' by informed observers.[142] In the face of FATF restrictions, compounded by its debilitating economic problems, Pakistan could do little but protest, aware that its freedom to operate while grey-listed was severely constrained.[143]

Given the regional geopolitical constraints, General Bajwa, stated at the Islamabad Security Dialogue in 2021 that 'national security [was] not only about protecting a country from internal and external threats but also providing [a] conducive environment in which aspirations of human security, national progress and development could be realized. Surely, *it is not solely a function of armed forces anymore*.'[144] Any country could not have national security alone, as '*a nation at peace and a region in harmony*' remained 'essential prerequisites' for the attainment of one's national security.[145] For him, 'it

[139] See 'There and Back Again: A Timeline of Pakistan's Journey Out of the FATF "Grey List"', *Dawn*, 21 October 2022.
[140] 'Hafiz Saeed: Will Pakistan's "Terror Cleric" Stay in Jail?' *BBC News*, 13 February 2020.
[141] The Article placed limits on the Indian parliament's legislative powers to defence, finance and foreign affairs. It also restricted non-residents from owning property in the regions. The revocation brought the territories under full control of the central government.
[142] Interview, Journalist 3.
[143] Taj M. Khattak, 'Hybrid War and Pakistan's National Security', *Hilal*, January 2021.
[144] See 'Full Text of Gen Bajwa's Speech at the Islamabad Security Dialogue', *Dawn*, 18 March 2021, emphasis added.
[145] Ibid., emphasis added.

[was] naive to apply the failed solutions of yesteryears to the challenges of today and tomorrow'. Bajwa went on to assert:

> In this environment, developing countries like Pakistan, face[d] multi-dimensional challenges, which [could not] be navigated single-handedly. A similar situation [was] confronted by other countries in our region as well, therefore, we all require a multilateral global and regional approach and cooperation to overcome these challenges ... it [was] time that we in South Asia created synergy through connectivity, peaceful co-existence and resource sharing to fight hunger, illiteracy and disease instead of fighting each other.[146]

Bajwa outlined a vision of balanced relations with the great powers by asserting that 'while CPEC remain[ed] central to our vision, only seeing Pakistan through the CPEC prism [was] also misleading'. Pakistan's 'immensely vital geostrategic location and a transformed vision' made it a country of 'immense and diverse potential', which could make a positive contribution to 'regional development and prosperity'. He emphasized that Pakistan's primary foreign policy vision centred on 'geoeconomics', was composed of four pillars:

> One: Moving towards a lasting and enduring peace within and outside
> Two: Non-interference of any kind in the internal affairs of our neighbouring and regional countries
> Three: Boosting intra-regional trade and connectivity
> Four: Bringing sustainable development and prosperity through establishment of investment and economic hubs within the region.[147]

Although the Bajwa Doctrine was praised for signalling a move 'from chauvinism to realism',[148] it emerged only after Pakistan had pushed itself into a corner towards the end of the 2010s. After the American withdrawal from Afghanistan in August 2021, Pakistan's significance for the United States decreased even more. The country could have benefited from the international community's engagement and goodwill during the twenty-year occupation of Afghanistan, but that opportunity was not taken because the interests of the military were prioritized over the Pakistani state and its people. The continuous support of the Afghan Taliban might have kept the military dominant domestically, but it did not serve the interests of the country in the short or long term.

[146] Ibid.
[147] Ibid.
[148] Warraich, 'The Bajwa Doctrine'.

At the time when Pakistan was double-crossing the international community, India was joining the list of the world's foremost economic powers. It is no wonder that when the Pakistani military finally began discussing peace by the late 2010s, India had neither the desire nor the need to reciprocate.[149] By that time, an assertive India under Modi wanted the 'complete isolation' of Pakistan 'from the international community', as described by the Indian Finance Minister Arun Jaitly in February 2019.[150] Pakistan had, for decades, lost opportunities to discuss peace with India, and when it finally changed its mind due to circumstances beyond its control, the administration of Narendra Modi had its fist tightly clenched.[151] Prime Minister Modi's Hindu-centred politics depended on anti-Pakistan rhetoric and there was little incentive for him to improve relations with his neighbour.[152] In fact, hard-line rhetoric against Pakistan had major dividends for Modi's political future and he, understandably, was in no hurry to move away from that.

The Bajwa Doctrine envisaged an independent foreign policy for Pakistan, which did not correspond with Pakistan's clientelist and rent-based economic reality. The country's import-heavy economy lacked the industrial base to trade with the rest of the world. Economic growth is dependent on continuity of robust policies, which in turn requires political stability. Since the return to civilian rule in 2008, the Pakistani military, often in collaboration with the judiciary, paralysed the governments of both the Pakistan People's Party and the Pakistan Muslim League (Nawaz) between 2008 and 2018. The efforts to derail the political process included ventures like manufactured sit-ins to paralyse the capital city and summary disqualifications of elected prime ministers through the courts. These, coupled with the flawed military policy of targeting only the 'bad Taliban' and keeping contacts alive with the 'good Taliban', worsened the security situation in the country.[153] It was no surprise that these schemes brought the country's economy to its knees by 2022, a year described as Pakistan's '*annus miserabilis*'.[154] This poor economic outlook precluded a reorientation of the foreign policy. For example, when the Pakistani military attempted to reorient the country's foreign policy towards Muslim states in December 2020, shifting from traditional patrons like Saudi Arabia towards Turkey and Malaysia through their 'selected'[155] Prime Minister Imran

[149] Interview, Columnist, Islamabad, Pakistan, January 2024.
[150] 'India Withdraws Most Favoured Nation Status from Pakistan'.
[151] Interview, Politician 1, Islamabad, Pakistan, January 2024.
[152] Interview, Columnist.
[153] Zia Ur Rehman, 'Good Taliban, Bad Taliban: The Case of Hafiz Gul Bahadur', *Dawn*, 21 March 2024.
[154] Sushant Sareen, 'Pakistan: From Annus Horribilis into Annus Miserabilis', *Observer Research Foundation*, 5 January 2023.
[155] Interview, Politician 1.

Khan, Saudi Arabia asked Pakistan to 'return $1 billion of a $3 billion loan that was part of an overall aid package that included $3.2 billion oil credit facility'.[156] The Saudis demand quickly put an end to that ambition, with Pakistan stumbling back into the Saudi bloc very quickly.[157]

Instead of letting the civilian government devise Pakistan's economic and security policies, the Bajwa Doctrine was also symptomatic of the deeper malaise: the military's involvement in policy and decision-making in Pakistan's hybrid-state system. The Doctrine was the military's vision for Pakistan, and it indirectly contributed to ensuring the military's dominance over the civilian dispensation in the country. The military lacked the wherewithal to run the entire state without competent civilian partners and without an educated and skilled workforce. Its longstanding policy had both undermined civilian politicians and kept the masses impoverished, proving conducive to neither political stability nor economic development by the time General Asim Munir took charge of the Pakistan Army in November 2022. It is therefore unsurprising that Munir reverted to the nationalist and militant policies of earlier times, resulting in further curbs on civil liberties at home and a more confrontational posture towards Pakistan's neighbours, India and Afghanistan.[158] The mid-2025 Trump–Munir 'bonhomie', driven by Washington's aim to counter China and Iran, further intensified this trend.[159]

6.6 Conclusion

Pakistan's foreign policy since the start of the current century consistently followed the path of ensuring the military's dominance of the policy at the cost of the nation's prosperity. It reflected reactive engagement with circumstances rather than a well-considered strategy for moving ahead with the times. The Pakistani state's engagements with the United States was based on duplicity without an eye on the result. A government official asserted during my interview that Pakistan's double game was something 'we should be proud of', as it made the West 'shriek'. When I observed that this strategy primarily benefited a small civil–military elite while imposing negative economic consequences on ordinary Pakistanis, the official responded that such disparities were common worldwide, arguing that in any country it is the elite who benefit, and

[156] Khaled Ahmed, 'Pakistan After Hostile Cartography', *Newsweek Pakistan*, 13 August 2020.
[157] Interview, Academic 2.
[158] Omer Farooq Khan, 'Asim Munir's Grip on Power Deepens Pakistan's Militarisation', *The Times of India*, 7 August 2025.
[159] Abid Hussain, 'Amid US-Pakistan Thaw, Two Key Challenges: Iran and China', *Al Jazeera*, 20 June 2025.

that their interests should be prioritized over those of ordinary Pakistanis.[160] That exchange encapsulated the core issue of policymaking in Pakistan: the ruling elite's inability to prioritize the nation's collective benefit over the interests of a select few. The decades-long double-crossing of the United States was done without giving any thought to the fact that a day might come when the world would no longer tolerate that approach. As another interviewee put it aptly, 'how long can you take a folly and a lie'.[161]

Pakistan's relationship with India also told the story of a dominant military showing little interest in the welfare of the rest of the country. As long as the West was financially supporting Pakistan, the latter continued to use militant proxies to launch attacks inside India. Additionally, the rivalry with India was used as an excuse whenever American officials requested Pakistan to play a helpful role in securing the Afghanistan-Pakistan border. The Pakistani military consistently argued that it could not shift the focus away from the eastern border to help against the Afghan militants. It was only after the threat of FATF blacklisting that Pakistan realized that it could no longer use that option.[162] That, combined with the dire state of the economy,[163] led Pakistan to talk about peace with India by 2020/2021. But by this point it was too late as India was no longer interested in that option.[164] This was ironic because in the past, New Delhi had extended its offer of talks several times, but Pakistan rebuffed it, asserting that the offers were not genuine. However, when Pakistan finally realized that jihad through militant proxies was no longer an option, and it had no choice but to trade with India,[165] its neighbour had moved on.

Pakistan transitioned towards a civilian government in 2008 with the departure of General Pervez Musharraf. However, despite civilians exercising comparatively greater control, the country continued to operate under a hybrid system of governance.[166] The military, for its part, opposed a return to a two-party system and ultimately engineered Imran Khan's rise as a disruptive force, facilitating his election in 2018 through alleged rigging before later orchestrating his removal in 2022.[167] The country's economic and

[160] Interview, Government official.
[161] Interview, Journalist 3.
[162] Interview, Academic 2.
[163] Interview, Government official.
[164] Interview, Columnist.
[165] Interview, Government official.
[166] Interview, Politician 1.
[167] On Imran Khan's views regarding the US security involvement in Pakistan, see Wali Aslam, 'The Political Appropriation of Casualties in Threat Construction: The Case of US Drone Strikes in Pakistan', in *Precision Strike Warfare and International Intervention: Strategic, Ethico-Legal and Decisional Implications*, edited by Michel Aaronson, Wali Aslam, Tom Dyson, and Regina Rauxloh (Abingdon: Routledge, 2015): pp. 73–92.

political institutions are yet to recover from those experiments by the military. An interviewee described the decade of 2014–2024 as the 'lost decade'[168] due to the military's 'Project Imran Khan'. After his removal, the military openly admitted that launching him was a mistake[169] as the two sides did not get along. The military's internal political interference and external double games left the country's economy in tatters.[170]

It has been asserted by analysts that a situation of 'controlled chaos'[171] in the country suited the Pakistani military in the past. Perpetual tensions with India and the use of militants as instruments of foreign policy created instability which ensured the military's dominance within the state. However, by the time of writing in 2026, that approach had become untenable, as external rents had diminished and the civilian side had been weakened to the point of posing no threat to the military's dominance. An enhanced role for the military in governance within a hybrid system meant that it was now directly accountable for the country's underdeveloped economy. Consequently, it became even more imperative for the military to bolster its image by all means at its disposal.[172] By 2026, the case for direct trade with India had become even stronger, given that 'informal trade' between the two countries had reached an estimated $10 billion by 2024–2025.[173] India and China might have several disagreements, yet their trade was thriving at the time of writing: Indian trade with China during the fiscal year 2023–2024 grew to $118.4 billion, with imports standing at $101.7 billion and exports at $16.67 billion.[174] Pakistan's trade with China, on the other hand, was only $1.4 billion, with imports counting for $1.162 billion and exports amounting to a mere $246 million.[175] Pakistan's inability to export high-tech products to China explained the low trade volume between the two, as its exports primarily consisted of low-value raw materials and commodities.[176] That, together with the limited 'people-to-people contacts', underscored the shallowness of the 'high-sounding epithets of the political and military relationship'.[177]

[168] Interview, Government official.
[169] Interview, Columnist.
[170] Interview, Journalist 3.
[171] Abbas Nasir, 'Controlled Chaos? Think Not', *Dawn*, 18 February 2024.
[172] Interview, Academic 2.
[173] Priyanka Shankar and Yashraj Sharma, 'The $10bn India-Pakistan Trade Secret Hidden by Official Data', *Al Jazeera*, 1 May 2025.
[174] Ravi Dutta Mishra, 'India's Top Trade Partner: China Regains Spot on Higher Imports', *The Indian Express*, 13 May 2024.
[175] Figures from http://www.ceicdata.com
[176] Devasher, *Pakistan: Courting the Abyss*, p. 313. See also Saeed Shafqat, 'CPEC and Changing Dynamics of China–India Relations', in *China and South Asia: Changing Regional Dynamics, Development and Power Play*, edited by Rajiv Ranjan and Guo Changgang (Abingdon: Routledge, 2021), pp. 17–34.
[177] Devasher, *Pakistan: Courting the Abyss*, p. 313.

Building a strong economy could also benefit Pakistan in its approach of bringing a balance into its dealings with the great powers. Pakistan could not become a client state of either China or the United States. The United States was happy for Pakistan to trade with China and other countries as long as its certain 'red lines' were observed—such as giving exclusive control of its defence to China or only involving China in its critical defence projects—yet neither of these were likely to happen.[178]

Political stability is a major requirement for a stable economy and growth of foreign direct investment. It also requires human capital, which can participate in building economic momentum. As mentioned, Pakistan's security elites have historically targeted both of these ideals in order to perpetuate their control over the state. Pakistan has not been able to develop a democratic culture, since its military has targeted and divided the civilian leadership whenever the country was not under military rule. The result of that steadfast hammering at the political system was that there did not exist civilian partners to share the burden of national rejuvenation envisaged by General Bajwa in the Islamabad Security Dialogue. The masses were kept impoverished to pay for defence against 'threats' from all sides, with the country scoring poorly on all human development indicators. For instance, Pakistan stood in the 164th position out of 193 countries, as ranked by the United Nations Development Programme in 2024.[179]

However, moving towards trade and relative peace with India would be a tough target for Pakistan, the mid-2025 tensions notwithstanding. First, achieving this goal would create an internal problem for the military, who used the 'India-threat theory' for nearly eighty years; persuading the entire army to change its threat perception so rapidly would have institutional implications and would surely encounter resistance.[180] Second, casting India as an enemy was important for the military's dominance internally; doing away with describing that country as an existential threat would lead to a recalibration of the civil–military balance in the years to come, for which the military is not prepared. Third, neither the current military leadership nor the civilian government in Pakistan would risk taking such a step, with 'Imran Khan breathing down their necks' and ready to exploit any opportunity for populist gains.[181] Fourth, there exists a divide between senior officers in the military

[178] Interview, Academic 2.
[179] Anwar Iqbal, 'Pakistan's Human Development Ranking Declines', *Dawn*, 14 March 2024.
[180] Interview, Academic 2.
[181] Ibid.

regarding India. While some may be persuaded of the need for a different approach towards the country's neighbour, a significant number appear to retain a more hardline, ideologically driven 'jihadi' orientation and would be resistant to any such shift in focus.[182] Regardless, the lack of willingness on the Indian side to engage constituted a separate obstacle altogether.

[182] Interview, Politician 1.

7
The fortress of Islam

7.1 The argument

In this book, I argue that the roots of Pakistan's suboptimal foreign policy choices over nearly eight decades are directly linked to the country's civil–military imbalance. Apart from two brief periods (the 1950s and early 1970s), the Pakistani military has exercised a tight control over the country's foreign policy by deploying a narrative of the country facing threats from all directions. Such a narrative is then used to justify the military's control domestically. According to my interpretation, the internal and external dimensions of Pakistani politics are intrinsically linked. Contrary to the two most frequently cited books on Pakistan's foreign policy,[1] I argue that the roots of Pakistan's external behaviour cannot be attributed solely to a hostile regional and global geopolitical environment. Rather they lie in the longstanding dominance of the civil and military bureaucracy over the state since its creation. The military has harboured revisionist agendas towards India and Afghanistan and has actively worked against civilian leaders' visions to normalize relations with these countries. Whereas the examples of Pakistan's revisionism towards India abound (e.g., Operation Gibraltar, Operation Desert Hawk), one would be hard-pressed to find similar examples from the Indian side until the 2010s.

Although this book primarily examines the military's role in shaping Pakistan's foreign policy, this should not be taken to discount the ineptitude of civilian leadership at certain critical junctures, particularly during the first decade after independence. That period was marked by persistent infighting among civilian political elites, which distracted from state-building and effective governance. This fragmentation empowered the civil bureaucracy and laid the institutional foundations for the military's subsequent dominance over both domestic and foreign policy.

Civilian leaders have also, at various points, championed an aggressive policy towards India. In some instances, they undermined peace initiatives

[1] Abdul Sattar, *Pakistan's Foreign Policy 1947–2019* (Karachi: Oxford University Press Pakistan, 2020); Shahid Amin, *Pakistan's Foreign Policy: A Reappraisal* (Karachi: Oxford University Press Pakistan, 2000).

Pakistan's Foreign Policy. Wali Aslam, Oxford University Press. © Wali Aslam (2026).
DOI: 10.1093/9780198909552.003.0007

in pursuit of short-term political gains. For example, after the civil war of 1971, the military was on the back foot and suffered from low morale following its defeat in what became the newly independent state of Bangladesh. Prime Minister Zulfikar Ali Bhutto had a significant opportunity to correct the civil–military balance and to re-shape the contours of Pakistani foreign policy. However, he appeared more interested in consolidating his position as a populist leader by rekindling anti-India rhetoric, thereby squandering an opportunity to effect meaningful change.

In some cases, Pakistan's civilian leaders showed excessive loyalty to the military's vision instead of standing by their own principles. Such instances of loyalty included belligerent and threatening statements to Afghanistan, India, and the United States, which only strengthened the military's hand in Pakistani foreign policy.[2] They were often willing to be co-opted by the military, as this route offered a shortcut to higher office in the country. Given the military's pervasive control, the civilian leaders believed that their political survival depended on operating within the constraints of the existing system—a system that granted the military veto power over foreign policy decision-making. More recently, the civilian leadership have avoided talking peace for fear of losing popularity, given the strength of anti-India narratives prevalent in the country. Those narratives have been pushed hard by the country's security establishment from the very early years and are found everywhere from school textbooks[3] to the mainstream media channels of Pakistan.[4] At certain times, civilian leaders have also feared that they would lose the military's backing were they to extend an olive branch to India. Taken together, these dynamics suggest that civilian leaders' approaches to the country's external relations can only be properly understood within the broader context of Pakistan's civil–military imbalance.

Although India has not historically mirrored Pakistan's revisionist tendencies, some of its policy missteps have, nonetheless adversely affected the civil–military imbalance in Pakistan, which, in turn, has negatively shaped the country's relations with its neighbours. These include the initial Indian reluctance to acknowledge Pakistan's independence, its military exercises on the border with Pakistan in 1950 and 1987, and the start of the nuclear programme in the 1960s. These developments contributed to the feeling of

[2] '"Blood will Flow if India Stops River Water", Threatens Bilawal Bhutto', *The Hindu*, 26 April 2025.

[3] Ahsan Butt, 'Nationalistic Narratives in Pakistani Textbooks', *United States Institute of Peace Brief 210*, 2016.

[4] Tehmina Firdous and Mehrajudin Aslam, 'Contesting Narratives: A Comparative Analysis of Indian, Pakistani and Kashmiri Print Media Framing of the Abrogation of Article 370', *International Journal for Multidisciplinary Research* 6, no. 2 (2024): pp. 1–18.

insecurity in Pakistan, which helped the latter's military to claim more space in shaping its foreign policy.

Similarly, certain strategic mistakes by the United States also deepened Pakistani insecurity, creating greater room for the military's dominance over the decision-making process in the country. Examples of those mistakes include the American inability to address Pakistan's security concerns after the fall of Kabul in the winter of 2001–2002. The bonhomie between the Pakistani military ruler General Musharraf and US President George W. Bush in the latter's first term strengthened the military tremendously at the expense of the two main civilian leaders, Benazir Bhutto and Nawaz Sharif, who were sent into exile by Musharraf after the takeover of 1999. The close defence cooperation between China and Pakistan[5] has also had negative consequences for the civil–military imbalance in Pakistan and has contributed to the erratic trend in the country's foreign policy for several decades.

7.2 The contextual roots of the 'siege mentality'

After summarizing the core arguments of the book, I will now examine the narrative of pervasive external threats that fosters a siege mentality within the psyche of the Pakistani populace. Throughout this book, I have analysed the discourse of the country's elite to demonstrate how a siege mentality is continually constructed and mobilized to generate public anxiety in Pakistan. That analysis has been conducted using a theoretical framework built upon ontological security theory, post-structuralist discourse theory, and the concept of 'fantasy'. The country's creation, and its continued existence, has always been justified through such a narrative of all-encompassing and ever-present dangers, such that alternative rationales for Pakistan's ontological existence have been seriously discounted. I will conclude this book by analysing that narrative in the context of Pakistan's foundational mythology.

Every country has its own national myths that both shape and are shaped by the collective national psyche.[6] Such myths perform an important role in providing comfort in fantasies of greatness at times when reality is hard to endure. The 'shining city on the hill' comes to mind when picturing the United States as a country that will lead the rest of the world in charting

[5] Sameer Lalwani, *A Threshold Alliance: The China-Pakistan Military Relationship*, United States Institute of Peace, March 2023, available at https://www.usip.org/sites/default/files/2023-03/sr-517_threshold-alliance-china-pakistan-military-relationship.pdf

[6] See, for example, Wilber Caldwell, *American Narcissism: The Myth of National Superiority* (Sanford, NC: Algora, 2008); Stephanie Barczewski, *Myth and National Identity in Nineteenth-Century Britain* (Oxford: Oxford University Press, 2000).

the path for democracy, given its own democratic credentials.[7] In the case of Pakistan, the country is often described as '*Islam ka Qila*' or the 'fortress of Islam'.[8] Being the first modern nation state to have come into being in the name of a religion, the myth provides not only an identity to Pakistan, it also defines a mission for it: the country is an impregnable and robust bastion of Islam that will be the religion's last stronghold. Defending Islam is the nation's primary and, if required, sole objective. Schoolchildren are taught from an early age that they are special to have been chosen for this project and should feel proud of it. This national myth is kept permanently alive through discursive means, used by the ruling elites to portray both Pakistan and Islam in perpetual danger. In a typical Orwellian sense, the masses are told of the existential threats the state faces at the hands of internal and external enemies—and these threats call for a formidable response.

Within the context of Turkey, Umut Adisonmez and Recep Onursal discuss 'dislocatory moments' that countries may face as a result of upheavals in the domestic or regional political scene (such as the 2016 coup attempt against Turkish President Erdoğan).[9] Such moments create ontological anxiety among the masses, and those in powerful positions exploit them to further extend their control over state and society alike. In the case of Pakistan, the state uses such dislocatory moments to strengthen the country's identity as the fortress of Islam—yet promoting that identity requires presenting Pakistan as a nation under siege, facing threats from all sides. Defending the state emerges as the primary objective. That process invariably places the armed forces, the de facto primary defenders of the state, in the exalted position of the saviour of the nation.

Daniel Bar-Tal and Dikla Antebi have talked about a 'siege mentality' with reference to the Israeli national psyche, which receives nourishment from the historical persecution faced by the Jewish people for millennia.[10] Seen through the lens of the discipline of 'group psychology', they define this mentality as 'a belief held by group members that the rest of the world has negative behavioural intentions towards them'.[11] It is the 'perception by group members that the out-groups have intentions to do wrong or inflict harm on their

[7] Rogers M. Smith, 'From the Shining City on a Hill to a Great Metropolis on a Plain? Stories of Immigration and Peoplehood', *Social Research* 77, no. 1 (2010): pp. 21–44.

[8] Khaled Ahmed, 'The "Castle of Islam" Concept', *The Express Tribune*, 3 March 2012.

[9] Umut Adisonmez and Recep Onursal, '"Strong, but Anxious State": The Fantasmatic Narratives on Ontological Insecurity and Anxiety in Turkey', *Uluslararası İlişkiler/International Relations* 19, no. 73 (2022): p. 65.

[10] Daniel Bar-Tal and Dikla Antebi, 'Siege Mentality in Israel', *International Journal of Intercultural Relations* 16, no. 3 (1992): pp. 251–275.

[11] Ibid., p. 251.

(in)group'.[12] The 'crucial focus' of this belief is directed towards the rest of the world, with the nation 'standing alone against the hostile world'.[13] The implication, according to Bar-Tal and Antebi, is that there is a threat to a group's very existence, and this group cannot expect help from anyone in times of difficulty.

The scholars of post-structural discourse theory will tell us that it is important to study the beliefs of a group of people about the rest of the world in order to understand how they react to different developments in their lives. That is because 'people act and react to the world they believe exists'.[14] The question of whether the 'world is actually hostile' towards said group is 'irrelevant' to understanding the behaviour of that group, because if the group believes (or is manipulated to believe) that the rest of the world has 'negative behavioural intentions towards them, they act accordingly'.[15] In the case of Israel, this siege mentality is rooted in the suffering that Jewish communities around the world faced for hundreds, if not thousands, of years.[16]

The concept of the 'fortress of Islam' is used to promote a siege mentality within Pakistani society in a slightly different way. Where the people lack a single coherent identity (unlike that of the Jewish people), they are presented as one group brought together under the umbrella of Islam (implying that no cleavages or differences exist within Islam). Given its critical role in defining the Pakistani identity, religion assumes a central importance for the matters of the polity, and thus a threat to Pakistan is often equated with a threat to Islam itself. Whereas the siege mentality among the Jews across the world is sustained by a long history of persecution, the same mentality is kept alive in Pakistan through assertions that the precious gift of Islam, which gives meaning to the country's people, might be in danger—and they need to unite behind its military to defend it. Subsequently, defending the religion takes priority over everything else. Doing so, therefore, calls for a central role for the armed forces in the state's affairs, given their primary role as the defenders of the state.

The fear among the masses of being under siege is invaluable in helping the Pakistan military to keep tight control over the state and its people. As we have seen throughout the book, the military has adopted a variety of means to stoke these fears throughout the country's existence. The most recent imaginary onslaught facing the country, as popularized by the military, is the

[12] Ibid., p. 251.
[13] Ibid., p. 251.
[14] Ibid., p. 252.
[15] Ibid., p. 252.
[16] Ibid., p. 253.

notion of 'fifth-generation warfare', or 'hybrid warfare', a concept that will be defined and examined in detail below. This particular instrument for cultivating a siege mentality is especially suited to the twenty-first-century context, in which the internet, digital technologies, and social media platforms have emerged as formidable social forces. The 'fifth-generation war' narrative conveniently identifies these domains as the principal sources of the existential threat said to confront the country. It is also important to remember that the 'fifth-generation war' narrative is not an entirely new phenomenon; rather, it seeks to reproduce and reinforce the myths propagated among the populace since the country's inception. In this sense, it represents an adaptation of those enduring narratives to contemporary political and technological realities. Given the significant role this latest instrument of control has played over the past decade—and is likely to continue playing in the near future—I examine it at length in this concluding chapter. I will do so by critically analysing the written pieces in the military's flagship magazine, *Hilal*, on this subject, examining them in the context of the contemporary sociopolitical milieu.

7.3 How does the Pakistani military define the 'fifth-generation war'?

The obvious source of the fifth-generation war can be traced to the launch of Imran Khan (and his political party, the Pakistan Tehreek-e-Insaf or PTI) as a spoiler in Pakistan's two-party system, at a point when the two main political parties, the Pakistan Muslim League (Nawaz) or PML(N) and the Pakistan People's Party (PPP), signed a Charter of Democracy in 2006 to coordinate their efforts against the military rule of then President General Pervez Musharraf. That was also the time when Pakistani relations with the West, especially the United States, were coming under increasing strain.

The military invested in Imran Khan to serve two purposes.[17] Domestically, he could counter the force of the two most popular political forces in the country; and externally, Khan could be deployed to send a message of resistance to the West and to champion the Pakistani nationalist perspective on the War against Terrorism.[18] Khan was presented as the change that Pakistan needed

[17] Interview, Academic 2, London, March 2024.

[18] Imran Khan led a popular campaign against the use of American drones in Pakistan and called for the release of Aafia Siddiqui, a Pakistani scientist and al-Qaeda operative detained in the United States. See 'Embassy Legal Attaché Named in Case Involving Alleged Bagram Detainee', US Embassy Cable, EO12958, 31 July 2008, available at https://www.theguardian.com/world/us-embassy-cables-documents/164310; see also 'Codel Lynch's Meeting with Pakistan Tehreek-E-Insaf (PTI) Party Leader Imran Khan', US Embassy Cable, EO12958, 6 February 2010, available at https://www.theguardian.com/world/us-embassy-cables-documents/247596

against the forces 'bent upon saving the status quo'.[19] The mantra of the fifth-generation war was part of 'Project Imran Khan', and its propagation heavily relied on social media.

According to the authors writing in *Hilal*, the fifth-generation war is an 'undeclared' war that 'aims to influence public opinion and narratives and spread manipulated public messages that are fabricated and promoted by adversaries'.[20] This 'multi-modal or multi-variant conflict' involves 'the use of disinformation and propaganda to sow internal disorder and dissent within a country' as a form of psychological warfare.[21] In this war, 'print, electronic and in particular social media' are used for 'effective manipulation'.[22] Accordingly, 'the attacker's military operations are conducted covertly by irregular forces against the whole society and in particular, against its political structure, state authorities and local government, the economy, the morale of the population and against the Armed Forces'.[23] The purpose is to target the 'entire nation' and its 'ideological foundations' by creating divisions on the basis of ethnicity, religion, and culture.[24] According to Maria Khalid, the enemy wants to weaken the 'self-belief' of the sovereign Pakistani nation with a view to 'divide and weaken' it and to penetrate its ranks to affect its 'foreign policy course' and its 'strategic orientation'.[25] As Pakistan's army chief said in 2020:

> The enemies who plotted our doom and destruction are watching us with disdain. Having failed and feeling frustrated, they have now subjected us to a 24/7 hybrid war. This war is not fought in the battlefields but in the minds. In previous wars, soldiers were taking the brunt on the frontline; in this new war, however, leadership at all levels is the target. It is a huge challenge that you will be confronted with from the day one.[26]

In this war, according to the military narrative articulated in *Hilal*, a bigger threat than the external enemy is the enemy within Pakistani society, who might appear in the form of 'pseudo liberals' or 'religious extremists'.[27] These forces are 'zealously pushing' their narratives in educational institutions, ranging from schools to universities to seminaries, to 'capture young

[19] Farrukh Saleem, 'Blind Alleys', *Hilal*, October 2014.
[20] Ehsan Mehmood Khan, 'Hybrid Warfare: A Conceptual Perspective', *Hilal*, February 2018.
[21] Maria Khalid, 'In the Age of Hybrid Warfare', *Hilal*, February 2018.
[22] Ibid.
[23] Ibid.
[24] Ibid.
[25] Ibid.
[26] Maria Khalid, 'Standing Together as One Amidst Challenges of Hybrid War', *Hilal*, November 2020.
[27] Amir Zia, 'War of Narratives', *Hilal*, September 2015.

minds'[28] and create 'strategic bewilderness'[29] among them. This point of view continues: these liberals would naturally consider the security forces to be the main obstruction to efforts to unite the peoples of India and Pakistan, who are essentially the same given their history, values, and culture. These extremists question the Partition of British India as 'a slip of history'[30] and turn a blind eye to the 'ancient biases, irrationality and fanaticism of the extremist Hindu-mindset'.[31]

> [T]he so-called liberals freely run propaganda campaigns against the core national interests from undermining the Kashmir cause to that of the criticism on the country's nuclear programme mainly in the English-language press, which serves as the main vehicle to highlight their small protests, conferences and other activities. They resort to propaganda against the country's security agencies and the armed forces and try to portray them as rouge institutions. They also indulge in foreign-inspired advocacy campaigns in the name of human rights—by which they basically mean rights of criminals and terrorists.[32]

Another component of this war, we are told, is religious extremists' opposition to the security forces for aligning with the West and for waging a war on al-Qaeda-linked militants.[33] According to *Hilal*, these groups might appear different, but they share the same purpose, which is to 'weaken and destroy' Pakistan.[34]

We are also told by *Hilal* that the main battlefields of this new war include media, educational institutions,[35] and the NGO sector. That is where the enemies within the country make coalitions with their foreign sponsors. The NGOs initiate 'strategic chaos' at the ground level of society by penetrating it in the guise of 'youth groups, human rights activists, corruption watch dogs, social media platforms and academia', all of whom are tasked with the job of 'perception management' for various rewards, including 'monetary benefits, scholarships for foreign degrees, internships inside and outside Pakistan and jobs in opinion making platforms'.[36]

[28] Ibid.
[29] Waqar Kauravi, 'Building Capacity in Non Kinetic Warfare', *Hilal*, December 2015.
[30] Zia, 'War of Narratives'.
[31] Amir Zia, 'Brand India', *Hilal*, November 2015.
[32] Zia, 'War of Narratives'.
[33] Ibid.
[34] Ibid.
[35] Amir Zia, 'Hybrid Warfare Manifested', *Hilal*, February 2018.
[36] Kauravi, 'Building Capacity in Non Kinetic Warfare'.

> Foreign-funded social media training sessions focus on tarnishing the image of the state and its institutions in the name of citizen journalism. Similarly, in the name of culture and literature, such events are funded which promote narrow ethnic agendas or provide platforms to promote anti-Pakistan sentiment and propaganda against its institutions ... [A]t their sponsored or funded media programs, future and working journalists are fed with warped interpretation of history that paints the Pakistan Movement as a reactionary and politically incorrect struggle. Pakistan Army is ... described as the mother of all ills and held responsible for tensions with India and Afghanistan. They are asked to cover those themes and stories that supplement their biases and propaganda against Pakistan.[37]

According to this narrative, the alleged fifth-generation war against Pakistan has been facilitated by the ineptitude of successive '"elected" leaders', who are said to have failed to inspire hope among citizens by neglecting to provide 'socio-economic justice, rule of law and fair administration to all Pakistani citizens'.[38]

> Thanks to weak policies, the enemy is no longer knocking at the door. The enemy is already within our ranks. Pakistan is being subverted from within. As a country and as a nation, at this critical juncture in our history we cannot leave ourselves to the vagaries of time or at the mercy of our systemic aberrations.[39]

In keeping with the military's historical stance on the matter, it is argued in *Hilal*, that Pakistani politics has been run by 'elite classes' who have been 'inimical to the promotion of genuine nationhood in the country'.[40] With the help of 'civil bureaucracy', the civilian leaders rule 'in the decadent Mughal style'.[41] Pakistan's failures are primarily presented as political failures, such as the dismemberment of the country in 1971 when corrupt civilian leaders failed to 'resolve the internal contradictions of the state, providing the enemy an opportunity to damage us from within'.[42] In that situation, we are told, the armed forces of Pakistan were placed in a 'no-win situation' long before the outbreak of actual hostilities as they faced a 'hostile local population':

> Our political failure led to the military failure. India manipulated the anger and resentment among Bengalis over their real or perceived exploitation by the

[37] Zia, 'Hybrid Warfare Manifested'.
[38] Shamshad Ahmed, 'National Unity and Security', *Hilal*, November 2015.
[39] Ibid.
[40] Ibid.
[41] Ibid.
[42] Zia, 'Hybrid Warfare Manifested'.

> [W]estern wing. Indians trained and armed the rebels and infiltrated agents in the former East Pakistan. However, much before fanning insurgency and declaring an all-out war, the Indian propaganda machine had penetrated various segments of the society, especially among students, teachers, opinion-makers, journalists, writers and intellectuals. In a nutshell, our mistakes offered a perfect dream victory to India.[43]

We are told that in the fifth-generation war, the minds of Pakistani people are targeted through 'intrigue, conspiracies, secrecy, espionage, mystery and media wars; subversion of [their] religious, sectarian and ethnic fabric as well as social issues; and economic strangulation.'[44] This confusion and division results in 'violence, chaos, anarchy and instability within the country.'[45] The common woman and man who are facing economic challenges in their daily lives are in an 'emotionally unstable state', which makes them especially 'vulnerable to enemy propaganda.'[46] They are apparently unable to differentiate between right and wrong in the face of 'such bombardment of conflicting news and unverified information.'[47]

> It is a victim-driven warfare wherein a victim often becomes an actor against itself … Everyone seems to be a warrior and a victim concomitantly. However, complexity, combination and simultaneity remain to be the hallmark of hybridity … This means implosion of states or their governments from within rather than explosion [*sic*].[48]

In this way, we are told, the external enemies

> create confusion, discord and division within the … country; nurture, exploit and stoke its ethnic, sectarian, religious and class contradictions; sponsor terrorists and subversive activities; target and damage its economy, and erode the national unity and cohesion of the enemy state by striking at its ideological foundations.[49]

We are informed that one of the objectives behind these hostile activities is to demoralize the armed forces and portray 'Pakistan's military leadership and sections of intelligence agencies as hurdles to international peace efforts.'[50]

[43] Ibid.

[44] Taj M. Khattak, 'Hybrid War and Pakistan's National Security', *Hilal*, January 2021.

[45] Amir Zia, 'Religious Harmony, Sectarian Cleavages and the World of Social Media', *Hilal*, November 2020.

[46] Abdul Rauf, 'Fake News in the Post-Truth Era', *Hilal*, November 2021.

[47] Maria Khalid, 'ITAI & CySec: Progression in Simultaneity', *Hilal*, July 2018.

[48] Khan, 'Hybrid Warfare: A Conceptual Perspective'.

[49] Zia, 'Hybrid Warfare Manifested'.

[50] Ather Javed, 'Espionage, Illegal Intelligence Networks and Growing Vulnerabilities for Pakistan', *Hilal*, February 2018.

Those involved in such activities are 'fringe dissident elements' who range from 'the shadowy nationalist militant groups to handful of social media activists, elements from academia, media and the NGOs', and their objective is 'to echo the enemy's propaganda line'.[51] The supporters of the military narrative argue that, given that the army is the 'the only institution which stands between ... [these] fringe dissident elements... and their designs against Pakistan',[52] it is no surprise that [the army] is subjected to the 'musketry volleys of criticism where quivers of fabricated and concocted figures are emptied throughout the year'.[53] *Hilal* argues that the fifth-generation war is an Indian-Afghan-American assault to weaken the world's 'lone Muslim nuclear power', and that it aims to tarnish 'Pakistan's image globally' and 'create division and discord among Pakistanis'.[54] In an apparent reference to the West, it is mentioned that Pakistan's 'frenemies [portmanteau combining "friends" and "enemies"] spend millions of dollars annually on perception management in Pakistan in the name of promoting education, traditional and new media, culture and social work. The aim is to win over, influence and brainwash educated Pakistanis so that they can promote their narrative in the country'.[55]

We are also told that in this war, the role of the Ministry of Information is 'more important than [the] Ministry of Defence ... as the ... first firewall in national security'.[56] At the individual level, there is a need for the entire society to 'remain vigilant' and the 'leaders and strategic thinkers must keep the nation hopeful and must not let the divisive elements or hostile identities to strengthen and encourage dissent and division'.[57] It is, thus, 'incumbent upon the academicians, religious figures, politicians, artists and the entire nation to not let anyone discredit our ideology, identity and [its] very fabric' and show 'extreme vigilance' against 'ploys guised beautifully'.[58]

Displaying unity is said to be the first requirement to combat the onslaught of the fifth-generation war, since a 'politically polarized and divided Pakistan remains an easy target of this modern undeclared warfare that aims to widen and sharpen these very cracks to weaken and destroy us from within'.[59] It is a 'long-term low-intensity game, which can be transformed into

[51] Zia, 'Hybrid Warfare Manifested'.
[52] Ibid.
[53] Bilal Naseer Mahmood, 'An Unceasing Propaganda and the Myth of Three Million', *Hilal*, May 2021.
[54] Zia, 'Hybrid Warfare Manifested'.
[55] Ibid.
[56] Kauravi, 'Building Capacity in Non-Kinetic Warfare'.
[57] Khalid, 'In the Age of Hybrid Warfare'.
[58] Ibid.
[59] Zia, 'Hybrid Warfare Manifested'.

a high-intensity and disastrous conflict for the enemy if not forcefully and effectively managed'.[60]

> [H]ybrid threat merits hybrid response. This calls for integration of society in state's efforts to maintain peace and ensure security in face of multiplicity, complexity and simultaneity of threat posed by the prevailing environment [*sic*]... [N]o single agency, organization or institution can combat the hybrid threat. Synergetic application of the entire national potential and all elements of national power is the only way to be victorious in a hybrid war.[61]

Contributors to *Hilal* repeatedly emphasize that, along with the geographical boundaries, the nation's ideological frontiers are also under attack in the fifth-generation war, and because the ideologies relate to the 'subjective domain' of the nation at large,

> every citizen has to act as a soldier to counter the hostile narratives ... [which aim] to discredit the nation and its moorings. ... Since Pakistan is amongst the few countries created on ideology instead of common culture, language or race [*sic*], therefore it is imperative to counter these 'disinformation' endeavors targeted towards our very mooring and symbol of 'national integration'.[62]

In the face of a fifth-generation war, the people of Pakistan can have faith in the armed forces of the country that have been protecting the country's borders with 'great resolve and determination ... giving sacrifices for [its] survival and existence as an independent state'.[63] The civilian rulers on the other hand, we are told, 'continue to disappoint the nation by failing to rise above their vested interests'. As a consequence, the country stumbles from one crisis to another, thereby impacting the 'nation's common sense of purpose, unity and security'.[64]

7.4 Interpreting the Pakistani military's narrative of the 'fifth-generation war'

Although the narrative of the fifth-generation war coincided with the political promotion of Imran Khan, interpreting its significance only through that

[60] Ibid.
[61] Khan, 'Hybrid Warfare: A Conceptual Perspective'.
[62] Mahmood, 'An Unceasing Propaganda'.
[63] Ahmed, 'National Unity and Security'.
[64] Ibid.

lens would obscure its wider significance in shaping contemporary Pakistani society for the foreseeable future. Its longevity has already outlasted Khan's tenure as prime minister; hence, a more in-depth analysis of the narrative is warranted.

Bar-Tal and Antebi, the Israeli authors mentioned above, assert that a siege mentality has a far-reaching influence on the members of any group and can be studied through five effects.[65] First, the threatened group members develop negative attitudes towards other groups or the rest of the world. Second, the group develops mistrust of the loosely defined 'other'. Third, 'the threatened group becomes sensitive to information that may indicate negative intentions'. Fourth, the members of the threatened group develop 'internal mechanisms to cope with the potential threat by increasing pressures among group members toward conformity and unity'. Finally, in view of the perceived threat, the besieged group employs 'all necessary means to protect itself'.[66]

A rapid review of the concept of so-called 'fifth-generation warfare' suggests that it is not a coherent or empirically grounded form of warfare at all, but rather a discursive construct deployed by the military establishment to discredit progressive elements within Pakistani society and cultivate a siege mentality among the public, thereby provoking the kinds of reactions described by Bar-Tal and Antebi. By constantly pointing towards an all-front and permanent war, a certain group mentality is created that displays negative attitudes and mistrust towards the rest of the world. The idea of the presence of the enemy hiding within society lays the foundations of a suspicious public, in which neighbour distrusts neighbour and the masses remain divided.[67] The general public is told, for the reasons discussed below, that they are emotionally vulnerable and unable to distinguish right from wrong, thereby sowing doubt in their minds regarding their own capabilities and decision-making powers.

It is important to remember that those in power do not always prefer a resilient and confident society because a citizenry lacking in confidence is easy to manipulate and to rule over.[68] Critical thinkers are particularly targeted, with universities and civil society organizations portrayed as hotbeds of anti-state activities. This is because evidence shows that higher levels of education at all levels are 'robustly associated with

[65] Bar-Tal and Antebi, 'Siege Mentality in Israel', p. 264.
[66] Ibid., pp. 264–265.
[67] Mark Webb, 'The Epistemology of Trust and the Politics of Suspicion', *Pacific Philosophical Quarterly* 73, no. 4 (1992): pp. 390–400.
[68] Roland Benedikter and Karim Fathi, 'What is a Resilient Society?' *International Policy Digest*, 17 September 2017.

democratization'.[69] Culture and literature thus become primary battlefields.[70] All of this serves to undermine the role of reflection and innovation, and presents knowledge as a threat. The purpose is to curb individuality and to encourage a uniformity of thinking. As Khaled Ahmed notes, a siege mentality is

> a narrowness dictated by an indoctrinated mind, deaf to words of reason and new thinking. It is paranoia mixed with fear, a recession from reality that rejects the breeze blowing in from the desert outside, pregnant with new ideas helpful in resolving old problems.[71]

The narrative of the fifth-generation war instructs that, in keeping with the motto of Pakistan ('unity, faith and discipline'[72]), the nation can only combat the wide range of hybrid threats by uniting behind their armed forces and having faith in them. The idea that every citizen is a soldier in this war serves to reward conformity and promote the regimentation of society, wherein militaristic discipline leaves little room for dissent and disagreement.[73] The defence of the state through 'all necessary means' becomes the primary objective and the nation is asked to place its faith in the armed forces because they are experts in all matters of defence. Paternalistically coercing the masses to place their faith in 'experts' undermines the foundations of a vigilant society in which citizens remain alert to defend their rights and to check the abuses of their rulers—abuses which the latter might want to conceal behind their 'expertise'.[74] The fifth-generation war narrative aims to present Pakistan as a fortress so that the 'people can be easily oppressed and the ruler can be certain he will be obeyed'.

> The ruler cannot control the invader besieging his castle but he can control the people living inside the castle. If the people want to escape his authority, there is nothing they can do. In a state without castles, if you can't stand the oppressive authority of the ruler, you exile yourself. If you live inside a castle, there is no place to run.[75]

[69] Howard Sanborn and Clayton Thyne, 'Learning Democracy: Education and the Fall of Authoritarian Regimes', *British Journal of Political Science* 44, no. 4 (2014): pp. 773–797; see also Nadim Mirshak, 'Education as Resistance: Egyptian Civil Society and Rethinking Political Education Under Authoritarian Contexts', *Critical Sociology* 46, no. 6 (2020): pp. 915–930.

[70] Edward Said and David Barsamian, *Culture and Resistance* (Chicago, IL: Haymarket, 2019).

[71] Ahmed, 'The "Castle of Islam" Concept'.

[72] Khaled Ahmed, 'The Sad Story of "Unity-Faith-Discipline"', *Newsweek Pakistan*, 21 March 2022.

[73] Karen Evans, 'War and Peace: Is Militarisation the New Norm?' in *Social Censure and Critical Criminology*, edited by Anthony Amatrudo (London: Palgrave Macmillan, 2017), pp. 275–302.

[74] Stuart Ritchie, 'Don't Trust the Psychologists on Coronavirus', *UnHerd*, 31 March 2020.

[75] Khaled Ahmed, 'The "Castle of Islam" Concept'.

A siege mentality, built through concepts like 'the fifth-generation war', encourages an isolationist mindset. Ideology plays an important role in promoting that frame of mind, as it helps create 'the uniformity of thinking that the castle demands ... It is only by thinking the same way that you can live inside a castle without conflict. Hence the importance of an agreed ideology protected by a Penal Code prescribing the measure of punishment that a breach of ideology, will entail.'[76]

Following on from the paternalistic notion presented above, the idea of the fifth-generation war has been deployed by the Pakistani elite to sow ontological insecurity and anxiety among the masses: a beautiful future is envisioned only if the masses support their military, whereas a horrific vision is the price they will pay for not extending that support. The beatific scenario does not tolerate 'difference of opinion and tolerance of opinion opposed to the indoctrination of the state'.[77]

The Pakistani people may indeed be living through a period of fifth-generation war, but it is not committed by the hand of an unknown, ambiguous, and a vaguely defined external enemy. It is being waged on them by their own ruling elite who, like a protection racket or a mafia, ask the people to support them so that they can save the masses from the enemy next door. However, it is obvious to any critical observer that, in the absence of such support, it is the protection racket itself that further threatens the well-being of the people and not some vaguely envisaged foreign enemy. Only by supporting the ruling elite will the people be left alone, but the price of this will be the permanent dominance by the elite through this protection racket at the cost of the entire nation.

Today, the Pakistani public are facing severe curbs on both mainstream[78] and social media at the hands of the country's security and intelligence agencies.[79] The NGO sector is under greater scrutiny than ever before, with numerous foreign organizations told to shut shop and leave.[80] Universities[81] and civil society groups[82] are under pressure not to voice opinions that are

[76] Ibid.
[77] Ibid.
[78] Raksha Kumar, 'As Pakistan Approaches a Crucial Election, its Media Watchdog Bans Critical Voices from TV', *Reuters Institute*, 3 October 2023.
[79] Asif Shahzad, 'Social Media Platform X Blocked in Pakistan over National Security, Ministry Says', *Reuters*, 17 April 2024.
[80] Sabrina Toppa, 'As Pakistan Cracks Down on NGOs, Civil Society Questions Next Steps', *DevEx*, 10 January 2018.
[81] Zoha Fareed Chishti, '(No) Memory of War and Violence: Conference Series on 1971 War Cancelled', *The LUMS Post*, 13 September 2021.
[82] Munawer Azeem, 'Discussion in Islamabad on Afghan Refugees' Issue Abandoned after Police "Raid"', *Dawn*, 25 October 2023.

critical of the military or of the ideology of Pakistan.[83] The country is swiftly moving from a 'hybrid regime' to an 'authoritarian state' in which civil rights and freedoms are in serious danger.[84] Regardless of whether the concept of a fifth-generation war continues to be used, the control and oppression is ever increasing and the mechanisms for exerting that control are constantly being updated to adapt to the requirements of contemporary digital society. This control is exercised to enable the military to dominate both external and domestic policy, given their inherent linkages. The 'fortress of Islam' continues to gain strength.

In *The Prince*, Niccolò Machiavelli asserts that the usefulness of fortresses depends on circumstances and that, 'if they do you good in one way, they injure you in another'.

> [T]his question can be reasoned thus: the prince who has more to fear from the people than from foreigners ought to build fortresses, but he who has more to fear from foreigners than from the people ought to leave them alone.[85]

For that reason, Machiavelli's advice might be worth paying attention to: the 'best possible fortress is not to be hated by the people, because, although you may hold the fortresses, yet they will not save you if the people hate you, for there will never be wanting foreigners to assist a people who have taken arms against you'.[86] Whereas the mechanisms of control are ever tightening, it is nevertheless important to note that civil resistance and an 'all-spectrum' pushback that the military faces from other Pakistani institutions,[87] civil society, and the media provides hope.[88] Though the people may not be able to immediately shake the yoke that has been around their necks for decades, the scene is set for a resilient society to make yet another stand.

7.5 The future of Pakistan's foreign policy

This book tells the story of imprudent decision-making by the Pakistani policy elite, which led to the country losing several opportunities to strengthen its

[83] Durdana Najam, 'The Vagaries of Pakistan Ideology', *The Express Tribune*, 18 April 2024.
[84] Veengas, 'Media and Social Media Platforms in Pakistan Face Increasing Restrictions', *The Wire*, 26 February 2024.
[85] Niccolò Machiavelli, *The Prince*, trans. W. K. Marriott (The Gutenberg Project, 2006), Chapter 20, Paragraph 6, available at https://www.gutenberg.org/cache/epub/1232/pg1232-images.html#chap20
[86] Ibid.
[87] Shah Meer Baloch, 'Pakistani Judges Say Intelligence Agency Threatened Them over Imran Khan', *The Guardian*, 3 April 2024.
[88] 'Pakistan's Civil Society Feels the Pressure', *CIVICUS*, 26 July 2023.

position in the world. Yet despite those missed opportunities, Pakistan is not completely isolated in the world. It remains relevant in the context of the 'rise' of China, the Taliban rule in Afghanistan, the US–Iran rivalry, and the American pivot to Asia. The changing geopolitical environment and the increased international focus on Asia opens potential avenues for the country to rectify the mistakes of the past and to better its economy. However, the country can only benefit from these opportunities if the decision-making is conducted with the interest of the broader nation in mind, instead of the interests of a small elite. Ensuring a greater civil–military balance will be essential to end this perpetual siege mentality and move towards policies intended to engage with the wider world, in order to enhance the country's prosperity and to reap the benefits of the population dividend. That project would also necessitate articulating alternative rationales for Pakistan's existence instead of merely serving as the fortress of Islam. Such an ideal is possible to achieve once the Pakistani military loosens its grip on political decision-making—that might appear to be an impossible ideal, but it remains the only path forward for the country.

Bibliography

Primary sources

Primary sources 1: United States Foreign Relations archives

Office of the Historian. National Intelligence Estimate, Foreign Relations of the United States (1952–1954), Volume XI, Part II: 1083–1089, 30 June 1953a.

Office of the Historian. Memorandum of Conversation, by the Ambassador in Pakistan, Foreign Relations of the United States (1952–1954), Volume XI, Part II, 7 December 1953b.

Office of the Historian. The Ambassador in Pakistan (Hildreth) to the Department of State, Foreign Relations of the United States (1952–1954), Volume XI, Part II, 8 December 1953c.

Office of the Historian. Memorandum of Conversation, by the Officer in Charge, Economic Affairs, Office of South Asian Affairs (Fluker), Foreign Relations of the United States (1952–1954), Volume XI, Part II, 22 June 1954a.

Office of the Historian. The Memorandum by the Secretary of State to the Embassy in Pakistan, Washington, D.C., Foreign Relations of the United States (1952–1954), Volume XI, Part II: 1865–1867, 20 August 1954b.

Office of the Historian. Memorandum of Conversation, by the Acting Officer in Charge of Pakistan-Afghanistan Affairs (Thacher), Foreign Relations of the United States (1952–1954), Volume XI, Part II, 18 October 1954c.

Office of the Historian. Memorandum by the Assistant Secretary of Defense for International Security Affairs (Hensel), Foreign Relations of the United States (1955–1957), Volume VIII, 17 February 1955a.

Office of the Historian. National Intelligence Estimate, Foreign Relations of the United States (1955–1957), Volume VIII, 15 March 1955b.

Office of the Historian. Memorandum from the Operations Coordinator (Bishop) to the Assistant Secretary of State for Near Eastern, South Asian, and African Affairs (Allen), Foreign Relations of the United States (1955–1957), Volume VIII, 28 July 1955c.

Office of the Historian. Telegram from the Embassy in Pakistan to the Department of State, Foreign Relations of the United States (1955–1957), Volume VIII, 30 March 1957a.

Office of the Historian. Letter from the Ambassador in Pakistan (Langley) to the Assistant Secretary of State for Near Eastern, South Asian, and African Affairs (Rountree), Foreign Relations of the United States (1955–1957), Volume VIII, 27 December 1957b.

Office of the Historian. Telegram from the Embassy in Pakistan to the Department of State, Foreign Relations of the United States (1958–1960), Volume XV, 14 March 1958.

Office of the Historian. Editorial Note, Foreign Relations of the United States (1958–1960), South and Southeast Asia, Volume XV.

Office of the Historian. National Intelligence Estimate, Foreign Relations of the United States (1958–1960), Volume XV, 5 May 1959a.

Office of the Historian. Memorandum of a Conversation, Department of State, Washington, 3 p.m., Foreign Relations of the United States (1958–1960), Volume XV, 8 May 1959b.

Office of the Historian. Memorandum of a Conversation, Washington, Foreign Relations of the United States (1958–1960), Volume XV, 17 July 1959c.

Office of the Historian. Telegram from the Embassy in Pakistan to the Department of State, Foreign Relations of the United States (1958–1960), Volume XV, 17 August 1959d.

Office of the Historian. Telegram from the Embassy in Pakistan to the Department of State, Foreign Relations of the United States (1958–1960), Volume XV, 16 September 1959e.

Office of the Historian. Airgram from the Embassy in Pakistan to the Department of State, Foreign Relations of the United States (1958–1960), Volume XV, 23 September 1959f.

Office of the Historian. Memorandum of a Conversation, Karachi, 9 a.m., Foreign Relations of the United States (1958–1960), Volume XV, 8 December 1959g.

Office of the Historian. Telegram from the Department of State to the Embassy in Pakistan, Foreign Relations of the United States (1958–1960), Volume XV, 26 January 1960a.

Office of the Historian. Letter from the Officer in Charge of Pakistan-Afghanistan Affairs (Poullada) to the Special Assistant at the Embassy in Pakistan for Mutual Security Affairs (Linebaugh), Foreign Relations of the United States (1958–1960), Volume XV, 14 March 1960b.

Office of the Historian. Memorandum from the Assistant Secretary of State for Near Eastern and South Asian Affairs (Jones) to the Deputy Under Secretary of State for Political Affairs (Hare), Foreign Relations of the United States (1958–1960), Volume XV, 2 December 1960c.

Office of the Historian. Airgram from the Embassy in Pakistan to the Department of State, Foreign Relations of the United States (1958–1960), Volume XV, 3 December 1960d.

Office of the Historian. Memorandum of Conversation, Foreign Relations of the United States (1961–1963), Volume XIX, 11 July 1961.

Office of the Historian. Telegram from the Department of State to the Embassy in Pakistan, Foreign Relations of the United States (1961–1963), Volume XIX, 13 November 1962.

Office of the Historian. Memorandum from Robert Komer of the National Security Council Staff to the President's Special Assistant for National Security Affairs (Bundy), Washington, Foreign Relations of the United States (1964–1968), Volume XXV, 2 June 1964a.

Office of the Historian. Telegram from the Embassy in Pakistan to the Department of State, Foreign Relations of the United States (1964–1968), Volume XXV, 11 August 1964b.

Office of the Historian. Memorandum of Conversation, Lahore, Foreign Relations of the United States (1969–1976), Volume E-7, Documents on South Asia (1969–1972), 1 August 1969.

Office of the Historian. Memorandum of Conversation Between Henry Kissinger and Indira Gandhi, Foreign Relations of the United States (1969–1976), Volume XI: 407–409, 7 July 1971.

Office of the Historian. Telegram 26 from the Consulate General in Karachi to the Department of State, Foreign Relations of the United States (1969–1976), Volume E-7, 1154Z, 3 January 1972a.

Office of the Historian. Memorandum from Secretary of State Rogers to President Nixon, Washington, Foreign Relations of the United States (1969–1976), Volume E-7, 17 January 1972b.

Office of the Historian. Letter from the Pakistani Ambassador (Raza) to President Nixon, Washington, Foreign Relations of the United States (1969–1976), Volume E-7, 18 January 1972c.

Office of the Historian. Telegram 1581 from the Embassy in Pakistan to the Department of State, Foreign Relations of the United States (1969–1976), Volume E-7, 1330Z, 17 February 1972d.

Office of the Historian. Telegram 1622 from the Embassy in Pakistan to the Department of State, Foreign Relations of the United States (1969–1976), Volume E-7, 1800Z, 18 February 1972e.

Office of the Historian. Telegram 1834 from the Embassy in Pakistan to the Department of State, Foreign Relations of the United States (1969–1976), Volume E-7, 1117Z, 25 February 1972f.

Office of the Historian. Telegram 2213 from the Embassy in Pakistan to the Department of State, Foreign Relations of the United States (1969–1976), Volume E-7, 0725Z, 10 March 1972g.

Office of the Historian. Memorandum from Secretary of State Rogers to President Nixon, Washington, Foreign Relations of the United States (1969–1976), Volume E-7, 17 March 1972h.

Office of the Historian. Telegram 32 from the Embassy in Pakistan to the Department of State, Foreign Relations of the United States (1969–1976), Volume E-8, 1048Z, 2 January 1973a.

Office of the Historian. Memorandum of Conversation, Washington, 1:15–2:20 p.m., Foreign Relations of the United States (1969–1976), Volume E-8, 29 January 1973b.

Office of the Historian. Intelligence Note RNAN-9 Prepared in the Bureau of Intelligence and Research, Washington, Foreign Relations of the United States (1969–1976), Volume E-8, 15 February 1973c.

Office of the Historian. Memorandum from the President's Assistant for National Security Affairs (Kissinger) to President Nixon, Washington, Foreign Relations of the United States (1969–1976), Volume E-8, 7 March 1973d.

Office of the Historian. Memorandum of Conversation, Washington, 11 a.m.–12:45 p.m., Foreign Relations of the United States (1969–1976), Volume E-8, 18 September 1973e.

Office of the Historian. Memorandum of Conversation, Islamabad, Foreign Relations of the United States (1973–1976), Volume E-8, 31 October 1974.

Primary sources 2: Pakistan military publications

Abbasi, Saadia. 'Pakistan and the United States: Where Do We Go from Here?' *Hilal*, August 2011, pp. 7–9.

Ahmed, Shamshad. 'National Unity and Security,' *Hilal*, November 2015.

Akram, Zamir. 'No U.S. Exit Strategy from Afghanistan,' *Hilal*, January 2018.

Anees, Atia. 'National Integration,' *Hilal*, June 2009, pp. 17–18.

Baqai, Huma. 'Pakistan's Foreign Policy Turns a New Leaf,' *Hilal*, July 2015.

Baqai, Huma. 'Pakistan-India-U.S.: The Troubled Triangle,' *Hilal*, April 2018.

Baqai, Huma. 'Towards Conflict Resolution and Lasting Peace,' *Hilal*, April 2021.

Bokhari, Sayed G. B. 'Unjust Maligning of ISI,' *Hilal*, May 2009, p. 13.

Cloughley, Brian. 'Not Too Far,' *Hilal*, October 2013, pp. 4–8.

Editorial. 'National Unity,' *Hilal*, January 2009, p. 3.

Editorial. 'Reflection, Not Mere Reaction,' *Hilal*, June 2011, p. 3.

Javed, Ather. 'Espionage, Illegal Intelligence Networks and Growing Vulnerabilities for Pakistan,' *Hilal*, February 2018.

Hilaly, Zafar. 'The Challenges Ahead,' *Hilal*, April 2010, pp. 5–6.

Hilaly, Zafar. 'Vibrant Armed Forces,' *Hilal*, September 2010, pp. 16–17.

Hilaly, Zafar. 'No; to Targeting Allies,' *Hilal*, November 2010, pp. 5–6.

Hussain, Nazir. 'The Contemporary Dynamics of Middle Asia and Pakistan's Role in Peace and Development,' *Hilal*, May 2018.

Iqbal, Khalid. 'Likely Fallout of Indian Posturing,' *Hilal*, March 2009, pp. 15–16.

Jaspal, Zafar. 'U.S. National Security Strategy and Pakistan: An Appraisal,' *Hilal*, January 2018.

Khalid, Maria. 'In the Age of Hybrid Warfare,' *Hilal*, February 2018.

Kauravi, Waqar. 'Building Capacity in Non Kinetic Warfare,' *Hilal*, December 2015.

Khalid, Maria. 'ITAI & Cysec: Progression in Simultaneity,' *Hilal*, July 2018.

Khalid, Maria. 'Standing Together as One Amidst Challenges of Hybrid War,' *Hilal*, November 2020.

Khan, Ehsan Mehmood. 'A View from the Other Side of the Planet,' *Hilal*, October 2009, p. 25.
Khan, Ehsan Mehmood. 'Hybrid Warfare: A Conceptual Perspective,' *Hilal*, February 2018.
Khan, Muhammad. 'Convolution of Military Operations in Tribal Areas,' *Hilal*, June 2009, pp. 14–16.
Khan, Muhammad. 'Curbing Militancy: The Responsibility of Allies,' *Hilal*, July 2009, pp. 22–24.
Khan, Muhammad. 'Pakistani Nukes Are Well Guarded,' *Hilal*, September 2009, pp. 19–21.
Khan, Muhammad. 'Towards a Better Pak-Iran Relationship,' *Hilal*, December 2009, pp. 15–17.
Khan, Muhammad. 'Limited War Under Nuclear Overhang,' *Hilal*, January 2010, pp. 9–11.
Khan, Muhammad. 'Robert Gates' Visit in the Perspective,' *Hilal*, March 2010, pp. 31–33.
Khan, Muhammad. 'Towards Pak-Afghan Harmonization,' *Hilal*, April 2010, pp. 9–10.
Khan, Muhammad. 'While Managing the Crises Situation,' *Hilal*, July 2011, pp. 23–24.
Khan, Muhammad. 'Security Environment in South Asia,' *Hilal*, September 2011, pp. 18–19.
Khan, Muhammad. 'Misperceiving Pakistan's Connections with Haqqanis,' *Hilal*, November 2011, pp. 13–14.
Khan, Muhammad. 'A Much Awaited Pakistani Stance,' *Hilal*, January 2012, pp. 11–12.
Khan, Muhammad. 'Balochistan Needs Attention,' *Hilal*, March 2012, pp. 7–9.
Khan, Muhammad Khalid. 'A Resolve to Defend,' *Hilal*, January 2012, p. 10.
Khattak, Taj M. 'Hybrid War and Pakistan's National Security,' *Hilal*, January 2021.
Mahmood, Bilal Naseer. 'An Unceasing Propaganda and the Myth of Three Million,' *Hilal*, May 2021.
Mahmood, Zafar. 'Pakistan Conferring MFN Treatment to India,' *Hilal*, January 2012, pp. 20–21.
Malik, Ahmad Rashid. 'Afpak Strategy–An Obituary,' *Hilal*, April 2010, pp. 15–16.
Malik, Ahmad Rashid. 'Pakistan-United States Strategic Dialogue: Shaping a New Relationship,' *Hilal*, May 2010, pp. 22–23.
Malik, Ahmad Rashid. 'Comparative Defence Perspectives,' *Hilal*, September 2010, pp. 20–21.
Malik, Ahmad Rashid. 'An Alliance in Troubled Waters,' *Hilal*, November 2011, pp. 11–12.
Malik, Ahmad Rashid. 'Nuclear Discrimination,' *Hilal*, January 2012, pp. 13–14.
Malik, Ahmad Rashid. 'The New Engagement with United States,' *Hilal*, February 2012, pp. 11–12.
Masood, Talat. 'Pakistan's Membership of Shanghai Cooperation Organisation (SCO),' *Hilal*, July 2017.
Qadir, Shaukat. 'In Search of a Foreign Policy,' *Hilal*, June 2012, pp. 5–6.
Rafaqat, Syed. 'Security Without Starvation,' *Hilal*, November 2008, pp. 10–11.
Rais, Rasul Bakhsh. 'Will the U.S. Walk Away Again?' *Hilal*, June 2011, pp. 5, 10.
Rauf, Abdul. 'Fake News in the Post-Truth Era,' *Hilal*, November 2021.
Rizvi, Hasan Askari. 'Complexities of Pakistan-India Relations,' *Hilal*, June 2010, pp. 5–6.
Rizvi, Hasan Askari. 'The Imperatives of Changing World Order,' *Hilal*, March 2014.
Saleem, Farrukh. 'Blind Alleys,' *Hilal*, October 2014.
Shabbir, Saima. 'We Stand Together,' *Hilal*, June 2011, p. 11.
Shahid, Bilal. 'The World Should Acknowledge Our Sacrifices,' *Hilal*, January 2009, p. 24.
Shahid, Bilal. 'Scourge of Extremism: Do More with What?' *Hilal*, March 2009, pp. 24–25.
Yamin, Tughral. 'Pakistan–US Relationship,' *Hilal*, November 2011, pp. 9–10.
Yamin, Tughral. 'The Massacre at Salala and Its Aftermath,' *Hilal*, January 2012, pp. 9–10.
Yamin, Tughral. 'Sri Lanka in the Cross-Hair,' *Hilal*, April 2012, pp. 13–14.
Zia, Amir. 'War of Narratives,' *Hilal*, September 2015.
Zia, Amir. 'Brand India,' *Hilal*, November 2015.
Zia, Amir. 'Unrealistic Expectations,' *Hilal*, February 2016.

Zia, Amir. 'Hybrid Warfare Manifested,' *Hilal*, February 2018.
Zia, Amir. 'Religious Harmony, Sectarian Cleavages and the World of Social Media,' *Hilal*, November 2020.

Primary sources 3: United States embassy cables (by date)

'Terror Finance: Embassy Presses GOP on UN 1267 Committee Cases,' US Embassy Cable, EO12958, 27 November 2006, available at: https://www.theguardian.com/world/us-embassy-cables-documents/87074
'President Musharraf Briefs Gen Schoomaker and Ambassador on Strategy for Border Security,' US Embassy Cable, EO12958, 3 January 2007, available at: https://www.theguardian.com/world/2010/nov/30/pervez-musharraf-taliban
'Musharraf Tells Boucher About Pakistan's Plans for Border Security,' US Embassy Cable, EO12958, 17 January 2007, available at: https://www.theguardian.com/world/us-embassy-cables-documents/92731
'President Musharraf Meets Speaker Pelosi and 110th Congressional Delegation,' US Embassy Cable, EO12958, 31 January 2007, available at: https://www.theguardian.com/world/us-embassy-cables-documents/94672
'Musharraf Tells McCain: Don't Pull Out of Iraq Without Tripartite Political Settlement,' US Embassy Cable, EO12958, 10 April 2007, available at: https://www.theguardian.com/world/us-embassy-cables-documents/103788
'Musharraf Tells Deputy Secretary Pakistan Is Committed to Fight Al Qaeda, Taliban,' US Embassy Cable, EO12958, 21 September 2007, available at: https://www.theguardian.com/world/us-embassy-cables-documents/123103
'CODEL Schiff Meets with PPP Leader Zardari: Political Moves and Balancing Aid,' US Embassy Cable, EO12958, 30 May 2008, available at: https://www.theguardian.com/world/us-embassy-cables-documents/156183
'Embassy Legal Attaché Named in Case Involving Alleged Bagram Detainee,' US Embassy Cable, EO12958, 31 July 2008, available at: https://www.theguardian.com/world/us-embassy-cables-documents/164310
'USDP Edelman's October 15 Meeting in London,' US Embassy Cable, EO12958, 21 October 2008, available at: https://www.theguardian.com/world/us-embassy-cables-documents/174627
'Pakistan Reactions to Terror in Mumbai,' US Embassy Cable, EO12958, 28 November 2008, available at: https://www.theguardian.com/world/us-embassy-cables-documents/180496
'CentCom General Petraeus Meets with Pakistan COAS Kayani,' US Embassy Cable, EO12958, 24 January 2009, available at: https://www.theguardian.com/world/us-embassy-cables-documents/188670
'Scenesetter for Special Envoy Holbrooke,' US Embassy Cable, EO12958, 4 February 2009, available at: https://www.theguardian.com/world/us-embassy-cables-documents/190330
'Indian Foreign Secretary: "Huge Stake" in Special Representative Holbrooke's success,' US Embassy Cable, EO12958, 17 February 2009, available at: https://www.theguardian.com/world/us-embassy-cables-documents/192309
'Scenesetter for General Kayani's Visit to Washington,' US Embassy Cable, EO12958, 19 February 2009, available at: https://www.theguardian.com/world/us-embassy-cables-documents/192895
'Focusing the U.S.-Pakistan Strategic Dialogue,' US Embassy Cable, EO12958, 21 February 2009, available at: https://www.theguardian.com/world/us-embassy-cables-documents/193196
'Special Advisor Holbrooke's Meeting with Saudi Assistant Interior Minister Prince Mohammed Bin Nayef,' US Embassy Cable, EO12958, 17 May 2009, available at: https://www.theguardian.com/world/us-embassy-cables-documents/207396

'Reviewing Our Afghanistan-Pakistan Strategy,' US Embassy Cable, EO12958, 23 September 2009, available at: https://www.theguardian.com/world/us-embassy-cables-documents/226531

'Ambassador Meets with Kayani and Pasha About Kerry-Lugar,' US Embassy Cable, EO12958, 7 October 2009, available at: https://www.theguardian.com/world/us-embassy-cables-documents/228747

'Pakistani Ambassador on Military Operations in FATA; Saudi Relations,' US Embassy Cable, EO12958, 26 October 2009, available at: https://www.theguardian.com/world/us-embassy-cables-documents/231326

'Current Indian Government Thinking on Afghanistan and Pakistan,' US Embassy Cable, EO12958, 27 November 2009, available at: https://www.theguardian.com/world/us-embassy-cables-documents/236969

'SRAP Holbrooke Discusses Afghanistan and Pakistan,' US Embassy Cable, EO12958, 28 January 2010, available at: https://www.theguardian.com/world/us-embassy-cables-documents/245980

'Codel Lynch's Meeting with Pakistan Tehreek-E-Insaf (PTI) Party Leader Imran Khan,' US Embassy Cable, EO12958, 6 February 2010, available at: https://www.theguardian.com/world/us-embassy-cables-documents/247596

Primary sources 4: Interviews

(2019–2020)

Politician 1, Islamabad, Pakistan
Close confidante of Zulfikar Ali Bhutto, Lahore, Pakistan
Security officials, location withheld

(January 2024)

Politician 1, Islamabad, Pakistan (second interview, first in April 2019)
Politician 2, Islamabad, Pakistan
Politician 3, Islamabad, Pakistan
Journalist 1, Islamabad, Pakistan
Columnist, Islamabad, Pakistan
Government official, Islamabad, Pakistan
Civil society activist, Islamabad, Pakistan
Journalist 2, Lahore, Pakistan
Journalist 3, Lahore, Pakistan
Academic 1, Lahore, Pakistan
Academic 3, Lahore, Pakistan

(March 2024)

Academic 2, London, United Kingdom

(August 2024)

Academic 4, Europe (Exact location withheld)

Secondary sources

Abhyanker, Rajendra. *West Asia and the Region: Defining India's Role* (New Delhi: Academic Foundation, 2007).

Adeney, Katharine. 'How to Understand Pakistan's Hybrid Regime: The Importance of a Multidimensional Continuum,' *Democratization* 24, no. 1 (2017): pp. 119–137.

Adil, Mohammad. 'Why Obama Won't Visit Pakistan,' *Foreign Policy*, 10 April 2015.

Adisonmez, Umut Can and Recep Onursal. 'Governing Anxiety, Trauma and Crisis: The Political Discourse on Ontological (In)Security After the July 15 Coup Attempt in Turkey,' *Middle East Critique* 29, no. 3 (2020): pp. 291–306.

Adisonmez, Umut Can and Recep Onursal. '"Strong, but Anxious State": The Fantasmatic Narratives on Ontological Insecurity and Anxiety in Turkey,' *Uluslararası İlişkiler/International Relations* 19, no. 73 (2022): pp. 61–75.

Ahall, Linda. 'Poststructuralism,' in *Security Studies: An Introduction*, edited by Paul Williams and Matt McDonald (Abingdon: Routledge, 2023), pp. 86–99.

Ahmad, Bashir and Maawra Salam. 'Bureaucracy in Pakistan Between Delivery of Civil Services and Politics,' *Journal of Political Studies* 23, no. 2 (2016): pp. 413–438.

Ahmad, Naveed. 'The Non-Aligned Movement and Pakistan,' *Pakistan Horizon* 32, no. 4 (1979): pp. 79–91.

Ahmad, Naveed. 'Sino-Pakistan Relations (1971–1981),' *Pakistan Horizon* 34, no. 3 (1981): pp. 56–81.

Ahmed, Ishtiaq. *The Punjab: Bloodied, Partitioned and Cleansed* (Karachi: Oxford University Press Pakistan, 2012).

Ahmed, Ishtiaq. *Pakistan: The Garrison State* (Karachi: Oxford University Press Pakistan, 2013).

Ahmed, Ishtiaq. *Jinnah: His Successes, Failures and Role in History* (London: Penguin, 2019).

Ahmed, Ishtiaq. 'Not Easily Forgotten: Reading an Indian Diplomat's View of Kashmir,' *The Friday Times*, 2 August 2022, available at: https://thefridaytimes.com/02-Aug-2022/not-easily-forgotten-reading-an-indian-diplomat-s-view-of-kashmir

Ahmed, Khaled. 'The "Castle of Islam" Concept,' *The Express Tribune*, 3 March 2012.

Ahmed, Khaled. 'Homework for Nawaz,' *The Indian Express*, 24 May 2013.

Ahmed, Khaled. 'Bullying Pakistan Won't Work,' *Newsweek Pakistan*, 22 June 2013.

Ahmed, Khaled. 'Pakistan After Hostile Cartography,' *Newsweek Pakistan*, 13 August 2020.

Ahmed, Khaled. 'The Sad Story of "Unity-Faith-Discipline",' *Newsweek Pakistan*, 21 March 2022.

Alam, Kemal. 'The Bajwa Doctrine: The Pakistani Military Has Done More Than Enough,' *RUSI Newsbrief* 38, no. 1 (16 February 2018).

Alavi, Hamza. 'The State in Post-colonial Societies: Pakistan and Bangladesh,' *New Left Review* 1, no. 74 (1972): pp. 59–81.

Ali, Mehrunnisa. 'Jinnah's Perception of Pakistan's Foreign Policy and the Kashmir Issue,' *Pakistan Horizon* 43, no. 2 (April 1990): pp. 57–70.

Al Jazeera. 'Pakistan Taliban Chief Mullah Fazlullah "Killed in Drone Attack",' 15 June 2018.

Al Jazeera. 'India's Narendra Modi Gets Top UAE Honour Amid Kashmir Crisis,' 24 August 2019.

Amin, Shahid. *Pakistan's Foreign Policy: A Reappraisal* (Karachi: Oxford University Press Pakistan, 2000).

Amin, Shahid. *Pakistan's Foreign Policy: A Reappraisal* (Karachi: Oxford University Press Pakistan, 2021).

Angermuller, Johannes. 'Poststructuralist Discourse Studies: From Structure to Practice,' in *The Cambridge Handbook of Discourse Studies*, edited by Anna de Finna and Alexandra Georgakopoulou (Cambridge: Cambridge, 2020), pp. 235–254.

Antreou, Magdalene. *The Ailing Turkish Democracy: The Transformation and Perpetuation of a Hybrid Competitive System* (Cambridge: Cambridge Scholars, 2019).

Aslam, Wali. 'A Critical Evaluation of American Drone Strikes in Pakistan: Legality, Legitimacy and Prudence,' *Critical Studies on Terrorism* 4, no. 3 (2011): pp. 313–329.

Aslam, Wali. 'Understanding the "Pak" in "AfPak": the Obama Administration's Security Policy for Pakistan at the Mid-term,' *Journal of Policing, Intelligence and Counter Terrorism* 7, no. 1 (2012): pp. 2–21.

Aslam, Wali. *The United States and Great Power Responsibility in International Society: Drones, Invasion and Rendition* (London: Routledge, 2013).

Aslam, Wali. 'Drones and the Issue of Continuity in America's Pakistan Policy Under Obama,' in *Obama's Foreign Policy: Ending the War on Terror*, edited by Michelle Bentley and Jack Holland (Abingdon: Routledge, 2013), pp. 139–161.

Aslam, Wali. 'Drone Opposition in Pakistan and the Issues of Post-colonial Identity,' *E-International Relations*, 26 February 2014.

Aslam, Wali. 'The Political Appropriation of Casualties in Threat Construction: The Case of US Drone Strikes in Pakistan,' in *Precision Strike Warfare and International Intervention: Strategic, Ethico-Legal and Decisional Implications*, edited by Michel Aaronson, Wali Aslam, Tom Dyson, and Regina Rauxloh (Abingdon: Routledge, 2015), pp. 73–92.

Aslam, Wali. 'Great-Power Responsibility, Side-Effect Harms and American Drone Strikes in Pakistan,' *Journal of Military Ethics* 15, no. 2 (2016): pp. 143–162.

Aslam, Wali. *Terrorist Relocation and the Societal Consequences of US Drone Strikes in Pakistan* (London: Remote Control Project, 2018), available at https://researchportal.bath.ac.uk/en/publications/terrorist-relocation-and-the-societal-consequences-of-us-drone-st/

Aslam, Wali. 'Post-9/11 Higher Education Reforms and the Instruction of International Relations in Pakistan,' *International Studies Perspectives* 19, no. 3 (2018): pp. 267–287.

Aslam, Wali, Leslie Wehner, Kei Koga, Janis van der Westhuizen, Cameron Thies, and Feliciano de Sá Guimarães. 'Misplaced States and the Politics of Regional Identity: Towards a Theoretical Framework,' *Cambridge Review of International Affairs* 33, no. 4 (2020): pp. 505–526.

Aslam, Wali. 'Pakistan's Hard Misplacement and the Politics of Regional Identity,' *Cambridge Review of International Affairs* 33, no. 4 (2020): pp. 542–554.

Aslam, Wali. 'International Interventions and Normative Prudence as a "Forgotten" Virtue of Statecraft,' *Journal of Intervention and Statebuilding* 15, no. 2 (2021): pp. 181–200.

Aslam, Wali and Alex Neads. 'Renegotiating Societal–Military Relations in Pakistan: The Case of the Pashtun Tahafuz Movement,' *Democratization* 28, no. 2 (2021): pp. 265–284.

Aslam, Wali. 'Misaligned Expectations and Broken Promises: United States Foreign Policy Approach Towards Pakistan (1947–2021),' in *The Routledge Handbook of US Foreign Policy in the Indo-Pacific*, edited by Oliver Turner, Nicola Nymalm, and Wali Aslam (Abingdon: Routledge, 2022), pp. 330–342.

Aslam, Wali. 'Whose Interest? Whose Peace? An Evaluation of Pakistan's Peace Agreements with the Taliban in the Tribal Areas (2004–2015),' *Studies in Conflict and Terrorism* 46, no. 1 (2023): pp. 21–45.

Ayoob, Mohammed. 'The Taliban and the Changing Nature of Pashtun Nationalism,' *The National Interest*, 10 January 2019.

Azeem, Munawer. 'Discussion in Islamabad on Afghan Refugees' Issue Abandoned After Police "Raid",' *Dawn*, 25 October 2023.

Aziz, Mazhar. *Military Control in Pakistan: The Parallel State* (Abingdon: Routledge, 2008).

Aziz, Saba. 'Pakistan-US War of Words Over Donald Trump's Tweet,' *Al Jazeera*, 2 June 2018.

Azmi, Raziullah. 'Pakistan-United States Relations: An Appraisal,' *Pakistan Horizon* 36, no. 3 (1983): pp. 37–50.

Baciu, Cornelia. *Civil-Military Relations and Global Security Governance: Strategy, Hybrid Orders and the Case of Pakistan* (Abingdon: Routledge, 2021).

Bacon, Tricia. 'Slipping the Leash? Pakistan's Relationship with the Afghan Taliban,' *Survival* 60, no. 5 (2018): pp. 159–180.

Bajwa, Qamar Javed. 'General Bajwa at the Munich Security Conference,' 17 February 2018, available at: https://securityconference.org/assets/02_Dokumente/03_Materialien/MSC2018_speech_QamarJavedBajwa.pdf

Bajwa, Qamar Javed. 'General Bajwa's Address at the Islamabad Security Dialogue,' *Dawn*, 18 March 2021.

Bajwa, Qamar Javed. 'Full Text of Gen Bajwa's Speech at the Islamabad Security Dialogue,' *Dawn*, 18 March 2021.

Baloch, Shah Meer. 'Pakistani Judges Say Intelligence Agency Threatened Them Over Imran Khan,' *The Guardian*, 3 April 2024.

Barczewski, Stephanie. *Myth and National Identity in Nineteenth-Century Britain* (Oxford: Oxford University Press, 2000).

Barrett, Roby. *The Greater Middle East and the Cold War: US Foreign Policy Under Eisenhower and Kennedy* (London: I.B. Tauris, 2009).

Bar-Tal, Daniel and Dikla Antebi. 'Siege Mentality in Israel,' *International Journal of Intercultural Relations* 16, no. 3 (1992): pp. 251–275.

BBC Media Centre. 'BBC Documentary Reveals New Evidence of Pakistan Double Cross,' 26 October 2011, available at: https://www.bbc.co.uk/mediacentre/latestnews/261011pakistan

BBC News. 'US "Threatened" to Bomb Pakistan,' 22 September 2006.

BBC News. 'Musharraf Admits US Aid Diverted,' 14 September 2009.

BBC News. 'Adm Mike Mullen's Address,' 22 September 2011.

BBC News. 'Secret Pakistan: Double Cross,' 26 October 2011.

BBC News. 'Pakistan Outrage After "NATO Attack Kills Soldiers",' 26 November 2011.

BBC News. 'US Cuts Pakistan Security Assistance Over Terror Groups,' 4 January 2018.

BBC News. 'Hafiz Saeed: Will Pakistan's "Terror Cleric" Stay in Jail?' 13 February 2020.

Behagel, Jelle and Ayşem Mert. 'The Political Nature of Fantasy and Political Fantasies of Nature,' *Journal of Language and Politics* 20, no. 1 (2021): pp. 79–94.

Behuria, Ashok. 'Fighting the Taliban: Pakistan at War with Itself,' *Australian Journal of International Affairs* 61, no. 4 (2007): pp. 529–543.

Benedikter, Roland and Karim Fathi. 'What Is a Resilient Society?' *International Policy Digest*, 17 September 2017.

Berenskötter, Felix. 'Anxiety, Time, and Agency,' *International Theory* 12, no. 2 (2020): pp. 273–290.

Bhagat, Gawdat. 'Pakistan and Saudi Arabia Relations: An Assessment,' in *Pakistan: The US, Geopolitics and Grand Strategies*, edited by Usama Butt and Julian Schofield (London: Pluto, 2012), pp. 188–205.

Bhatt, Sheela. 'Advani Destroyed Agra Summit with Pakistan: A S Dulat,' *Rediff*, 6 July 2015.

Bhutto, Benazir. *Daughter of Destiny: An Autobiography* (London: HarperCollins, 2008).

Boni, Filippo. *Sino-Pakistani Relations: Politics, Military and Regional Dynamics* (Abingdon: Routledge, 2019).

Bose, Sarmila. 'Anatomy of Violence: Analysis of Civil War in East Pakistan in 1971,' *Economic and Political Weekly* 40, no. 41 (2005): pp. 4463–4471.

Bourke-White, Margaret. *Halfway to Freedom* (New York: Simon & Schuster, 1949).

Bracher, Karl. *Stages of Seizing Power* (Berlin: Ullstein, 1960).

Bray, John. 'New Directions in Pakistan's Foreign Policy,' *The World Today* 48, no. 4 (1992): pp. 65–68.

Bredi, Daniela. 'Notes on Saudi Arabia's Economic and Ideological Expansionism—The Case of Pakistan,' in *Pakistan in the 80s: Ideology, Regionalism, Economy, Foreign Policy*, edited by Wolfgang Zingel and Stephanie Lallemant (Lahore: Vanguard, 1985), pp. 537–549.

Brookes, Adam. 'Mullen Mauling Shows Patience Wearing Thin with Pakistan,' *BBC News*, 23 September 2011.

Burki, Shahid. 'Pakistan Ends 2010 as Sick Man of South Asia,' *Gulf News*, 24 December 2010.

Business Standard. 'Anti-Pakistan Protests Break Out Across Afghanistan: The Protesters Asked Afghanistan President Ashraf Ghani to Declare War on Pakistan,' 8 April 2017.

Butt, Ahsan. 'Nationalistic Narratives in Pakistani Textbooks,' *United States Institute of Peace Brief 210*, 2016.

Butt, Usama and Julian Schofield. (eds.) *Pakistan: The US, Geopolitics and Grand Strategies* (London: Pluto, 2012).

Calabrese, John. 'Balancing on "the Fulcrum of Asia": China's Pakistan Strategy,' *Indian Journal of Asian Affairs* 27/28, no. 1/2 (2014–2015): pp. 1–20.

Calamur, Krishnadev. 'Pakistan's Week Keeps Getting Worse,' *The Atlantic*, 4 January 2018.

Campbell, David. *Writing Security: United States Foreign Policy and the Politics of Identity* (Saint Paul, MN: University of Minnesota Press, 1998).

Cassani, Andrea. 'Hybrid What? Partial Consensus and Persistent Divergences in the Analysis of Hybrid Regimes,' *International Political Science Review* 35, no. 5 (2014): pp. 542–558.

Central Intelligence Agency. 'Iran-Pakistan Border Dispute: An Evaluation of the Iranian Claims.' Intelligence Report number 6564, 16 September 1954, Office of Intelligence Research, Department of State, United States of America, available at: https://www.cia.gov/library/readingroom/docs/CIA-RDP08C01297R000500130001-4.pdf

Chacko, Priya. *Indian Foreign Policy: The Politics of Postcolonial Identity from 1947 to 2004* (Florence: Routledge, 2012).

Chaudhri, Mohammed Ahsen. 'Geopolitical Factors in Pakistan India Relations,' in *Pakistan and South Asian Environment*, edited by Ejaz Aslam Qureshi (Lahore: Sang-e-Meel, 1991), pp. 105–129.

Chaudhry, Rabia. *The Changing Dynamics of Civil Military Relations in Pakistan: Soldiers of Development* (Abingdon: Routledge, 2024).

Cheema, Pervaiz. *Pakistan's Defence Policy 1947–58* (London: Palgrave, 1990).

Chishti, Zoha Freed. '(No) Memory of War and Violence: Conference Series on 1971 War Cancelled,' *The LUMS Post*, 13 September 2021.

Ciftci, Sabri. 'Military Might: A Domestic Economy Explanation of Turkish Foreign Policy,' *Turkish Studies* 24, no. 5 (2023): pp. 764–787.

CIVICUS. 'Pakistan's Civil Society Feels the Pressure,' 26 July 2023.

Cloughley, Brian. *A History of Pakistan Army: Wars and Insurrections* (New York: Carrell Books, 2016).

Coll, Steve. *Ghost Wars: The Secret History of the CIA, Afghanistan and Bin Laden* (London: Penguin, 2005).

Conroy, Scott. 'Pakistanis React to "Stone Age" Claim,' *CBS News*, 23 September 2006.

Corera, Gordon. *Shopping for Bombs: Nuclear Proliferation, Global Insecurity, and the Rise and Fall of the A.Q. Khan Network* (London: Hurst, 2009).

Crikey. 'Pakistan-US Relations: Annus Horribilis,' 9 December 2011.

Croissant, Aurel, David Kuehn, Paul Chambers, and Siegfried O. Wolf. 'Beyond the Fallacy of Coup-Ism: Conceptualizing Civilian Control of the Military in Emerging Democracies,' *Democratization* 17, no. 5 (2009): pp. 950–975.

Curtis, Adam. *Bitter Lake* (London: BBC, 2015), available at: https://www.bbc.co.uk/programmes/p02gyz6b

Curtis, Lisa. 'The Reorientation of Pakistan's Foreign Policy Towards Its Region,' *Contemporary South Asia* 20, no. 2 (2012): pp. 256–269.

Curtis, Lisa and Derek Scissors. 'The Limits of the Pakistan-China Alliance,' *The Heritage Foundation Report*, 19 January 2012.

Dawn. 'Nur Khan Reminisces '65 War,' 6 September 2005.

Dawn. 'DCC to Be Reconstituted as Committee on National Security,' 22 August 2013.

Dawn. '"Bajwa Doctrine" Will Bring Durable Peace: Ghafoor,' 12 January 2018.

Dawn. 'India Withdraws Most Favoured Nation Status for Pakistan,' 15 February 2019.

Dawn. 'India Revokes Occupied Kashmir's Special Autonomy Through Rushed Presidential Decree,' 5 August 2019.

Dawn. 'FATF Meeting,' 18 February 2020.

Dawn. 'There and Back Again: A Timeline of Pakistan's Journey Out of the FATF "Grey List",' 21 October 2022.

Delvoie, Louis. 'The Islamization of Pakistan's Foreign Policy,' *International Journal* 51, no. 1 (1995/6): pp. 126–147.

Devasher, Tilak. *Pakistan: Courting the Abyss* (New Delhi: HarperCollins, 2016).

Diamond, Larry. 'Elections Without Democracy: Thinking About Hybrid Regimes,' *Journal of Democracy* 13, no. 2 (2002): pp. 21–35.

Dunbar, David. 'Bhutto—Two Years On,' *The World Today* 30, no. 1 (1974): pp. 23–24.

Eberle, Jakub. 'Narrative, Desire, Ontological Security, Transgression: Fantasy as a Factor in International Politics,' *Journal of International Relations and Development* 22, no. 1 (2019): pp. 243–248.

Eng, Pierre van der. 'Foreign Debt Will be Pakistan's Economic Touchstone in 2024,' *East Asia Forum Report*, 5 March 2024, available at: https://eastasiaforum.org/2024/03/05/foreign-debt-will-be-pakistans-economic-touchstone-in-2024/

Evans, Karen. 'War and Peace: Is Militarisation the New Norm?' in *Social Censure and Critical Criminology*, edited by Anthony Amatrudo (London: Palgrave Macmillan, 2017), pp. 275–302.

Fair, Christine. 'Islam and Politics in Pakistan,' in *The Muslim World After 9/11*, edited by Angel Rabasa, Cheryl Benard, Peter Chalk, Christine Fair, Theodore Karasik, Rollie Lal, Ian Lesser, and David Thaler (Santa Monica, CA: Rand, 2004), pp. 247–294.

Fair, Christine. 'Pakistan's Ability to Mitigate Sources of Insecurity,' in *Pakistan: Can the United States Secure an Insecure State?* edited by Christine Fair et al. (Santa Monica, CA: Rand: 2010), pp. 75–136.

Fair, Christine. 'The US-Pakistan Relations After a Decade of the War on Terror,' *Contemporary South Asia* 20, no. 2 (2012): pp. 243–253.

Fair, Christine. *Fighting to the End: The Pakistan Army's Way of War* (Oxford: Oxford University Press, 2014).

Fair, Christine. 'The China-Pakistan Axis: Asia's New Geopolitics. By Andrew Small. New York: Oxford University Press, 2015. xvi, 319,' Book Review, *The Journal of Asian Studies* 75, no. 3 (2016): pp. 797–798.

Fair, Christine. *In Their Own Words: Understanding Lashkar-e-Tayyaba* (Oxford: Oxford University Press, 2018).

Fair, Christine. 'The Milli Muslim League: The Domestic Politics of Pakistan's Lashkar-e-Taiba,' *The Hudson Institute Commentary*, 3 June 2018, available at: https://www.hudson.org/national-security-defense/the-milli-muslim-league-the-domestic-politics-of-pakistan-s-lashkar-e-taiba#footNote25

FATF. 'Who We Are,' Financial Action Task Force, 2020, available at: https://www.fatf-gafi.org/about

Fazl-e-Haider, Syed. 'China's Big Gamble in Pakistan: A 10-Year Scorecard For CPEC,' *The Interpreter*, 1 August 2023.

Filkins, Dexter. 'Pakistanis Tell of Motive in Taliban Leader's Arrest,' *The New York Times*, 22 August 2010.

Firdous, Tehmina and Mehrajudin Aslam. 'Contesting Narratives: A Comparative Analysis of Indian, Pakistani and Kashmiri Print Media Framing of the Abrogation of Article 370,' *International Journal for Multidisciplinary Research* 6, no. 2 (2024): pp. 1–18.

Firstpost. 'Afghan Taliban Dismisses Fear of Attacks on India After US Troop Withdrawal, Says Neighbourhood's Support Vital to Rebuild Country,' 15 October 2019.

Friedrich, Carl Joachim and Zbigniew Brzezinski. *Totalitarian Dictatorship and Autocracy* (Cambridge, MA: Harvard, 1965).

Gall, Sandy. *War Against the Taliban: Why It All Went Wrong in Afghanistan* (London: Bloomsbury, 2012).

Garlick, Jeremy. 'Deconstructing the China-Pakistan Economic Corridor: Pipe Dreams Versus Geopolitical Realities,' *Journal of Contemporary China* 27, no. 112 (2018): pp. 519–533.

Garner, George. 'Chechnya and Kashmir: The Jihadist Evolution of Nationalism to Jihad and Beyond,' *Terrorism and Political Violence* 25, no. 3 (2013): pp. 419–434.

Garver, John. 'China and South Asia,' *The Annals of the American Academy of Political and Social Science* 519 (1992): pp. 67–85.

Garver, John. 'Sino-Indian Rapproachement and the Sino-Pakistan Entente,' *Political Science Quarterly* 111, no. 2 (1996): pp. 323–347.

Garver, John. 'The Future of the Sino-Pakistan Entente Cordiale,' in *South Asia in 2020: Future Strategic Balances and Alliances*, edited by Michael Chambers (Walcott, NY: Scholar's Choice Publishers, 2002), pp. 385–449.

Gauhar, Altaf. 'Pakistan: Ayub Khan's Abdication,' *Third World Quarterly* 7, no. 1 (1985): pp. 102–131.

Ghori, Karamatullah. 'Sixty Years of Pakistan's Foreign Policy,' *Pakistan Horizon* 60, no. 2 (2007): pp. 9–24.

Gibel, Bryan. 'U.S. Aid to Pakistan: The Kerry-Lugar Bill,' *PBS News*, available at: https://www.pbs.org/frontlineworld/stories/pakistan901/aid.html

Gilbert, Leah and Payam Mohseni. 'Beyond Authoritarianism: The Conceptualization of Hybrid Regimes,' *Studies in Comparative International Development* 46, no. 3 (2011): pp. 270–297.

Gill, John. *An Atlas of the 1971 India-Pakistan War: The Creation of Bangladesh* (Washington, D.C.: NDU Press, 2003).

Glynos, Jason. 'Ideological Fantasy at Work,' *Journal of Political Ideologies* 13, no. 3 (2008): pp. 275–296.

Glynos, Jason and David Howarth. *Logics of Critical Explanation in Social and Political Theory* (Abingdon: Routledge, 2007).

Grare, Frédéric. 'Pakistan's Double Dealing,' *Foreign Policy*, 1 March 2006.

Grare, Frédéric. 'The Evolution of Sectarian Conflicts in Pakistan and the Ever-Changing Face of Islamic Violence,' *South Asia: Journal of South Asian Studies* 30, no. 1 (2007): pp. 127–143.

Goldberg, Jeffrey and Marc Ambinder. 'The Ally from Hell,' *The Atlantic*, December 2011.

Goldenberg, Suzanne. 'Bush Threatened to Bomb Pakistan, Says Musharraf,' *The Guardian*, 22 September 2006.

Hagan, Joe. 'Regime Type, Foreign Policy, and International Relations,' *The International Studies Encyclopedia* (2010): pp. 6153–6178.

Haider, Zeeshan. 'Pakistan Grants India Most Favoured Nation Trade Status,' *Reuters*, 2 November 2011.

Haider, Ziad. *The Ideological Struggle for Pakistan* (Stanford, CA: Hoover Press, 2010).

Haqqani, Husain. *Pakistan: Between Mosque and Military* (Washington, D.C.: Brookings, 2005).

Haqqani, Husain. *Magnificent Delusions: Pakistan, the United States and an Epic History of Misunderstanding* (New York: PublicAffairs, 2013).

Harrison, Selig. 'Fanning Flames in South Asia,' in *Pakistan and South Asian Environment*, edited by Ejaz Aslam Qureshi (Lahore: Sang-e-Meel, 1991), pp. 13–34.

Hasan, Khwaja. 'The Foreign Policy of Mr. Liaquat Ali Khan,' *Pakistan Horizon* 4, no. 4 (1951): pp. 181–199.

Hersh, Seymour. 'The Getaway: Questions Surround a Secret Pakistani Airlift,' *The New Yorker*, 21 January 2002.

Hindustan Times. 'Transcript of Speech and Q&A Session with President of Pakistan Asif Ali Zardari,' 26 November 2008.

Hintz, Lisel. *Identity Politics Inside Out: National Identity Contestation and Foreign Policy in Turkey* (New York: Oxford University Press, 2018).

Hiro, Dilip. *The Longest August: The Unflinching Rivalry Between India and Pakistan* (New York: Nation Books, 2015).

Hoffman, Michael. 'Military Extrication and Temporary Democracy: The Case of Pakistan,' *Democratization* 18, no. 1 (2011): pp. 75–99.

Hoodbhoy, Pervez. 'Saudizing Pakistan: How Pakistan Is Changing and What This Means for South Asia and the World,' in *Routledge Handbook of Contemporary Pakistan*, edited by Aparna Pande (Abingdon: Routledge, 2016), pp. 466–478.

Hopf, Ted. *Reconstructing the Cold War: The Early Years, 1945–1958* (New York: Oxford University Press, 2012).

Hudson, Valerie. *Foreign Policy Analysis: Classic and Contemporary Theory* (Lanham, MD: Rowman & Littlefield, 2013).

Hussain, Abid. 'Tensions High After Pakistan Launches Cross-Border Attacks into Afghanistan,' *Al Jazeera*, 18 March 2024.

Hussain, Abid. 'Amid US-Pakistan Thaw, Two Key Challenges: Iran and China,' *Al Jazeera*, 20 June 2025.

Hussain, Zahid. *Frontline Pakistan: The Struggle with Militant Islam* (New York: Columbia University Press, 2007).

Hussain, Zahid. 'Ironics of thc Afghan War,' *Dawn*, 29 March 2019.

Hyde, Susan and Elizabeth Saunders. 'Recapturing Regime Type in International Relations: Leaders, Institutions, and Agency Space,' *International Organization* 74, no. 2 (2020): pp. 363–395.

Ibrahimi, S. Yaqub. 'The Taliban's Islamic Emirate of Afghanistan (1996–2001): "War-Making and State-Making" as an Insurgency Strategy,' *Small Wars & Insurgencies* 28, no. 6 (2017): pp. 947–972.

Imran, Kashif. 'Pakistan Says Taliban Deputy FM's Rejection of Durand Line "Fanciful", Can't Change Facts,' *Arab News*, 17 February 2024.

Imtiaz, Saba. '"Taliban Architect" Naseerullah Babar Passes Away,' *The Express Tribune*, 11 January 2011.

International Crisis Group. 'Pakistan: The Mullahs and the Military,' Asia Report 49, 20 March 2003, available at: https://www.crisisgroup.org/asia/south-asia/pakistan/pakistan-mullahs-and-military.

International Crisis Group. 'The State of Sectarianism in Pakistan,' Asia Report 95, 18 April 2005, available at: https://www.crisisgroup.org/asia/south-asia/pakistan/state-sectarianism-pakistan

Iqbal, Anwar. 'Terror War Killed 80,000 During 2005–13: Report,' *Dawn*, 31 March 2015.

Iqbal, Anwar. 'Pakistan Placed on FATF "Grey List" Despite Diplomatic Efforts to Avert Decision,' *Dawn*, 28 June 2018.

Iqbal, Anwar. 'Pakistan's Human Development Ranking Declines,' *Dawn*, 14 March 2024.

Iqbal, Khalid. 'Fallacy of Two Front Squeeze,' *Defence Journal* 20, no. 10 (2017): pp. 70–72.

Iqbal, Nosheen. 'Barack Obama and His Wild, Drug-Taking Roomie,' *The Guardian*, 3 October 2012.
Islam, Nasir. 'Islam and National Identity: The Case of Pakistan and Bangladesh,' *International Journal of Middle East Studies* 13, no. 1 (1981): pp. 55–73.
Ispahani, Abul Hassan. 'The Alternative to the India-Pakistan Arms Race,' *Pakistan Horizon* 2, no. 2 (1967): pp. 109–119.
Jabeen, Mussarat and Muhammad Mazhar. 'Security Game: SEATO and CENTO as Instrument of Economic and Military Assistance to Encircle Pakistan.' *Pakistan Economic and Social Review* 49, no. 1 (2011): pp. 109–132.
Jacobs, Thomas. 'The Dislocated Universe of Laclau and Mouffe: An Introduction to Post-Structuralist Discourse Theory,' *Critical Review* 30, nos. 3–4 (2018): pp. 294–315.
Jamal, Umair. 'What the Apparent Collapse of Afghan Peace Talks Means for Pakistan,' *The Diplomat*, 10 September 2019.
Jamal, Umair. 'Can a Negative Decision at the FATF Bolster Hardliners in Pakistan?' *The Diplomat*, 16 October 2019.
Jamal, Umair. 'The Kuala Lumpur Summit 2019 Shows Pakistan's Diplomatic Subservience to Saudi Arabia,' *The Diplomat*, 20 December 2019.
Janjua, Haroon. 'Is China's Investment in Pakistan Slowing Down?' *DW*, 10 November 2023.
Jinnah, Mohammad Ali. *Quaid-i-Azam Mohammad Ali Jinnah: Speeches as Governor-General of Pakistan, 1947–1948* (Karachi: Ferozsons, 1949).
Jones, Seth. 'Pakistan's Dangerous Game,' *Survival* 49, no. 1 (2007): pp. 15–32.
Jones, Seth. 'The Rise of Afghanistan's Insurgency: State Failure and Jihad,' *International Security* 32, no. 4 (2008): pp. 7–40.
Kaarbo, Juliet. 'A Foreign Policy Analysis Perspective on the Domestic Politics Turn in IR Theory,' *International Studies Review* 17, no. 2 (2015): pp. 189–216.
Kapur, S. Paul. *Jihad as Grand Strategy: Islamist Militancy, National Security, and the Pakistani State* (New York: Oxford University Press, 2017).
Kapur, S. Paul and Sumit Ganguly. 'The Jihad Paradox: Pakistan and Islamist Militancy in South Asia,' *International Security* 37, no. 1 (2012): pp. 111–141.
Khan, Asghar. *Generals in Politics* (New Delhi: Vikas, 1983).
Khan, Asghar. *We've Learnt Nothing from History: Pakistan: Politics and Military Power* (Karachi: Oxford University Press Pakistan, 2005).
Khan, Hasan Naser. 'CPEC for Punjab or Pakistan: Myth and Reality,' *The Express Tribune*, 17 January 2016.
Khan, Jamil. 'CPEC—A Game Changer for Pakistan,' *The Business Recorder*, 12 August 2022.
Khan, Obaid Abrar. 'Pak Political Parties Must Hammer Out Rifts: China,' *The News*, 10 January 2016.
Khan, Omer Farooq. 'Asim Munir's Grip on Power Deepens Pakistan's Militarisation,' *The Times of India*, 7 August 2025.
Khan, Raza. 'CPEC Agreement Is Confidential, Ahsan Iqbal Tells Senate,' *Dawn*, 28 July 2016.
Khan, Sahar. 'Double Game: Why Pakistan Supports Militants and Resists U.S. Pressure to Stop,' *Cato Institute*, 20 September 2018.
Khan, Seema and Zahid Shahab Ahmed. 'Impact of The China-Pakistan Economic Corridor on Nation-Building in Pakistan: A Case Study of Balochistan,' *Journal of Contemporary China* 33, no. 150 (2024): pp. 1–15.
Khattak, Afrasiab. 'Change or Isolation?' *The Nation*, 9 September 2017.
Khattak, Daud. 'Will Pakistan Part Ways with Its Proxies?' *The Diplomat*, 13 September 2017.
Khattak, Daud. 'The Taliban's Diplomatic Reemergence,' *The Diplomat*, 25 October 2019.
Kinne, Brendon. 'Decision Making in Autocratic Regimes: A Poliheuristic Perspective,' *International Studies Perspectives* 6, no. 1 (2005): pp. 114–128.

Kinnvall, Catarina. 'Globalization and Religious Nationalism: Self, Identity, and the Search for Ontological Security,' *Political Psychology* 25, no. 5 (2004): pp. 741–767.

Kivimäki, Timo. 'When Ideologies Became Dangerous: An Analysis of the Transformation of the Relationship Between Security and Oppositional Ideologies in US Presidential Discourse,' *Global Society* 37, no. 2 (2023): pp. 225–244.

Kneuer, Marianne. 'Autocratic Regimes and Foreign Policy,' *Oxford Research Encyclopedia of Politics*, 26 September 2017.

Knopf, Jeffrey. 'Domestic Sources of Preferences for Arms Cooperation: The Impact of Protest,' *Journal of Peace Research* 35, no. 6 (1998): pp. 677–695.

Koshy, Ninan. 'On Re-hyphenating India and Pakistan,' *Mainstream Weekly*, 21 December 2008.

Kristof, Nicholas. 'A Girl, a School and Hope,' *The New York Times*, 10 November 2010.

Kugelman, Michael. 'The Haqqani Threat to the US-Pakistan Détente,' *The Diplomat*, 31 July 2014.

Kugelman, Michael. 'How Pakistan Can Help with the Afghan Peace Process,' *Herald*, 21 January 2019.

Kumar, Aishwarya. 'India Should Hold Backchannel Talks with Taliban, Its Return to Power in Afghanistan Is Inevitable,' *News 18*, 16 January 2019.

Kumar, Ashish. *Missteps and Missed Opportunities for Peace in Afghanistan*, United States Institute of Peace, 27 October 2022.

Kumar, Radha. 'Untying the Kashmir Knot,' *World Policy Journal* 19, no. 1 (2002): pp. 11–24.

Kumar, Raksha. 'As Pakistan Approaches a Crucial Election, Its Media Watchdog Bans Critical Voices from TV,' *Reuters Institute*, 3 October 2023.

Kurita, Masahiro. 'The "Pakistan Factor" in China-Indian Relations,' *The National Institute of Defence Studies Commentary Number 48*, Ministry of Defence, Japan, 29 July 2015, available at: https://www.nids.mod.go.jp/english/publication/commentary/pdf/commentary048e.pdf

Kux, Dennis. *Disenchanted Allies: The United States and Pakistan, 1947–2000* (Baltimore, MD: The Johns Hopkins University Press, 2001).

Lakhani, Nina. '"We Couldn't Fail Them": How Pakistan's Floods Spurred Fight at COP for Loss and Damage Fund,' *The Guardian*, 20 November 2022.

Lalwani, Sameer. *A Threshold Alliance: The China-Pakistan Military Relationship*, United States Institute of Peace, March 2023, available at: https://www.usip.org/sites/default/files/2023-03/sr-517_threshold-alliance-china-pakistan-military-relationship.pdf

Lamb, Alastair. *Incomplete Partition: The Genesis of the Kashmir Dispute* (Oxford: Oxford University Press, 2002).

Lambah, Satinder Singh. *In Pursuit of Peace: India-Pakistan Relations Under Six Prime Ministers* (Gurugram: Penguin India Publishers, 2023).

Landler, Mark and Alissa Rubin. 'War Plan for Karzai: Reach Out to Taliban,' *The New York Times*, 28 January 2010.

La Porte, Jr. Robert. 'Pakistan in 1972: Picking Up the Pieces,' *Asian Survey* 13, no. 2 (1973): pp. 187–198.

Laufer, Michael. 'A. Q. Khan Nuclear Chronology,' *Carnegie Endowment for International Peace Report*, 7 September 2005.

Lehman, Howard and Jennifer McCoy. 'The Dynamics of the Two-Level Bargaining Game: The 1988 Brazilian Debt Negotiations,' *World Politics* 44, no. 4 (1992): pp. 600–644.

Levitsky, Steven. *Competitive Authoritarianism: Hybrid Regimes After the Cold War* (New York: Cambridge University Press, 2010).

Levy, Jack. 'Domestic Politics and War,' *The Journal of Interdisciplinary History* 18, no. 4 (1988): pp. 653–673.

Linton, Suzannah. 'Completing the Circle: Accountability for the Crimes of the 1971 Bangladesh War of Liberation,' *Criminal Law Forum* 21, no. 2 (2010): pp. 191–311.

Lipson, Charles. *Reliable Partners: How Democracies Have Made a Separate Peace* (Princeton, NJ: Princeton University Press, 2003).

Macander, Michelle. 'Examining Extremism: Lashkar-e-Taiba,' *Centre for Strategic and International Studies, 28 October* 2021.

Machiavelli, Niccolò. *The Prince*, trans. W. K. Marriott (The Gutenberg Project, 2006), available at: https://www.gutenberg.org/cache/epub/1232/pg1232-images.html#chap20.

Mahdi, Niloufer. *Pakistan's Foreign Policy, 1971–1981: The Search for Security* (Lahore: Ferozsons, 1999).

Malik, Mohan. 'South Asia in China's Foreign Relations,' *Pacifica Review: Peace, Security & Global Change* 13, no. 1 (2001): pp. 73–90.

Malik, Mohan. 'The China Factor in the India-Pakistan Conflict,' *Parameters* 33, no. 1 (2003): pp. 35–50, available at: https://press.armywarcollege.edu/cgi/viewcontent.cgi?article=2134&context=parameters

Manish and Prashant Kaushik. 'CPEC, Afghanistan and India's Concerns,' *India Quarterly* 75, no. 2 (2019): pp. 253–261.

Mattoo, Amitabh. 'Shadow of the Dragon: Indo-US relations and China,' in *Engaging India: U.S. Strategic Relations with the World's Largest Democracy*, edited by Gary K. Bertsch, Seema Gahlaut, and Anupam Srivastava (Abingdon: Routledge, 1999), pp. 219–220.

Mazzucco, Leonardo Jacopo Maria and Kristian P. Alexander. 'Saudi Arabia and Qatar are Cooperating with the Taliban. But Their Approaches to Afghanistan are Different,' *The Atlantic Council*, 12 January 2023.

Meyerle, Jerry. 'Afghanistan-Pakistan Border: Back to Politics as Usual?' *The Diplomat*, 29 November 2013.

Mirshak, Nadim. 'Education as Resistance: Egyptian Civil Society and Rethinking Political Education Under Authoritarian Contexts,' *Critical Sociology* 46, no. 6 (2020): pp. 915–930.

Mishra, Abhinandan. 'No Question of Taliban Fighters Targeting India,' *The Guardian*, 5 October 2019.

Mishra, Manjo. 'Engaging Taliban and India's Afghan Strategy,' *Asia Times*, 22 October 2019.

Mishra, Ravi Dutta. 'India's Top Trade Partner: China Regains Spot on Higher Imports,' *The Indian Express*, 13 May 2024.

Mitzen, Jennifer. 'Ontological Security in World Politics: State Identity and the Security Dilemma,' *European Journal of International Relations* 12, no. 3 (2006): pp. 341–370.

Mohan, C. Raja. 'The Great Game Folio: Obama and Pakistan,' *The Indian Express*, 21 January 2015.

Moini, Qasim. 'Carrot-and-Stick Policy Needed with Afghan Taliban, Says Ahmed Rashid,' *Dawn*, 12 March 2016.

Mookherjee, Nayanika. '"Remembering to Forget": Public Secrecy and Memory of Sexual Violence in the Bangladesh War of 1971,' *Journal of the Royal Anthropological Institute* 12, no. 2 (2006): pp. 433–450.

Morlino, Leonardo. 'Are There Hybrid Regimes? Or Are They Just An Optical Illusion?' *European Political Science Review* 1, no. 2 (2009): pp. 273–296.

Moskalenko, Vladimir. 'Pakistan's Foreign Policy,' *Asian Survey* 14, no. 3 (1974): pp. 267–278.

Mufti, Mariam. 'The Influence of Domestic Politics on the Making of US-Pakistan Foreign Policy,' in *Pakistan: The US, Geopolitics and Grand Strategies*, edited by Usama Butt and Julian Schofield (London: Pluto, 2012), pp. 64–86.

Mufti, Mariam. 'What Do We Know About Hybrid Regimes After Two Decades of Scholarship? *Politics and Governance* 6, no. 2 (2018): pp. 112–119.

Musharraf, Pervez. 'General Pervez Musharraf's Address to the Nation,' 13 October 1999, available at: https://presidentmusharraf.wordpress.com/2007/07/10/address-nation-13-oct-1999/

Musharraf, Pervez. 'President's Address to Nation: 7-Point Agenda and Future Course,' 17 October 1999, available at: https://presidentmusharraf.wordpress.com/2007/07/10/address-nation-7-point-agenda/

Musharraf, Pervez. 'President's Address to the Nation,' 19 September 2001, available at: https://presidentmusharraf.wordpress.com/2006/07/13/address-19-september-2001/

Musharraf, Pervez. 'Remarks by Secretary of State Colin L. Powell and President Pervez Musharraf in Islamabad,' 15 October 2001, available at: https://presidentmusharraf.wordpress.com/2007/05/12/colin-powell-press-conference-2001/

Musharraf, Pervez. 'President Pervez Musharraf's Speech at the United Nations General Assembly's 59th Session,' 22 September 2004, https://presidentmusharraf.wordpress.com/2007/03/17/united-nations-59th-session/

Mustafa, Zubeida. 'Recent Trends in Pakistan's Policy Towards the Middle East,' *Pakistan Horizon* 28, no. 4 (1975): pp. 1–17.

Najam, Durdana. 'The Vagaries of Pakistan Ideology,' *The Express Tribune*, 18 April 2024.

Naqvi, Muhammad. 'South Asian Cooperation,' in *Pakistan and South Asian Environment*, edited by Ejaz Aslam Qureshi (Lahore: Sang-e-Meel, 1991), pp. 176–195.

Narilkar, Amrita. 'Is India a Responsible Great Power?' *Third World Quarterly* 32, no. 9 (2011): pp. 1607–1621.

Nasir, Abbas. 'Controlled Chaos? Think Not,' *Dawn*, 18 February 2024.

Nawaz, Shuja. 'The First Kashmir War Revisited,' *India Review* 7, no. 2 (2008): pp. 115–154.

Nawaz, Shuja. *Crossed Swords: Pakistan, Its Army and the Wars Within* (Karachi: Oxford University Press Pakistan, 2008).

NDTV. 'India's Ex Diplomat Details What Led to Agra Summit Collapse in 2001,' 8 January 2024.

New America. *Drone Strikes: Pakistan*. New America Foundation, 2020, available at: https://www.newamerica.org/in-depth/americas-counterterrorism-wars/pakistan/.

Nissenbaum, Dion, Tom Wright, Owais Tohid, and Adam Entous. 'Airstrike Ravages U.S.-Pakistan Ties,' *The Wall Street Journal*, 28 November 2011.

Noorani, Abdul. 'Bhutto and Nonalignment,' *Economic and Political Weekly* 11, no. 51 (1976): pp. 1957–1958.

Norval, Mitchell. *Sir George Cunningham: A Memoir* (Edinburgh: Blackwood Publishers, 1968).

Notezai, Muhammad. 'The Dilemma of Pakistan's Foreign Policy,' *The Diplomat*, 12 August 2016.

Notezai, Muhammad. 'Trouble at the Pakistan-Iran Border,' *The Diplomat*, 30 October 2018.

Nye, Joseph. 'Energy and Security in the 1980s,' *World Politics* 35, no. 1 (1982): pp. 121–134.

Oğuzlu, Tarik. 'Middle Easternization of Turkey's Foreign Policy: Does Turkey Dissociate from the West?' *Turkish Studies* 9, no. 1 (2008), pp. 3–20.

Omrani, Bijan. 'The Durand Line: History and Problems of the Afghan-Pakistan Border,' *Asian Affairs* 40, no. 2 (2009): pp. 177–195.

Ozkececi-Taner, Binnur. 'Domestic Politics and Foreign Policy,' *Oxford Research Encyclopedia of Politics*, 26 September 2017.

Padmanabhan, Keshav. 'How Much Debt Is Too Much Debt? Curious Case of Pakistan's $69 Billion Chinese Debt Trap,' *The Print*, 13 November 2023.

Paliwal, Avinash. 'The "India Question" in Afghanistan,' *Lawfare*, 6 October 2019, available at: https://www.lawfareblog.com/india-question-afghanistan

Panda, Ankit. 'Pakistan to Grant India Most-Favored Nation Status,' *The Diplomat*, 19 March 2014.

Pande, Aparna. *Explaining Pakistan's Foreign Policy: Escaping India* (Abingdon: Routledge, 2011).

Pande, Aparna. 'Friends of Last Resort: Pakistan's Relations with China and Saudi Arabia,' in *Pakistan's Enduring Challenges*, edited by Christine Fair and Sarah Watson (Philadelphia, PA: University of Pennsylvania Press, 2015), pp. 256–277.

Pant, Harsh and Avinash Paliwal. 'India's Afghan Dilemma Is Tougher Than Ever,' *The Diplomat*, 19 February 2019.

Paracha, Nadeem F. 'The Pakistan Zeitgeists: A Nation Through the Ages,' *Dawn*, 29 May 2014.

Paul, T. V. *The Warrior State: Pakistan in the Contemporary World* (New York: Oxford University Press, 2014).

Paul, T. V. *The Unfinished Quest: India's Search for Major Power Status from Nehru to Modi* (New York: Oxford University Press, 2024).

PBS News. 'Frontline interview: Anne Patterson,' August 2009, available at: https://www.pbs.org/wgbh/pages/frontline/obamaswar/interviews/patterson.html

Pentz, Peter. 'The Mujahidin Middleman: Pakistan's Role in the Afghan Crisis and the International Rule of Non-Intervention,' *Dickinson Journal of International Law* 6, no. 3 (1988): pp. 377–401.

Perlez, Jane. 'Rebuffed by China, Pakistan May Seek I.M.F. Aid,' *The New York Times*, 18 October 2008.

Perlez, Jane. 'Onetime Taliban Handler Dies in Their Hands,' *The New York Times*, 24 January 2011.

Putnam, Robert. 'Diplomacy and Domestic Politics: The Logic of Two-Level Games,' *International Organization* 42, no. 3 (1988): pp. 427–460.

Qazi, Ashraf. 'Our Afghanistan Policy,' *Dawn*, 9 September 2017.

Qureshi, Khalida. 'Pakistan and Iran: A Study in Neighbourly Diplomacy,' *Pakistan Horizon* 21, no. 1 (1968): pp. 33–39.

Qureshi, Saqib. '*US Foreign Policy to Pakistan, 1947–1960: Re-Constructing Strategy*,' PhD Thesis, Department of International Relations, London School of Economics and Political Science, United Kingdom, September 2001, pp. 278–281, available at: http://etheses.lse.ac.uk/3055/

Radio Free Europe/Radio Liberty. 'Pakistani PM Says China Ties "Sweeter Than Honey",' 5 July 2013.

Radio Free Europe/Radio Liberty. 'Intense Border Clashes Between Taliban, Pakistan Cause Deaths, Destruction,' 17 May 2024.

Rafiq, Arif. 'Pakistan's Roller Coaster 2012,' *The Diplomat*, 8 January 2012.

Rahi, Arwin. 'Do the Afghan Taliban and Pakistan Get on as Well as Some Think?' *Dawn*, 27 April 2019.

Rais, Rasul Bakhsh. 'Pakistan in the Regional and Global Power Structure,' *Asian Survey* 31, no. 4 (1991): pp. 378–392.

Rana, Shahbaz. 'Eastern CPEC Route Unfeasible: Report,' *The Express Tribune*, 25 July 2015.

Rana, Uday. 'When Vajpayee and Musharraf "Almost Resolved" the Kashmir Dispute,' *News18*, 13 February 2018.

Rashid, Ahmed. 'Pakistan Undermines UN in Peace Process,' *The Nation*, 23 January 1998.

Rashid, Ahmed. *Taliban: Militant Islam, Oil and Fundamentalism in Central Asia* (London: I.B. Tauris, 2002).

Rashid, Ahmed. 'Musharraf's Double Game Unravels,' *The New York Times*, 10 August 2005.

Rashid, Ahmed. 'The Void Left Behind,' *The Washington Post*, 27 December 2007.

Rashid, Ahmed. 'Afghan Taliban Don't See India as Their Enemy,' Talk delivered at Asia Society, New York, 27 March 2012, available at: https://asiasociety.org/blog/asia/ahmed-rashid-afghan-taliban-dont-see-india-their-enemy

Rath, Saroj. 'Pakistan's Double Game: Supporting and Opposing Terrorism,' *World Affairs: The Journal of International Issues* 15, no. 2 (2011): pp. 92–105.

Rehman, Zia Ur. 'Good Taliban, Bad Taliban: The Case of Hafiz Gul Bahadur,' *Dawn*, 21 March 2024.

Riaz, Ali. *Voting in a Hybrid Regime: Explaining the 2018 Bangladeshi election* (Singapore: Palgrave, 2019).

Ritchie, Stuart. 'Don't Trust the Psychologists on Coronavirus,' *UnHerd*, 31 March 2020.

Rizvi, Hasan Askari. 'Pakistan: Ideology and Foreign Policy,' *Asian Affairs* 10, no. 1 (1983): pp. 48–59.

Rizvi, Hasan Askari. *Military and Politics in Pakistan* (New Delhi: Konark, 1989).

Rizvi, Hasan Askari. *Pakistan and the Geostrategic Environment: A Study of Foreign Policy* (New York: Palgrave Macmillan, 1993).

Rizvi, Hasan Askari. *Military, State and Society in Pakistan* (Basingstoke: Palgrave Macmillan, 2000).

Rizvi, Hasan Askari. *The Military and Politics in Pakistan 1947–1997* (Lahore: Sang-e-Meel, 2000).

Roy, Olivier. 'The Taliban: A Strategic Tool for Pakistan,' in *Pakistan: Nationalism Without a Nation*, edited by Christophe Jaffrelot (London: Zed, 2002), pp. 149–160.

Rubin, Barnett. 'Is Afghanistan Ready For Peace? How Great Powers Can End the War,' *Foreign Affairs*, 30 July 2018.

Rubin, Barnett and Ahmed Rashid. 'From Great Game to Grand Bargain: Ending Chaos in Afghanistan and Pakistan,' *Foreign Affairs* 87, no. 6 (2008): pp. 30–44.

Rubin, Michael. 'When Realities Collide: Differing US-Pakistan Threat Perceptions,' in *Pakistan: The US, Geopolitics and Grand Strategies*, edited by Usama Butt and Julian Schofield (London: Pluto, 2012), pp. 45–60.

Saberin, Zeenat and Shereena Qazi. 'Afghan Peace Conference: India Shares Table with Taliban,' *Al Jazeera*, 9 November 2018.

Said, Edward and David Barsamian. *Culture and Resistance* (Chicago, IL: Haymarket, 2019).

Sajid, Islamuddin. 'Pakistan: PM Recalls Sacrifices Made in US War on Terror,' *Anadolu Agency*, 19 November 2018, available at: https://www.aa.com.tr/en/asia-pacific/pakistanpm-recalls-sacrifices-made-in-us-war-on-terror/1315757

Saleem, Farrukh. 'Gwadar,' *The News*, 31 July 2016.

Samad, Saleem. 'What Led to The War,' *Dhaka Tribune*, 16 December 2019.

Samanta, Pranab. 'Pakistan on FATF's Grey List: How India Convinced China & Pak Shot Itself in the Foot,' *The Print*, 23 February 2018.

Sanborn, Howard and Clayton Thyne. 'Learning Democracy: Education and the Fall of Authoritarian Regimes,' *British Journal of Political Science* 44, no. 4 (2014): pp. 773–797.

Sareen, Sushant. *The Jihad Factory: Pakistan's Islamic Revolution in the Making* (New Delhi: Hindustan Publishing, 2005).

Sareen, Sushant. 'Pakistan: From Annus Horribilis into Annus Miserabilis,' *Observer Research Foundation*, 5 January 2023.

Sattar, Abdul. *Pakistan's Foreign Policy 1947–2005: A Concise History* (Karachi: Oxford University Press Pakistan, 2007).

Sattar, Abdul. *Pakistan's Foreign Policy 1947–2019: A Concise History* (Karachi: Oxford University Press Pakistan, 2020).

Seale, Patrick. 'The US Is Losing Pakistan,' *The Diplomat*, 4 June 2011.

Sengupta, Anwesha. 'Breaking Up: Dividing Assets Between India and Pakistan in Times of Partition,' *The Indian Economic & Social History Review* 51, no. 4 (2014): pp. 529–548.

Schadewitz, Nicole, and Timothy Jachna. 'Comparing Inductive and Deductive Methodologies for Design Patterns Identification and Articulation,' in *International Design Research Conference IADSR 2007: Emerging Trends in Design Research*, Hong Kong, 12–15 November 2007, available at: https://oro.open.ac.uk/15257/

Schofield, Julian and Reeta Tremblay. 'Why Pakistan Failed: Tribal Focoism in Kashmir,' *Small Wars & Insurgencies* 19, no. 1 (2008): pp. 23–38.

Schoppa, Leonard. 'Two-level Games and Bargaining Outcomes: Why *Gaiatsu* Succeeds in Japan in Some Cases but not Others,' *International Organization* 47, no. 3 (1993): pp. 353–386.

Schweller, Randall. 'Unanswered Threats: A Neoclassical Realist Theory of Underbalancing,' *International Security* 29, no. 2 (2004): pp. 159–201.

Shafqat, Saeed. *Civil-Military Relations in Pakistan: From Zulfikar Ali Bhutto to Benazir Bhutto* (Boulder, CO: Westview, 1997).

Shafqat, Saeed. 'CPEC and Changing Dynamics of China-India Relations,' in *China and South Asia: Changing Regional Dynamics, Development and Power Play*, edited by Rajiv Ranjan and Guo Changgang (Abingdon: Routledge, 2021), pp. 17–34.

Shah, Aqil. *The Army and Democracy: Military Politics in Pakistan* (Cambridge, MA: Harvard, 2014).

Shah, Saeed. 'Pakistan's New Plight in Kashmir: What to Do About the Jihadists,' *The Wall Street Journal*, 11 August 2018.

Shahid, Saleem. 'Kabul Playing "Double Game", Says Minister,' *Dawn*, 3 December 2023.

Shahzad, Asif. 'Social Media Platform X Blocked in Pakistan Over National Security, Ministry Says,' *Reuters*, 17 April 2024.

Shamim, Syed. 'Nuclear Proliferation in South Asia: Towards World War III,' *Social Sciences Research Network*, 27 January 2018, available at: SSRN: https://ssrn.com/abstract=3111513

Shankar, Priyanka and Yashraj Sharma. 'The $10bn India-Pakistan Trade Secret Hidden by Official Data,' *Al Jazeera*, 1 May 2025.

Sharma, Rajeev. 'US Must Wise Up on Pak Aid,' *The Diplomat*, 27 April 2010.

Sherazee, Mahnoor and Hammad Abbasi. 'Health Indicators Project Bleak Future for Pakistani Children,' *Dawn*, 26 February 2014.

Siddiqa, Ayesha. *Military Inc.: Inside Pakistan's Military Economy* (London: Pluto, 2007).

Sikand, Yoginder. 'Changing Course of Kashmiri Struggle: From National Liberation to Islamist Jihad?' *Economic and Political Weekly* 36, no. 3 (2001): pp. 218–227.

Small, Andrew. *The China-Pakistan Axis: Asia's New Geopolitics* (New York: Oxford University Press, 2015).

Smith, Paul J. 'The China-Pakistan-United States Strategic Triangle: From Cold War to the "War on Terrorism",' *Asian Affairs* 38, no. 4 (2011): pp. 197–220.

Smith, Jeff. 'US Takes Gloves Off with Pakistan,' *The Diplomat*, 25 September 2011.

Smith, Rogers M. 'From the Shining City on a Hill to a Great Metropolis on a Plain? Stories of Immigration and Peoplehood,' *Social Research* 77, no. 1 (2010): pp. 21–44.

Solomon, Ty. *The Politics of Subjectivity in American Foreign Policy Discourses* (Ann Arbor, MI: University of Michigan Press, 2015).

Steele, Brent. *Ontological Security in International Relations* (Abingdon: Routledge, 2008).

Stel, Nora. *Hybrid Political Order and the Politics of Uncertainty: Refugee Governance in Lebanon* (Abingdon: Routledge, 2020).

Stern, Jessica. 'Pakistan's Jihad Culture,' *Foreign Affairs* 79, no. 6 (2000): pp. 115–116.

Tahir-Kheli, Shirin. 'Iran and Pakistan: Cooperation in an Area of Conflict,' *Asian Survey* 17, no. 5 (1977): pp. 474–490.

Tahir-Kheli, Shirin. *The United States and Pakistan: The Evolution of an Influence Relationship* (New York: Praeger, 1982).

Tahir-Kheli, Shirin. 'In Search of an Identity: Islam and Pakistan's Foreign Policy,' in *Islam in Foreign Policy*, edited by Adeed Dawisha (Cambridge: Cambridge University Press, 1983), pp. 68–83.

Talbot, Ian. 'Pakistan's Hybrid Regime: Growing Democratization, or Increased Authoritarianism?' in *Routledge Handbook of Autocratization in South Asia*, edited by Sten Widmalm (Abingdon: Routledge, 2021), pp. 141–150.

Tankel, Stephen. 'Lashkar-e-Taiba: from 9/11 to Mumbai,' *The International Centre for the Study of Radicalisation and Political Violence*, April/May 2009, available at: https://icsr.info/wp-content/uploads/2009/04/Tankel_01.pdf

The Diplomat. 'US Must Grow Up on Pakistan,' 26 July 2010.

The Guardian. 'India and US Reach Nuclear Agreement,' 2 March 2006.

The Guardian. 'US Worried Taliban Battle Funds Being Wasted on Barbed Wire,' 30 November 2010.

The Guardian. 'Taliban Refuse to Take Part in Afghanistan Peace Talks,' 5 March 2016.

The Hindu. '"Blood Will Flow if India Stops River Water,' Threatens Bilawal Bhutto,' 26 April 2025.

The Nation. 'Pakistan Thrice Tried to Formalize Durand Line: Taliban Official,' 29 February 2016.

The Telegraph. 'Donald Trump "Approves Conditional Peace Deal with Taliban",' 12 February 2020.

The Times of India. 'N-deal for Pak? US denies, India Remains Anxious,' 23 March 2010.

The Times of India. 'We Didn't Think About Consequences of "Kabul Tea": Bilawal Bhutto-Zardari,' 16 January 2024.

The Tribune. 'Don't Link J&K Issue with Kabul: Taliban,' 9 August 2019, available at: https://www.tribuneindia.com/news/archive/nation/don-t-link-j-k-issue-with-kabul-taliban-815019

The Wire. 'Kashmir: Taliban Urges India, Pakistan to Refrain from Violence, Use "Rational Pathways",' 9 August 2019.

Time. 'Pakistan: Death in the Chair,' 6 October 1958, available at: https://time.com/archive/6888304/pakistan-death-in-the-chair/

Toppa, Sabrina. 'As Pakistan Cracks Down on NGOs, Civil Society Questions Next Steps,' *DevEx*, 10 January 2018.

Tripp, Aili Mari. *Museveni's Uganda: Paradoxes of Power in a Hybrid Regime* (Boulder, CO: Lynne Rienner, 2010).

United States Central Command. 'Gates: Pakistan Must Do Better Job Near Border,' Press Release, 18 July 2008, available at: https://www.centcom.mil/MEDIA/NEWS-ARTICLES/News-Article-View/Article/883676/gates-pakistan-must-do-better-job-near-border/

United States Department of State. 'Joint Statement: U.S.-Pakistan Strategic Dialogue at the Ministerial Level,' Washington, D.C., United States, 22 October 2010, available at: https://2009-2017.state.gov/r/pa/prs/ps/2010/10/149854.htm

United States Senate Armed Services Committee. 'The Statement of Admiral Michael Mullen, Chairman Joint Chiefs of Staff, Before the Senate Armed Services Committee on Afghanistan and Iraq,' 22 September 2011, available at: https://www.armed-services.senate.gov/imo/media/doc/Mullen%2009-22-11.pdf

United States Institute for Peace. *How to Deal with Pakistan*, Washington, D.C., Seminar, 18 October 2017.

Urinboyev, Rustamjon. *Migration and Hybrid Political Regimes: Navigating the Legal Landscape in Russia* (Oakland, CA: University of California Press, 2020).

Van Hollen, Christopher. 'The Tilt Policy Revisited: Nixon-Kissinger Geopolitics and South Asia,' *Asian Survey* 20, no. 4 (1980): pp. 339–361.

Veengas. 'Media and Social Media Platforms in Pakistan Face Increasing Restrictions,' *The Wire*, 26 February 2024.

Verma, Raj. 'China's New Security Concept: India, Terrorism, China's Geostrategic Interests and Domestic Stability in Pakistan,' *The Pacific Review* 33, no. 6 (2019): pp. 991–1021.

Verwey, Duncan. 'How Do Orientalist Narratives Inhibit Western Understanding of Terrorism in Pakistan?' Unpublished Dissertation, MA International Security, University of Bath, United Kingdom, 2019.

Vucetic, Srdjan. 'Identity and Foreign Policy,' *Oxford Research Encyclopedia of Politics*, 26 April 2017.

Waldman, Matt. 'The Sun in the Sky: The Relationship Between Pakistan's ISI and Afghan Insurgents,' *Carr Center for Human Rights Policy*, Kennedy School of Government, Harvard University, June 2010, available at: https://assets.publishing.service.gov.uk/media/57a08b0c40f0b652dd000a78/DP18.pdf

Walsh, Declan. 'A CIA Spy, a Hail of Bullets, Three Killed and a US-Pakistan Diplomatic Row,' *The Guardian*, 20 February 2011.

Wani, Ayjaz. 'Pakistan: Why Has China Halted CPEC Projects,' *Observer Research Foundation*, 25 November 2020.

Warraich, Suhail. 'The Bajwa Doctrine: from Chauvinism to Realism,' *The News*, 18 March 2018.

Webb, Mark. 'The Epistemology of Trust and the Politics of Suspicion,' *Pacific Philosophical Quarterly* 73, no. 4 (1992): pp. 390–400.

Weinberg, G. L. *The Foreign Policy of Hitler's Germany: Diplomatic Revolution in Europe, 1933–36* (Chicago, IL: University of Chicago Press, 1971).

Weitz, Richard. 'How China Can Fix Pakistan,' *The Diplomat*, 17 May 2011.

Weitz, Richard. 'How Pakistan Kids Itself on China,' *The Diplomat*, 28 December 2011.

Wilber, Caldwell. *American Narcissism: The Myth of National Superiority* (Sanford, NC: Algora, 2008).

Wilde, Gabriele, et al. (eds.) *Civil Society and Gender Relations in Authoritarian and Hybrid Regimes: New Theoretical Approaches and Empirical Case Studies* (New York: Columbia University Press, 2018).

Williams, Brian. 'The CIA's Covert Predator Drone War in Pakistan, 2004–2010: The History of an Assassination Campaign,' *Studies in Conflict and Terrorism* 33, no. 10 (2010): pp. 871–892.

Wilkinson, Steven. 'Democratic Consolidation and Failure: Lessons from Bangladesh and Pakistan,' *Democratization* 7, no. 3 (2007): pp. 203–226.

Wojczewski, Thorsten. 'Identity and World Order in India's Post-Cold War Foreign Policy Discourse,' *Third World Quarterly* 40, no. 1 (2019): pp. 180–198.

Wolf, Siegfried. 'China's Role in Pakistan: International and Domestic Implications,' in *Studies on Pakistan*, edited by Sylvia Mishra (New Delhi: Paragon, 2014), pp. 155–181.

Wolf, Siegfried. *The China-Pakistan Economic Corridor of the Belt and Road Initiative: Concept, Context and Assessment* (London: Springer, 2020).

Wriggins, W. Howard. 'Pakistan's Search for a Foreign Policy After the Invasion of Afghanistan,' *Pacific Affairs* 57, no. 2 (1984): pp. 284–303.

Yavuz, M. Hakan. *Nostalgia for the Empire: The Politics of Neo-Ottomanism* (New York: Oxford University Press, 2020).

Yavuz, M. Hakan. 'The Motives Behind the AKP's Foreign Policy: Neo-Ottomanism and Strategic Autonomy,' *Turkish Studies* 23, no. 5 (2022): pp. 659–680.

Yuan, Jingdong. 'Beijing's Balancing Act: Courting New Delhi, Reassuring Islamabad,' *Journal of International Affairs* 64, no. 2 (2011): pp. 37–54.

Yusufzai, Rhimullah. 'The Taliban and India Can Be Reconciled: Taliban Spokesman Zabihullah Mujahid Spoke to Rahimullah Yusufzai,' *Outlook*, 15 February 2020.

Zaeef, Abdul Salam. *My Life with the Taliban* (London: Hurst, 2010).

Zaheer, Hasan. *The Separation of East Pakistan: The Rise and Realization of Bengali Muslim Nationalism* (Oxford: Oxford University Press, 1994).

Zaidi, S. Akbar. 'Who Benefits from Aid to Pakistan,' *Economic and Political Weekly* 46, no. 32 (2011): pp. 103–109.

Zakaria, Muhammad, Wen Jun, and Haseeb Ahmed. 'Effect of Terrorism on Economic Growth in Pakistan: An Empirical Analysis,' *Economic Research-Ekonomska Istraživanja* 32, no. 1 (2019): pp. 1794–1812.

Zehra, Naseem. *From Kargil to the Coup: Events That Shook Pakistan* (Lahore: Sang-e-Meel, 2018).

Zengerle, Patricia. 'With Eye on Afghanistan Talks, Trump Vows to Stop "Endless Wars",' *Reuters*, 6 February 2019.

Ziring, Lawrence. 'Recent Trends in Pakistan's Foreign Policy,' *Asian Affairs* 2, no. 5 (1975): pp. 295–307.

Ziring, Lawrence. 'Pakistan and India: Politics, Personalities, and Foreign Policy,' *Asian Survey* 18, no. 7 (1978): pp. 706–730.

Ziring, Lawrence. *Pakistan at the Crosscurrent of History* (London: Oneworld, 2003).

Index

For the benefit of digital users, indexed terms that span two pages (e.g., 52–53) may, on occasion, appear on only one of those pages.